Felix Mitelman, M.D., Ph.D., D.Sc., **Lund**

Catalogue of Chromosome Aberrations in Cancer

Editor:
Harold P. Klinger, M.D., Ph.D.

KARGER

Reprinted from
Cytogenetics and Cell Genetics

Vol. 36, Nos. 1–2 (1983)

S. Karger · Medical and Scientific Publishers · Basel · München · Paris · London · New York · Tokyo · Sydney

Printed in Switzerland by Thür AG Offsetdruck, Pratteln
ISBN 3–8055–3813–8

Contents

Cytogenet. Cell Genet. *36:* 4 (1983)

Preface

Immense amounts of data about cancer-associated chromosome abberations have been collected during the last ten years, and the systematic evaluation of these data has disclosed a number of correlations between chromosome change and neoplastic disease. These results are of practical clinical significance for the evaluation of diagnosis, treatment, and prognosis; they are also of theoretical interest in cancer research and cell biology.

New data, widely scattered in the literature, appear at a rapid rate, and there is an urgent need for a collection and survey of all the relevant material. Since the introduction of chromosome banding techniques we have tried to collect and periodically summarize the information available. By 1981 the complexity of the information prompted adoption of computer methods for assembling, revising, and indexing these data.

Many cytogeneticists, clinicians and cell biologists have indicated a need for a systematic, concise, and uniform presentation of the material included in this registry. In evaluating the chromosome abnormality in a neoplastic disorder, it is becoming difficult to ascertain from the vast body of literature, what, if anything, has been published before on the same type of chromosome abnormality. It is hoped that this Catalogue will fulfill this need and make the information readily accessible.

The material for the present compilation, comprising 3,844 cases, has been collected from three main sources: published cases, unpublished cases kindly communicated by numerous colleagues, and unpublished cases from our laboratory. I have aimed at completeness up to August, 1983.

Every effort was made to avoid duplicate recording of a single case. In most papers it is clearly stated if data on a patient have been published previously. Unfortunately, this is not always the case. In many instances it has been possible to identify such patients, although I have probably not always been successful.

Considerable effort was also spent in checking the manuscript; I apologize for any errors that may remain. Since I intend to continue to update this Catalogue, any suggestions concerning errors, omissions, or other shortcomings are very much encouraged. Published or unpublished data on new cases of human neoplasms analyzed with banding techniques will also be extremely welcome. Data on consecutive series of patients with any type of premalignant or malignant disorders are especially valuable. Such information will be included, with due reference, in the next edition of the Catalogue.

I would like to express my sincere gratitude to a great number of colleagues around the world who over the years have contributed preprints, reprints, and/or unpublished cases. They will, I hope, forgive my failure to mention them here by name; all contributors may be found in the reference list.

Computerization was made possible by the patient and efficient assistance of Claes Westrup at the Oncologic Centre, University Hospital, Lund. Mrs. Marianne Lagergren and Miss Eva Hjalmarsson deserve special mention for excellent secretarial help. I also thank my scientific tutor, Professor Albert Levan, who twenty years ago introduced me to tumor cytogenetics, and his son Professor Göran Levan, my co-worker for many years, for encouragement and advice.

Financial support for this project was generously provided by a special grant from the John and Augusta Persson Foundation for Medical Research. I also gratefully acknowledge support from the Swedish Cancer Society and the International Agency for Research on Cancer, Lyon.

Felix Mitelman[1]
Lund,
August, 1983

[1] Prof. Felix Mitelman, Department of Clinical Genetics, University Hospital, S–221 85 Lund (Sweden)

Cytogenet. Cell Genet. *36:* 5–516 (1983)

0301–0171/83/0362–0005$2.75/0

On the use of the Catalogue

This Catalogue lists karyotype changes in 3,844 cases of neoplastic disorders studied by means of chromosome banding techniques. An additional 593 cases with normal karyotypes and 212 cases of chronic myeloid leukemia (CML) with the Philadelphia chromosome (Ph^1), t(9;22), as the sole abnormality are registered in our computer files but not included in the Catalogue.

Only clonal aberrations have been recorded. A clone was defined as at least two cells with the same extra chromosome or structural aberration or three cells with the same missing chromosome. Only the aberrant clones within a tumor have been listed in the Catalogue; thus, normal diploid stem- or sidelines have been excluded.

In general, the nomenclature for chromosome aberrations follows the recommendations proposed by ISCN (1978) with the following slight modifications:

? (question mark) indicates questionable identification of a chromosome or chromosome structure, as well as uncertainty regarding the modal/stem- or sideline chromosome number and sex. Thus, the designation X? indicates that the sex is unknown; ??,XX,... indicates a female karyotype with uncertain or unknown chromosome number.

cx (complex) indicates that unidentified, usually complex, structural changes were present in addition to the karyotype presented.

der (derivative chromosome) designates a structurally rearranged chromosome that cannot be fully characterized. Identified break points in such marker chromosomes are specified within parenthesis immediately following the chromosome involved.

t (translocation) indicates all types of translocations, Robertsonian, reciprocal, or tandem. In complex translocations, the order of the break points specified for the chromosomes involved does not necessarily reflect the mechanism of origin of the rearrangement. For example, a four-way translocation affecting bands p11, p22, and q11 in chromosome 7, bands p13 and p24 in chromosome 9, and band q13 in chromosome 12, is given as t(7;9;12)(p11p22q11;p13p24;q13)

Constitutional aberrations, unbalanced and balanced, are included in the description of the tumor karyotype. The presence of a constitutional aberration is indicated by the designation 'c' immediately after the karyotype. Thus, an extra chromosome 8 in the bone marrow of a leukemic female patient with Down's syndrome is given as 48,XX,+8,+21'c'.

Secondary neoplasms, that is any neoplasm in a patient with a previous benign or malignant neoplasm, treated or untreated, are indicated by the symbol 's' immediately after the karyotype.

The morphologic diagnoses have been collected from various sources such as the International Classification of Diseases for Oncology (ICD-O), the Systematized Nomenclature of Medicine (SNOMED), the French-American-British (FAB) proposals for the classification of acute leukemias and myelodysplastic syndromes, the classification systems for non-Hodgkin lymphomas proposed by Rappaport, Lukes and Collins, Lennert (Kiel-classification), and the US National Cancer Institute Working Formulation. All morphologic diagnoses used in this Catalogue are presented in table I. NOS designates 'not otherwise specified'.

In the Catalogue the karyotypes of all cases are arranged according to:

1. The chromosome involved in aberrations (chromosomes 1–22, the X and the Y). The distribution of the 3,844 cases among the 24 chromosomes is presented in table II. Obviously the same case may be listed under several chromosomes depending on the number of chro-

Table I. Number of cases according to histopathologic diagnosis

Histopathology	Number of cases
Hematological disorders	
Unclassified leukemias	
Leukemia, NOS	2
Acute leukemia, NOS	12
Chronic leukemia, NOS	2
Non-lymphocytic leukemias	
Acute non-lymphocytic leukemia (ANLL)	271
ANLL, FAB type M1	77
ANLL, FAB type M2	187
ANLL, FAB type M1 + M2	212
ANLL, FAB type M3	139
ANLL, FAB type M4	137
ANLL, FAB type M5	64
ANLL, FAB type M5a	22
ANLL, FAB type M5b	4
ANLL, FAB type M4 + M5	27
ANLL, FAB type M6	70
ANLL, special type	3
Chronic myeloid leukemia, t(9;22)	616
Chronic myeloid leukemia, aberrant translocation	205
Chronic myeloid leukemia, Ph^1 negative	42
Eosinophilic leukemia	12
Megakaryocytic leukemia	2
Myeloproliferative disorders	
Polycythemia vera	114
Myelosclerosis/Myelofibrosis	47
Idiopathic thrombocythemia	10
Chronic myeloproliferative disease, NOS	24
Myeloproliferative disorder, special type	19
Dysmyelopoietic syndromes	
Preleukemia, NOS	98
Chronic monocytic leukemia	2
Chronic myelomonocytic leukemia	10
Refractory anemia, NOS	44
Sideroblastic anemia, NOS	10
Acquired idiopathic sideroblastic anemia	16
Refractory anemia without excess of blasts	12
Refractory anemia with excess of blasts	27
Erythrocytopenia	2
Granulocytopenia	3
Thrombocytopenia	8
Pancytopenia	17
Aplastic anemia	6
Paroxysmal nocturnal hemoglobinuria	2
Dysmyelopoietic syndrome, special type	9
Hematopoietic proliferation disorder, NOS	2
Special leukemias	
Hairy cell leukemia	4
Lymphocytic leukemias	
Acute lymphocytic leukemia (ALL)	163
ALL, FAB type L1	32
ALL, FAB type L2	44
ALL, FAB type L3	36
Chronic lymphocytic leukemia	71

Table I (continued)

Histopathology	Number of cases
Prolymphocytic leukemia	9
Lymphocytic leukemia, special type	14
Monoclonal gammopathies	
Multiple myeloma	37
Macroglobulinemia	4
Plasma cell leukemia/Plasmocytoma	12
Solid tumors	
Unclassified neoplasms	
Malignant neoplasm, NOS	1
Epithelial neoplasms	
Epithelial tumor NOS, uncertain, benign, or malignant	2
Carcinoma, NOS	99
Carcinoma, NOS, metastatic	37
Large cell carcinoma	2
Small cell carcinoma	30
Papillary carcinoma	14
Squamous cell carcinoma	22
Squamous cell carcinoma, metastatic	1
Transitional cell carcinoma	6
Adenocarcinoma	38
Adenocarcinoma, metastatic	21
Adenoma	14
Adenomatous polyp	8
Mesenchymal neoplasms	
Sarcoma, NOS	3
Liposarcoma	1
Leiomyosarcoma	1
Rhabdomyosarcoma	1
Mesothelioma, malignant	4
Melanomas	
Malignant melanoma	11
Malignant melanoma, metastatic	5
Neurogenic neoplasms	
Astrocytoma, grade I, II	1
Astrocytoma, grade III, IV	1
Oligodendroglioma	1
Neuroblastoma	24
Retinoblastoma	25
Meningioma	65
Embryonal and miscellaneous neoplasms	
Embryonal carcinoma, NOS	1
Blastoma, NOS	2
Teratoma, benign	1
Teratoma, malignant	3
Seminoma	5
Hydatiform mole	1
Lymphomas	
1. Unclassified and miscellaneous lymphomas	
Benign lymphomatous tumor, NOS	1
Malignant lymphoma, NOS	4
Non-Hodgkin's lymphoma, NOS	9

Table I (continued)

Histopathology	Number of cases
Non-Hodgkin's lymphoma, lymphocytic, NOS	25
Non-Hodgkin's lymphoma, histiocytic, NOS	22
Non-Hodgkin's lymphoma, T-cell, NOS	2
Malignant histiocytosis	4
Angioimmunoblastic lymphadenopathy	7
Lennert's lymphoma	1
2. Hodgkin's lymphomas	
Hodgkin's disease, NOS	4
Hodgkin's disease, lymphocytic predominance	1
Hodgkin's disease, mixed cellularity	6
Hodgkin's disease, lymphocytic depletion	5
Hodgkin's disease, nodular sclerosis	2
3a. Non-Hodgkin's lymphomas: Rappaport classification	
Lymphocytic, well differentiated, diffuse	1
Lymphocytic, poorly differentiated, NOS	2
Lymphocytic, poorly differentiated, diffuse	17
Lymphocytic, poorly differentiated, nodular	7
Mixed cell, diffuse	4
Mixed cell, nodular	1
Histiocytic, diffuse	27
Histiocytic, nodular	2
3b. Non-Hodgkin's lymphomas: Kiel classification	
Lymphocytic, T-zone	5
Immunocytoma	7
Centrocytic	2
Centroblastic-centrocytic, diffuse	5
Centroblastic-centrocytic, follicular	12
Centroblastic, diffuse	3
Centroblastic, follicular	1
Lymphoblastic, unclassified	2
Lymphoblastic, Burkitt's type	94
Lymphoblastic, convoluted type	9
Immunoblastic	13
3c. Non-Hodgkin's lymphomas: Lukes and Collins classification	
Mycosis fungoides	12
Sezary's syndrome	11
Follicular center cell, follicular, small cleaved	1
Folicular center cell, diffuse, large cleaved	3
Follicular center cell, diffuse, small non-cleaved	1
Immunoblastic type (B-cell)	2
3d. Non-Hodgkin's lymphomas: Working formulation	
Small lymphocytic	11
Follicular, small cleaved cell	14
Follicular, mixed small cleaved and large cell	2
Follicular, large cell	3
Diffuse, small cleaved cell	1
Diffuse, mixed small and large cell	2
Diffuse, large cell	3
Large cell, immunoblastic	3
Small non-cleaved cell	3
Total number	3, 844

Table II. Distribution of 3,844 cases of neoplastic disorders among the 24 chromosomes

Chromosome	Number of cases
1	678
2	289
3	462
4	276
5	520
6	409
7	632
8	1,083
9	651
10	258
11	463
12	378
13	335
14	521
15	362
16	246
17	806
18	287
19	311
20	280
21	543
22	806
X	186
Y	211
Total	10,993

mosomes involved. It should be noted that cases of CML with additional aberrations, superimposed on t(9;22), are only entered under the chromosomes involved in additional changes; such cases are not listed under chromosomes 9 and 22.

2. Within each chromosome entry, the cases are categorized according to the morphologic diagnosis, and these are listed in the same order for each chromosome (see table I). The main headings are repeated in the same order for each chromosome, even if there is no case entered under them.

3. The karyotypes within each morphologic diagnosis are arranged according to increasing chromosome numbers. Identical aberrations are therefore not necessarily all found together.

Karoytoypes not separated by a slash (/), but listed consecutively for the same case, indicate that the tumor was examined either on different occasions or at different locations. Whenever a tumor was studied at different times, the karyotypes are presented chronologically. It should be noted that the relevant chromosome aberration may be present in only one of several karyotypes presented for the patient.

In compiling this Catalogue some arbitrary decisions were necessary:

1. No attempt is made to interpret karyotypic changes, e.g. inferred breakpoints were not catalogued. Neither have obvious mistakes been corrected. Thus, breakpoints localized to non-existing band are listed as given in the original report. I of course take full responsibility for any errors that I may have introduced.

2. Only positively identified clonal aberrations are catalogued. Therefore, the sum of of the catalogued abnormalities may not necessarily add up to the given modal chromosome number.

3. In general, the order of the chromosome changes in a karyotype follows the description given by the author. This means that in some karyotypes numerical changes may precede structural aberrations, or vice versa, irrespective of the order of the chromosomes involved.

4. In polyploid tumors only structural aberrations are recorded; the sex chromosomes in these tumors are given as XX or XY only to indicate the sex of the patient and not to reflect the actual sex chromosome complement.

5. The morphologic classification of the non-Hodgkin lymphomas presents a special problem since at least four different classification systems are used in the literature. If more than one classification was attempted in a report, the case has in general been entered in the Catalogue on the basis of the following priority system: (1) Working Formulation, (2) Kiel, (3) Lukes and Collins, and (4) Rappaport. Among the possible equivalent classifications of a single tumor, Mycosis fungoides and Sezary's syndrome are entered according to the Lukes and Collins classification, while the convoluted and plasmacytoid lymphocytic (immunocytoma) types are entered according to the Kiel classification. Also, all Burkitt's lymphomas are entered under the Kiel classification. Unclassifiable lymphomas and Histiocytic lymphomas in all classification systems are presented as Non-Hodgkin's lymphoma, NOS and Non-Hodgkin's lymphoma, histiocytic, NOS, respectively, under the heading *Unclassified and miscellaneous lymphomas.* Certain hematologic lymphoid disorders, such as chronic lymphocytic leukemia, plasma cell leukemia, and hairy cell leukemia, are included in the section *Hematological disorders.*

6. In principle, only cases studied in direct preparations or after short term in vitro culture have been included. However, in a few instances, cell lines from malignant melanomas, neuroblastomas, and Burkitt's lymphomas are included.

7. When the same case is dealt with in more than one paper, only the most recent reference is generally given in the tables but all of the original papers, which may contain additional important information, are included in the reference list as well as references to cases with normal karyotypes that are not listed in the Catalogue but are recorded in the registry. The reference list should, therefore, contain most of the literature on the results of chromosome analyses obtained by means of banding techniques in neoplastic disorders.

The data contained in the computer is coded for by a number of parameters:

Mode of ascertainment, i.e. whether or not a case belongs to an unselected, consecutive series of patients studied in a laboratory;

Age, sex, ethnic group, and geographic region;

Previous neoplasm – morphologic diagnosis. topography and type of treatment used;

Hereditary disorder, including constitutional chromosome aberration in the patient and/or in relatives;

Obvious environmental or occupational exposure to potentially mutagenic or carcinogenic agents;

Clinical and hematologic data, including immunology, tumor site, clinical state, and survival period;

Type of tissue studied, technique used for chromosome preparation, and time of culture.

All this information can easily be retrieved and used for scientific purposes. Workers active in the field may, upon request, obtain any information directly from the author. Any suggestions on how to improve the usefulness of this Catalogue will be gratefully appreciated.

Chromosome 1

HEMATOLOGICAL DISORDERS

UNCLASSIFIED LEUKEMIAS

Acute leukemia, NOS

45,X,-X,+t(X;1)(q13;p12),-7,+8,-9	Mamaeva et al 1983
45,XY,-1	Shiloh et al 1979
46,XX,+t(1;19)(q25;q13),der(8),19q+	Prigogina et al 1979
46,XX,dup(1)(p31p36),dup(7)(p15p21),dup(3)(q12q29), ins(12)(q12)"s"	Berger et al 1977
46,XY,+del(1)(q32),-9,del(9)(q13),del(22)(q11)	Prigogina et al 1979

NONLYMPHOCYTIC LEUKEMIAS

Acute nonlymphocytic leukemia (ANLL)

??,X?,+t(1;10)(q21;q22)	Anglani et al 1981
??,X?,+t(1;3)(q21;p11)	Anglani et al 1981
43,X,-Y,t(1;8)(p22;q24),-1,-11,-11,-13,-16,-17,-22, t(1;22)(q11;p11),t(11;13)(q23q25;q12q14),t(17;?)(p11;?), +mar	Oshimura et al 1976b
43,XX,-1,3p+,3q-,4q+,-5,-5,-7,+9,+9,10p+,+10,+10,10p-,10q-, +11q-,-13,-15,-16,-17,-18,-21"s"	Whang-Peng et al 1979
43,XX,t(1;20),-5,-6,7q-,11q-,-12,17q+,-18,19q+,-21	Mitelman 1983
44,XX,t(8;11)(p23;p13 or p15),-20/46,XX,del(7)(q11), t(1;3)(q2?;q2?),-20,-21,-22,+3mar	Philip et al 1978a
44-45,XX,-3,-5,-6,-18,-19,t(1;2)(q2?;q11),t(5;?)(q13;?)	Philip et al 1978a
45,XY,-1	Bernard et al 1982a
45,XY,t(1;14)(p22;p11),t(1;15)(p22;p11),-2,del(5)(p13),-17, +mar	Philip et al 1978a
46,X?,t(1;13)(p?;q?)	Wurster-Hill 1983
46,XX,-1,+2,del(1)(q31)	Zuelzer et al 1976
46,XX,del(1)(p32),del(5)(q13),+8,t(9;?)(p11;?), t(11;22)(q11;q13),-16,-17,-19,+3mar	Muir et al 1977
46,XX,del(7)(q22)/46,XX,del(7)(q22),del(5)(q22q31)/46,XX, del(7)(q22),-9,t(1;9)(q22;p24)/46,XX,t(2;3)(q31;q27), del(7)(q22),del(5)(q22q31orq33)"s"	Oshimura et al 1976b
46,XX,t(1;1)(q42;q21),t(3;7)(q2?;q22)	Philip et al 1978a
46,XX,t(1;17)(q2?;q21)	Philip et al 1978a
46,XX,t(1;22)(p36;q11),del(5)(q14q34),del(7)(q22q36),r(22)	Yunis 1982
46,XX,t(1;8)(q?;q?)	Nordenson 1983

46,XY,+t(1;12)(q21;p13),-12	Trent et al 1983a
46,XY,1p+,12p-,?t(1;12),?11p+,inv(12)(p?),-2,+mar	Mitelman 1983
46,XY,1q+,5p-,-7,+mar	Chrz et al 1983
46,XY,del(1)(p22),t(1;?)(q32;?)	Zuelzer et al 1976
46,XY,del(2)(q31),-7,del(9)(p13),t(2;21)(q21;q22), t(2;?)(p21;?)/46,XY,t(1;?)(p22;?),t(17;?)(q25;?)	Philip et al 1978a
46,XY,dup(1)(q25q44),t(11;17)(p15;q21)/46,XY, t(11;17)(p15;q21)/47,XY,t(11;17)(p15;q21),+del(1)(p11)	Mitelman 1983
46,XY,dup(1)(q25q44),t(11;17)(p15;q21)/46,XY, t(11;17)(p15;q21)	
46,XY,dup(1),t(11;17)/46,XY,t(11;17)/47,XY,t(11;17), +del(1)	
46,Y,t(X;11)(q22.3;q23.3),ins(X;11)(q22.3;p13p15), dup(1)(q11q21),t(4;6)(p16;p21),t(7;12)(p13;q15)	Nacheva et al 1982
47,XX,+t(1;16),-16,-16,+13,+20	Prieto et al 1978b
47,XX,-4,+13,t(1;4)(q22q25;p14p16)	Oshimura et al 1976b
47,XX,del(1)(p22),+10"s"	Geraedts et al 1980a
47,XX,t(3;5)(q21;q31),t(1;13)(p36;q14),+8,+22	Oshimura et al 1976b
47,XX,t(9;22)(q34;q11),+22q-/48,XX,t(9;22),+8,+22q-	Mitelman 1983
49,XX,t(9;22)(q34;q11),+8,+13,+22q-	
49,XX,t(9;22)(q34;q11),+8,+22q-,+mar	
50,XX,t(9;22)(q34;q11),dup(1)(q25q44),+del(3)(p11),+8, +22q-,+mar	
49,X,t(X;10)(p11;p11),+del(1)(p22),+del(1)(p22), del(5)(q23),+7/49,X,t(X;10)(p11;p11),+del(1)(p22),+7,+8	Mitelman 1983
49-51,XX,-1,+8,+1-3mar"s"	Weinfeld et al 1977a
51,X,-Y,+1,+18,+20,+22,+inv(4),+mar	Zuelzer et al 1976
52,XX,1p+,+10,+13,+19,+21,+21,+22"c"	Berger et al 1973
56,XY,+Y,+6,+14,+21,+t(5;6)(q?;p?),+del(1)(p22q25), +del(2)(q12),+del(3)(p12),+del(4)(q13),+del(5)(p14q14), +mar	Zuelzer et al 1976

ANLL, FAB type M1

41,XY,-1,-2,-4,-5,-12,-15,-17,-17,-18,-22,+5mar	Sessarego et al 1981a
46,XX,-1,inv(1),3q+,5q-,t(13;17),-18,+mar/51,XX,-1,inv(1), 3q+,5q-,t(13;17),-18,+6,+9,+9,+11,+11,+mar	Hagemeijer et al 1981a
46,XX,dup(1)(q21q33),7q+,18p+	Alimena 1983
47-48,XY,+8,t(9;22)(q34;q11),-20,+t(1;20)(q21;q13), t(10;21)(p11;q22),+22q-	Sasaki et al 1983
49,XX,+X,-1,+der(1)(q21q31),+del(1)(p22),+t(1;5)(p11;q35), -5,t(9;22)(q34;q11),+22q-	Sasaki et al 1983
50,XX,+3,+19,+t(1;6)(q21;p22)	Yunis et al 1981
58,XY,+1,+4,+5,+6,+8,+11,+13,+13,+14,+14,+15,+21,+21, t(9;22)(q34;q11)	Alimena 1983

ANLL, FAB type M2

44,X,t(X;8),del(5)(q12q21),t(1;6),t(7;19),-7,-8,t(7;8),-21	Kerkhofs et al 1982
45,X,-X,t(8;21)(q21;q21)	Hagemeijer et al 1981a
45,X,-X,t(8;21)(q21;q21),1p+,17p-	
45,X,-X,t(8;21)(q22;q22)/46,X,-X,t(8;21),1p+,7q+,+19/47,X, -X,t(8;21),1p+,+19,+der(21)	Sakurai et al 1982b

45,X,-Y,t(8;21)(q22;q22),t(1;5)(q43;q11)	Berger et al 1982b
45,X,-Y,t(8;21)(q22;q22),+1,-8/46,X,-Y,t(8;21),+8	Berger et al 1982b
46,XX,+1p-,+9,-16,-16/46,XX,+1p-,+13,-16,-16	Prieto et al 1981
46,XX,-7,+8,inv(1)(p36q13)"s"	Rowley et al 1981a
46,XX,1q+,2q+,-3,-11,+r,+mar	Berger et al 1981a
46,XX,t(1;8;21)	Tricot et al 1981
46,XY,+1,5q-,-7"s"	Berger et al 1981a
46,XY,del(1)(p22p32),+19,-21	Hagemeijer et al 1981a
46,XY,t(1;17)(p36;q21)"s"	Rowley et al 1977a
47,XX,1p+,der(5),-6,12p+,-13,-14,-16,-17,-17,-21,+22,+7mar, dmin	Cooperman and Klinger 1981
48,XY,+1,+15	Lessard and Le Prise 1983
53,XX,+1,-4,-5,+6,-7,+8,+11,+12,+21,+21,+3mar"s"	Zaccaria et al 1983b

<u>ANLL, FAB type M1+M2</u>

43,XX,1p+,-4,-5,-6,-7,-13,-17,del(18)(q21),+3mar	Rowley and Potter 1976
45,XY,+t(1;10)(q23;q26),t(8;21)(q22;q22),-10/45,X,-Y, +t(1;10),t(8;21),-10	Fitzgerald et al 1983
45,XY,t(19;22)(p13;q11),t(1;7)(q21;q22),t(7;9),+del(8)(p?), -15	Oshimura and Sandberg 1977
47,XX,+19,t(1;11)	Benedict et al 1979
46,XX,del(1)(q32)/46,XX,del(1)(q32),del(13)(q22)	Fitzgerald et al 1983
46,XY,t(1;11)(q21;p15),del(1)(p22)"s"	Sandberg et al 1982a
46,XY,t(1;6;11)(q12;q23;p15)	Yamada and Furusawa 1976
46-47,XX,del(1)(q32),+8,t(15;17)(q25or26;q22or23),+18	Mitelman et al 1981
46-47,XX,t(1;?)(p36;?),del(18)(q12),del(20)(q11), +1-2mar"s"	Mitelman 1983
47,X,-X,+del(1)(p11),del(5)(q13),t(5;12)(q21;q24), t(11;12)(q25;q13),t(13;14)(q11;p11),+16,+21	Mitelman et al 1981
47,X,-Y,t(1;13)(q32p32q11;q11),del(3)(q21), t(8;21)(q22;q22),del(9)(q22),+8,+18	Sakurai et al 1982b
47,XX,del(1)(q?),+18	Benedict et al 1979
48,X,-Y,1p-,+3p-,+15,+mar"s"	Sandberg et al 1982a
48,XX,+1,del(5)(q13orq15),+11"s"	Rowley and Potter 1976
48,XX,+t(1;22)(q11;p12),+del(19),-22,+mar	Fitzgerald et al 1983
48,XX,t(1;3)(q44;p14),del(3)(q21p14),-5,+10,+11,+21	Mitelman et al 1981
49,XX,+1,-5,+7,+8	Nordenson 1983
51,XY,+1,+4,+5,+8,+20	Nordenson 1983

<u>ANLL, FAB type M3</u>

45-51,XY,t(3;12)(p14;q24),t(13;18)(q11;p11),+1,+6,+16,+19, +21,+22	Mitelman et al 1981
46,XX,t(15;17)(q22;q21)/46,XX,t(15;17),t(1;4)(p35;q21), del(X)(q25)	Berger et al 1981e
46,XY,t(15;17)(q26;q22)/47,XY,t(15;17),+8	Van Den Berghe et al 1979c
46,XY,t(15;17)/46,XY,del(1)(q12)/47,XY,+8/47,XY, del(1)(q12),+8/47,XY,t(15;17),+8	
46,XY,t(15;17)/46,XY,t(15;17),t(1;9)(p36;q22)	Hurd et al 1982
46,XY,t(1;17)(p36;q21)	Yamada et al 1983a
47,XY,t(15;17),del(1)(q42),+del(1)(p22q42)	Sakurai et al 1982a

ANLL, FAB type M4

42,XY,-5,-9,-10,-15,-16,-17,-20,-21,-22,+5mar	Testa et al 1979
43,XY,+1,-5,+6,-9,-10,-15,-16,-17,-20,-21,-22,+6mar	
44-46,XY,+1,-5,-17,-18,-20,del(6)(q21),-15,t(1;6)(q11;q24), +3mar	Mitelman et al 1981
44-46,XY,+1,t(1;6)(q11;q27),-5,del(6)(q21),-15,-17,-18,-20, +mar	Alimena 1983
45,X,-X,t(8;17;21)(q22;q23;q22)	Testa et al 1979
45,X,-X,t(8;17;21)(q22;q23;q22)/45,X,-X, t(1;11)(p36q32;q13),t(8;17;21)	
46,XX,1p+	Testa et al 1979
47,XX,+6,-17,+t(1;17)(q21;p13)	Kaneko et al 1982b
46,XX,-1,-3,-5,+10,+11,-13,-14,+21,+del(3?)(p?q?), ?t(1;3;5)(p?;p?q?;q?)	Mitelman 1983
46,XX,t(3;13)(q29;q12),del(5)(q13q31)/46,XX, t(1;18)(q44;q21),t(3;13),del(5)	Alimena 1983
46,XY,1p+,4q+,del(7)(p11)	Morse et al 1979a
46,XY,t(1;9)(p34;q34),inv(8)(p11q12)	Rowley and Potter 1976
47,XX,+1p-"s"	Berger et al 1981a

ANLL, FAB type M5

46,XY,del(11)(q13q23),t(10;11)(p25;q23),t(1;7)(p21;q36)	Yunis et al 1981
46,Y,-X,-1,-7,-8,-14,-15,-17,-20,inv(6)(p23q13), del(11)(q23),+t(X;?)(p11;?),+t(1;?)(p24;?), +t(12;?)(q22;?),+t(17;?)(p11;?),+4mar	Kaneko et al 1982b
47,XX,+del(1)(p11)/48,XX,+del(1)(p11),+6/48,XX, +del(1)(p11),+8	Garson 1980
47,XY,+8,t(1;13)(p31;q11)/48,XY,+8,+18q-	Hagemeijer et al 1981a
47,XY,1p+,1q+,2p+,-3,4q+,+13,+mar	Kaneko et al 1982b
51-52,XX,+3,+6,+9,-10,+18,+19,del(1)(p22),del(1)(p22), dup(11)(q11q21),+t(10;?)(p13;?)	Kaneko et al 1982b
57-58,XX,+1,+3,-4,-5,+6,-8,+11,+12,-17,+19,-20, del(10)(q24),+8-9mar	Testa et al 1979
60-61,XX,Xp+,+1,+2,-4,+6,-7,+19,+20,del(2)(q31), del(10)(q24),+12-13mar	

ANLL, FAB type M5a

46,XX,t(1;12)"c"	Berger et al 1982a
46,XY,t(1;9)(q11;q34),t(10;11)(p14;q14)	Berger et al 1982a

ANLL, FAB type M5b

47,XXX,t(1;11)(q21;q24)"c"	Berger et al 1982a

ANLL, FAB type M4+M5

48,XY,+i(1q),+3,+12,-14	Shiraishi et al 1982
48,XY,t(1;6)(q21;q27),+del(8)(q22),+22	Shiraishi et al 1982
66,XY,+1,+1,+2,+4,+6,+6,+7,+8,+8,+10,+11,+11,+12,+12,+18, +19,+21,+21,+22,+22	Li et al 1983

ANLL, FAB type M6

43-45,XX,der(1),inv(2)(p25q14),del(3)(q25),-4,-5,-6,6q-,-7, 7p+,11p+,-12,-17,-21,+r,+1-4mar"s"	Smadja et al 1983b
44,XX,-1,-2	Nordenson 1983
46,X?,+1,-5,-7,+mar	Wurster-Hill 1983
46,XY,dup(1)(p32p34),dup(3)(q12q27),del(7)(p11)"s"	Berger et al 1980a
48,XX,+t(1;12)(p22;p13),+21,del(22)(q11)	Brodeur et al 1983
48,XX,t(1;8)(p11;q11),+8,+19	Morse et al 1979a
51,XY,+1,+2,+6,-7,+8,-14,+15,+21,+mar"s"	Rowley et al 1981a

Chronic myeloid leukemia, t(9;22)

??,X?,t(9;22),1p+,+22q-,cx	Carbonell et al 1982a
44,X,-X,t(9;22),1q+,-3	Sadamori et al 1980
45,X,-Y,t(9;22)/45,X,-Y,t(9;22),t(1;11)(p?;p?)"s"	Hagemeijer et al 1980b
45,XY,t(9;22),+del(1)(q12),-3,-7,del(11)(q21),-19,+mar	Lyall and Garson 1978
46,X?,t(9;22),1p+,-7,+mar/46,X?,t(9;22),1p+,-7,8q+,+mar	Carbonell et al 1982a
46,XX,t(9;22)(q34;q11)/46,XX,t(9;22),t(1;?)(q44;?)	Mitelman 1983
46,XX,t(9;22)(q34;q11)/47,XX,t(9;22),+mar	
46,XX,t(9;22),i(17q)/47,XX,t(9;22),i(17q),+20/46,XX, t(9;22),-1,+20	Olah and Rak 1981
46,XX,t(9;22),t(1;17)(p?;q?)	Borgström et al 1982
46,XX,t(9;22),t(1;4)(q22;q31),del(7)(q11)	Miyamoto 1980
46,XY,t(9;22)(q34;q11)/50,XY,t(9;22),t(1;15)(p36;q22),+8, +10,+19,+22q-	Misawa 1978
46,XY,t(9;22)(q34;q11),+t(1;17)(q21;p11)/47,XY,t(9;22), +t(1;17),+18/48,XY,t(9;22),+t(1;17),+8,+19	Mamaeva et al 1983
46,XY,t(9;22)(q34;q11) 27,X,-Y,t(9;22),-1,-2,-3,-4,-5,-6,-7,-9,-11,-13,-14,-15, -16,-17,-18,-19,-20,-22	Ishihara et al 1982
46,XY,t(9;22),dup(1)(q12q31)	Kohno and Sandberg 1980
46,XY,t(9;22),t(12;19)(q13;q13)/46,XY,t(9;22), t(1;6)(p36;p21)	Seabright 1983
46,XY,t(9;22),t(1;17)(q21;q25)	Prigogina et al 1978
46,XY,t(9;22),t(1;6)(q25;q25)/46,XY,t(9;22), t(1;15)(q12;p11)	Miyamoto 1980
46,XY,t(9;22),t(7;8),1q+	Lilleyman et al 1978
47,X?,t(9;22),r(19),+22q-/48,X?,t(9;22),+17,+22q-/49,X?, t(9;22),+8,+17,+22q-/50,X?,t(9;22),+8,+17,+r(19), +22q-/49,X?,t(9;22),1q+,+8,+17,+22q-	Carbonell et al 1982a
47,XY,t(9;22),+22q-/56,XY,t(9;22),+1,+2,+5,+6,+7,+8,+10, +12,+17,+22q-	Stoll and Oberling 1978
47,XY,t(9;22),i(17q)/49,XY,t(9;22),i(1q),+8,+21,+22q-	Sonta and Sandberg 1977b
47,XY,t(9;22),inv(1)(p13q44),+22q-	Seabright 1983
49,XY,t(9;22),+6,+8,1p+,14q+,+22q-	Sadamori et al 1980
50-54,XY,t(9;22),+1,+5,+8,+9,-16,+17,+18,+19,+21,+22q-	Alimena et al 1980
52,XY,t(9;22),+1,+8,+10,+11,+21,+22q-	Sadamori et al 1980

Chronic myeloid leukemia, aberrant translocation

??,X?,t(1;4;20;22)	Geraedts et al 1977
45,XY,t(18;22)(q23;q11),+del(1)(q11),6q-,-9,-9	Nowell 1983
46,X,t(X;9;22)(q27;q34;q12)	Dallapiccola and Alimena 1979
46,X,t(X;9;22)/47-51,X,t(X;9;22),+1,+4,+8,+9,-14,i(17q), -19,+22q-	
46,X?,t(1;9;22)(p31;q34;q11)	Ishihara 1983
46,X?,t(1;9;22)(p32;q34;q11)	Hagemeijer et al 1980b
46,XX,t(1;11;22)(q24;p11;q11)	Najfeld et al 1983
46,XX,t(1;9;22)(q22;q34;q11q13)	Fleischman et al 1981
46,XX,t(1;9;22)(q32;q34;q11)	Borgström 1981
46,XX,t(1;9;22)(q32;q34;q11)	Pasquali 1983
46,XX,t(6;22)(q26;q11),t(1;9)(q21;q24)	Berger et al 1976
46,XY,t(1;7;19;22)(q?;q?;p?;q11)	Borgström et al 1982
46,XY,t(1;7;19;22),i(17q)	
46,XY,t(1;9;22)(p11;q34;q11)	Borgström 1981
46,XY,t(1;9;22)(q21;q34;q11)/46,XY,t(9;22)(q34;q11)	Berger et al 1981d
46,XY,t(1;9;22)(q21;q34;q11)	Lessard and Le Prise 1982
46,XY,t(1;9;22)(q23;q34;q12)	Berger and Bernheim 1978
46,XY,t(1;9;22)(q23;q22;q12)	Verma and Dosik 1977
46,XY,t(1;9;22)(q32;q34;q11)	Fleischman et al 1981
46,XY,t(9;11;22)(q34;q13;q11)/52,XY,t(9;11;22),+6,+9q+, +11q-,+18,+19,+del(22q),1q+	Lawler et al 1976

Chronic myeloid leukemia, Ph1 negative

46,XY,t(1;6)(p36;q15),del(3)(q25),del(17)(p11)	Srivastava et al 1981
46,XY,t(2;8)(q21;q24),del(1)(p33),dup(1)(p21p32)	Brodeur et al 1979

MYELOPROLIFERATIVE DISORDERS

Polycythemia vera

45,X,-Y	Berger and Bernheim 1979
54-57,X,-Y,+6,+8,+11,+17,+20,+21,t(1;22)(p12;q13)	
46,XX,-15,t(1;15)(q11;q11)	Wurster-Hill et al 1976
46,XX,dup(1)(q21q44),inv(1),inv(9)	Mamaeva et al 1983
46,XX,t(1;9)(q11;q13)	Kirkland et al 1980
46,XX,t(2;11)(p13;q21)/46,XX,t(1;15)(p1?;q1?), t(2;11)(p13;q21),del(16)(p12)/51,XX,+3,+8,+8,+19,t(1;15), t(2;11),del(16)	Testa et al 1981b
46,XY,t(1;5)(p36;q31)	Westin 1976
46,XY,1p+	Kirkland et al 1980
46,XY,dup(1)(q21q31)	Carbonell et al 1983
46,XY,t(12;17)(q13;p11)/47,X,-Y,+t(Y;1)(q12;q21),+9	Testa et al 1981b
46,XY,t(1;13)(q12;p12)	Hsu et al 1977
47,XX,+del(1)(p11)	Wurster-Hill et al 1976
47,XX,t(1;9)(p2?;p2?),t(1;9),del(7)(q2?),del(20)(q11)"s"	Nowell and Finan 1978
47,XX,t(1;9)(q22;q13)	Westin et al 1976
47,XY,+9	Van Den Berghe et al 1979g

Karyotype	Reference
47,XY,del(1)(q12),t(1;18)(q12;q23),+9,del(20)(q12)/47,XY,+9,del(11)(q21),del(20)(q12)/47,XY,+9,del(20)(q12)	
47,XY,+9/47,XY,del(5)(q14q32),del(8)(q?),+9,del(11)(q21),del(13)(q21)/47,XY,del(5)(q14q32),del(6)(q?),+9,del(11)(q21),del(13)(q21),del(1)(q12),del(20)(q12)	
47,XY,+del(1)(p21),del(20)(q11)/47,XY,+9,del(20)(q11)	Westin et al 1976
47,XY,+del(1)(p21),del(5)(q31),del(20)(q11)/47,XY,+9,del(20)(q11)	Swolin et al 1981
47,XY,+del(1)(p?),-20,+mar	Nowell and Finan 1978

Myelosclerosis / Myelofibrosis

Karyotype	Reference
??,XY,t(1;12)(q11;q24)	Miyamoto et al 1981b
45,XY,t(1;4;7)(p32;q28;q11q32)	Nowell and Finan 1978
46,XX,-7,+t(1;7)(p1?;p11)"s"	Geraedts et al 1980b
46,XX,1q+,1q-	Nowell and Finan 1978
46,XX,t(1;6)(q25;p22)	Hsu et al 1977
46,XY,-7,t(1;7)(p1?;p11),17p+	Geraedts et al 1980b
46,XY,t(1;2)(q?;q?),-8,inv(13)(q12q31),i(17q)	Whang-Peng et al 1978
46,XY,t(1;6)(q23;p21)	Gahrton et al 1978a
47,XX,del(22)(q?),t(1;2),t(6;7),+r,+mar	Page et al 1979
47,XY,-7,+21,+t(1;7)(p1?;p11)"s"	Geraedts et al 1980b
50,XY,+del(1)(p?),+8,+9,+21	Nowell and Finan 1978
51,XX,+1,del(2)(q33),+6,+9,+11,+17,-19,+20q+	Najfeld et al 1978b

Idiopathic thrombocythemia

Karyotype	Reference
46,XY,dup(1)(q21q32)	Knuutila et al 1983

Chronic myeloproliferative disease, NOS

Karyotype	Reference
46,X?,t(1;2)(q?;q?)	Wurster-Hill 1983
46,XX,6p+,t(1;6)(q25;p25)	Najfeld 1983
46,XX,t(1;22)(q?;q13)	Lessard and Le Prise 1983
47,XX,+1p-	Lessard and Le Prise 1983

Myeloproliferative disorder, special type

Karyotype	Reference
45,XX,t(1;?)(p36;?),-5,-7,-12,t(13;?)(q34;?),-17,-22,+4mar/45,XX,t(1;?),-2,-3,-5,-7,-12,-17,-22,+8mar"s"	Rowley et al 1977a
46,XY,t(1;13)(p36;q13),del(3)(p21),-5,inv(7)(p11q22),t(16;16)(q22;q24),+mar	Clare et al 1982

DYSMYELOPOIETIC SYNDROMES

Preleukemia, NOS

Karyotype	Reference
45,XX,-1,-4,5q-,-7,-12,-18,14q+,+4mar"s"	Pedersen-Bjergaard et al 1981
45,XX,-7/48,XX,+t(1;7),-7,+11,+13"s"	Pedersen-Bjergaard et al 1981
46,X,-Y,+t(1;?)(p?;?)	Geraedts et al 1980a
46,XX,1p+,9q-,18q+,r(20)	Geraedts et al 1980a
46,XX,dup(1)(q23q44)	Anderson and Bagby 1982

46,XX,t(1;16)(q11;p11),t(1;20)(p?;p?)	Anderson and Bagby 1982
46,XX,t(1;3;11)(p1?;q2?;q?)	Ruutu et al 1977b
46,XY,-1,+r	Knuutila et al 1981
46,XY,-7,+t(1;7)(q?;p?)"s"	Pedersen-Bjergaard et al 1982
46,XY,t(1;16)	Knuutila et al 1981
46,XY,t(1;17)(q24;p13),-2,-5,-7,+r,+2mar"s"	Anderson and Bagby 1982
47,X,Xp-,1p+,5q-,6q+,11q-,11p+,+11,12p-	Swansbury and Lawler 1980
50,XY,-5,-21,+1,+9,+11,17p+,+1-3mar	Borgström et al 1982
52,XY,+1,+3,+21,+3mar	Panani et al 1980
53,XX,+1,+2,t(4;19),+6,i(17q),+21,+21,+22	Seabright 1983

<u>Chronic myelomonocytic leukemia</u>

46,XY,t(1;7)(p36;q22)	Roush 1981

<u>Refractory anemia, NOS</u>

47,X,del(Y)(q11),+1	Warburton and Bluming 1973

<u>Sideroblastic anemia, NOS</u>

44,X,-Y,-21/45,XY,-21/46,XY,1q+	Yamada and Furusawa 1976

<u>Refractory anemia without excess of blasts</u>

46,XY,1p-,3q-,5q-	Teerenhovi et al 1981

<u>Refractory anemia with excess of blasts</u>

44-47,XY,-3,-5,-7,-12,-16,-20,del(1)(q32),3p+,del(11)(q22), 3-6mar,+r	Streuli et al 1980
46,XX,del(5)(q15)/49-51,XX,+1,del(5)(q15),+11,-14,+22,+mar	Swolin et al 1981
46,XY,t(1;1)(p?;q?)"s"	Berger et al 1981a
49,XX,+X,del(1)(p21?),+9,+mar	Mitelman 1983
56-64,XY,+1,-2,+6,+9,+13,+15,+19,+21,10-13mar	Streuli et al 1980

<u>Granulocytopenia</u>

46,XX,?1q+	Mitelman 1983

<u>Pancytopenia</u>

46,XY,t(1;11)(q11orq12;q25)	Najfeld et al 1978a

<u>Dysmyelopoietic syndrome, special type</u>

45,XX,t(13;14)(p13;q11),del(5)(q14)/50,XX,t(13;14),+1,+6, -7,+8,+10,del(5),+mar"c""s"	Rowley et al 1981a

SPECIAL LEUKEMIAS

LYMPHOCYTIC LEUKEMIAS

Acute lymphocytic leukemia (ALL)

??,X?,t(1;22)(p36;q11)	Pittman et al 1979
28,XX,-1,-2,-3,-4,-5,-7,-8,-9,-11,-12,-13,-14,-15,-16,-17, -19,-20,-22	Kaneko and Sakurai 1980
45,X,-X,dup(1)(q31q41),del(5)(q22),del(17)(p11)	Humbert et al 1978
45,XY,1q+,-6	Oshimura et al 1977a
46,X,-Y,+1,-7,+20,t(1;22)(q24;q12)	Vercherat et al 1980
46,X,Xp+,del(1)(q21),del(6)(q21),i(17q)	Oshimura et al 1977a
46,XX,-11,+t(1;11)(q21;q23)	Mamaeva et al 1983
46,XX,1q-,7q-,t(1;19)	Goldstone et al 1979
55,XX,+1q-,+7q-,+t(1;19),+19,+22,+4mar	
46,XY,-8,del(3)(q12q25),t(8;14)(q24;q32),+t(1;8)(p11;q11)	Kaneko et al 1980
46,XY,1q+	Yamada and Furusawa 1976
46,XY,1q+	Hustinx and Rutten 1983
46,XY,del(1)(p?)	Zuelzer et al 1976
46,XY,dup(1)(q21q32),4q+	Morse et al 1979a
46,XY,dup(1)(q31q41)	Humbert et al 1978
46,XY,t(4;11)(q21;q23)	Oshimura et al 1977a
46-49,XY,t(4;11)(q21;q23),+6,+8,+13,+17,t(1;13)(p22;q12), t(7;9)(q11;q34)	
46,XY,t(8;22)(q24;q12),t(3;11)(p12p21;q23),inv(2)(p11q13), del(2)(p21)/46,XY,t(8;22)(q24;q12),dup(1)(q23q44)"s"	Fonatsch et al 1982
47,X,del(Y)(q11),del(3)(q11),t(Y;12)/50,X,del(Y)(q11), del(3)(q11),del(1)(p31),t(Y;12),+4mar	Zuelzer et al 1976
47,XX,t(1;?)(q25;?),+13,del(1)(q32)	Oshimura et al 1977a
47,XY,del(1)(q22),+mar	Chaganti et al 1979
47,XY,del(1)(q22),+mar	Chaganti et al 1979
47,XY,t(4;11)(q21;q23),1p+,del(2)(p16),del(6)(q21),i(7q), del(8)(q21),+mar	Arthur et al 1982
48,XY,t(1;22)(q21;p11),+5,+8,19p+	Humbert et al 1978
54,XX,7p+,+10,+10,+14,+14,+18,+18,+21,+21/27,X,-X,-1,-2,-3, -4,-5,-6,-7,7p+,-8,-9,-11,-12,-13,-15,-16,-17,-19,-20, -22	Oshimura et al 1977a
61,XX,+X,+1,+2,+3,+4,+5,+6,+8,del(6)(q21),+13,+14,+15,+16, +18,+21,+22	Oshimura et al 1977a
62-64,XY,1q-,cx	Prigogina et al 1979

ALL, FAB type L1

26,XX,-1,-2,-3,-4,-5,-6,-7,-8,-9,-10,-11,-12,-13,-14,-15, -16,-17,-19,-20,-22	Hoeltge et al 1982
28,XX,-1,-2,-3,-4,-5,-6,-7,-8,-9,-11,-12,-13,-15,-16,-17, -19,-20,-22/56,XX,+X,+X,+10,+10,+14,+14,+18,+18,+21,+21	Brodeur et al 1981a
46,XX,t(1;19)(q21;q13),del(3)(q23),del(11)(q21)	Kaneko et al 1982e
46,XY,-11,+t(1;11)(q21;q14)	Kaneko et al 1982e
46,XY,del(3)(q12q25),t(8;14)(q24;q32),-8,+t(1;8)(p11;q11)	Kaneko et al 1982e
47,XY,del(17)(q11q21),t(1;13)(p34;q14),+mar	Kaneko et al 1981b

47,XY,t(22;?),-1,-1,-2,-3,-9,+6mar	Sandberg et al 1980
53,XX,+1,+5,+14,+15,+19,+20,+22	Kaneko et al 1981b

ALL, FAB type L2

26,XX,-1,-2,-3,-4,-5,-6,-7,-8,-9,-10,-11,-12,-13,-15,-16,-17,-18,-19,-20,-22	Brodeur et al 1981a
45,X,-X,-1,-14,-17,+21,t(11;12)(p15;p11),+i(17q),+t(1;?)(p34;?)	Kaneko et al 1981b
45,XX,1q-,7q-,-?	Borgström et al 1981
46,XY,+1,-8,t(14;?)(q32;?)	Mitelman 1983
46,XY,1q+/46,XY,1q+,14q+	Borgström et al 1981
49,XY,+7,+12,-13,t(6;18)(p25;q21),t(11;14)(q23;q32),+t(9;?)(p24;?),+t(1;13)(q12;p13)	Kaneko et al 1982e
55-65,XX,+1,+1,+6	Lessard and Le Prise 1983
56,XX,+1,+6,+10,+13,+15,+16,+17,+18,+20,+21	Mitelman 1983

ALL, FAB type L3

??,XY,t(8;14)(q?;q?),t(Y;1)	Prieto et al 1981
46,XX,dup(1)(q21q32),t(8;14)(q24;q32)	Rossi et al 1982
46,XY,dup(1)(q23q32),t(1;6)(q21;q13),t(14;?)(q32;?),del(6)(q13),del(8)(p21)	Slater et al 1979
46,XY,t(8;14)	Borgström et al 1981
46,XY,t(8;14),1q+	
46,XY,t(8;14)(q24;q32)/46,XY,t(8;14)(q24;q32),dup(1)(q21q24)	Berger and Bernheim 1982
46,XY,t(8;14)(q24;q32)/46,XY,t(8;14)(q24;q32),dup(1)(q21q24)	Berger and Bernheim 1982
46,XY,t(8;14)(q24;q32)/46,XY,t(1;8;14)(q23;q24;q32)	Berger and Bernheim 1982
46,XY,t(8;22)(q24;q12)/46,XY,1q+,+del(1)(p22),-5,t(8;22)(q24;q12)/47,XY,+del(1)(p22),t(8;22)(q24;q12)	Abe et al 1982a
46-47,XX,t(14;?)(q32),t(13;?)(q31;?),1q+,+7	Slater et al 1979
47,XY,dup(1)(q21q32),del(6)(q23),t(8;14)(q23;q32),+9	Roos et al 1982
47,XY,ins(1),t(3;22)(q?;q?),+7	Penchansky et al 1981

Chronic lymphocytic leukemia

??,XY,1q-,t(1;10)(q11;p15),t(1;12)(q11;p12)	Miyamoto et al 1981b
41,XX,t(14;14)(q11;q32),t(1;13)(q44;q11),t(15;18)(q11;q23),10q+,-16,-20	Kaiser McCaw et al 1975
44-48,XY,t(Y;9)(q12;q13),del(1)(p22p32),t(4;6)(p16;q15),del(6)(q15),t(12;14)(q15;q32),+mar	Robert et al 1982
45,XX,-8,-14,-15,-22,+2,+9,+1p+,+mar	Finan et al 1978
46,XX,t(1;8)(q?;q?)	Schröder et al 1981
46,XY,del(11)(q22),+12,-13,-21,-22,+3mar/45,XY,del(11),+12,-13,-15,-21,-22,+2mar/46,XY,del(11),+12,+12,-13,-21,-22,+2mar/46,XY,t(1;8)(p22;q24)	Robert et al 1982
46,XY,t(1;10)(q22;q26)	Fleischman and Prigogina 1977
46,XY,t(1;7;9)(p36;q21;q12),t(12;14)(q13;q24)	Miyoshi et al 1979a
48,XY,+12,+19/48,XY,t(1;11)(q31;q22),+12,+19	Vahdati et al 1983a

Lymphocytic leukemia, special type

45,XX,-1,2q+,14q+	Ueshima et al 1981
46,X,-Y,-1,1p+,del(9)(q32),12p+,14q+,+2mar	Ueshima et al 1981
46,XY,t(1;7)(p36;q22),del(9)(q12),t(12;14)(q13;q32)	Miyoshi et al 1981a
47,X,-Y,-4,+7,del(1)(q32),i(18q),21q+,i(22q),+2mar	Ueshima et al 1981
47,X,-Y,dup(Y),1q-,4q+,6q-,7q+,+18q+	Miyamoto et al 1982b

MONOCLONAL GAMMOPATHIES

Multiple myeloma

45,X,-Y,-1,+t(1;3)(q3?;q2?),-3,+11,14q+/45,X,-Y,-1,-1, +t(1;1)(q2?;q3?),14q+	Philip et al 1980
45,XY,del(1),+del(1),-10,-11,-11,+t(11;?),-13,-14,-20, +3mar	Philip et al 1980
46,XX,del(1)(p35)	Nordenson 1983
46,XX,dup(1)(q21q31)	Spriggs et al 1976
46,XY,+t(1;?)(p1?;?),+t(1;12)(q21;q24),del(6)(p11),-8,+9, -12,-13,-16,-19,22q+,+mar	Philip et al 1980
46,XY,ins(1)(p36q21),t(11;14)(q13;q32)	Liang et al 1979
47,X,-X,inv(1)(p22q12),t(7;?)(q11;?),+9,t(11;?)(p15;?),-13, -14,t(14;?)(q11;?),+15,17q+	Philip et al 1980
47,X,-Y,+t(1;15)(q12;p11),+3,+11,-15,+19,-22/47,X,-Y,+3, +11,+19,-22	Philip et al 1980
47,XY,dup(1)(q11q32),t(1;10)(p22;p15),inv(6)(p21q13), t(11;14)(q23;q32),21p+,+mar	Liang et al 1979
48,X,-X,+1,+7,-8,+9,+21,del(1)(p36),4p+,del(5)(q22),9p+, 14q+	Liang et al 1979
52,X,-X,+t(1;?)(q12;?),+3,+7,+8,+9,+11,-13,+18,-21,+2mar	Philip et al 1980
52,XY,-1,+t(1;15)(q12;p1?),+t(1;16)(p36;p13), t(3;5)(q2?;q13),+3,+5,+7,+8,+9,+11,-15,+18,-20	Philip et al 1980
52,XY,-1,+t(1;?)(p31;?),+3,+2mar	Philip et al 1980
53,XY,+3,+5,-7,-8,+9,+11,-14,+20,del(1)(p34),del(6)(q25), +5mar	Liang et al 1979
55,XY,+1,+5,+9,+11,-12,+13,+15,del(1)(q32),del(6)(q25), t(9;19)(q13;q13),14q+,16p+,+4mar	Liang et al 1979
45,XY,-2,-2,-3,+5,-8,-10,+14,-15,-16,del(6)(q25),+4mar	

Macroglobulinemia

46,Y,-X,del(1)(q21),del(5)(q23),+del(5)(q23), +t(7;19)(p22;q13),del(8)(p12),+t(1;16)(q12;p13), +t(1;18)(q12;p11),-7,-16,-18,-19,+mar	Ueshima et al 1983

Plasma cell leukemia / Plasmocytoma

43,X,-X,-1,+t(1;8)(q11;q11),+t(1;9)(q21;p24), +t(1;10)(p31;q26),+t(1;?)(q12;?),-8,-9,-10,-13,-22	Ueshima et al 1983
44,X,-Y,+t(1;9)(q12q44;q34),t(6;8)(q13;p11),-9,-13	Ueshima et al 1983
44,X,-Y,del(1)(p11),+del(1)(q11),t(11;14)(q11;q32),14q+, -21,-22	Ueshima et al 1983
46,XX,t(1;6)(p22p34;p25),t(11;14)(q13;q32)	Gahrton et al 1980b

46,XY,+t(1;16)(q21;p13),del(6)(q21),-16,-18,+mar	Ueshima et al 1983
47,XY,+t(1;16),del(6)(q21),+t(8;15)(p23;q12),-8,14q+,-15, -16,-18,+3mar	
78,XY,cx,del(1)(q21),t(1;17)(q12;p13),3q+,3p-,9p+,11p+, 14q+,16p+	Ueshima et al 1983
86,XX,cx,t(1;?)(q11;?),2q+,dup(7)(q22q36),t(8;?)(p11;?), t(9;13)(p23;q21)	Ueshima et al 1983

SOLID TUMORS

UNCLASSIFIED NEOPLASMS

Malignant neoplasm, NOS

Karyotype	Reference
47,XX,-2,-5,+6,-10,11q+,-13,+18,-21,t(1;?)(q21;?), t(1;13)(q21;p11),t(1;6)(q11;q23)	Kovacs 1978a

EPITHELIAL NEOPLASMS

Carcinoma, NOS

Karyotype	Reference
??,XX,+1	Atkin and Baker 1977a
??,XX,+1	Atkin and Baker 1977a
??,XX,+1	Atkin and Baker 1977a
??,XX,+1	Atkin and Baker 1977a
??,XX,+1	Atkin and Baker 1977a
??,XX,del(1)(q31)	Riet-Fox et al 1979
32-35,XX,cx,del(1)(p12),dup(1)(q21q32),2q+,5q+,7p+,11p+, hsr(10),t(11;12)(q13;q24),12p+,13q+	Kusyk et al 1982
39-41,XX,del(1)(q32),1p+,t(1;14)(q21;p11),cx	Kovacs 1978a
40,XX,1q+,cx	Atkin and Baker 1977a
41,XX,1p+,1q-,cx	Atkin and Baker 1977a
41,XX,t(1;1;?)(p36q21;q12;?),t(1;?)(q12;?),del(6)(q21), t(5;12)(q?;q?),del(5)(q12q32)	Atkin and Baker 1982a
43,X,-X,t(1;11)(q21;q23),3q+,4q+,5q+,-11,-13,-16,17p+	Atkin and Baker 1982a
43-44,XX,del(1)(p?),cx	Atkin and Baker 1977a
43-49,XX,-1,-8,+10,-11,t(1;11)(q11;q12)	Kusyk et al 1982
44,XX,ins(1;1)(q32;q12q31),i(2q),+3,t(9;11)(q13;p15),-12, -15,17p+,-18	Atkin and Baker 1982a
44,XY,+5,-8,-11,-13,-13,-14,-14,-15,+19,-22,-22,t(1;11), 5q-,9q+,+5mar	Reichmann et al 1981
44-45,XX,-1,dup(1)(q21q32)	Riet-Fox et al 1979
44-45,XX,1p+,cx	Atkin and Baker 1977a
44-45,XX,t(1;4)(q11;q35),+3,5q-,-14,i(17q),-21	Atkin and Baker 1982a
44-50,XX,+3,dup(1)(q21q32),t(10;11)(q26;q25),t(2;17), t(6;13),t(9;15),del(10)(q24)	Riet-Fox et al 1979
45,XX,1q+,cx	Atkin and Baker 1977a
45,XY,-1,-6,-14,+21,+2mar/45,XY,t(1;14)(q11?;q24?),-6,-16, +21	Sonta and Sandberg 1978a
46,XX,i(1q)	Kovacs 1978a
46,XX,i(1q),t(1;4)(p?;q?)	Atkin and Baker 1982a
46,XX,t(1;?)(p32;?),-2,3q-,5q-,6q+,-18,-21,+mar	Atkin and Baker 1982a
46,XY,-4,-5,-11,-13,-17,+20,+21,1p+,9p-,+3mar	Reichmann et al 1981
46-47,XX,ab(1)(q?),cx	Atkin and Baker 1977a
47,XX,+1,+3,t(11;14)(q?;q?),-12,-21,+r	Atkin and Baker 1982a
47,XX,+i(1q),cx	Atkin and Baker 1977a
48,XX,+3,t(1;20)(p13;p13),+i(1q)	Kovacs 1978a
48,XX,+del(1)(p32),-10,-21,+mar	Atkin and Baker 1982a
48,XX,i(1q),cx	Atkin and Baker 1977a
48,XY,+1,-5,+8,+17	Sonta and Sandberg 1978a
49,X,-Y,-3,-5,-14,+15,-16,-17,+19,+21,+22,1q+,5q-,6q-,8p-, 9p-,11p-,18p+,20q-,22q-,t(Y;14),+3mar	Reichmann et al 1981

50,XX,i(1p),cx	Atkin and Baker 1977a
52,XY,+X,+Y,-2,+7,+8,+9,+10,+13,-17,+i(1q)	Reichmann et al 1981
52-54,X?,cx,t(1;1)(q23;p32),del(1)(q21)	Kovacs 1978a
54,XX,+1,+2,+8,+11,+13,+14,+21,+21	Sonta et al 1977
54,XY,+1,+5,+9,+17,+19,+20,+21,+21	Sonta and Sandberg 1978a
55,XY,t(1;17)(p36;q21),+2,+2,+4,+8,+8,-10,+15,+18,+20,+21, +2mar	Sonta and Sandberg 1978a
58,XX,+X,+1,+2,+3,5p-,+6,+7,+9,11q+,-13,-14,-15,+16,+19, +20,+6mar	Kovacs 1978a
63,XX,-1,+3,+5,+6,+7,+8,+10,+11,+12,-13,-15,+20,+21,+22, +9mar	Sandberg 1977
63,XX,cx,t(1;4)(q21;q35),7p+,i(17q)	Kovacs 1978a
63-66,XX,cx,1p+,16q+,del(3)(p13)	Riet-Fox et al 1979
66,XX,t(Y;13),1q+,i(4q),i(9p),i(9q),cx	Martin et al 1979
68-70,XX,cx,i(1q)	Kovacs 1978a
69-70,XX,cx,t(13;13),t(1;10),1p+,del(1)(p22)	Riet-Fox et al 1979
71-75,XX,cx,11p+,17q+,del(1)(p31),del(3)(p13),del(4)(p1?)	Riet-Fox et al 1979
73,XY,+1,+7,+11,+13,-16,+17,+18,+19	Sonta and Sandberg 1978a
73,XY,cx,i(1q)	Atkin and Baker 1977b
76-79,XY,cx,1q-,8p-,dmin	Reichmann et al 1981
79,XX,cx,i(1q),t(1;3)(p11;q29),12q+,15p+,i(5p)	Kovacs 1978a
79,XX,cx,i(1q),t(1;3)(p13;q29),t(1;14)(p13;p13), t(12;?)(q24;?),t(15;21)(p11;q11),i(5p),r	Kovacs 1981

<u>Carcinoma NOS, metastatic</u>

??,X,-X,t(1;?),hsr(1),6q-,t(9;?),dic(13),dmin,cx	Bartnitzke and Bullerdiek 1983
??,XX,1p-,t(3;?)(q?;?),cx	Bartnitzke and Bullerdiek 1983
??,XX,t(1;9),t(2;?),cx	Bartnitzke and Bullerdiek 1983
??,XY,t(1;?),cx	Bartnitzke and Bullerdiek 1983
35,XX,-1,-3,-4,-5,-7,-8,-10,-11,-12,-13,-14,-15,-16,-17, -18,-19,-20,-21,-22,+8mar	Pathak et al 1979
40,XX,cx,t(1;11)(q?;q?)	Jones Cruciger et al 1976
41-42,XX,t(1;18)(p13orp21;q11),3p-,t(3;?)(p25;?), t(8;?)(p2?;?),t(9;?)(q34;?),cx	Bartnitzke and Bullerdiek 1983
43,XX,cx,t(1;11)(q?;q?)	Jones Cruciger et al 1976
43,XX,cx,t(1;3)(q?;q?)	Jones Cruciger et al 1976
45,X,Xq-,t(1;?)(p11;?),t(1;?)(q23;?),t(3;?),t(5;?)(q2?;?), del(9)(p2?),cx	Bartnitzke and Bullerdiek 1983
45,X?,+16,-18,1p-,3p-,16p-	Wake et al 1981
45-50,XX,-2,+7,-22,t(1;12),t(13;14),inv(7)(q?)	Ayraud 1975
47,XX,+t(1;1)(p11;q21)	Atkin and Baker 1978
49,XX,cx,t(1;7)(q?;q?)	Jones Cruciger et al 1976
51,XX,cx,t(1;5)(q?;q?)	Jones Cruciger et al 1976
59,XX,cx,t(1;12)(q?;q?)	Jones Cruciger et al 1976
60,XX,t(1;?),t(7;?)(q3?;?),i(10q),der(11),t(14;21),cx	Bartnitzke and Bullerdiek 1983
60-62,XX,del(1)(q?),t(1;7),2q+,del(3)(q?),der(5),dmin,cx	Bartnitzke and Bullerdiek 1983
60-70,XY,i(1)(q11),i(17q),cx	Kakati et al 1976b

64-100,XX,t(6;?)(p11q27;?),t(4;10)(q?;q?),t(3;?)(q25;?), del(1)(q2?),t(6;11)(p21;q13)	Kakati et al 1975
65,XX,i(1q),t(1;9),der(5),t(6;10)(p?;q?),t(14;?),der(21), cx	Bartnitzke and Bullerdiek 1983
66,XX,cx,t(1;12)(q?;q?)	Jones Cruciger et al 1976
79-80,XX,t(1;?)(q12;?),t(1;2),t(3;?),t(7;?)(q36;?),t(8;?), t(9;?)(q34;?),cx	Bartnitzke and Bullerdiek 1983
81-83,XX,t(1;3)(p22;p25),del(1)(p13),t(5;13)(q13;q14), del(4)(p12),i(2)(p11),i(5)(p11),t(4;5)(q12;q11q13)	Kakati et al 1976a
85-90,XY,i(9q),del(1)(q2?),t(7;?)(q35;?),t(3;?)(p11;?), t(21;?)(q11;?)	Kakati et al 1975
100,XY,t(12;16),inv(3)(q?),del(1)(q?),del(1)(p?),cx	Ayraud 1975

Large cell carcinoma

55-70,XY,t(4;5;14),del(1)(q11q21),t(7;11)(p22;q13), t(7;21)(q11;q22),t(3;8)(p?;q?),t(6;11)(p11;p11),i(16q), t(3;?)(q11;?),t(21;?)(q11;?),t(13;?)	Kakati et al 1975

Small cell carcinoma

40,XY,cx,t(1;16)(q21;q24),t(1;19)(q23;q13),del(3)(p21p24), del(3)(p23q26),del(10)(q22),del(11)(q23), t(5;13)(q13;q32),del(17)(p12),dmin	Whang-Peng et al 1982b
44,XY,cx,del(1)(q31),t(2;3)(p?;q?),del(3)(p14p23), t(9;13)(p11;q11),del(11)(p15),12q+,del(X)(q22)	Whang-Peng et al 1982b
44-45,XX,1p+,del(3)(p?q?)	Wurster-Hill and Maurer 1978
44-46,X?,1p+	Wurster-Hill and Maurer 1978
45-46,X?,1p+,cx	Wurster-Hill and Maurer 1978
46,X?,del(1)(p?q?),cx	Wurster-Hill and Maurer 1978
46,XX,del(1)(p?),del(4)(p?),del(1)(q32)	Wurster-Hill and Maurer 1978
47,XY,+del(1)(p?),+del(3)(q?),+del(3)(p?),-4,+8,-15,+18, -22	Wurster-Hill and Maurer 1978
49,XY,cx,t(1;1)(q32;p36),del(3)(p14)	Whang-Peng et al 1982b
50,X?,del(1)(p?),cx	Wurster-Hill and Maurer 1978
58,XY,cx,del(1)(p13),del(1)(q21),del(3)(p21), inv(3)(p14p23),hsr(15)(p?)	Whang-Peng et al 1982b
67,XY,cx,inv(1)(q32p36),dup(1)(q32q44),del(3)(p14q13), del(3)(p14p23),t(7;13)(p11;q11),del(9)(q11),del(11)(p11), t(12;19)(q24;p13)	Whang-Peng et al 1982b
69-70,X?,+del(1)(q?),cx	Wurster-Hill and Maurer 1978
71-87,X?,1p+,cx	Wurster-Hill and Maurer 1978
77-78,XX,del(1)(q?),del(1)(p?),del(3)(p?),del(3)(q?), del(4)(p?)	Wurster-Hill and Maurer 1978
78-98,X?,del(1)(q21),cx	Wurster-Hill and Maurer 1978

80,XX,cx,del(1)(q32),del(3)(p14p21),t(3;4)(p23q11;q11), t(5;20)(q11;q13),dmin	Whang-Peng et al 1982b
80,XY,+del(1)(q11),cx	Wurster-Hill and Maurer 1978
83-86,X?,del(1)(q?),cx	Wurster-Hill and Maurer 1978
83-92,X?,del(1)(q32),cx	Wurster-Hill and Maurer 1978
106,XY,cx,del(1)(p32),del(3)(p13),del(3)(p14p23)	Whang-Peng et al 1982b
110,XY,cx,del(1)(q41),del(3)(p13p23),del(3)(p13), t(3;16)(q24;p13),del(6)(q24)	Whang-Peng et al 1982b

Papillary carcinoma

34-36,XX,cx,1p-,1q-,6q-,14q+	Wake et al 1980
41,XX,cx,1p-,1q-,3p+,3q+,6q-,11q-	Wake et al 1980
41,XX,cx,t(6;14)(q21;q24),del(1)(p?q?),13q+	Wake et al 1980
42-54,X,-X,-14,-16,-16,+1,+3,+12,+20,1p-,6q-	Wake et al 1981
46,XX,t(6;14)(q21;q24),1p-,1q-,3p+,3q+	Wake et al 1980
50,X,-X,+1,+3,+12,+20,-14,-16,1p-,6q-	Wake et al 1980
62,XX,cx,1p-,1q-,t(6;14)(q21;q24),10q-	Wake et al 1980
68,XX,cx,t(6;14)(q21;q24),del(1)(p?q?),1p+,4q-,8q-,10q+	Wake et al 1980
69,XX,cx,1p+,1q-,1p-,1q-,2q+,3q+,4q+,6p+,12q+,14q+	Wake et al 1980

Squamous cell carcinoma

40,XX,1q+	Atkin and Baker 1979
41,XX,1p+	Atkin and Baker 1979
43,XX,1p-	Atkin and Baker 1979
44,XX,+1	Atkin and Baker 1979
45,XX,1q+	Atkin and Baker 1979
46,XX,der(1)	Atkin and Baker 1979
47,XX,der(1)	Atkin and Baker 1979
47,XX,i(1q)	Atkin and Baker 1979
48,XX,i(1q)	Atkin and Baker 1979
51,XX,cx,+1	Atkin and Baker 1979
51,XX,cx,i(1p)	Atkin and Baker 1979
62-68,XX,cx,i(1q)	Atkin and Baker 1979
64-70,XX,cx,1p+,del(1)(p?q?)	Atkin and Baker 1979
70,XX,cx,i(1q-),1p-	Atkin and Baker 1979
72-82,XX,cx,1p+,1q-	Atkin and Baker 1979
74-76,XX,cx,1p-	Atkin and Baker 1979
77-81,XX,cx,der(1)	Atkin and Baker 1979
77-84,XX,cx,1p-	Atkin and Baker 1979
81-87,XX,cx,i(1q)	Atkin and Baker 1979
87-91,XX,cx,der(1)	Atkin and Baker 1979
99-107,XX,cx,1p-	Atkin and Baker 1979

Squamous cell carcinoma, metastatic

80-88,XX,cx,i(1q),del(1)(p?q?)	Atkin and Baker 1979

Transitional cell carcinoma

43,XY,cx,1p-	Atkin and Baker 1977b
44,XY,cx,1q+	Atkin and Baker 1977b
61-65,XY,cx,i(1q)	Atkin and Baker 1977b
62-68,XX,cx,der(1)	Atkin and Baker 1977b
65,XY,cx,1q-	Atkin and Baker 1977b
80-85,XX,cx,der(1)	Atkin and Baker 1977b

Adenoma

46,XX,t(1;3)(p21;p21),t(1;13)(p36;q14),t(3;8)(p21;q12), t(1;5;20)(p21;p14q12;p13),t(5;8)(p14;q12), t(X;9)(q27;q12)	Mark et al 1982a
46,XY,t(6;21)(q13;q22)/46,XY,t(1;14)(q42;p12)/45,X,-Y 45,X,-Y 46,Y,t(X;15)(q24;q15)/45,X,-Y	Mark et al 1981b

Adenocarcinoma

??,XX,cx,del(1)(q42),del(7)(q2?),del(18)(p11),i(8p),i(8q)	Riet-Fox et al 1979
??,XX,t(1;15)(q32;q25)	Trent and Davis 1979
38,X,-X,-2,-3,-10,-14,-22,del(1)(q25),del(6)(q15), del(14)(p11),dmin	Trent and Salmon 1981
38,X,-X,cx,-17,-22,del(1)(q25),del(2)(q23),del(6)(q15)	Trent and Salmon 1980
40,XX,1q-,cx	Atkin and Pickthall 1977
40,XX,cx,1q+	Atkin and Pickthall 1977
40-42,X,-X,t(1;?)(p13;?),del(1)(p12),2q+,del(3)(q21), del(6)(q13),del(7)(q11),t(11;11)(p15;q13), t(12;12)(q24;?),13p+,15p+,i(22q)	Kusyk et al 1981
42,XX,cx,1p+,1q-,1p-	Atkin and Pickthall 1977
44,XX,-1,+5,-15,-18,2q+/44,XX,-1,+5,-6,-15,-18,2q+,10p-, +mar	Martin et al 1979
44,XX,-1,t(1;8),del(1)(p22)	Riet-Fox et al 1979
44-50,X,-X,-1,t(1;?)(q21;?),del(2)(q33),4p+,-5,-7,+8, del(10)(q12),t(10;13)(q12;q34),del(17)(q25),-20	Kusyk et al 1982
47,XY,-1,+2mar	Sonta et al 1977
48-48,XX,+3,+7,+8,-1,-5,-14,-17,5q+,6q+,del(1)(q42), del(5)(q3?),del(10)(q24),del(14)(q2?)	Riet-Fox et al 1979
54,XX,+5,+6,+7,+9,+11,+21,+21,+del(1)(q32)	Sonta et al 1977
57,XX,cx,1p+,1q-,1p-	Atkin and Pickthall 1977
67,XX,cx,der(1)	Atkin and Pickthall 1977
68,XX,cx,der(1)	Atkin and Pickthall 1977
68-74,XX,cx,t(13;12),17q+,t(16;16),1q-,i(12p)	Riet-Fox et al 1979
71,XX,cx,1q-,4q+,6q-,9q+,10q-,12q+,17q+,14q+	Wake et al 1981
74-75,XY,cx,t(18;19),del(1)(p22),del(6)(q21)	Riet-Fox et al 1979
76,XY,cx,1p+	Riet-Fox et al 1979
77,XX,cx,1p+	Atkin and Pickthall 1977
83-86,XX,cx,1p-	Atkin and Pickthall 1977

Adenocarcinoma, metastatic

Karyotype	Reference
??,X,del(X)(q21),del(1)(q12),i(7q),del(9)(q31), del(16)(q23)	Ayraud et al 1977
??,X,del(X)(q21),t(1;16)(q?;p?),del(12)(p12),del(16)(q21)	Ayraud et al 1977
??,XX,t(1;16),2q-,del(3)(p14),del(4)(q22),del(6)(q22), i(7q),del(9)(q31),t(13;13)	Ayraud et al 1977
??,XX,t(1;16),del(1)(q12),del(4)(q22),del(16)(p11)	Ayraud et al 1977
39-41,XX,-3,-7,-8,der(1),der(2),der(4),der(12),cx	Tiepolo and Zuffardi 1973
40,XX,t(4;5;8),t(3;?)(q21;?),t(1;11)(q21;p11), t(9;11)(q11;p11),del(1)(p36)	Kakati et al 1975
43-46,XX,t(1;16),t(1;22),+r	Bertrand et al 1979
48,XX,+1,+3,+6,-16,1p-	Wake et al 1981
49,XX,1q+,+3,4q+,+7,+8,12q+,17q+	Granberg et al 1973
56-64,XX,del(X)(q22),del(1)(p12),del(5)(p21),del(7)(q31), t(14;18)(q11;q11),18q+,i(22q)	Berger and Lacour 1974
60-65,XX,del(1)(p13),i(1)(q11),i(17q),cx	Kakati et al 1976b
66,XX,cx,i(1q),inv(1)(p32q12),inv(1)(q12q43), del(6)(q15orq16),dmin	Trent and Salmon 1981

Adenomatous polyp

Karyotype	Reference
48-52,X,-Y,+7,+8,+13,+14,+16,+18,-20,+12q-,1q+	Reichmann et al 1982b
50,XX,+1,+5,+17,+19	Sonta and Sandberg 1978a

MESENCHYMAL NEOPLASMS

Sarcoma, NOS

Karyotype	Reference
68-69,XX,cx,inv(1)	Atkin and Pickthall 1977

Leiomyosarcoma

Karyotype	Reference
42,XX,del(1)(p12orp13),del(11)(q13orq14),+7,-9,-13,-14,-15, -18,+19,-22	Mark 1976

Mesothelioma, malignant

Karyotype	Reference
41,XY,del(1)(p13),t(1;3)(p13;q11),t(7;7)(q11;q36), t(9;12)(p24;q13),del(12)(q13),del(13)(q12q14),-14,-15, t(17;?)(p13;?),i(19p),-20,-21,-22	Mark 1978
43,XY,t(1;1)(q42;q32p22),t(2;6)(p25;p21),inv(3)(p24q13), t(5;18)(p15;q11),-6,+9,del(13)(q24q14),-14,-18,-22	Mark 1978
86,X?,cx,del(1)(p?q?),2p-,2p+,6q-,14q+	Wake et al 1981

MELANOMAS

Malignant melanoma

Karyotype	Reference
??,X?,t(1;6)(q15;q23),cx	Trent et al 1983b
50-152,XX,del(1)(p22)	McCulloch et al 1976

Malignant melanoma, metastatic

Karyotype	Reference
24,X,cx,t(1;15),t(1;14)(q12;p11),del(7)(p13), t(6;7)(p11;q11),t(13;21)(p11;q11),+r	Atkin and Baker 1981
41,XY,dup(1)(q21q32),t(1;7)(q25;q36),t(7;9)(q11;q11), del(9)(q13),-3,-4,-5,-8,-10,-11,-16,+18,-20,-22	Kakati et al 1977
41-43,XY,del(1)(q21q25),i(1q),t(1;9)(p22;q34), t(11;12)(q11;p11),t(10;12)(q26;q13),t(4;?)(q35;?), del(5)(p13),t(14;?)(p12;?),i(17q),t(18;?)(p11;?), del(6)(q21)	Kakati et al 1977
45,XY,t(1;9;13)(q?;q?;q?),r(2),t(1;9)(p?;q?),i(8q), t(11;?)(q?;?),-10,-13,+mar	Chen and Shaw 1973
80,XY,t(11;12)(q23;q13),del(5)(q13),del(5)(p11), del(9)(q13),t(5;9),t(14;?)(q11;?),t(7;12)(q22;q24), del(1)(p13),del(11)(q23)	Kakati et al 1977

NEUROGENIC NEOPLASMS

Astrocytoma, grade III, IV

Karyotype	Reference
44,XY,-4,-17,-22,1q-,5p-,-11q+,+mar	Yamada et al 1980

Neuroblastoma

Karyotype	Reference
44,X?,del(1)(p22),-4,-7,del(7)(q22),t(7;16)(q22q32;q24)	Trent 1983
45,XX,+14,-15,-18,del(1)(p32),dup(1)(p13p31),del(6)(q23), del(11)(q23)	Brodeur et al 1981b
45,XY,-21,t(1;14)(p32;q31),t(8;?)(q24;?),dup(14)(q11q24)	Brodeur et al 1981b
45,XY,-3,del(1)(p13),t(1;9)(q11;p11),t(9;15)(p11;p11), hsr(16p)	Brodeur et al 1981b
46,XX,del(1)(p31)	Brodeur et al 1977
46,XX,del(1)(p31),22p+	Brodeur et al 1977
46,XX,t(1;?)(p32;?),11p+,17q+,dmin	Gilbert et al 1982
46,XX,t(1;?)(p32;?),2q+,7p+,13q+,hsr(5q),dmin	Gilbert et al 1982
46,XY,del(1)(p32),6q+,13q+,16q+,dmin	Gilbert et al 1982
46,XY,t(1;11)(p?;p?),dmin	Gilbert et al 1982
46,XY,t(1;?)(p32;?),6p+,hsr(6q),7q-,hsr(7p),19q+	Gilbert et al 1982
46,XY,t(1;?)(p?;?),19q+,20q+,-22	Gilbert et al 1982
47,X,-Y,del(1)(p?),+6,16q+	Brodeur et al 1977
47,XY,+1,11q-,16q+,19q+	Gilbert et al 1982
47,XY,1p+,4p+,6q+,10q+,12q+,+19	Brodeur et al 1977
53,XX,+1,+4,+15,+20,dup(1)(p13p32),t(3;15)(q13;q26), del(4)(q31),t(7;11)(q36;q13)	Brodeur et al 1981b
82,XXY,cx,del(1)(p22),del(10)(q22),del(17)(p11)	Brodeur et al 1981b
83,XX,cx,hsr(1)(p34),hsr(13p)	Gilbert et al 1982

Retinoblastoma

Karyotype	Reference
44,XY,1p+,-3,t(3;12)(q?;p?),+6q-,der(9),13p+,-14,-16	Kusnetsova et al 1982
45,X,-X,+1p+,+i(6p),-16,17q+,-19,22p+	Kusnetsova et al 1982
46,X,-Y,1p+,+1p-,5q+,7p+,10q-,16q-,17q+,21p+	Kusnetsova et al 1982
46,X,t(X;1)(q28;q11)	Gardner et al 1982
46,XX,+del(1)(p?),-13	Balaban et al 1982
46,XX,1p+,18p+	Kusnetsova et al 1982
46,XX,inv(1),inv(2),inv(3),inv(4),t(6;7),inv(9),t(11;?), 11p-,inv(12),i(17q),-20	Gardner et al 1982
46,XX,t(1;15)(q11q32;q12),t(1;16)(q21;q11),dup(17)(p11q25)	Gardner et al 1982
46,XY,+4,t(5;13)(q35;q21),t(7;17)(p22;q12),t(1;9)(q22;p24), -10,del(16)(q11)	Gardner et al 1982
46,XY,+del(1)(p13),del(16)(p11),-19	Gardner et al 1982
46,XY,dup(1)(q25q44),t(7;10)(q36;q11),-8,-10, t(12;13)(q11;p11),dup(13)(q22q34),+18,+22	Gardner et al 1982
47,XX,+t(1;22)(q?;q?),+i(6p),-22	Kusnetsova et al 1982
47,XY,dup(1)(q12q44),del(2)(q13),+del(6)(q16), t(14;?)(q32;?),i(17q),+20,-22	Gardner et al 1982
47,XY,t(X;1)(q28;p11),t(2;?)(q11;?),t(2;15)(p25;q11), t(3;?)(q27;?),4p+,t(5;?)(q35;?),i(17q),+mar	Gardner et al 1982

Meningioma, benign

Karyotype	Reference
38,X,-Y,-22,-21,-17,-14,-13,-10,-9,-4,-1p-,+mar	Zankl et al 1975a
41,X,-Y,-22,-18,-15,-10,1p-	Zankl et al 1975a
41,XY,+del(1)(p?),-5,-10,-17,-20,-22	Zang and Zankl 1983
41-42,XY,-1,-5,-8,9q+,-14,-18,-22,+r	Mark 1973b
43,XX,-1,-6,-7,-11,-22,+2mar/40-45,XX,-1,-6,-7,-9,-11,-19, -22,+2mar	Mark et al 1972b
43,XY,-1,-8,-17,-19,-22,+2mar/44,XY,-1,-17,-22,+mar	Mark et al 1972a
45,XX,-15,-22,1p-,2p-,4p-,+mar	Zankl et al 1979
45,XX,-22,+13,-14/44,XX,-22,+13,-14,-1	Mark 1973a
45,XY,1q+,-4	Zang and Zankl 1983

EMBRYONAL AND MISCELLANEOUS NEOPLASMS

Teratoma, malignant

Karyotype	Reference
48,X,-Y,+1,del(5)(q12q32),+8,+10	Tricot et al 1983
67,XY,+1,+3,+4,+5,+6,+7,+7,+8,+10,+11,+13,+13,+14,+16,+17, +18,+20,+20,+21,+21	Sonta et al 1977

LYMPHOMAS

1. UNCLASSIFIED AND MISCELLANEOUS LYMPHOMAS

Malignant lymphoma, NOS

Karyotype	Reference
47,XX,t(1;14)(q23;q32),+3,t(6;21)(p25;q11),+8, t(7;11)(q11;q25),-21	Yamada et al 1980

Non-Hodgkin's lymphoma, NOS

Karyotype	Reference
46,X,-X,del(1)(q22),t(2;11)(p21;q21),t(9;15)(q11;p12), del(9)(q11),-14,t(14;17)(q23;q23),-16,+mar	Kristoffersson 1983
46-48,XX,del(1)(q32),+1,+3,r(7),14q+,-20,+mar	Kristoffersson 1983
46-48,XX,t(1;?)(q21orq22;?),t(2;?)(q37;?),-8,+9,+12, 16q+"s"	Mark et al 1976
50,XX,+X,t(1;11)(q25;q23),+3,+7,-13,t(14;18)(p?;p?),-15, t(19;?)(q13;?),+3mar	Reeves and Pickup 1980

Non-Hodgkin's lymphoma, lymphocytic, NOS

Karyotype	Reference
45-46,XY,-1,3q+,del(9)(q22),+14,+17	Reeves 1973
46,XY,del(1)(p22),del(9)(q22),del(10)(p13)	Reeves 1973
46-47,XX,t(1;1)(q25;p36),dup(1)(q25q44),t(14;?)(q32;?),+9	Slavutsky et al 1981
47,XY,+7,-9,-11,1p+,6q-,10p-,14q-, t(14;?;14)(q24orq32;?;q13),15q-,22q-	Fukuhara et al 1979b
47-49,XX,+X,-14,i(17q),t(1;17),1p+	Fleischman and Prigogina 1977
48,XX,del(1)(q22q25),i(mar1),del(12)(p12),dup(14)(q24q32)	Mark 1975a
48-49,XY,1q-,+3,+del(3)(p?q?),-4,-6,+7,-9,-17,-18,-19, +5mar	Fleischmann et al 1976
80,XY,cx,4p+,t(1;4)(p22;p15),t(1;6)(p11;p11)	Kakati et al 1980

Non-Hodgkin's lymphoma, histiocytic, NOS

Karyotype	Reference
45,X,-X,del(1)(q22),t(1;17)(q?;q?),del(2)(q33), del(3)(p21p25),+ins(3)(p14p21p25),+5,del(6)(q14),t(1;7)	Mark et al 1978
45,XY,del(6)(q13),del(6)(q21),15q+,i(1q)	Reeves 1973
46,XX,del(1)(p31p35),+3,del(9)(p11),t(10;?)(q26;?), del(11)(q13),-13,t(11;14)(q13;q32),t(17;21), t(9;22)(q11;p13),+r	Mark et al 1977
46,XY,1p-,6q-,t(11;14)(q21;q32)	Kakati et al 1980
48,XY,1p-,+3,6q-,+7,9p-,t(11;14)(q21;q32)	
46,XY,1q+,t(14;?)(q32;?),18p-	Kakati et al 1980
47,XX,t(1;18)(p36;q21),+7,t(8;12)(q11;q15), t(8;12;13)(q11q22;q15;q14),del(13)(q14),t(8;14)(q22;q32)	Mark et al 1978
47,XY,-6,-14,1q-,+1p-,2q+,3q+,9p+,10p-,10p+,12p-,18q+,22p+, t(2;14)(q37?;q13?)	Fukuhara et al 1979b
47-50,XY,t(1;11)(q?;p?),1q-,1p-,3q+,11p+,t(11;17)	Fleischman and Prigogina 1977
50,XY,dup(1)(q12q32),+2,4p+,del(4)(q21),t(8;14)(q24;q32), +17p+,18q+,+20	Brynes et al 1978

80-100,XY,cx,del(1)(p13),dup(1)(q32q43),inv(1)(p13q12)	Kakati et al 1980
89,XX,i(1q),del(3)(p15p23),del(10)(q24),del(12)(p12), i(17q)	Mark 1977

Malignant histiocytosis

48,XX,1p+?,del(2)(p14?),+del(3)(q12?),+mar	Kristoffersson 1983
48,XY,+i(1q),t(2;5)(p21;q35),+mar	Kristoffersson 1983

2. HODGKIN'S LYMPHOMAS

Hodgkin's disease, NOS

52,XX,+1,-9,-15,+7mar	Fleischmann and Krizsa 1977

Hodgkin's disease, mixed cellularity

61-80,XX,del(1)(q32),i(1q),del(2)(p13),del(6)(q13or15), del(6)(q23or25),15q+,12q+,del(5)(q31),del(7)(q11), del(10)(q26),del(3)(p13)	Reeves 1973

Hodgkin's disease, lymphocytic depletion

45,X,t(X;1)(q28;q11),t(3;5)(q29;q33),del(8)(q22),-9, del(11)(p13),-17	Hossfeld and Schmidt 1978
48,X,-Y,t(1;?)(q44;?),i(1q),del(1)(q21),del(3)(q21),-10, t(3;11)(q23;q21),+14,t(14;?)(q32;?),+15,-17,+19,+20,-22	Hossfeld and Schmidt 1978
60,XY,+del(1)(p21),t(1;5)(q44;q33),dup(5)(q12), t(1;5)(p21q23;q13),del(6)(q22),t(9;?)(q12;?), t(3;12)(q11;q24),t(14;?)(p11;?),i(17q)	Hossfeld and Schmidt 1978

Hodgkin's disease, nodular sclerosis

47,XY,t(1;?)(p36;?),+2,t(2;6)(q31;q27),+8,del(12)(p11), t(17;?)(p13;?),-14,-20,+mar	Reeves and Pickup 1980

3A. NON-HODGKIN'S LYMPHOMAS - RAPPAPORT CLASSIFICATION

Lymphocytic, well differentiated, diffuse

48,XY,1q+,3q+,+i(3p),-6,-10,t(14;?)(q12;?),+16,17p-,+18, +mar	Miyamoto et al 1981a

Lymphocytic, poorly differentiated, diffuse

44,XX,t(11;14)(q13;q32),-3,-9,-13,1p+,+2mar	Fukuhara et al 1979a
46,XY,t(14;18)(q32;q21),t(10;14)(q24;q32),-1,-17,-22,+3mar	Fukuhara and Rowley 1978
47,XX,dup(1)(q21q32),+3	Slavutsky et al 1981
47,XY,-1,+7,-9,+9q+,+9p+,+10,14q+,-17,t(17;18)(p11;q11)	Kristoffersson et al 1981
47,XY,t(14;14)(q32;q?),+7,-9,-11,1p+,6q-,10p-,15q-,22q-, +2mar	Fukuhara and Rowley 1978

Karyotype	Reference
47,XY,t(1;4)(q23;q33),+3,t(10;15),t(11;14)(q13;q32),r(13), r(13)	San Roman et al 1982
48,XX,1q-,+3,+4,6q-,+7,10q-,-15	Kakati et al 1980
48,XX,t(14;18)(q32;q21),1p+,12q+,+mar	Fukuhara and Rowley 1978
87,XY,del(1)(q23),del(3)(p14),del(6)(q21),del(7)(p11), del(11)(q21),t(17;?)(p13;?)	Mark et al 1979
88,XX,del(1)(p22),del(3)(q21),t(3;9)(q?;q?),del(6)(q21), t(22;?)(q13;?)	Mark et al 1979

Lymphocytic, poorly differentiated, nodular

Karyotype	Reference
45-47,XX,14q+,del(18)(q21),+del(1)(p34q32)	Reeves 1973
48,XY,t(14;18)(q32;q21),-6,-11,+12,+del(1)(p?),3q+,+2mar	Fukuhara et al 1979a
52,X,-X,+2,-4,+5,+7,+8,-14,-15,-16,+20,+21, +t(1;15)(p11;q11),+t(12;17)(q24;q11),+t(14;?)(q32;?), +mar	Kaneko et al 1982a
92,XY,t(2;9)(q11;p11),t(5;?)(p15;?),t(17;?)(q25;?),17q+, del(1)(p22),del(3)(p21),del(X)(q24),del(11)(q23)	Reeves 1973

Mixed cell, diffuse

Karyotype	Reference
50,XY,+11,+12,-15,+del(3)(q11),+del(3)(q11), +t(X;1)(p22;q11)	Kaneko et al 1982a
50,XY,+X,-1,+5,+7,+12,-14,-17,del(13)(q12q14), ins(1;1)(q21;q21q32),+t(14;?)(q32;?),+t(1;17)(q21;q25)	Kaneko et al 1982a
53,XY,t(8;14)(q24;q32),t(14;18)(q32;q21),+20,1q-,+1p-,2q-, 4q+,5q-,5p+,6q-,+7q+,10q+,11q+,11q-,13p+,+4mar	Fukuhara and Rowley 1978

Histiocytic, diffuse

Karyotype	Reference
45,X,-Y,t(14;15)(q32;q21orq22),-15,+18,-22,1q+,3p+,5p-,7p-, 9p+,12p+,16q+,17p-,+2mar	Fukuhara and Rowley 1978
45,X,del(X)(q24),del(1)(q21),+del(1)(p11),del(2)(q22), del(3)(q21),del(6)(p11),del(6)(q15),t(7;9)(p22;q13), t(8;10)(p23;q21),del(9)(q13),del(10)(q21), t(2;11)(q22;q23),-13,del(14)(q24),del(15)(q22), del(16)(q22),+17,-19	Mark et al 1979
45-47,XX,cx,t(1;2)?	Kristoffersson 1983
45-47,XX,cx,t(1;2)?	
46,X,-X,-2,-10,+11,-13,-14,-17,del(1)(p22p32), del(3)(p13p21),del(6)(q21),del(9)(p13),+t(2;?)(p23;?), +t(14;?)(q32;?),t(18;?)(q23;?),+t(X;13)(q26;q12),+mar	Kaneko et al 1982a
47,X,-X,t(4;14)(q?;q32),-18,1q+,2p+,6q-,7q+,9q-,12q+,+3mar	Fukuhara and Rowley 1978
47,XX,-1,-2,-6,+8,-14,+16,-19,del(11)(q23), t(8;14)(q22;q32),+t(2;6)(q21;p23),del(6)(q21), +t(1;14)(p11;q11),+t(19;?)(q13;?),+t(14;?)(q11;?)	Kaneko et al 1982a
47,XY,-11,-22,del(1)(q42),del(2)(q33),del(5)(q22q31), del(7)(q32),del(9)(p22),t(21;22)(q22;q11), +t(11;?)(q23;?),+t(21;22)(q22;q11),+mar	Kaneko et al 1982a
50,XX,del(1)(p22),+3,+7,+12,t(1;14)(p22;q32),+18	Mark et al 1979
50,XY,t(8;14)(q24;q32),+2,+20,1q+,4p+,+4q+,17p+,18q+	Fukuhara and Rowley 1978
61,XY,cx,t(1;16)(q23;p11),dup(1)(q23q32),t(1;6)(q23;p21), t(3;6)(q21;p21),t(7;7)(q32;p22),13p+	Fitzgerald et al 1980

82,XX,del(X)(q13),i(1q),i(1p),del(7)(p11),i(9p),
del(14)(q22),t(1;14)(q23;q32),i(15q),i(17q) — Mark et al 1979
84,XY,t(1;14)(q2?;q24orq32),cx — Fukuhara and Rowley 1978
89,XY,del(1)(p22),t(5;13)(q15;q34),t(14;20)(q32;q12),
t(1;17)(p22;q25),t(1;19)(q21;p13),del(20)(q12) — Mark et al 1979

Histiocytic, nodular

46,XX,-6,+t(1;6)(q11orq21;q25orq27) — Kaneko et al 1982a

3B. NON-HODGKIN'S LYMPHOMAS - KIEL CLASSIFICATION

Lymphocytic, T-zone

45,XX,-1,-2,-3,-4,+t(1;2)(p32q32;p13),+t(2;4)(p13;p15),
+t(3;4)(q11;?) — Gödde-Salz et al 1981a

Immunocytoma

46,XX,1p+,+2,del(17)(p12),17p+ — Kristoffersson 1983
46,XX,1p+
47,XX,1p+,+12,del(17)(p12)

Centroblastic-centrocytic, diffuse

48-50,XY,-1,+7,9q+,+9p+,+10,+14q+,-17,-18,
+t(17;18)(p11;q11),-19,+20,+22,+mar — Kristoffersson 1983
48-51,XY,del(1)(p33q23q25),-2,+del(3)(p22),+4,4p+,+5,+6p+,
+7,+8,+9,+21,+22 — Kristoffersson 1983
45-51,XY,del(1)(p33q23q25),del(2)(q14),+del(3)(p22),+4,+5,
+7,+8,-12
49,XY,+1,+20,+r — Kristoffersson 1983

Centroblastic-centrocytic, follicular

46,XY,t(1;?)(p36;?),t(1;1)(p36;q1?),del(7)(q11),
del(7)(q22),del(9)(p11),t(14;?)(q32;?) — Reeves and Pickup 1980
46-47,XX,del(1)(q11q22),+11 — Kristoffersson 1983
46-49,XX,1p+,3q-,4q+,+6,+13,-14,14q+,17p+ — Kristoffersson 1983
47,XY,t(1;14)(q21;q32),+12,-16,+19,t(14;?)(q32;?) — Reeves and Pickup 1980
48,XX,+X,t(1;?)(p36;?),del(2)(p21),t(3;?)(q27orq29;?),
t(14;?)(q32;?),t(17;18) — Reeves and Pickup 1980

Centroblastic, diffuse

42-45,XY,1p+,-6,-8,14q+,14q-,-16,i(17q),+18,-20,+mar — Kristoffersson 1983
46-49,XY,1p+,+2,-6,+del(9)(p21),+12,+13,14q+?,i(17q),+18 — Kristoffersson 1983
47,XY,14q+,+22

Lymphoblastic, Burkitt's type

??,XX,dup(1)(q12orq21q31),t(8;14),+10q- — Douglass et al 1980a
??,XY,-5,+6 — Douglass et al 1980a

Karyotype	Reference
??,XY,1p+,t(8;14)	
??,XY,dup(1)(q12orq21q31),3p-,3q-,+6p-,+8,+14,+15,+18,-21	Douglass et al 1980a
??,XY,dup(1)(q?q31),t(8;14)	Douglass et al 1980a
43,XY,+1,+2,+3,-7,-8,-9,-12,-16,-22,del(1)(q25), del(2)(p11),del(10)(q22),14q+	Biggar et al 1981
45,X,-Y,t(1;14)(q22;q32),t(8;22)(q24;q11),3q+,6q-,6q+,7q+, 9p-,14q-,18q-	Berger et al 1981c
46,X,t(X;1)(p11;q25),del(1)(q25),7p+,17p-,t(2;8)(p11;q24)	Slater et al 1982
46,XX,1p+,14q+,+mar,+r	Kristoffersson 1983
46,XX,dup(1)(q12orq21q31),t(8;14)	Douglass et al 1980a
46,XX,dup(1)(q12orq21q31),t(8;14)	Douglass et al 1980a
46,XX,dup(1)(q12q32),t(8;14)(q24;q32)	Miyamoto et al 1982a
46,XX,t(8;14)(q24;q32),dup(1)(q23q32),del(3)(p25)	Slater et al 1982
46,XY,dup(1)(q21q44),del(2)(q32)	Biggar et al 1981
46,XY,dup(1)(q23q42)	Slater et al 1982
46,XY,t(8;22)(q23;q12)/46,XY,t(8;22)(q23;q12), t(1;6)(q21;q27)	Berger et al 1981c
46,XY,t(8;22)(q23;q12)/46,XY,t(8;22)(q23;q12), +t(1;6)(q23;q26)	Berger et al 1981c
46,XY,t(8;22)(q23;q11)/46,XY,t(8;22)(q23;q11), t(1;4)(p32;q32)	Berger et al 1981c
46-47,XY,t(14;?)(q32;?),dup(1)(q21q32),t(2;3)(q13;q29),+5, 6p-,-8,del(9)(q?),+12,+18,+mar	Miyoshi et al 1981c
46-52,X?,t(8;14)(q24;q32),+1,+3,+5,+7,i(17q),+X	Zech et al 1976a
47,XX,t(8;14)(q24;q32),dup(1)(q24q32),del(2)(q34), del(5)(q31q34),+8"s"	Berger and Bernheim 1982
47,XY,+11,-14,+21,inv(1)(p11p31),11q+,21p+	Biggar et al 1981
47,XY,t(8;22)(q24;q11),+22/47,XY,t(8;22)(q24;q11),+22, dup(1)(q23q24)	Berger and Bernheim 1982
47-49,XY,t(8;14)(q23;q32),+7,+t(X;1)(p21;q21)	Kakati et al 1979
50,XX,+1,+2,+3,+11,del(2)(p14),11p+,t(8;14)(q24;q32)	Biggar et al 1981

Lymphoblastic, convoluted type

Karyotype	Reference
43-46,XY,del(1)(p32),del(2)(p21),-6,t(7;9)(p11;q11),-9,-10, t(11;15)(p11;q11),-11,del(12)(q14),-13,-14,i(17q),-17, +18,del(19)(p11),-22,+2mar	Kristoffersson 1983
46,X,-Y,t(1;?)(q21orq23;?),+mar/46,X,-Y,-1,-2,+der(1), +t(2;?)(q33orq35;?),+mar/48,X,-Y,-5,-14,+18,+18,+20, del(9)(p13),t(1;14),+dup(5)(q13q35)	Kaneko et al 1982d
47,XY,-7,-9,del(6)(q23?),t(14;18)(p11;q11),+t(9;?)(q34;?), +2mar/47,XY,-7,-9,del(6),del(1)(q32),t(14;18),t(9;?), +t(1;?)(q23orq25;?)	Kaneko et al 1982d
47,XY,t(1;2)(p34;p21),+4,t(13;18)(p13;p11)/47,XY,+3	Kaneko et al 1982d
48,XX,+13,+t(1;5)(q21;p15)/48,XX,-5,+8,+13,del(1)(q32), +t(1;5)(q21;p15)	Kaneko et al 1982d

Immunoblastic

Karyotype	Reference
45,XY,del(1)(p22orp23),del(3)(q11),del(3)(q21),del(8)(p11), del(9)(p11),del(9)(q12),t(12;?)(q24;?)	Reeves and Stathopoulos 1976
47-48,XX,+del(3)(q11),+17/49,XX,+del(3)(q11),+16,+17	Kristoffersson 1983
45-47,XX,1p+,-14,-15,+16,+17,-19,+mar	

49,XY,del(1)(p32),del(1)(q32),+5,del(6)(q21),+2mar	Kristoffersson 1983
51,XY,t(1;6)(p22;q13orq15),t(2;13)(q21;q14), t(8;16)(q24;p11),+11,+12,del(15)(q22),+16,+20,+mar	Reeves and Pickup 1980
78-83,XY,+del(1)(q11),+5,+7,+12,+13q+,+14q+,+14q+,+20,+mar	Kristoffersson 1983
80-86,XY,del(1)(q11),+3,+7,+9,+12,+13q+,+14q+,+14q+,+15q+, +18,+20,+mar	

3C. NON-HODGKIN'S LYMPHOMAS - LUKES AND COLLINS CLASSIFICATION

Mycosis fungoides

46-48,XX,del(1)(q42),+3,-11,-12,+2-6mar	Edelson et al 1979
48,XX,del(1)(p22),t(2;8;14)(q37;q24;q24),+5,del(7)(p13), del(9)(q22),t(9;18)(q11;q24),del(10)(p13), del(13)(q12q14),del(5)(q15q22),del(5)(q13),del(5)(q15)	Fukuhara et al 1978a
46,XX,t(1;14)(q32;q32)	

Sezary's syndrome

42,XY,-13,-21,-22,1p+,17q+	Nowell et al 1982
46,XY,-1,-10,+t(1;9),+t(10;11),9q-,11q-	Nowell et al 1982
50,XX,1q+,2p-,+3q-,+6q-,+9,-15,+21,+mar	Nowell et al 1982

Follicular center cell, diffuse, large cleaved

47,XY,-6,-14,1q-,+1p-,2q+,3q+,9p+,10p-,10p+,12p-,18q+,19p+, 19q+,+22p+,+mar	Fukuhara et al 1978b

Follicular center cell, diffuse, small non-cleaved

44,X,-Y,t(1;14)(p21;q13),5p+,del(6)(q13),del(8)(p21),9p+, 9p-,t(11;21)(q23;q22),13q+,-10,+12,-17p+,der(19),22q+	Fukuhara et al 1979a

Immunoblastic type (B-cell)

49,XX,+i(1q),+5,del(8)(q24),t(3;22)(q29;q11),+13	Berger et al 1983a

3D. NON-HODGKIN'S LYMPHOMAS - WORKING FORMULATION

Small lymphocytic

46,XX,t(1;8),t(8;9)	Yunis et al 1982

Follicular, small cleaved cell

44,XY,+der(1),+2,t(2;13),-3,5q-,-12,t(14;18)(q32;q21), i(17q)	Yunis et al 1982
46,XY,t(1;2),t(4;10),t(14;18)(q32;q21),t(1;22)	Yunis et al 1982
47,XX,t(X;1),ins(1;8),t(2;8),t(14;18)(q32;q21),+17	Yunis et al 1982
48,X,-X,inv(1),1q-,+3,4q-,ins(5),+7,+mar	Yunis et al 1982
48,X,r(Y),+2,+3,t(14;18)(q32;q21),t(1;22)	Yunis et al 1982

Follicular, mixed small cleaved and large cell

79,XY,cx,t(1;1),9p-,t(14;18)(q32;q21) Yunis et al 1982

Follicular, large cell

50,XY,1q-,t(2;3),+3,+10,t(1;12),13q-,+15,-19,t(3;19),+21 Yunis et al 1982

Diffuse, small cleaved cell

49,XY,dup(1)(q?),8p-,-9,+14,+17,+20,+mar Yunis et al 1982

Diffuse, large cell

54,XY,dup(1)(q?),inv(5),+3,+3,+9,+13,+18,+18,+20,+mar Yunis et al 1982

Large cell, immunoblastic

48,Y,t(X;?),t(1;?),t(3;6),+5,-6,t(8;14)(q24;q32),i(17q), +mar Yunis et al 1982

Small non-cleaved cell

46,X,-X,t(8;14)(q24;q32),dup(13),+r/45,X,-X,t(8;14),-12, t(1;15),der(22),+r Yunis et al 1982
46,XX,dup(1)(q?),t(8;14)(q24;q32) Yunis et al 1982

Chromosome 2

HEMATOLOGICAL DISORDERS

UNCLASSIFIED LEUKEMIAS

Acute leukemia, NOS

46,XY,2p+	Chrz et al 1983
46,XY,t(2;19)(p23;p13)	Prigogina et al 1979

NONLYMPHOCYTIC LEUKEMIAS

Acute nonlymphocytic leukemia (ANLL)

43,XX,2q-,inv(3),-4,del(5)(q?),-7,17p+,-19	Rowley 1976
44-45,XX,-3,-5,-6,-18,-19,t(1;2)(q2?;q11),t(5;?)(q13;?)	Philip et al 1978a
45,XY,t(1;14)(p22;p11),t(1;15)(p22;p11),-2,del(5)(p13),-17, +mar	Philip et al 1978a
45-47,XX,-2,-7,7q-,-8,+15,+22,+2mar"s"	Whang-Peng et al 1979
46,XX,-1,+2,del(1)(q31)	Zuelzer et al 1976
46,XX,2q+"s"	Mitelman 1983
46,XX,2q+"s"	
46,XX,2q+"s"	
46,XX,2q+"s"	
46,XX,2q+"s"	
46,XX,2q+"s"	
46,XX,del(7)(q22)/46,XX,del(7)(q22),del(5)(q22q31)/46,XX, del(7)(q22),-9,t(1;9)(q22;p24)/46,XX,t(2;3)(q31;q27), del(7)(q22),del(5)(q22q31orq33)"s"	Oshimura et al 1976b
46,XY,1p+,12p-,?t(1;12),?11p+,inv(12)(p?),-2,+mar	Mitelman 1983
46,XY,del(2)(p11)	Bernard et al 1982a
46,XY,del(2)(q31),-7,del(9)(p13),t(2;21)(q21;q22), t(2;?)(p21;?)/46,XY,t(1;?)(p22;?),t(17;?)(q25;?)	Philip et al 1978a
56,XY,+Y,+6,+14,+21,+t(5;6)(q?;p?),+del(1)(p22q25), +del(2)(q12),+del(3)(p12),+del(4)(q13),+del(5)(p14q14), +mar	Zuelzer et al 1976

ANLL, FAB type M1

Karyotype	Reference
41,XY,-1,-2,-4,-5,-12,-15,-17,-17,-18,-22,+5mar	Sessarego et al 1981a
45,XY,t(2;5)(q22orq23;q14q15),6p-,-7,t(6;7)(p21;p22q21), t(7;10)(p21;q21),t(14;19)(q11;q13)"s"	Hagemeijer et al 1981a
46,XY,+del(2)(p13),i(9q),-16,del(17)(p11)	Brodeur et al 1983
46,XY,-2,-5,-7,-13,+16,-21,-21,+5mar	Sessarego et al 1981a

ANLL, FAB type M2

Karyotype	Reference
43,XX,+2,3q+,-5,-20,-22	Berger et al 1981a
43,XX,-3,-4,-7,-19,del(5)(q14),17p+	Testa et al 1979
43,XX,-3,-4,-7,-19,del(2)(q31?),del(5)(q14)	
46,XX,1q+,2q+,-3,-11,+r,+mar	Berger et al 1981a
46,XX,2q+,5q-,7q-,17q+"s"	Berger et al 1981a
46,XX,t(2;8;21)(p11q13;q22;q22)	Pasquali and Casalone 1981

ANLL, FAB type M1+M2

Karyotype	Reference
42-44,XX,+8,-12,-13,-14,t(2;5)(q12;q35)	Mitelman et al 1981
44-45,XY,+2,-5,-7	Mitelman et al 1981
45,XX,-2,5q+	Yamada and Furusawa 1976
45,XY,t(2;2)(q12;p25),del(7)(q32)	Prigogina et al 1979
45-46,XY,t(2;6)(q37;q23),-7,del(11)(q21),del(22)(q11)	Mitelman et al 1981
46,XY,t(9;22)(q34;q11),t(2;15)(q37;q12)	Mitelman et al 1981

ANLL, FAB type M3

Karyotype	Reference
46,XX,2q-,6p-,7q-,10q+,14q+,-16,+mar	Hagemeijer et al 1981a
46,XX,del(2)(q32q34),t(15;17)(q25;q22)	Webber and Garson 1983
46,XX,t(2;15;17)(q21?;q25?;q21?;),9p+	Bernstein et al 1982b

ANLL, FAB type M4

Karyotype	Reference
43,XY,-2,-3,-6,-7,-12,-15,-21,-22,del(2)(p11),del(11)(p14), t(14;?)(q32;?),t(17;?)(p13;?),del(22)(q11),+5mar"s"	Rowley et al 1981a
45,XX,2p-,-5,-9,-17,+r,+mar	Berger et al 1981a
45,XX,del(2)(q31),del(7)(p13),-13	Rowley and Potter 1976
45-46,XY,inv(2),-7	Mitelman et al 1981
46,XX,2p+	Mitelman et al 1981
46,XX,t(2;11;12;17)(q31;p11;p13;q23)	Fitzgerald et al 1983
46,XY,2q+,del(18)(p11),del(22)(q11)/47,XY,2q+,+8, del(18)(p11),del(22)(q11)"s"	Alimena et al 1981
45-47,XY,2q+,+8,del(18)(p11),-21,del(22)(q11)"s"	
46,X,-Y,2q+,+8,del(21)(q21)"s"	
46,XY,inv(2)	Hustinx and Rutten 1983

ANLL, FAB type M5

44,X,-Y,del(2)(p11),-15,-21,+mar Weber et al 1979
46,XX,-18,-22,+2r/46,XX,t(2;9)(q23;p22),-3,-22,+2r/49,XX, t(2;9),-22,+4r Alimena 1983
46,XY,t(2;11)(q37;q12),t(2;12)(q24;q24),10p+ Mitelman 1983
46,XY,t(2;11)(q37;q12),t(2;12)(q24;q24),10p+
46-51,XX,-3,-18,-22,t(2;9)(q23;p22),+2-4r Mitelman et al 1981
47,XY,1p+,1q+,2p+,-3,4q+,+13,+mar Kaneko et al 1982b
57-58,XX,+1,+3,-4,-5,+6,-8,+11,+12,-17,+19,-20, del(10)(q24),+8-9mar Testa et al 1979
60-61,XX,Xp+,+1,+2,-4,+6,-7,+19,+20,del(2)(q31), del(10)(q24),+12-13mar

ANLL, FAB type M4+M5

66,XY,+1,+1,+2,+4,+6,+6,+7,+8,+8,+10,+11,+11,+12,+12,+18, +19,+21,+21,+22,+22 Li et al 1983

ANLL, FAB type M6

43,XY,-2,4q-,-9,11q-,-15,-19,-20,+2mar/46,XY,-15,-20,+2mar Li et al 1983
43-45,XX,der(1),inv(2)(p25q14),del(3)(q25),-4,-5,-6,6q-,-7, 7p+,11p+,-12,-17,-21,+r,+1-4mar"s" Smadja et al 1983b
44,XX,-1,-2 Nordenson 1983
44,XY,3p-,t(2;6),t(5;6),t(17;20)"s" Sandberg et al 1982a
44,XY,t(2;13)(q37;q12?),t(13;?)(p11;?),-15"s" Weinfeld et al 1977a
44-45,XY,-2,-4,-7,-10,-13,-16,-17,-22,+3mar Mitelman et al 1981
44-47,XY,+2,-3,-5,-7,+8,-14,-19,+2mar Mitelman et al 1981
46,XX,-7,-16,-21,+t(2;7)(p11;q11),del(2)(q33),del(11)(q22), +11p+,14p+,+t(16;21;21;21)(p13;q22;q11;q22)"s" Rowley et al 1981a
51,XY,+1,+2,+6,-7,+8,-14,+15,+21,+mar"s" Rowley et al 1981a
72,XY,cx,t(2;15)(q31;q15),del(6)(p22q15),t(8;13)(q23?;q13), t(11;14)(p12;q13),i(16p),t(15;18)(q15;p11) Douglass and Freeman 1983

Chronic myeloid leukemia, t(9;22)

26,XX,t(9;22)(q34;q11),-1,-2,-3,-4,-5,-6,-7,-9,-10,-11,-12, -13,-14,-15,-16,-17,-18,-19,-20 Hartley and McBeath 1981
45,XX,t(9;22),-2,t(2;14)(q21;q32) Sessarego et al 1979
46,X?,t(9;22)(q34;q11),t(2;2)(q21;p25) Lawler et al 1976
46,XX,t(9;22)(q34;q11),2p-,7p+,-8,+mar Alimena et al 1982b
46,XX,t(9;22)(q34;q11)/46,XX,t(9;22),del(6)(q16)/46,XX, t(9;22),del(20)(q11)/45,XX,t(9;22),del(20)(q11), t(1;2)(p31;q37) Mitelman 1983
46,XX,t(9;22)/46,XX,t(9;22),del(6)(q16)/46,XX,t(9;22), del(20)(q11)
46,XX,t(9;22),2p+/48,XX,t(9;22),2p+,+19,+22q- Sonta and Sandberg 1977b
49,XX,t(9;22),2p+,+19,+22q-,+22q-
46,XX,t(9;22),2q+,i(17q),19p- Olah and Rak 1981
46,XX,t(9;22),t(2;11)(p25;p12) Miyamoto 1980
46,XX,t(9;22),t(2;4)(q?;q?) Sadamori et al 1980
46,XY,t(9;22)(q34;q11) Ishihara et al 1982

27,X,-Y,t(9;22),-1,-2,-3,-4,-5,-6,-7,-9,-11,-13,-14,-15, -16,-17,-18,-19,-20,-22	
47,XY,t(9;22),+22q-/56,XY,t(9;22),+1,+2,+5,+6,+7,+8,+10, +12,+17,+22q-	Stoll and Oberling 1978
48-49,XX,t(9;22)(q34;q11),t(2;4),+19,+del(22q)	Sonta and Sandberg 1978b
50,XX,t(9;22),t(2;12)(p13;q24),+8,+8,+12q+,+12q+,+del(22q)	Sharp et al 1976

Chronic myeloid leukemia, aberrant translocation

??,X?,t(2;9;22)(q14;q34;q11)	Smadja et al 1980
46,X?,t(2;22)(q37;q11)	Van Den Berghe 1983
46,XX,t(2;9;22)(p11;q34;q11)	Rochon and Vaillancourt 1980
46,XX,t(2;9;22)(p21;q34;q11)	Pasquali et al 1979b
46,XX,t(2;9;22)(p24q24;q34;q11)	Pasquali 1983
46,XX,t(2;9;22)(q11;q34;q11),t(2;7)(q11;q11)	Fleischman et al 1981
46,XX,t(2;9;22)(q11;q34;q11)	Smadja et al 1983a
46,XX,t(2;9;22)(q14q37;q34;q11)	Van Den Akker et al 1980
46,XX,t(2;9;22)(q23;q34;q11)	Pasquali et al 1979b
46,XX,t(2;9;22)/50,XX,t(2;9;22),+6,+10,+19,+der(22)	
46,XY,t(2;22)(q37;q11)	Hayata et al 1975b
46,XY,t(2;22;22)(q11;q11;q13)	Fraisse et al 1980
46,XY,t(2;9;22)(q11;q34;q11)	Alimena et al 1982b
46,XY,t(2;9;22)(q24orq31;q34;q11)	Tanzer et al 1977
46,XY,t(2;9;22)(q32;q34;q11)	Nowell 1983
46,XY,t(2;9;22)(q37;q22q34;q11)	Testa et al 1982

Chronic myeloid leukemia, Ph1 negative

45,XX,-7,inv(2)	Engel et al 1977
46,XY,t(2;8)(q21;q24),del(1)(p33),dup(1)(p21p32)	Brodeur et al 1979

MYELOPROLIFERATIVE DISORDERS

Polycythemia vera

??,X?,+2,+del(13)	Vykoupil et al 1980
??,X?,+2,-10	Vykoupil et al 1980
??,X?,+2,-10	Vykoupil et al 1980
??,X?,+2,-8,-10	Vykoupil et al 1980
??,X?,2q+,6q-	Vykoupil et al 1980
46,XX,t(2;11)(p13;q21)/46,XX,t(1;15)(p1?;q1?), t(2;11)(p13;q21),del(16)(p12)/51,XX,+3,+8,+8,+19,t(1;15), t(2;11),del(16)	Testa et al 1981b

Myelosclerosis / Myelofibrosis

44,XY,5p-,-7,-10,11q-,i(17q)/45,XY,-7,18q-/45,XY,2p-,2q-, 5p-,6q-,9q-,10q-,-11,+i(17q)	Whang-Peng et al 1978
46,XY,t(1;2)(q?;q?),-8,inv(13)(q12q31),i(17q)	Whang-Peng et al 1978
47,XX,del(22)(q?),t(1;2),t(6;7),+r,+mar	Page et al 1979
51,XX,+1,del(2)(q33),+6,+9,+11,+17,-19,+20q+	Najfeld et al 1978b

Idiopathic thrombocythemia

46,XX,t(2;9)(q?;p?) Köpf et al 1982

Chronic myeloproliferative disease, NOS

46,X?,t(1;2)(q?;q?) Wurster-Hill 1983

Myeloproliferative disorder, special type

45,XX,t(1;?)(p36;?),-5,-7,-12,t(13;?)(q34;?),-17,-22,
+4mar/45,XX,t(1;?),-2,-3,-5,-7,-12,-17,-22,+8mar"s" Rowley et al 1977a

DYSMYELOPOIETIC SYNDROMES

Preleukemia, NOS

45,XX,-2,-3,5q-,-6,-9,16p-,17p+,+3mar Swansbury and Lawler 1980
45,XY,-2,-4,-6,-8,-10,-10,+17,+18,+3mar Panani et al 1980
45,XY,-2,-5,-7,-8,-11,-12,-13,-14,+t(2;5),+t(11;12),
+t(16;17),+17,+3mar Watt et al 1982
46,XX,dup(2)(q34q36)"s" Berger et al 1980a
46,XY,t(1;17)(q24;p13),-2,-5,-7,+r,+2mar"s" Anderson and Bagby 1982
53,XX,+1,+2,t(4;19),+6,i(17q),+21,+21,+22 Seabright 1983

Refractory anemia, NOS

46,XX,t(2;4;13)(p11;q12;q34),del(5)(q15q31) Kerkhofs et al 1982

Sideroblastic anemia, NOS

46,XY,t(2;5)(q37;q?) Kamihira et al 1977

Acquired idiopathic sideroblastic anemia

46,XY,t(2;11)(p21;q25),del(5)(q13q31)/46,XY,
t(2;11)(p21;q25),del(11)(q14) Kardon et al 1982a

Refractory anemia with excess of blasts

56-64,XY,+1,-2,+6,+9,+13,+15,+19,+21,10-13mar Streuli et al 1980

Pancytopenia

46,XX,-4,-5,-6,-7,-11,-13,-16,+19,+20,t(7;13)(q?;q?),
+4mar/43,XX,-2,-3,-5,-7,-16,+19,15q+,+mar Nowell and Finan 1978

Paroxysmal nocturnal hemoglobinuria

47,XY,+2/47,XY,+9	Cohen et al 1979

Dysmyelopoietic syndrome, special type

45,XY,-7,t(2;3)(p21;q28)"s"	Rowley et al 1981a
45,XY,del(2)(q33),-5,del(7)(q22q32),-12,-15, t(12;15)(q11;q26),del(19)(q12),+r"s"	Albain et al 1983

SPECIAL LEUKEMIAS

Hairy cell leukemia

47,XY,-2,3q+,14q-	Khalid et al 1981

LYMPHOCYTIC LEUKEMIAS

Acute lymphocytic leukemia (ALL)

28,XX,-1,-2,-3,-4,-5,-7,-8,-9,-11,-12,-13,-14,-15,-16,-17, -19,-20,-22	Kaneko and Sakurai 1980
32,XY,-2,-3,-4,-5,-7,-8,-11,-12,-13,-14,-15,-16,-17,-20	Shabtai and Halbrecht 1981b
46,XX,2q+,4q-,12p+,13q+,14q+,-20,+21	Whang-Peng et al 1976c
46,XX,t(9;22)(q34;q11),t(2;9)(q21;p13),del(9)(p13)	Morse et al 1982a
46,XY,+2,t(6;?)(p21;?),+7,-12,-13,14q+,17p+	Oshimura et al 1977a
46,XY,t(8;22)(q24;q12),t(3;11)(p12p21;q23),inv(2)(p11q13), del(2)(p21)/46,XY,t(8;22)(q24;q12),dup(1)(q23q44)"s"	Fonatsch et al 1982
47,XY,t(4;11)(q21;q23),1p+,del(2)(p16),del(6)(q21),i(7q), del(8)(q21),+mar	Arthur et al 1982
53,XX,+2,+5,i(7q),+8,+13,+21,+21,+22,+17q+	Oshimura et al 1977a
54,XX,7p+,+10,+10,+14,+14,+18,+18,+21,+21/27,X,-X,-1,-2,-3, -4,-5,-6,-7,7p+,-8,-9,-11,-12,-13,-15,-16,-17,-19,-20, -22	Oshimura et al 1977a
55,XX,+2,+4,+9,+14,+15,+19,+21,+21	Oshimura et al 1977a
61,XX,+X,+1,+2,+3,+4,+5,+6,+8,del(6)(q21),+13,+14,+15,+16, +18,+21,+22	Oshimura et al 1977a

ALL, FAB type L1

26,XX,-1,-2,-3,-4,-5,-6,-7,-8,-9,-10,-11,-12,-13,-14,-15, -16,-17,-19,-20,-22	Hoeltge et al 1982
28,XX,-1,-2,-3,-4,-5,-6,-7,-8,-9,-11,-12,-13,-15,-16,-17, -19,-20,-22/56,XX,+X,+X,+10,+10,+14,+14,+18,+18,+21,+21	Brodeur et al 1981a
34-36,XY,-2,-3,-4,-7,-9,-12,-13,-14,-16,-17,-20,+22	Sandberg et al 1982c
46,XY,del(2)(q32),-11,-12,+2mar/46,X,-Y,del(2),-4,-5,-11, -12,-21,+6	Kaneko et al 1981b
47,XY,t(22;?),-1,-1,-2,-3,-9,+6mar	Sandberg et al 1980

ALL, FAB type L2

Karyotype	Reference
26,XX,-1,-2,-3,-4,-5,-6,-7,-8,-9,-10,-11,-12,-13,-15,-16, -17,-18,-19,-20,-22	Brodeur et al 1981a
46,XY,t(2;14)(p?;q?)	Prieto et al 1981
47,XY,+18	Kaneko et al 1982e
46,XY,-2,-5,-6,-7,+8,-9,-10,-11,-12,-16,+18,+t(2;?)(q35;?), +t(11;?)(q21;?),+5mar	

ALL, FAB type L3

Karyotype	Reference
46,XX,t(2;8)(p12;q24)	Kaneko et al 1982e
46,XX,t(2;8;14)(p11orp12;q24;q32),t(21;21)(p11;q11), +t(21;21)	Ekblom et al 1982
46,XY,t(2;8)(p13;q24)	Rowley et al 1981b

Chronic lymphocytic leukemia

Karyotype	Reference
45,XX,-2,t(8;12)(q?;q?)	Pittman et al 1982
45,XX,-8,-14,-15,-22,+2,+9,+1p+,+mar	Finan et al 1978
46,XX,-3,t(2;3),t(14;18)(q32;q?),2q-,17q+	Finan et al 1978
46,XX,del(6)(q13q16),t(2;7)(p11;q22)	Vahdati et al 1983a
46,XY,2p+,del(11)(q22)	Fleischman and Prigogina 1977
46,XY,2q+	Finan et al 1978
46,XY,del(2)(q31)	Pittman et al 1982
47,XX,+12,del(2)(q31)	Gahrton et al 1980c
47,XX,+12,del(2)(q31)	Robert et al 1982
47,XY,+2,t(9;13),t(9;19),i(17q)	Finan et al 1978

Prolymphocytic leukemia

Karyotype	Reference
44,XX,-2,-9,-10,11p+,+mar	Pittman et al 1982
46,XX,-2,-2,-8,+3mar	Pittman et al 1982

Lymphocytic leukemia, special type

Karyotype	Reference
45,XX,-1,2q+,14q+	Ueshima et al 1981
45,Y,-X,t(Y;14)(q12;q32),-13,-17,del(2)(q33), t(3;13)(q21;q34),del(6)(q16orq22),9p+,9q+,10q+,16q+,18q+	Miyoshi et al 1981a

MONOCLONAL GAMMOPATHIES

Multiple myeloma

Karyotype	Reference
55,XY,+1,+5,+9,+11,-12,+13,+15,del(1)(q32),del(6)(q25), t(9;19)(q13;q13),14q+,16p+,+4mar	Liang et al 1979
45,XY,-2,-2,-3,+5,-8,-10,+14,-15,-16,del(6)(q25),+4mar	

Plasma cell leukemia / Plasmocytoma

Karyotype	Reference
86,XX,cx,t(1;?)(q11;?),2q+,dup(7)(q22q36),t(8;?)(p11;?), t(9;13)(p23;q21)	Ueshima et al 1983

SOLID TUMORS

UNCLASSIFIED NEOPLASMS

Malignant neoplasm, NOS

Karyotype	Reference
47,XX,-2,-5,+6,-10,11q+,-13,+18,-21,t(1;?)(q21;?), t(1;13)(q21;p11),t(1;6)(q11;q23)	Kovacs 1978a

EPITHELIAL NEOPLASMS

Carcinoma, NOS

Karyotype	Reference
32-35,XX,cx,del(1)(p12),dup(1)(q21q32),2q+,5q+,7p+,11p+, hsr(10),t(11;12)(q13;q24),12p+,13q+	Kusyk et al 1982
44,XX,ins(1;1)(q32;q12q31),i(2q),+3,t(9;11)(q13;p15),-12, -15,17p+,-18	Atkin and Baker 1982a
44-50,XX,+3,dup(1)(q21q32),t(10;11)(q26;q25),t(2;17), t(6;13),t(9;15),del(10)(q24)	Riet-Fox et al 1979
46,X,-X,+mar/61,XX,cx,der(2)(q23)	Vang Nielsen et al 1982
46,XX,+2,-7,-10,+13,-16,-17,+20,+5q-,15p+,15p+,18p+	Reichmann et al 1981
46,XX,+8,+12,+21,-2,-15,-16	Wake et al 1981
46,XX,del(2)(p24),+5,+8,t(9;17)(p?;p?),del(16)(q21),-20, -21	Mark 1975b
46,XX,t(1;?)(p32;?),-2,3q-,5q-,6q+,-18,-21,+mar	Atkin and Baker 1982a
46,XX,t(3;11)(q12;q14),-6,+7,-8,-15,t(16;?)(q23;?),+2r	Mark 1975b
47,XX,+2	Sonta et al 1977
48-52,XX,+2,+11,+12,i(13q)	Riet-Fox et al 1979
52,XY,+X,+Y,-2,+7,+8,+9,+10,+13,-17,+i(1q)	Reichmann et al 1981
54,XX,+1,+2,+8,+11,+13,+14,+21,+21	Sonta et al 1977
55,XY,t(1;17)(p36;q21),+2,+2,+4,+8,+8,-10,+15,+18,+20,+21, +2mar	Sonta and Sandberg 1978a
58,XX,+X,+1,+2,+3,5p-,+6,+7,+9,11q+,-13,-14,-15,+16,+19, +20,+6mar	Kovacs 1978a
67,XY,cx,2q+	Reichmann et al 1981
73-81,XX,cx,2q+,i(13q)	Reichmann et al 1981
80-83,XX,cx,t(2;5;16),t(2;5;18),t(11;13),11q+,t(8;10), del(2)(q24),del(6)(q21),i(14q),dup(9)(q11q12)	Riet-Fox et al 1979
82,XX,cx,2q+,5q-,9q+,i(17q)	Reichmann et al 1981

Carcinoma NOS, metastatic

Karyotype	Reference
??,XX,t(1;9),t(2;?),cx	Bartnitzke and Bullerdiek 1983
45-50,XX,-2,+7,-22,t(1;12),t(13;14),inv(7)(q?)	Ayraud 1975
46,XX,i(2q),6q-,cx	Bartnitzke and Bullerdiek 1983
46,XY,+2,-6,-9,del(22)(q11),+mar	Hansson and Korsgaard 1974
60-62,XX,del(1)(q?),t(1;7),2q+,del(3)(q?),der(5),dmin,cx	Bartnitzke and Bullerdiek 1983
62-64,XY,del(2)(q34),del(6)(q14),ins(7)(q?),20q+,t(5;18), i(5p)	Ayraud 1975

79-80,XX,t(1;?)(q12;?),t(1;2),t(3;?),t(7;?)(q36;?),t(8;?), t(9;?)(q34;?),cx	Bartnitzke and Bullerdiek 1983
81-83,XX,t(1;3)(p22;p25),del(1)(p13),t(5;13)(q13;q14), del(4)(p12),i(2)(p11),i(5)(p11),t(4;5)(q12;q11q13)	Kakati et al 1976a

Large cell carcinoma

60,XY,del(5)(p11),t(6;10)(q11;p11),t(7;11)(p22;q13), t(14;18)(q11;q11),del(2)(q21),t(6;?)(p11;?)	Pickthall 1976

Small cell carcinoma

44,XY,cx,del(1)(q31),t(2;3)(p?;q?),del(3)(p14p23), t(9;13)(p11;q11),del(11)(p15),12q+,del(X)(q22)	Whang-Peng et al 1982b
75,XY,cx,t(2;3)(p?;q?),del(3)(p11),del(3)(p13;q13)	Whang-Peng et al 1982b

Papillary carcinoma

54,XY,+del(2)(q11),t(2;14)(q11;q32),t(3;11)(p13;p15), +i(5p),+8,+9,+12,+16,+17,+21,+2mar	Pathak et al 1982
69,XX,cx,1p+,1q-,1p-,1q-,2q+,3q+,4q+,6p+,12q+,14q+	Wake et al 1980

Adenoma

46,X,t(X;4)(q25;p16),del(2)(p13p21),del(5)(q15q31),-7, t(9;12)(p13;q13),del(11)(q11q14), t(7;9;12)(p11p22q11;p13p24;q13)/45,X,-X	Mark et al 1980

Adenocarcinoma

37,X,-X,-3,-8,-15,del(2)(p23),del(6)(q15),t(11;?)(p12;?)	Trent and Salmon 1981
38,X,-X,-2,-3,-10,-14,-22,del(1)(q25),del(6)(q15), del(14)(p11),dmin	Trent and Salmon 1981
38,X,-X,cx,-17,-22,del(1)(q25),del(2)(q23),del(6)(q15)	Trent and Salmon 1980
40-42,X,-X,t(1;?)(p13;?),del(1)(p12),2q+,del(3)(q21), del(6)(q13),del(7)(q11),t(11;11)(p15;q13), t(12;12)(q24;?),13p+,15p+,i(22q)	Kusyk et al 1981
44,XX,-1,+5,-15,-18,2q+/44,XX,-1,+5,-6,-15,-18,2q+,10p-, +mar	Martin et al 1979
44-50,X,-X,-1,t(1;?)(q21;?),del(2)(q33),4p+,-5,-7,+8, del(10)(q12),t(10;13)(q12;q34),del(17)(q25),-20	Kusyk et al 1982
45-47,XX,2q+,-3,-18,+mar	Couturier-Turpin et al 1982
47,XX,2q+,+16	Couturier-Turpin et al 1982
50,XY,2q+,+5,+10,+11,-16,-18,+20,+20,+mar	Couturier-Turpin et al 1982
72,XX,cx,del(2)(p23),hsr(3)(q13q28),del(6)(q21), t(10;11)(p12;p13)	Trent and Salmon 1981

Adenocarcinoma, metastatic

Karyotype	Reference
??,XX,t(1;16),2q-,del(3)(p14),del(4)(q22),del(6)(q22), i(7q),del(9)(q31),t(13;13)	Ayraud et al 1977
39-41,XX,-3,-7,-8,der(1),der(2),der(4),der(12),cx	Tiepolo and Zuffardi 1973
67,XX,cx,del(2)(p21),dmin	Trent and Salmon 1981

MESENCHYMAL NEOPLASMS

Rhabdomyosarcoma

Karyotype	Reference
83-87,XY,t(2;13)(q37;q14),cx	Seidal et al 1982

Mesothelioma, malignant

Karyotype	Reference
40-46,XY,+16,+20,-22,t(14;15),del(2)(p?),i(5p)	Ayraud 1975
43,XY,t(1;1)(q42;q32p22),t(2;6)(p25;p21),inv(3)(p24q13), t(5;18)(p15;q11),-6,+9,del(13)(q24q14),-14,-18,-22	Mark 1978
86,X?,cx,del(1)(p?q?),2p-,2p+,6q-,14q+	Wake et al 1981

MELANOMAS

Malignant melanoma

Karyotype	Reference
44,X?,+2,+9,-6,-8,-15,-21,9p-,19q+	Wake et al 1981
78-82,XY,t(4;9)(q21;q21),del(9)(p13),t(14;14)(p11;q12), del(2)(q24),t(14;?)(q11;?),t(21;?)(q11;?),del(17)(q11)	Kakati et al 1977

Malignant melanoma, metastatic

Karyotype	Reference
45,XY,t(1;9;13)(q?;q?;q?),r(2),t(1;9)(p?;q?),i(8q), t(11;?)(q?;?),-10,-13,+mar	Chen and Shaw 1973

NEUROGENIC NEOPLASMS

Neuroblastoma

Karyotype	Reference
46,XX,t(1;?)(p32;?),2q+,7p+,13q+,hsr(5q),dmin	Gilbert et al 1982

Retinoblastoma

Karyotype	Reference
46,XX,inv(1),inv(2),inv(3),inv(4),t(6;7),inv(9),t(11;?), 11p-,inv(12),i(17q),-20	Gardner et al 1982
47,XY,del(13)(q14),t(20;?)(q13;?),2q+,i(17q),dmin	Balaban et al 1982
47,XY,dup(1)(q12q44),del(2)(q13),+del(6)(q16), t(14;?)(q32;?),i(17q),+20,-22	Gardner et al 1982
47,XY,t(X;1)(q28;p11),t(2;?)(q11;?),t(2;15)(p25;q11), t(3;?)(q27;?),4p+,t(5;?)(q35;?),i(17q),+mar	Gardner et al 1982

Meningioma, benign

45,XX,-15,-22,1p-,2p-,4p-,+mar	Zankl et al 1979
45,XX,-22/43-44,XX,-2,-16,-22	Mark 1973a
46,XY,2q+	Zang and Zankl 1983

EMBRYONAL AND MISCELLANEOUS NEOPLASMS

Hydatiform mole

47,XX,+2	Honoré et al 1974

LYMPHOMAS

1. UNCLASSIFIED AND MISCELLANEOUS LYMPHOMAS

Non-Hodgkin's lymphoma, NOS

46,X,-X,del(1)(q22),t(2;11)(p21;q21),t(9;15)(q11;p12), del(9)(q11),-14,t(14;17)(q23;q23),-16,+mar	Kristoffersson 1983
46-48,XX,t(1;?)(q21orq22;?),t(2;?)(q37;?),-8,+9,+12, 16q+"s"	Mark et al 1976
52,XY,+2,+3,t(5;9)(q22;q32),+6,+12,+14,+20	Reeves and Pickup 1980

Non-Hodgkin's lymphoma, histiocytic, NOS

45,X,-X,del(1)(q22),t(1;17)(q?;q?),del(2)(q33), del(3)(p21p25),+ins(3)(p14p21p25),+5,del(6)(q14),t(1;7)	Mark et al 1978
46,XX,2p+,-15,del(16)(q13)	Reeves 1973
47,XY,-6,-14,1q-,+1p-,2q+,3q+,9p+,10p-,10p+,12p-,18q+,22p+, t(2;14)(q37?;q13?)	Fukuhara et al 1979b
50,XY,dup(1)(q12q32),+2,4p+,del(4)(q21),t(8;14)(q24;q32), +17p+,18q+,+20	Brynes et al 1978
50,XY,t(2;?)(p25;?),del(2)(p13),+i(3p),del(6)(p21),+7,+8, +9,+9,-11,-13,+14,-15,+20,+21,-22	Mark et al 1978
44,XY,del(2)(p13),t(6;?)(p21;?),del(11)(q13), t(11;14)(q13;q31),i(15q),-20,-22	

Non-Hodgkin's lymphoma, T-cell, NOS

47,XX,t(2;?),inv(3),4p-,del(5)(q14q32),6p+,6q+,t(7;7),9q+, 10p-,13q-,t(14;14),-15,-15,16q-,17p+,+21,+2mar	Gaeke et al 1981

Malignant histiocytosis

48,XX,1p+?,del(2)(p14?),+del(3)(q12?),+mar	Kristoffersson 1983
48,XY,+i(1q),t(2;5)(p21;q35),+mar	Kristoffersson 1983

2. HODGKIN'S LYMPHOMAS

Hodgkin's disease, NOS

47,XY,-2,-5,+16,-17,+3mar — Fleischmann et al 1976

Hodgkin's disease, mixed cellularity

61-80,XX,del(1)(q32),i(1q),del(2)(p13),del(6)(q13or15), del(6)(q23or25),15q+,12q+,del(5)(q31),del(7)(q11), del(10)(q26),del(3)(p13) — Reeves 1973

Hodgkin's disease, lymphocytic depletion

46,XY,+2,t(2;?)(q25;?),t(3;11)(q29;p13),t(4;11)(q33;q13), -13 — Hossfeld and Schmidt 1978

Hodgkin's disease, nodular sclerosis

47,XY,t(1;?)(p36;?),+2,t(2;6)(q31;q27),+8,del(12)(p11), t(17;?)(p13;?),-14,-20,+mar — Reeves and Pickup 1980

3A. NON-HODGKIN'S LYMPHOMAS - RAPPAPORT CLASSIFICATION

Lymphocytic, poorly differentiated, diffuse

45,XY,t(14;18)(q32;q21),+2,-5,-6,-11 — Fukuhara and Rowley 1978
48-49,XX,2p-,+7,+i(8q),+12,14q+,15q-,18q- — Kakati et al 1980

Lymphocytic, poorly differentiated, nodular

52,X,-X,+2,-4,+5,+7,+8,-14,-15,-16,+20,+21, +t(1;15)(p11;q11),+t(12;17)(q24;q11),+t(14;?)(q32;?), +mar — Kaneko et al 1982a
92,XY,t(2;9)(q11;p11),t(5;?)(p15;?),t(17;?)(q25;?),17q+, del(1)(p22),del(3)(p21),del(X)(q24),del(11)(q23) — Reeves 1973

Mixed cell, diffuse

53,XY,t(8;14)(q24;q32),t(14;18)(q32;q21),+20,1q-,+1p-,2q-, 4q+,5q-,5p+,6q-,+7q+,10q+,11q+,11q-,13p+,+4mar — Fukuhara and Rowley 1978

Histiocytic, diffuse

45,X,del(X)(q24),del(1)(q21),+del(1)(p11),del(2)(q22), del(3)(q21),del(6)(p11),del(6)(q15),t(7;9)(p22;q13), t(8;10)(p23;q21),del(9)(q13),del(10)(q21), t(2;11)(q22;q23),-13,del(14)(q24),del(15)(q22), del(16)(q22),+17,-19 — Mark et al 1979
45-47,XX,cx,t(1;2)? — Kristoffersson 1983
45-47,XX,cx,t(1;2)?

46,X,-X,-2,-10,+11,-13,-14,-17,del(1)(p22p32), del(3)(p13p21),del(6)(q21),del(9)(p13),+t(2;?)(p23;?), +t(14;?)(q32;?),t(18;?)(q23;?),+t(X;13)(q26;q12),+mar — Kaneko et al 1982a
47,X,-X,t(4;14)(q?;q32),-18,1q+,2p+,6q-,7q+,9q-,12q+,+3mar — Fukuhara and Rowley 1978
47,XX,-1,-2,-6,+8,-14,+16,-19,del(11)(q23), t(8;14)(q22;q32),+t(2;6)(q21;p23),del(6)(q21), +t(1;14)(p11;q11),+t(19;?)(q13;?),+t(14;?)(q11;?) — Kaneko et al 1982a
47,XY,-10,-13,+15,-16,+19,+20,del(6)(q21),del(8)(q22), t(2;14)(q21;q32),t(4;4)(q31;q35),+t(10;?)(p11;?) — Kaneko et al 1982a
47,XY,-11,-22,del(1)(q42),del(2)(q33),del(5)(q22q31), del(7)(q32),del(9)(p22),t(21;22)(q22;q11), +t(11;?)(q23;?),+t(21;22)(q22;q11),+mar — Kaneko et al 1982a
49,XY,+del(2)(q31),+3,del(10)(q22),+17,+t(10;17)(q22;p13), -19 — Mark et al 1979
50,XY,t(8;14)(q24;q32),+2,+20,1q+,4p+,+4q+,17p+,18q+ — Fukuhara and Rowley 1978
51,XY,+X,-2,+5,+11,-12,+21,del(6)(q21),+t(2;?)(q33orq35;?), +t(12;?)(p11;?),+mar — Kaneko et al 1982a

Histiocytic, nodular

46,XY,-4,inv(16)(p11q24),+t(2;4)(p11;p11) — Kaneko et al 1982a

3B. NON-HODGKIN'S LYMPHOMAS - KIEL CLASSIFICATION

Lymphocytic, T-zone

45,XX,-1,-2,-3,-4,+t(1;2)(p32q32;p13),+t(2;4)(p13;p15), +t(3;4)(q11;?) — Gödde-Salz et al 1981a

Immunocytoma

46,XX,1p+,+2,del(17)(p12),17p+ — Kristoffersson 1983
46,XX,1p+
47,XX,1p+,+12,del(17)(p12)

Centroblastic-centrocytic, diffuse

48-51,XY,del(1)(p33q23q25),-2,+del(3)(p22),+4,4p+,+5,+6p+, +7,+8,+9,+21,+22 — Kristoffersson 1983
45-51,XY,del(1)(p33q23q25),del(2)(q14),+del(3)(p22),+4,+5, +7,+8,-12

Centroblastic-centrocytic, follicular

48,XX,+X,t(1;?)(p36;?),del(2)(p21),t(3;?)(q27orq29;?), t(14;?)(q32;?),t(17;18) — Reeves and Pickup 1980

Centroblastic, diffuse

46-49,XY,1p+,+2,-6,+del(9)(p21),+12,+13,14q+?,i(17q),+18 — Kristoffersson 1983
47,XY,14q+,+22

Lymphoblastic, Burkitt's type

Karyotype	Reference
43,XY,+1,+2,+3,-7,-8,-9,-12,-16,-22,del(1)(q25), del(2)(p11),del(10)(q22),14q+	Biggar et al 1981
45,X,-X,t(2;8)(p11;q24)	Abe et al 1982b
45,XY,-2,5q+,11q+	Biggar et al 1981
46,X,t(X;1)(p11;q25),del(1)(q25),7p+,17p-,t(2;8)(p11;q24)	Slater et al 1982
46,XX,t(2;8)(p12;q23)	Van Den Berghe et al 1979d
46,XX,t(2;8)(p11;q24)	Bornkamm et al 1980
46,XY,del(2)(p12),t(8;14)(q24;q32)	Biggar et al 1981
46,XY,dup(1)(q21q44),del(2)(q32)	Biggar et al 1981
46,XY,t(2;8)(p12;q24)	Miyoshi et al 1979b
46,XY,t(2;8)(p12;q24)	Berger and Bernheim 1982
46,XY,t(2;8;9)(p11;q23;q21q31)	Philip et al 1981
46-47,XY,t(14;?)(q32;?),dup(1)(q21q32),t(2;3)(q13;q29),+5, 6p-,-8,del(9)(q?),+12,+18,+mar	Miyoshi et al 1981c
47,XX,t(8;14)(q24;q32),dup(1)(q24q32),del(2)(q34), del(5)(q31q34),+8"s"	Berger and Bernheim 1982
48-53,X?,t(8;14)(q24;q32),+X,+2,+3,+9,+10,+11,+12,+20	Zech et al 1976a
50,XX,+1,+2,+3,+11,del(2)(p14),11p+,t(8;14)(q24;q32)	Biggar et al 1981

Lymphoblastic, convoluted type

Karyotype	Reference
43-46,XY,del(1)(p32),del(2)(p21),-6,t(7;9)(p11;q11),-9,-10, t(11;15)(p11;q11),-11,del(12)(q14),-13,-14,i(17q),-17, +18,del(19)(p11),-22,+2mar	Kristoffersson 1983
46,X,-Y,t(1;?)(q21orq23;?),+mar/46,X,-Y,-1,-2,+der(1), +t(2;?)(q33orq35;?),+mar/48,X,-Y,-5,-14,+18,+18,+20, del(9)(p13),t(1;14),+dup(5)(q13q35)	Kaneko et al 1982d
46,XY,t(2;9)(q33;q34)	Kaneko et al 1982d
47,XY,t(1;2)(p34;p21),+4,t(13;18)(p13;p11)/47,XY,+3	Kaneko et al 1982d

Immunoblastic

Karyotype	Reference
51,XY,t(1;6)(p22;q13orq15),t(2;13)(q21;q14), t(8;16)(q24;p11),+11,+12,del(15)(q22),+16,+20,+mar	Reeves and Pickup 1980

3C. NON-HODGKIN'S LYMPHOMAS - LUKES AND COLLINS CLASSIFICATION

Mycosis fungoides

Karyotype	Reference
48,XX,del(1)(p22),t(2;8;14)(q37;q24;q24),+5,del(7)(p13), del(9)(q22),t(9;18)(q11;q24),del(10)(p13), del(13)(q12q14),del(5)(q15q22),del(5)(q13),del(5)(q15)	Fukuhara et al 1978a
46,XX,t(1;14)(q32;q32)	

Sezary's syndrome

44-46,XX,t(2;9),t(2;13),17p+,19p+,-21,-22	Edelson et al 1979
46,XY,-2,+mar	Liang et al 1980
46,XY,2q+,der(12),15q+	Nowell et al 1982
50,XX,1q+,2p-,+3q-,+6q-,+9,-15,+21,+mar	Nowell et al 1982

Follicular center cell, diffuse, large cleaved

45,XY,-6,-9,-17,2q-,3p-,3q-,4q-,5q-,6p+,7p+,7q+,+7q-,9p-, 12q+,13p+,20p-,+mar	Fukuhara et al 1978b
47,XY,+Y,t(2;3)(q11;q29)	Fukuhara et al 1978b
47,XY,-6,-14,1q-,+1p-,2q+,3q+,9p+,10p-,10p+,12p-,18q+,19p+, 19q+,+22p+,+mar	Fukuhara et al 1978b

3D. NON-HODGKIN'S LYMPHOMAS - WORKING FORMULATION

Small lymphocytic

47,XY,+2,2q-,17p-,t(18;22)	Yunis et al 1982

Follicular, small cleaved cell

44,XY,+der(1),+2,t(2;13),-3,5q-,-12,t(14;18)(q32;q21), i(17q)	Yunis et al 1982
46,XX,t(2;3),i(6p),t(14;18)(q32;q21)	Yunis et al 1982
46,XY,t(1;2),t(4;10),t(14;18)(q32;q21),t(1;22)	Yunis et al 1982
47,XX,t(X;1),ins(1;8),t(2;8),t(14;18)(q32;q21),+17	Yunis et al 1982
47,XY,+X,t(2;3;17),6q-,t(14;18)(q32;q21)	Yunis et al 1982
48,X,r(Y),+2,+3,t(14;18)(q32;q21),t(1;22)	Yunis et al 1982

Follicular, large cell

47,XX,t(2;13),t(14;18)(q32;q21),+21	Yunis et al 1982
50,XY,1q-,t(2;3),+3,+10,t(1;12),13q-,+15,-19,t(3;19),+21	Yunis et al 1982

Large cell, immunoblastic

45,XY,ins(2;21),t(8;14)(q24;q32),t(9;19),t(17;21),-21	Yunis et al 1982

Chromosome 3

HEMATOLOGICAL DISORDERS

UNCLASSIFIED LEUKEMIAS

Acute leukemia, NOS

Karyotype	Reference
46,XX,dup(1)(p31p36),dup(7)(p15p21),dup(3)(q12q29), ins(12)(q12)"s"	Berger et al 1977
54,XX,+X,+3,+6,+8,+8,+13,+19,+22	Muir et al 1977

NONLYMPHOCYTIC LEUKEMIAS

Acute nonlymphocytic leukemia (ANLL)

Karyotype	Reference
??,X?,+t(1;3)(q21;p11)	Anglani et al 1981
39,X,-X,-3,-5,-13,-14,-15,-16,-17,-21,-22,del(7)(p11), t(12;?)(p13;?),t(16;?)(p13;?),+3mar	Mitelman 1983
39,X,-X,-3,-5,-13,-14,-15,-16,-17,-21,-22,del(7)(p11), t(12;?)(p13;?),t(16;?)(p13;?),+3mar	
43,XX,-1,3p+,3q-,4q+,-5,-5,-7,+9,+9,10p+,+10,+10,10p-,10q-, +11q-,-13,-15,-16,-17,-18,-21"s"	Whang-Peng et al 1979
43,XX,2q-,inv(3),-4,del(5)(q?),-7,17p+,-19	Rowley 1976
44,X,-Y,-3,+4,-12,+15,-20	Bernard et al 1982a
44,XX,t(8;11)(p23;p13 or p15),-20/46,XX,del(7)(q11), t(1;3)(q2?;q2?),-20,-21,-22,+3mar	Philip et al 1978a
44,XY,del(5)(q21),del(7)(q11),del(11)(q11), t(7;11;12)(q11;q11;p13),del(13)(q11),-20,-21/45,XY, t(3;12)(q11;p13),del(5)(q21),del(7)(q11),del(13)(q11), -20	Mitelman 1983
44-45,XX,-3,-5,-6,-18,-19,t(1;2)(q2?;q11),t(5;?)(q13;?)	Philip et al 1978a
45,XX,del(3)(p11),t(5;7;12)(q31;q22;q13-24),t(5;?)(q13;?)	Philip et al 1978a
46,XX,3p-	Bernard et al 1982a
46,XX,del(7)(q22)/46,XX,del(7)(q22),del(5)(q22q31)/46,XX, del(7)(q22),-9,t(1;9)(q22;p24)/46,XX,t(2;3)(q31;q27), del(7)(q22),del(5)(q22q31orq33)"s"	Oshimura et al 1976b
46,XX,ins(3)(p21;q21q29)	Hustinx and Rutten 1983
46,XX,t(1;1)(q42;q21),t(3;7)(q2?;q22)	Philip et al 1978a
46,XX,t(3;11)(q12;p13)"s"	Geraedts et al 1980a
46,XY,3p-,4q-	Nordenson 1983
46,XY,del(3)(q21)	Sasaki et al 1976

Karyotype	Reference
46,XY,t(3;?)(q26;?)"s"	Geraedts et al 1980a
46,XY,t(9;10)(p24;p12),t(3;16),t(4;15)(p13;q14)"c"	Van Den Berghe et al 1979a
47,X,Xq-,t(3;5),-5,t(11;?),t(12;17),-18,-21,+6mar	Pedersen-Bjergaard et al 1980
47,XX,3p-,+mar	Chrz et al 1983
47,XX,t(3;5)(q21;q31),t(1;13)(p36;q14),+8,+22	Oshimura et al 1976b
47,XX,t(9;22)(q34;q11),+22q-/48,XX,t(9;22),+8,+22q- 49,XX,t(9;22)(q34;q11),+8,+13,+22q- 49,XX,t(9;22)(q34;q11),+8,+22q-,+mar 50,XX,t(9;22)(q34;q11),dup(1)(q25q44),+del(3)(p11),+8, +22q-,+mar	Mitelman 1983
48,XX,3q+,-5,+9,+16,+19	Nordenson 1983
48,XY,+del(3)(q11),+7,t(6;19)(q15;q13)/50,XY,+del(3)(q11), +7,t(6;19)(q15;q13),+2mar/51,XY,+del(3),+7,+9,t(6;19), +2mar	Mitelman 1983
48,XY,del(3)(q13),-5,+8,+r(?),+mar	Mitelman 1983
51,XY,+3,+5,-9,+19,+20,+21,t(6;14)(p25;q11),t(9;?)(q22;?)	Philip et al 1978a
56,XY,+Y,+6,+14,+21,+t(5;6)(q?;p?),+del(1)(p22q25), +del(2)(q12),+del(3)(p12),+del(4)(q13),+del(5)(p14q14), +mar	Zuelzer et al 1976

ANLL, FAB type M1

Karyotype	Reference
44,XX,-5,-7,-20,-22,+2mar/44,XX,-3,-11,-11p+,-22	Prieto et al 1981
45,XY,-3,3p+	Prieto et al 1981
45,XY,3p-,5q-,-7"s"	Pedersen-Bjergaard et al 1981
46,XX,-1,inv(1),3q+,5q-,t(13;17),-18,+mar/51,XX,-1,inv(1), 3q+,5q-,t(13;17),-18,+6,+9,+9,+11,+11,+mar	Hagemeijer et al 1981a
46,XX,inv(3)(p11q11)	Prieto et al 1981
46,XY,del(3)(p13)	Yunis et al 1981
46,XY,inv(3)(q21q26)	Bernstein et al 1982a
46,XY,t(3;3)(q26;q29)	Bernstein et al 1982b
46,XY,t(3;5;10)(p?;q?;p?)	Lessard and Le Prise 1983
47,XY,+3	Prieto et al 1981
50,XX,+3,+19,+t(1;6)(q21;p22)	Yunis et al 1981

ANLL, FAB type M2

Karyotype	Reference
43,XX,+2,3q+,-5,-20,-22	Berger et al 1981a
43,XX,-3,-4,-7,-19,del(5)(q14),17p+ 43,XX,-3,-4,-7,-19,del(2)(q31?),del(5)(q14)	Testa et al 1979
43,XY,-3,-4,-7,-7,-8,-8,-17,-20,+5mar	Berger et al 1981a
45,XX,-7,ins(3;3)(q21;q21q26)"s"	Rowley et al 1981a
45,XX,-7,t(3;9)(q29;p21)/44,X,-X,-7,t(3;9)"s"	Rowley et al 1981a
45,XY,-21/46,XY,+3,-21	Prieto et al 1981
45,XY,-5,-7,del(3)(p13),+t(5;17)(q11;q11)"s"	Rowley et al 1981a
45,XY,-7,del(3)(p13),del(8)(q22),del(16)(p12)"s"	Rowley et al 1981a
46,XX,1q+,2q+,-3,-11,+r,+mar	Berger et al 1981a
46,XX,t(3;13)(p14;q34)	Lessard and Le Prise 1983
46,XY,del(3)(p21)	Alimena 1983
46,XY,inv(3)(q21q26)	Bernstein et al 1982a
47,XY,+3,t(7;8)(p11;q11)	Kaneko et al 1982b

ANLL, FAB type M1+M2

Karyotype	Reference
43,XY,-3,-5,-21	Rowley and Potter 1976
44,X,-X,-5,t(3;7)(q13;p12)/44,XX,-7,-8/44,XX,-5,-8"s"	Sandberg et al 1982a
44,X,-Y,-3,6q-,+mar	Shiraishi et al 1982
44,XX,-3,-4,-5,+9,-17,-18,-21,-22,+5mar	Garson 1980
44,XX,del(3)(p21),del(5)(q12),-7,-20,i(22)(q11)	Mitelman et al 1981
45,XX,ins(3;3)(q26;q21q26),-7"s"	Sweet et al 1979
45,XX,t(3;6)(p23;p25),del(5)(q13),-12,i(17q)	Mitelman et al 1981
46,XX,r(3)(p26q29)	Shiraishi et al 1982
47,X,-Y,t(1;13)(q32p32q11;q11),del(3)(q21), t(8;21)(q22;q22),del(9)(q22),+8,+18	Sakurai et al 1982b
47,XX,del(3)(q21),+8	Fitzgerald et al 1983
47,XY,del(3)(q12),del(5)(q12),t(6;?)(q11;?),-13,+2mar"s"	Testa et al 1981b
47,XY,i(3q),+7	Nordenson 1983
48,X,-Y,1p-,+3p-,+15,+mar"s"	Sandberg et al 1982a
48,XX,+3,+8	Mitelman et al 1981
48,XX,t(1;3)(q44;p14),del(3)(q21p14),-5,+10,+11,+21	Mitelman et al 1981

ANLL, FAB type M3

Karyotype	Reference
45-51,XY,t(3;12)(p14;q24),t(13;18)(q11;p11),+1,+6,+16,+19, +21,+22	Mitelman et al 1981
46,XY,t(15;17)(q?;q?),t(3;11)(q?;q?),t(7;19)(q11;p13)	Trujillo and Cork 1981
46,XY,t(3;18),t(3;12),3p-	Teerenhovi et al 1978
46,XY,t(3;5;17)(p21;q25?;q21?),4p+	Bernstein et al 1982b

ANLL, FAB type M4

Karyotype	Reference
43,XY,-2,-3,-6,-7,-12,-15,-21,-22,del(2)(p11),del(11)(p14), t(14;?)(q32;?),t(17;?)(p13;?),del(22)(q11),+5mar"s"	Rowley et al 1981a
44,XY,del(3)(p22),-9,-12,t(17;?)(p13;?)"s"	Mitelman 1983
44,XY,t(3;5)(q29;q13),t(7;12)(q12;q14),t(7;10)(q12;q21),-8, -17,-18,-20	Mitelman et al 1981
46,XX,-1,-3,-5,+10,+11,-13,-14,+21,+del(3?)(p?q?), ?t(1;3;5)(p?;p?q?;q?)	Mitelman 1983
46,XX,ins(3;3)(q21;q21q26)	Rowley and Potter 1976
46,XX,t(3;13)(q29;q12),del(5)(q13q31)/46,XX, t(1;18)(q44;q21),t(3;13),del(5)	Alimena 1983
46,XY,t(3;12)(p14;q24)	Sandberg et al 1982b
46,XY,t(3;12)(p21;q34)	Prieto et al 1981
46,XY,t(3;12)(q21;q13),del(5)(q14),ins(8;5)(q22;q14q22), -17,+2mar	Mitelman 1983
46,XY,t(3;22)(q21;q13)	Alimena 1983
46,XY,t(3;5)(q25;q33)	Rowley and Potter 1976
49,XX,del(3)(q25),del(7)(q22),+8,+17,+19	Alimena 1983

ANLL, FAB type M5

Karyotype	Reference
45,XX,der(3),-7"s"	Prigogina et al 1979
46,XX,-18,-22,+2r/46,XX,t(2;9)(q23;p22),-3,-22,+2r/49,XX, t(2;9),-22,+4r	Alimena 1983
46,XY,t(3;9;13)(p21;q22;q14)	Benedict et al 1979

46,XY,t(6;11)(q26;q22)/52-53,XY,t(6;11)(q26;q22),+der(6), +3,+10,+13,+18,+19,+21,dmin — Hagemeijer et al 1981a
46,XY,t(6;11)(q27;q23)/51-52,XY,t(6;11),+der(6),+3,+19,+19, +21,dmin/53,XY,t(6;11),+der(6),+4,+10,+13,+18,+19,+21 — Löwenberg et al 1982
46-51,XX,-3,-18,-22,t(2;9)(q23;p22),+2-4r — Mitelman et al 1981
47,XY,1p+,1q+,2p+,-3,4q+,+13,+mar — Kaneko et al 1982b
51-52,XX,+3,+6,+9,-10,+18,+19,del(1)(p22),del(1)(p22), dup(11)(q11q21),+t(10;?)(p13;?) — Kaneko et al 1982b
57-58,XX,+1,+3,-4,-5,+6,-8,+11,+12,-17,+19,-20, del(10)(q24),+8-9mar — Testa et al 1979
60-61,XX,Xp+,+1,+2,-4,+6,-7,+19,+20,del(2)(q31), del(10)(q24),+12-13mar

ANLL, FAB type M4+M5

48,XY,+i(1q),+3,+12,-14 — Shiraishi et al 1982

ANLL, FAB type M6

43,X,-Y,-4,-13,-14,-15,+2mar/46,X,-Y,-3,+9,+mar — Li et al 1983
43,XY,-3,-15,-15"s" — Sandberg et al 1982a
43-45,XX,der(1),inv(2)(p25q14),del(3)(q25),-4,-5,-6,6q-,-7, 7p+,11p+,-12,-17,-21,+r,+1-4mar"s" — Smadja et al 1983b
44,X,del(Y)(q?),+Xq-,3p-,3q-,4q-,-6,+7q-,11q+,-15,-18"s" — Bradley et al 1982
44,XY,-3,-18 — Yamada and Furusawa 1976
44,XY,-3,5q-,-7,-D,+20 — Li et al 1983
44,XY,3p-,t(2;6),t(5;6),t(17;20)"s" — Sandberg et al 1982a
44,XY,t(3;5;11),17p+,-22 — Garson 1980
44-47,XY,+2,-3,-5,-7,+8,-14,-19,+2mar — Mitelman et al 1981
45,XY,del(5)(q11),-7/45,XY,del(5)(q11), ins(3;3)(q12;q21q27)"s" — Swolin et al 1981
46,XY,dup(1)(p32p34),dup(3)(q12q27),del(7)(p11)"s" — Berger et al 1980a
46,XY,t(3;5)(q13;q28) — Hagemeijer et al 1981a
47,XY,+Y,3p-,5q-,10q- — Garson 1980

Chronic myeloid leukemia, t(9;22)

44,X,-X,t(9;22),1q+,-3 — Sadamori et al 1980
45,XY,t(9;22),+del(1)(q12),-3,-7,del(11)(q21),-19,+mar — Lyall and Garson 1978
46,X?,t(9;22),t(3;20) — Carbonell et al 1982a
46,XX,t(9;22) — Mitelman 1983
46,XX,t(9;22)(q34;q11)/48,XX,t(9;22),+3,+8
44,XX,-5,del(6)(q16),-8,-14,-15,-18,-18,-20,+21,-22,-22, +6mar
44,XX,-5,del(6)(q16),-8,-14,-15,-18,-18,-20,+21,-22,-22, +6mar
46,XX,t(9;22),del(3)(p23)/46,XX,t(9;22) — Hartley and McBeath 1981
46,XX,t(9;22),der(3),del(15)(q22) — Fleischman et al 1981
46,XX,t(9;22),i(17q)/47-49,XX,t(9;22),i(17q),+3,+8 — Prigogina et al 1978
46,XY,t(9;22)(q34;q11) — Ishihara et al 1982
27,X,-Y,t(9;22),-1,-2,-3,-4,-5,-6,-7,-9,-11,-13,-14,-15, -16,-17,-18,-19,-20,-22
46,XY,t(9;22),t(X;9)(p22;q22),t(2;3)(p22;q27) — Fleischman et al 1981
47,XY,t(9;22),+22q-/46,XY,t(9;22),-9,+22q-/46,XY,t(9;22), t(3;5)(p?;q?),t(6;13)(p?;q?) — Kohno and Sandberg 1980

48,XX,t(9;22)(q34;q11),+8,+19/49,XX,t(9;22),+3,+8,+19/50, XX,t(9;22),+3,+8,+19,+22q- Mitelman 1983
48,XY,t(9;22),+3,+17 Stoll and Oberling 1978
48,XY,t(9;22),t(3;4)(p?;p?),+21,+22q- Sadamori et al 1980
49,XX,t(9;22),+3,+8,+22q- Seabright 1983
53,XY,+Y,t(9;22)(q34;q11),+3,+6,+8,+9,+10,+12,-15,+22q-/56, XY,+Y,t(9;22),+3,+6,+8,+8,+9,+10,+16,+19,+22q- Stoll and Oberling 1982
56,XY,+Y,t(9;22),+3,+6,+8,+8,+9,+10,+12,+19,+22q-
55,XX,t(9;22),+3,+8,+12,+14,+15,+19,+20,+22,+22q- Lilleyman et al 1978
55,XY,t(9;22),+3,+6,+8,+8,+9,+10,+16,+19,+22q-/53,XY, t(9;22),-15,+3,+6,+8,+8,+9,+10,+12,+22q- Stoll and Oberling 1978

Chronic myeloid leukemia, aberrant translocation

??,X?,t(3;22)(p21;q11) Potter et al 1981
??,X?,t(3;4;9;11;22) Potter et al 1981
??,X?,t(3;9;22)(p21;q34;q11) Potter et al 1981
??,X?,t(3;9;22)(q11orq12;q34;q11) Tanzer et al 1980a
46,X?,t(3;9;22) Nowell 1983
46,X?,t(3;9;22)(p21;q34;q12) Rozynkowa 1983
46,X?,t(3;9;22)(p21;q34;q12) Rozynkowa 1983
46,X?,t(3;9;22)(q22;q34;q11) Van Den Berghe 1983
46,XX,t(3;22)(p21;q11)/48,XX,t(3;22),+19,+21 Pravtcheva et al 1976
46,XX,t(3;4;9;22)(q21;q31;q34;q11) Chessells et al 1979
46,XX,t(3;9;22)(p21;q34;q11) Nowell et al 1975
46,XX,t(3;9;22)(p21;q34;q11) Anderson et al 1978
46,XX,t(3;9;22)(p21;q34;q12) Rozynkowa et al 1977
46,XX,t(3;9;22)(p21;q34;q11) Mohandas et al 1980
46,XX,t(3;9;22)(p21;q34;q11) Seabright 1983
46,XX,t(3;9;22)(q21;q34;q11) Pasquali 1983
46,XX,t(3;9;22)(q23;q34;q11) Borgström 1981
46,XY,t(19;22)(q13;q11),t(3;21)(p11;p11) Seabright 1983
46,XY,t(3;9)(q21;q34),t(17;22)(q21;q11) Oshimura et al 1982a
46,XY,t(3;9;10;22)(p13;q34;q22;q11),17q+ Seabright 1983
46,XY,t(3;9;22)(p21;q34;q11) Lessard and Le Prise 1982
46,XY,t(3;9;22)(q21;q34;q11) Pasquali et al 1979b
46,XY,t(3;9;22)(q21;q34;q11) Pasquali 1983
46,XY,t(3;9;22)(q23;q34;q11) Seabright 1983
46,XY,t(4;22)(q35;q21),t(3;5)(q27;q22) Sessarego et al 1981b
46,XY,t(4;22)(q35;q21),t(3;5)(q27;q22),i(17q)/50,XY, t(4;22),t(3;5),i(17q),+8,+15,+21,+22q-

Chronic myeloid leukemia, Ph1 negative

46,X,t(Y;3;9)(q12;q25;q34) Verhest et al 1980
46,X?,t(3;9)(q12;q34) Verhest et al 1980
46,XX,t(3;11)(q21;p15) Mitelman 1983
46,XX,t(3;11)(q21;p15)/49,XX,t(3;11)(q21;p15),+8,+19,+21
46,XX,t(3;11)/49,XX,t(3;11),+8,+19,+21
46,XY,t(1;6)(p36;q15),del(3)(q25),del(17)(p11) Srivastava et al 1981
46,XY,t(3;6)(q?;p?),t(17;21) Borgström et al 1982
46,XY,t(3;7)(q29;q11) Altman et al 1974
47,XY,t(3;4)(p14;q11q31),+8 Kohno et al 1979a

Eosinophilic leukemia

Karyotype	Reference
49,XY,+10,+15,+19,del(3)(q?)	Huang et al 1979

MYELOPROLIFERATIVE DISORDERS

Polycythemia vera

Karyotype	Reference
44,XY,-3,-5	Nowell and Finan 1978
46,X?,3q+,8p+,8q+	Wurster-Hill 1983
46,XX,t(2;11)(p13;q21)/46,XX,t(1;15)(p1?;q1?), t(2;11)(p13;q21),del(16)(p12)/51,XX,+3,+8,+8,+19,t(1;15), t(2;11),del(16)	Testa et al 1981b
46,XY,t(3;7)(p?;q?)	Lessard and Le Prise 1983
48,XY,+3,-5,t(13;14)(q?;q?),+2mar	Nowell and Finan 1978

Myelosclerosis / Myelofibrosis

Karyotype	Reference
45,XY,-7,t(3;3)(q?;q?)	Nowell and Finan 1978

Myeloproliferative disorder, special type

Karyotype	Reference
45,XX,t(1;?)(p36;?),-5,-7,-12,t(13;?)(q34;?),-17,-22, +4mar/45,XX,t(1;?),-2,-3,-5,-7,-12,-17,-22,+8mar"s"	Rowley et al 1977a
46,XX,t(3;12)(q29;q24)	Sandberg et al 1982b
46,XY,t(1;13)(p36;q13),del(3)(p21),-5,inv(7)(p11q22), t(16;16)(q22;q24),+mar	Clare et al 1982

DYSMYELOPOIETIC SYNDROMES

Preleukemia, NOS

Karyotype	Reference
43,X,-X,-3,-13,4q-,8q-,12q+	Geraedts et al 1980a
43,X,-Y,-3,5q-,-7,-10,-16,-22,+3mar"s"	Pedersen-Bjergaard et al 1981
44,XY,+Y,-3,-5,+del(6)(q21),-8,t(17;?)(p13;?),+18,-20,+22	Mitelman 1983
44,XY,t(3;17),-5,-20,+mar"s"	Pedersen-Bjergaard et al 1981
45,XX,-2,-3,5q-,-6,-9,16p-,17p+,+3mar	Swansbury and Lawler 1980
46,XX,t(1;3;11)(p1?;q2?;q?)	Ruutu et al 1977b
46,XX,t(3;12)(q21;p13),del(7)(q11),del(11)(q11), t(7;11)(q11;q11),t(11;12)(q22;p13)	Mitelman 1983
46,XY,t(3;12)(q?;p?)	Anderson and Bagby 1982
47,XY,+8,t(3;6)(p21;p12),+del(3)(p21)	Panani et al 1977
47-48,XY,+3,+9,+12	Panani et al 1977
47-48,XY,+3,+9	Panani et al 1977
52,XY,+1,+3,+21,+3mar	Panani et al 1980

Chronic myelomonocytic leukemia

46,X?,t(3;16)(p13;q22)	Geary et al 1975

Acquired idiopathic sideroblastic anemia

46,XX,del(5)(q12q23)/45,XX,3p-,4q+,6q-,-22"s"	Kerkhofs et al 1982

Refractory anemia without excess of blasts

45,XY,3q-,5q-,-7	Teerenhovi et al 1981
46,XY,1p-,3q-,5q-	Teerenhovi et al 1981

Refractory anemia with excess of blasts

43,XX,3q-,del(5)(q13q33),7q-,-12,14p+,16q+,-17,-18,18q+	Kerkhofs et al 1982
44-45,XX,del(3)(q?),-7,-16,-18,+1-2mar	Mitelman 1983
44-47,XY,-3,-5,-7,-12,-16,-20,del(1)(q32),3p+,del(11)(q22), 3-6mar,+r	Streuli et al 1980
44-50,XY,-3,-5,del(3)(p23),+mar	Streuli et al 1980

Pancytopenia

46,XX,-4,-5,-6,-7,-11,-13,-16,+19,+20,t(7;13)(q?;q?), +4mar/43,XX,-2,-3,-5,-7,-16,+19,15q+,+mar	Nowell and Finan 1978

Dysmyelopoietic syndrome, special type

44,XY,-5,-7,-17,del(3)(q13),+del(3)(p13),del(6)(q13)"s"	Rowley et al 1981a
45,XY,-7,t(2;3)(p21;q28)"s"	Rowley et al 1981a

Thrombocytosis

46,XX,ins(3;3)(q27;q21q27)	Norrby et al 1982
46,XX,ins(3;3)(q27;q21q27)/46,XX,ins(3;3),del(7)(q31)	

SPECIAL LEUKEMIAS

Hairy cell leukemia

47,XY,-2,3q+,14q-	Khalid et al 1981

LYMPHOCYTIC LEUKEMIAS

Acute lymphocytic leukemia (ALL)

28,XX,-1,-2,-3,-4,-5,-7,-8,-9,-11,-12,-13,-14,-15,-16,-17,-19,-20,-22	Kaneko and Sakurai 1980
32,XY,-2,-3,-4,-5,-7,-8,-11,-12,-13,-14,-15,-16,-17,-20	Shabtai and Halbrecht 1981b
46,XX,t(3;11)(p11;p1?)	Inoue et al 1977
46,XY,-8,del(3)(q12q25),t(8;14)(q24;q32),+t(1;8)(p11;q11)	Kaneko et al 1980
46,XY,3q-,6q-,19q+	Dewald et al 1978
46,XY,del(3)(p?),t(9;22)(q34;q11)	Mitelman 1983
46,XY,del(3)(p?),t(9;22)(q34;q11)	
46,XY,t(8;22)(q24;q12),t(3;11)(p12p21;q23),inv(2)(p11q13),del(2)(p21)/46,XY,t(8;22)(q24;q12),dup(1)(q23q44)"s"	Fonatsch et al 1982
47,X,del(Y)(q11),del(3)(q11),t(Y;12)/50,X,del(Y)(q11),del(3)(q11),del(1)(p31),t(Y;12),+4mar	Zuelzer et al 1976
48-53,XY,3q+	Lessard and Le Prise 1983
49,XX,+10,-21,+3r"c"	Stern et al 1979b
54,XX,7p+,+10,+10,+14,+14,+18,+18,+21,+21/27,X,-X,-1,-2,-3,-4,-5,-6,-7,7p+,-8,-9,-11,-12,-13,-15,-16,-17,-19,-20,-22	Oshimura et al 1977a
54-57,XY,+3,+5,+6,+7,+12,+13,+14,+21,+2mar	Prigogina et al 1979
55,XY,+X,+3,+6,+10,+13,+14,+16,+21,+21	Oshimura et al 1977a
56,XX,+3,+6,+12,+13,+14,+15,+18,+21,+r	Prigogina et al 1979
57,XX,+3,+4,+8,11p+,+14,+16,+17,+18,+21,+22,+2mar/47,XX,+8	Morse et al 1978
61,XX,+X,+1,+2,+3,+4,+5,+6,+8,del(6)(q21),+13,+14,+15,+16,+18,+21,+22	Oshimura et al 1977a

ALL, FAB type L1

26,XX,-1,-2,-3,-4,-5,-6,-7,-8,-9,-10,-11,-12,-13,-14,-15,-16,-17,-19,-20,-22	Hoeltge et al 1982
28,XX,-1,-2,-3,-4,-5,-6,-7,-8,-9,-11,-12,-13,-15,-16,-17,-19,-20,-22/56,XX,+X,+X,+10,+10,+14,+14,+18,+18,+21,+21	Brodeur et al 1981a
34-36,XY,-2,-3,-4,-7,-9,-12,-13,-14,-16,-17,-20,+22	Sandberg et al 1982c
46,XX,t(1;19)(q21;q13),del(3)(q23),del(11)(q21)	Kaneko et al 1982e
46,XY,del(3)(q12q25),t(8;14)(q24;q32),-8,+t(1;8)(p11;q11)	Kaneko et al 1982e
47,XY,t(22;?),-1,-1,-2,-3,-9,+6mar	Sandberg et al 1980

ALL, FAB type L2

26,XX,-1,-2,-3,-4,-5,-6,-7,-8,-9,-10,-11,-12,-13,-15,-16,-17,-18,-19,-20,-22	Brodeur et al 1981a
46,XX,3q+	Borgström et al 1981

ALL, FAB type L3

46,XY,t(8;14)(q23;q32),del(3)(p21)	Lessard and Le Prise 1983
46,XY,t(8;14)(q24;q32),del(3)(p24),18p+	Berger and Bernheim 1982
46,XY,t(8;14)(q24;q32)/46,XY,t(3;8;14)(q14;q24;q32)	Berger and Bernheim 1982
47,XY,ins(1),t(3;22)(q?;q?),+7	Penchansky et al 1981

Chronic lymphocytic leukemia

Karyotype	Reference
44,XX,-3,14q+,-17	Najfeld et al 1980
46,XX,-3,t(2;3),t(14;18)(q32;q?),2q-,17q+	Finan et al 1978
46,XY,t(11;14)(q13;q32)/46,XY,t(3;4;9)(p12;q21q31;q31)	Robert et al 1982
46,XY,t(3;6)(q12;q26),t(11;14)(q13;q32)	San Roman et al 1982
47,XXY,t(3;8),t(11;14)(q11;q32),+i(18q)"c"	Finan et al 1978
47,XY,+3,+12,-20,/47,XY,+9,+12,-20/45,XY,-20/46,XY,+12,-20	Morita et al 1981
48,XX,+3,+12/47,XX,+12/46,XX,-11,+12/45,XX,-11	Morita et al 1981
48,XX,+3,+18,t(7;10)(p?;q?)	Hurley et al 1980

Lymphocytic leukemia, special type

Karyotype	Reference
45,Y,-X,t(Y;14)(q12;q32),-13,-17,del(2)(q33), t(3;13)(q21;q34),del(6)(q16orq22),9p+,9q+,10q+,16q+,18q+	Miyoshi et al 1981a
48,XX,+3,-6,+7,del(10)(p13q24),10q+,+mar	Ueshima et al 1981
50,XY,+3,+6,+2mar	Miyamoto et al 1982b

MONOCLONAL GAMMOPATHIES

Multiple myeloma

Karyotype	Reference
44,XY,del(3)(q25),t(5;12)(q15;q14),del(7)(q22), del(11)(q24),-14,-18,+mar	Mitelman 1983
45,X,-Y,-1,+t(1;3)(q3?;q2?),-3,+11,14q+/45,X,-Y,-1,-1, +t(1;1)(q2?;q3?),14q+	Philip et al 1980
45,X,del(X)(q25),t(3;8;22)(q13;q34;q11),-5,-7,-10,-11, del(13)(q13?),+16,19q+,+3mar	Van Den Berghe et al 1979f
46,XX,3p-	Chrz et al 1983
47,X,-Y,+t(1;15)(q12;p11),+3,+11,-15,+19,-22/47,X,-Y,+3, +11,+19,-22	Philip et al 1980
51,XY,t(3;?),+4mar	Philip et al 1980
52,X,-X,+t(1;?)(q12;?),+3,+7,+8,+9,+11,-13,+18,-21,+2mar	Philip et al 1980
52,XY,-1,+t(1;15)(q12;p1?),+t(1;16)(p36;p13), t(3;5)(q2?;q13),+3,+5,+7,+8,+9,+11,-15,+18,-20	Philip et al 1980
52,XY,-1,+t(1;?)(p31;?),+3,+2mar	Philip et al 1980
53,XY,+3,+5,-7,-8,+9,+11,-14,+20,del(1)(p34),del(6)(q25), +5mar	Liang et al 1979
55,XY,+1,+5,+9,+11,-12,+13,+15,del(1)(q32),del(6)(q25), t(9;19)(q13;q13),14q+,16p+,+4mar	Liang et al 1979
45,XY,-2,-2,-3,+5,-8,-10,+14,-15,-16,del(6)(q25),+4mar	
55,XY,+3,+5,+7,+9,+11,+14,+15,+21,+21/53,XY,+3,+5,-8,+9, +11,+14,+15,-19,+21,14q+,+3mar	Liang et al 1979

Macroglobulinemia

Karyotype	Reference
47,XX,+del(3)(p25)	Contrafatto 1977
47,XY,+del(3)(p25)	Contrafatto 1977

Plasma cell leukemia / Plasmocytoma

Karyotype	Reference
78,XY,cx,del(1)(q21),t(1;17)(q12;p13),3q+,3p-,9p+,11p+, 14q+,16p+	Ueshima et al 1983

SOLID TUMORS

UNCLASSIFIED NEOPLASMS

EPITHELIAL NEOPLASMS

Carcinoma, NOS

Karyotype	Reference
43,X,-X,t(1;11)(q21;q23),3q+,4q+,5q+,-11,-13,-16,17p+	Atkin and Baker 1982a
44,XX,ins(1;1)(q32;q12q31),i(2q),+3,t(9;11)(q13;p15),-12,-15,17p+,-18	Atkin and Baker 1982a
44-45,XX,t(1;4)(q11;q35),+3,5q-,-14,i(17q),-21	Atkin and Baker 1982a
44-50,XX,+3,dup(1)(q21q32),t(10;11)(q26;q25),t(2;17),t(6;13),t(9;15),del(10)(q24)	Riet-Fox et al 1979
46,XX,t(1;?)(p32;?),-2,3q-,5q-,6q+,-18,-21,+mar	Atkin and Baker 1982a
46,XX,t(3;11)(q12;q14),-6,+7,-8,-15,t(16;?)(q23;?),+2r	Mark 1975b
46,XX,t(3;17)(p24;p13)/46,XX,t(5;17)(q22;p13)/46,XX,t(9;18)(q13;q21)/46,XX,t(12;15)(q13;q22)	Stenman et al 1982
47,XX,+1,+3,t(11;14)(q?;q?),-12,-21,+r	Atkin and Baker 1982a
48,XX,+3,t(1;20)(p13;p13),+i(1q)	Kovacs 1978a
49,X,-Y,-3,-5,-14,+15,-16,-17,+19,+21,+22,1q+,5q-,6q-,8p-,9p-,11p-,18p+,20q-,22q-,t(Y;14),+3mar	Reichmann et al 1981
49,XX,+3,+11,-7,i(8q)	Riet-Fox et al 1979
49,XX,t(3;11)(q?;q?),+5q-,+6q-,+i(11q)	Atkin and Baker 1982a
49,XY,+3,+8,+21	Sonta and Sandberg 1978a
50,XX,inv(3)(p13q25),+8,+14,+21/63,XX,inv(3),del(11)(p13),cx	Sonta and Sandberg 1978a
57-61,XX,cx,t(11;16)(q?;q?),inv(3)(p14q29)	Riet-Fox et al 1979
58,XX,+X,+1,+2,+3,5p-,+6,+7,+9,11q+,-13,-14,-15,+16,+19,+20,+6mar	Kovacs 1978a
58,XY,del(3)(p14),+4,+7,+8,+10,+11,+12,+14,+15,+17,+18,+19,+21	Kovacs 1978a
60,XY,cx,3p+,6q+,8p-	Reichmann et al 1981
60-63,XX,cx,t(5;18),t(3;6),t(3;22),del(10)(q24),del(4)(p1?)	Riet-Fox et al 1979
63,XX,-1,+3,+5,+6,+7,+8,+10,+11,+12,-13,-15,+20,+21,+22,+9mar	Sandberg 1977
63-66,XX,cx,1p+,16q+,del(3)(p13)	Riet-Fox et al 1979
71-75,XX,cx,11p+,17q+,del(1)(p31),del(3)(p13),del(4)(p1?)	Riet-Fox et al 1979
79,XX,cx,i(1q),t(1;3)(p11;q29),12q+,15p+,i(5p)	Kovacs 1978a
79,XX,cx,i(1q),t(1;3)(p13;q29),t(1;14)(p13;p13),t(12;?)(q24;?),t(15;21)(p11;q11),i(5p),r	Kovacs 1981
81,XY,cx,3p-,dmin	Reichmann et al 1981

Carcinoma NOS, metastatic

Karyotype	Reference
??,XX,1p-,t(3;?)(q?;?),cx	Bartnitzke and Bullerdiek 1983
35,XX,-1,-3,-4,-5,-7,-8,-10,-11,-12,-13,-14,-15,-16,-17,-18,-19,-20,-21,-22,+8mar	Pathak et al 1979
41-42,XX,t(1;18)(p13orp21;q11),3p-,t(3;?)(p25;?),t(8;?)(p2?;?),t(9;?)(q34;?),cx	Bartnitzke and Bullerdiek 1983
43,XX,cx,t(1;3)(q?;q?)	Jones Cruciger et al 1976

45,X,Xq-,t(1;?)(p11;?),t(1;?)(q23;?),t(3;?),t(5;?)(q2?;?), del(9)(p2?),cx	Bartnitzke and Bullerdiek 1983
45,X?,+16,-18,1p-,3p-,16p-	Wake et al 1981
46,XX,t(9;22)(q34;q11)"s"	Togawa et al 1981
52,XX,+3,+5,+7,+8,+9,+11,12q+,13q-,-14,+16,i(17q)"s"	
48,XX,t(3;?)(q2?;?),t(7;14)(p21;q1?),cx	Bartnitzke and Bullerdiek 1983
60-62,XX,del(1)(q?),t(1;7),2q+,del(3)(q?),der(5),dmin,cx	Bartnitzke and Bullerdiek 1983
64-100,XX,t(6;?)(p11q27;?),t(4;10)(q?;q?),t(3;?)(q25;?), del(1)(q2?),t(6;11)(p21;q13)	Kakati et al 1975
70,XX,del(5)(p21),del(3)(p?),cx	Ayraud 1975
79-80,XX,t(1;?)(q12;?),t(1;2),t(3;?),t(7;?)(q36;?),t(8;?), t(9;?)(q34;?),cx	Bartnitzke and Bullerdiek 1983
81-83,XX,t(1;3)(p22;p25),del(1)(p13),t(5;13)(q13;q14), del(4)(p12),i(2)(p11),i(5)(p11),t(4;5)(q12;q11q13)	Kakati et al 1976a
85-90,XY,i(9q),del(1)(q2?),t(7;?)(q35;?),t(3;?)(p11;?), t(21;?)(q11;?)	Kakati et al 1975
100,XY,t(12;16),inv(3)(q?),del(1)(q?),del(1)(p?),cx	Ayraud 1975

Large cell carcinoma

55-70,XY,t(4;5;14),del(1)(q11q21),t(7;11)(p22;q13), t(7;21)(q11;q22),t(3;8)(p?;q?),t(6;11)(p11;p11),i(16q), t(3;?)(q11;?),t(21;?)(q11;?),t(13;?)	Kakati et al 1975

Small cell carcinoma

40,XY,cx,t(1;16)(q21;q24),t(1;19)(q23;q13),del(3)(p21p24), del(3)(p23q26),del(10)(q22),del(11)(q23), t(5;13)(q13;q32),del(17)(p12),dmin	Whang-Peng et al 1982b
44,XX,cx,del(3)(p14p23),t(14;14)(p11;q11)	Whang-Peng et al 1982b
44,XX,cx,del(3)(p14p23),t(3;19)(p13;p11)	Whang-Peng et al 1982b
44,XY,cx,del(1)(q31),t(2;3)(p?;q?),del(3)(p14p23), t(9;13)(p11;q11),del(11)(p15),12q+,del(X)(q22)	Whang-Peng et al 1982b
44-45,XX,1p+,del(3)(p?q?)	Wurster-Hill and Maurer 1978
47,XY,+del(1)(p?),+del(3)(q?),+del(3)(p?),-4,+8,-15,+18, -22	Wurster-Hill and Maurer 1978
49,XY,cx,t(1;1)(q32;p36),del(3)(p14)	Whang-Peng et al 1982b
58,XY,cx,del(1)(p13),del(1)(q21),del(3)(p21), inv(3)(p14p23),hsr(15)(p?)	Whang-Peng et al 1982b
60,XY,cx,del(3)(p14q23),del(3)(p14),12q+	Whang-Peng et al 1982b
66,XY,cx,del(3)(p23q26)	Whang-Peng et al 1982b
67,XY,cx,inv(1)(q32p36),dup(1)(q32q44),del(3)(p14q13), del(3)(p14p23),t(7;13)(p11;q11),del(9)(q11),del(11)(p11), t(12;19)(q24;p13)	Whang-Peng et al 1982b
68,XY,cx,del(3)(p14p23),del(22)(q11)	Whang-Peng et al 1982b
69,XY,cx,del(3)(p14q24),t(11;14)(p11;p11), t(11;13)(p11;p11),del(X)(q23)	Whang-Peng et al 1982b
75,XY,cx,t(2;3)(p?;q?),del(3)(p11),del(3)(p13;q13)	Whang-Peng et al 1982b
77-78,XX,del(1)(q?),del(1)(p?),del(3)(p?),del(3)(q?), del(4)(p?)	Wurster-Hill and Maurer 1978
80,XX,cx,del(1)(q32),del(3)(p14p21),t(3;4)(p23q11;q11), t(5;20)(q11;q13),dmin	Whang-Peng et al 1982b

Karyotype	Reference
82,XY,cx,del(3)(p14q13)	Whang-Peng et al 1982b
106,XY,cx,del(1)(p32),del(3)(p13),del(3)(p14p23)	Whang-Peng et al 1982b
110,XY,cx,del(1)(q41),del(3)(p13p23),del(3)(p13), t(3;16)(q24;p13),del(6)(q24)	Whang-Peng et al 1982b

Papillary carcinoma

Karyotype	Reference
41,XX,cx,1p-,1q-,3p+,3q+,6q-,11q-	Wake et al 1980
42-54,X,-X,-14,-16,-16,+1,+3,+12,+20,1p-,6q-	Wake et al 1981
46,XX,t(6;14)(q21;q24),1p-,1q-,3p+,3q+	Wake et al 1980
50,X,-X,+1,+3,+12,+20,-14,-16,1p-,6q-	Wake et al 1980
54,XY,+del(2)(q11),t(2;14)(q11;q32),t(3;11)(p13;p15), +i(5p),+8,+9,+12,+16,+17,+21,+2mar	Pathak et al 1982
69,XX,cx,1p+,1q-,1p-,1q-,2q+,3q+,4q+,6p+,12q+,14q+	Wake et al 1980

Adenoma

Karyotype	Reference
46,XX,t(1;3)(p21;p21),t(1;13)(p36;q14),t(3;8)(p21;q12), t(1;5;20)(p21;p14q12;p13),t(5;8)(p14;q12), t(X;9)(q27;q12)	Mark et al 1982a
46,XX,t(3;5)(p21;p15),del(8)(p12q12),t(8;9)(q12;q34)	Mark et al 1980
46,XX,t(3;8)(p21;q12)	Mark et al 1981a
46,XX,t(3;8)(p21;q12)	Mark et al 1982a
46,XX,t(3;8)(p21;q12)	Mark et al 1982a
46,XX,t(3;8)(p25;q21)	Mark et al 1980
46,XY,t(3;8)(p25;q21)	Mark et al 1980
46,Y,-X,ins(X;3)(q22;q25q27),del(3)(q25),t(8;14)(q12;q22)	Mark et al 1981a

Adenocarcinoma

Karyotype	Reference
37,X,-X,-3,-8,-15,del(2)(p23),del(6)(q15),t(11;?)(p12;?)	Trent and Salmon 1981
38,X,-X,-2,-3,-10,-14,-22,del(1)(q25),del(6)(q15), del(14)(p11),dmin	Trent and Salmon 1981
40-42,X,-X,t(1;?)(p13;?),del(1)(p12),2q+,del(3)(q21), del(6)(q13),del(7)(q11),t(11;11)(p15;q13), t(12;12)(q24;?),13p+,15p+,i(22q)	Kusyk et al 1981
45-47,XX,2q+,-3,-18,+mar	Couturier-Turpin et al 1982
48-48,XX,+3,+7,+8,-1,-5,-14,-17,5q+,6q+,del(1)(q42), del(5)(q3?),del(10)(q24),del(14)(q2?)	Riet-Fox et al 1979
60-61,XX,cx,3p-	Martin et al 1979
62-74,XX,cx,3q-,6q-,t(13;14),dmin	Hecht et al 1983
72,XX,cx,del(2)(p23),hsr(3)(q13q28),del(6)(q21), t(10;11)(p12;p13)	Trent and Salmon 1981

Adenocarcinoma, metastatic

Karyotype	Reference
??,XX,del(3)(p14),del(4)(q22),del(4)(q13),del(16)(q21)	Ayraud et al 1977
??,XX,del(3)(p14),del(4)(q13),del(16)(q23)	Ayraud et al 1977
??,XX,t(1;16),2q-,del(3)(p14),del(4)(q22),del(6)(q22), i(7q),del(9)(q31),t(13;13)	Ayraud et al 1977
39-41,XX,-3,-7,-8,der(1),der(2),der(4),der(12),cx	Tiepolo and Zuffardi 1973
40,XX,t(4;5;8),t(3;?)(q21;?),t(1;11)(q21;p11), t(9;11)(q11;p11),del(1)(p36)	Kakati et al 1975
40-45,XY,-3,-8,-18,+5,+10,11p+,14q+,del(4)(q32)	Ayraud 1975

Karyotype	Reference
47-48,X,Xq+,t(3;9)(q12;p11),t(3;3)(q29;p25),del(3)(p25), t(4;?)(q35;?),+8	Kakati et al 1975
48,XX,+1,+3,+6,-16,1p-	Wake et al 1981
49,XX,1q+,+3,4q+,+7,+8,12q+,17q+	Granberg et al 1973

Adenomatous polyp

48,XX,+8,+14/48,XX,-3,+8,+9,+14 — Mitelman et al 1974b

MESENCHYMAL NEOPLASMS

Sarcoma, NOS

49,XX,+3,+13,+14 — Sonta and Sandberg 1978a

Liposarcoma

50,XY,+3,+6,+8,+11,+14,-22,17p+ — Sonta et al 1977

Mesothelioma, malignant

Karyotype	Reference
41,XY,del(1)(p13),t(1;3)(p13;q11),t(7;7)(q11;q36), t(9;12)(p24;q13),del(12)(q13),del(13)(q12q14),-14,-15, t(17;?)(p13;?),i(19p),-20,-21,-22	Mark 1978
43,XY,t(1;1)(q42;q32p22),t(2;6)(p25;p21),inv(3)(p24q13), t(5;18)(p15;q11),-6,+9,del(13)(q24q14),-14,-18,-22	Mark 1978

MELANOMAS

Malignant melanoma

??,X?,t(3;6)(p21;q23),cx — Trent et al 1983b

Malignant melanoma, metastatic

41,XY,dup(1)(q21q32),t(1;7)(q25;q36),t(7;9)(q11;q11), del(9)(q13),-3,-4,-5,-8,-10,-11,-16,+18,-20,-22 — Kakati et al 1977

NEUROGENIC NEOPLASMS

Neuroblastoma

Karyotype	Reference
45,XY,-3,del(1)(p13),t(1;9)(q11;p11),t(9;15)(p11;p11), hsr(16p)	Brodeur et al 1981b
53,XX,+1,+4,+15,+20,dup(1)(p13p32),t(3;15)(q13;q26), del(4)(q31),t(7;11)(q36;q13)	Brodeur et al 1981b

Retinoblastoma

44,XY,1p+,-3,t(3;12)(q?;p?),+6q-,der(9),13p+,-14,-16	Kusnetsova et al 1982
46,XX,3q+,del(13)(q12q14)	Balaban et al 1982
46,XX,inv(1),inv(2),inv(3),inv(4),t(6;7),inv(9),t(11;?), 11p-,inv(12),i(17q),-20	Gardner et al 1982
47,XY,t(X;1)(q28;p11),t(2;?)(q11;?),t(2;15)(p25;q11), t(3;?)(q27;?),4p+,t(5;?)(q35;?),i(17q),+mar	Gardner et al 1982

Meningioma, benign

45,XX,-22/45,XX,3p-,-22	Zang and Zankl 1983

EMBRYONAL AND MISCELLANEOUS NEOPLASMS

Teratoma, malignant

67,XY,+1,+3,+4,+5,+6,+7,+7,+8,+10,+11,+13,+13,+14,+16,+17, +18,+20,+20,+21,+21	Sonta et al 1977

Seminoma

53,XY,+3,+4,+6,+7,+11,+13,+19	Sonta et al 1977

LYMPHOMAS

1. UNCLASSIFIED AND MISCELLANEOUS LYMPHOMAS

Malignant lymphoma, NOS

47,XX,t(1;14)(q23;q32),+3,t(6;21)(p25;q11),+8, t(7;11)(q11;q25),-21	Yamada et al 1980

Non-Hodgkin's lymphoma, NOS

44-47,XY,3q+,del(4)(p13),del(6)(p23),+7,del(9)(p13)	Kristoffersson 1983
46-48,XX,del(1)(q32),+1,+3,r(7),14q+,-20,+mar	Kristoffersson 1983
48-54,XX,+X,3p+,-4,-6,-8,-9,-10,+12,-15,+7-10mar	Pearson et al 1982
50,XX,+X,t(1;11)(q25;q23),+3,+7,-13,t(14;18)(p?;p?),-15, t(19;?)(q13;?),+3mar	Reeves and Pickup 1980
52,XY,+2,+3,t(5;9)(q22;q32),+6,+12,+14,+20	Reeves and Pickup 1980

Non-Hodgkin's lymphoma, lymphocytic, NOS

45-46,XY,-1,3q+,del(9)(q22),+14,+17	Reeves 1973
46,XX,+3,t(13;22)(p11;p11)	Fleischman and Prigogina 1977
47,XY,t(3;4)(q29;p12),del(6)(p22),+7,del(9)(p13)	Kristoffersson et al 1981
47-48,XX,+3,+18	Fleischman and Prigogina 1977
47-50,XX,+3,+18,7q-	Fleischman and Prigogina 1977

Karyotype	Reference
48,XY,+3,+3,-15,+mar	Reeves and Pickup 1980
48-49,XY,1q-,+3,+del(3)(p?q?),-4,-6,+7,-9,-17,-18,-19,+5mar	Fleischmann et al 1976
50,XY,+3,+3,+12q-,+18,+14q+	Fleischman and Prigogina 1977

Non-Hodgkin's lymphoma, histiocytic, NOS

Karyotype	Reference
45,X,-X,del(1)(q22),t(1;17)(q?;q?),del(2)(q33),del(3)(p21p25),+ins(3)(p14p21p25),+5,del(6)(q14),t(1;7)	Mark et al 1978
46,XX,del(1)(p31p35),+3,del(9)(p11),t(10;?)(q26;?),del(11)(q13),-13,t(11;14)(q13;q32),t(17;21),t(9;22)(q11;p13),+r	Mark et al 1977
46,XY,1p-,6q-,t(11;14)(q21;q32)	Kakati et al 1980
48,XY,1p-,+3,6q-,+7,9p-,t(11;14)(q21;q32)	
47,XX,del(3)(q11),14q+	Nordenson 1983
47,XY,-6,-14,1q-,+1p-,2q+,3q+,9p+,10p-,10p+,12p-,18q+,22p+,t(2;14)(q37?;q13?)	Fukuhara et al 1979b
47-50,XY,t(1;11)(q?;p?),1q-,1p-,3q+,11p+,t(11;17)	Fleischman and Prigogina 1977
48,XY,del(3)(p21),del(9)(q22),11q+,12q+,del(6)(q23orq25),15q+,del(12)(q15)	Reeves 1973
48,XY,t(3;8)(q22;q11),t(5;10)(q33;q22),del(7)(q21),t(4;9)(q11;p24),t(10;13)(q22;q21),r(12),del(19)(p13),del(22)(q11),+7	Mark et al 1978
50,XY,t(2;?)(p25;?),del(2)(p13),+i(3p),del(6)(p21),+7,+8,+9,+9,-11,-13,+14,-15,+20,+21,-22	Mark et al 1978
44,XY,del(2)(p13),t(6;?)(p21;?),del(11)(q13),t(11;14)(q13;q31),i(15q),-20,-22	
82,XXX,t(3;12)(q13;q15),t(3;8)(q22;q11),del(3)(q11),del(5)(q31),i(7q),del(10)(q24),t(10;10)(q24;q26),inv(11)(p15q21),del(12)(q13),del(13)(q14)	Mark et al 1978
89,XX,i(1q),del(3)(p15p23),del(10)(q24),del(12)(p12),i(17q)	Mark 1977

Non-Hodgkin's lymphoma, T-cell, NOS

Karyotype	Reference
47,XX,t(2;?),inv(3),4p-,del(5)(q14q32),6p+,6q+,t(7;7),9q+,10p-,13q-,t(14;14),-15,-15,16q-,17p+,+21,+2mar	Gaeke et al 1981

Malignant histiocytosis

Karyotype	Reference
48,XX,1p+?,del(2)(p14?),+del(3)(q12?),+mar	Kristoffersson 1983

Angioimmunoblastic lymphadenopathy

Karyotype	Reference
47,XX,+3,-5,+mar	Kaneko et al 1982c
47,XY,+3/47,XY,+3,-9,+21/48,XY,+3,+8,+9,-20	Hossfeld et al 1976
48,XY,+3,+18/51,XY,+5,+15,+19,+21,+22	Kaneko et al 1982c

2. HODGKIN'S LYMPHOMAS

Hodgkin's disease, mixed cellularity

Karyotype	Reference
61-80,XX,del(1)(q32),i(1q),del(2)(p13),del(6)(q13or15), del(6)(q23or25),15q+,12q+,del(5)(q31),del(7)(q11), del(10)(q26),del(3)(p13)	Reeves 1973
90-102,XY,t(3;11)(p14;q21)	Reeves 1973

Hodgkin's disease, lymphocytic depletion

Karyotype	Reference
45,X,t(X;1)(q28;q11),t(3;5)(q29;q33),del(8)(q22),-9, del(11)(p13),-17	Hossfeld and Schmidt 1978
46,XY,+2,t(2;?)(q25;?),t(3;11)(q29;p13),t(4;11)(q33;q13), -13	Hossfeld and Schmidt 1978
48,X,-Y,t(1;?)(q44;?),i(1q),del(1)(q21),del(3)(q21),-10, t(3;11)(q23;q21),+14,t(14;?)(q32;?),+15,-17,+19,+20,-22	Hossfeld and Schmidt 1978
60,XY,+del(1)(p21),t(1;5)(q44;q33),dup(5)(q12), t(1;5)(p21q23;q13),del(6)(q22),t(9;?)(q12;?), t(3;12)(q11;q24),t(14;?)(p11;?),i(17q)	Hossfeld and Schmidt 1978

3A. NON-HODGKIN'S LYMPHOMAS - RAPPAPORT CLASSIFICATION

Lymphocytic, well differentiated, diffuse

Karyotype	Reference
48,XY,1q+,3q+,+i(3p),-6,-10,t(14;?)(q12;?),+16,17p-,+18, +mar	Miyamoto et al 1981a

Lymphocytic, poorly differentiated, NOS

Karyotype	Reference
46,XY,+3,-8	Mark et al 1979

Lymphocytic, poorly differentiated, diffuse

Karyotype	Reference
44,XX,t(11;14)(q13;q32),-3,-9,-13,1p+,+2mar	Fukuhara et al 1979a
47,XX,dup(1)(q21q32),+3	Slavutsky et al 1981
47,XY,+3	Mark et al 1979
47,XY,t(1;4)(q23;q33),+3,t(10;15),t(11;14)(q13;q32),r(13), r(13)	San Roman et al 1982
48,XX,1q-,+3,+4,6q-,+7,10q-,-15	Kakati et al 1980
50,XX,+3,+4,+10,+18/51,XX,+3,+4,+9,+10,+18	Mark et al 1979
87,XY,del(1)(q23),del(3)(p14),del(6)(q21),del(7)(p11), del(11)(q21),t(17;?)(p13;?)	Mark et al 1979
88,XX,del(1)(p22),del(3)(q21),t(3;9)(q?;q?),del(6)(q21), t(22;?)(q13;?)	Mark et al 1979

Lymphocytic, poorly differentiated, nodular

48,XY,t(14;18)(q32;q21),-6,-11,+12,+del(1)(p?),3q+,+2mar — Fukuhara et al 1979a
92,XY,t(2;9)(q11;p11),t(5;?)(p15;?),t(17;?)(q25;?),17q+, del(1)(p22),del(3)(p21),del(X)(q24),del(11)(q23) — Reeves 1973

Mixed cell, diffuse

50,XY,+11,+12,-15,+del(3)(q11),+del(3)(q11), +t(X;1)(p22;q11) — Kaneko et al 1982a

Mixed cell, nodular

50,XX,t(3;6)(q29;p21),del(9)(p13q22),+11,+12,+21 — Mark et al 1976

Histiocytic, diffuse

45,X,-Y,t(14;15)(q32;q21orq22),-15,+18,-22,1q+,3p+,5p-,7p-, 9p+,12p+,16q+,17p-,+2mar — Fukuhara and Rowley 1978
45,X,del(X)(q24),del(1)(q21),+del(1)(p11),del(2)(q22), del(3)(q21),del(6)(p11),del(6)(q15),t(7;9)(p22;q13), t(8;10)(p23;q21),del(9)(q13),del(10)(q21), t(2;11)(q22;q23),-13,del(14)(q24),del(15)(q22), del(16)(q22),+17,-19 — Mark et al 1979
45-49,XX,+3,+7p+,-18 — Kakati et al 1980
46,X,-X,-2,-10,+11,-13,-14,-17,del(1)(p22p32), del(3)(p13p21),del(6)(q21),del(9)(p13),+t(2;?)(p23;?), +t(14;?)(q32;?),t(18;?)(q23;?),+t(X;13)(q26;q12),+mar — Kaneko et al 1982a
49,XY,+del(2)(q31),+3,del(10)(q22),+17,+t(10;17)(q22;p13), -19 — Mark et al 1979
50,XX,+3,+3,-4,-6,-9,+12,-17,+18,+i(6p),+3mar — Kaneko et al 1982a
50,XX,del(1)(p22),+3,+7,+12,t(1;14)(p22;q32),+18 — Mark et al 1979
50,XY,del(3)(p24),+7,+12,+14,+20 — Kristoffersson et al 1981
61,XY,cx,t(1;16)(q23;p11),dup(1)(q23q32),t(1;6)(q23;p21), t(3;6)(q21;p21),t(7;7)(q32;p22),13p+ — Fitzgerald et al 1980

3B. NON-HODGKIN'S LYMPHOMAS - KIEL CLASSIFICATION

Lymphocytic, T-zone

44-46,X,+t(Y;14)(q24;q12),-3,del(3)(q21),+9,14q+,+mar — Gödde-Salz et al 1981a
45,XX,-1,-2,-3,-4,+t(1;2)(p32q32;p13),+t(2;4)(p13;p15), +t(3;4)(q11;?) — Gödde-Salz et al 1981a
46,XY,+t(Y;14)(q12;q24),3q-,-5,-13,+mar — Gödde-Salz et al 1981b
47,XY,+3 — Gödde-Salz et al 1981a
47,XY,+3 — Gödde-Salz et al 1981a

Immunocytoma

45,X,-Y,t(Y;18)(q11;p11),t(3;3)(q21;q29),5q+ Kristoffersson 1983
47,XX,+3,+del(3)(q11),del(10)(q11),-21 Kristoffersson 1983
46-47,XY,+3,+mar Kristoffersson 1983
49,XX,+3,del(6)(q14),+7,+18/49,XX,+3,del(6)(q14), del(9)(q22),+12,+18 Kristoffersson 1983

Centroblastic-centrocytic, diffuse

48-51,XY,del(1)(p33q23q25),-2,+del(3)(p22),+4,4p+,+5,+6p+, +7,+8,+9,+21,+22 Kristoffersson 1983
45-51,XY,del(1)(p33q23q25),del(2)(q14),+del(3)(p22),+4,+5, +7,+8,-12

Centroblastic-centrocytic, follicular

46-49,XX,1p+,3q-,4q+,+6,+13,-14,14q+,17p+ Kristoffersson 1983
48,XX,+X,t(1;?)(p36;?),del(2)(p21),t(3;?)(q27orq29;?), t(14;?)(q32;?),t(17;18) Reeves and Pickup 1980

Centroblastic, diffuse

46-51,XX,+3,del(6)(q21),+del(6)(q21),+7,-8,del(9)(q11),+13, -16,-17,+19,-20,-21,+1-5mar Kristoffersson 1983

Lymphoblastic, unclassified

45,XY,3p-,14q+,-C Kristoffersson 1983

Lymphoblastic, Burkitt's type

??,XY,dup(1)(q12orq21q31),3p-,3q-,+6p-,+8,+14,+15,+18,-21 Douglass et al 1980a
43,XY,+1,+2,+3,-7,-8,-9,-12,-16,-22,del(1)(q25), del(2)(p11),del(10)(q22),14q+ Biggar et al 1981
45,X,-Y,t(1;14)(q22;q32),t(8;22)(q24;q11),3q+,6q-,6q+,7q+, 9p-,14q-,18q- Berger et al 1981c
45-47,XY,t(14;?)(q32;?),del(3)(p?),-9,-22 Philip et al 1977b
46,XX,t(8;14)(q24;q32),dup(1)(q23q32),del(3)(p25) Slater et al 1982
46-47,XY,t(14;?)(q32;?),dup(1)(q21q32),t(2;3)(q13;q29),+5, 6p-,-8,del(9)(q?),+12,+18,+mar Miyoshi et al 1981c
46-52,X?,t(8;14)(q24;q32),+1,+3,+5,+7,i(17q),+X Zech et al 1976a
48-53,X?,t(8;14)(q24;q32),+X,+2,+3,+9,+10,+11,+12,+20 Zech et al 1976a
49,XY,+3,+7,+12,14q+ Biggar et al 1981
50,XX,+1,+2,+3,+11,del(2)(p14),11p+,t(8;14)(q24;q32) Biggar et al 1981

Lymphoblastic, convoluted type

45,XY,t(3;?)(q29;?),t(8;?)(q2?;?),t(12;17)(p13;q11) Kaneko et al 1982d
47,XY,t(1;2)(p34;p21),+4,t(13;18)(p13;p11)/47,XY,+3 Kaneko et al 1982d

Immunoblastic

Karyotype	Reference
45,XY,del(1)(p22orp23),del(3)(q11),del(3)(q21),del(8)(p11), del(9)(p11),del(9)(q12),t(12;?)(q24;?)	Reeves and Stathopoulos 1976
45-46,XX,+3,del(6)(q21),+4,+16,+21,-C,-C	Kristoffersson 1983
47-48,XX,+del(3)(q11),+17/49,XX,+del(3)(q11),+16,+17	Kristoffersson 1983
45-47,XX,1p+,-14,-15,+16,+17,-19,+mar	
48-49,XY,+3,+7,+11,t(12;?)(p13;?),t(13;?)(p13;?),-16,-20, +21,+22	Reeves and Pickup 1980
48-49,XY,del(3)(p23),+7,+12,+14,+20	Kristoffersson 1983
51,XY,+X,t(3;3)(q11;q27),t(8;?)(q2?;?),+7,+12,+21,+21	Reeves and Pickup 1980
78-83,XY,+del(1)(q11),+5,+7,+12,+13q+,+14q+,+14q+,+20,+mar	Kristoffersson 1983
80-86,XY,del(1)(q11),+3,+7,+9,+12,+13q+,+14q+,+14q+,+15q+, +18,+20,+mar	

3C. NON-HODGKIN'S LYMPHOMAS - LUKES AND COLLINS CLASSIFICATION

Mycosis fungoides

Karyotype	Reference
46-48,XX,del(1)(q42),+3,-11,-12,+2-6mar	Edelson et al 1979

Sezary's syndrome

Karyotype	Reference
50,XX,1q+,2p-,+3q-,+6q-,+9,-15,+21,+mar	Nowell et al 1982

Follicular center cell, diffuse, large cleaved

Karyotype	Reference
45,XY,-6,-9,-17,2q-,3p-,3q-,4q-,5q-,6p+,7p+,7q+,+7q-,9p-, 12q+,13p+,20p-,+mar	Fukuhara et al 1978b
47,XY,+Y,t(2;3)(q11;q29)	Fukuhara et al 1978b
47,XY,-6,-14,1q-,+1p-,2q+,3q+,9p+,10p-,10p+,12p-,18q+,19p+, 19q+,+22p+,+mar	Fukuhara et al 1978b

Immunoblastic type (B-cell)

Karyotype	Reference
49,XX,+i(1q),+5,del(8)(q24),t(3;22)(q29;q11),+13	Berger et al 1983a

3D. NON-HODGKIN'S LYMPHOMAS - WORKING FORMULATION

Follicular, small cleaved cell

Karyotype	Reference
44,XY,+der(1),+2,t(2;13),-3,5q-,-12,t(14;18)(q32;q21), i(17q)	Yunis et al 1982
46,XX,t(2;3),i(6p),t(14;18)(q32;q21)	Yunis et al 1982
47,XY,+X,t(2;3;17),6q-,t(14;18)(q32;q21)	Yunis et al 1982
48,X,-X,inv(1),1q-,+3,4q-,ins(5),+7,+mar	Yunis et al 1982
48,X,r(Y),+2,+3,t(14;18)(q32;q21),t(1;22)	Yunis et al 1982
49,XY,+X,+3,ins(11;3),t(14;18)(q32;q21),+18	Yunis et al 1982

Follicular, mixed small cleaved and large cell

50,Y,inv(X),+3,+7,+10,t(12;12),+12,t(14;18)(q32;q21),+18, 18q-,-21 — Yunis et al 1982

Follicular, large cell

50,XY,1q-,t(2;3),+3,+10,t(1;12),13q-,+15,-19,t(3;19),+21 — Yunis et al 1982

Diffuse, mixed small and large cell

47,XX,+9,t(9;15)(p13;q11),t(12;14)/49,XX,+3,+5,+9,t(9;15), t(12;14) — Yunis et al 1982

Diffuse, large cell

46,XY,t(11;14)/47,XY,t(11;14),+3 — Yunis et al 1982
54,XY,dup(1)(q?),inv(5),+3,+3,+9,+13,+18,+18,+20,+mar — Yunis et al 1982

Large cell, immunoblastic

48,Y,t(X;?),t(1;?),t(3;6),+5,-6,t(8;14)(q24;q32),i(17q), +mar — Yunis et al 1982

Chromosome 4

HEMATOLOGICAL DISORDERS

UNCLASSIFIED LEUKEMIAS

NONLYMPHOCYTIC LEUKEMIAS

Acute nonlymphocytic leukemia (ANLL)

Karyotype	Reference
43,XX,-1,3p+,3q-,4q+,-5,-5,-7,+9,+9,10p+,+10,+10,10p-,10q-, +11q-,-13,-15,-16,-17,-18,-21"s"	Whang-Peng et al 1979
43,XX,-4,-5,-21,t(8;21)(q22;q22)	Oshimura et al 1976b
43,XX,2q-,inv(3),-4,del(5)(q?),-7,17p+,-19	Rowley 1976
44,X,-Y,-3,+4,-12,+15,-20	Bernard et al 1982a
46,XX,4q-,5q-,-16,-17,+2mar	Rowley 1976
46,XX,t(4;8),dic(21)	Seabright 1983
46,XY,3p-,4q-	Nordenson 1983
46,XY,t(9;10)(p24;p12),t(3;16),t(4;15)(p13;q14)"c"	Van Den Berghe et al 1979a
46,Y,t(X;11)(q22.3;q23.3),ins(X;11)(q22.3;p13p15), dup(1)(q11q21),t(4;6)(p16;p21),t(7;12)(p13;q15)	Nacheva et al 1982
47,XX,-4,+13,t(1;4)(q22q25;p14p16)	Oshimura et al 1976b
50,XX,4q-,-11,+5mar	Bernard et al 1982a
51,X,-Y,+1,+18,+20,+22,+inv(4),+mar	Zuelzer et al 1976
56,XY,+Y,+6,+14,+21,+t(5;6)(q?;p?),+del(1)(p22q25), +del(2)(q12),+del(3)(p12),+del(4)(q13),+del(5)(p14q14), +mar	Zuelzer et al 1976

ANLL, FAB type M1

Karyotype	Reference
41,XY,-1,-2,-4,-5,-12,-15,-17,-17,-18,-22,+5mar	Sessarego et al 1981a
46,XX,t(4;19)(p15;p13)	Yunis et al 1981
46,XY,der(4)(q?)	Bernstein et al 1982b
58,XY,+1,+4,+5,+6,+8,+11,+13,+13,+14,+14,+15,+21,+21, t(9;22)(q34;q11)	Alimena 1983

ANLL, FAB type M2

43,XX,-3,-4,-7,-19,del(5)(q14),17p+	Testa et al 1979
43,XX,-3,-4,-7,-19,del(2)(q31?),del(5)(q14)	
43,XY,-3,-4,-7,-7,-8,-8,-17,-20,+5mar	Berger et al 1981a
43,XY,-5,-7,-12,-16,4p+,17p+,20q-,+mar"s"	Rowley et al 1981a
45-47,XX,-9,-17,-20,del(4)(q21),del(5)(q12q31), t(9;17)(p13;p11),+mar	Testa et al 1979
46,XX,-4,-6,-7,-20,+mar	Golomb et al 1978c
46,XX,-4,5q-"s"	Pedersen-Bjergaard et al 1981
46,XX,4q+,del(7)(q31),t(8;21)(q21;q21),del(9)(q12q21)	Hagemeijer et al 1981a
46,XX,del(4)(q31q34),t(7;12)(q21;q13)	Yunis et al 1981
46,XX,t(4;10)(q31;q22),t(6;9)(p23;q34)	Kaneko et al 1982b
46,XY,4q+	Brodeur et al 1983
49,XY,+4,+r(7),+19	Kaneko et al 1982b
53,XX,+1,-4,-5,+6,-7,+8,+11,+12,+21,+21,+3mar"s"	Zaccaria et al 1983b

ANLL, FAB type M1+M2

45,XY,-21,t(4;9)(p16;q22)	Yamada and Furusawa 1976
37-45,XY,-7,-12,-13,-14,-16,-18,-19,-21,del(4)(q29), del(5)(q24),der(9),der(17),11q+,12p+,t(14;?)	Fitzgerald et al 1983
42,X,-Y,4q+,-8,-16,17p+,-21,-22,+Yp+"s"	Sandberg et al 1982a
43,XX,1p+,-4,-5,-6,-7,-13,-17,del(18)(q21),+3mar	Rowley and Potter 1976
44,XX,-3,-4,-5,+9,-17,-18,-21,-22,+5mar	Garson 1980
45-46,XX,4q+,t(5;11)(q35;p13)	Mitelman et al 1981
45-46,XX,t(5;11),4q+,-21	Alimena et al 1977
46,XX,4q+	Prigogina et al 1979
46,XY,4q+,del(5)(q15)	Mitelman et al 1981
47,XY,+19,t(7;12)(q21;p13),del(4)(q26),del(5)(p12)	Mitelman et al 1981
51,XY,+1,+4,+5,+8,+20	Nordenson 1983

ANLL, FAB type M3

45,XX,-4,-5,-14,t(21;?)(q22;?),+2mar"s"	Rowley et al 1977a
46,XX,4q+,t(15;17)	Rochon and Vaillancourt 1981
46,XX,t(15;17)(q22;q21)/46,XX,t(15;17),t(1;4)(p35;q21), del(X)(q25)	Berger et al 1981e
46,XY,t(15;17)(q22;q12)/46,XY,t(15;17),t(21;?)(p11;?)/46, XY,del(4)(q22),t(15;17)	Kaneko et al 1982b
46,XY,t(3;5;17)(p21;q25?;q21?),4p+	Bernstein et al 1982b

ANLL, FAB type M4

42,XY,-4,-15,-16,-22	Prieto et al 1981
46,XX,4q+,-20,+mar	Takeuchi et al 1981
46,XY,1p+,4q+,del(7)(p11)	Morse et al 1979a
47,XY,+i(4p)	Hagemeijer et al 1981a

ANLL, FAB type M5

46,XX,-18,-22,+2r/46,XX,t(2;9)(q23;p22),-3,-22,+2r/49,XX, t(2;9),-22,+4r	Alimena 1983
46,XX,ins(10;11)(p11;q23q24)/52,XX,+4,+8,ins(10;11),+12, +16,+19,+20	Kaneko et al 1982b
46,XY,4p+	Nordenson 1983
46,XY,t(6;11)(q27;q23)/51-52,XY,t(6;11),+der(6),+3,+19,+19, +21,dmin/53,XY,t(6;11),+der(6),+4,+10,+13,+18,+19,+21	Löwenberg et al 1982
47,XY,1p+,1q+,2p+,-3,4q+,+13,+mar	Kaneko et al 1982b
49,XX,+4,+8,+8,t(9;11)(p22;q23)	Kaneko et al 1982b
57-58,XX,+1,+3,-4,-5,+6,-8,+11,+12,-17,+19,-20, del(10)(q24),+8-9mar	Testa et al 1979
60-61,XX,Xp+,+1,+2,-4,+6,-7,+19,+20,del(2)(q31), del(10)(q24),+12-13mar	

ANLL, FAB type M5a

48,XY,+4,+8,t(11;17)(q24;q21)"s"	Dewald et al 1983

ANLL, FAB type M4+M5

66,XY,+1,+1,+2,+4,+6,+6,+7,+8,+8,+10,+11,+11,+12,+12,+18, +19,+21,+21,+22,+22	Li et al 1983

ANLL, FAB type M6

??,XY,t(4;?)(q23;?),t(5;17),t(8;9),+8,+11, t(12;16)(p12;p12)	Hustinx and Rutten 1983
40-45,XY,-4,-6,-7,-10,-11,-20,+1-3mar	Mitelman et al 1981
43,X,-Y,-4,-13,-14,-15,+2mar/46,X,-Y,-3,+9,+mar	Li et al 1983
43,XY,-2,4q-,-9,11q-,-15,-19,-20,+2mar/46,XY,-15,-20,+2mar	Li et al 1983
43,XY,-4,5q-,t(4;6),-7,dic(12),-15"s"	Hagemeijer et al 1981a
43-45,XX,der(1),inv(2)(p25q14),del(3)(q25),-4,-5,-6,6q-,-7, 7p+,11p+,-12,-17,-21,+r,+1-4mar"s"	Smadja et al 1983b
44,X,del(Y)(q?),+Xq-,3p-,3q-,4q-,-6,+7q-,11q+,-15,-18"s"	Bradley et al 1982
44-45,XY,-2,-4,-7,-10,-13,-16,-17,-22,+3mar	Mitelman et al 1981
45,XX,t(4;22)(p16;q11),-7/46,XX,t(4;22)(p16;q11)	Oshimura and Sandberg 1977
45,XY,del(4)(q23),del(5)(q13),-6,del(6)(q21),del(7)(p14), del(8)(p21),del(9)(q31),16q+	Rowley and Potter 1976
45-49,XY,-4,-7,-8,-10,+14,+16,-21,-22,+2mar	Mitelman et al 1981
46,XX,t(4;16)(q11;q23)	Hagemeijer et al 1981a
46,XX,t(4;9)(q21;q34),t(4;13)(q12;q12),del(13)(q12orq13), del(12)(p11)	Fitzgerald et al 1983
48,XX,+8,+9,+9,-11,del(4)(q23)"s"	Weinfeld et al 1977a

Chronic myeloid leukemia, t(9;22)

46,X?,t(9;22),19q+/46,X?,t(9;22),4p+,10q+,19q+	Fleischman et al 1981
46,XX,t(9;22)(q34;q11)/46,X,-X,t(9;22),4q+,+mar	McGlave et al 1982
46,XX,t(9;22),del(4)(q31)	Sharp et al 1976
46,XX,t(9;22),t(1;4)(q22;q31),del(7)(q11)	Miyamoto 1980
46,XX,t(9;22),t(2;4)(q?;q?)	Sadamori et al 1980

Karyotype	Reference
46,XX,t(9;22),t(4;9)(p16;q32)	Sadamori et al 1980
45,XX,t(9;22),t(4;9),-7	
46,XX,t(9;22)/45,XX,t(9;22),-4	Miyamoto 1980
46,XY,t(9;22)(q34;q11)	Ishihara et al 1982
27,X,-Y,t(9;22),-1,-2,-3,-4,-5,-6,-7,-9,-11,-13,-14,-15, -16,-17,-18,-19,-20,-22	
46,XY,t(9;22),t(4;12;15)(q21;q24;q21)	Hartley and McBeath 1981
47,XXY,t(9;22)(q34;q11)/53,XXY,+X,t(9;22),+4,+8,+18,+18, +22q-"c"	Mitelman 1983
47,XXY,t(9;22)/53,XXY,+X,t(9;22),+4,+8,+18,+18,+22q-"c"	
47,XXY,t(9;22)"c"	
47,XY,t(9;22),4p+,+22q-	Sadamori et al 1980
48,XY,t(9;22),+22q-,+22q-/47,XY,t(9;22),+17/46,XY,t(9;22), -4,-7,+6,+17	Stoll and Oberling 1978
45,XY,t(9;22),-21/46,XY,t(9;22),-10,+17	
48,XY,t(9;22),+8,+22q-/61,XY,+Y,t(9;22),+4,+5,+6,+8,+8,+8, +8,+15,+17,+18,+19,+20,+22q-,+22q-	Wurster-Hill 1983
48,XY,t(9;22),t(3;4)(p?;p?),+21,+22q-	Sadamori et al 1980
51,XX,t(9;22),+4,+17,+19,+del(22q),+mar	Hossfeld 1975b
59,X,-X,t(9;22),+4,+6,+6,t(7;?)(p15;?),+der(7),+8,+8,+12q+, +17,+19,+19,+21,+22,+22q-,+22q-	Miyamoto 1980

Chronic myeloid leukemia, aberrant translocation

Karyotype	Reference
??,X?,t(1;4;20;22)	Geraedts et al 1977
??,X?,t(3;4;9;11;22)	Potter et al 1981
46,X,t(X;9;22)(q27;q34;q12)	Dallapiccola and Alimena 1979
46,X,t(X;9;22)/47-51,X,t(X;9;22),+1,+4,+8,+9,-14,i(17q), -19,+22q-	
46,X?,t(4;22)(p16;q11)	Van Den Berghe 1983
46,XX,t(3;4;9;22)(q21;q31;q34;q11)	Chessells et al 1979
46,XX,t(4;22)(p?;q11)	Sekhon et al 1978
46,XX,t(4;9;22)(p14;q34;q11)	Tomiyasu et al 1982
46,XX,t(4;9;22)(q23;q34;q11)	Ciric and Rolovic 1980
46,XX,t(4;9;22)(q31;q34;q11)	Sudries et al 1980
46,XX,t(4;9;22)(q31;q34;q11)	Kessous et al 1980
46,XX,t(4;9;22)(q32;q34;q12)	Fraisse et al 1980
46,XY,t(4;22)(q35;q21),t(3;5)(q27;q22)	Sessarego et al 1981b
46,XY,t(4;22)(q35;q21),t(3;5)(q27;q22),i(17q)/50,XY, t(4;22),t(3;5),i(17q),+8,+15,+21,+22q-	
46,XY,t(4;9;22)(q21;q34;q11)	Pasquali et al 1979b
46,XY,t(4;9;22)(q23;q34;q11)	Rowley et al 1976
46,XY,t(4;9;22)(q31;q34;q11)	Seabright 1983

Chronic myeloid leukemia, Ph1 negative

Karyotype	Reference
47,XY,t(3;4)(p14;q11q31),+8	Kohno et al 1979a

Eosinophilic leukemia

46,XY,4q+ Ellman et al 1979

MYELOPROLIFERATIVE DISORDERS

Polycythemia vera

??,X?,4q+,+13,15q-,20q- Vykoupil et al 1980
47,XX,+9/49,XX,+9,+20,+20,t(4;5),-7,-11,-12,+3mar Zech et al 1976b

Myelosclerosis / Myelofibrosis

45,XY,t(1;4;7)(p32;q28;q11q32) Nowell and Finan 1978

Chronic myeloproliferative disease, NOS

46,XY,t(4;12)(q35;q11),i(17q),del(20)(q11) Berger et al 1975

DYSMYELOPOIETIC SYNDROMES

Preleukemia, NOS

43,X,-X,-3,-13,4q-,8q-,12q+ Geraedts et al 1980a
45,XX,-1,-4,5q-,-7,-12,-18,14q+,+4mar"s" Pedersen-Bjergaard et al 1981
45,XY,-2,-4,-6,-8,-10,-10,+17,+18,+3mar Panani et al 1980
53,XX,+1,+2,t(4;19),+6,i(17q),+21,+21,+22 Seabright 1983

Refractory anemia, NOS

46,XX,t(2;4;13)(p11;q12;q34),del(5)(q15q31) Kerkhofs et al 1982

Acquired idiopathic sideroblastic anemia

46,XX,del(5)(q12q23)/45,XX,3p-,4q+,6q-,-22"s" Kerkhofs et al 1982

Refractory anemia with excess of blasts

45,XX,-9,-17,-20,del(4)(q21),del(5)(q12q31), +t(9;17)(p13?;p11) Streuli et al 1980

Pancytopenia

46,XX,-4,-5,-6,-7,-11,-13,-16,+19,+20,t(7;13)(q?;q?), +4mar/43,XX,-2,-3,-5,-7,-16,+19,15q+,+mar Nowell and Finan 1978

Dysmyelopoietic syndrome, special type

47,XX,del(4)(q25),+8 Mitelman 1983
47,XX,del(4)(q?),+8

SPECIAL LEUKEMIAS

LYMPHOCYTIC LEUKEMIAS

Acute lymphocytic leukemia (ALL)

Karyotype	Reference
28,XX,-1,-2,-3,-4,-5,-7,-8,-9,-11,-12,-13,-14,-15,-16,-17,-19,-20,-22	Kaneko and Sakurai 1980
32,XY,-2,-3,-4,-5,-7,-8,-11,-12,-13,-14,-15,-16,-17,-20	Shabtai and Halbrecht 1981b
46,X?,t(4;11)(q21;q23)	Parkin et al 1982
46,X?,t(4;11)(q21;q23)	Parkin et al 1982
46,XX,2q+,4q-,12p+,13q+,14q+,-20,+21	Whang-Peng et al 1976c
46,XX,t(4;11)(q13;q22)	Van Den Berghe et al 1979b
46,XX,t(4;11)(q13;q22)	Van Den Berghe et al 1979b
46,XX,t(4;11)(q21;q23)	Prigogina et al 1979
46,XX,t(4;11)(q21;q23)/49-50,XX,t(4;11),+8,+13,+19,+4q-	Prigogina et al 1979
46,XX,t(4;11)(q21;q23)	Morse et al 1982b
46,XX,t(4;11)(q21;q23)	Arthur et al 1982
46,XX,t(4;11)(q21;q23),del(17)(p11)	
46,XX,t(4;11)(q21;q23)	Arthur et al 1982
46,XX,t(4;11)(q21;q23)	Arthur et al 1982
46,XX,t(4;11)(q21;q23)	Arthur et al 1982
46,XX,t(4;11)(q21;q23),i(Xq)	Arthur et al 1982
46,XX,t(4;11)(q21;q23)	Esseltine et al 1982
46,XX,t(4;12)(q12;p13)	Mitelman 1983
46,XX,t(4;14)(q35;q12)	Prigogina et al 1979
46,XY,4q+,del(4)(q12)	Secker-Walker et al 1979
46,XY,dup(1)(q21q32),4q+	Morse et al 1979a
46,XY,t(4;11)(q13;q22)	Van Den Berghe et al 1979b
46,XY,t(4;11)(q21;q23)	Prigogina et al 1979
46,XY,t(4;11)(q21;q23)	Oshimura et al 1977a
46-49,XY,t(4;11)(q21;q23),+6,+8,+13,+17,t(1;13)(p22;q12),t(7;9)(q11;q34)	
46,XY,t(4;11)(q21;q23),i(7q)	Morse et al 1982b
46,XY,t(4;11)(q21;q23)	Arthur et al 1982
46,XY,t(4;11)(q22;q25)	Mitelman 1983
47,XY,4q-,+17q-	Whang-Peng et al 1976c
47,XY,t(4;11)(q21;q23),1p+,del(2)(p16),del(6)(q21),i(7q),del(8)(q21),+mar	Arthur et al 1982
48,XY,t(4;12)(q?;q?),t(6;15)(q?;q?),t(9;22)(q34;q11),11q-,+20,+21	Nordenson 1983
48-53,XX,t(4;11)(q21;q23),+6,+7,+9,+13,+21	Prigogina et al 1979
49,XX,del(4)(p15),t(4;11)(q21;q23),+5,+8,del(9)(p13),del(9)(p13),t(12;?)(q24;?),+mar	Morse et al 1982b
53,XY,+X,+4,+8,+14,+15,+17,+21	Oshimura et al 1977a
53-55,XY,+4,+5,+6,+20,+22	Lessard and Le Prise 1983
54,XX,7p+,+10,+10,+14,+14,+18,+18,+21,+21/27,X,-X,-1,-2,-3,-4,-5,-6,-7,7p+,-8,-9,-11,-12,-13,-15,-16,-17,-19,-20,-22	Oshimura et al 1977a

55,XX,+2,+4,+9,+14,+15,+19,+21,+21 Oshimura et al 1977a
55,XX,+4,+5,+6,+10,+17,+21,+3mar Humbert et al 1978
57,XX,+3,+4,+8,11p+,+14,+16,+17,+18,+21,+22,+2mar/47,XX,+8 Morse et al 1978
61,XX,+X,+1,+2,+3,+4,+5,+6,+8,del(6)(q21),+13,+14,+15,+16,+18,+21,+22 Oshimura et al 1977a

ALL, FAB type L1

26,XX,-1,-2,-3,-4,-5,-6,-7,-8,-9,-10,-11,-12,-13,-14,-15,-16,-17,-19,-20,-22 Hoeltge et al 1982
28,XX,-1,-2,-3,-4,-5,-6,-7,-8,-9,-11,-12,-13,-15,-16,-17,-19,-20,-22/56,XX,+X,+X,+10,+10,+14,+14,+18,+18,+21,+21 Brodeur et al 1981a
34-36,XY,-2,-3,-4,-7,-9,-12,-13,-14,-16,-17,-20,+22 Sandberg et al 1982c
45,XX,4p-,-18,t(22;?)(q11;?) Priest et al 1980
46,XY,del(2)(q32),-11,-12,+2mar/46,X,-Y,del(2),-4,-5,-11,-12,-21,+6 Kaneko et al 1981b
52-56,XY,t(9;22),+4,+6,+9,+10,+14,+15,+20,+21,+22q- Sandberg et al 1980
55,XY,+4,+6,+7mar Kaneko et al 1981b
56,XY,+X,+4,+6,+10,+15,+17,+18,+21,+21,+mar Kaneko et al 1982e
58,XX,+4,+6,+7,+10,+14,+14,+21,+del(18)(q21),+4mar Kaneko et al 1982e

ALL, FAB type L2

26,XX,-1,-2,-3,-4,-5,-6,-7,-8,-9,-10,-11,-12,-13,-15,-16,-17,-18,-19,-20,-22 Brodeur et al 1981a
46,XY,-4,+6,-8,-10,-11,-15,-17,+22,del(6)(q21),+i(17q),+3mar Kaneko et al 1981b
46,XY,t(4;11)(q21;q23) Weh and Hossfeld 1982
47,XX,+X,t(4;11)(q21;q23) Weh and Hossfeld 1982
53,XY,t(9;22),+4,+5,+6,+7,+8,+17,+21 Sandberg et al 1980
56,XY,t(9;22)(q34;q11),+4,+5,+8,+9,+11,+16,+17,+18,+19,+20 Mitelman 1983

Chronic lymphocytic leukemia

44-48,XY,t(Y;9)(q12;q13),del(1)(p22p32),t(4;6)(p16;q15),del(6)(q15),t(12;14)(q15;q32),+mar Robert et al 1982
46,XY,inv(4)(p14q12),del(7)(q11) Pittman et al 1982
46,XY,t(11;14)(q13;q32)/46,XY,t(3;4;9)(p12;q21q31;q31) Robert et al 1982

Lymphocytic leukemia, special type

45,X,-X,-4,+t(4;7)(p16;q11),15q+ Ueshima et al 1981
47,X,-Y,-4,+7,del(1)(q32),i(18q),21q+,i(22q),+2mar Ueshima et al 1981
47,X,-Y,dup(Y),1q-,4q+,6q-,7q+,+18q+ Miyamoto et al 1982b

MONOCLONAL GAMMOPATHIES

Multiple myeloma

48,X,-X,+1,+7,-8,+9,+21,del(1)(p36),4p+,del(5)(q22),9p+,14q+ Liang et al 1979

SOLID TUMORS

UNCLASSIFIED NEOPLASMS

EPITHELIAL NEOPLASMS

Carcinoma, NOS

43,X,-X,t(1;11)(q21;q23),3q+,4q+,5q+,-11,-13,-16,17p+ Atkin and Baker 1982a
44-45,XX,t(1;4)(q11;q35),+3,5q-,-14,i(17q),-21 Atkin and Baker 1982a
45,XY,-4,-7,+8,-9,-12,+13,-17,-18,-19,-20,+21,+3mar Reichmann et al 1981
46,XX,i(1q),t(1;4)(p?;q?) Atkin and Baker 1982a
46,XY,-4,-5,-11,-13,-17,+20,+21,1p+,9p-,+3mar Reichmann et al 1981
47,XY,-5,+7,-10,-13,-14,4q-,5q-,+4mar Reichmann et al 1981
49,XY,+4,+5,+8 Sonta et al 1977
55,XY,t(1;17)(p36;q21),+2,+2,+4,+8,+8,-10,+15,+18,+20,+21, +2mar Sonta and Sandberg 1978a
58,XY,del(3)(p14),+4,+7,+8,+10,+11,+12,+14,+15,+17,+18,+19, +21 Kovacs 1978a
60-63,XX,cx,t(5;18),t(3;6),t(3;22),del(10)(q24), del(4)(p1?) Riet-Fox et al 1979
63,XX,cx,t(1;4)(q21;q35),7p+,i(17q) Kovacs 1978a
66,XX,t(Y;13),1q+,i(4q),i(9p),i(9q),cx Martin et al 1979
71-75,XX,cx,11p+,17q+,del(1)(p31),del(3)(p13),del(4)(p1?) Riet-Fox et al 1979

Carcinoma NOS, metastatic

35,XX,-1,-3,-4,-5,-7,-8,-10,-11,-12,-13,-14,-15,-16,-17, -18,-19,-20,-21,-22,+8mar Pathak et al 1979
47,XX,+4,+16,-21 Musilova et al 1981
64-100,XX,t(6;?)(p11q27;?),t(4;10)(q?;q?),t(3;?)(q25;?), del(1)(q2?),t(6;11)(p21;q13) Kakati et al 1975
81-83,XX,t(1;3)(p22;p25),del(1)(p13),t(5;13)(q13;q14), del(4)(p12),i(2)(p11),i(5)(p11),t(4;5)(q12;q11q13) Kakati et al 1976a

Large cell carcinoma

55-70,XY,t(4;5;14),del(1)(q11q21),t(7;11)(p22;q13), t(7;21)(q11;q22),t(3;8)(p?;q?),t(6;11)(p11;p11),i(16q), t(3;?)(q11;?),t(21;?)(q11;?),t(13;?) Kakati et al 1975

Small cell carcinoma

46,XX,del(1)(p?),del(4)(p?),del(1)(q32) Wurster-Hill and Maurer 1978
47,XY,+del(1)(p?),+del(3)(q?),+del(3)(p?),-4,+8,-15,+18, -22 Wurster-Hill and Maurer 1978
77-78,XX,del(1)(q?),del(1)(p?),del(3)(p?),del(3)(q?), del(4)(p?) Wurster-Hill and Maurer 1978
80,XX,cx,del(1)(q32),del(3)(p14p21),t(3;4)(p23q11;q11), t(5;20)(q11;q13),dmin Whang-Peng et al 1982b

Papillary carcinoma

Karyotype	Reference
68,XX,cx,t(6;14)(q21;q24),del(1)(p?q?),1p+,4q-,8q-,10q+	Wake et al 1980
69,XX,cx,1p+,1q-,1p-,1q-,2q+,3q+,4q+,6p+,12q+,14q+	Wake et al 1980

Adenoma

Karyotype	Reference
46,X,t(X;4)(q25;p16),del(2)(p13p21),del(5)(q15q31),-7, t(9;12)(p13;q13),del(11)(q11q14), t(7;9;12)(p11p22q11;p13p24;q13)/45,X,-X	Mark et al 1980

Adenocarcinoma

Karyotype	Reference
44-50,X,-X,-1,t(1;?)(q21;?),del(2)(q33),4p+,-5,-7,+8, del(10)(q12),t(10;13)(q12;q34),del(17)(q25),-20	Kusyk et al 1982
71,XX,cx,1q-,4q+,6q-,9q+,10q-,12q+,17q+,14q+	Wake et al 1981

Adenocarcinoma, metastatic

Karyotype	Reference
??,XX,del(3)(p14),del(4)(q22),del(4)(q13),del(16)(q21)	Ayraud et al 1977
??,XX,del(3)(p14),del(4)(q13),del(16)(q23)	Ayraud et al 1977
??,XX,t(1;16),2q-,del(3)(p14),del(4)(q22),del(6)(q22), i(7q),del(9)(q31),t(13;13)	Ayraud et al 1977
??,XX,t(1;16),del(1)(q12),del(4)(q22),del(16)(p11)	Ayraud et al 1977
39-41,XX,-3,-7,-8,der(1),der(2),der(4),der(12),cx	Tiepolo and Zuffardi 1973
40,XX,t(4;5;8),t(3;?)(q21;?),t(1;11)(q21;p11), t(9;11)(q11;p11),del(1)(p36)	Kakati et al 1975
40-45,XY,-3,-8,-18,+5,+10,11p+,14q+,del(4)(q32)	Ayraud 1975
47-48,X,Xq+,t(3;9)(q12;p11),t(3;3)(q29;p25),del(3)(p25), t(4;?)(q35;?),+8	Kakati et al 1975
49,XX,1q+,+3,4q+,+7,+8,12q+,17q+	Granberg et al 1973

MESENCHYMAL NEOPLASMS

Sarcoma, NOS

Karyotype	Reference
48,XX,+4,+5	Sonta et al 1977

MELANOMAS

Malignant melanoma

Karyotype	Reference
78-82,XY,t(4;9)(q21;q21),del(9)(p13),t(14;14)(p11;q12), del(2)(q24),t(14;?)(q11;?),t(21;?)(q11;?),del(17)(q11)	Kakati et al 1977

Malignant melanoma, metastatic

Karyotype	Reference
41,XY,dup(1)(q21q32),t(1;7)(q25;q36),t(7;9)(q11;q11), del(9)(q13),-3,-4,-5,-8,-10,-11,-16,+18,-20,-22	Kakati et al 1977

41-43,XY,del(1)(q21q25),i(1q),t(1;9)(p22;q34), t(11;12)(q11;p11),t(10;12)(q26;q13),t(4;?)(q35;?), del(5)(p13),t(14;?)(p12;?),i(17q),t(18;?)(p11;?), del(6)(q21) — Kakati et al 1977

NEUROGENIC NEOPLASMS

Astrocytoma, grade III, IV

44,XY,-4,-17,-22,1q-,5p-,-11q+,+mar — Yamada et al 1980

Neuroblastoma

44,X?,del(1)(p22),-4,-7,del(7)(q22),t(7;16)(q22q32;q24) — Trent 1983
47,XY,1p+,4p+,6q+,10q+,12q+,+19 — Brodeur et al 1977
53,XX,+1,+4,+15,+20,dup(1)(p13p32),t(3;15)(q13;q26), del(4)(q31),t(7;11)(q36;q13) — Brodeur et al 1981b

Retinoblastoma

46,XX,inv(1),inv(2),inv(3),inv(4),t(6;7),inv(9),t(11;?), 11p-,inv(12),i(17q),-20 — Gardner et al 1982
46,XY,+4,t(5;13)(q35;q21),t(7;17)(p22;q12),t(1;9)(q22;p24), -10,del(16)(q11) — Gardner et al 1982
47,XX,12p+,12q+,+i(17q),t(4;16)(q35;q24),-16 — Gardner et al 1982
47,XX,i(6p),+4q+,-16 — Kusnetsova et al 1982
47,XY,t(X;1)(q28;p11),t(2;?)(q11;?),t(2;15)(p25;q11), t(3;?)(q27;?),4p+,t(5;?)(q35;?),i(17q),+mar — Gardner et al 1982

Meningioma, benign

38,X,-Y,-22,-21,-17,-14,-13,-10,-9,-4,-1p-,+mar — Zankl et al 1975a
45,XX,-15,-22,1p-,2p-,4p-,+mar — Zankl et al 1979
45,XY,-22/44,XY,-4,-22 — Mark 1973a
45,XY,1q+,-4 — Zang and Zankl 1983

EMBRYONAL AND MISCELLANEOUS NEOPLASMS

Blastoma, NOS

46,XY,t(4;?)(p14;?),t(9;11)(q22;p14),del(11)(p13p15), del(11)(q21q23) — Slater and Kraker 1982

Teratoma, malignant

67,XY,+1,+3,+4,+5,+6,+7,+7,+8,+10,+11,+13,+13,+14,+16,+17, +18,+20,+20,+21,+21 — Sonta et al 1977

Seminoma

53,XY,+3,+4,+6,+7,+11,+13,+19	Sonta et al 1977

LYMPHOMAS

1. UNCLASSIFIED AND MISCELLANEOUS LYMPHOMAS

Non-Hodgkin's lymphoma, NOS

44-47,XY,3q+,del(4)(p13),del(6)(p23),+7,del(9)(p13)	Kristoffersson 1983
48-54,XX,+X,3p+,-4,-6,-8,-9,-10,+12,-15,+7-10mar	Pearson et al 1982

Non-Hodgkin's lymphoma, lymphocytic, NOS

42-45,XX,-4,-5,-9,-16,+mar	Fleischmann et al 1976
46-47,XY,-4,-15,-16,+3mar	Fleischmann et al 1976
47,XY,t(3;4)(q29;p12),del(6)(p22),+7,del(9)(p13)	Kristoffersson et al 1981
47-48,XY,+7,-11,-14,-15,4q+,9p-,14q+	Fleischman and Prigogina 1977
48-49,XY,1q-,+3,+del(3)(p?q?),-4,-6,+7,-9,-17,-18,-19, +5mar	Fleischmann et al 1976
80,XY,cx,4p+,t(1;4)(p22;p15),t(1;6)(p11;p11)	Kakati et al 1980

Non-Hodgkin's lymphoma, histiocytic, NOS

45,X,del(Y)(q12),t(4;10)(q12;q24),t(6;7)(q21;q32), del(8)(p21)	Mark et al 1978
48,XY,t(3;8)(q22;q11),t(5;10)(q33;q22),del(7)(q21), t(4;9)(q11;p24),t(10;13)(q22;q21),r(12),del(19)(p13), del(22)(q11),+7	Mark et al 1978
50,XY,dup(1)(q12q32),+2,4p+,del(4)(q21),t(8;14)(q24;q32), +17p+,18q+,+20	Brynes et al 1978

Non-Hodgkin's lymphoma, T-cell, NOS

47,XX,t(2;?),inv(3),4p-,del(5)(q14q32),6p+,6q+,t(7;7),9q+, 10p-,13q-,t(14;14),-15,-15,16q-,17p+,+21,+2mar	Gaeke et al 1981

2. HODGKIN'S LYMPHOMAS

Hodgkin's disease, lymphocytic depletion

46,XY,+2,t(2;?)(q25;?),t(3;11)(q29;p13),t(4;11)(q33;q13), -13	Hossfeld and Schmidt 1978

3A. NON-HODGKIN'S LYMPHOMAS - RAPPAPORT CLASSIFICATION

Lymphocytic, poorly differentiated, diffuse

43,XX,-4,-5,-9,i(17q),-18,-21,+2mar — Goh et al 1980
47,XY,t(1;4)(q23;q33),+3,t(10;15),t(11;14)(q13;q32),r(13), r(13) — San Roman et al 1982
48,XX,1q-,+3,+4,6q-,+7,10q-,-15 — Kakati et al 1980
50,XX,+3,+4,+10,+18/51,XX,+3,+4,+9,+10,+18 — Mark et al 1979

Lymphocytic, poorly differentiated, nodular

52,X,-X,+2,-4,+5,+7,+8,-14,-15,-16,+20,+21, +t(1;15)(p11;q11),+t(12;17)(q24;q11),+t(14;?)(q32;?), +mar — Kaneko et al 1982a

Mixed cell, diffuse

46,XY,-4,-9,del(6)(p23),+2mar — Kaneko et al 1982a
53,XY,t(8;14)(q24;q32),t(14;18)(q32;q21),+20,1q-,+1p-,2q-, 4q+,5q-,5p+,6q-,+7q+,10q+,11q+,11q-,13p+,+4mar — Fukuhara and Rowley 1978

Histiocytic, diffuse

46,X,-X,-4,-11,-17,+18,-19,del(7)(p15),del(8)(q22),+i(17q), +t(4;19)(p11;q13),t(11;?)(q23;?) — Kaneko et al 1982a
47,X,-X,t(4;14)(q?;q32),-18,1q+,2p+,6q-,7q+,9q-,12q+,+3mar — Fukuhara and Rowley 1978
47,XY,-10,-13,+15,-16,+19,+20,del(6)(q21),del(8)(q22), t(2;14)(q21;q32),t(4;4)(q31;q35),+t(10;?)(p11;?) — Kaneko et al 1982a
50,XX,+3,+3,-4,-6,-9,+12,-17,+18,+i(6p),+3mar — Kaneko et al 1982a
50,XY,t(8;14)(q24;q32),+2,+20,1q+,4p+,+4q+,17p+,18q+ — Fukuhara and Rowley 1978

Histiocytic, nodular

46,XY,-4,inv(16)(p11q24),+t(2;4)(p11;p11) — Kaneko et al 1982a

3B. NON-HODGKIN'S LYMPHOMAS - KIEL CLASSIFICATION

Lymphocytic, T-zone

45,XX,-1,-2,-3,-4,+t(1;2)(p32q32;p13),+t(2;4)(p13;p15), +t(3;4)(q11;?) — Gödde-Salz et al 1981a

Centrocytic

44,XY,del(4)(p13),-11,-13,t(13;14)(q11;q32) — Kristoffersson 1983

Centroblastic-centrocytic, diffuse

Karyotype	Reference
48-51,XY,del(1)(p33q23q25),-2,+del(3)(p22),+4,4p+,+5,+6p+, +7,+8,+9,+21,+22	Kristoffersson 1983
45-51,XY,del(1)(p33q23q25),del(2)(q14),+del(3)(p22),+4,+5, +7,+8,-12	

Centroblastic-centrocytic, follicular

Karyotype	Reference
46-47,XY,del(4)(q28),-10,+12,-18,del(18)(q22)	Kristoffersson 1983
46-49,XX,1p+,3q-,4q+,+6,+13,-14,14q+,17p+	Kristoffersson 1983

Lymphoblastic, Burkitt's type

Karyotype	Reference
46,XY,t(4;5;7)(q13;p13;p22),t(8;14)(q23;q32)	Kaiser McCaw et al 1977
46,XY,t(8;22)(q23;q11)/46,XY,t(8;22)(q23;q11), t(1;4)(p32;q32)	Berger et al 1981c

Lymphoblastic, convoluted type

Karyotype	Reference
47,XY,t(1;2)(p34;p21),+4,t(13;18)(p13;p11)/47,XY,+3	Kaneko et al 1982d

Immunoblastic

Karyotype	Reference
45-46,XX,+3,del(6)(q21),+4,+16,+21,-C,-C	Kristoffersson 1983
46,XY,del(4)(q33),del(6)(q23),t(5;7)(q33;q22)	Kaneko et al 1982a
47-48,XY,+4,del(4)(q21),+5,t(15;?)(q25;?),+19,-21	Reeves and Stathopoulos 1976

3C. NON-HODGKIN'S LYMPHOMAS - LUKES AND COLLINS CLASSIFICATION

Mycosis fungoides

Karyotype	Reference
47,XY,+Xp-,t(4;14),+14	Edelson et al 1979

Follicular center cell, diffuse, large cleaved

Karyotype	Reference
45,XY,-6,-9,-17,2q-,3p-,3q-,4q-,5q-,6p+,7p+,7q+,+7q-,9p-, 12q+,13p+,20p-,+mar	Fukuhara et al 1978b

Immunoblastic type (B-cell)

Karyotype	Reference
48,XY,t(4;12)(q?;q?),t(6;15)(q?;q?),t(9;22)(q34;q11),11q-, +20,+21	Nordenson et al 1983

3D. NON-HODGKIN'S LYMPHOMAS - WORKING FORMULATION

Follicular, small cleaved cell

Karyotype	Reference
46,XY,t(1;2),t(4;10),t(14;18)(q32;q21),t(1;22)	Yunis et al 1982
48,X,-X,inv(1),1q-,+3,4q-,ins(5),+7,+mar	Yunis et al 1982

Chromosome 5

HEMATOLOGICAL DISORDERS

UNCLASSIFIED LEUKEMIAS

Acute leukemia, NOS

41-44,XX,5q-,i(8q),11q+,+mar	Prigogina et al 1979

NONLYMPHOCYTIC LEUKEMIAS

Acute nonlymphocytic leukemia (ANLL)

??,XX,del(5)(q?),cx	Van Den Berghe et al 1976
??,XY,del(5)(q?),cx	Van Den Berghe et al 1976
39,X,-X,-3,-5,-13,-14,-15,-16,-17,-21,-22,del(7)(p11), t(12;?)(p13;?),t(16;?)(p13;?),+3mar	Mitelman 1983
39,X,-X,-3,-5,-13,-14,-15,-16,-17,-21,-22,del(7)(p11), t(12;?)(p13;?),t(16;?)(p13;?),+3mar	
42,XX,-5,-12,-14,-16,-18,t(12;16)(q11;q22q24)"s"	Oshimura et al 1976b
42,XY,-5,t(12;?)(p11;?),-18,-22,-22	Hustinx and Rutten 1983
43,XX,-1,3p+,3q-,4q+,-5,-5,-7,+9,+9,10p+,+10,+10,10p-,10q-, +11q-,-13,-15,-16,-17,-18,-21"s"	Whang-Peng et al 1979
43,XX,-4,-5,-21,t(8;21)(q22;q22)	Oshimura et al 1976b
43,XX,2q-,inv(3),-4,del(5)(q?),-7,17p+,-19	Rowley 1976
43,XX,t(1;20),-5,-6,7q-,11q-,-12,17q+,-18,19q+,-21	Mitelman 1983
43-44,XX,-5,-7,-14,+mar,dmin	Marinello and Levan 1982
43-44,XY,-5,-5,-14,-15,-16,-17,-18,-20,+5mar	Mitelman 1983
43-45,XY,-5,5q-,-7,-19"s"	Whang-Peng et al 1979
44,XY,5q-,-7,12p-,-22"s"	Whang-Peng et al 1979
44,XY,del(5)(q21),del(7)(q11),del(11)(q11), t(7;11;12)(q11;q11;p13),del(13)(q11),-20,-21/45,XY, t(3;12)(q11;p13),del(5)(q21),del(7)(q11),del(13)(q11), -20	Mitelman 1983
44-45,XX,-3,-5,-6,-18,-19,t(1;2)(q2?;q11),t(5;?)(q13;?)	Philip et al 1978a
45,XX,-5	Pedersen-Bjergaard et al 1980
45,XX,del(3)(p11),t(5;7;12)(q31;q22;q13-24),t(5;?)(q13;?)	Philip et al 1978a
45,XX,t(13;14),5q-"c""s"	Rowley 1976
45,XY,-5,-15,+r"s"	Berger et al 1981a
45,XY,-5,t(20;?)(q13;?)	Hustinx and Rutten 1983

45,XY,5q+,-20	Bernard et al 1982a
45,XY,t(1;14)(p22;p11),t(1;15)(p22;p11),-2,del(5)(p13),-17, +mar	Philip et al 1978a
46,XX,4q-,5q-,-16,-17,+2mar	Rowley 1976
46,XX,5q-	Nordenson 1983
46,XX,5q-,"s"	Berger et al 1981a
46,XX,5q-/47,XX,5q-,+21	Rowley 1976
46,XX,del(1)(p32),del(5)(q13),+8,t(9;?)(p11;?), t(11;22)(q11;q13),-16,-17,-19,+3mar	Muir et al 1977
46,XX,del(5)(q12q31),del(11)(q23)	Sadamori et al 1981a
46,XX,del(5)(q12q31),-9,15p+,18p-,-21,+2mar	Kerkhofs et al 1982
46,XX,del(5)(q?)/47,XX,del(5)(q?),+9"s"	Verhest et al 1977
46,XX,del(5),del(6),del(7),21q+"s"	Geraedts et al 1980a
46,XX,del(7)(q22)/46,XX,del(7)(q22),del(5)(q22q31)/46,XX, del(7)(q22),-9,t(1;9)(q22;p24)/46,XX,t(2;3)(q31;q27), del(7)(q22),del(5)(q22q31orq33)"s"	Oshimura et al 1976b
46,XX,t(1;22)(p36;q11),del(5)(q14q34),del(7)(q22q36),r(22)	Yunis 1982
46,XY,-5,dmin	Bernard et al 1982a
46,XY,1q+,5p-,-7,+mar	Chrz et al 1983
46,XY,5q-	Bernard et al 1982a
46,XY,del(5)(q12q31)"s"	Cabrol and Abele 1978
46,XY,del(5)(q13q31)"s"	Geraedts et al 1980a
46,XY,del(5)(q15)	Mitelman 1983
47,X,Xq-,t(3;5),-5,t(11;?),t(12;17),-18,-21,+6mar	Pedersen-Bjergaard et al 1980
47,XX,del(5)(q15q23),+8	Oshimura et al 1976b
47,XX,del(5)(q?),+21	Van Den Berghe et al 1976
47,XX,t(3;5)(q21;q31),t(1;13)(p36;q14),+8,+22	Oshimura et al 1976b
47,XY,del(5)(q?),+8	Van Den Berghe et al 1976
48,XX,3q+,-5,+9,+16,+19	Nordenson 1983
48,XY,del(3)(q13),-5,+8,+r(?),+mar	Mitelman 1983
48-49,XX,-5,del(15)(q?),+19,t(5;?)(q13;?)	Philip et al 1978a
49,X,t(X;10)(p11;p11),+del(1)(p22),+del(1)(p22), del(5)(q23),+7/49,X,t(X;10)(p11;p11),+del(1)(p22),+7,+8	Mitelman 1983
49,XY,t(5;?)(q35;?),+6,+13,+14	Philip et al 1978a
51,XX,del(5)(q?),+7,+8,+13,+17,+22	Van Den Berghe et al 1976
51,XY,+3,+5,-9,+19,+20,+21,t(6;14)(p25;q11),t(9;?)(q22;?)	Philip et al 1978a
56,XY,+Y,+6,+14,+21,+t(5;6)(q?;p?),+del(1)(p22q25), +del(2)(q12),+del(3)(p12),+del(4)(q13),+del(5)(p14q14), +mar	Zuelzer et al 1976

<u>ANLL, FAB type M1</u>

39,XY,-5,-7,-12,-16,-17,-20,-21	Berger et al 1981a
41,XY,-1,-2,-4,-5,-12,-15,-17,-17,-18,-22,+5mar	Sessarego et al 1981a
43,XX,del(5)(q12q21),-7,11p+,-15,-17	Kerkhofs et al 1982
44,XX,-5,-7,-20,-22,+2mar/44,XX,-3,-11,-11p+,-22	Prieto et al 1981
45,X,-X,-5,-7,-11,-17,+21,+3mar"s"	Zaccaria et al 1983b
45,XX,del(5)(q12q31),-7,12p-"s"	Kerkhofs et al 1982
45,XY,-5	Alimena 1983
45,XY,3p-,5q-,-7"s"	Pedersen-Bjergaard et al 1981
45,XY,5q-,-7"s"	Pedersen-Bjergaard et al 1981

Karyotype	Reference
45,XY,t(2;5)(q22orq23;q14q15),6p-,-7,t(6;7)(p21;p22q21), t(7;10)(p21;q21),t(14;19)(q11;q13)"s"	Hagemeijer et al 1981a
46,XX,-1,inv(1),3q+,5q-,t(13;17),-18,+mar/51,XX,-1,inv(1), 3q+,5q-,t(13;17),-18,+6,+9,+9,+11,+11,+mar	Hagemeijer et al 1981a
46,XY,-2,-5,-7,-13,+16,-21,-21,+5mar	Sessarego et al 1981a
46,XY,del(5)(q13q31)"s"	Kerkhofs et al 1982
46,XY,t(3;5;10)(p?;q?;p?)	Lessard and Le Prise 1983
49,XX,+X,-1,+der(1)(q21q31),+del(1)(p22),+t(1;5)(p11;q35), -5,t(9;22)(q34;q11),+22q-	Sasaki et al 1983
58,XY,+1,+4,+5,+6,+8,+11,+13,+13,+14,+14,+15,+21,+21, t(9;22)(q34;q11)	Alimena 1983

ANLL, FAB type M2

Karyotype	Reference
42,XY,5q-,7q-,-8,-12,-17,-21,-22,+i(17q),+2mar	Berger et al 1981a
43,XX,+2,3q+,-5,-20,-22	Berger et al 1981a
43,XX,-3,-4,-7,-19,del(5)(q14),17p+	Testa et al 1979
43,XX,-3,-4,-7,-19,del(2)(q31?),del(5)(q14)	
43,XY,-5,-7,-12,-16,4p+,17p+,20q-,+mar"s"	Rowley et al 1981a
44,X,t(X;8),del(5)(q12q21),t(1;6),t(7;19),-7,-8,t(7;8),-21	Kerkhofs et al 1982
44,XX,-5,-7,-18,+mar"s"	Rowley et al 1977a
45,X,-Y,t(8;21)(q22;q22),t(1;5)(q43;q11)	Berger et al 1982b
45,XX,5q-,-7,-18,+mar	Teerenhovi et al 1981
45,XX,5q-,-C,-C,+G"s"	Pedersen-Bjergaard et al 1981
45,XY,-5,-7,+mar"s"	Rowley et al 1977a
45,XY,-5,-7,-8,+16,-17,+r,+mar	Berger et al 1981a
45,XY,-5,-7,del(3)(p13),+t(5;17)(q11;q11)"s"	Rowley et al 1981a
45,XY,-7,del(5)(q13),t(12;17)(q13?;p12?)"s"	Rowley et al 1981a
45-47,XX,-9,-17,-20,del(4)(q21),del(5)(q12q31), t(9;17)(p13;p11),+mar	Testa et al 1979
46,XX,-4,5q-"s"	Pedersen-Bjergaard et al 1981
46,XX,2q+,5q-,7q-,17q+"s"	Berger et al 1981a
46,XX,5q-	Teerenhovi et al 1981
46,XX,del(5)(p13)	Yunis et al 1981
46,XX,del(5)(q14q34)	Yunis et al 1981
46,XX,t(8;21)(q22;q22),del(5)(q11?;q13?)	Bernstein et al 1982b
46,XY,+1,5q-,-7"s"	Berger et al 1981a
46,XY,-7,-12,-17,del(5)(q14),8q+,del(17)(p11),+mar	Testa et al 1979
45,XY,-7,+8,-12,-17,del(5)(q14),8q+,t(12;17)	
46,XY,del(5)(q13?q22?),t(9;22)(q34;q11)	Sasaki et al 1983
46,XY,del(5)(q14q33),t(6;14)(q23;q22)	Kerkhofs et al 1982
47,XX,1p+,der(5),-6,12p+,-13,-14,-16,-17,-17,-21,+22,+7mar, dmin	Cooperman and Klinger 1981
47,XY,del(5)(q15),+21	Swolin et al 1981
48,XX,-5,t(8;?)(p23;?),del(11)(q23),t(15;?)(p11;?),-21, +4mar"s"	Rowley et al 1977a
48,XX,5q-,8p-,+2mar"s"	Pedersen-Bjergaard et al 1981
53,XX,+1,-4,-5,+6,-7,+8,+11,+12,+21,+21,+3mar"s"	Zaccaria et al 1983b

ANLL, FAB type M1+M2

Karyotype	Reference
37-45,XY,-7,-12,-13,-14,-16,-18,-19,-21,del(4)(q29), del(5)(q24),der(9),der(17),11q+,12p+,t(14;?)	Fitzgerald et al 1983
42-44,XX,+8,-12,-13,-14,t(2;5)(q12;q35)	Mitelman et al 1981
43,XX,1p+,-4,-5,-6,-7,-13,-17,del(18)(q21),+3mar	Rowley and Potter 1976
43,XX,del(5)(q12q32),-7,-12,-18,t(7;12;21)(q11;q12;q22)	Petit and Van Den Berghe 1979b
43,XY,-3,-5,-21	Rowley and Potter 1976
44,X,-X,-5,t(3;7)(q13;p12)/44,XX,-7,-8/44,XX,-5,-8"s"	Sandberg et al 1982a
44,XX,-3,-4,-5,+9,-17,-18,-21,-22,+5mar	Garson 1980
44,XX,del(3)(p21),del(5)(q12),-7,-20,i(22)(q11)	Mitelman et al 1981
44,XY,-5,-12,-18,+mar,+dmin	Prigogina et al 1979
44,XY,-5,t(5;12)(q12;q24),i(18)(q11),-19	Mitelman et al 1981
44-45,XX,-5,-7	Mitelman et al 1981
44-45,XY,+2,-5,-7	Mitelman et al 1981
45,X,-Y,+del(5)(q11),del(11)(q21),-14,-16,-18,-20,+2mar,+r	Mitelman et al 1981
45,X,-Y,del(5)(q22)/45,X,-Y,5q+	Shiraishi et al 1982
45,XX,-2,5q+	Yamada and Furusawa 1976
45,XX,-5,-11,-13,-18,-21,+4mar,dmin"s"	Sandberg et al 1982a
45,XX,t(3;6)(p23;p25),del(5)(q13),-12,i(17q)	Mitelman et al 1981
45,XY,-5,-19,-20,-22,+t(7;12)(q11;q13),+r,+3mar	Fitzgerald et al 1983
45,XY,t(5;11),t(5;17)	Garson 1980
45-46,XX,4q+,t(5;11)(q35;p13)	Mitelman et al 1981
45-46,XX,t(5;11),4q+,-21	Alimena et al 1977
46,X,del(X)(q21),t(5;21)(q13;q22)	Mitelman et al 1981
46,XX,-7,del(5)(q22),+r	Prigogina et al 1979
46,XX,del(5)(q12q32),del(12)(p11)	Petit and Van Den Berghe 1979b
46,XX,del(5)(q14)	Mitelman et al 1981
46,XY,4q+,del(5)(q15)	Mitelman et al 1981
46,XY,del(5)(q15)	Prigogina et al 1979
46,XY,t(5;8;21)(q31;q22;q22)/47,XY,t(5;8;21),+mar	Fitzgerald et al 1983
47,X,-X,+del(1)(p11),del(5)(q13),t(5;12)(q21;q24), t(11;12)(q25;q13),t(13;14)(q11;p11),+16,+21	Mitelman et al 1981
47,XY,+19,t(7;12)(q21;p13),del(4)(q26),del(5)(p12)	Mitelman et al 1981
47,XY,+del(5)(q?)	Benedict et al 1979
47,XY,del(3)(q12),del(5)(q12),t(6;?)(q11;?),-13,+2mar"s"	Testa et al 1981b
47-49,XY,-5,del(5)(q15),-7,i(8q),dic(9),-13,-16,-22, +6-9mar"s"	Testa et al 1981b
48,XX,+1,del(5)(q13orq15),+11"s"	Rowley and Potter 1976
48,XX,-5,del(5)(q13q31),+8,+t(11;?)(p14;?),-17,-19,+21, +2mar,dmin"s"	Testa et al 1981b
48,XX,t(1;3)(q44;p14),del(3)(q21p14),-5,+10,+11,+21	Mitelman et al 1981
49,XX,+1,-5,+7,+8	Nordenson 1983
50-52,XY,-5,-7,+9,+14,+17,+19,+20,+21,+mar	Mitelman et al 1981
51,XY,+1,+4,+5,+8,+20	Nordenson 1983

ANLL, FAB type M3

Karyotype	Reference
44,XX,-5,-8,-17,-18,+2mar	Gallagher et al 1979
45,XX,-4,-5,-14,t(21;?)(q22;?),+2mar"s"	Rowley et al 1977a
46,XY,t(3;5;17)(p21;q25?;q21?),4p+	Bernstein et al 1982b

ANLL, FAB type M4

Karyotype	Reference
42,XY,-5,-9,-10,-15,-16,-17,-20,-21,-22,+5mar	Testa et al 1979
43,XY,+1,-5,+6,-9,-10,-15,-16,-17,-20,-21,-22,+6mar	
43-46,X,-Y,del(5)(q31),-7,+18,-20	Swolin et al 1981
44,XY,del(5)(q15q23),del(7)(q22),-9,-12,-13,del(15)(q22), +mar/46,X,-Y,del(5),del(7),+8,-12,-13,del(15),+mar,dmin	Alimena 1983
44,XY,t(3;5)(q29;q13),t(7;12)(q12;q14),t(7;10)(q12;q21),-8, -17,-18,-20	Mitelman et al 1981
44-46,XY,+1,-5,-17,-18,-20,del(6)(q21),-15,t(1;6)(q11;q24), +3mar	Mitelman et al 1981
44-46,XY,+1,t(1;6)(q11;q27),-5,del(6)(q21),-15,-17,-18,-20, +mar	Alimena 1983
45,XX,2p-,-5,-9,-17,+r,+mar	Berger et al 1981a
45,XY,del(5)(q13q31),del(7)(q22),del(15)(q22q24)	Alimena et al 1982a
46,XX,-1,-3,-5,+10,+11,-13,-14,+21,+del(3?)(p?q?), ?t(1;3;5)(p?;p?q?;q?)	Mitelman 1983
46,XX,t(3;13)(q29;q12),del(5)(q13q31)/46,XX, t(1;18)(q44;q21),t(3;13),del(5)	Alimena 1983
46,XX,t(5;10)(q35;q23)	Kaneko et al 1982b
47,XY,+5/47,XY,+5,del(5)(q13)	Testa et al 1979
46,XY,t(3;12)(q21;q13),del(5)(q14),ins(8;5)(q22;q14q22), -17,+2mar	Mitelman 1983
46,XY,t(3;5)(q25;q33)	Rowley and Potter 1976
46,XY,t(5;19),16p+,19p+	Golomb et al 1978c
47,XX,+19,20p+/50,XX,+5,+8,+12,+mar	Morse et al 1978
47,XX,+5	Fitzgerald et al 1983
49,XX,+5,+8,+15	Fitzgerald et al 1983

ANLL, FAB type M5

Karyotype	Reference
45,XY,-5,11p+	Brodeur et al 1983
46,XY,t(17;21)(q11;q11orq21),+i(8q),del(5)(q11q13),9p+, 11q+	Fitzgerald et al 1983
46,XY,t(5;9)(p11;p11)	Mitelman 1983
46,XY,t(5;9)(p11;p11)	
57-58,XX,+1,+3,-4,-5,+6,-8,+11,+12,-17,+19,-20, del(10)(q24),+8-9mar	Testa et al 1979
60-61,XX,Xp+,+1,+2,-4,+6,-7,+19,+20,del(2)(q31), del(10)(q24),+12-13mar	

ANLL, FAB type M5a

Karyotype	Reference
44,Y,-X,-5,del(11)(q22),17p+,+r	Berger et al 1982a

ANLL, FAB type M4+M5

Karyotype	Reference
42,XX,-5,-7,-12,-15,-16,-18,+2mar	Li et al 1983

ANLL, FAB type M6

Karyotype	Reference
??,XY,t(4;?)(q23;?),t(5;17),t(8;9),+8,+11, t(12;16)(p12;p12)	Hustinx and Rutten 1983
43,X,-X,del(5)(q14q34),-7,11q-,-12,12p+,t(12;17),t(20;21)	Kerkhofs et al 1982
43,XX,5q-,-7,-17,-18	Garson 1980
43,XY,-4,5q-,t(4;6),-7,dic(12),-15"s"	Hagemeijer et al 1981a
43-45,XX,der(1),inv(2)(p25q14),del(3)(q25),-4,-5,-6,6q-,-7, 7p+,11p+,-12,-17,-21,+r,+1-4mar"s"	Smadja et al 1983b
44,XX,del(5)(q13q31orq33),del(11)(q21),-17,-21"c""s"	Papa et al 1979
44,XY,-3,5q-,-7,-D,+20	Li et al 1983
44,XY,3p-,t(2;6),t(5;6),t(17;20)"s"	Sandberg et al 1982a
44,XY,t(3;5;11),17p+,-22	Garson 1980
44-46,XY,-5,-17,-19,+mar	Mitelman et al 1981
44-47,XY,+2,-3,-5,-7,+8,-14,-19,+2mar	Mitelman et al 1981
45,X,-X,-5,-21,22q-,+3mar"s"	Sandberg et al 1982a
45,XY,del(4)(q23),del(5)(q13),-6,del(6)(q21),del(7)(p14), del(8)(p21),del(9)(q31),16q+	Rowley and Potter 1976
45,XY,del(5)(q11),-7/45,XY,del(5)(q11), ins(3;3)(q12;q21q27)"s"	Swolin et al 1981
45,XY,del(5)(q13q31),r(8),-17,-21,+mar	Kerkhofs et al 1982
46,X?,+1,-5,-7,+mar	Wurster-Hill 1983
46,XY,5p-,-9,-12,13q-,+2mar	Nordenson 1983
46,XY,t(3;5)(q13;q28)	Hagemeijer et al 1981a
47,XX,-5,i(17q),-19,+3mar	Najfeld 1976
47,XY,+Y,3p-,5q-,10q-	Garson 1980

ANLL, special type

Karyotype	Reference
48,XY,-5,-17,+idic(17q),+idic(17q),+idic(17q),+mar	Atkin et al 1981

Chronic myeloid leukemia, t(9;22)

Karyotype	Reference
26,XX,t(9;22)(q34;q11),-1,-2,-3,-4,-5,-6,-7,-9,-10,-11,-12, -13,-14,-15,-16,-17,-18,-19,-20	Hartley and McBeath 1981
35,XX,t(9;22)(q34;q11),-3,-4,-5,+8,-9,-10,-11,-12,-13,-14, -15,-16,-17/66-72,XX,t(9;22)	Pedersen and Boesen 1983
43-46,XX,t(9;22),-5,-7,-10,-17,-19,5p+,14q+,+r(?),+mar	Sadamori et al 1981b
46,X,del(X)(q?),t(9;22),5q-,inv(11),16q+,20q+	Hagemeijer et al 1980b
46,XX,t(9;22)(q34;q11)/51,XX,t(9;22),+5,+8,+17,+19,+22q-	Misawa 1978
46,XX,t(9;22)(q34;q11),5q+,t(8;19),17p+	Alimena et al 1982b
46,XX,t(9;22)(q34;q11),del(5)(q14q23)	Tomiyasu et al 1982
46,XX,t(9;22),t(5;21)(q?;q?)	Sadamori et al 1980
46,XY,t(9;22)	Hagemeijer et al 1981b
46,XY,t(9;22),r(18)	
44,XY,t(9;22),r(18),t(5;17)(q11;p11)	
46,XY,t(9;22)	Mitelman 1983
50,XY,t(9;22)(q34;q11),del(3)(q11),+4,+5,+12,+22q-	
46,XY,t(9;22)(q34;q11)	Como and Graze 1979

Karyotype	Reference
35,XY,t(9;22)(q34;q11),-3,-4,-5,-7,-9,-11,-12,-13,-15,-16, +3mar	
46,XY,t(9;22)(q34;q11)	Ishihara et al 1982
27,X,-Y,t(9;22),-1,-2,-3,-4,-5,-6,-7,-9,-11,-13,-14,-15, -16,-17,-18,-19,-20,-22	
46,XY,t(9;22),-5,-17,t(5;17),+22q-	Borgström et al 1982
46,XY,t(9;22),r(18)/45,XY,t(9;22),-5,-17,t(5;17)(q11;p11), r(18)	Hagemeijer et al 1980b
46,XY,t(9;22),t(X;6)(q28?;q22?),t(5;17)(q13?;q25)	Bernstein et al 1980b
46,XY,t(9;22)/46,XY,t(9;22),del(5)(q12),del(13)(q21), t(17;21)(p11;q11)	Bernstein et al 1980b
46-49,XX,t(9;22),+del(22q),+5,+6,+9	Prigogina et al 1978
47,XX,t(9;22)/47,XX,t(9;22),t(5;11)(p15;q22), del(5)(q21q31),+del(8)(q21q23)	Hagemeijer et al 1980b
47,XY,t(9;22)(q34;q11),t(3;5),-8,-10,+17,+22q-,+mar	Alimena et al 1982b
47,XY,t(9;22)(q34;q11),-5,+8,+mar	Tomiyasu et al 1982
47,XY,t(9;22),+17/47,XY,t(9;22),-22,+5,+22q-/49,XY,t(9;22), +5,+21,+22q-	Stoll and Oberling 1978
47,XY,t(9;22),+22q-/56,XY,t(9;22),+1,+2,+5,+6,+7,+8,+10, +12,+17,+22q-	Stoll and Oberling 1978
47,XY,t(9;22),+22q-/46,XY,t(9;22),-9,+22q-/46,XY,t(9;22), t(3;5)(p?;q?),t(6;13)(p?;q?)	Kohno and Sandberg 1980
48,XY,t(9;22),+8,+13,-15,t(9;15),+i(22q)	Stoll and Oberling 1978
50,XY,t(9;22),+5,+8,+13,-15,t(9;15),+i(22q),+22q-	
48,XY,t(9;22),+8,+22q-/61,XY,+Y,t(9;22),+4,+5,+6,+8,+8,+8, +8,+15,+17,+18,+19,+20,+22q-,+22q-	Wurster-Hill 1983
50-54,XY,t(9;22),+1,+5,+8,+9,-16,+17,+18,+19,+21,+22q-	Alimena et al 1980
50-55,XY,t(9;22)(q34;q11),+1,+5,+8,+9,-16,+17,+18,+19,+21, +22q-,+mar	Alimena et al 1982b

Chronic myeloid leukemia, aberrant translocation

Karyotype	Reference
??,X?,t(5;9;22)(q13;q34;q11)	Potter et al 1981
46,X?,t(5;9;22)(q35;q11;q11)	Ishihara 1983
46,XX,inv(5)(p14q12),t(9;13;17;22)(q31;q11;q12;q12)	Fraisse et al 1980
46,XX,t(17;22),inv(5),t(9;13)	Freycon et al 1982
46,XX,t(5;22)(q35;q11)/46,XX,t(5;22),i(17q)/47,XX,t(5;22), +8,i(17q)	Pasquali et al 1979b
46,XX,t(5;9;22)(q13;q34;q11)	Alimena 1983
46,XX,t(9;15;22)(q34;q11;q11)	Hays et al 1981
54-57,XX,t(9;15;22),+5,+7,+8,+10,+11,+12,+17,+22q-	
46,XY,t(4;22)(q35;q21),t(3;5)(q27;q22)	Sessarego et al 1981b
46,XY,t(4;22)(q35;q21),t(3;5)(q27;q22),i(17q)/50,XY, t(4;22),t(3;5),i(17q),+8,+15,+21,+22q-	
46,XY,t(5;13)(q33;q14),del(22)(q11)	Rolovic et al 1983
46,XY,t(5;9;22)(q13;q34;q11)	Oshimura et al 1982a
47,XY,inv(9)(p11q13),t(5;9;22)(q13;q34;q11), +del(22)(q11)"c"	Nowell et al 1975

MYELOPROLIFERATIVE DISORDERS

Polycythemia vera

Karyotype	Reference
??,X?,+5,+8	Vykoupil et al 1980
44,XY,-3,-5	Nqwell and Finan 1978
46,X?,5q-,11q-	Lessard and Le Prise 1983
44,X,-X,del(5)(q14q32),-7	Van Den Berghe et al 1979g
44,XX,del(5)(q14q32),-6,-7,-8,t(12;?)(p12;?), t(16;18)(q11;q12)	
46,XY,t(1;5)(p36;q31)	Westin 1976
46,XY,del(5)(q14q32),del(7)(q21q31)	Van Den Berghe et al 1979g
46,XY,del(20)(q11),t(5;20)(q35;q11)	Zech et al 1976b
46,XY,del(5)(q15),del(20)(q11)/46,XY,del(11)(q22)/46,XY, del(13)(q14q32)	Testa et al 1981b
47,XX,+9/49,XX,+9,+20,+20,t(4;5),-7,-11,-12,+3mar	Zech et al 1976b
47,XY,+9	Van Den Berghe et al 1979g
47,XY,del(1)(q12),t(1;18)(q12;q23),+9,del(20)(q12)/47,XY, +9,del(11)(q21),del(20)(q12)/47,XY,+9,del(20)(q12)	
47,XY,+9/47,XY,del(5)(q14q32),del(8)(q?),+9,del(11)(q21), del(13)(q21)/47,XY,del(5)(q14q32),del(6)(q?),+9, del(11)(q21),del(13)(q21),del(1)(q12),del(20)(q12)	
47,XY,+del(1)(p21),del(5)(q31),del(20)(q11)/47,XY,+9, del(20)(q11)	Swolin et al 1981
48,XX,t(9;22),+5,+8	Armenta et al 1976
48,XY,+3,-5,t(13;14)(q?;q?),+2mar	Nowell and Finan 1978

Myelosclerosis / Myelofibrosis

Karyotype	Reference
44,XY,-5,-7,-11,-12,t(7;12)(p?;q?),t(17;18)(p11;p11), +mar"s"	Nowell and Finan 1978
44,XY,5p-,-7,-10,11q-,i(17q)/45,XY,-7,18q-/45,XY,2p-,2q-, 5p-,6q-,9q-,10q-,-11,+i(17q)	Whang-Peng et al 1978
46,X,del(X)(p?q?),del(5)(q?),13q+,15q-"s"	Whang-Peng et al 1978
46,XX,t(5;12)(q14;q13)/45,XX,-16	Whang-Peng et al 1978

Idiopathic thrombocythemia

Karyotype	Reference
46,XX,5q-	Nowell and Finan 1978
46,XY,del(5)(q14q22)"s"	Kerkhofs et al 1982

Chronic myeloproliferative disease, NOS

Karyotype	Reference
45,X?,-5	Lessard and Le Prise 1983

Myeloproliferative disorder, special type

43,XX,-5,-13,-18/44,XX,-5,t(13;?)(p11;?),-18"s"	Rowley et al 1977a
45,XX,-7/45,XX,-5,-7,+mar"s"	Rowley et al 1977a
45,XX,t(1;?)(p36;?),-5,-7,-12,t(13;?)(q34;?),-17,-22, +4mar/45,XX,t(1;?),-2,-3,-5,-7,-12,-17,-22,+8mar"s"	Rowley et al 1977a
46,XX,-5,del(6)(q13),-7,+8,-17,+2mar"s"	Rowley et al 1977a
46,XX,del(5)(q15)/47,XX,+8	Swolin et al 1981
46,XY,t(1;13)(p36;q13),del(3)(p21),-5,inv(7)(p11q22), t(16;16)(q22;q24),+mar	Clare et al 1982

DYSMYELOPOIETIC SYNDROMES

Preleukemia, NOS

43,X,-Y,-3,5q-,-7,-10,-16,-22,+3mar"s"	Pedersen-Bjergaard et al 1981
44,XY,+Y,-3,-5,+del(6)(q21),-8,t(17;?)(p13;?),+18,-20,+22	Mitelman 1983
44,XY,-5,-6,-7,-8,t(17;?),+2mar"s"	Pedersen-Bjergaard et al 1982
44,XY,t(3;17),-5,-20,+mar"s"	Pedersen-Bjergaard et al 1981
45,XX,-1,-4,5q-,-7,-12,-18,14q+,+4mar"s"	Pedersen-Bjergaard et al 1981
45,XX,-2,-3,5q-,-6,-9,16p-,17p+,+3mar	Swansbury and Lawler 1980
45,XY,-2,-5,-7,-8,-11,-12,-13,-14,+t(2;5),+t(11;12), +t(16;17),+17,+3mar	Watt et al 1982
45,XY,-5,-17,t(5;17)(p?;p?)	Borgström et al 1982
46,XX,5q-	Teerenhovi et al 1981
46,XX,5q-,22q-	Teerenhovi et al 1981
46,XX,5q-/47,XX,+8"s"	Teerenhovi et al 1981
46,XX,del(5)(q13orq14),del(22)(q11)	Ruutu et al 1977b
46,XX,del(5)(q13q23)	Geraedts et al 1980a
46,XX,del(5)(q13q31)	Geraedts et al 1980a
46,XX,del(5)(q15q31),del(11)(q14)	Swolin et al 1981
46,XX,t(5;12)(?q15;q13),-17,+mar"s"	Anderson and Bagby 1982
46,XY,-5,+mar"s"	Anderson and Bagby 1982
46,XY,5q-"s"	Pedersen-Bjergaard et al 1981
46,XY,5q-,-11,+18,18p+,18p+	Swansbury and Lawler 1980
46,XY,5q-,del(11)(q14)	Swansbury and Lawler 1980
46,XY,del(5)(q13q31)/47,XY,+8	Swolin et al 1981
46,XY,del(5)(q13q33)	Swolin et al 1981
46,XY,t(1;17)(q24;p13),-2,-5,-7,+r,+2mar"s"	Anderson and Bagby 1982
47,X,Xp-,1p+,5q-,6q+,11q-,11p+,+11,12p-	Swansbury and Lawler 1980
50,XY,-5,-21,+1,+9,+11,17p+,+1-3mar	Borgström et al 1982

Refractory anemia, NOS

Karyotype	Reference
45,XY,del(5)(q14q22),-7	Kerkhofs et al 1982
46,XX,5q-	Lessard and Le Prise 1983
46,XX,5q-	Kerkhofs et al 1982
46,XX,del(5)(q12q31)	Verhest et al 1976
46,XX,del(5)(q12q31),del(8)(q22)	Watt et al 1983
46,XX,del(5)(q13)	Mitelman 1983
46,XX,del(5)(q13)	
46,XX,del(5)(q13q31)	Hartley and McCallum 1981
46,XX,del(5)(q13q31)	Kerkhofs et al 1982
46,XX,del(5)(q13q31)	Kerkhofs et al 1982
46,XX,del(5)(q13q33)	Kerkhofs et al 1982
46,XX,del(5)(q13q34)/46,XX,t(9;11)/47,XX,+8	Swolin et al 1981
46,XX,del(5)(q13q34)	Kerkhofs et al 1982
46,XX,del(5)(q14q32),t(11;21)(q25;q21)	Van Den Berghe et al 1979e
46,XX,del(5)(q14q32)	Kerkhofs et al 1982
46,XX,del(5)(q15)	Kaffe et al 1978
46,XX,del(5)(q15)	Prieto et al 1976
46,XX,del(5)(q15),inv(12)(p13q21)	Mitelman 1983
46,XX,del(5)(q21q33)	Kerkhofs et al 1982
46,XX,del(5)(q?)	Verhest et al 1977
46,XX,del(5)(q?)	Mahmood et al 1979
46,XX,del(5)(q?)	Mahmood et al 1979
46,XX,del(5)(q?)	Mahmood et al 1979
46,XX,del(5)(q?)	Mahmood et al 1979
46,XX,del(5)(q?)	Mahmood et al 1979
46,XX,del(5)(q?)	Sokal et al 1975
46,XX,del(5)(q?),del(10)(p13)/47,XX,del(5),del(10),+mar	Mitelman 1983
46,XX,del(5)(q?)	Sokal et al 1975
46,XX,del(5)(q?)	Sokal et al 1975
46,XX,del(5)(q?)	Sokal et al 1975
46,XX,t(2;4;13)(p11;q12;q34),del(5)(q15q31)	Kerkhofs et al 1982
46,XY,del(5)(q12q34)	Kerkhofs et al 1982
46,XY,del(5)(q14q33)	Mecucci et al 1982
46,XY,del(5)(q?)	Mahmood et al 1979
46,XY,del(5)(q?)	DiBenedetto et al 1979
46,XY,del(5)(q?)	Sokal et al 1975

Sideroblastic anemia, NOS

Karyotype	Reference
46,XX,-5,+16	Seabright 1983
46,XX,del(5)(q12)	Mitelman 1983
46,XY,t(2;5)(q37;q?)	Kamihira et al 1977

Acquired idiopathic sideroblastic anemia

Karyotype	Reference
46,XX,del(5)(q12q23)/45,XX,3p-,4q+,6q-,-22"s"	Kerkhofs et al 1982
46,XX,del(5)(q13q31)	Kerkhofs et al 1982
46,XX,del(5)(q13q34)/46,X,Xq-,5q-,13q-"s"	Kerkhofs et al 1982
46,XY,del(5)(q13q31)	Kerkhofs et al 1982
46,XY,t(2;11)(p21;q25),del(5)(q13q31)/46,XY,t(2;11)(p21;q25),del(11)(q14)	Kardon et al 1982a

Refractory anemia without excess of blasts

44,XX,5q-,-7,+8,-12,-13,-17,-20,+2mar	Teerenhovi et al 1981
45,XY,3q-,5q-,-7	Teerenhovi et al 1981
46,XX,5q-	Teerenhovi et al 1981
46,XX,5q-/47,XX,+8/47,XX,+14/46,X,-X,+8	Teerenhovi et al 1981
46,XX,del(5)(q14)	Mitelman 1983
46,XX,del(5)(q14)	
46,XX,del(5)(q21)	Mitelman 1983
46,XX,del(5)(q?)	Mitelman 1983
46,XY,1p-,3q-,5q-	Teerenhovi et al 1981
46,XY,del(5)(q13)	Mitelman 1983
47,XY,5q-,-7,+2mar/48,XY,-8,-9,+11,+r,+2mar	Teerenhovi et al 1981

Refractory anemia with excess of blasts

43,XX,3q-,del(5)(q13q33),7q-,-12,14p+,16q+,-17,-18,18q+	Kerkhofs et al 1982
44,XX,5q-,-13,-17,-20,-22,+2mar	Teerenhovi et al 1981
44-47,XY,-3,-5,-7,-12,-16,-20,del(1)(q32),3p+,del(11)(q22), 3-6mar,+r	Streuli et al 1980
44-50,XY,-3,-5,del(3)(p23),+mar	Streuli et al 1980
45,XX,-9,-17,-20,del(4)(q21),del(5)(q12q31), +t(9;17)(p13?;p11)	Streuli et al 1980
45,XY,-5,11q+,-11,-18,+mar	Berger et al 1981a
46,XX,del(5)(q13)"s"	Mitelman 1983
46,XX,del(5)(q14q34)	Kerkhofs et al 1982
46,XX,del(5)(q15)/49-51,XX,+1,del(5)(q15),+11,-14,+22,+mar	Swolin et al 1981
46,XY,del(5)(q12)/47,XY,del(5)(q21),+21/49-50,XY, del(5)(q12),+9,+21,+2mar	Kardon et al 1982a
46,XY,del(5)(q12q31),-7,20q+,+22	Kerkhofs et al 1982
47,XXX,5q-"c"	Teerenhovi et al 1981

Thrombocytopenia

46,XY,del(5)(q13q31)/46,XY,inv(7)(q13q32)	Kerkhofs et al 1982

Pancytopenia

46,XX,-4,-5,-6,-7,-11,-13,-16,+19,+20,t(7;13)(q?;q?), +4mar/43,XX,-2,-3,-5,-7,-16,+19,15q+,+mar	Nowell and Finan 1978
46,XX,-7,+8,-12,-16,-17,t(6;12)(q?;q?),5q-,6q-,+mar	Nowell and Finan 1978

Dysmyelopoietic syndrome, special type

44,XY,-5,-7,-17,del(3)(q13),+del(3)(p13),del(6)(q13)"s"	Rowley et al 1981a
45,XX,t(13;14)(p13;q11),del(5)(q14)/50,XX,t(13;14),+1,+6, -7,+8,+10,del(5),+mar"c""s"	Rowley et al 1981a
45,XY,del(2)(q33),-5,del(7)(q22q32),-12,-15, t(12;15)(q11;q26),del(19)(q12),+r"s"	Albain et al 1983

SPECIAL LEUKEMIAS

LYMPHOCYTIC LEUKEMIAS

Acute lymphocytic leukemia (ALL)

Karyotype	Reference
28,XX,-1,-2,-3,-4,-5,-7,-8,-9,-11,-12,-13,-14,-15,-16,-17,-19,-20,-22	Kaneko and Sakurai 1980
32,XY,-2,-3,-4,-5,-7,-8,-11,-12,-13,-14,-15,-16,-17,-20	Shabtai and Halbrecht 1981b
45,X,-X,dup(1)(q31q41),del(5)(q22),del(17)(p11)	Humbert et al 1978
45,XX,-5,19q+	Prigogina et al 1979
45,XX,5q+,14q+,-15	Chrz et al 1983
46,XY,del(5)(q12q23),del(9)(p21)	Abe et al 1979
46,XY,del(5)(q13),del(15)(q15)/46,XY,del(7)(q32)/46,XY,del(6)(q21)	Secker-Walker et al 1979
46-47,XY,del(5)(q15),del(12)(p12),del(17)(p11),+8	Prigogina et al 1979
47,X,-X,del(5)(p23),+t(17;?)(p13;?),+del(18)(p11)	Oshimura et al 1977a
48,XX,+5,-19,+21,+mar	Seabright 1983
48,XY,t(1;22)(q21;p11),+5,+8,19p+	Humbert et al 1978
49,XX,del(4)(p15),t(4;11)(q21;q23),+5,+8,del(9)(p13),del(9)(p13),t(12;?)(q24;?),+mar	Morse et al 1982b
52-56,XY,+5,+8,+11,+19,+20,+21,del(6)(q15),+mar	Prigogina et al 1979
53,XX,+2,+5,i(7q),+8,+13,+21,+21,+22,+17q+	Oshimura et al 1977a
53-55,XY,+4,+5,+6,+20,+22	Lessard and Le Prise 1983
54,XX,7p+,+10,+10,+14,+14,+18,+18,+21,+21/27,X,-X,-1,-2,-3,-4,-5,-6,-7,7p+,-8,-9,-11,-12,-13,-15,-16,-17,-19,-20,-22	Oshimura et al 1977a
54-55,XX,+5,+12,+14,+16,+17,+18,+21	Prigogina et al 1979
54-55,XY,+5,+8,+9,+14,+21,+mar	Prigogina et al 1979
54-57,XY,+3,+5,+6,+7,+12,+13,+14,+21,+2mar	Prigogina et al 1979
55,XX,+4,+5,+6,+10,+17,+21,+3mar	Humbert et al 1978
57,XY,+X,+5,+6,i(7q),+10,+12,+13,+15,+18,+21,+21,+22	Oshimura et al 1977a
61,XX,+X,+1,+2,+3,+4,+5,+6,+8,del(6)(q21),+13,+14,+15,+16,+18,+21,+22	Oshimura et al 1977a

ALL, FAB type L1

Karyotype	Reference
26,XX,-1,-2,-3,-4,-5,-6,-7,-8,-9,-10,-11,-12,-13,-14,-15,-16,-17,-19,-20,-22	Hoeltge et al 1982
28,XX,-1,-2,-3,-4,-5,-6,-7,-8,-9,-11,-12,-13,-15,-16,-17,-19,-20,-22/56,XX,+X,+X,+10,+10,+14,+14,+18,+18,+21,+21	Brodeur et al 1981a
46,XY,del(2)(q32),-11,-12,+2mar/46,X,-Y,del(2),-4,-5,-11,-12,-21,+6	Kaneko et al 1981b
53,XX,+1,+5,+14,+15,+19,+20,+22	Kaneko et al 1981b

ALL, FAB type L2

Karyotype	Reference
26,XX,-1,-2,-3,-4,-5,-6,-7,-8,-9,-10,-11,-12,-13,-15,-16,-17,-18,-19,-20,-22	Brodeur et al 1981a
46,XX,t(5;8;14)(q11;q24;q32),t(12;22)(p13;q11),-18,+t(18;?)(q23;?)	Kaneko et al 1982e

47,XY,+18	Kaneko et al 1982e
46,XY,-2,-5,-6,-7,+8,-9,-10,-11,-12,-16,+18,+t(2;?)(q35;?), +t(11;?)(q21;?),+5mar	
53,XY,t(9;22),+4,+5,+6,+7,+8,+17,+21	Sandberg et al 1980
56,XY,t(9;22)(q34;q11),+4,+5,+8,+9,+11,+16,+17,+18,+19,+20	Mitelman 1983

ALL, FAB type L3

46,XY,t(8;22)(q24;q12)/46,XY,1q+,+del(1)(p22),-5, t(8;22)(q24;q12)/47,XY,+del(1)(p22),t(8;22)(q24;q12)	Abe et al 1982a

Chronic lymphocytic leukemia

50,XX,+5,+5,t(13;?)(q34;?),+22	Nordenson 1983

Prolymphocytic leukemia

44,XY,-5,-8,-9,+mar	Pittman et al 1982

Lymphocytic leukemia, special type

46,XY,-10,-13,5p+,+t(10;13)(p13;q11q32),21p+	Ueshima et al 1981

MONOCLONAL GAMMOPATHIES

Multiple myeloma

44,XY,del(3)(q25),t(5;12)(q15;q14),del(7)(q22), del(11)(q24),-14,-18,+mar	Mitelman 1983
45,X,del(X)(q25),t(3;8;22)(q13;q34;q11),-5,-7,-10,-11, del(13)(q13?),+16,19q+,+3mar	Van Den Berghe et al 1979f
48,X,-X,+1,+7,-8,+9,+21,del(1)(p36),4p+,del(5)(q22),9p+, 14q+	Liang et al 1979
52,XY,-1,+t(1;15)(q12;p1?),+t(1;16)(p36;p13), t(3;5)(q2?;q13),+3,+5,+7,+8,+9,+11,-15,+18,-20	Philip et al 1980
53,XY,+3,+5,-7,-8,+9,+11,-14,+20,del(1)(p34),del(6)(q25), +5mar	Liang et al 1979
55,XY,+1,+5,+9,+11,-12,+13,+15,del(1)(q32),del(6)(q25), t(9;19)(q13;q13),14q+,16p+,+4mar	Liang et al 1979
45,XY,-2,-2,-3,+5,-8,-10,+14,-15,-16,del(6)(q25),+4mar	
55,XY,+3,+5,+7,+9,+11,+14,+15,+21,+21/53,XY,+3,+5,-8,+9, +11,+14,+15,-19,+21,14q+,+3mar	Liang et al 1979

Macroglobulinemia

46,Y,-X,del(1)(q21),del(5)(q23),+del(5)(q23), +t(7;19)(p22;q13),del(8)(p12),+t(1;16)(q12;p13), +t(1;18)(q12;p11),-7,-16,-18,-19,+mar	Ueshima et al 1983

Plasma cell leukemia / Plasmocytoma

45,XY,-5,del(6)(q11),t(6;7)(q?;q?),+9,-12,-13,-16,-17, t(9;22)(q34;q11),+2mar	Karpas et al 1982

SOLID TUMORS

UNCLASSIFIED NEOPLASMS

Malignant neoplasm, NOS

Karyotype	Reference
47,XX,-2,-5,+6,-10,11q+,-13,+18,-21,t(1;?)(q21;?), t(1;13)(q21;p11),t(1;6)(q11;q23)	Kovacs 1978a

EPITHELIAL NEOPLASMS

Carcinoma, NOS

Karyotype	Reference
32-35,XX,cx,del(1)(p12),dup(1)(q21q32),2q+,5q+,7p+,11p+, hsr(10),t(11;12)(q13;q24),12p+,13q+	Kusyk et al 1982
41,XX,t(1;1;?)(p36q21;q12;?),t(1;?)(q12;?),del(6)(q21), t(5;12)(q?;q?),del(5)(q12q32)	Atkin and Baker 1982a
43,X,-X,t(1;11)(q21;q23),3q+,4q+,5q+,-11,-13,-16,17p+	Atkin and Baker 1982a
44,XY,+5,-8,-11,-13,-13,-14,-14,-15,+19,-22,-22,t(1;11), 5q-,9q+,+5mar	Reichmann et al 1981
44-45,XX,t(1;4)(q11;q35),+3,5q-,-14,i(17q),-21	Atkin and Baker 1982a
46,XX,+2,-7,-10,+13,-16,-17,+20,+5q-,15p+,15p+,18p+	Reichmann et al 1981
46,XX,del(2)(p24),+5,+8,t(9;17)(p?;p?),del(16)(q21),-20, -21	Mark 1975b
46,XX,t(1;?)(p32;?),-2,3q-,5q-,6q+,-18,-21,+mar	Atkin and Baker 1982a
46,XX,t(3;17)(p24;p13)/46,XX,t(5;17)(q22;p13)/46,XX, t(9;18)(q13;q21)/46,XX,t(12;15)(q13;q22)	Stenman et al 1982
46,XY,-4,-5,-11,-13,-17,+20,+21,1p+,9p-,+3mar	Reichmann et al 1981
47,XY,-5,+7,-10,-13,-14,4q-,5q-,+4mar	Reichmann et al 1981
47-48,Y,-X,-5,-8,+13,-14,-16,-18,-22,+7mar	Reichmann et al 1981
48,XY,+1,-5,+8,+17	Sonta and Sandberg 1978a
49,X,-Y,-3,-5,-14,+15,-16,-17,+19,+21,+22,1q+,5q-,6q-,8p-, 9p-,11p-,18p+,20q-,22q-,t(Y;14),+3mar	Reichmann et al 1981
49,XX,t(3;11)(q?;q?),+5q-,+6q-,+i(11q)	Atkin and Baker 1982a
49,XY,+4,+5,+8	Sonta et al 1977
54,XY,+1,+5,+9,+17,+19,+20,+21,+21	Sonta and Sandberg 1978a
58,XX,+X,+1,+2,+3,5p-,+6,+7,+9,11q+,-13,-14,-15,+16,+19, +20,+6mar	Kovacs 1978a
60-63,XX,cx,t(5;18),t(3;6),t(3;22),del(10)(q24), del(4)(p1?)	Riet-Fox et al 1979
63,XX,-1,+3,+5,+6,+7,+8,+10,+11,+12,-13,-15,+20,+21,+22, +9mar	Sandberg 1977
71-77,XY,cx,5q-	Reichmann et al 1981
79,XX,cx,i(1q),t(1;3)(p11;q29),12q+,15p+,i(5p)	Kovacs 1978a
79,XX,cx,i(1q),t(1;3)(p13;q29),t(1;14)(p13;p13), t(12;?)(q24;?),t(15;21)(p11;q11),i(5p),r	Kovacs 1981
80-83,XX,cx,t(2;5;16),t(2;5;18),t(11;13),11q+,t(8;10), del(2)(q24),del(6)(q21),i(14q),dup(9)(q11q12)	Riet-Fox et al 1979
82,XX,cx,2q+,5q-,9q+,i(17q)	Reichmann et al 1981
91,XX,cx,5q-	Reichmann et al 1981

Carcinoma NOS, metastatic

35,XX,-1,-3,-4,-5,-7,-8,-10,-11,-12,-13,-14,-15,-16,-17,-18,-19,-20,-21,-22,+8mar	Pathak et al 1979
45,X,Xq-,t(1;?)(p11;?),t(1;?)(q23;?),t(3;?),t(5;?)(q2?;?),del(9)(p2?),cx	Bartnitzke and Bullerdiek 1983
46,XX,t(9;22)(q34;q11)"s"	Togawa et al 1981
52,XX,+3,+5,+7,+8,+9,+11,12q+,13q-,-14,+16,i(17q)"s"	
51,XX,cx,t(1;5)(q?;q?)	Jones Cruciger et al 1976
60-62,XX,del(1)(q?),t(1;7),2q+,del(3)(q?),der(5),dmin,cx	Bartnitzke and Bullerdiek 1983
62-64,XY,del(2)(q34),del(6)(q14),ins(7)(q?),20q+,t(5;18),i(5p)	Ayraud 1975
65,XX,i(1q),t(1;9),der(5),t(6;10)(p?;q?),t(14;?),der(21),cx	Bartnitzke and Bullerdiek 1983
70,XX,del(5)(p21),del(3)(p?),cx	Ayraud 1975
81-83,XX,t(1;3)(p22;p25),del(1)(p13),t(5;13)(q13;q14),del(4)(p12),i(2)(p11),i(5)(p11),t(4;5)(q12;q11q13)	Kakati et al 1976a

Large cell carcinoma

55-70,XY,t(4;5;14),del(1)(q11q21),t(7;11)(p22;q13),t(7;21)(q11;q22),t(3;8)(p?;q?),t(6;11)(p11;p11),i(16q),t(3;?)(q11;?),t(21;?)(q11;?),t(13;?)	Kakati et al 1975
60,XY,del(5)(p11),t(6;10)(q11;p11),t(7;11)(p22;q13),t(14;18)(q11;q11),del(2)(q21),t(6;?)(p11;?)	Pickthall 1976

Small cell carcinoma

40,XY,cx,t(1;16)(q21;q24),t(1;19)(q23;q13),del(3)(p21p24),del(3)(p23q26),del(10)(q22),del(11)(q23),t(5;13)(q13;q32),del(17)(p12),dmin	Whang-Peng et al 1982b
80,XX,cx,del(1)(q32),del(3)(p14p21),t(3;4)(p23q11;q11),t(5;20)(q11;q13),dmin	Whang-Peng et al 1982b

Papillary carcinoma

54,XY,+del(2)(q11),t(2;14)(q11;q32),t(3;11)(p13;p15),+i(5p),+8,+9,+12,+16,+17,+21,+2mar	Pathak et al 1982

Adenoma

46,X,t(X;4)(q25;p16),del(2)(p13p21),del(5)(q15q31),-7,t(9;12)(p13;q13),del(11)(q11q14),t(7;9;12)(p11p22q11;p13p24;q13)/45,X,-X	Mark et al 1980
46,XX,t(1;3)(p21;p21),t(1;13)(p36;q14),t(3;8)(p21;q12),t(1;5;20)(p21;p14q12;p13),t(5;8)(p14;q12),t(X;9)(q27;q12)	Mark et al 1982a
46,XX,t(3;5)(p21;p15),del(8)(p12q12),t(8;9)(q12;q34)	Mark et al 1980
46,XY,del(5)(q14q31),ins(15;5)(q22;q14q31)	Mark et al 1982a

Adenocarcinoma

Karyotype	Reference
44,XX,+5,-9,-13,-13,-15,-17,-18,+22,+2mar	Martin et al 1979
44,XX,-1,+5,-15,-18,2q+/44,XX,-1,+5,-6,-15,-18,2q+,10p-,+mar	Martin et al 1979
44-50,X,-X,-1,t(1;?)(q21;?),del(2)(q33),4p+,-5,-7,+8,del(10)(q12),t(10;13)(q12;q34),del(17)(q25),-20	Kusyk et al 1982
48-48,XX,+3,+7,+8,-1,-5,-14,-17,5q+,6q+,del(1)(q42),del(5)(q3?),del(10)(q24),del(14)(q2?)	Riet-Fox et al 1979
50,XY,2q+,+5,+10,+11,-16,-18,+20,+20,+mar	Couturier-Turpin et al 1982
54,XX,+5,+6,+7,+9,+11,+21,+21,+del(1)(q32)	Sonta et al 1977

Adenocarcinoma, metastatic

Karyotype	Reference
40,XX,t(4;5;8),t(3;?)(q21;?),t(1;11)(q21;p11),t(9;11)(q11;p11),del(1)(p36)	Kakati et al 1975
40-45,XY,-3,-8,-18,+5,+10,11p+,14q+,del(4)(q32)	Ayraud 1975
56-64,XX,del(X)(q22),del(1)(p12),del(5)(p21),del(7)(q31),t(14;18)(q11;q11),18q+,i(22q)	Berger and Lacour 1974

Adenomatous polyp

Karyotype	Reference
50,XX,+1,+5,+17,+19	Sonta and Sandberg 1978a

MESENCHYMAL NEOPLASMS

Sarcoma, NOS

Karyotype	Reference
48,XX,+4,+5	Sonta et al 1977

Mesothelioma, malignant

Karyotype	Reference
40-46,XY,+16,+20,-22,t(14;15),del(2)(p?),i(5p)	Ayraud 1975
43,XY,t(1;1)(q42;q32p22),t(2;6)(p25;p21),inv(3)(p24q13),t(5;18)(p15;q11),-6,+9,del(13)(q24q14),-14,-18,-22	Mark 1978

MELANOMAS

Malignant melanoma

Karyotype	Reference
46,XY,5q+,+7,-8,-9,-10,-13,-20,+21,+3mar	McCulloch et al 1976

Malignant melanoma, metastatic

Karyotype	Reference
41,XY,dup(1)(q21q32),t(1;7)(q25;q36),t(7;9)(q11;q11),del(9)(q13),-3,-4,-5,-8,-10,-11,-16,+18,-20,-22	Kakati et al 1977
41-43,XY,del(1)(q21q25),i(1q),t(1;9)(p22;q34),t(11;12)(q11;p11),t(10;12)(q26;q13),t(4;?)(q35;?),del(5)(p13),t(14;?)(p12;?),i(17q),t(18;?)(p11;?),del(6)(q21)	Kakati et al 1977

80,XY,t(11;12)(q23;q13),del(5)(q13),del(5)(p11), del(9)(q13),t(5;9),t(14;?)(q11;?),t(7;12)(q22;q24), del(1)(p13),del(11)(q23) — Kakati et al 1977

NEUROGENIC NEOPLASMS

Astrocytoma, grade III, IV

44,XY,-4,-17,-22,1q-,5p-,-11q+,+mar — Yamada et al 1980

Neuroblastoma

46,XX,t(1;?)(p32;?),2q+,7p+,13q+,hsr(5q),dmin — Gilbert et al 1982

Retinoblastoma

46,X,-Y,1p+,+1p-,5q+,7p+,10q-,16q-,17q+,21p+ — Kusnetsova et al 1982
46,XY,+4,t(5;13)(q35;q21),t(7;17)(p22;q12),t(1;9)(q22;p24), -10,del(16)(q11) — Gardner et al 1982
47,XY,5q+,+i(7q)/47,XY,+i(6p),12q+ — Kusnetsova et al 1982
47,XY,t(X;1)(q28;p11),t(2;?)(q11;?),t(2;15)(p25;q11), t(3;?)(q27;?),4p+,t(5;?)(q35;?),i(17q),+mar — Gardner et al 1982

Meningioma, benign

41,XY,+del(1)(p?),-5,-10,-17,-20,-22 — Zang and Zankl 1983
41-42,XY,-1,-5,-8,9q+,-14,-18,-22,+r — Mark 1973b
51,XY,+5,6q+,+12,+16,+17,+18,+20,-22/52,XY,+5,+8,+10,+12, +17,+18,+20,-22 — Zang and Zankl 1983
52,X,-Y,-22,+20,+19,+17,+15,+11,+9,+7,+5 — Zankl et al 1975a

EMBRYONAL AND MISCELLANEOUS NEOPLASMS

Teratoma, malignant

48,X,-Y,+1,del(5)(q12q32),+8,+10 — Tricot et al 1983
67,XY,+1,+3,+4,+5,+6,+7,+7,+8,+10,+11,+13,+13,+14,+16,+17, +18,+20,+20,+21,+21 — Sonta et al 1977

LYMPHOMAS

1. UNCLASSIFIED AND MISCELLANEOUS LYMPHOMAS

Malignant lymphoma, NOS

46-47,XY,-5,-14,+1-3mar Knuutila et al 1981

Non-Hodgkin's lymphoma, NOS

52,XY,+2,+3,t(5;9)(q22;q32),+6,+12,+14,+20 Reeves and Pickup 1980

Non-Hodgkin's lymphoma, lymphocytic, NOS

42-45,XX,-4,-5,-9,-16,+mar Fleischmann et al 1976
51,XX,+5,+9,+11,+del(11)(q?),+del(12)(q?),14q+,i(17q) Prieto et al 1978a

Non-Hodgkin's lymphoma, histiocytic, NOS

45,X,-X,del(1)(q22),t(1;17)(q?;q?),del(2)(q33), del(3)(p21p25),+ins(3)(p14p21p25),+5,del(6)(q14),t(1;7) Mark et al 1978
48,XY,t(3;8)(q22;q11),t(5;10)(q33;q22),del(7)(q21), t(4;9)(q11;p24),t(10;13)(q22;q21),r(12),del(19)(p13), del(22)(q11),+7 Mark et al 1978
80-90,XX,cx,del(7)(p15),t(5;8)(q13;q11) Kakati et al 1980
82,XXX,t(3;12)(q13;q15),t(3;8)(q22;q11),del(3)(q11), del(5)(q31),i(7q),del(10)(q24),t(10;10)(q24;q26), inv(11)(p15q21),del(12)(q13),del(13)(q14) Mark et al 1978

Non-Hodgkin's lymphoma, T-cell, NOS

47,XX,t(2;?),inv(3),4p-,del(5)(q14q32),6p+,6q+,t(7;7),9q+, 10p-,13q-,t(14;14),-15,-15,16q-,17p+,+21,+2mar Gaeke et al 1981

Malignant histiocytosis

48,XY,+i(1q),t(2;5)(p21;q35),+mar Kristoffersson 1983
50,XY,+5,+12,+13,+21 Sonta et al 1977

Angioimmunoblastic lymphadenopathy

47,XX,+3,-5,+mar Kaneko et al 1982c
48,XY,+3,+18/51,XY,+5,+15,+19,+21,+22 Kaneko et al 1982c
56,XY,+5,+5,+6,+10,+15,+15,+19,+20,+21,+22 Kaneko et al 1982c

2. HODGKIN'S LYMPHOMAS

Hodgkin's disease, NOS

47,XY,-2,-5,+16,-17,+3mar — Fleischmann et al 1976

Hodgkin's disease, mixed cellularity

61-80,XX,del(1)(q32),i(1q),del(2)(p13),del(6)(q13or15), del(6)(q23or25),15q+,12q+,del(5)(q31),del(7)(q11), del(10)(q26),del(3)(p13) — Reeves 1973

Hodgkin's disease, lymphocytic depletion

45,X,t(X;1)(q28;q11),t(3;5)(q29;q33),del(8)(q22),-9, del(11)(p13),-17 — Hossfeld and Schmidt 1978

60,XY,+del(1)(p21),t(1;5)(q44;q33),dup(5)(q12), t(1;5)(p21q23;q13),del(6)(q22),t(9;?)(q12;?), t(3;12)(q11;q24),t(14;?)(p11;?),i(17q) — Hossfeld and Schmidt 1978

3A. NON-HODGKIN'S LYMPHOMAS - RAPPAPORT CLASSIFICATION

Lymphocytic, poorly differentiated, diffuse

43,XX,-4,-5,-9,i(17q),-18,-21,+2mar — Goh et al 1980

45,XY,t(14;18)(q32;q21),+2,-5,-6,-11 — Fukuhara and Rowley 1978

Lymphocytic, poorly differentiated, nodular

45,XY,-5,-8,+i(8q) — Kakati et al 1980

52,X,-X,+2,-4,+5,+7,+8,-14,-15,-16,+20,+21, +t(1;15)(p11;q11),+t(12;17)(q24;q11),+t(14;?)(q32;?), +mar — Kaneko et al 1982a

92,XY,t(2;9)(q11;p11),t(5;?)(p15;?),t(17;?)(q25;?),17q+, del(1)(p22),del(3)(p21),del(X)(q24),del(11)(q23) — Reeves 1973

Mixed cell, diffuse

50,XY,+X,-1,+5,+7,+12,-14,-17,del(13)(q12q14), ins(1;1)(q21;q21q32),+t(14;?)(q32;?),+t(1;17)(q21;q25) — Kaneko et al 1982a

53,XY,t(8;14)(q24;q32),t(14;18)(q32;q21),+20,1q-,+1p-,2q-, 4q+,5q-,5p+,6q-,+7q+,10q+,11q+,11q-,13p+,+4mar — Fukuhara and Rowley 1978

Histiocytic, diffuse

45,X,-Y,t(14;15)(q32;q21orq22),-15,+18,-22,1q+,3p+,5p-,7p-, 9p+,12p+,16q+,17p-,+2mar — Fukuhara and Rowley 1978

47,XY,-11,-22,del(1)(q42),del(2)(q33),del(5)(q22q31), del(7)(q32),del(9)(p22),t(21;22)(q22;q11), +t(11;?)(q23;?),+t(21;22)(q22;q11),+mar — Kaneko et al 1982a

51,XY,+X,-2,+5,+11,-12,+21,del(6)(q21),+t(2;?)(q33orq35;?), +t(12;?)(p11;?),+mar — Kaneko et al 1982a

89,XY,del(1)(p22),t(5;13)(q15;q34),t(14;20)(q32;q12), t(1;17)(p22;q25),t(1;19)(q21;p13),del(20)(q12) — Mark et al 1979

3B. NON-HODGKIN'S LYMPHOMAS - KIEL CLASSIFICATION

Lymphocytic, T-zone

46,XY,+t(Y;14)(q12;q24),3q-,-5,-13,+mar — Gödde-Salz et al 1981b

Immunocytoma

45,X,-Y,t(Y;18)(q11;p11),t(3;3)(q21;q29),5q+ — Kristoffersson 1983
45-46,XY,-8 — Kristoffersson 1983
47,XY,15p+,+20/46,XY,del(5)(q23),del(6)(p22),+del(7)(p11), -21

Centroblastic-centrocytic, diffuse

48-51,XY,del(1)(p33q23q25),-2,+del(3)(p22),+4,4p+,+5,+6p+, +7,+8,+9,+21,+22 — Kristoffersson 1983
45-51,XY,del(1)(p33q23q25),del(2)(q14),+del(3)(p22),+4,+5, +7,+8,-12

Lymphoblastic, Burkitt's type

??,XY,-5,+6 — Douglass et al 1980a
??,XY,1p+,t(8;14)
45,XY,-2,5q+,11q+ — Biggar et al 1981
46,XY,t(4;5;7)(q13;p13;p22),t(8;14)(q23;q32) — Kaiser McCaw et al 1977
46-47,XY,t(14;?)(q32;?),dup(1)(q21q32),t(2;3)(q13;q29),+5, 6p-,-8,del(9)(q?),+12,+18,+mar — Miyoshi et al 1981c
46-52,X?,t(8;14)(q24;q32),+1,+3,+5,+7,i(17q),+X — Zech et al 1976a
47,XX,t(8;14)(q24;q32),dup(1)(q24q32),del(2)(q34), del(5)(q31q34),+8"s" — Berger and Bernheim 1982
48,XX,+5,+7,-9,+10,+11,-16,del(5)(p13q31),del(10)(p11q25), del(11)(q12) — Biggar et al 1981

Lymphoblastic, convoluted type

46,X,-Y,t(1;?)(q21orq23;?),+mar/46,X,-Y,-1,-2,+der(1), +t(2;?)(q33orq35;?),+mar/48,X,-Y,-5,-14,+18,+18,+20, del(9)(p13),t(1;14),+dup(5)(q13q35) — Kaneko et al 1982d
48,XX,+13,+t(1;5)(q21;p15)/48,XX,-5,+8,+13,del(1)(q32), +t(1;5)(q21;p15) — Kaneko et al 1982d

Immunoblastic

46,XY,del(4)(q33),del(6)(q23),t(5;7)(q33;q22) — Kaneko et al 1982a
47-48,XY,+4,del(4)(q21),+5,t(15;?)(q25;?),+19,-21 — Reeves and Stathopoulos 1976
49,XX,+5,+19,+22 — Kaneko et al 1982a
49,XY,del(1)(p32),del(1)(q32),+5,del(6)(q21),+2mar — Kristoffersson 1983
78-83,XY,+del(1)(q11),+5,+7,+12,+13q+,+14q+,+14q+,+20,+mar — Kristoffersson 1983

80-86,XY,del(1)(q11),+3,+7,+9,+12,+13q+,+14q+,+14q+,+15q+, +18,+20,+mar

3C. NON-HODGKIN'S LYMPHOMAS - LUKES AND COLLINS CLASSIFICATION

Mycosis fungoides

Karyotype	Reference
48,XX,del(1)(p22),t(2;8;14)(q37;q24;q24),+5,del(7)(p13), del(9)(q22),t(9;18)(q11;q24),del(10)(p13), del(13)(q12q14),del(5)(q15q22),del(5)(q13),del(5)(q15)	Fukuhara et al 1978a
46,XX,t(1;14)(q32;q32)	

Follicular center cell, diffuse, large cleaved

Karyotype	Reference
45,XY,-6,-9,-17,2q-,3p-,3q-,4q-,5q-,6p+,7p+,7q+,+7q-,9p-, 12q+,13p+,20p-,+mar	Fukuhara et al 1978b

Follicular center cell, diffuse, small non-cleaved

Karyotype	Reference
44,X,-Y,t(1;14)(p21;q13),5p+,del(6)(q13),del(8)(p21),9p+, 9p-,t(11;21)(q23;q22),13q+,-10,+12,-17p+,der(19),22q+	Fukuhara et al 1979a

Immunoblastic type (B-cell)

Karyotype	Reference
49,XX,+i(1q),+5,del(8)(q24),t(3;22)(q29;q11),+13	Berger et al 1983a

3D. NON-HODGKIN'S LYMPHOMAS - WORKING FORMULATION

Follicular, small cleaved cell

Karyotype	Reference
44,XY,+der(1),+2,t(2;13),-3,5q-,-12,t(14;18)(q32;q21), i(17q)	Yunis et al 1982
48,X,-X,inv(1),1q-,+3,4q-,ins(5),+7,+mar	Yunis et al 1982

Diffuse, mixed small and large cell

Karyotype	Reference
47,XX,+9,t(9;15)(p13;q11),t(12;14)/49,XX,+3,+5,+9,t(9;15), t(12;14)	Yunis et al 1982

Diffuse, large cell

Karyotype	Reference
54,XY,dup(1)(q?),inv(5),+3,+3,+9,+13,+18,+18,+20,+mar	Yunis et al 1982

Large cell, immunoblastic

Karyotype	Reference
48,Y,t(X;?),t(1;?),t(3;6),+5,-6,t(8;14)(q24;q32),i(17q), +mar	Yunis et al 1982
99,XY,cx,5p-,5q-,6q-,7p-,i(12q)	Yunis et al 1982

Chromosome 6

HEMATOLOGICAL DISORDERS

UNCLASSIFIED LEUKEMIAS

Acute leukemia, NOS

54,XX,+X,+3,+6,+8,+8,+13,+19,+22	Muir et al 1977

NONLYMPHOCYTIC LEUKEMIAS

Acute nonlymphocytic leukemia (ANLL)

??,XY,+6,+10,+11,+13	Philip et al 1978a
43,XX,t(1;20),-5,-6,7q-,11q-,-12,17q+,-18,19q+,-21	Mitelman 1983
44-45,XX,-3,-5,-6,-18,-19,t(1;2)(q2?;q11),t(5;?)(q13;?)	Philip et al 1978a
45,X,-Y,t(6;11)(p2?;q13),t(8;11)(q22;q13),del(8)(q22)	Philip et al 1978a
46,XX,del(5),del(6),del(7),21q+"s"	Geraedts et al 1980a
46,Y,t(X;11)(q22.3;q23.3),ins(X;11)(q22.3;p13p15), dup(1)(q11q21),t(4;6)(p16;p21),t(7;12)(p13;q15)	Nacheva et al 1982
47,XX,+t(6;?)(q12;?)"s"	Geraedts et al 1980a
47,XY,+6	Philip et al 1978a
48,XY,+del(3)(q11),+7,t(6;19)(q15;q13)/50,XY,+del(3)(q11), +7,t(6;19)(q15;q13),+2mar/51,XY,+del(3),+7,+9,t(6;19), +2mar	Mitelman 1983
49,XY,t(5;?)(q35;?),+6,+13,+14	Philip et al 1978a
50,XX,t(9;22)(q34;q11),+6,+8,+10,+del(9)(p13), +del(22)(q11)	Abe and Sandberg 1979
50,XY,+6,+13,+16,+mar	Chrz et al 1983
51,XY,+3,+5,-9,+19,+20,+21,t(6;14)(p25;q11),t(9;?)(q22;?)	Philip et al 1978a
56,XY,+Y,+6,+14,+21,+t(5;6)(q?;p?),+del(1)(p22q25), +del(2)(q12),+del(3)(p12),+del(4)(q13),+del(5)(p14q14), +mar	Zuelzer et al 1976
57,XY,+6,+11,+17,+20,+21"s"	Berger et al 1981a

ANLL, FAB type M1

45,XY,t(2;5)(q22orq23;q14q15),6p-,-7,t(6;7)(p21;p22q21), t(7;10)(p21;q21),t(14;19)(q11;q13)"s"	Hagemeijer et al 1981a
46,XX,-1,inv(1),3q+,5q-,t(13;17),-18,+mar/51,XX,-1,inv(1), 3q+,5q-,t(13;17),-18,+6,+9,+9,+11,+11,+mar	Hagemeijer et al 1981a
46,XY,t(9;22)(q34;q11)	Sasaki et al 1983

46,XY,t(9;20;22)(q34;q12;q11)/46,XY,t(9;22)(q34;q11),-6,-7,
-7,-8,10q+,+t(7;8),i(8q)
50,XX,+3,+19,+t(1;6)(q21;p22) — Yunis et al 1981
50,XX,+6,+19,+21"c" — Kaneko et al 1982b
58,XY,+1,+4,+5,+6,+8,+11,+13,+13,+14,+14,+15,+21,+21,
t(9;22)(q34;q11) — Alimena 1983

ANLL, FAB type M2

44,X,t(X;8),del(5)(q12q21),t(1;6),t(7;19),-7,-8,t(7;8),-21 — Kerkhofs et al 1982
46,XX,-4,-6,-7,-20,+mar — Golomb et al 1978c
46,XX,6p- — Golomb et al 1978c
46,XX,t(4;10)(q31;q22),t(6;9)(p23;q34) — Kaneko et al 1982b
46,XY,del(5)(q14q33),t(6;14)(q23;q22) — Kerkhofs et al 1982
47,XX,1p+,der(5),-6,12p+,-13,-14,-16,-17,-17,-21,+22,+7mar,
dmin — Cooperman and Klinger 1981
48-50,XX,+del(6)(q16),+9,-18,+mar"s" — Mitelman 1983
47,XX,del(6)(q16),+mar"s"
53,XX,+1,-4,-5,+6,-7,+8,+11,+12,+21,+21,+3mar"s" — Zaccaria et al 1983b

ANLL, FAB type M1+M2

43,XX,1p+,-4,-5,-6,-7,-13,-17,del(18)(q21),+3mar — Rowley and Potter 1976
44,X,-Y,-3,6q-,+mar — Shiraishi et al 1982
45,XX,t(3;6)(p23;p25),del(5)(q13),-12,i(17q) — Mitelman et al 1981
45-46,XY,t(2;6)(q37;q23),-7,del(11)(q21),del(22)(q11) — Mitelman et al 1981
46,XX,t(6;9)(p23;q34) — Rowley and Potter 1976
46,XX,t(6;9)(p23;q34) — Testa et al 1979
48,XX,t(6;9)(p23;q34),+8,+13
46,XY,t(1;6;11)(q12;q23;p15) — Yamada and Furusawa 1976
46,XY,t(6;9)(p23;q34) — Fitzgerald et al 1983
47,XX,+6 — Benedict et al 1979
47,XY,del(3)(q12),del(5)(q12),t(6;?)(q11;?),-13,+2mar"s" — Testa et al 1981b
49,XX,+6,t(9;22)(q34;q11),+17,+21 — Nordenson 1983
57,X,-Y,-6,+8,+15,+15,+19,+19,+20,+20,+21,+21,+22,+3mar — Fitzgerald et al 1983
57,XY,+6,+11,+13,+15,+21,+21,+21,+21,+21,+21,+22"c" — Shiraishi et al 1982

ANLL, FAB type M3

45-51,XY,t(3;12)(p14;q24),t(13;18)(q11;p11),+1,+6,+16,+19,
+21,+22 — Mitelman et al 1981
46,XX,2q-,6p-,7q-,10q+,14q+,-16,+mar — Hagemeijer et al 1981a

ANLL, FAB type M4

42,XY,-5,-9,-10,-15,-16,-17,-20,-21,-22,+5mar — Testa et al 1979
43,XY,+1,-5,+6,-9,-10,-15,-16,-17,-20,-21,-22,+6mar
43,XY,-2,-3,-6,-7,-12,-15,-21,-22,del(2)(p11),del(11)(p14),
t(14;?)(q32;?),t(17;?)(p13;?),del(22)(q11),+5mar"s" — Rowley et al 1981a
44,XY,-6,-10 — Larson et al 1982
44-46,XY,+1,-5,-17,-18,-20,del(6)(q21),-15,t(1;6)(q11;q24),
+3mar — Mitelman et al 1981

44-46,XY,+1,t(1;6)(q11;q27),-5,del(6)(q21),-15,-17,-18,-20, +mar — Alimena 1983
47,XX,+6,-17,+t(1;17)(q21;p13) — Kaneko et al 1982b
46,XY,t(6;11)(q27;q23) — Yunis et al 1981

ANLL, FAB type M5

46,XX,6q- — Chrz et al 1983
46,XY,t(6;11)(q26;q22)/52-53,XY,t(6;11)(q26;q22),+der(6), +3,+10,+13,+18,+19,+21,dmin — Hagemeijer et al 1981a
46,XY,t(6;11)(q27;q23)/51-52,XY,t(6;11),+der(6),+3,+19,+19, +21,dmin/53,XY,t(6;11),+der(6),+4,+10,+13,+18,+19,+21 — Löwenberg et al 1982
46,Y,-X,-1,-7,-8,-14,-15,-17,-20,inv(6)(p23q13), del(11)(q23),+t(X;?)(p11;?),+t(1;?)(p24;?), +t(12;?)(q22;?),+t(17;?)(p11;?),+4mar — Kaneko et al 1982b
47,XX,+del(1)(p11)/48,XX,+del(1)(p11),+6/48,XX, +del(1)(p11),+8 — Garson 1980
50,XX,+6,+8,10p-,+12,+13 — Nordenson 1983
51,XX,+6,+8,+14,+19,+21,t(11;?)(q21orq24;?) — Soukup and Neely 1981
51-52,XX,+3,+6,+9,-10,+18,+19,del(1)(p22),del(1)(p22), dup(11)(q11q21),+t(10;?)(p13;?) — Kaneko et al 1982b
57-58,XX,+1,+3,-4,-5,+6,-8,+11,+12,-17,+19,-20, del(10)(q24),+8-9mar — Testa et al 1979
60-61,XX,Xp+,+1,+2,-4,+6,-7,+19,+20,del(2)(q31), del(10)(q24),+12-13mar

ANLL, FAB type M5a

46,XY,t(6;10;11)(p22;p14;q14) — Berger et al 1982a
47,XY,+6,t(11;19)(q24;p13) — Berger et al 1982a

ANLL, FAB type M4+M5

48,XY,+6,+8 — Li et al 1983
48,XY,t(1;6)(q21;q27),+del(8)(q22),+22 — Shiraishi et al 1982
66,XY,+1,+1,+2,+4,+6,+6,+7,+8,+8,+10,+11,+11,+12,+12,+18, +19,+21,+21,+22,+22 — Li et al 1983

ANLL, FAB type M6

40-45,XY,-4,-6,-7,-10,-11,-20,+1-3mar — Mitelman et al 1981
43,XX,del(6)(q15orq21),-7,-12"s" — Rowley et al 1977a
43,XY,-4,5q-,t(4;6),-7,dic(12),-15"s" — Hagemeijer et al 1981a
43-45,XX,der(1),inv(2)(p25q14),del(3)(q25),-4,-5,-6,6q-,-7, 7p+,11p+,-12,-17,-21,+r,+1-4mar"s" — Smadja et al 1983b
44,X,del(Y)(q?),+Xq-,3p-,3q-,4q-,-6,+7q-,11q+,-15,-18"s" — Bradley et al 1982
44,XY,3p-,t(2;6),t(5;6),t(17;20)"s" — Sandberg et al 1982a
45,XY,del(4)(q23),del(5)(q13),-6,del(6)(q21),del(7)(p14), del(8)(p21),del(9)(q31),16q+ — Rowley and Potter 1976
45-47,XY,del(6)(q22),-9,+18,+21 — Mitelman et al 1981
48,XX,del(6)(q21orq22q24orq25),+21,+21"c" — Hagemeijer et al 1981a
49,XX,+6,+6,+21,der(13)(q12q21) — Bernstein et al 1982b
51,XY,+1,+2,+6,-7,+8,-14,+15,+21,+mar"s" — Rowley et al 1981a
72,XY,cx,t(2;15)(q31;q15),del(6)(p22q15),t(8;13)(q23?;q13), t(11;14)(p12;q13),i(16p),t(15;18)(q15;p11) — Douglass and Freeman 1983

Chronic myeloid leukemia, t(9;22)

Karyotype	Reference
26,XX,t(9;22)(q34;q11),-1,-2,-3,-4,-5,-6,-7,-9,-10,-11,-12,-13,-14,-15,-16,-17,-18,-19,-20	Hartley and McBeath 1981
45,XX,t(9;22),-6	Kohn et al 1975
46,X?,t(9;22),6q-/46,X?,t(9;22),6q-,i(17q)	Carbonell et al 1982a
46,XX,t(9;22)(q34;q11)/51,XX,t(9;22),+6,+8,+10,+19,+22q-	Mitelman 1983
46,XX,t(9;22)(q34;q11)/46,XX,t(9;22),del(6)(q16)/46,XX,t(9;22),del(20)(q11)/45,XX,t(9;22),del(20)(q11),t(1;2)(p31;q37)	Mitelman 1983
46,XX,t(9;22)/46,XX,t(9;22),del(6)(q16)/46,XX,t(9;22),del(20)(q11)	
46,XY,t(9;22)(q34;q11),-19,3q+,6q-,+17	Greenberg et al 1978
46,XY,t(9;22)(q34;q11),del(6)(q13)	Srivastava et al 1981
46,XY,t(9;22),del(6)(q13),i(17q)	
46,XY,t(9;22)(q34;q11),t(6;13)"c"	Tomiyasu et al 1982
46,XY,t(9;22)(q34;q11)	Ishihara et al 1982
27,X,-Y,t(9;22),-1,-2,-3,-4,-5,-6,-7,-9,-11,-13,-14,-15,-16,-17,-18,-19,-20,-22	
46,XY,t(9;22),t(12;19)(q13;q13)/46,XY,t(9;22),t(1;6)(p36;p21)	Seabright 1983
46,XY,t(9;22),t(1;6)(q25;q25)/46,XY,t(9;22),t(1;15)(q12;p11)	Miyamoto 1980
46,XY,t(9;22),t(X;6)(q28?;q22?),t(5;17)(q13?;q25)	Bernstein et al 1980b
46-49,XX,t(9;22),+del(22q),+5,+6,+9	Prigogina et al 1978
47,X?,t(9;22),del(6)(q14),+8,i(17q)	Fleischman et al 1981
47,XX,t(9;22)(q34;q12)/62,XX,cx,t(9;22)(q34;q12),del(1)(q25),del(1)(q11),del(1)(q21),del(1)(q11q25),del(2)(p12),del(3)(q21),del(6)(q23),i(13q),+22q-,+22q-,+22q-,+22q-	Hjorth et al 1980
47,XX,t(9;22),i(17q),+8,6q-	Prigogina et al 1978
47,XY,t(9;22)(q34;q11),+8,i(17)(q11)/50,XY,t(9;22)(q34;q11),i(17)(q11),+4,+6,+8,+8/47,XY,t(9;22)(q34;q11),t(3;17)(q21;p1?),del(10)(q24),i(17)(q11),+mar	Hayata et al 1975b
47,XY,t(9;22)(q34;q11),del(6)(q11),+8,i(17q)	Mitelman 1983
47,XY,t(9;22)(q34;q11),del(6)(q11),+8,i(17q)	
47,XY,t(9;22),+22q-/56,XY,t(9;22),+1,+2,+5,+6,+7,+8,+10,+12,+17,+22q-	Stoll and Oberling 1978
47,XY,t(9;22),+22q-/46,XY,t(9;22),-9,+22q-/46,XY,t(9;22),t(3;5)(p?;q?),t(6;13)(p?;q?)	Kohno and Sandberg 1980
48,XY,t(9;22),+22q-,+22q-/47,XY,t(9;22),+17/46,XY,t(9;22),-4,-7,+6,+17	Stoll and Oberling 1978
45,XY,t(9;22),-21/46,XY,t(9;22),-10,+17	
48,XY,t(9;22),+8,+22q-/61,XY,+Y,t(9;22),+4,+5,+6,+8,+8,+8,+8,+15,+17,+18,+19,+20,+22q-,+22q-	Wurster-Hill 1983
49,XY,t(9;22)(q34;q11),+6,+8,t(22;?)(p11;?),+der(22)	Oshimura et al 1982a
49,XY,t(9;22),+6,+8,1p+,14q+,+22q-	Sadamori et al 1980
50,XY,+Y,t(9;22)(q34;q11),t(6;8)(q34;q11),+8,i(17q),+19,+22q-	Alimena et al 1982b
50,XY,t(9;22),+del(22q),+6,+10,+19	Prigogina et al 1978
51,XX,t(9;22)(q34;q11),+6,+8,i(17q),+19,+20,+22q-	Alimena et al 1982b
51,XY,t(9;22)(q34;q11),+6,+8,+19,+21,+22q-	Tomiyasu et al 1982

Karyotype	Reference
52-54,XX,t(9;22)(q34;q11),+6,+8,+8,+19,+19,+del(22)(q11), +del(22)(q11),+X	Hayata et al 1975b
52-54,XY,t(9;22)(q34;q11),+3,+6,+7,+8,+19,+21,+22q-,+22q-, +1-3mar	Ondreyco et al 1981
53,XY,+Y,t(9;22)(q34;q11),+3,+6,+8,+9,+10,+12,-15,+22q-/56, XY,+Y,t(9;22),+3,+6,+8,+8,+9,+10,+16,+19,+22q-	Stoll and Oberling 1982
56,XY,+Y,t(9;22),+3,+6,+8,+8,+9,+10,+12,+19,+22q-	
53,XY,t(9;22),+6,+9q+,+10,+12,+19,+21,+22q-/54,XY,t(9;22), +6,+8,+9q+,+10,+12,+19,+21,+22q-	Sonta and Sandberg 1977b
53-56,XY,+Y,t(9;22),+del(22q),+6,+10,+19,+8,+11,+21	Prigogina et al 1978
55,XY,t(9;22),+3,+6,+8,+8,+9,+10,+16,+19,+22q-/53,XY, t(9;22),-15,+3,+6,+8,+8,+9,+10,+12,+22q-	Stoll and Oberling 1978
55,XY,t(9;22),+6,+10,+10,+11,+13,+14,+14,+19,+21	Seabright 1983
59,X,-X,t(9;22),+4,+6,+6,t(7;?)(p15;?),+der(7),+8,+8,+12q+, +17,+19,+19,+21,+22,+22q-,+22q-	Miyamoto 1980

Chronic myeloid leukemia, aberrant translocation

Karyotype	Reference
??,X?,t(6;22)(p25;q11)	Potter et al 1981
45,XY,t(18;22)(q23;q11),+del(1)(q11),6q-,-9,-9	Nowell 1983
46,XX,t(2;9;22)(q23;q34;q11)	Pasquali et al 1979b
46,XX,t(2;9;22)/50,XX,t(2;9;22),+6,+10,+19,+der(22)	
46,XX,t(6;22)(p25;q12)	Mammon et al 1976
46,XX,t(6;22)(q26;q11),t(1;9)(q21;q24)	Berger et al 1976
46,XX,t(6;9;11;22)(p21;q34;q13;q11)	Carbonell et al 1980
46,XX,t(6;9;22)(p21;q34;q11)	Pasquali 1983
46,XX,t(6;9;22)(p21;q34;q11)	Pasquali 1983
46,XY,t(6;9;22)(p21;q34;q11)	Potter et al 1975
46,XY,t(6;9;22)(p21;q34;q11)	Potter et al 1975
46,XY,t(6;9;22)(q21;q34;q11)	Chessells et al 1979
46,XY,t(6;9;22)(q24;q34;q11)	Sudries et al 1980
46,XY,t(9;11;22)(q34;q13;q11)/52,XY,t(9;11;22),+6,+9q+, +11q-,+18,+19,+del(22q),1q+	Lawler et al 1976

Chronic myeloid leukemia, Ph1 negative

Karyotype	Reference
46,XY,t(1;6)(p36;q15),del(3)(q25),del(17)(p11)	Srivastava et al 1981
46,XY,t(3;6)(q?;p?),t(17;21)	Borgström et al 1982
46,XY,t(6;14)(p21;q32)	Mintz et al 1979

MYELOPROLIFERATIVE DISORDERS

Polycythemia vera

Karyotype	Reference
??,X?,+6	Vykoupil et al 1980
??,X?,+6,+11,+14,+22	Vykoupil et al 1980
??,X?,2q+,6q-	Vykoupil et al 1980
45,X,-Y	Berger and Bernheim 1979
54-57,X,-Y,+6,+8,+11,+17,+20,+21,t(1;22)(p12;q13)	
44,X,-X,del(5)(q14q32),-7	Van Den Berghe et al 1979g

44,XX,del(5)(q14q32),-6,-7,-8,t(12;?)(p12;?), t(16;18)(q11;q12)	
46,XY,del(6)(p22),del(11)(q26),del(12)(p12),del(13)(q13)	Kohno et al 1979b
47,XY,+9	Van Den Berghe et al 1979g
47,XY,del(1)(q12),t(1;18)(q12;q23),+9,del(20)(q12)/47,XY, +9,del(11)(q21),del(20)(q12)/47,XY,+9,del(20)(q12)	
47,XY,+9/47,XY,del(5)(q14q32),del(8)(q?),+9,del(11)(q21), del(13)(q21)/47,XY,del(5)(q14q32),del(6)(q?),+9, del(11)(q21),del(13)(q21),del(1)(q12),del(20)(q12)	

Myelosclerosis / Myelofibrosis

44,XY,5p-,-7,-10,11q-,i(17q)/45,XY,-7,18q-/45,XY,2p-,2q-, 5p-,6q-,9q-,10q-,-11,+i(17q)	Whang-Peng et al 1978
46,XX,t(1;6)(q25;p22)	Hsu et al 1977
46,XY,6p+	Seabright 1983
46,XY,t(1;6)(q23;p21)	Gahrton et al 1978a
47,XX,del(22)(q?),t(1;2),t(6;7),+r,+mar	Page et al 1979
51,XX,+1,del(2)(q33),+6,+9,+11,+17,-19,+20q+	Najfeld et al 1978b

Chronic myeloproliferative disease, NOS

46,XX,6p+,t(1;6)(q25;p25)	Najfeld 1983

Myeloproliferative disorder, special type

46,XX,-5,del(6)(q13),-7,+8,-17,+2mar"s"	Rowley et al 1977a

DYSMYELOPOIETIC SYNDROMES

Preleukemia, NOS

44,XY,+Y,-3,-5,+del(6)(q21),-8,t(17;?)(p13;?),+18,-20,+22	Mitelman 1983
44,XY,-5,-6,-7,-8,t(17;?),+2mar"s"	Pedersen-Bjergaard et al 1982
45,XX,-2,-3,5q-,-6,-9,16p-,17p+,+3mar	Swansbury and Lawler 1980
45,XY,-2,-4,-6,-8,-10,-10,+17,+18,+3mar	Panani et al 1980
47,X,Xp-,1p+,5q-,6q+,11q-,11p+,+11,12p-	Swansbury and Lawler 1980
47,XX,+6,t(18;?)(q23;?)	Geraedts et al 1980a
47,XY,+6	Panani et al 1980
47,XY,+8,t(3;6)(p21;p12),+del(3)(p21)	Panani et al 1977
53,XX,+1,+2,t(4;19),+6,i(17q),+21,+21,+22	Seabright 1983

Acquired idiopathic sideroblastic anemia

46,XX,del(5)(q12q23)/45,XX,3p-,4q+,6q-,-22"s"	Kerkhofs et al 1982

Refractory anemia with excess of blasts

56-64,XY,+1,-2,+6,+9,+13,+15,+19,+21,10-13mar	Streuli et al 1980

Pancytopenia

Karyotype	Reference
44,XY,-7,6p+	Nowell and Finan 1978
46,XX,-4,-5,-6,-7,-11,-13,-16,+19,+20,t(7;13)(q?;q?),+4mar/43,XX,-2,-3,-5,-7,-16,+19,15q+,+mar	Nowell and Finan 1978
46,XX,-7,+8,-12,-16,-17,t(6;12)(q?;q?),5q-,6q-,+mar	Nowell and Finan 1978
46,XX,t(6;16;16)(p?;p?;p?)	Nowell and Finan 1978

Aplastic anemia

Karyotype	Reference
47,XX,+6	Geraedts and Haak 1976

Dysmyelopoietic syndrome, special type

Karyotype	Reference
44,XY,-5,-7,-17,del(3)(q13),+del(3)(p13),del(6)(q13)"s"	Rowley et al 1981a
45,XX,t(13;14)(p13;q11),del(5)(q14)/50,XX,t(13;14),+1,+6,-7,+8,+10,del(5),+mar"c""s"	Rowley et al 1981a

SPECIAL LEUKEMIAS

Hairy cell leukemia

Karyotype	Reference
44,X,del(X)(q22),del(6)(q23),-7,-8,-10,del(11)(p15q21),-21,14q+,+r,+mar	Sadamori and Sandberg 1983a

LYMPHOCYTIC LEUKEMIAS

Acute lymphocytic leukemia (ALL)

Karyotype	Reference
45,XX,del(6)(q11),-8	Secker-Walker et al 1979
45,XY,1q+,-6	Oshimura et al 1977a
46,X,Xp+,del(1)(q21),del(6)(q21),i(17q)	Oshimura et al 1977a
46,X?,del(6)(q13)	Lawler et al 1975
46,XX,-8,-10,del(6)(q23),+2mar	Secker-Walker et al 1979
46,XY,+2,t(6;?)(p21;?),+7,-12,-13,14q+,17p+	Oshimura et al 1977a
46,XY,3q-,6q-,19q+	Dewald et al 1978
46,XY,del(5)(q13),del(15)(q15)/46,XY,del(7)(q32)/46,XY,del(6)(q21)	Secker-Walker et al 1979
46,XY,del(6)(q15)	Prigogina et al 1979
46,XY,del(6)(q15),-7,+11	Prigogina et al 1979
46,XY,del(6)(q23orq25)	Oshimura et al 1977a
46,XY,del(6)(q23orq25)/47,XY,del(6)(q23orq25),+8	Oshimura et al 1977a
46,XY,t(4;11)(q21;q23)	Oshimura et al 1977a
46-49,XY,t(4;11)(q21;q23),+6,+8,+13,+17,t(1;13)(p22;q12),t(7;9)(q11;q34)	
47,XY,t(4;11)(q21;q23),1p+,del(2)(p16),del(6)(q21),i(7q),del(8)(q21),+mar	Arthur et al 1982
48,XY,t(4;12)(q?;q?),t(6;15)(q?;q?),t(9;22)(q34;q11),11q-,+20,+21	Nordenson 1983
48-53,XX,t(4;11)(q21;q23),+6,+7,+9,+13,+21	Prigogina et al 1979
52-56,XY,+5,+8,+11,+19,+20,+21,del(6)(q15),+mar	Prigogina et al 1979

53-55,XY,+4,+5,+6,+20,+22	Lessard and Le Prise 1983
54,XX,+6,+7,+8,+14,+17,+18,+21,+22	Prigogina et al 1979
54,XX,7p+,+10,+10,+14,+14,+18,+18,+21,+21/27,X,-X,-1,-2,-3, -4,-5,-6,-7,7p+,-8,-9,-11,-12,-13,-15,-16,-17,-19,-20, -22	Oshimura et al 1977a
54-57,XY,+3,+5,+6,+7,+12,+13,+14,+21,+2mar	Prigogina et al 1979
55,XX,+4,+5,+6,+10,+17,+21,+3mar	Humbert et al 1978
55,XY,+X,+3,+6,+10,+13,+14,+16,+21,+21	Oshimura et al 1977a
56,XX,+3,+6,+12,+13,+14,+15,+18,+21,+r	Prigogina et al 1979
57,XY,+X,+5,+6,i(7q),+10,+12,+13,+15,+18,+21,+21,+22	Oshimura et al 1977a
61,XX,+X,+1,+2,+3,+4,+5,+6,+8,del(6)(q21),+13,+14,+15,+16, +18,+21,+22	Oshimura et al 1977a

ALL, FAB type L1

26,XX,-1,-2,-3,-4,-5,-6,-7,-8,-9,-10,-11,-12,-13,-14,-15, -16,-17,-19,-20,-22	Hoeltge et al 1982
28,XX,-1,-2,-3,-4,-5,-6,-7,-8,-9,-11,-12,-13,-15,-16,-17, -19,-20,-22/56,XX,+X,+X,+10,+10,+14,+14,+18,+18,+21,+21	Brodeur et al 1981a
46,XY,del(2)(q32),-11,-12,+2mar/46,X,-Y,del(2),-4,-5,-11, -12,-21,+6	Kaneko et al 1981b
46,XY,del(6)(q23q25)	Shabtai and Halbrecht 1981b
52-56,XY,t(9;22),+4,+6,+9,+10,+14,+15,+20,+21,+22q-	Sandberg et al 1980
55,XY,+4,+6,+7mar	Kaneko et al 1981b
56,XY,+X,+4,+6,+10,+15,+17,+18,+21,+21,+mar	Kaneko et al 1982e
58,XX,+4,+6,+7,+10,+14,+14,+21,+del(18)(q21),+4mar	Kaneko et al 1982e

ALL, FAB type L2

26,XX,-1,-2,-3,-4,-5,-6,-7,-8,-9,-10,-11,-12,-13,-15,-16, -17,-18,-19,-20,-22	Brodeur et al 1981a
46,XY,-4,+6,-8,-10,-11,-15,-17,+22,del(6)(q21),+i(17q), +3mar	Kaneko et al 1981b
46,XY,6q-	Prieto et al 1981
46,XY,del(6)(q21orq23),del(9)(p22),del(11)(q13)/46,XY, del(6),i(6p),del(7)(p15),del(9)(p22),del(11)(q13)	Kaneko et al 1982e
46,XY,del(6)(q24)	Shabtai and Halbrecht 1981b
47,XY,+18	Kaneko et al 1982e
46,XY,-2,-5,-6,-7,+8,-9,-10,-11,-12,-16,+18,+t(2;?)(q35;?), +t(11;?)(q21;?),+5mar	
49,XY,+7,+12,-13,t(6;18)(p25;q21),t(11;14)(q23;q32), +t(9;?)(p24;?),+t(1;13)(q12;p13)	Kaneko et al 1982e
52-54,XX,+6,+7,+9,+12,+3mar	Mitelman 1983
53,XY,t(9;22),+4,+5,+6,+7,+8,+17,+21	Sandberg et al 1980
55-65,XX,+1,+1,+6	Lessard and Le Prise 1983
56,XX,+1,+6,+10,+13,+15,+16,+17,+18,+20,+21	Mitelman 1983

ALL, FAB type L3

Karyotype	Reference
46,XY,del(6)(q23q25)	Shabtai and Halbrecht 1981b
46,XY,dup(1)(q23q32),t(1;6)(q21;q13),t(14;?)(q32;?), del(6)(q13),del(8)(p21)	Slater et al 1979
47,XY,dup(1)(q21q32),del(6)(q23),t(8;14)(q23;q32),+9	Roos et al 1982

Chronic lymphocytic leukemia

Karyotype	Reference
44,XX,6q-,i(8q),12p-,-14,t(14;14)(q31;q12),-20,20p+	Sparkes et al 1980
44-48,XY,t(Y;9)(q12;q13),del(1)(p22p32),t(4;6)(p16;q15), del(6)(q15),t(12;14)(q15;q32),+mar	Robert et al 1982
46,XX,del(6)(q13q16),t(2;7)(p11;q22)	Vahdati et al 1983a
46,XX,t(6;21)(p21;q22)/46,XX,del(6)(q15)/46,XX,del(6)(q21)	Robert et al 1982
46,XX,t(6;7)(q?;q?),t(7;13)(q?;q?),t(11;14)(q11;q32),17p+	Schröder et al 1981
46,XY,del(6)(p12)/46,XY,t(14;18)(q32;q21)	Robert et al 1982
46,XY,t(3;6)(q12;q26),t(11;14)(q13;q32)	San Roman et al 1982
46,XY,t(6;20)(q?;q?)	Schröder et al 1981

Lymphocytic leukemia, special type

Karyotype	Reference
45,X,-Y,6q-/46,XY,6q-	Miyamoto et al 1982b
45,Y,-X,t(Y;14)(q12;q32),-13,-17,del(2)(q33), t(3;13)(q21;q34),del(6)(q16orq22),9p+,9q+,10q+,16q+,18q+	Miyoshi et al 1981a
47,X,-Y,dup(Y),1q-,4q+,6q-,7q+,+18q+	Miyamoto et al 1982b
48,XX,+3,-6,+7,del(10)(p13q24),10q+,+mar	Ueshima et al 1981
48,XX,-6,-11,14q+,+18,+20,+22	Imamura et al 1981
50,XY,+3,+6,+2mar	Miyamoto et al 1982b

MONOCLONAL GAMMOPATHIES

Multiple myeloma

Karyotype	Reference
46,XY,+t(1;?)(p1?;?),+t(1;12)(q21;q24),del(6)(p11),-8,+9, -12,-13,-16,-19,22q+,+mar	Philip et al 1980
47,XY,dup(1)(q11q32),t(1;10)(p22;p15),inv(6)(p21q13), t(11;14)(q23;q32),21p+,+mar	Liang et al 1979
53,XY,+3,+5,-7,-8,+9,+11,-14,+20,del(1)(p34),del(6)(q25), +5mar	Liang et al 1979
55,XY,+1,+5,+9,+11,-12,+13,+15,del(1)(q32),del(6)(q25), t(9;19)(q13;q13),14q+,16p+,+4mar	Liang et al 1979
45,XY,-2,-2,-3,+5,-8,-10,+14,-15,-16,del(6)(q25),+4mar	

Plasma cell leukemia / Plasmocytoma

Karyotype	Reference
44,X,-Y,+t(1;9)(q12q44;q34),t(6;8)(q13;p11),-9,-13	Ueshima et al 1983
45,XY,-5,del(6)(q11),t(6;7)(q?;q?),+9,-12,-13,-16,-17, t(9;22)(q34;q11),+2mar	Karpas et al 1982
46,XX,t(1;6)(p22p34;p25),t(11;14)(q13;q32)	Gahrton et al 1980b
46,XY,+t(1;16)(q21;p13),del(6)(q21),-16,-18,+mar	Ueshima et al 1983
47,XY,+t(1;16),del(6)(q21),+t(8;15)(p23;q12),-8,14q+,-15, -16,-18,+3mar	

SOLID TUMORS

UNCLASSIFIED NEOPLASMS

Malignant neoplasm, NOS

Karyotype	Reference
47,XX,-2,-5,+6,-10,11q+,-13,+18,-21,t(1;?)(q21;?), t(1;13)(q21;p11),t(1;6)(q11;q23)	Kovacs 1978a

EPITHELIAL NEOPLASMS

Carcinoma, NOS

Karyotype	Reference
41,XX,t(1;1;?)(p36q21;q12;?),t(1;?)(q12;?),del(6)(q21), t(5;12)(q?;q?),del(5)(q12q32)	Atkin and Baker 1982a
44-50,XX,+3,dup(1)(q21q32),t(10;11)(q26;q25),t(2;17), t(6;13),t(9;15),del(10)(q24)	Riet-Fox et al 1979
45,XY,-1,-6,-14,+21,+2mar/45,XY,t(1;14)(q11?;q24?),-6,-16, +21	Sonta and Sandberg 1978a
46,XX,t(1;?)(p32;?),-2,3q-,5q-,6q+,-18,-21,+mar	Atkin and Baker 1982a
46,XX,t(3;11)(q12;q14),-6,+7,-8,-15,t(16;?)(q23;?),+2r	Mark 1975b
47,XY,6q+,-7,+2mar	Reichmann et al 1981
49,X,-Y,-3,-5,-14,+15,-16,-17,+19,+21,+22,1q+,5q-,6q-,8p-, 9p-,11p-,18p+,20q-,22q-,t(Y;14),+3mar	Reichmann et al 1981
49,XX,t(3;11)(q?;q?),+5q-,+6q-,+i(11q)	Atkin and Baker 1982a
58,XX,+X,+1,+2,+3,5p-,+6,+7,+9,11q+,-13,-14,-15,+16,+19, +20,+6mar	Kovacs 1978a
60,XY,cx,3p+,6q+,8p-	Reichmann et al 1981
60-63,XX,cx,t(5;18),t(3;6),t(3;22),del(10)(q24), del(4)(p1?)	Riet-Fox et al 1979
63,XX,-1,+3,+5,+6,+7,+8,+10,+11,+12,-13,-15,+20,+21,+22, +9mar	Sandberg 1977
80-83,XX,cx,t(2;5;16),t(2;5;18),t(11;13),11q+,t(8;10), del(2)(q24),del(6)(q21),i(14q),dup(9)(q11q12)	Riet-Fox et al 1979

Carcinoma NOS, metastatic

Karyotype	Reference
??,X,-X,t(1;?),hsr(1),6q-,t(9;?),dic(13),dmin,cx	Bartnitzke and Bullerdiek 1983
46,XX,i(2q),6q-,cx	Bartnitzke and Bullerdiek 1983
46,XY,+2,-6,-9,del(22)(q11),+mar	Hansson and Korsgaard 1974
46,XY,-6,+mar	Musilova et al 1981
62-64,XY,del(2)(q34),del(6)(q14),ins(7)(q?),20q+,t(5;18), i(5p)	Ayraud 1975
64-100,XX,t(6;?)(p11q27;?),t(4;10)(q?;q?),t(3;?)(q25;?), del(1)(q2?),t(6;11)(p21;q13)	Kakati et al 1975
65,XX,i(1q),t(1;9),der(5),t(6;10)(p?;q?),t(14;?),der(21), cx	Bartnitzke and Bullerdiek 1983
66,XX,6q-,cx	Bartnitzke and Bullerdiek 1983

Large cell carcinoma

Karyotype	Reference
55-70,XY,t(4;5;14),del(1)(q11q21),t(7;11)(p22;q13), t(7;21)(q11;q22),t(3;8)(p?;q?),t(6;11)(p11;p11),i(16q), t(3;?)(q11;?),t(21;?)(q11;?),t(13;?)	Kakati et al 1975
60,XY,del(5)(p11),t(6;10)(q11;p11),t(7;11)(p22;q13), t(14;18)(q11;q11),del(2)(q21),t(6;?)(p11;?)	Pickthall 1976

Small cell carcinoma

Karyotype	Reference
110,XY,cx,del(1)(q41),del(3)(p13p23),del(3)(p13), t(3;16)(q24;p13),del(6)(q24)	Whang-Peng et al 1982b

Papillary carcinoma

Karyotype	Reference
34-36,XX,cx,1p-,1q-,6q-,14q+	Wake et al 1980
41,XX,cx,1p-,1q-,3p+,3q+,6q-,11q-	Wake et al 1980
41,XX,cx,t(6;14)(q21;q24),del(1)(p?q?),13q+	Wake et al 1980
42-54,X,-X,-14,-16,-16,+1,+3,+12,+20,1p-,6q-	Wake et al 1981
46,XX,t(6;14)(q21;q24),1p-,1q-,3p+,3q+	Wake et al 1980
50,X,-X,+1,+3,+12,+20,-14,-16,1p-,6q-	Wake et al 1980
56-63,XX,cx,t(6;14)(q21;q24),6q-,6q+	Wake et al 1980
62,XX,cx,1p-,1q-,t(6;14)(q21;q24),10q-	Wake et al 1980
62,XX,cx,6q-,14q+	Wake et al 1980
68,XX,cx,t(6;14)(q21;q24)	Wake et al 1980
68,XX,cx,t(6;14)(q21;q24),del(1)(p?q?),1p+,4q-,8q-,10q+	Wake et al 1980
69,XX,cx,1p+,1q-,1p-,1q-,2q+,3q+,4q+,6p+,12q+,14q+	Wake et al 1980

Adenoma

Karyotype	Reference
46,XY,t(6;21)(q13;q22)/46,XY,t(1;14)(q42;p12)/45,X,-Y	Mark et al 1981b
45,X,-Y	
46,Y,t(X;15)(q24;q15)/45,X,-Y	

Adenocarcinoma

Karyotype	Reference
37,X,-X,-3,-8,-15,del(2)(p23),del(6)(q15),t(11;?)(p12;?)	Trent and Salmon 1981
38,X,-X,-2,-3,-10,-14,-22,del(1)(q25),del(6)(q15), del(14)(p11),dmin	Trent and Salmon 1981
38,X,-X,cx,-17,-22,del(1)(q25),del(2)(q23),del(6)(q15)	Trent and Salmon 1980
40-42,X,-X,t(1;?)(p13;?),del(1)(p12),2q+,del(3)(q21), del(6)(q13),del(7)(q11),t(11;11)(p15;q13), t(12;12)(q24;?),13p+,15p+,i(22q)	Kusyk et al 1981
44,XX,-1,+5,-15,-18,2q+/44,XX,-1,+5,-6,-15,-18,2q+,10p-, +mar	Martin et al 1979
48-48,XX,+3,+7,+8,-1,-5,-14,-17,5q+,6q+,del(1)(q42), del(5)(q3?),del(10)(q24),del(14)(q2?)	Riet-Fox et al 1979
54,XX,+5,+6,+7,+9,+11,+21,+21,+del(1)(q32)	Sonta et al 1977
62-74,XX,cx,3q-,6q-,t(13;14),dmin	Hecht et al 1983
71,XX,cx,1q-,4q+,6q-,9q+,10q-,12q+,17q+,14q+	Wake et al 1981
72,XX,cx,del(2)(p23),hsr(3)(q13q28),del(6)(q21), t(10;11)(p12;p13)	Trent and Salmon 1981
74-75,XY,cx,t(18;19),del(1)(p22),del(6)(q21)	Riet-Fox et al 1979

Adenocarcinoma, metastatic

Karyotype	Reference
??,XX,t(1;16),2q-,del(3)(p14),del(4)(q22),del(6)(q22), i(7q),del(9)(q31),t(13;13)	Ayraud et al 1977
48,XX,+1,+3,+6,-16,1p-	Wake et al 1981
66,XX,cx,i(1q),inv(1)(p32q12),inv(1)(q12q43), del(6)(q15orq16),dmin	Trent and Salmon 1981

MESENCHYMAL NEOPLASMS

Liposarcoma

Karyotype	Reference
50,XY,+3,+6,+8,+11,+14,-22,17p+	Sonta et al 1977

Mesothelioma, malignant

Karyotype	Reference
43,XY,t(1;1)(q42;q32p22),t(2;6)(p25;p21),inv(3)(p24q13), t(5;18)(p15;q11),-6,+9,del(13)(q24q14),-14,-18,-22	Mark 1978
86,X?,cx,del(1)(p?q?),2p-,2p+,6q-,14q+	Wake et al 1981

MELANOMAS

Malignant melanoma

Karyotype	Reference
??,X?,+del(6)(q21),cx	Becher et al 1983
??,X?,+i(6p),cx	Becher et al 1983
??,X?,-6,cx	Trent et al 1983b
??,X?,6q+,t(6;7)(p?;q?)	Becher et al 1983
??,X?,t(1;6)(q15;q23),cx	Trent et al 1983b
??,X?,t(3;6)(p21;q23),cx	Trent et al 1983b
??,X?,t(6;?)(q21;?),cx	Trent et al 1983b
44,X?,+2,+9,-6,-8,-15,-21,9p-,19q+	Wake et al 1981

Malignant melanoma, metastatic

Karyotype	Reference
24,X,cx,t(1;15),t(1;14)(q12;p11),del(7)(p13), t(6;7)(p11;q11),t(13;21)(p11;q11),+r	Atkin and Baker 1981
41-43,XY,del(1)(q21q25),i(1q),t(1;9)(p22;q34), t(11;12)(q11;p11),t(10;12)(q26;q13),t(4;?)(q35;?), del(5)(p13),t(14;?)(p12;?),i(17q),t(18;?)(p11;?), del(6)(q21)	Kakati et al 1977

NEUROGENIC NEOPLASMS

Neuroblastoma

Karyotype	Reference
45,XX,+14,-15,-18,del(1)(p32),dup(1)(p13p31),del(6)(q23), del(11)(q23)	Brodeur et al 1981b
46,XY,del(1)(p32),6q+,13q+,16q+,dmin	Gilbert et al 1982
46,XY,del(6)(q25),dmin	Brodeur et al 1981b
46,XY,t(1;?)(p32;?),6p+,hsr(6q),7q-,hsr(7p),19q+	Gilbert et al 1982
47,X,-Y,del(1)(p?),+6,16q+	Brodeur et al 1977
47,XY,1p+,4p+,6q+,10q+,12q+,+19	Brodeur et al 1977

Retinoblastoma

Karyotype	Reference
44,XY,1p+,-3,t(3;12)(q?;p?),+6q-,der(9),13p+,-14,-16	Kusnetsova et al 1982
45,X,-X,+1p+,+i(6p),-16,17q+,-19,22p+	Kusnetsova et al 1982
46,X,-X,+i(6p),17q+,20q+	Kusnetsova et al 1982
46,XX,+i(6p),12q+,19p+,-22	Kusnetsova et al 1982
46,XX,inv(1),inv(2),inv(3),inv(4),t(6;7),inv(9),t(11;?), 11p-,inv(12),i(17q),-20	Gardner et al 1982
47,XX,+6,del(13)(q12q14),dmin	Balaban et al 1982
47,XX,+t(1;22)(q?;q?),+i(6p),-22	Kusnetsova et al 1982
47,XX,i(6p),+4q+,-16	Kusnetsova et al 1982
47,XY,5q+,+i(7q)/47,XY,+i(6p),12q+	Kusnetsova et al 1982
47,XY,dup(1)(q12q44),del(2)(q13),+del(6)(q16), t(14;?)(q32;?),i(17q),+20,-22	Gardner et al 1982
51,XY,del(6)(q21),+8,+11,-12,t(13;18)(p11;q11),14q+,-16, +19,+22,+2mar	Hossfeld 1978

Meningioma, benign

Karyotype	Reference
43,XX,-1,-6,-7,-11,-22,+2mar/40-45,XX,-1,-6,-7,-9,-11,-19, -22,+2mar	Mark et al 1972b
44,XX,-6,-22/45,XX,-22	Zang and Zankl 1983
45,XX,-22/45-46,XX,-22,-9,+6,+14,+del(22q)	Mark 1973a
45,XX,t(6;15)(p12;q26),-22	Zang and Zankl 1983
51,XY,+5,6q+,+12,+16,+17,+18,+20,-22/52,XY,+5,+8,+10,+12, +17,+18,+20,-22	Zang and Zankl 1983

EMBRYONAL AND MISCELLANEOUS NEOPLASMS

Teratoma, malignant

Karyotype	Reference
67,XY,+1,+3,+4,+5,+6,+7,+7,+8,+10,+11,+13,+13,+14,+16,+17, +18,+20,+20,+21,+21	Sonta et al 1977

Seminoma

53,XY,+3,+4,+6,+7,+11,+13,+19	Sonta et al 1977

LYMPHOMAS

1. UNCLASSIFIED AND MISCELLANEOUS LYMPHOMAS

Malignant lymphoma, NOS

47,XX,t(1;14)(q23;q32),+3,t(6;21)(p25;q11),+8, t(7;11)(q11;q25),-21	Yamada et al 1980

Non-Hodgkin's lymphoma, NOS

44-47,XY,3q+,del(4)(p13),del(6)(p23),+7,del(9)(p13)	Kristoffersson 1983
48-54,XX,+X,3p+,-4,-6,-8,-9,-10,+12,-15,+7-10mar	Pearson et al 1982
52,XY,+2,+3,t(5;9)(q22;q32),+6,+12,+14,+20	Reeves and Pickup 1980

Non-Hodgkin's lymphoma, lymphocytic, NOS

47,XY,+7,-9,-11,1p+,6q-,10p-,14q-, t(14;?;14)(q24orq32;?;q13),15q-,22q-	Fukuhara et al 1979b
47,XY,t(3;4)(q29;p12),del(6)(p22),+7,del(9)(p13)	Kristoffersson et al 1981
48-49,XY,1q-,+3,+del(3)(p?q?),-4,-6,+7,-9,-17,-18,-19, +5mar	Fleischmann et al 1976
80,XY,cx,4p+,t(1;4)(p22;p15),t(1;6)(p11;p11)	Kakati et al 1980

Non-Hodgkin's lymphoma, histiocytic, NOS

45,X,-X,del(1)(q22),t(1;17)(q?;q?),del(2)(q33), del(3)(p21p25),+ins(3)(p14p21p25),+5,del(6)(q14),t(1;7)	Mark et al 1978
45,X,del(Y)(q12),t(4;10)(q12;q24),t(6;7)(q21;q32), del(8)(p21)	Mark et al 1978
45,XY,del(6)(q13),del(6)(q21),15q+,i(1q)	Reeves 1973
46,XY,1p-,6q-,t(11;14)(q21;q32)	Kakati et al 1980
48,XY,1p-,+3,6q-,+7,9p-,t(11;14)(q21;q32)	
47,XY,-6,-14,1q-,+1p-,2q+,3q+,9p+,10p-,10p+,12p-,18q+,22p+, t(2;14)(q37?;q13?)	Fukuhara et al 1979b
47-48,XY,+X,-13,-15,t(6;11)(p?;q?),t(13;14;?)	Fleischman and Prigogina 1977
48,XY,del(3)(p21),del(9)(q22),11q+,12q+,del(6)(q23orq25), 15q+,del(12)(q15)	Reeves 1973
50,XY,t(2;?)(p25;?),del(2)(p13),+i(3p),del(6)(p21),+7,+8, +9,+9,-11,-13,+14,-15,+20,+21,-22	Mark et al 1978
44,XY,del(2)(p13),t(6;?)(p21;?),del(11)(q13), t(11;14)(q13;q31),i(15q),-20,-22	

Non-Hodgkin's lymphoma, T-cell, NOS

46-47,XY,del(6)(q14),14q+,+mar	Kristoffersson 1983
47,XX,t(2;?),inv(3),4p-,del(5)(q14q32),6p+,6q+,t(7;7),9q+, 10p-,13q-,t(14;14),-15,-15,16q-,17p+,+21,+2mar	Gaeke et al 1981

Angioimmunoblastic lymphadenopathy

47,XY,del(6)(q21q23orq25),-9,+2mar	Kaneko et al 1982c
56,XY,+5,+5,+6,+10,+15,+15,+19,+20,+21,+22	Kaneko et al 1982c

2. HODGKIN'S LYMPHOMAS

Hodgkin's disease, mixed cellularity

45,XX,-14,del(6)(q21)	Bernstein et al 1981
47,XX,+X,t(6;14)(q15;q32)	Reeves and Pickup 1980
61-80,XX,del(1)(q32),i(1q),del(2)(p13),del(6)(q13or15), del(6)(q23or25),15q+,12q+,del(5)(q31),del(7)(q11), del(10)(q26),del(3)(p13)	Reeves 1973
81,XX,14q+,i(18q),i(18p),del(6)(q21),t(9;15)(q11;p12), del(16)(q13)	Reeves 1973

Hodgkin's disease, lymphocytic depletion

60,XY,+del(1)(p21),t(1;5)(q44;q33),dup(5)(q12), t(1;5)(p21q23;q13),del(6)(q22),t(9;?)(q12;?), t(3;12)(q11;q24),t(14;?)(p11;?),i(17q)	Hossfeld and Schmidt 1978

Hodgkin's disease, nodular sclerosis

47,XY,t(1;?)(p36;?),+2,t(2;6)(q31;q27),+8,del(12)(p11), t(17;?)(p13;?),-14,-20,+mar	Reeves and Pickup 1980

3A. NON-HODGKIN'S LYMPHOMAS - RAPPAPORT CLASSIFICATION

Lymphocytic, well differentiated, diffuse

48,XY,1q+,3q+,+i(3p),-6,-10,t(14;?)(q12;?),+16,17p-,+18, +mar	Miyamoto et al 1981a

Lymphocytic, poorly differentiated, NOS

46,XY,del(6)(q21),-8,-9,t(6;14)(q21;q32),+16,+20	Mark et al 1979

Lymphocytic, poorly differentiated, diffuse

45,XY,t(14;18)(q32;q21),+2,-5,-6,-11	Fukuhara and Rowley 1978
47,XY,t(14;14)(q32;q?),+7,-9,-11,1p+,6q-,10p-,15q-,22q-, +2mar	Fukuhara and Rowley 1978
48,XX,1q-,+3,+4,6q-,+7,10q-,-15	Kakati et al 1980

87,XY,del(1)(q23),del(3)(p14),del(6)(q21),del(7)(p11), del(11)(q21),t(17;?)(p13;?) — Mark et al 1979

88,XX,del(1)(p22),del(3)(q21),t(3;9)(q?;q?),del(6)(q21), t(22;?)(q13;?) — Mark et al 1979

Lymphocytic, poorly differentiated, nodular

48,XY,t(14;18)(q32;q21),-6,-11,+12,+del(1)(p?),3q+,+2mar — Fukuhara et al 1979a

Mixed cell, diffuse

46,XY,-4,-9,del(6)(p23),+2mar — Kaneko et al 1982a

53,XY,t(8;14)(q24;q32),t(14;18)(q32;q21),+20,1q-,+1p-,2q-, 4q+,5q-,5p+,6q-,+7q+,10q+,11q+,11q-,13p+,+4mar — Fukuhara and Rowley 1978

Mixed cell, nodular

50,XX,t(3;6)(q29;p21),del(9)(p13q22),+11,+12,+21 — Mark et al 1976

Histiocytic, diffuse

45,X,del(X)(q24),del(1)(q21),+del(1)(p11),del(2)(q22), del(3)(q21),del(6)(p11),del(6)(q15),t(7;9)(p22;q13), t(8;10)(p23;q21),del(9)(q13),del(10)(q21), t(2;11)(q22;q23),-13,del(14)(q24),del(15)(q22), del(16)(q22),+17,-19 — Mark et al 1979

46,X,-X,-2,-10,+11,-13,-14,-17,del(1)(p22p32), del(3)(p13p21),del(6)(q21),del(9)(p13),+t(2;?)(p23;?), +t(14;?)(q32;?),t(18;?)(q23;?),+t(X;13)(q26;q12),+mar — Kaneko et al 1982a

47,X,-X,t(4;14)(q?;q32),-18,1q+,2p+,6q-,7q+,9q-,12q+,+3mar — Fukuhara and Rowley 1978

47,XX,-1,-2,-6,+8,-14,+16,-19,del(11)(q23), t(8;14)(q22;q32),+t(2;6)(q21;p23),del(6)(q21), +t(1;14)(p11;q11),+t(19;?)(q13;?),+t(14;?)(q11;?) — Kaneko et al 1982a

47,XY,-10,-13,+15,-16,+19,+20,del(6)(q21),del(8)(q22), t(2;14)(q21;q32),t(4;4)(q31;q35),+t(10;?)(p11;?) — Kaneko et al 1982a

48,XY,-6,-8,+21,i(6p),+t(8;?)(q11;?),+t(19;?)(q13;?) — Kaneko et al 1982a

50,XX,+3,+3,-4,-6,-9,+12,-17,+18,+i(6p),+3mar — Kaneko et al 1982a

51,XY,+X,-2,+5,+11,-12,+21,del(6)(q21),+t(2;?)(q33orq35;?), +t(12;?)(p11;?),+mar — Kaneko et al 1982a

61,XY,cx,t(1;16)(q23;p11),dup(1)(q23q32),t(1;6)(q23;p21), t(3;6)(q21;p21),t(7;7)(q32;p22),13p+ — Fitzgerald et al 1980

Histiocytic, nodular

46,XX,-6,+t(1;6)(q11orq21;q25orq27) — Kaneko et al 1982a

3B. NON-HODGKIN'S LYMPHOMAS - KIEL CLASSIFICATION

Immunocytoma

45-46,XY,-8 — Kristoffersson 1983
47,XY,15p+,+20/46,XY,del(5)(q23),del(6)(p22),+del(7)(p11), -21
49,XX,+3,del(6)(q14),+7,+18/49,XX,+3,del(6)(q14), del(9)(q22),+12,+18 — Kristoffersson 1983

Centroblastic-centrocytic, diffuse

46,XY,del(6)(q12),t(7;11)(p11;q11) — Kristoffersson 1983
48-51,XY,del(1)(p33q23q25),-2,+del(3)(p22),+4,4p+,+5,+6p+, +7,+8,+9,+21,+22 — Kristoffersson 1983
45-51,XY,del(1)(p33q23q25),del(2)(q14),+del(3)(p22),+4,+5, +7,+8,-12

Centroblastic-centrocytic, follicular

46-49,XX,1p+,3q-,4q+,+6,+13,-14,14q+,17p+ — Kristoffersson 1983

Centroblastic, diffuse

42-45,XY,1p+,-6,-8,14q+,14q-,-16,i(17q),+18,-20,+mar — Kristoffersson 1983
46-49,XY,1p+,+2,-6,+del(9)(p21),+12,+13,14q+?,i(17q),+18 — Kristoffersson 1983
47,XY,14q+,+22
46-51,XX,+3,del(6)(q21),+del(6)(q21),+7,-8,del(9)(q11),+13, -16,-17,+19,-20,-21,+1-5mar — Kristoffersson 1983

Centroblastic, follicular

47,XY,del(6)(q23),+8,14q+?,i(17q) — Kristoffersson 1983

Lymphoblastic, Burkitt's type

??,XX,6p-,t(8;14) — Douglass et al 1980a
??,XY,-5,+6 — Douglass et al 1980a
??,XY,1p+,t(8;14)
??,XY,dup(1)(q12orq21q31),3p-,3q-,+6p-,+8,+14,+15,+18,-21 — Douglass et al 1980a
45,X,-Y,t(1;14)(q22;q32),t(8;22)(q24;q11),3q+,6q-,6q+,7q+, 9p-,14q-,18q- — Berger et al 1981c
45,X?,t(8;14)(q24;q32),t(7;16),t(6;17) — Zech et al 1976a
46,XY,t(8;22)(q23;q12)/46,XY,t(8;22)(q23;q12), t(1;6)(q21;q27) — Berger et al 1981c
46,XY,t(8;22)(q23;q12)/46,XY,t(8;22)(q23;q12), +t(1;6)(q23;q26) — Berger et al 1981c
46-47,XY,t(14;?)(q32;?),dup(1)(q21q32),t(2;3)(q13;q29),+5, 6p-,-8,del(9)(q?),+12,+18,+mar — Miyoshi et al 1981c
46-48,X?,t(8;14)(q24;q32),+7,15q-,+6,+13 — Zech et al 1976a
47,XY,+6,del(6)(p23q25) — Biggar et al 1981

Lymphoblastic, convoluted type

43-46,XY,del(1)(p32),del(2)(p21),-6,t(7;9)(p11;q11),-9,-10, t(11;15)(p11;q11),-11,del(12)(q14),-13,-14,i(17q),-17, +18,del(19)(p11),-22,+2mar	Kristoffersson 1983
46,XY,del(6)(q21q25)	Kaneko et al 1982d
47,XY,-7,-9,del(6)(q23?),t(14;18)(p11;q11),+t(9;?)(q34;?), +2mar/47,XY,-7,-9,del(6),del(1)(q32),t(14;18),t(9;?), +t(1;?)(q23orq25;?)	Kaneko et al 1982d

Immunoblastic

45-46,XX,+3,del(6)(q21),+4,+16,+21,-C,-C	Kristoffersson 1983
46,XY,del(4)(q33),del(6)(q23),t(5;7)(q33;q22)	Kaneko et al 1982a
49,XY,del(1)(p32),del(1)(q32),+5,del(6)(q21),+2mar	Kristoffersson 1983
51,XY,t(1;6)(p22;q13orq15),t(2;13)(q21;q14), t(8;16)(q24;p11),+11,+12,del(15)(q22),+16,+20,+mar	Reeves and Pickup 1980

3C. NON-HODGKIN'S LYMPHOMAS - LUKES AND COLLINS CLASSIFICATION

Mycosis fungoides

46,XY,6q-,der(12)	Nowell et al 1982

Sezary's syndrome

50,XX,1q+,2p-,+3q-,+6q-,+9,-15,+21,+mar	Nowell et al 1982

Follicular center cell, diffuse, large cleaved

45,XY,-6,-9,-17,2q-,3p-,3q-,4q-,5q-,6p+,7p+,7q+,+7q-,9p-, 12q+,13p+,20p-,+mar	Fukuhara et al 1978b
47,XY,-6,-14,1q-,+1p-,2q+,3q+,9p+,10p-,10p+,12p-,18q+,19p+, 19q+,+22p+,+mar	Fukuhara et al 1978b

Follicular center cell, diffuse, small non-cleaved

44,X,-Y,t(1;14)(p21;q13),5p+,del(6)(q13),del(8)(p21),9p+, 9p-,t(11;21)(q23;q22),13q+,-10,+12,-17p+,der(19),22q+	Fukuhara et al 1979a

Immunoblastic type (B-cell)

48,XY,t(4;12)(q?;q?),t(6;15)(q?;q?),t(9;22)(q34;q11),11q-, +20,+21	Nordenson et al 1983

3D. NON-HODGKIN'S LYMPHOMAS - WORKING FORMULATION

Follicular, small cleaved cell

46,XX,t(2;3),i(6p),t(14;18)(q32;q21)	Yunis et al 1982
46,XY,6q-,11q-	Yunis et al 1982
47,XY,+X,t(2;3;17),6q-,t(14;18)(q32;q21)	Yunis et al 1982

Follicular, large cell

75,XX,cx,i(6p),t(14;18)(q32;q21),i(17q)	Yunis et al 1982

Diffuse, large cell

47,XX,+X,6q-,t(14;17)	Yunis et al 1982

Large cell, immunoblastic

48,Y,t(X;?),t(1;?),t(3;6),+5,-6,t(8;14)(q24;q32),i(17q),+mar	Yunis et al 1982
99,XY,cx,5p-,5q-,6q-,7p-,i(12q)	Yunis et al 1982

Chromosome 7

HEMATOLOGICAL DISORDERS

UNCLASSIFIED LEUKEMIAS

Acute leukemia, NOS

45,X,-X,+t(X;1)(q13;p12),-7,+8,-9	Mamaeva et al 1983
46,XX,dup(1)(p31p36),dup(7)(p15p21),dup(3)(q12q29), ins(12)(q12)"s"	Berger et al 1977

NONLYMPHOCYTIC LEUKEMIAS

Acute nonlymphocytic leukemia (ANLL)

39,X,-X,-3,-5,-13,-14,-15,-16,-17,-21,-22,del(7)(p11), t(12;?)(p13;?),t(16;?)(p13;?),+3mar	Mitelman 1983
39,X,-X,-3,-5,-13,-14,-15,-16,-17,-21,-22,del(7)(p11), t(12;?)(p13;?),t(16;?)(p13;?),+3mar	
43,XX,-1,3p+,3q-,4q+,-5,-5,-7,+9,+9,10p+,+10,+10,10p-,10q-, +11q-,-13,-15,-16,-17,-18,-21"s"	Whang-Peng et al 1979
43,XX,2q-,inv(3),-4,del(5)(q?),-7,17p+,-19	Rowley 1976
43,XX,t(1;20),-5,-6,7q-,11q-,-12,17q+,-18,19q+,-21	Mitelman 1983
43-44,XX,-5,-7,-14,+mar,dmin	Marinello and Levan 1982
43-45,XY,-5,5q-,-7,-19"s"	Whang-Peng et al 1979
44,XX,t(8;11)(p23;p13 or p15),-20/46,XX,del(7)(q11), t(1;3)(q2?;q2?),-20,-21,-22,+3mar	Philip et al 1978a
44,XY,-7,-17,-20,-22,+2mar"s"	Pedersen-Bjergaard et al 1982
44,XY,5q-,-7,12p-,-22"s"	Whang-Peng et al 1979
44,XY,del(5)(q21),del(7)(q11),del(11)(q11), t(7;11;12)(q11;q11;p13),del(13)(q11),-20,-21/45,XY, t(3;12)(q11;p13),del(5)(q21),del(7)(q11),del(13)(q11), -20	Mitelman 1983
45,X,-X,-7,t(X;X)(q13;q13)	Philip et al 1978a
45,X,-X,7q-,-8,+i(17q)	Nordenson 1983
45,XX,-7	Kaufmann et al 1974
45,XX,-7	Philip et al 1978a
45,XX,-7"s"	Boetius et al 1977
45,XX,-7	Carbonell 1983
45,XX,-7"s"	Mitelman 1983

45,XX,-7"s"	
45,XX,-7,12q+"s"	Mitelman 1983
45,XX,-7"s"	Berger et al 1981a
45,XX,del(3)(p11),t(5;7;12)(q31;q22;q13-24),t(5;?)(q13;?)	Philip et al 1978a
45,XY,-7	Philip et al 1978a
45,XY,-7	Philip et al 1978a
45,XY,-7	Van Den Berghe et al 1979a
45,XY,-7"s"	Kross et al 1981
45,XY,-7	Trent et al 1983a
45,XY,-7	Bernard et al 1982a
45,XY,-7"s"	Berger et al 1981a
45,XY,-7/50,XY,+8,+15,+19,+20	Berger et al 1981a
45-47,XX,-2,-7,7q-,-8,+15,+22,+2mar"s"	Whang-Peng et al 1979
46,X?,del(7)(q22)	Lawler et al 1975
46,X?,del(7)(q22)	Lawler et al 1975
46,XX,-7,-8,+2mar	Auerbach et al 1982
46,XX,del(5),del(6),del(7),21q+"s"	Geraedts et al 1980a
46,XX,del(7)(q22)/46,XX,del(7)(q22),del(5)(q22q31)/46,XX, del(7)(q22),-9,t(1;9)(q22;p24)/46,XX,t(2;3)(q31;q27), del(7)(q22),del(5)(q22q31orq33)"s"	Oshimura et al 1976b
46,XX,r(7)	Slee et al 1981
46,XX,t(1;1)(q42;q21),t(3;7)(q2?;q22)	Philip et al 1978a
46,XX,t(1;22)(p36;q11),del(5)(q14q34),del(7)(q22q36),r(22)	Yunis 1982
46,XY,+del(7)(p11),-8,del(12)(p11),t(21;?)(q22;?)/50,XY,+Y, +del(7)(p11),del(12)(p11),+17,t(17;?)(q25;?), t(21;?)(q22;?),+22	Mitelman 1983
46,XY,+del(7)(p11),-8,del(12)(p11),t(21;?)(q22;?)/47,XY,+9	
46,XY,-7,+21/47,XY,+21	Philip et al 1978a
46,XY,1q+,5p-,-7,+mar	Chrz et al 1983
46,XY,7q-	Bernard et al 1982a
46,XY,del(2)(q31),-7,del(9)(p13),t(2;21)(q21;q22), t(2;?)(p21;?)/46,XY,t(1;?)(p22;?),t(17;?)(q25;?)	Philip et al 1978a
46,XY,del(20)(q11),del(7)(q22)	Philip et al 1978a
46,XY,del(7)(q11)	Slee et al 1981
46,XY,del(7)(q11)	Seabright 1983
46,XY,del(7)(q34)"s"	Davis et al 1981
46,XY,t(9;22)(q34;q11)	Mitelman 1983
46,XY,t(9;22)(q34;q11)	
46,XY,t(9;22)(q34;q11)	
46,XY,t(9;22)(q34;q11)/47,XY,t(9;22),7p+,+19	
46,XY,t(9;22)	
46,Y,t(X;11)(q22.3;q23.3),ins(X;11)(q22.3;p13p15), dup(1)(q11q21),t(4;6)(p16;p21),t(7;12)(p13;q15)	Nacheva et al 1982
47,XY,+21"c"	Mitelman 1983
47,XY,+21/47,XY,+21,r(7)"c"	
47,XY,+21,r(7)"c"	
47,XY,+21,r(7)"c"	
47,XY,t(7;20)(p13;p12),+8"c"	Riccardi et al 1978
48,XY,+del(3)(q11),+7,t(6;19)(q15;q13)/50,XY,+del(3)(q11), +7,t(6;19)(q15;q13),+2mar/51,XY,+del(3),+7,+9,t(6;19), +2mar	Mitelman 1983
49,X,t(X;10)(p11;p11),+del(1)(p22),+del(1)(p22), del(5)(q23),+7/49,X,t(X;10)(p11;p11),+del(1)(p22),+7,+8	Mitelman 1983
51,XX,del(5)(q?),+7,+8,+13,+17,+22	Van Den Berghe et al 1976

ANLL, FAB type M1

39,XY,-5,-7,-12,-16,-17,-20,-21	Berger et al 1981a
42,XX,-7,-17,-19,-21"s"	Papa et al 1979
43,XX,del(5)(q12q21),-7,11p+,-15,-17	Kerkhofs et al 1982
44,XX,-5,-7,-20,-22,+2mar/44,XX,-3,-11,-11p+,-22	Prieto et al 1981
45,X,-X,-5,-7,-11,-17,+21,+3mar"s"	Zaccaria et al 1983b
45,XX,-7,-9,+t(7;9)(q11;p11),22q-/46,XX,t(9;22)(q34;q11)	Sasaki et al 1983
45,XX,del(5)(q12q31),-7,12p-"s"	Kerkhofs et al 1982
45,XX,t(9;22)(q34;q11),-7	Sasaki et al 1983
45,XY,-7"s"	Pedersen-Bjergaard et al 1981
45,XY,-7	Yunis et al 1981
45,XY,-7"s"	Hagemeijer et al 1981a
45,XY,-7	Hagemeijer et al 1981a
45,XY,-7,t(9;22)(q34;q11)	Sasaki et al 1983
45,XY,3p-,5q-,-7"s"	Pedersen-Bjergaard et al 1981
45,XY,5q-,-7"s"	Pedersen-Bjergaard et al 1981
45,XY,t(2;5)(q22orq23;q14q15),6p-,-7,t(6;7)(p21;p22q21), t(7;10)(p21;q21),t(14;19)(q11;q13)"s"	Hagemeijer et al 1981a
46,XX,dup(1)(q21q33),7q+,18p+	Alimena 1983
46,XX,t(9;22)/45,XX,-7,t(9;22)	Hagemeijer et al 1981a
46,XY,-2,-5,-7,-13,+16,-21,-21,+5mar	Sessarego et al 1981a
46,XY,del(7)(q32)"s"	Mitelman 1983
46,XY,del(7)(q32)"s"	
46,XY,t(9;22)(q34;q11)	Sasaki et al 1983
46,XY,t(9;20;22)(q34;q12;q11)/46,XY,t(9;22)(q34;q11),-6,-7, -7,-8,10q+,+t(7;8),i(8q)	

ANLL, FAB type M2

42,XY,5q-,7q-,-8,-12,-17,-21,-22,+i(17q),+2mar	Berger et al 1981a
43,XX,-3,-4,-7,-19,del(5)(q14),17p+	Testa et al 1979
43,XX,-3,-4,-7,-19,del(2)(q31?),del(5)(q14)	
43,XY,-3,-4,-7,-7,-8,-8,-17,-20,+5mar	Berger et al 1981a
43,XY,-5,-7,-12,-16,4p+,17p+,20q-,+mar"s"	Rowley et al 1981a
44,X,t(X;8),del(5)(q12q21),t(1;6),t(7;19),-7,-8,t(7;8),-21	Kerkhofs et al 1982
44,XX,-5,-7,-18,+mar"s"	Rowley et al 1977a
44,XY,-7,17p+,-20,21q+"s"	Pedersen-Bjergaard et al 1981
45,X,-X,t(8;21)(q22;q22)/46,X,-X,t(8;21),1p+,7q+,+19/47,X, -X,t(8;21),1p+,+19,+der(21)	Sakurai et al 1982b
45,XX,-7	Pasquali et al 1982a
45,XX,-7	Lessard and Le Prise 198[illegible]
45,XX,-7,ins(3;3)(q21;q21q26)"s"	Rowley et al 1981a
45,XX,-7,t(3;9)(q29;p21)/44,X,-X,-7,t(3;9)"s"	Rowley et al 1981a
45,XX,5q-,-7,-18,+mar	Teerenhovi et al 1981
45,XY,-5,-7,+mar"s"	Rowley et al 1977a
45,XY,-5,-7,-8,+16,-17,+r,+mar	Berger et al 1981a
45,XY,-5,-7,del(3)(p13),+t(5;17)(q11;q11)"s"	Rowley et al 1981a
45,XY,-7,del(3)(p13),del(8)(q22),del(16)(p12)"s"	Rowley et al 1981a
45,XY,-7,del(5)(q13),t(12;17)(q13?;p12?)"s"	Rowley et al 1981a

Karyotype	Reference
45,XY,-7,t(9;22)(q34;q11)	Sasaki et al 1983
46,XX,-4,-6,-7,-20,+mar	Golomb et al 1978c
46,XX,-7,+8,inv(1)(p36q13)"s"	Rowley et al 1981a
46,XX,2q+,5q-,7q-,17q+"s"	Berger et al 1981a
46,XX,4q+,del(7)(q31),t(8;21)(q21;q21),del(9)(q12q21)	Hagemeijer et al 1981a
46,XX,7q-	Lessard and Le Prise 1983
46,XX,del(4)(q31q34),t(7;12)(q21;q13)	Yunis et al 1981
45,XY,-7"s"	Mitelman 1983
45,XY,-7"s"	
46,XY,+1,5q-,-7"s"	Berger et al 1981a
46,XY,-7,-12,-17,del(5)(q14),8q+,del(17)(p11),+mar	Testa et al 1979
45,XY,-7,+8,-12,-17,del(5)(q14),8q+,t(12;17)	
46,XY,inv(7)(p21q36)	Hagemeijer et al 1981a
47,XY,+3,t(7;8)(p11;q11)	Kaneko et al 1982b
49,XY,+4,+r(7),+19	Kaneko et al 1982b
53,XX,+1,-4,-5,+6,-7,+8,+11,+12,+21,+21,+3mar"s"	Zaccaria et al 1983b

ANLL, FAB type M1+M2

Karyotype	Reference
37-45,XY,-7,-12,-13,-14,-16,-18,-19,-21,del(4)(q29), del(5)(q24),der(9),der(17),11q+,12p+,t(14;?)	Fitzgerald et al 1983
43,XX,1p+,-4,-5,-6,-7,-13,-17,del(18)(q21),+3mar	Rowley and Potter 1976
43,XX,del(5)(q12q32),-7,-12,-18,t(7;12;21)(q11;q12;q22)	Petit and Van Den Berghe 1979b
44,X,-X,-5,t(3;7)(q13;p12)/44,XX,-7,-8/44,XX,-5,-8"s"	Sandberg et al 1982a
44,XX,del(3)(p21),del(5)(q12),-7,-20,i(22)(q11)	Mitelman et al 1981
44,XY,-7,-16,-19,-21,-22,+3mar"s"	Sandberg et al 1982a
44-45,XX,-5,-7	Mitelman et al 1981
44-45,XX,-7,-22	Mitelman et al 1981
44-45,XY,+2,-5,-7	Mitelman et al 1981
45,XX,-7	Garson 1980
45,XX,-7	Li et al 1983
45,XX,-7	Mitelman et al 1981
45,XX,-7	Mitelman et al 1981
45,XX,-7	Zech et al 1975
45,XX,-7	Shiraishi et al 1982
45,XX,-7	Mitelman 1983
45,XX,ins(3;3)(q26;q21q26),-7"s"	Sweet et al 1979
45,XY,-5,-19,-20,-22,+t(7;12)(q11;q13),+r,+3mar	Fitzgerald et al 1983
45,XY,-7	Mitelman et al 1981
45,XY,-7	Yamada and Furusawa 1976
45,XY,-7	Rowley and Potter 1976
45,XY,-7	Garson 1980
45,XY,-7"s"	Sandberg et al 1982a
45,XY,-7"s"	Sandberg et al 1982a
45,XY,-7	Li et al 1983
45,XY,-7,t(11;21)(q25;q11)	Fitzgerald et al 1983
45,XY,-7/47,XY,+11	Li et al 1983
45,XY,t(19;22)(p13;q11),t(1;7)(q21;q22),t(7;9),+del(8)(p?), -15	Oshimura and Sandberg 1977
45,XY,t(2;2)(q12;p25),del(7)(q32)	Prigogina et al 1979
45-46,XX,+14,-7,-22,+16,-20	Mitelman et al 1981
45-46,XY,t(2;6)(q37;q23),-7,del(11)(q21),del(22)(q11)	Mitelman et al 1981
46,XX,+14,-22/46,XX,-7,+14,+16,-20,-22	Alimena et al 1977
46,XX,-7,+19	Mitelman et al 1981

46,XX,-7,+19	Alimena et al 1977
46,XX,-7,del(5)(q22),+r	Prigogina et al 1979
46,XX,7q+,t(8;21)	Li et al 1983
46,XX,del(7)(q?)	Garson 1980
46,XY,-7,+mar	Rowley and Potter 1976
46,XY,7q-	Li et al 1983
46,XY,del(7)(q22),del(7)(p?),del(20)(q12)"s"	Whang-Peng et al 1977
46,XY,del(7)(q23)	Mitelman et al 1981
46,XY,del(7)(q31),del(11)(q14),del(12)(p12)	Mitelman et al 1981
46,XY,t(9;22),t(7;10)(q34;q22)	Wayne et al 1979
46-50,XY,+Y,+del(7)(p11),-8,+9,del(12)(p11),+17,21q+,+22	Mitelman et al 1981
47,XY,+19,t(7;12)(q21;p13),del(4)(q26),del(5)(p12)	Mitelman et al 1981
47,XY,i(3q),+7	Nordenson 1983
47-49,XY,-5,del(5)(q15),-7,i(8q),dic(9),-13,-16,-22, +6-9mar"s"	Testa et al 1981b
48,XY,+7,+8,+8,-17"s"	Sandberg et al 1982a
49,XX,+1,-5,+7,+8	Nordenson 1983
50-52,XY,-5,-7,+9,+14,+17,+19,+20,+21,+mar	Mitelman et al 1981

ANLL, FAB type M3

46,XX,2q-,6p-,7q-,10q+,14q+,-16,+mar	Hagemeijer et al 1981a
46,XX,t(15;17)(q26;q22),del(7)(q33orq35)	Van Den Berghe et al 1979c
46,XX,t(15;17)(q26;q22)/47,XX,t(15;17),-7,+8,-9,-10,+2mar	Van Den Berghe et al 1979c
46,XX,t(7;17)(q36;q22)	Yamada et al 1983a
46,XY,del(15)(q22q26)/45,XY,-7	Zahavi et al 1982
46,XY,t(15;17)(q25;q22)/46,XY,t(15;17),del(7)(q22), del(9)(q22)	Golomb et al 1979
46,XY,t(15;17)(q?;q?),t(3;11)(q?;q?),t(7;19)(q11;p13)	Trujillo and Cork 1981
47,XX,t(15;17)(q26;q22),+8,/47,XX,+8/46,XX, t(15;17)(q26;q22),-7,+8	Van Den Berghe et al 1979c

ANLL, FAB type M4

42,XY,-7,i(8)(q11),-16,-17,-18,t(21;21)(q12;q11),+mar	Rowley and Potter 1976
43,XY,-2,-3,-6,-7,-12,-15,-21,-22,del(2)(p11),del(11)(p14), t(14;?)(q32;?),t(17;?)(p13;?),del(22)(q11),+5mar"s"	Rowley et al 1981a
43-46,X,-Y,del(5)(q31),-7,+18,-20	Swolin et al 1981
44,XY,-7,+20,-21,-22"s"	Berger et al 1981a
44,XY,del(5)(q15q23),del(7)(q22),-9,-12,-13,del(15)(q22), +mar/46,X,-Y,del(5),del(7),+8,-12,-13,del(15),+mar,dmin	Alimena 1983
44,XY,t(3;5)(q29;q13),t(7;12)(q12;q14),t(7;10)(q12;q21),-8, -17,-18,-20	Mitelman et al 1981
45,X,-Y,del(7)(q32),t(8;21)(q22;q22)	Hustinx et al 1980
45,X,-Y/46,X,-Y,+7	Chrz et al 1983
45,XX,-7	Pasquali et al 1982a
45,XX,-7	Berger et al 1981a
45,XX,-7"s"	Hagemeijer et al 1981a
45,XX,-7"s"	Berger et al 1981a
45,XX,-7	Berger et al 1981a
45,XX,del(2)(q31),del(7)(p13),-13	Rowley and Potter 1976

45,XY,-7	Bernstein et al 1982b
45,XY,-7	Bernstein et al 1982b
45,XY,-7	Li et al 1981a
45,XY,-7"s"	Pasquali et al 1982a
45,XY,-7	Pasquali et al 1982a
45,XY,-7	Pasquali et al 1982a
45,XY,-7	Takeuchi et al 1981
45,XY,-7	Hagemeijer et al 1981a
45,XY,-7	Brodeur et al 1983
45,XY,7p-,21	Garson 1980
45,XY,del(5)(q13q31),del(7)(q22),del(15)(q22q24)	Alimena et al 1982a
45-46,XY,-7,+17	Mitelman et al 1981
45-46,XY,inv(2),-7	Mitelman et al 1981
46,XX,del(7)(q22)	Alimena 1983
46,XX,del(7)(q22)/45,XX,del(7)(q22),-21	Hagemeijer et al 1981a
46,XY,-7,+8	Hagemeijer et al 1981a
46,XY,1p+,4q+,del(7)(p11)	Morse et al 1979a
46,XY,7q-,-21,+i(?21q)"s"	Berger et al 1981a
47,XX,-7,+21,+22/45,XX,-7	Prieto et al 1981
47,XY,+7	Hagemeijer et al 1981a
47,XY,del(7)(q32),+22	Prigogina et al 1979
47,XY,t(7;12)(q36;p13),+8	Hagemeijer et al 1981a
49,XX,del(3)(q25),del(7)(q22),+8,+17,+19	Alimena 1983

ANLL, FAB type M5

45,XX,der(3),-7"s"	Prigogina et al 1979
45,XY,-7	Alimena 1983
46,XY,-7,+17/45,XY,-7	Alimena 1983
46,XY,del(11)(q13q23),t(10;11)(p25;q23),t(1;7)(p21;q36)	Yunis et al 1981
46,Y,-X,-1,-7,-8,-14,-15,-17,-20,inv(6)(p23q13), del(11)(q23),+t(X;?)(p11;?),+t(1;?)(p24;?), +t(12;?)(q22;?),+t(17;?)(p11;?),+4mar	Kaneko et al 1982b
46-47,XX,-7,+8,+20	Mitelman et al 1981
47,XY,del(7)(q22),+del(8)(q12)	Mitelman et al 1981
57-58,XX,+1,+3,-4,-5,+6,-8,+11,+12,-17,+19,-20, del(10)(q24),+8-9mar	Testa et al 1979
60-61,XX,Xp+,+1,+2,-4,+6,-7,+19,+20,del(2)(q31), del(10)(q24),+12-13mar	

ANLL, FAB type M5a

48,XX,t(7;8)(p11;q11),+i(8q),+i(8q),del(11)(q24)	Berger et al 1982a

ANLL, FAB type M4+M5

42,XX,-5,-7,-12,-15,-16,-18,+2mar	Li et al 1983
45,XX,-7	Zech et al 1975
45,XX,-7"s"	Secker Walker and Sandler 1978
45,XY,-7"s"	Sandberg et al 1982a
66,XY,+1,+1,+2,+4,+6,+6,+7,+8,+8,+10,+11,+11,+12,+12,+18, +19,+21,+21,+22,+22	Li et al 1983

ANLL, FAB type M6

Karyotype	Reference
40-45,XY,-4,-6,-7,-10,-11,-20,+1-3mar	Mitelman et al 1981
43,X,-X,del(5)(q14q34),-7,11q-,-12,12p+,t(12;17),t(20;21)	Kerkhofs et al 1982
43,XX,5q-,-7,-17,-18	Garson 1980
43,XX,del(6)(q15orq21),-7,-12"s"	Rowley et al 1977a
43,XY,-4,5q-,t(4;6),-7,dic(12),-15"s"	Hagemeijer et al 1981a
43-45,XX,der(1),inv(2)(p25q14),del(3)(q25),-4,-5,-6,6q-,-7, 7p+,11p+,-12,-17,-21,+r,+1-4mar"s"	Smadja et al 1983b
44,X,del(Y)(q?),+Xq-,3p-,3q-,4q-,-6,+7q-,11q+,-15,-18"s"	Bradley et al 1982
44,XY,-3,5q-,-7,-D,+20	Li et al 1983
44-45,XY,-2,-4,-7,-10,-13,-16,-17,-22,+3mar	Mitelman et al 1981
44-47,XY,+2,-3,-5,-7,+8,-14,-19,+2mar	Mitelman et al 1981
45,XX,-7	Yunis et al 1981
45,XX,-7	Pasquali et al 1982a
45,XX,t(4;22)(p16;q11),-7/46,XX,t(4;22)(p16;q11)	Oshimura and Sandberg 1977
45,XY,-7	Petit et al 1973
45,XY,-7	Pasquali et al 1982a
45,XY,del(4)(q23),del(5)(q13),-6,del(6)(q21),del(7)(p14), del(8)(p21),del(9)(q31),16q+	Rowley and Potter 1976
45,XY,del(5)(q11),-7/45,XY,del(5)(q11), ins(3;3)(q12;q21q27)"s"	Swolin et al 1981
45-49,XY,-4,-7,-8,-10,+14,+16,-21,-22,+2mar	Mitelman et al 1981
46,X?,+1,-5,-7,+mar	Wurster-Hill 1983
46,XX,-7,-16,-21,+t(2;7)(p11;q11),del(2)(q33),del(11)(q22), +11p+,14p+,+t(16;21;21;21)(p13;q22;q11;q22)"s"	Rowley et al 1981a
46,XY,dup(1)(p32p34),dup(3)(q12q27),del(7)(p11)"s"	Berger et al 1980a
46,XY,r(7)"s"	Mitelman 1983
51,XY,+1,+2,+6,-7,+8,-14,+15,+21,+mar"s"	Rowley et al 1981a

ANLL, special type

Karyotype	Reference
45,XX,-7	Berger et al 1981a
45,XY,-7	Berger et al 1981a

Chronic myeloid leukemia, t(9;22)

Karyotype	Reference
26,XX,t(9;22)(q34;q11),-1,-2,-3,-4,-5,-6,-7,-9,-10,-11,-12, -13,-14,-15,-16,-17,-18,-19,-20	Hartley and McBeath 1981
43-46,XX,t(9;22),-5,-7,-10,-17,-19,5p+,14q+,+r(?),+mar	Sadamori et al 1981b
44,XY,t(9;22),-17,-20	Stoll and Oberling 1978
45,XY,t(9;22),-8,-11,-18,+2mar	
47,XY,t(9;22),+7,i(17q)	
45,X,-Y,t(9;22)	Sonta and Sandberg 1977b
42,X,-Y,t(9;22),-7,-16,-17	
45,X?,t(9;22),-7,i(17q)	Carbonell et al 1982a
45,XY,t(9;22),+del(1)(q12),-3,-7,del(11)(q21),-19,+mar	Lyall and Garson 1978
45,XY,t(9;22),-20	Stoll and Oberling 1978
45,XY,t(9;22),-7,-8,+17	
45,XY,t(9;22),-7	Miyamoto 1980
45,XY,t(9;22),-7	Mitelman 1983
46,X,Yq-,t(9;22)	Miyamoto 1980
45,X,Yq-,t(9;22),-7	

46,X?,t(9;22),1p+,-7,+mar/46,X?,t(9;22),1p+,-7,8q+,+mar	Carbonell et al 1982a
46,X?,t(9;22),7q+	Fleischman et al 1981
46,X?,t(9;22),inv(7)/47,X?,t(9;22),+8,i(17q)	Carbonell et al 1982a
46,XX,t(9;22)(q34;q11)/46,XX,t(9;22),-7,-17,+13,+14	Sonta and Sandberg 1978b
46,XX,t(9;22)(q34;q11)/46,XX,t(9;22)(q34;q11), t(7;8)(q22;q13)	McGlave et al 1982
46,XX,t(9;22)(q34;q11),2p-,7p+,-8,+mar	Alimena et al 1982b
46,XX,t(9;22)(q34;q11),-7,+t(1;7)(q24;p11)	Tomiyasu et al 1982
46,XX,t(9;22),-7,-15,+20,+i(17q)	Stoll and Oberling 1978
46,XX,t(9;22),r(7)	Seabright 1983
46,XX,t(9;22),t(1;4)(q22;q31),del(7)(q11)	Miyamoto 1980
46,XX,t(9;22),t(4;9)(p16;q32)	Sadamori et al 1980
45,XX,t(9;22),t(4;9),-7	
46,XX,t(9;22)/45,XX,t(9;22),-7	Hagemeijer et al 1980b
46,XX,t(9;22)/45,XX,-7	Hartley and McBeath 1981
46,XX,t(9;22)/47,XX,t(9;22),+22q-/46,XX,t(9;22),-7, +22q-/47,XX,t(9;22),-7,+19,+22q-	Hagemeijer et al 1980b
46,XY,t(9;22)(q34;q12)	Whang-Peng et al 1976b
45,XY,t(9;22)(q34;q12),t(7;22)(p12;p13)	
46,XY,t(9;22)(q34;q12),del(8)(q23)	
46,XY,t(9;22)(q34;q11)	Como and Graze 1979
35,XY,t(9;22)(q34;q11),-3,-4,-5,-7,-9,-11,-12,-13,-15,-16, +3mar	
46,XY,t(9;22)(q34;q11),21q+/50,XY,t(9;22),+7,+8,+12,+15, 21q+	Tomiyasu et al 1982
46,XY,t(9;22)(q34;q11),t(7;11)(p14;p15)/47,XY,t(9;22), t(7;11),+22q-	Tomiyasu et al 1982
46,XY,t(9;22)(q34;q11)	Ishihara et al 1982
27,X,-Y,t(9;22),-1,-2,-3,-4,-5,-6,-7,-9,-11,-13,-14,-15, -16,-17,-18,-19,-20,-22	
46,XY,t(9;22),del(7)(q33?),del(15)(q?)	Kohno et al 1979b
46,XY,t(9;22),inv(7)	Stoll and Oberling 1978
46,XY,t(9;22),r(7)	Prigogina et al 1978
46,XY,t(9;22),t(7;10)(q34;q22)	Sharp et al 1976
46,XY,t(9;22),t(7;8),1q+	Lilleyman et al 1978
46,XY,t(9;22),t(7;?)(q36;?)	Fleischman et al 1981
47,XX,t(9;22)(q34;q11),t(7;11)(q21;p15),+22q-	Mitelman 1983
47,XX,t(9;22),+7	Kamada et al 1981
47,XX,t(9;22),+7	Chrz et al 1983
47,XY,t(9;22)(q34;q11),-7,13q+,20p+,+21,+mar	Alimena et al 1982b
47,XY,t(9;22),+22q-/56,XY,t(9;22),+1,+2,+5,+6,+7,+8,+10, +12,+17,+22q-	Stoll and Oberling 1978
48,XY,t(9;22)(q34;q11),+7,+22q-/47,XY,t(9;22),+22q-	Tomiyasu et al 1982
48,XY,t(9;22),+22q-,+22q-/47,XY,t(9;22),+17/46,XY,t(9;22), -4,-7,+6,+17	Stoll and Oberling 1978
45,XY,t(9;22),-21/46,XY,t(9;22),-10,+17	
48-51,XY,t(9;22)(q34;q?),+7,+12,+C,+G	Kaffe et al 1974
50-58,XY,+X,t(9;22)(q34;q11),+7,+8,+9,+10,+11,+12, +del(15)(q22),+16,+18,+19,+22q-	Kwan et al 1977
46,XY,t(9;22)	
51-52,XY,t(9;22),+del(22q),+7,+8,+12,+19,+21	Prigogina et al 1978
52-54,XY,t(9;22)(q34;q11),+3,+6,+7,+8,+19,+21,+22q-,+22q-, +1-3mar	Ondreyco et al 1981
59,X,-X,t(9;22),+4,+6,+6,t(7;?)(p15;?),+der(7),+8,+8,+12q+, +17,+19,+19,+21,+22,+22q-,+22q-	Miyamoto 1980

<u>Chronic myeloid leukemia, aberrant translocation</u>

??,X?,t(7;22)(p22;q11)	Potter et al 1981
46,X?,t(7;9;11;22)(p11;q34;q22;q11)	Ishihara 1983
46,X?,t(7;9;22)(q22;q34;q11)	Nowell 1983
46,XX,t(2;9;22)(q11;q34;q11),t(2;7)(q11;q11)	Fleischman et al 1981
46,XX,t(7;9)(q22;q34),t(17;22)(p13;q11)	Mitelman 1983
46,XX,t(7;9)(q22;q34),t(17;22)(p13;q11)	
46,XX,t(7;9;22)(p14;q34;q12)	Fraisse et al 1980
46,XX,t(7;9;22)(q11orq21;q34;q11)	Martin et al 1980
46,XX,t(7;9;22)(q22;q32q34;q11)	Pasquali et al 1979b
46,XX,t(9;15;22)(q34;q11;q11)	Hays et al 1981
54-57,XX,t(9;15;22),+5,+7,+8,+10,+11,+12,+17,+22q-	
46,XY,t(16;22)(q24;q11),t(9;17)(q34;q21)/46,XY,t(16;22), t(9;17),del(7)(q22)	Mitelman 1983
46,XY,t(16;22)(q24;q11),t(9;17)(q34;q21)	
46,XY,t(1;7;19;22)(q?;q?;p?;q11)	Borgström et al 1982
46,XY,t(1;7;19;22),i(17q)	
46,XY,t(7;22)(p22;q12)	Gahrton et al 1979
48,XY,t(7;22)(p22;q12),+8,+8,i(17q)	
46,XY,t(7;22)(p22;q11)	Adler et al 1978
46,XY,t(7;9;22)(q22;q34;q11)	Seabright 1983
46,XY,t(7;9;22)(q35;q31q34;q11)	Lessard and Le Prise 1982

<u>Chronic myeloid leukemia, Ph1 negative</u>

45,XX,-7,inv(2)	Engel et al 1977
45,XY,-7	Symann et al 1982
45,XY,-7	Symann et al 1982
45,XY,-7	Miyamoto 1980
46,XX,del(7)(p11)	Bernstein et al 1980b
46,XY,t(3;7)(q29;q11)	Altman et al 1974

MYELOPROLIFERATIVE DISORDERS

<u>Polycythemia vera</u>

45,X?,-7	Shabtai 1983
45,X?,-7	Wurster-Hill 1983
44,X,-X,del(5)(q14q32),-7	Van Den Berghe et al 1979g
44,XX,del(5)(q14q32),-6,-7,-8,t(12;?)(p12;?), t(16;18)(q11;q12)	
46,XY,del(5)(q14q32),del(7)(q21q31)	Van Den Berghe et al 1979g
46,XY,-7,+mar	Pedersen-Bjergaard et al 1980
46,XY,del(7)(q22)	Testa et al 1981b
46,XY,del(7)(q22)/46,XY,del(7)(q22),inv(14)(p11q24)	Tsuchimoto et al 1974
46,XY,t(3;7)(p?;q?)	Lessard and Le Prise 198?
47,XX,+9/49,XX,+9,+20,+20,t(4;5),-7,-11,-12,+3mar	Zech et al 1976b
47,XX,t(1;9)(p2?;p2?),t(1;9),del(7)(q2?),del(20)(q11)"s"	Nowell and Finan 1978

Myelosclerosis / Myelofibrosis

Karyotype	Reference
44,XY,-5,-7,-11,-12,t(7;12)(p?;q?),t(17;18)(p11;p11),+mar"s"	Nowell and Finan 1978
44,XY,5p-,-7,-10,11q-,i(17q)/45,XY,-7,18q-/45,XY,2p-,2q-,5p-,6q-,9q-,10q-,-11,+i(17q)	Whang-Peng et al 1978
45,XY,-7	Nowell and Finan 1978
45,XY,-7	Hagemeijer et al 1980a
45,XY,-7,t(3;3)(q?;q?)	Nowell and Finan 1978
45,XY,t(1;4;7)(p32;q28;q11q32)	Nowell and Finan 1978
46,XX,-7,+t(1;7)(p1?;p11)"s"	Geraedts et al 1980b
46,XX,7q-,-9,+mar	Carbonell et al 1983
46,XY,-16,+18	Whang-Peng et al 1978
46,XY,7q-,8q-,10q-,11p-,11q+,12q-,-16,+18	
46,XY,-7,t(1;7)(p1?;p11),17p+	Geraedts et al 1980b
46,XY,9q+,13q-,20q-/47,XY,-7,+2mar	Chrz et al 1983
47,XX,del(22)(q?),t(1;2),t(6;7),+r,+mar	Page et al 1979
47,XY,-7,+21,+t(1;7)(p1?;p11)"s"	Geraedts et al 1980b

Chronic myeloproliferative disease, NOS

Karyotype	Reference
46,XY,-7,+19/47,XY,+19/52,XY,+16,+19,+19,+21,+22,+22	Chrz et al 1983
46,XY,-7,+mar	Linch et al 1982

Myeloproliferative disorder, special type

Karyotype	Reference
45,XX,-7/45,XX,-5,-7,+mar"s"	Rowley et al 1977a
45,XX,t(1;?)(p36;?),-5,-7,-12,t(13;?)(q34;?),-17,-22,+4mar/45,XX,t(1;?),-2,-3,-5,-7,-12,-17,-22,+8mar"s"	Rowley et al 1977a
45,XY,-7	Humphrey et al 1981
45,XY,-7	Sieff et al 1981
45,XY,-7	Sieff et al 1981
45,XY,-7	Sieff et al 1981
45,XY,-7	Sieff et al 1981
45,XY,-7	Sieff et al 1981
46,XX,-5,del(6)(q13),-7,+8,-17,+2mar"s"	Rowley et al 1977a
46,XY,-7,+mar/45,XY,-7	Sieff et al 1981
46,XY,t(1;13)(p36;q13),del(3)(p21),-5,inv(7)(p11q22),t(16;16)(q22;q24),+mar	Clare et al 1982

DYSMYELOPOIETIC SYNDROMES

Preleukemia, NOS

Karyotype	Reference
43,X,-Y,-3,5q-,-7,-10,-16,-22,+3mar"s"	Pedersen-Bjergaard et al 1981
44,XY,-5,-6,-7,-8,t(17;?),+2mar"s"	Pedersen-Bjergaard et al 1982
45,XX,-1,-4,5q-,-7,-12,-18,14q+,+4mar"s"	Pedersen-Bjergaard et al 1981
45,XX,-7	MacDougall et al 1974
45,XX,-7"s"	Pedersen-Bjergaard et al 1981

45,XX,-7"s"	Pedersen-Bjergaard et al 1981
45,XX,-7"s"	Pedersen-Bjergaard et al 1981
45,XX,-7"s"	Pedersen-Bjergaard et al 1982
45,XX,-7"s"	Pedersen-Bjergaard et al 1982
45,XX,-7/45,X,-X	Panani et al 1980
45,XX,-7/48,XX,+t(1;7),-7,+11,+13"s"	Pedersen-Bjergaard et al 1981
45,XY,-2,-5,-7,-8,-11,-12,-13,-14,+t(2;5),+t(11;12), +t(16;17),+17,+3mar	Watt et al 1982
45,XY,-7	Ruutu et al 1977b
45,XY,-7	Shiloh et al 1979
45,XY,-7"s"	Pedersen-Bjergaard et al 1981
45,XY,-7"s"	Pedersen-Bjergaard et al 1982
45,XY,-7"s"	Pedersen-Bjergaard et al 1982
45,XY,-7	Hustinx and Rutten 1983
45,XY,-7	Gyger et al 1982
45,XY,-7,del(20)(q11)	Ruutu et al 1977b
46,XX,-7,+8,t(18;22)(p11;q11)"s"	Michalski et al 1982
46,XX,-7,+mar	Pedersen-Bjergaard et al 1980
46,XX,del(7)(q11),t(9;?)(p22;?)"s"	Anderson and Bagby 1982
46,XX,t(3;12)(q21;p13),del(7)(q11),del(11)(q11), t(7;11)(q11;q11),t(11;12)(q22;p13)	Mitelman 1983
46,XY,-7,+t(1;7)(q?;p?)"s"	Pedersen-Bjergaard et al 1982
46,XY,del(7)(p12),t(19;?)(q13;?)"s"	Anderson and Bagby 1982
46,XY,del(7)(q11)	Anderson and Bagby 1982
46,XY,del(7)(q11q22)	Geraedts et al 1980a
46,XY,t(1;17)(q24;p13),-2,-5,-7,+r,+2mar"s"	Anderson and Bagby 1982
47,XY,del(7)(q22),+8	Anderson and Bagby 1982

Chronic myelomonocytic leukemia

45,X,-Y	Streuli et al 1980
45,XY,-7,dmin/45,XY,-7	
45,XX,-7	Pasquali et al 1982a
45,XY,-7	Pasquali et al 1982a
45,XY,-7	Pasquali et al 1982a
46,XY,t(1;7)(p36;q22)	Roush 1981

Refractory anemia, NOS

45,XY,del(5)(q14q22),-7	Kerkhofs et al 1982
46,X?,t(7;14)(p?;?)	Lessard and Le Prise 1983

Refractory anemia without excess of blasts

Karyotype	Reference
44,XX,5q-,-7,+8,-12,-13,-17,-20,+2mar	Teerenhovi et al 1981
45,XY,3q-,5q-,-7	Teerenhovi et al 1981
47,XY,5q-,-7,+2mar/48,XY,-8,-9,+11,+r,+2mar	Teerenhovi et al 1981

Refractory anemia with excess of blasts

Karyotype	Reference
43,XX,3q-,del(5)(q13q33),7q-,-12,14p+,16q+,-17,-18,18q+	Kerkhofs et al 1982
44-45,XX,del(3)(q?),-7,-16,-18,+1-2mar	Mitelman 1983
44-47,XY,-3,-5,-7,-12,-16,-20,del(1)(q32),3p+,del(11)(q22), 3-6mar,+r	Streuli et al 1980
45,XX,-7	Pasquali et al 1982a
45,XX,-7"s"	Mitelman 1983
45,XY,-7	Pasquali et al 1982a
45,XY,-7	Pasquali et al 1982a
45,XY,-7	Pasquali et al 1982a
45,XY,-7	Kardon et al 1982a
45,XY,-7	Kardon et al 1982a
45,XY,-7	Mitelman 1983
45,XY,-7	
45,XY,-7	
46,XY,del(5)(q12q31),-7,20q+,+22	Kerkhofs et al 1982

Thrombocytopenia

Karyotype	Reference
46,XY,del(5)(q13q31)/46,XY,inv(7)(q13q32)	Kerkhofs et al 1982

Pancytopenia

Karyotype	Reference
44,XY,-7,6p+	Nowell and Finan 1978
45,XY,-7	Nowell and Finan 1978
45,XY,-7	Boetius et al 1977
45,XY,-7	Mitelman 1983
46,XX,-4,-5,-6,-7,-11,-13,-16,+19,+20,t(7;13)(q?;q?), +4mar/43,XX,-2,-3,-5,-7,-16,+19,15q+,+mar	Nowell and Finan 1978
46,XX,-7,+8,-12,-16,-17,t(6;12)(q?;q?),5q-,6q-,+mar	Nowell and Finan 1978

Dysmyelopoietic syndrome, special type

Karyotype	Reference
44,XY,-5,-7,-17,del(3)(q13),+del(3)(p13),del(6)(q13)"s"	Rowley et al 1981a
45,XX,t(13;14)(p13;q11),del(5)(q14)/50,XX,t(13;14),+1,+6, -7,+8,+10,del(5),+mar"c""s"	Rowley et al 1981a
45,XY,-7"s"	Rowley et al 1981a
45,XY,-7,t(2;3)(p21;q28)"s"	Rowley et al 1981a
45,XY,del(2)(q33),-5,del(7)(q22q32),-12,-15, t(12;15)(q11;q26),del(19)(q12),+r"s"	Albain et al 1983

Thrombocytosis

Karyotype	Reference
46,XX,ins(3;3)(q27;q21q27)	Norrby et al 1982
46,XX,ins(3;3)(q27;q21q27)/46,XX,ins(3;3),del(7)(q31)	

SPECIAL LEUKEMIAS

Hairy cell leukemia

Karyotype	Reference
44,X,del(X)(q22),del(6)(q23),-7,-8,-10,del(11)(p15q21),-21, 14q+,+r,+mar	Sadamori and Sandberg 1983a

LYMPHOCYTIC LEUKEMIAS

Acute lymphocytic leukemia (ALL)

Karyotype	Reference
28,XX,-1,-2,-3,-4,-5,-7,-8,-9,-11,-12,-13,-14,-15,-16,-17, -19,-20,-22	Kaneko and Sakurai 1980
32,XY,-2,-3,-4,-5,-7,-8,-11,-12,-13,-14,-15,-16,-17,-20	Shabtai and Halbrecht 1981b
43,X,-X,-7,-15,t(9;22)	Gibbs et al 1977
43,X,-X,-7,-8,t(9;22)(q34;q11)	Oshimura and Sandberg 1977
45,XY,-7,-18,+mar	Prigogina et al 1979
46,X,-Y,+1,-7,+20,t(1;22)(q24;q12)	Vercherat et al 1980
46,X?,del(7)(q22)	Lawler et al 1975
46,XX,1q-,7q-,t(1;19)	Goldstone et al 1979
55,XX,+1q-,+7q-,+t(1;19),+19,+22,+4mar	
46,XX,t(9;22)(q34;q11)/46,XX,t(9;22),del(7)(q11)/45,XX, t(9;22),-7	Shabtai and Halbrecht 1981b
46,XY,+2,t(6;?)(p21;?),+7,-12,-13,14q+,17p+	Oshimura et al 1977a
46,XY,del(5)(q13),del(15)(q15)/46,XY,del(7)(q32)/46,XY, del(6)(q21)	Secker-Walker et al 1979
46,XY,del(6)(q15),-7,+11	Prigogina et al 1979
46,XY,del(7)(q32)	Secker-Walker et al 1979
46,XY,t(4;11)(q21;q23)	Oshimura et al 1977a
46-49,XY,t(4;11)(q21;q23),+6,+8,+13,+17,t(1;13)(p22;q12), t(7;9)(q11;q34)	
46,XY,t(4;11)(q21;q23),i(7q)	Morse et al 1982b
47,XY,t(4;11)(q21;q23),1p+,del(2)(p16),del(6)(q21),i(7q), del(8)(q21),+mar	Arthur et al 1982
47,XY,t(7;12)(q22;p13),+19	Morse et al 1979b
48-53,XX,t(4;11)(q21;q23),+6,+7,+9,+13,+21	Prigogina et al 1979
49,XX,+14,+17,+21,7q-	Shabtai and Halbrecht 1981b
53,XX,+2,+5,i(7q),+8,+13,+21,+21,+22,+17q+	Oshimura et al 1977a
54,XX,+6,+7,+8,+14,+17,+18,+21,+22	Prigogina et al 1979
54,XX,7p+,+10,+10,+14,+14,+18,+18,+21,+21/27,X,-X,-1,-2,-3, -4,-5,-6,-7,7p+,-8,-9,-11,-12,-13,-15,-16,-17,-19,-20, -22	Oshimura et al 1977a
54-57,XY,+3,+5,+6,+7,+12,+13,+14,+21,+2mar	Prigogina et al 1979
57,XY,+X,+5,+6,i(7q),+10,+12,+13,+15,+18,+21,+21,+22	Oshimura et al 1977a

ALL FAB type L1

Karyotype	Reference
26,XX,-1,-2,-3,-4,-5,-6,-7,-8,-9,-10,-11,-12,-13,-14,-15, -16,-17,-19,-20,-22	Hoeltge et al 1982
28,XX,-1,-2,-3,-4,-5,-6,-7,-8,-9,-11,-12,-13,-15,-16,-17, -19,-20,-22/56,XX,+X,+X,+10,+10,+14,+14,+18,+18,+21,+21	Brodeur et al 1981a
34-36,XY,-2,-3,-4,-7,-9,-12,-13,-14,-16,-17,-20,+22	Sandberg et al 1982c
46,XX,del(7)(p15),del(9)(p22)	Kaneko et al 1982e
46,XY,t(7;19)(q11;q13)	Green et al 1982
47,XY,-7,-10,-12,+17,del(11)(p11),del(16)(q22), +t(10;?)(q22;?),+t(12;?)(q15;?),+mar	Kaneko et al 1982e
58,XX,+4,+6,+7,+10,+14,+14,+21,+del(18)(q21),+4mar	Kaneko et al 1982e

ALL, FAB type L2

Karyotype	Reference
26,XX,-1,-2,-3,-4,-5,-6,-7,-8,-9,-10,-11,-12,-13,-15,-16, -17,-18,-19,-20,-22	Brodeur et al 1981a
45,XX,1q-,7q-,-?	Borgström et al 1981
46,XY,del(6)(q21orq23),del(9)(p22),del(11)(q13)/46,XY, del(6),i(6p),del(7)(p15),del(9)(p22),del(11)(q13)	Kaneko et al 1982e
46,XY,del(7)(p15),del(8)(q22)	Kaneko et al 1981b
47,XY,+18	Kaneko et al 1982e
46,XY,-2,-5,-6,-7,+8,-9,-10,-11,-12,-16,+18,+t(2;?)(q35;?), +t(11;?)(q21;?),+5mar	
47,XY,t(9;22)(q34;q11),+del(7)(q11)/48,XY,t(9;22),+7, +del(7)	Mitelman 1983
47,XY,t(9;22)(q34;q11),+del(7)(q11)	
49,XY,+7,+12,-13,t(6;18)(p25;q21),t(11;14)(q23;q32), +t(9;?)(p24;?),+t(1;13)(q12;p13)	Kaneko et al 1982e
52-54,XX,+6,+7,+9,+12,+3mar	Mitelman 1983
53,XY,t(9;22),+4,+5,+6,+7,+8,+17,+21	Sandberg et al 1980

ALL, FAB type L3

Karyotype	Reference
46-47,XX,t(14;?)(q32),t(13;?)(q31;?),1q+,+7	Slater et al 1979
47,XY,ins(1),t(3;22)(q?;q?),+7	Penchansky et al 1981

Chronic lymphocytic leukemia

Karyotype	Reference
46,XX,del(6)(q13q16),t(2;7)(p11;q22)	Vahdati et al 1983a
46,XX,t(6;7)(q?;q?),t(7;13)(q?;q?),t(11;14)(q11;q32),17p+	Schröder et al 1981
46,XX,t(7;11)(q21;p11q14)	Vahdati et al 1983a
46,XY,inv(4)(p14q12),del(7)(q11)	Pittman et al 1982
46,XY,t(1;7;9)(p36;q21;q12),t(12;14)(q13;q24)	Miyoshi et al 1979a
47,XX,+7/49,XX,+14,+17,+21	Shabtai 1983
48,XX,+3,+18,t(7;10)(p?;q?)	Hurley et al 1980

Prolymphocytic leukemia

Karyotype	Reference
44,XX,-7,-8,t(9;17)(q22;q25),-11	Pittman et al 1982
47,XX,dup(7)(q11q36),+17	Pittman et al 1982

Lymphocytic leukemia, special type

45,X,-X,-4,+t(4;7)(p16;q11),15q+ Ueshima et al 1981
46,XY,t(1;7)(p36;q22),del(9)(q12),t(12;14)(q13;q32) Miyoshi et al 1981a
47,X,-Y,-4,+7,del(1)(q32),i(18q),21q+,i(22q),+2mar Ueshima et al 1981
47,X,-Y,dup(Y),1q-,4q+,6q-,7q+,+18q+ Miyamoto et al 1982b
48,XX,+3,-6,+7,del(10)(p13q24),10q+,+mar Ueshima et al 1981
49,XX,+7,+8,+12,8q+,8p+,del(10)(q23),21q+/49,XX,+8,+12, +del(7)(p13),8q+,8p+,10q-,21q+ Ueshima et al 1981

MONOCLONAL GAMMOPATHIES

Multiple myeloma

44,XY,del(3)(q25),t(5;12)(q15;q14),del(7)(q22), del(11)(q24),-14,-18,+mar Mitelman 1983
45,X,del(X)(q25),t(3;8;22)(q13;q34;q11),-5,-7,-10,-11, del(13)(q13?),+16,19q+,+3mar Van Den Berghe et al 1979f
45,XX,-7 Philip et al 1980
46,XY,-7,-12,+2mar Chrz et al 1983
47,X,-X,inv(1)(p22q12),t(7;?)(q11;?),+9,t(11;?)(p15;?),-13, -14,t(14;?)(q11;?),+15,17q+ Philip et al 1980
48,X,-X,+1,+7,-8,+9,+21,del(1)(p36),4p+,del(5)(q22),9p+, 14q+ Liang et al 1979
52,X,-X,+t(1;?)(q12;?),+3,+7,+8,+9,+11,-13,+18,-21,+2mar Philip et al 1980
52,XY,-1,+t(1;15)(q12;p1?),+t(1;16)(p36;p13), t(3;5)(q2?;q13),+3,+5,+7,+8,+9,+11,-15,+18,-20 Philip et al 1980
53,XY,+3,+5,-7,-8,+9,+11,-14,+20,del(1)(p34),del(6)(q25), +5mar Liang et al 1979
55,XY,+3,+5,+7,+9,+11,+14,+15,+21,+21/53,XY,+3,+5,-8,+9, +11,+14,+15,-19,+21,14q+,+3mar Liang et al 1979

Macroglobulinemia

46,Y,-X,del(1)(q21),del(5)(q23),+del(5)(q23), +t(7;19)(p22;q13),del(8)(p12),+t(1;16)(q12;p13), +t(1;18)(q12;p11),-7,-16,-18,-19,+mar Ueshima et al 1983

Plasma cell leukemia / Plasmocytoma

45,XY,-5,del(6)(q11),t(6;7)(q?;q?),+9,-12,-13,-16,-17, t(9;22)(q34;q11),+2mar Karpas et al 1982
46,XY,t(8;14)(q24;q32)/47,XY,t(8;14),+7/48,XY,t(8;14),+7, +18 Yamada et al 1983b
86,XX,cx,t(1;?)(q11;?),2q+,dup(7)(q22q36),t(8;?)(p11;?), t(9;13)(p23;q21) Ueshima et al 1983

SOLID TUMORS

UNCLASSIFIED NEOPLASMS

EPITHELIAL NEOPLASMS

Carcinoma, NOS

Karyotype	Reference
32-35,XX,cx,del(1)(p12),dup(1)(q21q32),2q+,5q+,7p+,11p+, hsr(10),t(11;12)(q13;q24),12p+,13q+	Kusyk et al 1982
45,XY,-4,-7,+8,-9,-12,+13,-17,-18,-19,-20,+21,+3mar	Reichmann et al 1981
45,XY,-7,+8,-14	Sandberg 1977
46,XX,+2,-7,-10,+13,-16,-17,+20,+5q-,15p+,15p+,18p+	Reichmann et al 1981
46,XX,t(3;11)(q12;q14),-6,+7,-8,-15,t(16;?)(q23;?),+2r	Mark 1975b
47,XX,-7,+15,+17,+mar	Sonta and Sandberg 1978a
47,XY,-5,+7,-10,-13,-14,4q-,5q-,+4mar	Reichmann et al 1981
47,XY,6q+,-7,+2mar	Reichmann et al 1981
48,XX,-7,+15,+17,+mar	Sandberg 1977
49,XX,+3,+11,-7,i(8q)	Riet-Fox et al 1979
49,XX,+7,+8,+13	Reichmann et al 1981
49,XX,+7,+8,+19	Reichmann et al 1981
50,XY,+X,+7,+8,+9	Reichmann et al 1981
52,XY,+X,+Y,-2,+7,+8,+9,+10,+13,-17,+i(1q)	Reichmann et al 1981
58,XX,+X,+1,+2,+3,5p-,+6,+7,+9,11q+,-13,-14,-15,+16,+19, +20,+6mar	Kovacs 1978a
58,XY,del(3)(p14),+4,+7,+8,+10,+11,+12,+14,+15,+17,+18,+19, +21	Kovacs 1978a
63,XX,-1,+3,+5,+6,+7,+8,+10,+11,+12,-13,-15,+20,+21,+22, +9mar	Sandberg 1977
63,XX,cx,t(1;4)(q21;q35),7p+,i(17q)	Kovacs 1978a
73,XY,+1,+7,+11,+13,-16,+17,+18,+19	Sonta and Sandberg 1978a

Carcinoma NOS, metastatic

Karyotype	Reference
??,XX,t(7;?)(p?;?),dmin,cx	Bartnitzke and Bullerdiek 1983
35,XX,-1,-3,-4,-5,-7,-8,-10,-11,-12,-13,-14,-15,-16,-17, -18,-19,-20,-21,-22,+8mar	Pathak et al 1979
45-50,XX,-2,+7,-22,t(1;12),t(13;14),inv(7)(q?)	Ayraud 1975
46,XX,t(9;22)(q34;q11)"s"	Togawa et al 1981
52,XX,+3,+5,+7,+8,+9,+11,12q+,13q-,-14,+16,i(17q)"s"	
48,XX,t(3;?)(q2?;?),t(7;14)(p21;q1?),cx	Bartnitzke and Bullerdiek 1983
49,XX,cx,t(1;7)(q?;q?)	Jones Cruciger et al 1976
60,XX,t(1;?),t(7;?)(q3?;?),i(10q),der(11),t(14;21),cx	Bartnitzke and Bullerdiek 1983
60-62,XX,del(1)(q?),t(1;7),2q+,del(3)(q?),der(5),dmin,cx	Bartnitzke and Bullerdiek 1983
62-64,XY,del(2)(q34),del(6)(q14),ins(7)(q?),20q+,t(5;18), i(5p)	Ayraud 1975
79-80,XX,t(1;?)(q12;?),t(1;2),t(3;?),t(7;?)(q36;?),t(8;?), t(9;?)(q34;?),cx	Bartnitzke and Bullerdiek 1983
85-90,XY,i(9q),del(1)(q2?),t(7;?)(q35;?),t(3;?)(p11;?), t(21;?)(q11;?)	Kakati et al 1975

Large cell carcinoma

Karyotype	Reference
55-70,XY,t(4;5;14),del(1)(q11q21),t(7;11)(p22;q13), t(7;21)(q11;q22),t(3;8)(p?;q?),t(6;11)(p11;p11),i(16q), t(3;?)(q11;?),t(21;?)(q11;?),t(13;?)	Kakati et al 1975
60,XY,del(5)(p11),t(6;10)(q11;p11),t(7;11)(p22;q13), t(14;18)(q11;q11),del(2)(q21),t(6;?)(p11;?)	Pickthall 1976

Small cell carcinoma

Karyotype	Reference
67,XY,cx,inv(1)(q32p36),dup(1)(q32q44),del(3)(p14q13), del(3)(p14p23),t(7;13)(p11;q11),del(9)(q11),del(11)(p11), t(12;19)(q24;p13)	Whang-Peng et al 1982b

Adenoma

Karyotype	Reference
46,X,t(X;4)(q25;p16),del(2)(p13p21),del(5)(q15q31),-7, t(9;12)(p13;q13),del(11)(q11q14), t(7;9;12)(p11p22q11;p13p24;q13)/45,X,-X	Mark et al 1980

Adenocarcinoma

Karyotype	Reference
??,XX,cx,del(1)(q42),del(7)(q2?),del(18)(p11),i(8p),i(8q)	Riet-Fox et al 1979
40-42,X,-X,t(1;?)(p13;?),del(1)(p12),2q+,del(3)(q21), del(6)(q13),del(7)(q11),t(11;11)(p15;q13), t(12;12)(q24;?),13p+,15p+,i(22q)	Kusyk et al 1981
44-50,X,-X,-1,t(1;?)(q21;?),del(2)(q33),4p+,-5,-7,+8, del(10)(q12),t(10;13)(q12;q34),del(17)(q25),-20	Kusyk et al 1982
48-48,XX,+3,+7,+8,-1,-5,-14,-17,5q+,6q+,del(1)(q42), del(5)(q3?),del(10)(q24),del(14)(q2?)	Riet-Fox et al 1979
49,XY,+7,-9,+11,+12,+mar	Couturier-Turpin et al 1982
54,XX,+5,+6,+7,+9,+11,+21,+21,+del(1)(q32)	Sonta et al 1977

Adenocarcinoma, metastatic

Karyotype	Reference
??,X,del(X)(q21),del(1)(q12),i(7q),del(9)(q31), del(16)(q23)	Ayraud et al 1977
??,XX,t(1;16),2q-,del(3)(p14),del(4)(q22),del(6)(q22), i(7q),del(9)(q31),t(13;13)	Ayraud et al 1977
39-41,XX,-3,-7,-8,der(1),der(2),der(4),der(12),cx	Tiepolo and Zuffardi 1973
49,XX,1q+,+3,4q+,+7,+8,12q+,17q+	Granberg et al 1973
56-64,XX,del(X)(q22),del(1)(p12),del(5)(p21),del(7)(q31), t(14;18)(q11;q11),18q+,i(22q)	Berger and Lacour 1974

Adenomatous polyp

Karyotype	Reference
45-47,XY,-7,+14	Mitelman et al 1974b
48-52,X,-Y,+7,+8,+13,+14,+16,+18,-20,+12q-,1q+	Reichmann et al 1982b

MESENCHYMAL NEOPLASMS

Leiomyosarcoma

42,XX,del(1)(p12orp13),del(11)(q13orq14),+7,-9,-13,-14,-15, -18,+19,-22 — Mark 1976

Mesothelioma, malignant

41,XY,del(1)(p13),t(1;3)(p13;q11),t(7;7)(q11;q36), t(9;12)(p24;q13),del(12)(q13),del(13)(q12q14),-14,-15, t(17;?)(p13;?),i(19p),-20,-21,-22 — Mark 1978

MELANOMAS

Malignant melanoma

??,X?,6q+,t(6;7)(p?;q?) — Becher et al 1983
46,XY,5q+,+7,-8,-9,-10,-13,-20,+21,+3mar — McCulloch et al 1976

Malignant melanoma, metastatic

24,X,cx,t(1;15),t(1;14)(q12;p11),del(7)(p13), t(6;7)(p11;q11),t(13;21)(p11;q11),+r — Atkin and Baker 1981
41,XY,dup(1)(q21q32),t(1;7)(q25;q36),t(7;9)(q11;q11), del(9)(q13),-3,-4,-5,-8,-10,-11,-16,+18,-20,-22 — Kakati et al 1977
80,XY,t(11;12)(q23;q13),del(5)(q13),del(5)(p11), del(9)(q13),t(5;9),t(14;?)(q11;?),t(7;12)(q22;q24), del(1)(p13),del(11)(q23) — Kakati et al 1977

NEUROGENIC NEOPLASMS

Oligodendroglioma

45,XY,t(21;22)(p11;q11),7q-,9q-,14q+ — Yamada et al 1980

Neuroblastoma

44,X?,del(1)(p22),-4,-7,del(7)(q22),t(7;16)(q22q32;q24) — Trent 1983
46,XX,t(1;?)(p32;?),2q+,7p+,13q+,hsr(5q),dmin — Gilbert et al 1982
46,XY,t(1;?)(p32;?),6p+,hsr(6q),7q-,hsr(7p),19q+ — Gilbert et al 1982
47,XX,+7,9q+,22q+ — Brodeur et al 1977
53,XX,+1,+4,+15,+20,dup(1)(p13p32),t(3;15)(q13;q26), del(4)(q31),t(7;11)(q36;q13) — Brodeur et al 1981b

Retinoblastoma

46,X,-Y,1p+,+1p-,5q+,7p+,10q-,16q-,17q+,21p+ — Kusnetsova et al 1982
46,XX,inv(1),inv(2),inv(3),inv(4),t(6;7),inv(9),t(11;?), 11p-,inv(12),i(17q),-20 — Gardner et al 1982

46,XY,+4,t(5;13)(q35;q21),t(7;17)(p22;q12),t(1;9)(q22;p24), -10,del(16)(q11)	Gardner et al 1982
46,XY,dup(1)(q25q44),t(7;10)(q36;q11),-8,-10, t(12;13)(q11;p11),dup(13)(q22q34),+18,+22	Gardner et al 1982
47,XY,5q+,+i(7q)/47,XY,+i(6p),12q+	Kusnetsova et al 1982

Meningioma, benign

43,XX,-1,-6,-7,-11,-22,+2mar/40-45,XX,-1,-6,-7,-9,-11,-19, -22,+2mar	Mark et al 1972b
47,XX,-22,+7,+9	Zankl et al 1975a
52,X,-Y,-22,+20,+19,+17,+15,+11,+9,+7,+5	Zankl et al 1975a

EMBRYONAL AND MISCELLANEOUS NEOPLASMS

Embryonal carcinoma, NOS

50,XXY,+7,+21,+mar"c"	Mann et al 1983

Blastoma, NOS

40-47,X,-X,cx,7q+,+del(10)(p13),+del(12)(p11), del(11)(p11p14)	Slater and Kraker 1982

Teratoma, malignant

67,XY,+1,+3,+4,+5,+6,+7,+7,+8,+10,+11,+13,+13,+14,+16,+17, +18,+20,+20,+21,+21	Sonta et al 1977

Seminoma

53,XY,+3,+4,+6,+7,+11,+13,+19	Sonta et al 1977

LYMPHOMAS

1. UNCLASSIFIED AND MISCELLANEOUS LYMPHOMAS

Malignant lymphoma, NOS

47,XX,t(1;14)(q23;q32),+3,t(6;21)(p25;q11),+8, t(7;11)(q11;q25),-21	Yamada et al 1980
48,XY,+7,+21"c"	Kristoffersson 1983

Non-Hodgkin's lymphoma, NOS

44-47,XY,3q+,del(4)(p13),del(6)(p23),+7,del(9)(p13)	Kristoffersson 1983
46-48,XX,del(1)(q32),+1,+3,r(7),14q+,-20,+mar	Kristoffersson 1983
47,X?,t(8;14)(q24;q32),+7	Zech et al 1976a
47-48,XY,t(14;?)(q32;?),16q-,-17,18q-,13q-,+7,+8,8q-	Catovsky et al 1977
50,XX,+X,t(1;11)(q25;q23),+3,+7,-13,t(14;18)(p?;p?),-15, t(19;?)(q13;?),+3mar	Reeves and Pickup 1980

Non-Hodgkin's lymphoma, lymphocytic, NOS

Karyotype	Reference
47,XY,+7,-9,-11,1p+,6q-,10p-,14q-, t(14;?;14)(q24orq32;?;q13),15q-,22q-	Fukuhara et al 1979b
47,XY,t(3;4)(q29;p12),del(6)(p22),+7,del(9)(p13)	Kristoffersson et al 1981
47-48,XY,+7,-11,-14,-15,4q+,9p-,14q+	Fleischman and Prigogina 1977
47-50,XX,+3,+18,7q-	Fleischman and Prigogina 1977
48-49,XY,1q-,+3,+del(3)(p?q?),-4,-6,+7,-9,-17,-18,-19, +5mar	Fleischmann et al 1976

Non-Hodgkin's lymphoma, histiocytic, NOS

Karyotype	Reference
45,X,-X,del(1)(q22),t(1;17)(q?;q?),del(2)(q33), del(3)(p21p25),+ins(3)(p14p21p25),+5,del(6)(q14),t(1;7)	Mark et al 1978
45,X,del(Y)(q12),t(4;10)(q12;q24),t(6;7)(q21;q32), del(8)(p21)	Mark et al 1978
46,XY,1p-,6q-,t(11;14)(q21;q32)	Kakati et al 1980
48,XY,1p-,+3,6q-,+7,9p-,t(11;14)(q21;q32)	
47,XX,t(1;18)(p36;q21),+7,t(8;12)(q11;q15), t(8;12;13)(q11q22;q15;q14),del(13)(q14),t(8;14)(q22;q32)	Mark et al 1978
48,XY,t(3;8)(q22;q11),t(5;10)(q33;q22),del(7)(q21), t(4;9)(q11;p24),t(10;13)(q22;q21),r(12),del(19)(p13), del(22)(q11),+7	Mark et al 1978
50,XY,t(2;?)(p25;?),del(2)(p13),+i(3p),del(6)(p21),+7,+8, +9,+9,-11,-13,+14,-15,+20,+21,-22	Mark et al 1978
44,XY,del(2)(p13),t(6;?)(p21;?),del(11)(q13), t(11;14)(q13;q31),i(15q),-20,-22	
80-90,XX,cx,del(7)(p15),t(5;8)(q13;q11)	Kakati et al 1980
82,XXX,t(3;12)(q13;q15),t(3;8)(q22;q11),del(3)(q11), del(5)(q31),i(7q),del(10)(q24),t(10;10)(q24;q26), inv(11)(p15q21),del(12)(q13),del(13)(q14)	Mark et al 1978

Non-Hodgkin's lymphoma, T-cell, NOS

Karyotype	Reference
47,XX,t(2;?),inv(3),4p-,del(5)(q14q32),6p+,6q+,t(7;7),9q+, 10p-,13q-,t(14;14),-15,-15,16q-,17p+,+21,+2mar	Gaeke et al 1981

2. HODGKIN'S LYMPHOMAS

Hodgkin's disease, mixed cellularity

Karyotype	Reference
61-80,XX,del(1)(q32),i(1q),del(2)(p13),del(6)(q13or15), del(6)(q23or25),15q+,12q+,del(5)(q31),del(7)(q11), del(10)(q26),del(3)(p13)	Reeves 1973

3A. NON-HODGKIN'S LYMPHOMAS - RAPPAPORT CLASSIFICATION

Lymphocytic, poorly differentiated, diffuse

47,XY,-1,+7,-9,+9q+,+9p+,+10,14q+,-17,t(17;18)(p11;q11)	Kristoffersson et al 1981
47,XY,t(14;14)(q32;q?),+7,-9,-11,1p+,6q-,10p-,15q-,22q-, +2mar	Fukuhara and Rowley 1978
48,XX,1q-,+3,+4,6q-,+7,10q-,-15	Kakati et al 1980
48-49,XX,2p-,+7,+i(8q),+12,14q+,15q-,18q-	Kakati et al 1980
87,XY,del(1)(q23),del(3)(p14),del(6)(q21),del(7)(p11), del(11)(q21),t(17;?)(p13;?)	Mark et al 1979

Lymphocytic, poorly differentiated, nodular

52,X,-X,+2,-4,+5,+7,+8,-14,-15,-16,+20,+21, +t(1;15)(p11;q11),+t(12;17)(q24;q11),+t(14;?)(q32;?), +mar	Kaneko et al 1982a

Mixed cell, diffuse

50,XY,+X,-1,+5,+7,+12,-14,-17,del(13)(q12q14), ins(1;1)(q21;q21q32),+t(14;?)(q32;?),+t(1;17)(q21;q25)	Kaneko et al 1982a
53,XY,t(8;14)(q24;q32),t(14;18)(q32;q21),+20,1q-,+1p-,2q-, 4q+,5q-,5p+,6q-,+7q+,10q+,11q+,11q-,13p+,+4mar	Fukuhara and Rowley 1978

Histiocytic, diffuse

45,X,-Y,t(14;15)(q32;q21orq22),-15,+18,-22,1q+,3p+,5p-,7p-, 9p+,12p+,16q+,17p-,+2mar	Fukuhara and Rowley 1978
45,X,del(X)(q24),del(1)(q21),+del(1)(p11),del(2)(q22), del(3)(q21),del(6)(p11),del(6)(q15),t(7;9)(p22;q13), t(8;10)(p23;q21),del(9)(q13),del(10)(q21), t(2;11)(q22;q23),-13,del(14)(q24),del(15)(q22), del(16)(q22),+17,-19	Mark et al 1979
45-49,XX,+3,+7p+,-18	Kakati et al 1980
46,X,-X,-4,-11,-17,+18,-19,del(7)(p15),del(8)(q22),+i(17q), +t(4;19)(p11;q13),t(11;?)(q23;?)	Kaneko et al 1982a
47,X,-X,-7,-17,+19,+i(7q),+i(17q),+mar	Kaneko et al 1982a
47,X,-X,t(4;14)(q?;q32),-18,1q+,2p+,6q-,7q+,9q-,12q+,+3mar	Fukuhara and Rowley 1978
47,XY,-11,-22,del(1)(q42),del(2)(q33),del(5)(q22q31), del(7)(q32),del(9)(p22),t(21;22)(q22;q11), +t(11;?)(q23;?),+t(21;22)(q22;q11),+mar	Kaneko et al 1982a
50,XX,del(1)(p22),+3,+7,+12,t(1;14)(p22;q32),+18	Mark et al 1979
50,XY,del(3)(p24),+7,+12,+14,+20	Kristoffersson et al 1981
61,XY,cx,t(1;16)(q23;p11),dup(1)(q23q32),t(1;6)(q23;p21), t(3;6)(q21;p21),t(7;7)(q32;p22),13p+	Fitzgerald et al 1980
73,XX,cx,del(7)(q32),del(9)(q22),del(10)(p14), t(7;?)(p22;?),t(10;?)(p15;?),t(11;?)(q25;?), t(17;?)(q25;?),t(19;?)(q13;?)	Kaneko et al 1982a
82,XX,del(X)(q13),i(1q),i(1p),del(7)(p11),i(9p), del(14)(q22),t(1;14)(q23;q32),i(15q),i(17q)	Mark et al 1979

3B. NON-HODGKIN'S LYMPHOMAS - KIEL CLASSIFICATION

Immunocytoma

Karyotype	Reference
45-46,XY,-8	Kristoffersson 1983
47,XY,15p+,+20/46,XY,del(5)(q23),del(6)(p22),+del(7)(p11), -21	
49,XX,+3,del(6)(q14),+7,+18/49,XX,+3,del(6)(q14), del(9)(q22),+12,+18	Kristoffersson 1983

Centroblastic-centrocytic, diffuse

Karyotype	Reference
46,XY,del(6)(q12),t(7;11)(p11;q11)	Kristoffersson 1983
48-50,XY,-1,+7,9q+,+9p+,+10,+14q+,-17,-18, +t(17;18)(p11;q11),-19,+20,+22,+mar	Kristoffersson 1983
48-51,XY,del(1)(p33q23q25),-2,+del(3)(p22),+4,4p+,+5,+6p+, +7,+8,+9,+21,+22	Kristoffersson 1983
45-51,XY,del(1)(p33q23q25),del(2)(q14),+del(3)(p22),+4,+5, +7,+8,-12	

Centroblastic-centrocytic, follicular

Karyotype	Reference
45-46,XY,-7,t(14;18)(q32;q21)	Kristoffersson 1983
46,XY,t(1;?)(p36;?),t(1;1)(p36;q1?),del(7)(q11), del(7)(q22),del(9)(p11),t(14;?)(q32;?)	Reeves and Pickup 1980
52,XX,+X,+7,+8,del(10)(q24),+12,t(13;?)(q22;?), t(14;?)(q32;?),del(18)(q21),+19,+mar	Reeves and Pickup 1980

Centroblastic, diffuse

Karyotype	Reference
46-51,XX,+3,del(6)(q21),+del(6)(q21),+7,-8,del(9)(q11),+13, -16,-17,+19,-20,-21,+1-5mar	Kristoffersson 1983

Lymphoblastic, Burkitt's type

Karyotype	Reference
43,XY,+1,+2,+3,-7,-8,-9,-12,-16,-22,del(1)(q25), del(2)(p11),del(10)(q22),14q+	Biggar et al 1981
45,X,-Y,t(1;14)(q22;q32),t(8;22)(q24;q11),3q+,6q-,6q+,7q+, 9p-,14q-,18q-	Berger et al 1981c
45,X?,t(8;14)(q24;q32),t(7;16),t(6;17)	Zech et al 1976a
46,X,t(X;1)(p11;q25),del(1)(q25),7p+,17p-,t(2;8)(p11;q24)	Slater et al 1982
46,XY,t(4;5;7)(q13;p13;p22),t(8;14)(q23;q32)	Kaiser McCaw et al 1977
46-48,X?,t(8;14)(q24;q32),+7,15q-,+6,+13	Zech et al 1976a
46-52,X?,t(8;14)(q24;q32),+1,+3,+5,+7,i(17q),+X	Zech et al 1976a
47,X?,t(8;14)(q24;q32),+7	Zech et al 1976a
47-49,XY,t(8;14)(q23;q32),+7,+t(X;1)(p21;q21)	Kakati et al 1979
48,XX,+5,+7,-9,+10,+11,-16,del(5)(p13q31),del(10)(p11q25), del(11)(q12)	Biggar et al 1981
49,XY,+3,+7,+12,14q+	Biggar et al 1981

Lymphoblastic, convoluted type

Karyotype	Reference
43-46,XY,del(1)(p32),del(2)(p21),-6,t(7;9)(p11;q11),-9,-10, t(11;15)(p11;q11),-11,del(12)(q14),-13,-14,i(17q),-17, +18,del(19)(p11),-22,+2mar	Kristoffersson 1983

46,XY,-7,+i(7q) Kaneko et al 1982d
47,XY,-7,-9,del(6)(q23?),t(14;18)(p11;q11),+t(9;?)(q34;?), +2mar/47,XY,-7,-9,del(6),del(1)(q32),t(14;18),t(9;?), +t(1;?)(q23orq25;?) Kaneko et al 1982d

Immunoblastic

46,XY,del(4)(q33),del(6)(q23),t(5;7)(q33;q22) Kaneko et al 1982a
48-49,XY,+3,+7,+11,t(12;?)(p13;?),t(13;?)(p13;?),-16,-20, +21,+22 Reeves and Pickup 1980
48-49,XY,del(3)(p23),+7,+12,+14,+20 Kristoffersson 1983
51,XY,+X,t(3;3)(q11;q27),t(8;?)(q2?;?),+7,+12,+21,+21 Reeves and Pickup 1980
78-83,XY,+del(1)(q11),+5,+7,+12,+13q+,+14q+,+14q+,+20,+mar Kristoffersson 1983
80-86,XY,del(1)(q11),+3,+7,+9,+12,+13q+,+14q+,+14q+,+15q+, +18,+20,+mar

3C. NON-HODGKIN'S LYMPHOMAS - LUKES AND COLLINS CLASSIFICATION

Mycosis fungoides

48,XX,del(1)(p22),t(2;8;14)(q37;q24;q24),+5,del(7)(p13), del(9)(q22),t(9;18)(q11;q24),del(10)(p13), del(13)(q12q14),del(5)(q15q22),del(5)(q13),del(5)(q15) Fukuhara et al 1978a
46,XX,t(1;14)(q32;q32)

Follicular center cell, diffuse, large cleaved

45,XY,-6,-9,-17,2q-,3p-,3q-,4q-,5q-,6p+,7p+,7q+,+7q-,9p-, 12q+,13p+,20p-,+mar Fukuhara et al 1978b

3D. NON-HODGKIN'S LYMPHOMAS - WORKING FORMULATION

Small lymphocytic

46,XY,r(7) Yunis et al 1982

Follicular, small cleaved cell

47,XY,+7,t(14;18)(q32;q21) Yunis et al 1982
48,X,-X,inv(1),1q-,+3,4q-,ins(5),+7,+mar Yunis et al 1982

Follicular, mixed small cleaved and large cell

50,Y,inv(X),+3,+7,+10,t(12;12),+12,t(14;18)(q32;q21),+18, 18q-,-21 Yunis et al 1982

Large cell, immunoblastic

99,XY,cx,5p-,5q-,6q-,7p-,i(12q) Yunis et al 1982

Chromosome 8

HEMATOLOGICAL DISORDERS

UNCLASSIFIED LEUKEMIAS

Acute leukemia, NOS

Karyotype	Reference
41-44,XX,5q-,i(8q),11q+,+mar	Prigogina et al 1979
45,X,-X,+t(X;1)(q13;p12),-7,+8,-9	Mamaeva et al 1983
46,XX,+t(1;19)(q25;q13),der(8),19q+	Prigogina et al 1979
47,XY,+8	Prigogina et al 1979
48,XX,+8,+21"c"	Debiec-Rychter et al 1982
54,XX,+X,+3,+6,+8,+8,+13,+19,+22	Muir et al 1977

NONLYMPHOCYTIC LEUKEMIAS

Acute nonlymphocytic leukemia (ANLL)

Karyotype	Reference
43,X,-Y,t(1;8)(p22;q24),-1,-11,-11,-13,-16,-17,-22, t(1;22)(q11;p11),t(11;13)(q23q25;q12q14),t(17;?)(p11;?), +mar	Oshimura et al 1976b
43,XX,-4,-5,-21,t(8;21)(q22;q22)	Oshimura et al 1976b
44,XX,t(8;11)(p23;p13 or p15),-20/46,XX,del(7)(q11), t(1;3)(q2?;q2?),-20,-21,-22,+3mar	Philip et al 1978a
45,X,-X,7q-,-8,+i(17q)	Nordenson 1983
45,X,-X,t(8;16)(q22;q24),+20,-21	Trent et al 1983a
45,X,-X,t(8;21)(q22;q22)"s"	Elfenbein et al 1978
45,X,-X,t(8;21)(q22;q22)	Sasaki et al 1976
45,X,-Y,t(6;11)(p2?;q13),t(8;11)(q22;q13),del(8)(q22)	Philip et al 1978a
45,X,-Y,t(8;21)(q22;q22)	Oshimura et al 1976b
45,X,-Y,t(8;21)(q22;q22)	Philip et al 1978a
45,X,-Y,t(8;21)(q22;q22),t(9;22)(q34;q11)	Francesconi and Pasquali 1978a
45,X,-Y,t(8;21)(q22;q22)	Sasaki et al 1976
45,X,-Y,t(8;21)(q22;q22)	Sasaki et al 1976
45,X,-Y,t(8;21)(q22;q22)	Seabright 1983
45,XY,-7/50,XY,+8,+15,+19,+20	Berger et al 1981a
45-47,XX,-2,-7,7q-,-8,+15,+22,+2mar"s"	Whang-Peng et al 1979
46,X,-X,+8	Mitelman 1983
46,X?,t(8;21)(q22;q22)	Hartley 1983
46,XX,-7,-8,+2mar	Auerbach et al 1982

46,XX,del(1)(p32),del(5)(q13),+8,t(9;?)(p11;?), Muir et al 1977
+(11;22)(q11;q13),-16,-17,-19,+3mar
46,XX,t(1;8)(q?;q?) Nordenson 1983
46,XX,t(4;8),dic(21) Seabright 1983
46,XX,t(8;21)(q22;q22) Oshimura et al 1976b
46,XX,t(8;21)(q22;q22) Philip et al 1978a
46,XX,t(8;21)(q22;q22) Philip et al 1978a
46,XX,t(8;21)(q22;q22),del(9)(q?) Rozynkowa et al 1977
46,XX,t(8;21)(q22;q22) Sasaki et al 1976
46,XX,t(8;21)(q22;q22) Seabright 1983
46,XY,+del(7)(p11),-8,del(12)(p11),t(21;?)(q22;?)/50,XY,+Y, Mitelman 1983
+del(7)(p11),del(12)(p11),+17,t(17;?)(q25;?),
t(21;?)(q22;?),+22
46,XY,+del(7)(p11),-8,del(12)(p11),t(21;?)(q22;?)/47,XY,+9
46,XY,t(8;21) Bernard et al 1982a
46,XY,t(8;21) Knuutila et al 1981
46,XY,t(8;21) Carbonell 1983
46,XY,t(8;21)(q22;q22) Sasaki et al 1976
46,XY,t(8;21)(q22;q22) Seabright 1983
46,XY,t(8;21)(q22;q22) Watt 1983
46,XY,t(8;21) Bernard et al 1982a
46,XY,t(8;21) Bernard et al 1982a
46,XY,t(9;22)(q34;q11),21q+/46,X,-Y,t(9;22),+8 Abe and Sandberg 1979
46-49,XY,+8,+14,-17,+18,-21"s" Cavallin-Ståhl et al 1977
47,X?,+8 Lawler et al 1975
47,XX,+21/48,XX,+8,+21 Sikand et al 1980
47,XX,+8 Philip et al 1978a
47,XX,+8 Philip et al 1978a
47,XX,+8 Philip et al 1978a
47,XX,+8 Philip et al 1978a
47,XX,+8 Philip et al 1978a
47,XX,+8 Philip et al 1978a
47,XX,+8 Mitelman et al 1976b
47,XX,+8 Sasaki et al 1976
47,XX,+8 Bernard et al 1982a
47,XX,+8 Petit 1983
47,XX,+8 Mitelman 1983
47,XX,+8 Bernard et al 1982a
47,XX,del(5)(q15q23),+8 Oshimura et al 1976b
47,XX,t(3;5)(q21;q31),t(1;13)(p36;q14),+8,+22 Oshimura et al 1976b
47,XX,t(9;22)(q34;q11),+22q-/48,XX,t(9;22),+8,+22q- Mitelman 1983
49,XX,t(9;22)(q34;q11),+8,+13,+22q-
49,XX,t(9;22)(q34;q11),+8,+22q-,+mar
50,XX,t(9;22)(q34;q11),dup(1)(q25q44),+del(3)(p11),+8,
+22q-,+mar
47,XY,+8 Philip et al 1978a
47,XY,+8 Philip et al 1978a
47,XY,+8 Philip et al 1978a
47,XY,+8 Mitelman et al 1976b
47,XY,+8 Mitelman et al 1976b
47,XY,+8 Carbonell 1983
47,XY,+8 Petit 1983
47,XY,+8 Mitelman et al 1976b
47,XY,+8 Bernard et al 1982a
47,XY,+8 Mitelman 1983

Karyotype	Reference
47,XY,+8/48,XY,+8,+14	Bernard et al 1982a
47,XY,+8/48,XY,+8,+9"s"	Mitelman 1983
47,XY,-8,10q+,+2mar	Bernard et al 1982a
47,XY,del(5)(q?),+8	Van Den Berghe et al 1976
47,XY,t(17;17),+8	Borgström et al 1982
47,XY,t(7;20)(p13;p12),+8"c"	Riccardi et al 1978
48,XX,+8,+21"c"	Mitelman 1983
48,XX,+8,+21"c"	
48,XX,+8,+9"s"	Geraedts et al 1980a
48,XX,+8,+mar	Sasaki et al 1976
48,XX,+8,-9,+t(X;9)(q13;q32)	Philip et al 1978a
48,XX,+8,t(9;22),+10	Chrz et al 1983
48,XY,+8,+8/47,XY,+8,+8,-15	Bernard et al 1982a
48,XY,del(3)(q13),-5,+8,+r(?),+mar	Mitelman 1983
48-50,XX,+8,+9,+del(12)(p11)"s"	Weinfeld et al 1977a
49,X,t(X;10)(p11;p11),+del(1)(p22),+del(1)(p22), del(5)(q23),+7/49,X,t(X;10)(p11;p11),+del(1)(p22),+7,+8	Mitelman 1983
49-51,XX,-1,+8,+1-3mar"s"	Weinfeld et al 1977a
50,XX,t(9;22)(q34;q11),+6,+8,+10,+del(9)(p13), +del(22)(q11)	Abe and Sandberg 1979
51,XX,del(5)(q?),+7,+8,+13,+17,+22	Van Den Berghe et al 1976
51,XY,+8,+9,+13,+14,+21	Oshimura et al 1976b

ANLL, FAB type M1

Karyotype	Reference
46,XY,del(8)(q21.3)	Yunis et al 1981
46,XY,t(8;21)(q22;q22)	Brodeur et al 1983
46,XY,t(9;22)(q34;q11)	Sasaki et al 1983
46,XY,t(9;20;22)(q34;q12;q11)/46,XY,t(9;22)(q34;q11),-6,-7, -7,-8,10q+,+t(7;8),i(8q)	
47,X,-X,+i(Xp)/48,X,-X,+i(Xp),+8/48,X,-X,+i(Xp),+20	Hagemeijer et al 1981a
47,XX,+8	Bernstein et al 1982b
47,XY,+8	Bernstein et al 1982b
47,XY,+8	Yunis et al 1981
47,XY,+8	Prieto et al 1981
47,XY,+8	Hagemeijer et al 1981a
47,XY,+8	Hagemeijer et al 1981a
47,XY,+8	Alimena 1983
47,XY,+8,-21,+dic(21;21)(p13;p11)"c"	Kaneko et al 1982b
47-48,XY,+8,t(9;22)(q34;q11),-20,+t(1;20)(q21;q13), t(10;21)(p11;q22),+22q-	Sasaki et al 1983
48,XX,+8,+21	Prieto et al 1981
48,XX,+8,+21"c"	Kaneko et al 1982b
48,XY,+8,+13,inv(9)"c"	Yunis et al 1981
48,XY,+8,+8,t(11;19)(q23;p12orp13)	Hagemeijer et al 1981a
58,XY,+1,+4,+5,+6,+8,+11,+13,+13,+14,+14,+15,+21,+21, t(9;22)(q34;q11)	Alimena 1983

ANLL, FAB type M2

Karyotype	Reference
??,X?,t(8;21)	Trujillo et al 1979
??,X?,t(8;21)	Trujillo et al 1979
??,X?,t(8;21)	Trujillo et al 1979
??,X?,t(8;21)	Trujillo et al 1979
??,X?,t(8;21)	Trujillo et al 1979

??,X?,t(8;21)	Trujillo et al 1979
??,X?,t(8;21)	Trujillo et al 1979
??,X?,t(8;21)	Trujillo et al 1979
??,X?,t(8;21)	Trujillo et al 1979
??,X?,t(8;21)	Trujillo et al 1979
??,X?,t(8;21)	Trujillo et al 1979
??,X?,t(8;21),t(Y;19)	Trujillo et al 1979
42,XY,5q-,7q-,-8,-12,-17,-21,-22,+i(17q),+2mar	Berger et al 1981a
43,XY,-3,-4,-7,-7,-8,-8,-17,-20,+5mar	Berger et al 1981a
44,X,t(X;8),del(5)(q12q21),t(1;6),t(7;19),-7,-8,t(7;8),-21	Kerkhofs et al 1982
45,,del(X)(q26),-Y,t(8;21)(q22;q22),del(9)(q24), del(18)(p11)	Brodeur et al 1983
45,X,-X,t(8;21)	Tricot et al 1981
45,X,-X,t(8;21)	Takeuchi et al 1981
45,X,-X,t(8;21)	Takeuchi et al 1981
45,X,-X,t(8;21)(q21;q21)	Hagemeijer et al 1981a
45,X,-X,t(8;21)(q21;q21)	Hagemeijer et al 1981a
45,X,-X,t(8;21)(q21;q21),1p+,17p-	
45,X,-X,t(8;21)(q22;q22)	Levan and Mitelman 1979
45,X,-X,t(8;21)(q22;q22)/46,X,-X,t(8;21),1p+,7q+,+19/47,X, -X,t(8;21),1p+,+19,+der(21)	Sakurai et al 1982b
45,X,-X,t(8;21)(q22;q22)	Oshimura et al 1982b
45,X,-X,t(8;21)(q22;q22)/46,XX,t(8;21)	Berger et al 1982b
45,X,-X,t(8;21)(q22;q22)/46,XX,t(8;21)	Sakurai et al 1982b
45,X,-X,t(8;21)(q22;q22)/46,XX,t(8;21)	Alimena 1983
45,X,-X,t(8;21)(q22;q22),del(9)(q31)	Brodeur et al 1983
45,X,-Y,del(8)(q22)/45,X,-Y,del(8)(q22),del(11)(q21)/46,XY, del(8)(q22)	Sakurai et al 1982b
45,X,-Y,t(8;21)	Tricot et al 1981
45,X,-Y,t(8;21)	Tricot et al 1981
45,X,-Y,t(8;21)	Tricot et al 1981
45,X,-Y,t(8;21)	Takeuchi et al 1981
45,X,-Y,t(8;21)	Hagemeijer et al 1981a
45,X,-Y,t(8;21)(q21;q22)	Prieto et al 1981
45,X,-Y,t(8;21)(q21;q21)	Hagemeijer et al 1981a
45,X,-Y,t(8;21)(q22;q22)	Kamada et al 1976
45,X,-Y,t(8;21)(q22;q22)	Bernstein et al 1982b
45,X,-Y,t(8;21)(q22;q22),del(9)(q13q22)	Hossfeld et al 1980
45,X,-Y,t(8;21)(q22;q22)/46,XY,t(8;21)	Oshimura et al 1982b
45,X,-Y,t(8;21)(q22;q22)	Oshimura et al 1982b
42,X,-Y,t(8;21)(q22;q22)	Berger et al 1982b
45,X,-Y,t(8;21)(q22;q22)/46,XY,t(8;21)	Berger et al 1982b
45,X,-Y,t(8;21)(q22;q22),t(1;5)(q43;q11)	Berger et al 1982b
45,X,-Y,t(8;21)(q22;q22)	Berger et al 1982b
45,X,-Y,t(8;21)(q22;q22),+1,-B/46,X,-Y,t(8;21),+8	Berger et al 1982b
45,X,-Y,t(8;21)(q22;q22)	Kamada et al 1981
45,X,-Y,t(8;21)(q22;q22)/46,X,-Y,t(8;21)(q22;q22), +t(11;12)(p11;q13)	Sakurai et al 1982b
45,X,-Y,t(8;21)(q22;q22)"s"	Mitelman 1983
45,X,-Y,t(8;21)(q22;q22)/46,X,-Y,t(8;21)(q22;q22),+der(21)	Mitelman 1983
45,X,-Y,t(8;21)	
45,X,-Y,t(8;21)	Golomb et al 1978c
45,XY,-5,-7,-8,+16,-17,+r,+mar	Berger et al 1981a
45,XY,-7,del(3)(p13),del(8)(q22),del(16)(p12)"s"	Rowley et al 1981a
46,X,-X,t(8;21)(q22;q22),+18p+	Berger et al 1982b

46,X,-Y,t(8;21)(q22;q22)	Kamada et al 1976
46,X,-Y,t(8;21)(q22;q22)	Bernstein et al 1982b
46,XX,-7,+8,inv(1)(p36q13)"s"	Rowley et al 1981a
46,XX,4q+,del(7)(q31),t(8;21)(q21;q21),del(9)(q12q21)	Hagemeijer et al 1981a
46,XX,dup(8)(q12q24)	Lessard and Le Prise 1983
46,XX,t(1;8;21)	Tricot et al 1981
46,XX,t(2;8;21)(p11q13;q22;q22)	Pasquali and Casalone 1981
46,XX,t(8;21)	Tricot et al 1981
46,XX,t(8;21)	Tricot et al 1981
46,XX,t(8;21)	Tricot et al 1981
46,XX,t(8;21)	Takeuchi et al 1981
46,XX,t(8;21)	Takeuchi et al 1981
46,XX,t(8;21)	Takeuchi et al 1981
46,XX,t(8;21)(q21;q21),del(9)(q11q13)	Hagemeijer et al 1981a
46,XX,t(8;21)(q22;q22),del(5)(q11?;q13?)	Bernstein et al 1982b
46,XX,t(8;21)(q22;q22),del(9)(q13q22)	Hossfeld et al 1980
46,XX,t(8;21)(q22;q22)	Oshimura et al 1982b
46,XX,t(8;21)(q22;q22)	Oshimura et al 1982b
46,XX,t(8;21)(q22;q22)	Berger et al 1982b
46,XX,t(8;21)(q22;q22)	Kamada et al 1981
46,XX,t(8;21)(q22;q22)	Kamada et al 1981
46,XX,t(8;21)(q22;q22)	Mitelman 1983
46,XX,t(8;21)(q22;q22)/45,X,-X,t(8;21)	Kaneko et al 1982b
46,XX,t(8;21)(q22;q22)	Brodeur et al 1983
46,XX,t(8;21)(q22;q22)	Brodeur et al 1983
46,XX,t(8;21),t(14;21)"c"	Tricot et al 1981
46,XY,-7,-12,-17,del(5)(q14),8q+,del(17)(p11),+mar	Testa et al 1979
45,XY,-7,+8,-12,-17,del(5)(q14),8q+,t(12;17)	
46,XY,t(8;21)	Tricot et al 1981
46,XY,t(8;21)	Hagemeijer et al 1981a
46,XY,t(8;21)(q22;q22),del(9)(q21)/45,X,-Y,t(8;21), del(9)/45,XY,t(8;21),-9	Bernstein et al 1982b
46,XY,t(8;21)(q22;q22)/45,X,-Y,t(8;21)/48,XY,t(8;21), +der(21),t(10;15)(p15;q22),+r	Sakurai et al 1982b
46,XY,t(8;21)(q22;q22)	Sakurai et al 1982b
46,XY,t(8;21)(q22;q22)	Oshimura et al 1982b
46,XY,t(8;21)(q22;q22)/46,XY,t(8;21),9q-/47,XY,t(8;21),+9	Berger et al 1982b
46,XY,t(8;21)(q22;q22)	Kamada et al 1981
46,XY,t(8;21)(q22;q22)	Alimena 1983
46,XY,t(8;21)(q22;q22)	Brodeur et al 1983
47,X,-X,+8,t(8;21)(q22;q22),+14	Brodeur et al 1983
47,X?,t(8;21),+21	Trujillo et al 1979
47,X?,t(8;21),+8	Trujillo et al 1979
47,X?,t(8;21),+8	Trujillo et al 1979
47,XX,+8	Bernstein et al 1982b
47,XX,+8	Lessard and Le Prise 1983
47,XX,+8	Lessard and Le Prise 1983
47,XX,+8,"s"	Berger et al 1981a
47,XX,+8,/48,XX,+8,+8/46,X,-X,+8"s"	Hagemeijer et al 1981a
47,XX,+8,t(12;22)(p13;q12)	Mitelman 1983
47,XX,+8,t(12;22)(p13;q12)	
47,XX,+8,t(12;22)(p13;q12)	
47,XX,t(8;11;21),+8	Golomb et al 1978c
47,XX,t(8;21)(q22;q22),+8	Sakurai et al 1982b

47,XY,+3,t(7;8)(p11;q11)	Kaneko et al 1982b
47,XY,+8	Prieto et al 1981
47,XY,+8	Prieto et al 1981
47,XY,+8"s"	Rowley et al 1981a
48,XX,-5,t(8;?)(p23;?),del(11)(q23),t(15;?)(p11;?),-21, +4mar"s"	Rowley et al 1977a
48,XX,5q-,8p-,+2mar"s"	Pedersen-Bjergaard et al 1981
53,XX,+1,-4,-5,+6,-7,+8,+11,+12,+21,+21,+3mar"s"	Zaccaria et al 1983b
90,XX,-Y,-Y,t(8;21)(q22;q22),t(8;21)(q22;q22)	Testa et al 1983

ANLL, FAB type M1+M2

42,X,-Y,4q+,-8,-16,17p+,-21,-22,+Yp+"s"	Sandberg et al 1982a
42-44,XX,+8,-12,-13,-14,t(2;5)(q12;q35)	Mitelman et al 1981
44,X,-X,-5,t(3;7)(q13;p12)/44,XX,-7,-8/44,XX,-5,-8"s"	Sandberg et al 1982a
45,X,-X,t(8;21)(q21;q22)	Mitelman et al 1981
45,X,-X,t(8;21)(q22;q22)	Mitelman et al 1981
45,X,-X,t(8;21)(q22;q22)	Rowley and Potter 1976
45,X,-X,t(8;21)(q22;q22)	Rowley and Potter 1976
45,X,-X,t(8;21)(q22;q22)	Prigogina et al 1979
45,X,-Y,t(8;21)	Li et al 1983
45,X,-Y,t(8;21)(q21;q22)	Mitelman et al 1981
45,X,-Y,t(8;21)(q22;q22)	Yamada and Furusawa 1976
45,X,-Y,t(8;21)(q22;q22)	Prigogina et al 1979
45,X,-Y,t(8;21)(q22;q22)	Prigogina et al 1979
45,X,-Y,t(8;21)(q22;q22)	Prigogina et al 1979
45,X,-Y,t(8;21)(q22;q22)	Prigogina et al 1979
45,X,-Y,t(8;21)(q22;q22),del(9)(q13)	Shiraishi et al 1982
45,X,-Y/46,X,-Y,+8	Alimena et al 1977
45,XX,-21/47,XX,+8	Li et al 1983
45,XY,+t(1;10)(q23;q26),t(8;21)(q22;q22),-10/45,X,-Y, +t(1;10),t(8;21),-10	Fitzgerald et al 1983
45,XY,-8,-10,+21/47,XY,+21	Alimena et al 1977
45,XY,t(19;22)(p13;q11),t(1;7)(q21;q22),t(7;9),+del(8)(p?), -15	Oshimura and Sandberg 1977
45-46,X,-Y,+8	Mitelman et al 1981
46,X,i(Xq),t(8;21)(q22;q22)	Shiraishi et al 1982
46,X,r(X),t(8;21)	Garson 1980
47,XX,+8/51,XX,+X,+8,+18,+20,+21	Testa et al 1979
46,XX,t(8;21)(q22;q22)	Rowley and Potter 1976
46,XX,7q+,t(8;21)	Li et al 1983
46,XX,t(6;9)(p23;q34)	Testa et al 1979
48,XX,t(6;9)(p23;q34),+8,+13	
46,XX,t(8;16)(p11;q13)	Mitelman et al 1981
46,XX,t(8;21)	Garson 1980
46,XX,t(8;21)	Li et al 1983
46,XX,t(8;21)(q22;q22)	Mitelman et al 1981
46,XX,t(8;21)(q22;q22)	Fitzgerald et al 1983
46,XX,t(8;9)(q23or24;q31),-17,+20	Mitelman et al 1981
48,XY,+8,+21/49,XY,+8,+18,+21	Testa et al 1979
46,XY,t(5;8;21)(q31;q22;q22)/47,XY,t(5;8;21),+mar	Fitzgerald et al 1983
46,XY,t(8;21)	Garson 1980
46,XY,t(8;21)(q22;q22)	Prigogina et al 1979
46,XY,t(9;22),del(10)(q22)	Wayne et al 1979

47,XY,t(9;22),+8,del(10)(q22)/48-50,XY,t(9;22),+8,+9, del(10)(q22),+10,+22q-	
46-47,XX,del(1)(q32),+8,t(15;17)(q25or26;q22or23),+18	Mitelman et al 1981
46-47,XY,+8,-18,21q+	Mitelman et al 1981
46-50,XY,+Y,+del(7)(p11),-8,+9,del(12)(p11),+17,21q+,+22	Mitelman et al 1981
47,X,-Y,t(1;13)(q32p32q11;q11),del(3)(q21), t(8;21)(q22;q22),del(9)(q22),+8,+18	Sakurai et al 1982b
47,XX,+8	Mitelman et al 1981
47,XX,+8	Jonasson et al 1974
47,XX,+8	Rowley and Potter 1976
47,XX,+8	Garson 1980
47,XX,+8	Garson 1980
47,XX,+8	Li et al 1983
47,XX,+8	Li et al 1983
47,XX,+8,+10,-19	Mitelman et al 1981
47,XX,+8	Alimena et al 1977
47,XX,del(3)(q21),+8	Fitzgerald et al 1983
47,XY,+8	Mitelman et al 1981
47,XY,+8	Mitelman et al 1981
47,XY,+8	Yamada and Furusawa 1976
47,XY,+8	Garson 1980
47,XY,+8	Garson 1980
47,XY,+8	Garson 1980
47,XY,+8	Fitzgerald et al 1983
47,XY,+8	Fitzgerald et al 1983
47,XY,+8	Nordenson 1983
47,XY,+8,r(11),t(8;16)(q?;q?)	Garson 1980
47,XY,+8	Sandberg et al 1982a
47-49,XY,-5,del(5)(q15),-7,i(8q),dic(9),-13,-16,-22, +6-9mar"s"	Testa et al 1981b
48,XX,+3,+8	Mitelman et al 1981
48,XX,+8,+9	Fitzgerald et al 1983
48,XX,-5,del(5)(q13q31),+8,+t(11;?)(p14;?),-17,-19,+21, +2mar,dmin"s"	Testa et al 1981b
48,XY,+7,+8,+8,-17"s"	Sandberg et al 1982a
48,XY,+8,+13	Fitzgerald et al 1983
49,XX,+1,-5,+7,+8	Nordenson 1983
50,XY,+8,+14,+19,+mar	Fitzgerald et al 1983
51,XY,+1,+4,+5,+8,+20	Nordenson 1983
57,X,-Y,-6,+8,+15,+15,+19,+19,+20,+20,+21,+21,+22,+3mar	Fitzgerald et al 1983

<u>**ANLL, FAB type M3**</u>

44,XX,-5,-8,-17,-18,+2mar	Gallagher et al 1979
47,XX,+8,i(17q)	Chapelle et al 1981
46,XX,t(15;17)(q22;q12)/46,XX,15q+,i(17q-)/47,XX,t(15;17), +8	Kondo and Sasaki 1982
46,XX,t(15;17)(q26;q22)/47,XX,t(15;17),-7,+8,-9,-10,+2mar	Van Den Berghe et al 1979c
46,XY,t(15;17)(q22;q21),8q+	Berger et al 1981e
46,XY,t(15;17)(q26;q22)/47,XY,t(15;17),+8	Van Den Berghe et al 1979c
46,XY,t(15;17)/46,XY,del(1)(q12)/47,XY,+8/47,XY, del(1)(q12),+8/47,XY,t(15;17),+8	
47,X?,t(15;17),+8	Sakurai et al 1982a

47,XX,t(15;17)(q22;q12),+8	Kondo and Sasaki 1982
47,XX,t(15;17)(q26;q22),+8,/47,XX,+8/46,XX, t(15;17)(q26;q22),-7,+8	Van Den Berghe et al 1979c
47,XX,t(15;17),+8/47,XX,+8	Sakurai et al 1982a
47,XY,+8	Van Den Berghe et al 1979c
47,XY,+8,t(15;17)(q24;q21)	Fitzgerald et al 1982b
47,XY,t(15;17)(q21?;q25?),+8/47,XY,t(15;17)(q21?;q25?), +9/48,XY,t(15;17),+8,+9	Van Den Berghe et al 1979c
47,XY,t(15;17)(q22;q21),+8	Berger et al 1981e

ANLL, FAB type M4

42,XY,-7,i(8)(q11),-16,-17,-18,t(21;21)(q12;q11),+mar	Rowley and Potter 1976
44,XY,del(5)(q15q23),del(7)(q22),-9,-12,-13,del(15)(q22), +mar/46,X,-Y,del(5),del(7),+8,-12,-13,del(15),+mar,dmin	Alimena 1983
44,XY,t(3;5)(q29;q13),t(7;12)(q12;q14),t(7;10)(q12;q21),-8, -17,-18,-20	Mitelman et al 1981
45,X,-X,t(8;17;21)(q22;q23;q22)	Testa et al 1979
45,X,-X,t(8;17;21)(q22;q23;q22)/45,X,-X, t(1;11)(p36q32;q13),t(8;17;21)	
45,X,-Y,del(7)(q32),t(8;21)(q22;q22)	Hustinx et al 1980
45,X,-Y,r(21)"c"	Testa et al 1979
45,X,-Y,r(21)/46,X,-Y,r(21),+8"c"	
46,XX,del(X)(q11),del(8)(q21)	Alimena 1983
46,XY,-7,+8	Hagemeijer et al 1981a
46,XY,2q+,del(18)(p11),del(22)(q11)/47,XY,2q+,+8, del(18)(p11),del(22)(q11)"s"	Alimena et al 1981
45-47,XY,2q+,+8,del(18)(p11),-21,del(22)(q11)"s"	
46,X,-Y,2q+,+8,del(21)(q21)"s"	
46,XY,t(1;9)(p34;q34),inv(8)(p11q12)	Rowley and Potter 1976
46,XY,t(3;12)(q21;q13),del(5)(q14),ins(8;5)(q22;q14q22), -17,+2mar	Mitelman 1983
46,XY,t(9;22)(q34;q11)/48,XY,t(9;22),+8,+22q-	Bernstein et al 1982b
46-47,XY,+8,-17	Mitelman et al 1981
47,XX,+19,20p+/50,XX,+5,+8,+12,+mar	Morse et al 1978
47,XX,+8	Fitzgerald et al 1983
47,XX,+8	Mitelman et al 1981
47,XX,+8	Shiloh et al 1979
47,XX,+8/48,XX,+8,+17	Garson 1980
47,XY,+13/49,XY,+8,+13,+mar	Muir et al 1977
47,XY,+8	Yamada and Furusawa 1976
47,XY,+8	Muir et al 1977
47,XY,+8	Garson 1980
47,XY,+8	Prieto et al 1981
47,XY,+8	Hagemeijer et al 1981a
47,XY,+8	Fitzgerald et al 1983
47,XY,+8,t(9;22)	Prieto et al 1981
47,XY,+8	Mitelman et al 1981
47,XY,+8/46,XY,i(17q)	Mitelman 1983
47,XY,+8/48,XY,+8,+9/49,XY,+8,+9,+11	Hagemeijer et al 1981a
47,XY,t(7;12)(q36;p13),+8	Hagemeijer et al 1981a
47-48,XY,+8,+10,+21,+22	Mitelman et al 1981

Karyotype	Reference
49,XX,+5,+8,+15	Fitzgerald et al 1983
49,XX,+8,+13,+21	Prigogina et al 1979
49,XX,del(3)(q25),del(7)(q22),+8,+17,+19	Alimena 1983

ANLL, FAB type M5

Karyotype	Reference
47,XX,+8	Testa et al 1979
46,XX,ins(10;11)(p11;q23q24)/52,XX,+4,+8,ins(10;11),+12, +16,+19,+20	Kaneko et al 1982b
46,XY,t(17;21)(q11;q11orq21),+i(8q),del(5)(q11q13),9p+, 11q+	Fitzgerald et al 1983
46,XY,t(9;22)(q34;q11)/47,XY,+8,t(9;22)	Hagemeijer et al 1981a
46,Y,-X,-1,-7,-8,-14,-15,-17,-20,inv(6)(p23q13), del(11)(q23),+t(X;?)(p11;?),+t(1;?)(p24;?), +t(12;?)(q22;?),+t(17;?)(p11;?),+4mar	Kaneko et al 1982b
46-47,XX,-7,+8,+20	Mitelman et al 1981
47,XX,+8	Shiloh et al 1979
47,XX,+8"s"	Hagemeijer et al 1981a
47,XX,+8	Petit 1983
47,XX,+8/48,XX,+8,+17/48,XX,+8,+20	Garson 1980
47,XX,+del(1)(p11)/48,XX,+del(1)(p11),+6/48,XX, +del(1)(p11),+8	Garson 1980
47,XY,+8	Garson 1980
47,XY,+8	Yunis et al 1981
47,XY,+8,t(1;13)(p31;q11)/48,XY,+8,+18q-	Hagemeijer et al 1981a
47,XY,+8/48,XY,+8,+17/48,XY,+8,+22	Garson 1980
47,XY,del(7)(q22),+del(8)(q12)	Mitelman et al 1981
48,XY,+8,+9	Kaneko et al 1982b
49,XX,+4,+8,+8,t(9;11)(p22;q23)	Kaneko et al 1982b
49,XY,t(8;9)(p21;p22),+del(8)(p21),+9p+	Brynes et al 1976
50,XX,+6,+8,10p-,+12,+13	Nordenson 1983
51,XX,+6,+8,+14,+19,+21,t(11;?)(q21orq24;?)	Soukup and Neely 1981
57-58,XX,+1,+3,-4,-5,+6,-8,+11,+12,-17,+19,-20, del(10)(q24),+8-9mar	Testa et al 1979
60-61,XX,Xp+,+1,+2,-4,+6,-7,+19,+20,del(2)(q31), del(10)(q24),+12-13mar	

ANLL, FAB type M5a

Karyotype	Reference
46,XY,+8,+8,-15,t(13;15)(p11;q11)	Berger et al 1982a
46,XY,del(10)(p12),del(11)(q23),der(8),ins(10)(p12)	Berger et al 1982a
46,XY,t(8;13)(q24;q12),del(11)(q23),t(11;13)(p12;q12),13p+	Berger et al 1982a
47,XX,+8	Berger et al 1982a
47,XX,+8	Berger et al 1982a
47,XX,+8,t(10;11)(p13;q14)	Berger et al 1982a
47,XX,+8,t(9;11)(p22;q24)	Dewald et al 1983
47,XX,+der(?8)	Berger et al 1982a
47,XY,+8,9q+,del(14)(q23),17p+	Berger et al 1982a
47,XY,+8,t(9;11)(p22;q24)	Dewald et al 1983
48,XX,t(7;8)(p11;q11),+i(8q),+i(8q),del(11)(q24)	Berger et al 1982a
48,XY,+4,+8,t(11;17)(q24;q21)"s"	Dewald et al 1983

ANLL, FAB type M4+M5

Karyotype	Reference
47,XX,+8	Chapelle et al 1976
47,XY,+8	Li et al 1983
47,XY,+8	Li et al 1983
47,XY,+8,dmin	Li et al 1983
47,XY,+8	Alimena et al 1977
47,XY,+8/47,XY,+8,t(17;17)	Chapelle et al 1976
47,XY,+8/48,XY,+8,+10/48,XY,+21,+22	Alimena et al 1977
48,XY,+6,+8	Li et al 1983
48,XY,t(1;6)(q21;q27),+del(8)(q22),+22	Shiraishi et al 1982
66,XY,+1,+1,+2,+4,+6,+6,+7,+8,+8,+10,+11,+11,+12,+12,+18, +19,+21,+21,+22,+22	Li et al 1983

ANLL, FAB type M6

Karyotype	Reference
??,XY,t(4;?)(q23;?),t(5;17),t(8;9),+8,+11, t(12;16)(p12;p12)	Hustinx and Rutten 1983
44-47,XY,+2,-3,-5,-7,+8,-14,-19,+2mar	Mitelman et al 1981
45,XY,del(4)(q23),del(5)(q13),-6,del(6)(q21),del(7)(p14), del(8)(p21),del(9)(q31),16q+	Rowley and Potter 1976
45,XY,del(5)(q13q31),r(8),-17,-21,+mar	Kerkhofs et al 1982
45-49,XY,-4,-7,-8,-10,+14,+16,-21,-22,+2mar	Mitelman et al 1981
47,XX,+8/48,XX,+15,+18	Testa et al 1979
46,XX,-8,+18	Li et al 1983
47,XX,+8	Larson et al 1982
47,XY,+8	Mitelman et al 1981
47,XY,+8	Alimena et al 1977
47,XY,+8	Hagemeijer et al 1981a
48,XX,+8,+9,+9,-11,del(4)(q23)"s"	Weinfeld et al 1977a
48,XX,t(1;8)(p11;q11),+8,+19	Morse et al 1979a
49,XY,+8,+12,+17,t(17;21)(p11;q22)"s"	Yamada and Furusawa 1976
49,XY,+8,+8,+9	Bernstein et al 1982b
51,XY,+1,+2,+6,-7,+8,-14,+15,+21,+mar"s"	Rowley et al 1981a
51,XY,+Y,+8,+15,+2mar	Hagemeijer et al 1981a
72,XY,cx,t(2;15)(q31;q15),del(6)(p22q15),t(8;13)(q23?;q13), t(11;14)(p12;q13),i(16p),t(15;18)(q15;p11)	Douglass and Freeman 1983

Chronic myeloid leukemia, t(9;22)

Karyotype	Reference
35,XX,t(9;22)(q34;q11),-3,-4,-5,+8,-9,-10,-11,-12,-13,-14, -15,-16,-17/66-72,XX,t(9;22)	Pedersen and Boesen 1983
44,XY,t(9;22),-17,-20 45,XY,t(9;22),-8,-11,-18,+2mar 47,XY,t(9;22),+7,i(17q)	Stoll and Oberling 1978
45,X,-Y,t(9;22)/47,XY,t(9;22),+8	Seabright 1983
45,X,-Y,t(9;22)/49,X,-Y,t(9;22),+t(9;22),+8,+12	Miyamoto 1980
45,X?,t(9;22),-8	Carbonell et al 1982a
45,XY,t(9;22)(q34;q11),+8,-18,-19	Alimena et al 1982b
45,XY,t(9;22),-17 47,XY,t(9;22),+8,-17,+22q-	Stoll and Oberling 1978
45,XY,t(9;22),-20 45,XY,t(9;22),-7,-8,+17	Stoll and Oberling 1978
45,XY,t(9;22),-20,-22,+22q-/46,XY,t(9;22),-20,-22,+8,+22q-	Stoll and Oberling 1978

Karyotype	Reference
44,XY,t(9;22),-17,-22,-22,+22q-/46,XY,t(9;22),-20,+22q-/47, XY,t(9;22),+11,-22,+22q-	
45,XY,t(9;22),-8	Stoll and Oberling 1978
45,X,-Y,t(9;22),i(17q)/45,XY,t(9;22),-8	
45,XY,t(9;22),-8	Stoll and Oberling 1978
46,X,-Y,t(9;22)(q34;q12),+13/47-48,X,-Y,t(9;22),+8,+13, +22q-/47,XY,t(9;22),+22q-	Izakovic et al 1982
46,X,-Y,t(9;22),+22q-/47,X,-Y,t(9;22),+8,i(17q),+22q-	Bernstein et al 1980b
46,X?,t(9;22),+8	Berger 1973
46,X?,t(9;22),1p+,-7,+mar/46,X?,t(9;22),1p+,-7,8q+,+mar	Carbonell et al 1982a
46,X?,t(9;22),i(17q)/47,X?,t(9;22),+8,i(17q)	Fleischman et al 1981
46,X?,t(9;22),i(17q)/47,X?,t(9;22),i(17q),+8	Fleischman et al 1981
46,X?,t(9;22),i(17q)/47,X?,t(9;22),+8,i(17q)	Carbonell et al 1982a
46,X?,t(9;22),i(17q)/47,X?,t(9;22),+8,i(17q)	Carbonell et al 1982a
46,X?,t(9;22),i(17q)/47,X?,t(9;22),+8,i(17q)	Carbonell et al 1982a
46,X?,t(9;22),inv(7)/47,X?,t(9;22),+8,i(17q)	Carbonell et al 1982a
46,X?,t(9;22)/47,X?,t(9;22),+8,i(17q)	Carbonell et al 1982a
46,XX,t(9;22)	Mitelman 1983
46,XX,t(9;22)(q34;q11)/48,XX,t(9;22),+3,+8	
44,XX,-5,del(6)(q16),-8,-14,-15,-18,-18,-20,+21,-22,-22, +6mar	
44,XX,-5,del(6)(q16),-8,-14,-15,-18,-18,-20,+21,-22,-22, +6mar	
46,XX,t(9;22)(q34;q11)/47,XX,t(9;22)(q34;q11),+8, i(17)(q11)	Rowley 1973a
46,XX,t(9;22)(q34;q11)/48,XX,t(9;22)(q34;q11),+8,+8	Hayata et al 1975b
46,XX,t(9;22)(q34;q11)/45,XX,t(9;22),t(8;14)(q11;q32), -der(8)	Hayata et al 1975b
46,XX,t(9;22)(q34;q11)/48,XX,+X,t(9;22),+8	Sonta and Sandberg 1978b
46,XX,t(9;22)(q34;q11)/48,XX,t(9;22),+8,+17	Misawa 1978
46,XX,t(9;22)(q34;q11)/51,XX,t(9;22),+5,+8,+17,+19,+22q-	Misawa 1978
46,XX,t(9;22)(q34;q11)/48,XX,t(9;22),+8,+19	Misawa 1978
46,XX,t(9;22)(q34;q11)/46,XX,t(9;22)(q34;q11), t(7;8)(q22;q13)	McGlave et al 1982
46,XX,t(9;22)(q34;q11)	Rajasekariah et al 1982
50,XX,t(9;22)(q34;q11),t(9;22)(q34;q11),+8,+11,+13, t(16;17)(p11;p12),+22	
46,XX,t(9;22)(q34;q11),t(8;17)(p12;q21)	Petit and Van Den Berghe 1981
46,XX,t(9;22)(q34;q11),+8,+22q-	Alimena et al 1982b
46,XX,t(9;22)(q34;q11),2p-,7p+,-8,+mar	Alimena et al 1982b
46,XX,t(9;22)(q34;q11),5q+,t(8;19),17p+	Alimena et al 1982b
46,XX,t(9;22)(q34;q11)/47,XX,t(9;22),+22q-/49,XX,t(9;22), +8,+17,+22q-/49,XX,t(9;22),+18,+19,+22q-/50,XX,t(9;22), +8,+17,+18,+22q-	Mitelman 1983
46,XX,t(9;22)(q34;q11)/51,XX,t(9;22),+6,+8,+10,+19,+22q-	Mitelman 1983
46,XX,t(9;22)(q34;q11),i(17q)/47,XX,t(9;22),+8,i(17q), +22q-	Tomiyasu et al 1982
46,XX,t(9;22),i(17q)/47-49,XX,t(9;22),i(17q),+3,+8	Prigogina et al 1978
46,XX,t(9;22),i(17q)/47,XX,t(9;22),i(17q),+8	Prigogina et al 1978
46,XX,t(9;22),i(17q)/47,XX,t(9;22),+8	Olah and Rak 1981
46,XX,t(9;22),t(8;13)(p12;q34),+der(8),-20	Lyall and Garson 1978
46,XX,t(9;22),t(8;?)(q24;?)	Fleischman et al 1981
46,XX,t(9;22)/47,XX,t(9;22),+8	Hagemeijer et al 1980b
46,XX,t(9;22)/47,XX,t(9;22),+22q-/48,XX,t(9;22),+8,+22q-	Bernstein et al 1980b

Karyotype	Reference
46,XX,t(9;22)/48,XX,t(9;22),+19,+22q-/50,XX,t(9;22),+8,+12, +19,+22q-/50,XX,t(9;22),+8,+14,+17,+19,+21	Hagemeijer et al 1980b
46,XX,t(9;22)/48,XX,t(9;22),+8,+19	Kohno and Sandberg 1980
46,XY,t(9;22)(q34;q11)	Mitelman 1983
46,XY,t(9;22)	
46,XY,t(9;22)	
46,XY,t(9;22)	
46,XY,t(9;22)	
46,XY,t(9;22)/49,XY,t(9;22),+8,+19,+22q-	
46,XY,t(9;22)/47,XY,t(9;22),+22q-/48,XY,t(9;22),+19, +22q-/49,XY,t(9;22),+8,+19,+22q-	
46,XY,t(9;22)(q34;q11)/47,XY,t(9;22)(q34;q11),+8	Hayata et al 1975b
46,XY,t(9;22)(q34;q12)	Whang-Peng et al 1976b
45,XY,t(9;22)(q34;q12),t(7;22)(p12;p13)	
46,XY,t(9;22)(q34;q12),del(8)(q23)	
46,XY,t(9;22)(q34;q11)	Mitelman et al 1975a
46,XY,t(9;22)(q34;q11)	
46,XY,t(9;22)(q34;q11)	
46,XY,t(9;22)(q34;q11)	
46,XY,t(9;22)(q34;q11)/49,XY,t(9;22),+8,+19,+del(22)(q11)	
46,XY,t(9;22)(q34;q11)/47,XY,t(9;22),+del(22)(q11)/48,XY, t(9;22),+19,+del(22)(q11)/49,XY,t(9;22),+8,+19, +del(22)(q11)	
46,XY,t(9;22)(q34;q11)/47,XY,t(9;22),i(17q),+del(22q)/49, XY,t(9;22),+8,+21,i(17q),+del(22q)	Sonta and Sandberg 1978b
46,XY,t(9;22)(q34;q11)/48,XY,t(9;22),+8,+8	Sonta and Sandberg 1978b
46,XY,t(9;22)(q34;q11),11p-/47-48,XY,t(9;22),+del(22)(q11), 11p-,+8/46,XY,t(9;22),11p-,dic(22q-)	Prigogina et al 1978
46,XY,t(9;22),dic(22q-)	
46,XY,t(9;22),dic(22q-)	
46,XY,t(9;22)(q34;q11)/47,XY,t(9;22),+22q-/47,XY,t(9;22), -20,t(11;?),+22q-/48,XY,t(9;22),+8,-20,+t(11;?),+22q-	Misawa 1978
46,XY,t(9;22)(q34;q11)/50,XY,t(9;22),t(1;15)(p36;q22),+8, +10,+19,+22q-	Misawa 1978
46,XY,t(9;22)(q34;q11)/47,XY,t(9;22),+8,i(17q)	Mitelman 1983
47,XY,t(9;22)(q34;q11),+8,i(17q)/46,XY,t(9;22)	
46,XY,t(9;22)(q34;q11),+t(1;17)(q21;p11)/47,XY,t(9;22), +t(1;17),+18/48,XY,t(9;22),+t(1;17),+8,+19	Mamaeva et al 1983
46,XY,t(9;22)(q34;q11),i(17q)/47,XY,t(9;22),+8,i(17q)/47, XY,t(9;22),+22q-/48,XY,t(9;22),+8,i(17q),+19	Tomiyasu et al 1982
46,XY,t(9;22)(q34;q11),21q+/50,XY,t(9;22),+7,+8,+12,+15, 21q+	Tomiyasu et al 1982
46,XY,t(9;22),-8,12q+,i(17q),+21/46,XY,t(9;22),-8,-8,12q+, +i(17q),+21	Olah and Rak 1981
46,XY,t(9;22),i(17q)/47,XY,t(9;22),+8/48,XY,t(9;22),+8,+19	Hagemeijer et al 1980b
46,XY,t(9;22),i(17q)/47,XY,t(9;22),i(17q),+8/48,XY,t(9;22), i(17q),+8,+8/49,XY,t(9;22),i(17q),+8,+8,+22q-/50,XY, t(9;22),i(17q),+8,+8,+22q-,+19	Hagemeijer et al 1980b
46,XY,t(9;22),i(17q)/47,XY,t(9;22),+8,i(17q)/48,XY,t(9;22), +8,+22q-	Bernstein et al 1980b
46,XY,t(9;22),i(17q)/47,XY,t(9;22),+8,i(17q)	Bernstein et al 1980b
46,XY,t(9;22),i(17q)/47,XY,t(9;22),+22q-/48,XY,t(9;22),+8, i(17q),+22q-	Miyamoto 1980
46,XY,t(9;22),t(7;8),1q+	Lilleyman et al 1978
46,XY,t(9;22),t(8;17)(q13;q25)	Fleischman et al 1981

46,XY,t(9;22)/46,XY,t(9;22),i(17q)	Prigogina et al 1978
46-47,XY,t(9;22),+8,i(17q)	
46,XY,t(9;22)/47,XY,t(9;22),i(17q),+8	Prigogina et al 1978
46,XY,t(9;22)/47,XY,t(9;22),+8	Olinici et al 1978b
46,XY,t(9;22)/47,XY,t(9;22),+8	
46,XY,t(9;22)/48,XY,t(9;22),+19,+22q-	
46,XY,t(9;22)/47,XY,t(9;22),+22q-/48,XY,t(9;22),+8, +22q-/48,XY,t(9;22),+22q-,+22q-	Sonta and Sandberg 1977b
46,XY,t(9;22)/47,XY,t(9;22),+22q-/48,XY,t(9;22),+8,+22q-	Bernstein et al 1980b
46,XY,t(9;22)/47,XY,t(9;22),+8	Kohno and Sandberg 1980
46,XY,t(9;22)/47,XY,t(9;22),+8	Kohno and Sandberg 1980
46,XY,t(9;22)/47,XY,t(9;22),+8/47,XY,t(9;22),+22q-	Kohno and Sandberg 1980
46,XY,t(9;22)/48,XY,t(9;22),+8,+22q-/49,XY,t(9;22),+8,13q-, +19,+22q-	Hagemeijer et al 1980b
46,XY,t(9;22)/48,XY,t(9;22),+8,+22q-	Bernstein et al 1980b
46,XY,t(9;22)/50,XY,t(9;22),+8,+17,+19,+22q-	Hagemeijer et al 1980b
46-47,XY,t(9;22),+8,i(17q)	Sharp et al 1976
47,X,-Y,t(9;22)(q34;q11),+8,i(17q),+22q-	Pasquali et al 1982b
47,X?,t(9;22)(q34;q11),+8,i(17)(q11)	Nowell et al 1975
47,X?,t(9;22)(q34;q11),+8	Lawler et al 1976
47,X?,t(9;22)(q34;q11),+8	
47,X?,t(9;22),+21/48,X?,t(9;22),+21,+21/50-52,X?,t(9;22), +8,+9,+12,+19,+21,+21	Carbonell et al 1982a
47,X?,t(9;22),+22q-/51,X?,t(9;22),+8,+8,+17,+19,+22q-	Fleischman et al 1981
47,X?,t(9;22),+22q-/48,X?,t(9;22),+8,21q+,+22q-/49,X?, t(9;22),+8,t(14;?),21q+,+22q-	Carbonell et al 1982a
47,X?,t(9;22),+22q-/49,X?,t(9;22),+8,+9,+22q-	Carbonell et al 1982a
47,X?,t(9;22),+8	Carbonell et al 1982a
47,X?,t(9;22),+8	Carbonell et al 1982a
47,X?,t(9;22),+8	Carbonell et al 1982a
47,X?,t(9;22),+8	Wurster-Hill 1983
47,X?,t(9;22),+8,i(17q)/48,X?,t(9;22),+8,i(17q),+22q-	Carbonell et al 1982a
47,X?,t(9;22),+8,i(17q+),19p-	Carbonell et al 1982a
47,X?,t(9;22),del(6)(q14),+8,i(17q)	Fleischman et al 1981
47,X?,t(9;22),r(19),+22q-/48,X?,t(9;22),+17,+22q-/49,X?, t(9;22),+8,+17,+22q-/50,X?,t(9;22),+8,+17,+r(19), +22q-/49,X?,t(9;22),1q+,+8,+17,+22q-	Carbonell et al 1982a
47,XX,t(9;22)(q34;q11),+8,i(17q)	Alimena et al 1979
47,XX,t(9;22)(q34;q11),+8,i(17q)/48,XX,t(9;22),+8,i(17q), +22q-	Tomiyasu et al 1982
47,XX,t(9;22)(q34;q11),+8	Sadamori and Sandberg 1983b
47,XX,t(9;22),+8	Lilleyman et al 1978
47,XX,t(9;22),+8	Stoll and Oberling 1978
47,XX,t(9;22),+8	Bernstein et al 1980b
47,XX,t(9;22),+8	Kohno and Sandberg 1980
47,XX,t(9;22),+8	Olah and Rak 1981
47,XX,t(9;22),+8	Fleischman et al 1981
47,XX,t(9;22),+8,16q+,i(17q)	Sadamori et al 1980
47,XX,t(9;22),+8,i(17q)	Lilleyman et al 1978
47,XX,t(9;22),+8,i(17q)	Lilleyman et al 1978
47,XX,t(9;22),+8,i(17q)	Stoll and Oberling 1978
47,XX,t(9;22),+8,i(17q)/48,XX,t(9;22),+i(17q),+22q-	
47,XX,t(9;22),+8,i(17q)	Lyall and Garson 1978
47,XX,t(9;22),+8,i(17q)	Borgström et al 1982

47,XX,t(9;22),+8,i(17q)	Borgström et al 1982
47,XX,t(9;22),+8,i(17q)	Borgström et al 1982
47,XX,t(9;22),i(17q),+8,6q-	Prigogina et al 1978
47,XX,t(9;22),i(17q),+8/48,XX,t(9;22),i(17q),+8,+19/49,XX, t(9;22),i(17q),+8,+19,+del(22)(q11)	Prigogina et al 1978
47,XX,t(9;22),i(17q),+8	Lessard and Le Prise 1982
47,XX,t(9;22)/47,XX,t(9;22),t(5;11)(p15;q22), del(5)(q21q31),+del(8)(q21q23)	Hagemeijer et al 1980b
47,XXY,t(9;22)(q34;q11)/53,XXY,+X,t(9;22),+4,+8,+18,+18, +22q-"c"	Mitelman 1983
47,XXY,t(9;22)/53,XXY,+X,t(9;22),+4,+8,+18,+18,+22q-"c"	
47,XXY,t(9;22)"c"	
47,XY,t(9;22)(q34;q11),+8,i(17)(q11)/50,XY, t(9;22)(q34;q11),i(17)(q11),+4,+6,+8,+8/47,XY, t(9;22)(q34;q11),t(3;17)(q21;p1?),del(10)(q24), i(17)(q11),+mar	Hayata et al 1975b
47,XY,t(9;22)(q34;q11),+8	Hayata et al 1975b
47,XY,t(9;22)(q34;q12),idic(17q),+8	Whang-Peng et al 1981
47,XY,t(9;22)(q34;q11),+8	Sadamori et al 1980
47,XY,t(9;22)(q34;q11),t(8;17)(q12;q22),+22q-	Petit and Van Den Berghe 1981
47,XY,t(9;22)(q34;q11),+8,i(17q)	Pasquali et al 1982b
47,XY,t(9;22)(q34;q11),del(1)(q21),dup(1)(q?),+8,+9,-14, -16,-21,+22q-,+mar	Alimena et al 1982b
47,XY,t(9;22)(q34;q11),t(3;5),-8,-10,+17,+22q-,+mar	Alimena et al 1982b
47,XY,t(9;22)(q34;q11),+8	Alimena et al 1982b
47,XY,t(9;22)(q34;q11),+8	Petit 1983
47,XY,t(9;22)(q34;q11),del(6)(q11),+8,i(17q)	Mitelman 1983
47,XY,t(9;22)(q34;q11),del(6)(q11),+8,i(17q)	
47,XY,t(9;22)(q34;q11),+8	Mitelman 1983
47,XY,t(9;22)(q34;q11),+8	
47,XY,t(9;22)(q34;q11),+8	
47,XY,t(9;22)(q34;q11),+8,i(17q)	Tomiyasu et al 1982
47,XY,t(9;22)(q34;q11),+8,i(17q)/46,XY,t(9;22),i(17q)	Tomiyasu et al 1982
47,XY,t(9;22)(q34;q11),-5,+8,+mar	Tomiyasu et al 1982
47,XY,t(9;22)(q34;q11),+8/47,XY,t(9;22),+19/47,XY,t(9;22), +21/47,XY,t(9;22),+22q-	Tomiyasu et al 1982
47,XY,t(9;22)(q34;q11),+8,i(17q)	Sadamori and Sandberg 1983b
47,XY,t(9;22),+22q-,+8,t(12;13)(q?;q?),-13,i(17q)	Rozynkowa et al 1977
47,XY,t(9;22),+22q-/56,XY,t(9;22),+1,+2,+5,+6,+7,+8,+10, +12,+17,+22q-	Stoll and Oberling 1978
47,XY,t(9;22),+22q-/48,XY,t(9;22),+8,+22q-/48,XY,t(9;22), +17,+22	Stoll and Oberling 1978
47,XY,t(9;22),+8	Lilleyman et al 1978
47,XY,t(9;22),+8	Lilleyman et al 1978
47,XY,t(9;22),+8	Stoll and Oberling 1978
47,XY,t(9;22),+8	Stoll and Oberling 1978
47,XY,t(9;22),+8	Bernstein et al 1980b
47,XY,t(9;22),+8	Sadamori et al 1980
47,XY,t(9;22),+8,i(17q)	
47,XY,t(9;22),+8	Sadamori et al 1980
47,XY,t(9;22),+8	
47,XY,t(9;22),+8	Lessard and Le Prise 1982
47,XY,t(9;22),+8,-17,-18,t(17;18)(p11;p11),+del(22q)	Sharp et al 1976

47,XY,t(9;22),+8,15q-/48,XY,t(9;22),+8,15q-,+19	Olah and Rak 1981
47,XY,t(9;22),+8,i(17q)/47,XY,t(9;22),+8	Chapelle et al 1976
47,XY,t(9;22),+8,i(17q)	Lilleyman et al 1978
47,XY,t(9;22),+8,i(17q)	McDermott et al 1978
47,XY,t(9;22),+8,i(17q)/47,XY,t(9;22),+16,i(17q)	Stoll and Oberling 1978
47,XY,t(9;22),+8,i(17q)	Lyall and Garson 1978
47,XY,t(9;22),+8,i(17q)	Lyall and Garson 1978
47,XY,t(9;22),+8,i(17q)/48,XY,t(9;22),+8,i(17q),+22q-	Lyall and Garson 1978
47,XY,t(9;22),+8,i(17q)	Miyamoto 1980
47,XY,t(9;22),+8,i(17q)	Borgström et al 1982
47,XY,t(9;22),+8,i(17q)	Borgström et al 1982
47,XY,t(9;22),+8,t(12;13)(q11;q34)	Fleischman et al 1981
47,XY,t(9;22),+8/48,XY,t(9;22),+8,+22q-	Stoll and Oberling 1978
47,XY,t(9;22),+8/48,XY,t(9;22),+8,+22q-	Olah and Rak 1981
47,XY,t(9;22),+8/50,XY,t(9;22),+8,+13,+i(17q),+22q-	Miyamoto 1980
47,XY,t(9;22),i(17q),+8	Prigogina et al 1978
47,XY,t(9;22),i(17q)/49,XY,t(9;22),i(1q),+8,+21,+22q-	Sonta and Sandberg 1977b
47-48,XY,t(9;22),+8,+22q-	Williams and Weiss 1982
48,X?,t(9;22)(q34;q11),+8,+del(22q)	Lawler et al 1976
48,XX,t(9;22)(q34;q11),+8,+19/49,XX,t(9;22),+3,+8,+19/50, XX,t(9;22),+3,+8,+19,+22q-	Mitelman 1983
48,XX,t(9;22)(q34;q11),+8,+22q-	Tomiyasu et al 1982
48,XX,t(9;22),+8,+19	Wurster-Hill 1983
48,XX,t(9;22),+8,+8,i(17q)	Borgström et al 1982
48,XX,t(9;22),+8,i(17q),+19/46,XX,t(9;22),i(17q)/47,XX, t(9;22),i(17q),+19	Hartley and McBeath 1981
48,XY,+X,t(9;22),+8,i(17q)	Borgström et al 1982
48,XY,t(9;22)(q34;q11),+8,+del(22)(q11)	Ishihara et al 1974
48,XY,t(9;22)(q34;q11),+8,i(17)(q11),+F	Rowley 1973a
48,XY,t(9;22)(q34;q11),+8,+22q-	Misawa 1978
48,XY,t(9;22)(q34;q11),+8,+22q-	Alimena et al 1982b
48,XY,t(9;22)(q34;q11),+8,i(17q),+22q-	Tomiyasu et al 1982
48,XY,t(9;22)(q34;q11),+8,+22q-	Tomiyasu et al 1982
48,XY,t(9;22)(q34;q11),+8,+22q-	Tomiyasu et al 1982
48,XY,t(9;22),+8,+13,-15,t(9;15),+i(22q)	Stoll and Oberling 1978
50,XY,t(9;22),+5,+8,+13,-15,t(9;15),+i(22q),+22q-	
48,XY,t(9;22),+8,+22q-	Lilleyman et al 1978
48,XY,t(9;22),+8,+22q-	Miyamoto 1980
48,XY,t(9;22),+8,+22q-/61,XY,+Y,t(9;22),+4,+5,+6,+8,+8,+8, +8,+15,+17,+18,+19,+20,+22q-,+22q-	Wurster-Hill 1983
48,XY,t(9;22),+8,+8,i(17q)	Lilleyman et al 1978
48,XY,t(9;22),+8,-9,-11,-12,-14,-19,+del(22q),+5mar	Sharp et al 1976
48,XY,t(9;22),+8,i(17q),+19	Lilleyman et al 1978
48,XY,t(9;22),+8,i(17q)	Lilleyman et al 1978
48,XY,t(9;22),+8,i(17q),+22q-	Stoll and Oberling 1978
48,XY,t(9;22),+8,i(17q),+22	Stoll and Oberling 1978
48-53,X?,t(9;22),+8,+8,+13,+14,+19,+21,+22q-	Carbonell et al 1982a
49,X?,t(9;22),+17,+19,+22q-/50,X?,t(9;22),+8,+17,+19, +22q-/51,X?,t(9;22),+8,+9,+17,+19,+22q-	Carbonell et al 1982a
49,XX,t(9;22)(q34;q11),8q-,+10,+14,+22q-	Alimena et al 1982b
49,XX,t(9;22),+3,+8,+22q-	Seabright 1983
49,XX,t(9;22),+8,+17,+21/49,XX,t(9;22),+8,+17,+17,-19,+21	Olah and Rak 1981
49,XX,t(9;22),+8,+19,+22	Lilleyman et al 1978
49,XX,t(9;22),+8,17q-,+19,+22q-/49,XX,t(9;22),+8,i(17q), +19,+22q-	Olah and Rak 1981

49,XY,t(9;22)(q34;q11),+6,+8,t(22;?)(p11;?),+der(22)	Oshimura et al 1982a
49,XY,t(9;22)(q34;q11),+8,dup(9)(q13q22),+17,+19	Tomiyasu et al 1982
49,XY,t(9;22),+17,+19,+22q-/50,XY,t(9;22),+8,+15,+19, +22q-/50,XY,t(9;22),+8,+9,+19,+22q-	Hartley and McBeath 1981
49,XY,t(9;22),+22q-,+8,+10	Rozynkowa et al 1977
49,XY,t(9;22),+6,+8,1p+,14q+,+22q-	Sadamori et al 1980
49,XY,t(9;22),+8,+17,+mar	Borgström et al 1982
49,XY,t(9;22),+8,+19,+22q-	Seabright 1983
49,XY,t(9;22),+8,+i(17q),+22q-/50,XY,t(9;22),+8,+10, +i(17q),+22q-	Olah and Rak 1981
50,XX,t(9;22),+8,+8,+19,+mar	Lilleyman et al 1978
50,XX,t(9;22),t(2;12)(p13;q24),+8,+8,+12q+,+12q+,+del(22q)	Sharp et al 1976
50,XY,+Y,t(9;22)(q34;q11),t(6;8)(q34;q11),+8,i(17q),+19, +22q-	Alimena et al 1982b
50,XY,t(9;22)(q34;q11),+8,+C,+del(22)(q11)+i(17)(q11)	Rowley 1973a
50,XY,t(9;22)(q34;q11),+8,+17,+21,+del(22)(q11)/51,XY, t(9;22),+8,+8,+17,+21,+del(22)(q11)/52,XY,t(9;22),+8,+8, +8,+17,+21,+del(22)(q11)	Sonta and Sandberg 1978b
50,XY,t(9;22),+8,+15,+17,+del(22q)	Sharp et al 1976
50,XY,t(9;22),+8,+17,+21,+22q-/51,XY,t(9;22),+8,+8,+17,+21, +22q-/54,XY,t(9;22),+8,+8,+8,+17,+21,+22q-,+22q-	Sonta and Sandberg 1977b
50,XY,t(9;22),+Y,+8,+19,+22q-	Sadamori et al 1980
50-54,XY,t(9;22),+1,+5,+8,+9,-16,+17,+18,+19,+21,+22q-	Alimena et al 1980
50-55,XY,t(9;22)(q34;q11),+1,+5,+8,+9,-16,+17,+18,+19,+21, +22q-,+mar	Alimena et al 1982b
50-58,XY,+X,t(9;22)(q34;q11),+7,+8,+9,+10,+11,+12, +del(15)(q22),+16,+18,+19,+22q- 46,XY,t(9;22)	Kwan et al 1977
51,XX,t(9;22)(q34;q11),+6,+8,i(17q),+19,+20,+22q-	Alimena et al 1982b
51,XX,t(9;22),+8,+9,+14,+19,+22q-	Miyamoto 1980
51,XX,t(9;22),t(8;22)(p12;p11),+del(8)(p12),+19	Lyall and Garson 1978
51,XY,t(9;22)(q34;q11),+6,+8,+19,+21,+22q-	Tomiyasu et al 1982
51,XY,t(9;22),+8,+14,+17,+19,+19	Borgström et al 1982
51-52,XY,t(9;22),+del(22q),+7,+8,+12,+19,+21	Prigogina et al 1978
52,XY,t(9;22)(q34;q11),1p+,+8,+9,i(17q),+19,+20,+21,+22q-	Alimena et al 1982b
52,XY,t(9;22),+1,+8,+10,+11,+21,+22q-	Sadamori et al 1980
52,XY,t(9;22),+8,+14,+15,+19,+21,+22q-	Sadamori et al 1980
52,XY,t(9;22),+8,+9,+12,+19,+21,+22q-	Nordenson 1983
52-54,XX,t(9;22)(q34;q11),+6,+8,+8,+19,+19,+del(22)(q11), +del(22)(q11),+X	Hayata et al 1975b
52-54,XY,t(9;22)(q34;q11),+3,+6,+7,+8,+19,+21,+22q-,+22q-, +1-3mar	Ondreyco et al 1981
53,XX,t(9;22)(q34;q11),+8,+10,+19,+20,+21,+22q-	Greenberg et al 1978
53,XX,t(9;22),+8,+11,+12,+16,+17,+19,+del(22q)/52,XX, t(9;22),+8,+11,+17,+19,+del(22q)	Gahrton et al 1974b
53,XY,+Y,t(9;22)(q34;q11),+3,+6,+8,+9,+10,+12,-15,+22q-/56, XY,+Y,t(9;22),+3,+6,+8,+8,+9,+10,+16,+19,+22q- 56,XY,+Y,t(9;22),+3,+6,+8,+8,+9,+10,+12,+19,+22q-	Stoll and Oberling 1982
53,XY,t(9;22),+6,+9q+,+10,+12,+19,+21,+22q-/54,XY,t(9;22), +6,+8,+9q+,+10,+12,+19,+21,+22q-	Sonta and Sandberg 1977b
53,XY,t(9;22),+8,+9,+19,+22q-,+mar	Lyall and Garson 1978
53-56,XY,+Y,t(9;22),+del(22q),+6,+10,+19,+8,+11,+21	Prigogina et al 1978
54,XY,t(9;22)(q34;q11),+8,+10,+11,+13,+14,+19,+21,+21	Hayata et al 1975b
55,XX,t(9;22),+3,+8,+12,+14,+15,+19,+20,+22,+22q-	Lilleyman et al 1978

Karyotype	Reference
55,XY,t(9;22),+3,+6,+8,+8,+9,+10,+16,+19,+22q-/53,XY, t(9;22),-15,+3,+6,+8,+8,+9,+10,+12,+22q-	Stoll and Oberling 1978
59,X,-X,t(9;22),+4,+6,+6,t(7;?)(p15;?),+der(7),+8,+8,+12q+, +17,+19,+19,+21,+22,+22q-,+22q-	Miyamoto 1980

Chronic myeloid leukemia, aberrant translocation

Karyotype	Reference
46,X,t(X;9;22)(q27;q34;q12)	Dallapiccola and Alimena 1979
46,X,t(X;9;22)/47-51,X,t(X;9;22),+1,+4,+8,+9,-14,i(17q), -19,+22q-	
46,X?,t(13;22)(q34;q11),t(8;9)(q22;q34)	Hossfeld 1983
46,XX,t(12;22)(p13;q11),t(8;17)(q13;q23)	Swolin et al 1983
46,XX,t(17;22)(q25;q11)/47,XX,t(17;22),+8	Bernstein et al 1980b
46,XX,t(5;22)(q35;q11)/46,XX,t(5;22),i(17q)/47,XX,t(5;22), +8,i(17q)	Pasquali et al 1979b
46,XX,t(9;15;22)(q34;q11;q11)	Hays et al 1981
54-57,XX,t(9;15;22),+5,+7,+8,+10,+11,+12,+17,+22q-	
46,XY,t(4;22)(q35;q21),t(3;5)(q27;q22)	Sessarego et al 1981b
46,XY,t(4;22)(q35;q21),t(3;5)(q27;q22),i(17q)/50,XY, t(4;22),t(3;5),i(17q),+8,+15,+21,+22q-	
46,XY,t(7;22)(p22;q12)	Gahrton et al 1979
48,XY,t(7;22)(p22;q12),+8,+8,i(17q)	
46,XY,t(8;9;22)(q22;q34;q11)/45,X,-Y, t(8;9;22)(q22;q34;q11)	Lawler et al 1976
47,XY,t(10;22)(q26;q11),+8,i(17q)	Prigogina et al 1978
47,XY,t(9;9;22)(p24;q34;q11),+8	Pasquali 1983
50,XY,del(22)(q11),+8,+11,+21,+21/51,XY,del(22)(q11),+8, +11,+13,+21,+21	Sonta and Sandberg 1978b
50,XY,del(22)(q11),+8,+11,+21,+21,/51,XY,del(22)(q11),+8, +11,+13,+21,+21	

Chronic myeloid leukemia, Ph1 negative

Karyotype	Reference
46,XX,t(3;11)(q21;p15)	Mitelman 1983
46,XX,t(3;11)(q21;p15)/49,XX,t(3;11)(q21;p15),+8,+19,+21	
46,XX,t(3;11)/49,XX,t(3;11),+8,+19,+21	
46,XY,t(12;13)(q11;p11)/46,XY,+8,-13,t(12;13)(q11;p11)	Labal de Vinuesa et al 1981
46,XY,t(2;8)(q21;q24),del(1)(p33),dup(1)(p21p32)	Brodeur et al 1979
47,XX,+8	Lindquist et al 1978
47,XX,+8	Kohno et al 1979a
47,XX,+8,17q-	Kohno et al 1979a
47,XX,+8	Engel et al 1977
47,XY,+8	Hsu et al 1974b
47,XY,+8	Mintz et al 1979
47,XY,+8/48,XY,+8,+8	Mintz et al 1979
47,XY,t(3;4)(p14;q11q31),+8	Kohno et al 1979a

Eosinophilic leukemia

Karyotype	Reference
45,X,-X,t(8;21)(q22;q22)	Kaneko et al 1983
47,XX,+8,14q+	Chilcote et al 1982
47,XY,+8	Weinfeld et al 1977b

MYELOPROLIFERATIVE DISORDERS

Polycythemia vera

Karyotype	Reference
??,X?,+2,-8,-10	Vykoupil et al 1980
??,X?,+5,+8	Vykoupil et al 1980
??,X?,+8,+11,+22	Vykoupil et al 1980
??,X?,t(9;22),+8,+10,+12,+16	Vykoupil et al 1980
45,X,-Y	Berger and Bernheim 1979
54-57,X,-Y,+6,+8,+11,+17,+20,+21,t(1;22)(p12;q13)	
46,X?,3q+,8p+,8q+	Wurster-Hill 1983
44,X,-X,del(5)(q14q32),-7	Van Den Berghe et al 1979g
44,XX,del(5)(q14q32),-6,-7,-8,t(12;?)(p12;?), t(16;18)(q11;q12)	
46,XX,t(2;11)(p13;q21)/46,XX,t(1;15)(p1?;q1?), t(2;11)(p13;q21),del(16)(p12)/51,XX,+3,+8,+8,+19,t(1;15), t(2;11),del(16)	Testa et al 1981b
47,X?,+9/48,X?,+8,+9	Shabtai 1983
47,XX,+8	Wurster-Hill et al 1976
47,XX,+8	Shabtai et al 1978
47,XX,+8	Shabtai et al 1978
47,XX,+8	Kirkland et al 1980
47,XX,+8	Lessard and Le Prise 1983
47,XX,+8	Carbonell et al 1983
47,XX,+8	Hsu et al 1974b
47,XX,+8	Zech et al 1976b
47,XY,+9	Van Den Berghe et al 1979g
47,XY,del(1)(q12),t(1;18)(q12;q23),+9,del(20)(q12)/47,XY, +9,del(11)(q21),del(20)(q12)/47,XY,+9,del(20)(q12)	
47,XY,+9/47,XY,del(5)(q14q32),del(8)(q?),+9,del(11)(q21), del(13)(q21)/47,XY,del(5)(q14q32),del(6)(q?),+9, del(11)(q21),del(13)(q21),del(1)(q12),del(20)(q12)	
48,XX,+8,+9	Testa et al 1981b
48,XX,+8,+9	Westin et al 1976
48,XX,+8,+mar	Nordenson 1983
48,XX,+8,del(11)(q21?),+mar	Testa et al 1981b
48,XX,t(9;22),+5,+8	Armenta et al 1976
48,XY,+8,+9	Wurster-Hill et al 1976
48,XY,+8,+9	Westin et al 1976
48,XY,+8,+9	Westin et al 1976

Myelosclerosis / Myelofibrosis

46,XX,8q-"s"	Whang-Peng et al 1978
46,XY,-16,+18	Whang-Peng et al 1978
46,XY,7q-,8q-,10q-,11p-,11q+,12q-,-16,+18	
46,XY,t(1;2)(q?;q?),-8,inv(13)(q12q31),i(17q)	Whang-Peng et al 1978
47,XX,+8	Nowell and Finan 1978
47,XX,+8	Jacobson et al 1978
47,XX,+8	Mitelman 1983
47,XX,+8	
47,XX,+8	
47,XX,+8	Nowell and Finan 1978
47,XX,+8/47,XX,+9	Bartoli et al 1979
47,XX,t(13;14),+8,+9"c"	Greef et al 1982
48,XX,+8,+21/49,XX,+8,+19,+21/50,XX,+8,+19,+19,+21"c"	Ueda et al 1981
50,XY,+del(1)(p?),+8,+9,+21	Nowell and Finan 1978

Chronic myeloproliferative disease, NOS

46,X?,t(8;13)(p?;q?)	Lessard and Le Prise 1983
47,XX,+8	Carbonell et al 1983
47,XY,+8	Carbonell et al 1983
47,XY,+8	Lessard and Le Prise 1983

Myeloproliferative disorder, special type

46,XX,-5,del(6)(q13),-7,+8,-17,+2mar"s"	Rowley et al 1977a
46,XX,del(5)(q15)/47,XX,+8	Swolin et al 1981
47,XX,+8,t(8;17)(q24.2;q22.1)	Hagemeijer et al 1980a
48,XY,+8,+21	Whaun et al 1981

DYSMYELOPOIETIC SYNDROMES

Preleukemia, NOS

43,X,-X,-3,-13,4q-,8q-,12q+	Geraedts et al 1980a
44,XY,+Y,-3,-5,+del(6)(q21),-8,t(17;?)(p13;?),+18,-20,+22	Mitelman 1983
44,XY,-5,-6,-7,-8,t(17;?),+2mar"s"	Pedersen-Bjergaard et al 1982
45,XY,-2,-4,-6,-8,-10,-10,+17,+18,+3mar	Panani et al 1980
45,XY,-2,-5,-7,-8,-11,-12,-13,-14,+t(2;5),+t(11;12),+t(16;17),+17,+3mar	Watt et al 1982
46,X,-Y,+8	Geraedts et al 1980a
46,XX,-7,+8,t(18;22)(p11;q11)"s"	Michalski et al 1982
46,XX,5q-/47,XX,+8"s"	Teerenhovi et al 1981
46,XY,del(5)(q13q31)/47,XY,+8	Swolin et al 1981
47,XX,+8	Ruutu et al 1977b
47,XX,+8	Geraedts et al 1980a
47,XX,+8	Anderson and Bagby 1982
47,XX,+8	Anderson and Bagby 1982
47,XX,+8	Panani et al 1980
47,XY,+8	Ruutu et al 1977b
47,XY,+8	Panani et al 1977

47,XY,+8	Geraedts et al 1980a
47,XY,+8	Geraedts et al 1980a
47,XY,+8,t(3;6)(p21;p12),+del(3)(p21)	Panani et al 1977
47,XY,+8	Panani et al 1980
47,XY,del(7)(q22),+8	Anderson and Bagby 1982

Chronic myelomonocytic leukemia

47,XX,+8/48,XX,+8,+21	Kardon et al 1982a

Refractory anemia, NOS

46,XX,del(5)(q12q31),del(8)(q22)	Watt et al 1983
46,XX,del(5)(q13q34)/46,XX,t(9;11)/47,XX,+8	Swolin et al 1981
47,XX,+8	Lessard and Le Prise 1983
47,XY,+8	Lessard and Le Prise 1983
48,XX,+8,+8	Lessard and Le Prise 1983

Sideroblastic anemia, NOS

47,XY,+8	Hellström et al 1971
47,XY,+8	Nordenson 1983
47,XY,+8	Shiloh et al 1979

Acquired idiopathic sideroblastic anemia

47,XX,+8"s"	Jonasson et al 1974
47,XY,+8	Chapelle et al 1976
47,XY,+8	Chapelle et al 1976
47,XY,+8	Bitran et al 1977

Refractory anemia without excess of blasts

44,XX,5q-,-7,+8,-12,-13,-17,-20,+2mar	Teerenhovi et al 1981
46,XX,5q-/47,XX,+8/47,XX,+14/46,X,-X,+8	Teerenhovi et al 1981
47,XY,+8	Mitelman 1983
47,XY,+8/46,X,-Y,+8	Mitelman 1983
47,XY,5q-,-7,+2mar/48,XY,-8,-9,+11,+r,+2mar	Teerenhovi et al 1981

Refractory anemia with excess of blasts

47,XX,+8	Streuli et al 1980

Granulocytopenia

47,XX,+8	Chapelle et al 1976
47,XX,+8	Chapelle et al 1976

Thrombocytopenia

47,XY,+8	Chapelle et al 1976

Pancytopenia

Karyotype	Reference
46,XX,-7,+8,-12,-16,-17,t(6;12)(q?;q?),5q-,6q-,+mar	Nowell and Finan 1978
47,XY,+8	Chapelle et al 1976
47,XY,+8	Chapelle et al 1976
47,XY,+8	Yamada and Furusawa 1976
47,XY,+8	Mitelman 1983
47,XY,+8	

Aplastic anemia

Karyotype	Reference
47,XX,+8	Yamada and Furusawa 1976
47,XX,+8	Hagemeijer et al 1980a
47,XY,+8	Yamada and Furusawa 1976
47,XY,+8	Mitelman 1983

Dysmyelopoietic syndrome, special type

Karyotype	Reference
45,XX,t(13;14)(p13;q11),del(5)(q14)/50,XX,t(13;14),+1,+6, -7,+8,+10,del(5),+mar"c""s"	Rowley et al 1981a
47,XX,del(4)(q25),+8	Mitelman 1983
47,XX,del(4)(q?),+8	
47,XY,+8	Mitelman 1983

SPECIAL LEUKEMIAS

Hairy cell leukemia

Karyotype	Reference
44,X,del(X)(q22),del(6)(q23),-7,-8,-10,del(11)(p15q21),-21, 14q+,+r,+mar	Sadamori and Sandberg 1983a

LYMPHOCYTIC LEUKEMIAS

Acute lymphocytic leukemia (ALL)

Karyotype	Reference
28,XX,-1,-2,-3,-4,-5,-7,-8,-9,-11,-12,-13,-14,-15,-16,-17, -19,-20,-22	Kaneko and Sakurai 1980
32,XY,-2,-3,-4,-5,-7,-8,-11,-12,-13,-14,-15,-16,-17,-20	Shabtai and Halbrecht 1981b
43,X,-X,-7,-8,t(9;22)(q34;q11)	Oshimura and Sandberg 1977
45,XX,-8,t(13;?)(q?;?),19q+	Prigogina et al 1979
45,XX,del(6)(q11),-8	Secker-Walker et al 1979
45,XY,-8	Whang-Peng et al 1976c
46,XX,-8,-10,del(6)(q23),+2mar	Secker-Walker et al 1979
46,XX,?inv(8)	Mitelman 1983
46,XX,t(4;11)(q21;q23)/49-50,XX,t(4;11),+8,+13,+19,+4q-	Prigogina et al 1979
46,XY,-8,del(3)(q12q25),t(8;14)(q24;q32),+t(1;8)(p11;q11)	Kaneko et al 1980
46,XY,del(6)(q23orq25)/47,XY,del(6)(q23orq25),+8	Oshimura et al 1977a
46,XY,t(4;11)(q21;q23)	Oshimura et al 1977a
46-49,XY,t(4;11)(q21;q23),+6,+8,+13,+17,t(1;13)(p22;q12), t(7;9)(q11;q34)	

46,XY,t(8;22)(q24;q12),t(3;11)(p12p21;q23),inv(2)(p11q13), del(2)(p21)/46,XY,t(8;22)(q24;q12),dup(1)(q23q44)"s" — Fonatsch et al 1982
46,XY,t(8;9)(q22;p24)/45,XY,t(8;9),-10,+12,-21 — Morse et al 1982a
46-47,XY,del(5)(q15),del(12)(p12),del(17)(p11),+8 — Prigogina et al 1979
47,XY,t(4;11)(q21;q23),1p+,del(2)(p16),del(6)(q21),i(7q), del(8)(q21),+mar — Arthur et al 1982
48,XY,t(1;22)(q21;p11),+5,+8,19p+ — Humbert et al 1978
49,XX,del(4)(p15),t(4;11)(q21;q23),+5,+8,del(9)(p13), del(9)(p13),t(12;?)(q24;?),+mar — Morse et al 1982b
52-56,XY,+5,+8,+11,+19,+20,+21,del(6)(q15),+mar — Prigogina et al 1979
53,XX,+2,+5,i(7q),+8,+13,+21,+21,+22,+17q+ — Oshimura et al 1977a
53,XY,+X,+4,+8,+14,+15,+17,+21 — Oshimura et al 1977a
54,XX,+6,+7,+8,+14,+17,+18,+21,+22 — Prigogina et al 1979
54,XX,7p+,+10,+10,+14,+14,+18,+18,+21,+21/27,X,-X,-1,-2,-3, -4,-5,-6,-7,7p+,-8,-9,-11,-12,-13,-15,-16,-17,-19,-20, -22 — Oshimura et al 1977a
54-55,XY,+5,+8,+9,+14,+21,+mar — Prigogina et al 1979
57,XX,+3,+4,+8,11p+,+14,+16,+17,+18,+21,+22,+2mar/47,XX,+8 — Morse et al 1978
61,XX,+X,+1,+2,+3,+4,+5,+6,+8,del(6)(q21),+13,+14,+15,+16, +18,+21,+22 — Oshimura et al 1977a

ALL, FAB type L1

26,XX,-1,-2,-3,-4,-5,-6,-7,-8,-9,-10,-11,-12,-13,-14,-15, -16,-17,-19,-20,-22 — Hoeltge et al 1982
28,XX,-1,-2,-3,-4,-5,-6,-7,-8,-9,-11,-12,-13,-15,-16,-17, -19,-20,-22/56,XX,+X,+X,+10,+10,+14,+14,+18,+18,+21,+21 — Brodeur et al 1981a
46,XY,del(3)(q12q25),t(8;14)(q24;q32),-8,+t(1;8)(p11;q11) — Kaneko et al 1982e
46,XY,t(8;14)(q11;q32) — Kardon et al 1982b

ALL, FAB type L2

26,XX,-1,-2,-3,-4,-5,-6,-7,-8,-9,-10,-11,-12,-13,-15,-16, -17,-18,-19,-20,-22 — Brodeur et al 1981a
44,XX,-8,t(13;14),t(9;22)"c" — Takeuchi et al 1981
46,XX,t(5;8;14)(q11;q24;q32),t(12;22)(p13;q11),-18, +t(18;?)(q23;?) — Kaneko et al 1982e
46,XY,+1,-8,t(14;?)(q32;?) — Mitelman 1983
46,XY,-4,+6,-8,-10,-11,-15,-17,+22,del(6)(q21),+i(17q), +3mar — Kaneko et al 1981b
46,XY,del(7)(p15),del(8)(q22) — Kaneko et al 1981b
47,XY,+18 — Kaneko et al 1982e
46,XY,-2,-5,-6,-7,+8,-9,-10,-11,-12,-16,+18,+t(2;?)(q35;?), +t(11;?)(q21;?),+5mar
48,XX,t(9;22),+8,+18,/46,XX,t(9;22) — Sandberg et al 1980
48,XX,t(9;22),i(17q),+18,+22q-
48,XY,+8,+21 — Prieto et al 1981
53,XY,t(9;22),+4,+5,+6,+7,+8,+17,+21 — Sandberg et al 1980
56,XY,t(9;22)(q34;q11),+4,+5,+8,+9,+11,+16,+17,+18,+19,+20 — Mitelman 1983

ALL, FAB type L3

??,XY,t(8;14)(q?;q?),t(Y;1) — Prieto et al 1981
46,XX,-16,-17,-22,+8,11q-,t(8;14),+1-2mar — Borgström et al 1981
46,XX,dup(1)(q21q32),t(8;14)(q24;q32) — Rossi et al 1982

46,XX,t(2;8)(p12;q24)	Kaneko et al 1982e
46,XX,t(2;8;14)(p11orp12;q24;q32),t(21;21)(p11;q11), +t(21;21)	Ekblom et al 1982
46,XX,t(8;14)(q24;q32)	Berger and Bernheim 1982
46,XY,dup(1)(q23q32),t(1;6)(q21;q13),t(14;?)(q32;?), del(6)(q13),del(8)(p21)	Slater et al 1979
46,XY,t(2;8)(p13;q24)	Rowley et al 1981b
46,XY,t(8;14)	Borgström et al 1981
46,XY,t(8;14),1q+	
46,XY,t(8;14)	Borgström et al 1981
46,XY,t(8;14)	Borgström et al 1981
46,XY,t(8;14)(q23;q32),del(3)(p21)	Lessard and Le Prise 1983
46,XY,t(8;14)(q24;q32)"s"	Mitelman et al 1979b
46,XY,t(8;14)(q24;q32),13q+	Berger and Bernheim 1982
46,XY,t(8;14)(q24;q32)	Berger and Bernheim 1982
46,XY,t(8;14)(q24;q32),-15,+16	Berger and Bernheim 1982
46,XY,t(8;14)(q24;q32)/46,XY,t(8;14)(q24;q32), dup(1)(q21q24)	Berger and Bernheim 1982
46,XY,t(8;14)(q24;q32),13q+	Berger and Bernheim 1982
46,XY,t(8;14)(q24;q32)	Berger and Bernheim 1982
46,XY,t(8;14)(q24;q32),+8,-22	Berger and Bernheim 1982
46,XY,t(8;14)(q24;q32)/46,XY,t(8;14)(q24;q32), dup(1)(q21q24)	Berger and Bernheim 1982
46,XY,t(8;14)(q24;q32)	Berger and Bernheim 1982
46,XY,t(8;14)(q24;q32),del(3)(p24),18p+	Berger and Bernheim 1982
46,XY,t(8;14)(q24;q32)/46,XY,t(3;8;14)(q14;q24;q32)	Berger and Bernheim 1982
46,XY,t(8;14)(q24;q32)/46,XY,t(1;8;14)(q23;q24;q32)	Berger and Bernheim 1982
46,XY,t(8;14)(q?;q?)	Prieto et al 1981
46,XY,t(8;22)(q24;q12)/46,XY,1q+,+del(1)(p22),-5, t(8;22)(q24;q12)/47,XY,+del(1)(p22),t(8;22)(q24;q12)	Abe et al 1982a
46,XY,t(8;22)(q24;q11)	Berger and Bernheim 1982
47,X?,t(8;14)(q24;q32),+8	Shabtai 1983
47,XX,t(8;14)(q22;q32),del(21)(q21),+mar	Alimena et al 1981
47,XX,t(8;14)(q24;q32),+mar/47,XX,t(8;14),del(21)(q21), +mar	Rossi et al 1982
47,XY,+8/47,XY,+8,14q+	Borgström et al 1981
47,XY,dup(1)(q21q32),del(6)(q23),t(8;14)(q23;q32),+9	Roos et al 1982

Chronic lymphocytic leukemia

44,XX,6q-,i(8q),12p-,-14,t(14;14)(q31;q12),-20,20p+	Sparkes et al 1980
44,XY,-8,del(16)(q21),-18	Pittman et al 1982
45,XX,-2,t(8;12)(q?;q?)	Pittman et al 1982
45,XX,-8,-14,-15,-22,+2,+9,+1p+,+mar	Finan et al 1978
45-46,XY,-8,9q+,-10,13p+,-14,-15,-19,-20,+mar	Vahdati et al 1983a
46,XX,t(1;8)(q?;q?)	Schröder et al 1981
46,XY,del(11)(q22),+12,-13,-21,-22,+3mar/45,XY,del(11),+12, -13,-15,-21,-22,+2mar/46,XY,del(11),+12,+12,-13,-21,-22, +2mar/46,XY,t(1;8)(p22;q24)	Robert et al 1982
46,XY,t(8;14)(q24;q32),19p-	Fleischman and Prigogina 1977
47,XXY,t(3;8),t(11;14)(q11;q32),+i(18q)"c"	Finan et al 1978
47,XY,+8	Nordenson 1983

Prolymphocytic leukemia

44,XX,-7,-8,t(9;17)(q22;q25),-11 — Pittman et al 1982
44,XY,-5,-8,-9,+mar — Pittman et al 1982
46,XX,-2,-2,-8,+3mar — Pittman et al 1982

Lymphocytic leukemia, special type

49,XX,+7,+8,+12,8q+,8p+,del(10)(q23),21q+/49,XX,+8,+12,
+del(7)(p13),8q+,8p+,10q-,21q+ — Ueshima et al 1981

MONOCLONAL GAMMOPATHIES

Multiple myeloma

45,X,del(X)(q25),t(3;8;22)(q13;q34;q11),-5,-7,-10,-11,
del(13)(q13?),+16,19q+,+3mar — Van Den Berghe et al 1979f
46,XY,+t(1;?)(p1?;?),+t(1;12)(q21;q24),del(6)(p11),-8,+9,
-12,-13,-16,-19,22q+,+mar — Philip et al 1980
47-48,XY,+8,+22,+mar — Shiloh et al 1979
48,X,-X,+1,+7,-8,+9,+21,del(1)(p36),4p+,del(5)(q22),9p+,
14q+ — Liang et al 1979
52,X,-X,+t(1;?)(q12;?),+3,+7,+8,+9,+11,-13,+18,-21,+2mar — Philip et al 1980
52,XY,-1,+t(1;15)(q12;p1?),+t(1;16)(p36;p13),
t(3;5)(q2?;q13),+3,+5,+7,+8,+9,+11,-15,+18,-20 — Philip et al 1980
53,XY,+3,+5,-7,-8,+9,+11,-14,+20,del(1)(p34),del(6)(q25),
+5mar — Liang et al 1979
55,XY,+1,+5,+9,+11,-12,+13,+15,del(1)(q32),del(6)(q25),
t(9;19)(q13;q13),14q+,16p+,+4mar — Liang et al 1979
45,XY,-2,-2,-3,+5,-8,-10,+14,-15,-16,del(6)(q25),+4mar
55,XY,+3,+5,+7,+9,+11,+14,+15,+21,+21/53,XY,+3,+5,-8,+9,
+11,+14,+15,-19,+21,14q+,+3mar — Liang et al 1979

Macroglobulinemia

46,Y,-X,del(1)(q21),del(5)(q23),+del(5)(q23),
+t(7;19)(p22;q13),del(8)(p12),+t(1;16)(q12;p13),
+t(1;18)(q12;p11),-7,-16,-18,-19,+mar — Ueshima et al 1983

Plasma cell leukemia / Plasmocytoma

43,X,-X,-1,+t(1;8)(q11;q11),+t(1;9)(q21;p24),
+t(1;10)(p31;q26),+t(1;?)(q12;?),-8,-9,-10,-13,-22 — Ueshima et al 1983
44,X,-Y,+t(1;9)(q12q44;q34),t(6;8)(q13;p11),-9,-13 — Ueshima et al 1983
46,XY,+t(1;16)(q21;p13),del(6)(q21),-16,-18,+mar — Ueshima et al 1983
47,XY,+t(1;16),del(6)(q21),+t(8;15)(p23;q12),-8,14q+,-15,
-16,-18,+3mar
46,XY,t(8;14)(q24;q32)/47,XY,t(8;14),+7/48,XY,t(8;14),+7,
+18 — Yamada et al 1983b
86,XX,cx,t(1;?)(q11;?),2q+,dup(7)(q22q36),t(8;?)(p11;?),
t(9;13)(p23;q21) — Ueshima et al 1983

SOLID TUMORS

UNCLASSIFIED NEOPLASMS

EPITHELIAL NEOPLASMS

Carcinoma, NOS

Karyotype	Reference
43-49,XX,-1,-8,+10,-11,t(1;11)(q11;q12)	Kusyk et al 1982
44,XY,+5,-8,-11,-13,-13,-14,-14,-15,+19,-22,-22,t(1;11), 5q-,9q+,+5mar	Reichmann et al 1981
45,XY,-4,-7,+8,-9,-12,+13,-17,-18,-19,-20,+21,+3mar	Reichmann et al 1981
45,XY,-7,+8,-14	Sandberg 1977
46,XX,+8,+12,+21,-2,-15,-16	Wake et al 1981
46,XX,del(2)(p24),+5,+8,t(9;17)(p?;p?),del(16)(q21),-20, -21	Mark 1975b
46,XX,t(3;11)(q12;q14),-6,+7,-8,-15,t(16;?)(q23;?),+2r	Mark 1975b
47,XX,+8,+13,-14	Sonta et al 1977
47-48,Y,-X,-5,-8,+13,-14,-16,-18,-22,+7mar	Reichmann et al 1981
48,XY,+1,-5,+8,+17	Sonta and Sandberg 1978a
48,XY,+8,-17,-18,+19,+20,t(10;18),t(15;17)	Reichmann et al 1981
49,X,-Y,-3,-5,-14,+15,-16,-17,+19,+21,+22,1q+,5q-,6q-,8p-, 9p-,11p-,18p+,20q-,22q-,t(Y;14),+3mar	Reichmann et al 1981
49,XX,+3,+11,-7,i(8q)	Riet-Fox et al 1979
49,XX,+7,+8,+13	Reichmann et al 1981
49,XX,+7,+8,+19	Reichmann et al 1981
49,XX,+8,+17,+20	Sonta and Sandberg 1978a
49,XY,+3,+8,+21	Sonta and Sandberg 1978a
49,XY,+4,+5,+8	Sonta et al 1977
50,XX,+X,+8,+12,+21	Sonta et al 1977
50,XX,+X,+8,+8,+13	Sonta and Sandberg 1978a
50,XX,inv(3)(p13q25),+8,+14,+21/63,XX,inv(3),del(11)(p13), cx	Sonta and Sandberg 1978a
50,XY,+X,+7,+8,+9	Reichmann et al 1981
52,XY,+X,+Y,-2,+7,+8,+9,+10,+13,-17,+i(1q)	Reichmann et al 1981
54,XX,+1,+2,+8,+11,+13,+14,+21,+21	Sonta et al 1977
55,XY,t(1;17)(p36;q21),+2,+2,+4,+8,+8,-10,+15,+18,+20,+21, +2mar	Sonta and Sandberg 1978a
58,XY,del(3)(p14),+4,+7,+8,+10,+11,+12,+14,+15,+17,+18,+19, +21	Kovacs 1978a
60,XY,cx,3p+,6q+,8p-	Reichmann et al 1981
63,XX,-1,+3,+5,+6,+7,+8,+10,+11,+12,-13,-15,+20,+21,+22, +9mar	Sandberg 1977
76-79,XY,cx,1q-,8p-,dmin	Reichmann et al 1981
80-83,XX,cx,t(2;5;16),t(2;5;18),t(11;13),11q+,t(8;10), del(2)(q24),del(6)(q21),i(14q),dup(9)(q11q12)	Riet-Fox et al 1979

Carcinoma NOS, metastatic

Karyotype	Reference
35,XX,-1,-3,-4,-5,-7,-8,-10,-11,-12,-13,-14,-15,-16,-17, -18,-19,-20,-21,-22,+8mar	Pathak et al 1979
41-42,XX,t(1;18)(p13orp21;q11),3p-,t(3;?)(p25;?), t(8;?)(p2?;?),t(9;?)(q34;?),cx	Bartnitzke and Bullerdiek 1983
46,XX,t(9;22)(q34;q11)"s"	Togawa et al 1981

52,XX,+3,+5,+7,+8,+9,+11,12q+,13q-,-14,+16,i(17q)"s"
79-80,XX,t(1;?)(q12;?),t(1;2),t(3;?),t(7;?)(q36;?),t(8;?), t(9;?)(q34;?),cx — Bartnitzke and Bullerdiek 1983

Large cell carcinoma

55-70,XY,t(4;5;14),del(1)(q11q21),t(7;11)(p22;q13), t(7;21)(q11;q22),t(3;8)(p?;q?),t(6;11)(p11;p11),i(16q), t(3;?)(q11;?),t(21;?)(q11;?),t(13;?) — Kakati et al 1975

Small cell carcinoma

47,XY,+del(1)(p?),+del(3)(q?),+del(3)(p?),-4,+8,-15,+18, -22 — Wurster-Hill and Maurer 1978

Papillary carcinoma

54,XY,+del(2)(q11),t(2;14)(q11;q32),t(3;11)(p13;p15), +i(5p),+8,+9,+12,+16,+17,+21,+2mar — Pathak et al 1982
68,XX,cx,t(6;14)(q21;q24),del(1)(p?q?),1p+,4q-,8q-,10q+ — Wake et al 1980

Adenoma

46,XX,del(8)(q12q22or23) — Mark et al 1982a
46,XX,t(1;3)(p21;p21),t(1;13)(p36;q14),t(3;8)(p21;q12), t(1;5;20)(p21;p14q12;p13),t(5;8)(p14;q12), t(X;9)(q27;q12) — Mark et al 1982a
46,XX,t(3;5)(p21;p15),del(8)(p12q12),t(8;9)(q12;q34) — Mark et al 1980
46,XX,t(3;8)(p21;q12) — Mark et al 1981a
46,XX,t(3;8)(p21;q12) — Mark et al 1982a
46,XX,t(3;8)(p21;q12) — Mark et al 1982a
46,XX,t(3;8)(p25;q21) — Mark et al 1980
46,XX,t(9;12)(p13p22;q13q15)/47,XX,+8,t(9;12) — Mark et al 1981a
46,XY,t(3;8)(p25;q21) — Mark et al 1980
46,Y,-X,ins(X;3)(q22;q25q27),del(3)(q25),t(8;14)(q12;q22) — Mark et al 1981a

Adenocarcinoma

??,XX,cx,del(1)(q42),del(7)(q2?),del(18)(p11),i(8p),i(8q) — Riet-Fox et al 1979
37,X,-X,-3,-8,-15,del(2)(p23),del(6)(q15),t(11;?)(p12;?) — Trent and Salmon 1981
38,X,-X,-8,-14,-20,+3-5mar — Trent and Salmon 1981
44,XX,-1,t(1;8),del(1)(p22) — Riet-Fox et al 1979
44-50,X,-X,-1,t(1;?)(q21;?),del(2)(q33),4p+,-5,-7,+8, del(10)(q12),t(10;13)(q12;q34),del(17)(q25),-20 — Kusyk et al 1982
48-48,XX,+3,+7,+8,-1,-5,-14,-17,5q+,6q+,del(1)(q42), del(5)(q3?),del(10)(q24),del(14)(q2?) — Riet-Fox et al 1979

Adenocarcinoma, metastatic

39-41,XX,-3,-7,-8,der(1),der(2),der(4),der(12),cx — Tiepolo and Zuffardi 1973
40,XX,t(4;5;8),t(3;?)(q21;?),t(1;11)(q21;p11), t(9;11)(q11;p11),del(1)(p36) — Kakati et al 1975
40-45,XY,-3,-8,-18,+5,+10,11p+,14q+,del(4)(q32) — Ayraud 1975
47-48,X,Xq+,t(3;9)(q12;p11),t(3;3)(q29;p25),del(3)(p25), t(4;?)(q35;?),+8 — Kakati et al 1975

49,XX,1q+,+3,4q+,+7,+8,12q+,17q+	Granberg et al 1973

Adenomatous polyp

47,XX,+8/48,XX,+8,+17	Mitelman et al 1974b
48,XX,+8,+14	Mitelman et al 1974b
48,XX,+8,+14/48,XX,-3,+8,+9,+14	Mitelman et al 1974b
48-52,X,-Y,+7,+8,+13,+14,+16,+18,-20,+12q-,1q+	Reichmann et al 1982b

MESENCHYMAL NEOPLASMS

Liposarcoma

50,XY,+3,+6,+8,+11,+14,-22,17p+	Sonta et al 1977

MELANOMAS

Malignant melanoma

44,X?,+2,+9,-6,-8,-15,-21,9p-,19q+	Wake et al 1981
46,XY,5q+,+7,-8,-9,-10,-13,-20,+21,+3mar	McCulloch et al 1976

Malignant melanoma, metastatic

41,XY,dup(1)(q21q32),t(1;7)(q25;q36),t(7;9)(q11;q11), del(9)(q13),-3,-4,-5,-8,-10,-11,-16,+18,-20,-22	Kakati et al 1977
45,XY,t(1;9;13)(q?;q?;q?),r(2),t(1;9)(p?;q?),i(8q), t(11;?)(q?;?),-10,-13,+mar	Chen and Shaw 1973

NEUROGENIC NEOPLASMS

Neuroblastoma

45,XY,-21,t(1;14)(p32;q31),t(8;?)(q24;?),dup(14)(q11q24)	Brodeur et al 1981b

Retinoblastoma

46,XY,dup(1)(q25q44),t(7;10)(q36;q11),-8,-10, t(12;13)(q11;p11),dup(13)(q22q34),+18,+22	Gardner et al 1982
51,XY,del(6)(q21),+8,+11,-12,t(13;18)(p11;q11),14q+,-16, +19,+22,+2mar	Hossfeld 1978

Meningioma, benign

41-42,XY,-1,-5,-8,9q+,-14,-18,-22,+r	Mark 1973b
43,XY,-1,-8,-17,-19,-22,+2mar/44,XY,-1,-17,-22,+mar	Mark et al 1972a
43-45,XX,-8,-16,-22	Mark et al 1972a
44,XX,-8,-22/40-44,XX,-8,-9,-19,-20,-21,-22	Mark et al 1972a
44-45,XX,-8,-22,+del(22q)	Mark 1973a
45,XX,-22/40-44,X,-X,-8,-9,-15,-21,-22	Mark et al 1972b
45,XX,-22/43-44,XX,-8,-16	Mark 1973a
45,XX,-22/43-44,XX,-8,-18,-22	Mark 1973a

45,XX,-22/44-45,XX,-8,-12,-21,-22	Mark et al 1972b
45;XY,-22/42-44,XY,-8,-16,-19,-22	Mark 1973a
45,XY,-22/42-45,XY,-8,-15,-17,-20,-22	Mark et al 1972b
46,XX,-22,+mar/44-45,XX,-8,-9,-22,+mar	Mark et al 1972b
47,XX,+12,del(22)(q11)/46,XX,-8,+12,-17,del(22)(q11)	Mark et al 1972a
51,XY,+5,6q+,+12,+16,+17,+18,+20,-22/52,XY,+5,+8,+10,+12, +17,+18,+20,-22	Zang and Zankl 1983

EMBRYONAL AND MISCELLANEOUS NEOPLASMS

Teratoma, malignant

48,X,-Y,+1,del(5)(q12q32),+8,+10	Tricot et al 1983
67,XY,+1,+3,+4,+5,+6,+7,+7,+8,+10,+11,+13,+13,+14,+16,+17, +18,+20,+20,+21,+21	Sonta et al 1977

LYMPHOMAS

1. UNCLASSIFIED AND MISCELLANEOUS LYMPHOMAS

Benign lymphomatous tumor, NOS

48,XY,t(8;14)(q24;q32),+17,+17q+"s"	Mitelman et al 1979a

Malignant lymphoma, NOS

47,XX,t(1;14)(q23;q32),+3,t(6;21)(p25;q11),+8, t(7;11)(q11;q25),-21	Yamada et al 1980

Non-Hodgkin's lymphoma, NOS

46-48,XX,t(1;?)(q21orq22;?),t(2;?)(q37;?),-8,+9,+12, 16q+"s"	Mark et al 1976
47,X?,t(8;14)(q24;q32),+7	Zech et al 1976a
47-48,XY,t(14;?)(q32;?),16q-,-17,18q-,13q-,+7,+8,8q-	Catovsky et al 1977
48-54,XX,+X,3p+,-4,-6,-8,-9,-10,+12,-15,+7-10mar	Pearson et al 1982

Non-Hodgkin's lymphoma, histiocytic, NOS

45,X,del(Y)(q12),t(4;10)(q12;q24),t(6;7)(q21;q32), del(8)(p21)	Mark et al 1978
45-47,XX,t(10;14)(q21;q32),+8,-18,-21	Mark et al 1977
47,XX,t(1;18)(p36;q21),+7,t(8;12)(q11;q15), t(8;12;13)(q11q22;q15;q14),del(13)(q14),t(8;14)(q22;q32)	Mark et al 1978
48,XY,t(3;8)(q22;q11),t(5;10)(q33;q22),del(7)(q21), t(4;9)(q11;p24),t(10;13)(q22;q21),r(12),del(19)(p13), del(22)(q11),+7	Mark et al 1978
50,XY,dup(1)(q12q32),+2,4p+,del(4)(q21),t(8;14)(q24;q32), +17p+,18q+,+20	Brynes et al 1978
50,XY,t(2;?)(p25;?),del(2)(p13),+i(3p),del(6)(p21),+7,+8, +9,+9,-11,-13,+14,-15,+20,+21,-22	Mark et al 1978

44,XY,del(2)(p13),t(6;?)(p21;?),del(11)(q13),
t(11;14)(q13;q31),i(15q),-20,-22
80-90,XX,cx,del(7)(p15),t(5;8)(q13;q11) Kakati et al 1980
82,XXX,t(3;12)(q13;q15),t(3;8)(q22;q11),del(3)(q11), Mark et al 1978
del(5)(q31),i(7q),del(10)(q24),t(10;10)(q24;q26),
inv(11)(p15q21),del(12)(q13),del(13)(q14)

Angioimmunoblastic lymphadenopathy

46,XX,t(8;14)(q?;q32) Mazur et al 1979
47,XY,+3/47,XY,+3,-9,+21/48,XY,+3,+8,+9,-20 Hossfeld et al 1976

2. HODGKIN'S LYMPHOMAS

Hodgkin's disease, lymphocytic depletion

45,X,t(X;1)(q28;q11),t(3;5)(q29;q33),del(8)(q22),-9, Hossfeld and Schmidt 1978
del(11)(p13),-17

Hodgkin's disease, nodular sclerosis

47,XY,t(1;?)(p36;?),+2,t(2;6)(q31;q27),+8,del(12)(p11), Reeves and Pickup 1980
t(17;?)(p13;?),-14,-20,+mar

3A. NON-HODGKIN'S LYMPHOMAS - RAPPAPORT CLASSIFICATION

Lymphocytic, poorly differentiated, NOS

46,XY,+3,-8 Mark et al 1979
46,XY,del(6)(q21),-8,-9,t(6;14)(q21;q32),+16,+20 Mark et al 1979

Lymphocytic, poorly differentiated, diffuse

47,XY,t(14;18)(q32;q21),-9,-10,+12,-16,8p-,13q+,+3mar Fukuhara and Rowley 1978
48-49,XX,2p-,+7,+i(8q),+12,14q+,15q-,18q- Kakati et al 1980

Lymphocytic, poorly differentiated, nodular

45,XY,-5,-8,+i(8q) Kakati et al 1980
52,X,-X,+2,-4,+5,+7,+8,-14,-15,-16,+20,+21, Kaneko et al 1982a
+t(1;15)(p11;q11),+t(12;17)(q24;q11),+t(14;?)(q32;?),
+mar

Mixed cell, diffuse

53,XY,t(8;14)(q24;q32),t(14;18)(q32;q21),+20,1q-,+1p-,2q-, Fukuhara and Rowley 1978
4q+,5q-,5p+,6q-,+7q+,10q+,11q+,11q-,13p+,+4mar

Histiocytic, diffuse

Karyotype	Reference
45,X,del(X)(q24),del(1)(q21),+del(1)(p11),del(2)(q22), del(3)(q21),del(6)(p11),del(6)(q15),t(7;9)(p22;q13), t(8;10)(p23;q21),del(9)(q13),del(10)(q21), t(2;11)(q22;q23),-13,del(14)(q24),del(15)(q22), del(16)(q22),+17,-19	Mark et al 1979
46,X,-X,-4,-11,-17,+18,-19,del(7)(p15),del(8)(q22),+i(17q), +t(4;19)(p11;q13),t(11;?)(q23;?)	Kaneko et al 1982a
47,XX,-1,-2,-6,+8,-14,+16,-19,del(11)(q23), t(8;14)(q22;q32),+t(2;6)(q21;p23),del(6)(q21), +t(1;14)(p11;q11),+t(19;?)(q13;?),+t(14;?)(q11;?)	Kaneko et al 1982a
47,XY,-10,-13,+15,-16,+19,+20,del(6)(q21),del(8)(q22), t(2;14)(q21;q32),t(4;4)(q31;q35),+t(10;?)(p11;?)	Kaneko et al 1982a
48,XY,-6,-8,+21,i(6p),+t(8;?)(q11;?),+t(19;?)(q13;?)	Kaneko et al 1982a
50,XY,t(8;14)(q24;q32),+2,+20,1q+,4p+,+4q+,17p+,18q+	Fukuhara and Rowley 1978

3B. NON-HODGKIN'S LYMPHOMAS - KIEL CLASSIFICATION

Immunocytoma

Karyotype	Reference
45-46,XY,-8	Kristoffersson 1983
47,XY,15p+,+20/46,XY,del(5)(q23),del(6)(p22),+del(7)(p11), -21	

Centroblastic-centrocytic, diffuse

Karyotype	Reference
48-51,XY,del(1)(p33q23q25),-2,+del(3)(p22),+4,4p+,+5,+6p+, +7,+8,+9,+21,+22	Kristoffersson 1983
45-51,XY,del(1)(p33q23q25),del(2)(q14),+del(3)(p22),+4,+5, +7,+8,-12	

Centroblastic-centrocytic, follicular

Karyotype	Reference
46-47,XX,t(8;9)(q13;p24),t(14;18)(q32;q21),+22	Kristoffersson 1983
52,XX,+X,+7,+8,del(10)(q24),+12,t(13;?)(q22;?), t(14;?)(q32;?),del(18)(q21),+19,+mar	Reeves and Pickup 1980

Centroblastic, diffuse

Karyotype	Reference
42-45,XY,1p+,-6,-8,14q+,14q-,-16,i(17q),+18,-20,+mar	Kristoffersson 1983
46-51,XX,+3,del(6)(q21),+del(6)(q21),+7,-8,del(9)(q11),+13, -16,-17,+19,-20,-21,+1-5mar	Kristoffersson 1983

Centroblastic, follicular

Karyotype	Reference
47,XY,del(6)(q23),+8,14q+?,i(17q)	Kristoffersson 1983

Lymphoblastic, Burkitt's type

Karyotype	Reference
??,X?,t(8;14)(q24.1;q32.5)	Manolova et al 1979b
??,X?,t(8;14)(q24.1;q32.5)	Manolova et al 1979b
??,X?,t(8;14)(q24.1;q32.5)	Manolova et al 1979b
??,X?,t(8;14)(q24.1;q32.5)	Manolova et al 1979b
??,X?,t(8;14)(q24.1;q32.5)	Manolova et al 1979b
??,XX,6p-,t(8;14)	Douglass et al 1980a
??,XX,dup(1)(q12orq21q31),t(8;14),+10q-	Douglass et al 1980a
??,XY,-5,+6	Douglass et al 1980a
??,XY,1p+,t(8;14)	
??,XY,dup(1)(q12orq21q31),3p-,3q-,+6p-,+8,+14,+15,+18,-21	Douglass et al 1980a
??,XY,dup(1)(q?q31),t(8;14)	Douglass et al 1980a
??,XY,t(8;14),+15,+21	Douglass et al 1980a
??,XY,t(8;14),12q-	Douglass et al 1980a
43,XY,+1,+2,+3,-7,-8,-9,-12,-16,-22,del(1)(q25), del(2)(p11),del(10)(q22),14q+	Biggar et al 1981
45,X,-X,t(2;8)(p11;q24)	Abe et al 1982b
45,X,-Y,t(1;14)(q22;q32),t(8;22)(q24;q11),3q+,6q-,6q+,7q+, 9p-,14q-,18q-	Berger et al 1981c
45,X?,t(8;14)(q24;q32),t(7;16),t(6;17)	Zech et al 1976a
46,X,t(X;1)(p11;q25),del(1)(q25),7p+,17p-,t(2;8)(p11;q24)	Slater et al 1982
46,X?,t(8;14)(q23;q32)	Kaiser McCaw et al 1977
46,X?,t(8;14)(q24;q32)	Zech et al 1976a
46,X?,t(8;14)(q24;q32)	Zech et al 1976a
46,X?,t(8;14)(q24;q32)	Zech et al 1976a
46,XX,t(2;8)(p12;q23)	Van Den Berghe et al 1979d
46,XX,dup(1)(q12orq21q31),t(8;14)	Douglass et al 1980a
46,XX,dup(1)(q12orq21q31),t(8;14)	Douglass et al 1980a
46,XX,dup(1)(q12q32),t(8;14)(q24;q32)	Miyamoto et al 1982a
46,XX,t(2;8)(p11;q24)	Bornkamm et al 1980
46,XX,t(8;14)(q24;q32),dup(1)(q23q32),del(3)(p25)	Slater et al 1982
46,XX,t(8;14)(q24;q32)	Biggar et al 1981
46,XX,t(8;14)(q24;q32)	Berger and Bernheim 1982
46,XX,t(8;14),16q-	Douglass et al 1980a
46,XX,t(8;22)(q24;q13)	Miyoshi et al 1981c
46,XY,del(2)(p12),t(8;14)(q24;q32)	Biggar et al 1981
46,XY,t(2;8)(p12;q24)	Miyoshi et al 1979b
46,XY,t(2;8)(p12;q24)	Berger and Bernheim 1982
46,XY,t(2;8;9)(p11;q23;q21q31)	Philip et al 1981
46,XY,t(4;5;7)(q13;p13;p22),t(8;14)(q23;q32)	Kaiser McCaw et al 1977
46,XY,t(8;14)	Miyoshi et al 1981c
46,XY,t(8;14)	Douglass et al 1980a
46,XY,t(8;14)	Douglass et al 1980a
46,XY,t(8;14)	Sakurai et al 1983
46,XY,t(8;14)	Sakurai et al 1983
46,XY,t(8;14)	Sakurai et al 1983
46,XY,t(8;14)(q23;q32)	Kaiser McCaw et al 1977
46,XY,t(8;14)(q24;q32),17p-,17q+	Miyoshi et al 1981c
46,XY,t(8;14)(q24;q32),-13,16p+,+mar	Slater et al 1982
46,XY,t(8;14)(q24;q32)	Biggar et al 1981
46,XY,t(8;14)(q24;q32)	Berger and Bernheim 1982
46,XY,t(8;14)(q24;q32)	Berger and Bernheim 1982

46,XY,t(8;14)(q24;q32)	Berger and Bernheim 1982
46,XY,t(8;14)(q24;q32)	Berger and Bernheim 1982
46,XY,t(8;14)(q24;q32)	Berger and Bernheim 1982
46,XY,t(8;22)(q23;q12)/46,XY,t(8;22)(q23;q12), t(1;6)(q21;q27)	Berger et al 1981c
46,XY,t(8;22)(q23;q12)/46,XY,t(8;22)(q23;q12), +t(1;6)(q23;q26)	Berger et al 1981c
46,XY,t(8;22)(q23;q11)/46,XY,t(8;22)(q23;q11), t(1;4)(p32;q32)	Berger et al 1981c
46-47,XY,t(14;?)(q32;?),dup(1)(q21q32),t(2;3)(q13;q29),+5, 6p-,-8,del(9)(q?),+12,+18,+mar	Miyoshi et al 1981c
46-48,X?,t(8;14)(q24;q32),+7,15q-,+6,+13	Zech et al 1976a
46-52,X?,t(8;14)(q24;q32),+1,+3,+5,+7,i(17q),+X	Zech et al 1976a
47,X?,t(8;14)(q23;q32),+mar	Kaiser McCaw et al 1977
47,X?,t(8;14)(q24;q32)	Zech et al 1976a
47,X?,t(8;14)(q24;q32)	Zech et al 1976a
47,X?,t(8;14)(q24;q32),+7	Zech et al 1976a
47,XX,+12,t(8;14)(q24;q32),del(12)(p11q22),del(17)(q21)	Biggar et al 1981
47,XX,+14,t(8;14)(q24;q32)	Biggar et al 1981
47,XX,t(8;14)(q24;q32),dup(1)(q24q32),del(2)(q34), del(5)(q31q34),+8"s"	Berger and Bernheim 1982
47,XX,t(8;22)(q24;q11),+8,t(11;21)(q14;q22)	Berger and Bernheim 1982
47,XY,t(8;22)(q24;q11),+22/47,XY,t(8;22)(q24;q11),+22, dup(1)(q23q24)	Berger and Bernheim 1982
47-49,XY,t(8;14)(q23;q32),+7,+t(X;1)(p21;q21)	Kakati et al 1979
48,X?,t(8;14)(q23;q32),+2mar	Kaiser McCaw et al 1977
48-53,X?,t(8;14)(q24;q32),+X,+2,+3,+9,+10,+11,+12,+20	Zech et al 1976a
50,XX,+1,+2,+3,+11,del(2)(p14),11p+,t(8;14)(q24;q32)	Biggar et al 1981

<u>Lymphoblastic, convoluted type</u>

45,XY,t(3;?)(q29;?),t(8;?)(q2?;?),t(12;17)(p13;q11)	Kaneko et al 1982d
48,XX,+13,+t(1;5)(q21;p15)/48,XX,-5,+8,+13,del(1)(q32), +t(1;5)(q21;p15)	Kaneko et al 1982d

<u>Immunoblastic</u>

45,XY,del(1)(p22orp23),del(3)(q11),del(3)(q21),del(8)(p11), del(9)(p11),del(9)(q12),t(12;?)(q24;?)	Reeves and Stathopoulos 1976
46,XY,r(8)	Reeves and Pickup 1980
51,XY,+X,t(3;3)(q11;q27),t(8;?)(q2?;?),+7,+12,+21,+21	Reeves and Pickup 1980
51,XY,t(1;6)(p22;q13orq15),t(2;13)(q21;q14), t(8;16)(q24;p11),+11,+12,del(15)(q22),+16,+20,+mar	Reeves and Pickup 1980

3C. NON-HODGKIN'S LYMPHOMAS - LUKES AND COLLINS CLASSIFICATION

Mycosis fungoides

48,XX,del(1)(p22),t(2;8;14)(q37;q24;q24),+5,del(7)(p13), del(9)(q22),t(9;18)(q11;q24),del(10)(p13), del(13)(q12q14),del(5)(q15q22),del(5)(q13),del(5)(q15) — Fukuhara et al 1978a
46,XX,t(1;14)(q32;q32)

Follicular center cell, diffuse, small non-cleaved

44,X,-Y,t(1;14)(p21;q13),5p+,del(6)(q13),del(8)(p21),9p+, 9p-,t(11;21)(q23;q22),13q+,-10,+12,-17p+,der(19),22q+ — Fukuhara et al 1979a

Immunoblastic type (B-cell)

49,XX,+i(1q),+5,del(8)(q24),t(3;22)(q29;q11),+13 — Berger et al 1983a

3D. NON-HODGKIN'S LYMPHOMAS - WORKING FORMULATION

Small lymphocytic

46,XX,t(1;8),t(8;9) — Yunis et al 1982

Follicular, small cleaved cell

47,XX,t(X;1),ins(1;8),t(2;8),t(14;18)(q32;q21),+17 — Yunis et al 1982

Diffuse, small cleaved cell

49,XY,dup(1)(q?),8p-,-9,+14,+17,+20,+mar — Yunis et al 1982

Large cell, immunoblastic

45,XY,ins(2;21),t(8;14)(q24;q32),t(9;19),t(17;21),-21 — Yunis et al 1982
48,Y,t(X;?),t(1;?),t(3;6),+5,-6,t(8;14)(q24;q32),i(17q), +mar — Yunis et al 1982

Small non-cleaved cell

46,X,-X,t(8;14)(q24;q32),dup(13),+r/45,X,-X,t(8;14),-12, t(1;15),der(22),+r — Yunis et al 1982
46,XX,dup(1)(q?),t(8;14)(q24;q32) — Yunis et al 1982
47,XY,t(8;14)(q24;q32),+mar — Yunis et al 1982

Chromosome 9

HEMATOLOGICAL DISORDERS

UNCLASSIFIED LEUKEMIAS

Acute leukemia, NOS

Karyotype	Reference
45,X,-X,+t(X;1)(q13;p12),-7,+8,-9	Mamaeva et al 1983
46,XY,+del(1)(q32),-9,del(9)(q13),del(22)(q11)	Prigogina et al 1979

NONLYMPHOCYTIC LEUKEMIAS

Acute nonlymphocytic leukemia (ANLL)

Karyotype	Reference
43,XX,-1,3p+,3q-,4q+,-5,-5,-7,+9,+9,10p+,+10,+10,10p-,10q-, +11q-,-13,-15,-16,-17,-18,-21"s"	Whang-Peng et al 1979
44,XX,-9,-10,-17,del(14)(q22),t(9;10)(q22orq34;q11orq22)	Oshimura et al 1976b
44-45,XX,?	Mitelman 1983
46,XX,-9,t(9;9)(p11;q11)	
45,X,-Y,t(8;21)(q22;q22),t(9;22)(q34;q11)	Francesconi and Pasquali 1978a
45,X,-Y,t(Y;9)(p11;p11)	Sasaki et al 1976
45,XX,-9	Lampert et al 1972
45,XY,-9,13q-"s"	Geraedts et al 1980a
46,X?,t(9;22)(q34;q12)	Rozynkowa et al 1977
46,XX,del(1)(p32),del(5)(q13),+8,t(9;?)(p11;?), t(11;22)(q11;q13),-16,-17,-19,+3mar	Muir et al 1977
46,XX,del(5)(q12q31),-9,15p+,18p-,-21,+2mar	Kerkhofs et al 1982
46,XX,del(5)(q?)/47,XX,del(5)(q?),+9"s"	Verhest et al 1977
46,XX,del(7)(q22)/46,XX,del(7)(q22),del(5)(q22q31)/46,XX, del(7)(q22),-9,t(1;9)(q22;p24)/46,XX,t(2;3)(q31;q27), del(7)(q22),del(5)(q22q31orq33)"s"	Oshimura et al 1976b
46,XX,del(9)(q22)	Morse et al 1982a
46,XX,t(8;21)(q22;q22),del(9)(q?)	Rozynkowa et al 1977
46,XX,t(9;22)(q34;q11)	Abe and Sandberg 1979
46,XY,+del(7)(p11),-8,del(12)(p11),t(21;?)(q22;?)/50,XY,+Y, +del(7)(p11),del(12)(p11),+17,t(17;?)(q25;?), t(21;?)(q22;?),+22	Mitelman 1983
46,XY,+del(7)(p11),-8,del(12)(p11),t(21;?)(q22;?)/47,XY,+9	
46,XY,-9,17p+,+mar	Borgström et al 1982

46,XY,del(2)(q31),-7,del(9)(p13),t(2;21)(q21;q22), t(2;?)(p21;?)/46,XY,t(1;?)(p22;?),t(17;?)(q25;?)	Philip et al 1978a
46,XY,del(9)(q22)	Sasaki et al 1976
46,XY,inv(9),t(11;15)(q11;q26),t(11;13)(q11;q34)"c"	Philip et al 1978a
46,XY,t(9;10)(p24;p12),t(3;16),t(4;15)(p13;q14)"c"	Van Den Berghe et al 1979a
46,XY,t(9;22)	Lessard and Le Prise 1982
46,XY,t(9;22)	Lessard and Le Prise 1982
46,XY,t(9;22)	Carbonell 1983
46,XY,t(9;22)(q34;q11),21q+/46,X,-Y,t(9;22),+8	Abe and Sandberg 1979
46,XY,t(9;22)(q34;q11)	Abe and Sandberg 1979
46,XY,t(9;22)(q34;q11)	Mitelman 1983
46,XY,t(9;22)(q34;q11)	
46,XY,t(9;22)(q34;q11)	
46,XY,t(9;22)(q34;q11)/47,XY,t(9;22),7p+,+19	
46,XY,t(9;22)	
47,XX,+9	Bernard et al 1982a
47,XX,t(9;22)(q34;q11),+22q-/48,XX,t(9;22),+8,+22q-	Mitelman 1983
49,XX,t(9;22)(q34;q11),+8,+13,+22q-	
49,XX,t(9;22)(q34;q11),+8,+22q-,+mar	
50,XX,t(9;22)(q34;q11),dup(1)(q25q44),+del(3)(p11),+8, +22q-,+mar	
47,XY,+8/48,XY,+8,+9"s"	Mitelman 1983
47,XY,+9	Mitelman et al 1976b
48,XX,+8,+9"s"	Geraedts et al 1980a
48,XX,+8,-9,+t(X;9)(q13;q32)	Philip et al 1978a
48,XX,+8,t(9;22),+10	Chrz et al 1983
48,XX,3q+,-5,+9,+16,+19	Nordenson 1983
48,XX,t(9;22)(q34;q11),+del(22)(q11),+21	Abe and Sandberg 1979
48,XY,+9,+21	Mitelman 1983
48,XY,+del(3)(q11),+7,t(6;19)(q15;q13)/50,XY,+del(3)(q11), +7,t(6;19)(q15;q13),+2mar/51,XY,+del(3),+7,+9,t(6;19), +2mar	Mitelman 1983
48-50,XX,+8,+9,+del(12)(p11)"s"	Weinfeld et al 1977a
50,XX,t(9;22)(q34;q11),+6,+8,+10,+del(9)(p13), +del(22)(q11)	Abe and Sandberg 1979
51,XY,+3,+5,-9,+19,+20,+21,t(6;14)(p25;q11),t(9;?)(q22;?)	Philip et al 1978a
51,XY,+8,+9,+13,+14,+21	Oshimura et al 1976b

<u>ANLL, FAB type M1</u>

45,XX,-7,-9,+t(7;9)(q11;p11),22q-/46,XX,t(9;22)(q34;q11)	Sasaki et al 1983
45,XX,t(9;22)(q34;q11),-7	Sasaki et al 1983
45,XY,-7,t(9;22)(q34;q11)	Sasaki et al 1983
45,XX,-9	Mitelman 1983
46,XX,-1,inv(1),3q+,5q-,t(13;17),-18,+mar/51,XX,-1,inv(1), 3q+,5q-,t(13;17),-18,+6,+9,+9,+11,+11,+mar	Hagemeijer et al 1981a
46,XX,t(9;11)(q34;q13)	Alimena 1983
46,XX,t(9;22)/45,XX,-7,t(9;22)	Hagemeijer et al 1981a
46,XY,+del(2)(p13),i(9q),-16,del(17)(p11)	Brodeur et al 1983
46,XY,t(9;22)(q34;q11)	Sasaki et al 1983
46,XY,t(9;20;22)(q34;q12;q11)/46,XY,t(9;22)(q34;q11),-6,-7, -7,-8,10q+,+t(7;8),i(8q)	
47,XY,t(9;22)(q34;q11),+18	Sasaki et al 1983

Karyotype	Reference
47-48,XY,+8,t(9;22)(q34;q11),-20,+t(1;20)(q21;q13), t(10;21)(p11;q22),+22q-	Sasaki et al 1983
48,XY,+8,+13,inv(9)"c"	Yunis et al 1981
49,XX,+X,-1,+der(1)(q21q31),+del(1)(p22),+t(1;5)(p11;q35), -5,t(9;22)(q34;q11),+22q-	Sasaki et al 1983
58,XY,+1,+4,+5,+6,+8,+11,+13,+13,+14,+14,+15,+21,+21, t(9;22)(q34;q11)	Alimena 1983

ANLL, FAB type M2

Karyotype	Reference
45,,del(X)(q26),-Y,t(8;21)(q22;q22),del(9)(q24), del(18)(p11)	Brodeur et al 1983
45,X,-X,-9,-18,+2mar	Prieto et al 1981
45,X,-X,t(8;21)(q22;q22),del(9)(q31)	Brodeur et al 1983
45,X,-Y,t(8;21)(q22;q22),del(9)(q13q22)	Hossfeld et al 1980
45,XX,-7,t(3;9)(q29;p21)/44,X,-X,-7,t(3;9)"s"	Rowley et al 1981a
45,XY,-7,t(9;22)(q34;q11)	Sasaki et al 1983
45-47,XX,-9,-17,-20,del(4)(q21),del(5)(q12q31), t(9;17)(p13;p11),+mar	Testa et al 1979
46,XX,+1p-,+9,-16,-16/46,XX,+1p-,+13,-16,-16	Prieto et al 1981
46,XX,4q+,del(7)(q31),t(8;21)(q21;q21),del(9)(q12q21)	Hagemeijer et al 1981a
46,XX,del(20)(q11)/48,XX,+9,del(20)(q11),+21	Golomb et al 1978c
46,XX,t(4;10)(q31;q22),t(6;9)(p23;q34)	Kaneko et al 1982b
46,XX,t(8;21)(q21;q21),del(9)(q11q13)	Hagemeijer et al 1981a
46,XX,t(8;21)(q22;q22),del(9)(q13q22)	Hossfeld et al 1980
46,XX,t(9;11)(p22;q24)	Yunis et al 1981
46,XX,t(9;22)(q34;q11),i(17q)	Sasaki et al 1983
46,XY,9q-	Larson et al 1982
46,XY,del(5)(q13?q22?),t(9;22)(q34;q11)	Sasaki et al 1983
46,XY,t(8;21)(q22;q22),del(9)(q21)/45,X,-Y,t(8;21), del(9)/45,XY,t(8;21),-9	Bernstein et al 1982b
46,XY,t(8;21)(q22;q22)/46,XY,t(8;21),9q-/47,XY,t(8;21),+9	Berger et al 1982b
46,XY,t(9;22)(q34;q11)	Bernstein et al 1982b
47,XY,del(9)(q22),+19	Kaneko et al 1982b
48-50,XX,+del(6)(q16),+9,-18,+mar"s"	Mitelman 1983
47,XX,del(6)(q16),+mar"s"	

ANLL, FAB type M1+M2

Karyotype	Reference
45,XY,-21,t(4;9)(p16;q22)	Yamada and Furusawa 1976
37-45,XY,-7,-12,-13,-14,-16,-18,-19,-21,del(4)(q29), del(5)(q24),der(9),der(17),11q+,12p+,t(14;?)	Fitzgerald et al 1983
42-43,XX,cx,t(11;15)(q11;q11),t(9;?)(p?;?),-15,-17,-18,-20, -21	Fitzgerald et al 1983
44,XX,-3,-4,-5,+9,-17,-18,-21,-22,+5mar	Garson 1980
45,X,-Y,t(8;21)(q22;q22),del(9)(q13)	Shiraishi et al 1982
45,XY,t(19;22)(p13;q11),t(1;7)(q21;q22),t(7;9),+del(8)(p?), -15	Oshimura and Sandberg 1977
45-46,XY,9p+,-14	Mitelman et al 1981
46,XX,t(6;9)(p23;q34)	Rowley and Potter 1976
46,XX,t(6;9)(p23;q34)	Testa et al 1979
48,XX,t(6;9)(p23;q34),+8,+13	
46,XX,t(8;9)(q23or24;q31),-17,+20	Mitelman et al 1981
46,XX,t(9;22)	Garson 1980
46,XX,t(9;22)	Garson 1980

46,XX,t(9;22)(q34;q11)	Sasaki et al 1975
46,XX,t(9;22)(q34;q11)	Sasaki et al 1975
46,XY,9q+	Nordenson 1983
46,XY,9q+	Nordenson 1983
46,XY,t(6;9)(p23;q34)	Fitzgerald et al 1983
46,XY,t(9;16)(p?;q?)	Garson 1980
46,XY,t(9;22)(q34;q11)	Mitelman et al 1981
46,XY,t(9;22)(q34;q11),t(2;15)(q37;q12)	Mitelman et al 1981
46,XY,t(9;22),del(10)(q22)	Wayne et al 1979
47,XY,t(9;22),+8,del(10)(q22)/48-50,XY,t(9;22),+8,+9, del(10)(q22),+10,+22q-	
46,XY,t(9;22),t(7;10)(q34;q22)	Wayne et al 1979
46,XY,t(9;22)/47,XY,t(9;22),+17/47,XY,t(9;22),+22q-	Nagao et al 1977
46-50,XY,+Y,+del(7)(p11),-8,+9,del(12)(p11),+17,21q+,+22	Mitelman et al 1981
47,X,-Y,t(1;13)(q32p32q11;q11),del(3)(q21), t(8;21)(q22;q22),del(9)(q22),+8,+18	Sakurai et al 1982b
47,XX,+9	Fitzgerald et al 1983
47,XX,+del(9)(q13)	Mitelman et al 1981
47,XY,+9	Prigogina et al 1979
47-49,XY,-5,del(5)(q15),-7,i(8q),dic(9),-13,-16,-22, +6-9mar"s"	Testa et al 1981b
48,XX,+8,+9	Fitzgerald et al 1983
49,XX,+6,t(9;22)(q34;q11),+17,+21	Nordenson 1983
50-52,XY,+9,+14,+17,+19,+20,+21	Alimena et al 1977
50-52,XY,-5,-7,+9,+14,+17,+19,+20,+21,+mar	Mitelman et al 1981

ANLL, FAB type M3

46,XX,t(15;17)(q26;q22)/47,XX,t(15;17),-7,+8,-9,-10,+2mar	Van Den Berghe et al 1979c
46,XX,t(2;15;17)(q21?;q25?;q21?;),9p+	Bernstein et al 1982b
46,XY,t(15;17)(q22;q21),t(X;9)(q28;q22)	Berger et al 1981e
46,XY,t(15;17)(q25;q22)/46,XY,t(15;17),del(7)(q22), del(9)(q22)	Golomb et al 1979
46,XY,t(15;17)/46,XY,t(15;17),t(1;9)(p36;q22)	Hurd et al 1982
47,XY,del(12)(p11),del(20)(q11),+9	Mitelman et al 1981
47,XY,t(15;17)(q21?;q25?),+8/47,XY,t(15;17)(q21?;q25?), +9/48,XY,t(15;17),+8,+9	Van Den Berghe et al 1979c

ANLL, FAB type M4

42,XY,-5,-9,-10,-15,-16,-17,-20,-21,-22,+5mar	Testa et al 1979
43,XY,+1,-5,+6,-9,-10,-15,-16,-17,-20,-21,-22,+6mar	
44,XY,del(3)(p22),-9,-12,t(17;?)(p13;?)"s"	Mitelman 1983
44,XY,del(5)(q15q23),del(7)(q22),-9,-12,-13,del(15)(q22), +mar/46,X,-Y,del(5),del(7),+8,-12,-13,del(15),+mar,dmin	Alimena 1983
45,X,-X,-9,-17,-18,-21,+5mar	Prieto et al 1981
45,XX,2p-,-5,-9,-17,+r,+mar	Berger et al 1981a
46,XY,del(9)(q12q31)	Hagemeijer et al 1981a
46,XY,del(9)(q?)	Garson 1980
46,XY,t(1;9)(p34;q34),inv(8)(p11q12)	Rowley and Potter 1976
46,XY,t(9;22)(q34;q11)/48,XY,t(9;22),+8,+22q-	Bernstein et al 1982b
47,XY,+8,t(9;22)	Prieto et al 1981
47,XY,+8/48,XY,+8,+9/49,XY,+8,+9,+11	Hagemeijer et al 1981a

47,XY,t(9;22)(q34;q11),t(16;?)(p13;?),-18,del(20)(p12q12), +22,+mar — Li et al 1981b
48,XY,+9,+21"s" — Berger et al 1981a

ANLL, FAB type M5

46,XX,-18,-22,+2r/46,XX,t(2;9)(q23;p22),-3,-22,+2r/49,XX, t(2;9),-22,+4r — Alimena 1983
46,XX,t(9;11)(p21;q23) — Hagemeijer et al 1982b
46,XX,t(9;11)(p21;q23) — Brodeur et al 1983
46,XY,t(17;21)(q11;q11orq21),+i(8q),del(5)(q11q13),9p+, 11q+ — Fitzgerald et al 1983
46,XY,t(3;9;13)(p21;q22;q14) — Benedict et al 1979
46,XY,t(5;9)(p11;p11) — Mitelman 1983
46,XY,t(5;9)(p11;p11)
46,XY,t(9;11)(p21;q23) — Hagemeijer et al 1982b
46,XY,t(9;11)(p21;q23) — Hagemeijer et al 1981a
46,XY,t(9;11)(p21;q23) — Brodeur et al 1983
46,XY,t(9;11)(p22;q13) — Michalski et al 1983
46,XY,t(9;22)(q34;q11)/47,XY,+8,t(9;22) — Hagemeijer et al 1981a
46-51,XX,-3,-18,-22,t(2;9)(q23;p22),+2-4r — Mitelman et al 1981
48,XY,+8,+9 — Kaneko et al 1982b
48,XY,+9,+13 — Yunis et al 1981
49,XX,+4,+8,+8,t(9;11)(p22;q23) — Kaneko et al 1982b
49,XY,t(8;9)(p21;p22),+del(8)(p21),+9p+ — Brynes et al 1976
50,XX,t(9;11)(p21;q23),+4mar — Hagemeijer et al 1982b
51-52,XX,+3,+6,+9,-10,+18,+19,del(1)(p22),del(1)(p22), dup(11)(q11q21),+t(10;?)(p13;?) — Kaneko et al 1982b

ANLL, FAB type M5a

46,XY,t(1;9)(q11;q34),t(10;11)(p14;q14) — Berger et al 1982a
47,XX,+8,t(9;11)(p22;q24) — Dewald et al 1983
47,XY,+8,9q+,del(14)(q23),17p+ — Berger et al 1982a
47,XY,+8,t(9;11)(p22;q24) — Dewald et al 1983

ANLL, FAB type M5b

46,XX,t(9;11)(p22;q24) — Dewald et al 1983
47,XY,+9 — Berger et al 1982a

ANLL, FAB type M4+M5

45,XY,+9,-15,-22 — Li et al 1983
46,XY,t(9;22) — Li et al 1983
47,XY,+9"s" — Rutten et al 1974

ANLL, FAB type M6

??,XY,t(4;?)(q23;?),t(5;17),t(8;9),+8,+11, t(12;16)(p12;p12) — Hustinx and Rutten 1983
43,X,-Y,-4,-13,-14,-15,+2mar/46,X,-Y,-3,+9,+mar — Li et al 1983
43,XY,-2,4q-,-9,11q-,-15,-19,-20,+2mar/46,XY,-15,-20,+2mar — Li et al 1983
45,XY,del(4)(q23),del(5)(q13),-6,del(6)(q21),del(7)(p14), del(8)(p21),del(9)(q31),16q+ — Rowley and Potter 1976

Karyotype	Reference
45-47,XY,del(6)(q22),-9,+18,+21	Mitelman et al 1981
46,XX,t(4;9)(q21;q34),t(4;13)(q12;q12),del(13)(q12orq13), del(12)(p11)	Fitzgerald et al 1983
46,XY,5p-,-9,-12,13q-,+2mar	Nordenson 1983
46,XY,del(9)(q22)	Alimena 1983
48,XX,+8,+9,+9,-11,del(4)(q23)"s"	Weinfeld et al 1977a
49,XY,+8,+8,+9	Bernstein et al 1982b

Chronic myeloid leukemia, t(9;22)

Karyotype	Reference
26,XX,t(9;22)(q34;q11),-1,-2,-3,-4,-5,-6,-7,-9,-10,-11,-12, -13,-14,-15,-16,-17,-18,-19,-20	Hartley and McBeath 1981
35,XX,t(9;22)(q34;q11),-3,-4,-5,+8,-9,-10,-11,-12,-13,-14, -15,-16,-17/66-72,XX,t(9;22)	Pedersen and Boesen 1983
45,XY,t(9;22),-9,t(20;21)(p13;q22)	Miyamoto 1980
46,X?,t(9;22)(q34;q11),t(1;9)(q11;p11-13),+1,-der(1)	Oshimura et al 1976a
46,XX,t(9;22)(q34;q11),t(9;12)(q34;q13)	Dor et al 1977
46,XX,t(9;22)(q34;q11),del(9)(p21)	Tomiyasu et al 1982
46,XX,t(9;22),t(4;9)(p16;q32)	Sadamori et al 1980
45,XX,t(9;22),t(4;9),-7	
46,XX,t(9;22),var(9)(q12)	Hossfeld 1975a
46,XY,t(9;22)(q34;q11),t(1;9)(q11;p11)	Sonta and Sandberg 1978b
46,XY,t(9;22)(q34;q11)	Como and Graze 1979
35,XY,t(9;22)(q34;q11),-3,-4,-5,-7,-9,-11,-12,-13,-15,-16, +3mar	
46,XY,t(9;22)(q34;q11)	Ishihara et al 1982
27,X,-Y,t(9;22),-1,-2,-3,-4,-5,-6,-7,-9,-11,-13,-14,-15, -16,-17,-18,-19,-20,-22	
46,XY,t(9;22),t(X;9)(p22;q22),t(2;3)(p22;q27)	Fleischman et al 1981
46,XY,t(9;22)/46,XY,t(9;22),i(9q),-22,+mar	Prigogina et al 1978
46,XY,t(9;22)/49,XY,t(9;22),+9,+19,+22q-	Gall et al 1976
46,XY,t(9;22)/49,XY,t(9;22),+9,+19,+22q-	
46-49,XX,t(9;22),+del(22q),+5,+6,+9	Prigogina et al 1978
47,X?,t(9;22),+21/48,X?,t(9;22),+21,+21/50-52,X?,t(9;22), +8,+9,+12,+19,+21,+21	Carbonell et al 1982a
47,X?,t(9;22),+22q-/49,X?,t(9;22),+8,+9,+22q-	Carbonell et al 1982a
47,XY,t(9;22)(q34;q11),del(1)(q21),dup(1)(q?),+8,+9,-14, -16,-21,+22q-,+mar	Alimena et al 1982b
47,XY,t(9;22),+22q-/46,XY,t(9;22),-9,+22q-/46,XY,t(9;22), t(3;5)(p?;q?),t(6;13)(p?;q?)	Kohno and Sandberg 1980
47,XY,t(9;22),+9q+,i(17q)	Kamada et al 1981
48,XY,t(9;22),+8,+13,-15,t(9;15),+i(22q)	Stoll and Oberling 1978
50,XY,t(9;22),+5,+8,+13,-15,t(9;15),+i(22q),+22q-	
48,XY,t(9;22),+8,-9,-11,-12,-14,-19,+del(22q),+5mar	Sharp et al 1976
49,X?,t(9;22),+17,+19,+22q-/50,X?,t(9;22),+8,+17,+19, +22q-/51,X?,t(9;22),+8,+9,+17,+19,+22q-	Carbonell et al 1982a
49,XY,t(9;22)(q34;q11),+8,dup(9)(q13q22),+17,+19	Tomiyasu et al 1982
49,XY,t(9;22),+17,+19,+22q-/50,XY,t(9;22),+8,+15,+19, +22q-/50,XY,t(9;22),+8,+9,+19,+22q-	Hartley and McBeath 1981
49,XY,t(9;22),+9,+10,+12/45,XY,t(5;17)(p11;q21)	Hagemeijer et al 1981b
49,XY,t(9;22),+9,+14,+22q-	Lyall and Garson 1978
50-54,XY,t(9;22),+1,+5,+8,+9,-16,+17,+18,+19,+21,+22q-	Alimena et al 1980
50-55,XY,t(9;22)(q34;q11),+1,+5,+8,+9,-16,+17,+18,+19,+21, +22q-,+mar	Alimena et al 1982b

Karyotype	Reference
50-58,XY,+X,t(9;22)(q34;q11),+7,+8,+9,+10,+11,+12, +del(15)(q22),+16,+18,+19,+22q-	Kwan et al 1977
46,XY,t(9;22)	
51,XX,t(9;22),+8,+9,+14,+19,+22q-	Miyamoto 1980
52,XY,t(9;22)(q34;q11),1p+,+8,+9,i(17q),+19,+20,+21,+22q-	Alimena et al 1982b
52,XY,t(9;22),+8,+9,+12,+19,+21,+22q-	Nordenson 1983
53,XY,+Y,t(9;22)(q34;q11),+3,+6,+8,+9,+10,+12,-15,+22q-/56, XY,+Y,t(9;22),+3,+6,+8,+8,+9,+10,+16,+19,+22q-	Stoll and Oberling 1982
56,XY,+Y,t(9;22),+3,+6,+8,+8,+9,+10,+12,+19,+22q-	
53,XY,t(9;22),+6,+9q+,+10,+12,+19,+21,+22q-/54,XY,t(9;22), +6,+8,+9q+,+10,+12,+19,+21,+22q-	Sonta and Sandberg 1977b
53,XY,t(9;22),+8,+9,+19,+22q-,+mar	Lyall and Garson 1978
55,XY,t(9;22),+3,+6,+8,+8,+9,+10,+16,+19,+22q-/53,XY, t(9;22),-15,+3,+6,+8,+8,+9,+10,+12,+22q-	Stoll and Oberling 1978

<u>Chronic myeloid leukemia, aberrant translocation</u>

Karyotype	Reference
??,X?,t(2;9;22)(q14;q34;q11)	Smadja et al 1980
??,X?,t(3;4;9;11;22)	Potter et al 1981
??,X?,t(3;9;22)(p21;q34;q11)	Potter et al 1981
??,X?,t(3;9;22)(q11orq12;q34;q11)	Tanzer et al 1980a
??,X?,t(5;9;22)(q13;q34;q11)	Potter et al 1981
??,X?,t(9;10;22)	Geraedts et al 1977
??,X?,t(9;13;22)(q34;q22;q11)	Potter et al 1981
??,X?,t(9;15;22)(q34;q15;q11)	Potter et al 1981
??,X?,t(9;15;22)(q34;q15;q11)	Potter et al 1981
44,Y,-X,t(9;17;22)(q34;p11;q11),t(X;9)(?;p24),i(17q)	Engel et al 1977
45,XY,t(18;22)(q23;q11),+del(1)(q11),6q-,-9,-9	Nowell 1983
46,X,t(X;9;22)(q27;q34;q12)	Dallapiccola and Alimena 1979
46,X,t(X;9;22)/47-51,X,t(X;9;22),+1,+4,+8,+9,-14,i(17q), -19,+22q-	
46,X?,t(13;22)(q34;q11),t(8;9)(q22;q34)	Hossfeld 1983
46,X?,t(1;9;22)(p31;q34;q11)	Ishihara 1983
46,X?,t(1;9;22)(p32;q34;q11)	Hagemeijer et al 1980b
46,X?,t(3;9;22)	Nowell 1983
46,X?,t(3;9;22)(p21;q34;q12)	Rozynkowa 1983
46,X?,t(3;9;22)(p21;q34;q12)	Rozynkowa 1983
46,X?,t(3;9;22)(q22;q34;q11)	Van Den Berghe 1983
46,X?,t(5;9;22)(q35;q11;q11)	Ishihara 1983
46,X?,t(7;9;11;22)(p11;q34;q22;q11)	Ishihara 1983
46,X?,t(7;9;22)(q22;q34;q11)	Nowell 1983
46,X?,t(9;11;22)(q34;q13;q11)	Carbonell et al 1982a
46,X?,t(9;11;22)(q34;q13;q11)	Ishihara 1983
46,X?,t(9;14;22)(q34;q2?;q11)	Nowell 1983
46,X?,t(9;17;22)(q34;p13;q11)	Ishihara and Minamihisamatsu 1981
46,X?,t(9;9;22)	Ishihara 1983
46,XX,inv(5)(p14q12),t(9;13;17;22)(q31;q11;q12;q12)	Fraisse et al 1980
46,XX,t(12;22)(q34;q11),t(9;11)(q34;q11)	Seabright 1983
46,XX,t(17;22),inv(5),t(9;13)	Freycon et al 1982
46,XX,t(1;9;22)(q22;q34;q11q13)	Fleischman et al 1981
46,XX,t(1;9;22)(q32;q34;q11)	Borgström 1981
46,XX,t(1;9;22)(q32;q34;q11)	Pasquali 1983
46,XX,t(2;9;22)(p11;q34;q11)	Rochon and Vaillancourt 1980

Karyotype	Reference
46,XX,t(2;9;22)(p21;q34;q11)	Pasquali et al 1979b
46,XX,t(2;9;22)(p24q24;q34;q11)	Pasquali 1983
46,XX,t(2;9;22)(q11;q34;q11),t(2;7)(q11;q11)	Fleischman et al 1981
46,XX,t(2;9;22)(q11;q34;q11)	Smadja et al 1983a
46,XX,t(2;9;22)(q14q37;q34;q11)	Van Den Akker et al 1980
46,XX,t(2;9;22)(q23;q34;q11)	Pasquali et al 1979b
46,XX,t(2;9;22)/50,XX,t(2;9;22),+6,+10,+19,+der(22)	
46,XX,t(3;4;9;22)(q21;q31;q34;q11)	Chessells et al 1979
46,XX,t(3;9;22)(p21;q34;q11)	Nowell et al 1975
46,XX,t(3;9;22)(p21;q34;q11)	Anderson et al 1978
46,XX,t(3;9;22)(p21;q34;q12)	Rozynkowa et al 1977
46,XX,t(3;9;22)(p21;q34;q11)	Mohandas et al 1980
46,XX,t(3;9;22)(p21;q34;q11)	Seabright 1983
46,XX,t(3;9;22)(q21;q34;q11)	Pasquali 1983
46,XX,t(3;9;22)(q23;q34;q11)	Borgström 1981
46,XX,t(4;9;22)(p14;q34;q11)	Tomiyasu et al 1982
46,XX,t(4;9;22)(q23;q34;q11)	Ciric and Rolovic 1980
46,XX,t(4;9;22)(q31;q34;q11)	Sudries et al 1980
46,XX,t(4;9;22)(q31;q34;q11)	Kessous et al 1980
46,XX,t(4;9;22)(q32;q34;q12)	Fraisse et al 1980
46,XX,t(5;9;22)(q13;q34;q11)	Alimena 1983
46,XX,t(6;22)(q26;q11),t(1;9)(q21;q24)	Berger et al 1976
46,XX,t(6;9;11;22)(p21;q34;q13;q11)	Carbonell et al 1980
46,XX,t(6;9;22)(p21;q34;q11)	Pasquali 1983
46,XX,t(6;9;22)(p21;q34;q11)	Pasquali 1983
46,XX,t(7;9)(q22;q34),t(17;22)(p13;q11)	Mitelman 1983
46,XX,t(7;9)(q22;q34),t(17;22)(p13;q11)	
46,XX,t(7;9;22)(p14;q34;q12)	Fraisse et al 1980
46,XX,t(7;9;22)(q11orq21;q34;q11)	Martin et al 1980
46,XX,t(7;9;22)(q22;q32q34;q11)	Pasquali et al 1979b
46,XX,t(9;11;22)(q34;q13;q11)	Sessarego et al 1982
46,XX,t(9;14;22)(q34;q24;q11)	Potter et al 1975
46,XX,t(9;14;22)(q34;q12orq13;q11)	Tanzer et al 1980a
46,XX,t(9;14;22)(q34;q11;q11)	Borgström 1981
46,XX,t(9;14;22)(q34;q13;q11)	Seabright 1983
46,XX,t(9;14;22)(q34;q21;q11)	Pasquali 1983
46,XX,t(9;15;22)(q34;q11;q11)	Hays et al 1981
54-57,XX,t(9;15;22),+5,+7,+8,+10,+11,+12,+17,+22q-	
46,XX,t(9;17;22)(q34;q21;q11)	Sonta and Sandberg 1978b
46,XX,t(9;17;22)(q34;q21;q11)	Rowley et al 1979
46,XX,t(9;17;22)(q34;q21;q11)	Sonta and Sandberg 1977a
46,XX,t(9;20;22)(q31orq33;q11;q11)	Oshimura et al 1982a
46,XX,t(9;22)(q22q34;q11)	Lessard et al 1981
46,XY,t(16;22)(q24;q11),t(9;17)(q34;q21)/46,XY,t(16;22), t(9;17),del(7)(q22)	Mitelman 1983
46,XY,t(16;22)(q24;q11),t(9;17)(q34;q21)	
46,XY,t(1;9;22)(p11;q34;q11)	Borgström 1981
46,XY,t(1;9;22)(q21;q34;q11)/46,XY,t(9;22)(q34;q11)	Berger et al 1981d
46,XY,t(1;9;22)(q21;q34;q11)	Lessard and Le Prise 1982
46,XY,t(1;9;22)(q23;q34;q12)	Berger and Bernheim 1978
46,XY,t(1;9;22)(q23;q22;q12)	Verma and Dosik 1977
46,XY,t(1;9;22)(q32;q34;q11)	Fleischman et al 1981
46,XY,t(2;9;22)(q11;q34;q11)	Alimena et al 1982b
46,XY,t(2;9;22)(q24orq31;q34;q11)	Tanzer et al 1977
46,XY,t(2;9;22)(q32;q34;q11)	Nowell 1983

46,XY,t(2;9;22)(q37;q22q34;q11)	Testa et al 1982
46,XY,t(3;9)(q21;q34),t(17;22)(q21;q11)	Oshimura et al 1982a
46,XY,t(3;9;10;22)(p13;q34;q22;q11),17q+	Seabright 1983
46,XY,t(3;9;22)(p21;q34;q11)	Lessard and Le Prise 1982
46,XY,t(3;9;22)(q21;q34;q11)	Pasquali et al 1979b
46,XY,t(3;9;22)(q21;q34;q11)	Pasquali 1983
46,XY,t(3;9;22)(q23;q34;q11)	Seabright 1983
46,XY,t(4;9;22)(q21;q34;q11)	Pasquali et al 1979b
46,XY,t(4;9;22)(q23;q34;q11)	Rowley et al 1976
46,XY,t(4;9;22)(q31;q34;q11)	Seabright 1983
46,XY,t(5;9;22)(q13;q34;q11)	Oshimura et al 1982a
46,XY,t(6;9;22)(p21;q34;q11)	Potter et al 1975
46,XY,t(6;9;22)(p21;q34;q11)	Potter et al 1975
46,XY,t(6;9;22)(q21;q34;q11)	Chessells et al 1979
46,XY,t(6;9;22)(q24;q34;q11)	Sudries et al 1980
46,XY,t(7;9;22)(q22;q34;q11)	Seabright 1983
46,XY,t(7;9;22)(q35;q31q34;q11)	Lessard and Le Prise 1982
46,XY,t(8;9;22)(q22;q34;q11)/45,X,-Y, t(8;9;22)(q22;q34;q11)	Lawler et al 1976
46,XY,t(9;10;15;19;22)(q34;q22;q21;q13;q12)	Tomiyasu et al 1982
46,XY,t(9;10;22)(q34;q11;q11)	Hayata et al 1975b
46,XY,t(9;11;22)(q34;q13;q11)/52,XY,t(9;11;22),+6,+9q+, +11q-,+18,+19,+del(22q),1q+	Lawler et al 1976
46,XY,t(9;11;22)(q34;q13;q11)	Fleischman et al 1981
46,XY,t(9;12;22)(q34;q15;q11)	Borgström 1981
46,XY,t(9;13;22)(q13-21;q13;q11)	Hayata et al 1975b
46,XY,t(9;13;22)(q34;q14;q11)	Seabright 1983
46,XY,t(9;14;22)(q34;q24;q11)	Borgström 1981
46,XY,t(9;14;22)(q34;q21;q11)	Pasquali 1983
46,XY,t(9;14;22)(q34;q24;q11)	Tomiyasu et al 1982
46,XY,t(9;15;22)(q34;q15;q11)	Seabright 1983
46,XY,t(9;17;22)(q34;q21;q11)	Pasquali 1983
46,XY,t(9;20;22)(q34;p11p13;q11),i(17q)	Pasquali et al 1982b
46,XY,t(9;20;22)(q34;p11;q11),i(17q)	Pasquali 1983
46,XY,t(9;20;22)(q34;p11p13;q11),i(17q)	Casalone et al 1983
46,XY,t(9;22;?)(q34;q11q13;?)	Fleischman et al 1981
46,XY,t(9;9;11;22)(q34;p24;p13;q11)	Gahrton et al 1977
46,XY,t(9;9;22)(q34;q22;q11)	Hayata et al 1975b
47,XY,+Y,t(9;13;15;22)(q34;q12;q22;q11)	Potter et al 1975
47,XY,inv(9)(p11q13),t(5;9;22)(q13;q34;q11), +del(22)(q11)"c"	Nowell et al 1975
47,XY,t(9;9;22)(p24;q34;q11),+8	Pasquali 1983

Chronic myeloid leukemia, Ph1 negative

46,X,t(Y;3;9)(q12;q25;q34)	Verhest et al 1980
46,X?,t(3;9)(q12;q34)	Verhest et al 1980
46,XX,t(9;11)(q34;q13)	Warburton and Shah 1976
48,XY,+9,+22	Seabright 1983

Eosinophilic leukemia

Karyotype	Reference
48,XY,t(9;22),r(12),+22,+22q-	Hartley 1983

MYELOPROLIFERATIVE DISORDERS

Polycythemia vera

Karyotype	Reference
43-45,XX,-9,-18,-22	Wurster-Hill et al 1976
??,X?,t(9;22),+8,+10,+12,+16	Vykoupil et al 1980
46,XX,dup(1)(q21q44),inv(1),inv(9)	Mamaeva et al 1983
46,XX,t(1;9)(q11;q13)	Kirkland et al 1980
46,XY,t(12;17)(q13;p11)/47,X,-Y,+t(Y;1)(q12;q21),+9	Testa et al 1981b
46,XY,t(9;22)(q34;q11)	Nicoara et al 1977
47,X?,+9,22p+	Shabtai 1983
47,X?,+9/48,X?,+8,+9	Shabtai 1983
47,XX,+9	Wurster-Hill et al 1976
47,XX,+9	Wurster-Hill et al 1976
47,XX,+9	Lessard and Le Prise 1983
47,XX,+9	Nowell and Finan 1978
47,XX,+9/49,XX,+9,+20,+20,t(4;5),-7,-11,-12,+3mar	Zech et al 1976b
47,XX,t(1;9)(p2?;p2?),t(1;9),del(7)(q2?),del(20)(q11)"s"	Nowell and Finan 1978
47,XX,t(1;9)(q22;q13)	Westin et al 1976
47,XY,+9	Shabtai et al 1978
47,XY,+9	Zech et al 1976b
47,XY,+9	Zech et al 1976b
47,XY,+9	Testa et al 1981b
47,XY,+9,del(20)(q11)	Mitelman 1983
47,XY,+9	Zech et al 1976b
47,XY,+9	Van Den Berghe et al 1979g
47,XY,del(1)(q12),t(1;18)(q12;q23),+9,del(20)(q12)/47,XY,+9,del(11)(q21),del(20)(q12)/47,XY,+9,del(20)(q12)	
47,XY,+9/47,XY,del(5)(q14q32),del(8)(q?),+9,del(11)(q21),del(13)(q21)/47,XY,del(5)(q14q32),del(6)(q?),+9,del(11)(q21),del(13)(q21),del(1)(q12),del(20)(q12)	
47,XY,+del(1)(p21),del(20)(q11)/47,XY,+9,del(20)(q11)	Westin et al 1976
47,XY,+del(1)(p21),del(5)(q31),del(20)(q11)/47,XY,+9,del(20)(q11)	Swolin et al 1981
48,XX,+8,+9	Testa et al 1981b
48,XX,+8,+9	Westin et al 1976
48,XX,t(9;22),+5,+8	Armenta et al 1976
48,XY,+8,+9	Wurster-Hill et al 1976
48,XY,+8,+9	Westin et al 1976
48,XY,+8,+9	Westin et al 1976

Myelosclerosis / Myelofibrosis

Karyotype	Reference
44,XY,5p-,-7,-10,11q-,i(17q)/45,XY,-7,18q-/45,XY,2p-,2q-,5p-,6q-,9q-,10q-,-11,+i(17q)	Whang-Peng et al 1978
46,XX,7q-,-9,+mar	Carbonell et al 1983
46,XY,9q+,13q-,20q-/47,XY,-7,+2mar	Chrz et al 1983
47,X?,+9	Davidson and Knight 1973

47,X?,+9	Davidson and Knight 1973
47,XX,+8/47,XX,+9	Bartoli et al 1979
47,XX,t(13;14),+8,+9"c"	Greef et al 1982
50,XY,+del(1)(p?),+8,+9,+21	Nowell and Finan 1978
51,XX,+1,del(2)(q33),+6,+9,+11,+17,-19,+20q+	Najfeld et al 1978b

Idiopathic thrombocythemia

46,X,t(X;9;22)(q11;q34;q11)	Fitzgerald et al 1981
46,XX,t(2;9)(q?;p?)	Köpf et al 1982
46,XX,t(9;22)(q34;q12)	Rajendra et al 1981

Chronic myeloproliferative disease, NOS

47,XX,+9	Knight et al 1974

DYSMYELOPOIETIC SYNDROMES

Preleukemia, NOS

45,XX,-2,-3,5q-,-6,-9,16p-,17p+,+3mar	Swansbury and Lawler 198[illegible]
46,XX,1p+,9q-,18q+,r(20)	Geraedts et al 1980a
46,XX,del(7)(q11),t(9;?)(p22;?)"s"	Anderson and Bagby 1982
46,XX,t(9;22)(q34;q11)	Roth et al 1980
47,XY,+9	Panani et al 1977
47-48,XY,+3,+9,+12	Panani et al 1977
47-48,XY,+3,+9	Panani et al 1977
50,XY,-5,-21,+1,+9,+11,17p+,+1-3mar	Borgström et al 1982

Refractory anemia, NOS

46,XX,del(5)(q13q34)/46,XX,t(9;11)/47,XX,+8	Swolin et al 1981

Refractory anemia without excess of blasts

47,XY,5q-,-7,+2mar/48,XY,-8,-9,+11,+r,+2mar	Teerenhovi et al 1981

Refractory anemia with excess of blasts

45,XX,-9,-17,-20,del(4)(q21),del(5)(q12q31), +t(9;17)(p13?;p11)	Streuli et al 1980
46,XY,del(5)(q12)/47,XY,del(5)(q21),+21/49-50,XY, del(5)(q12),+9,+21,+2mar	Kardon et al 1982a
49,XX,+X,del(1)(p21?),+9,+mar	Mitelman 1983
56-64,XY,+1,-2,+6,+9,+13,+15,+19,+21,10-13mar	Streuli et al 1980

Erythrocytopenia

46,XX,t(9;22)(q34;q11)	Shiraishi et al 1980

Pancytopenia

46,XY,del(?9)(p13p22)	Mitelman 1983

Paroxysmal nocturnal hemoglobinuria

47,XY,+2/47,XY,+9	Cohen et al 1979

SPECIAL LEUKEMIAS

LYMPHOCYTIC LEUKEMIAS

Acute lymphocytic leukemia (ALL)

??,X?,t(9;22)(q34;q11)	Pittman et al 1979
??,X?,t(9;22)(q34;q11)	Pittman et al 1979
??,X?,t(9;22)(q34;q11)	Pittman et al 1979
28,XX,-1,-2,-3,-4,-5,-7,-8,-9,-11,-12,-13,-14,-15,-16,-17, -19,-20,-22	Kaneko and Sakurai 1980
43,X,-X,-7,-15,t(9;22)	Gibbs et al 1977
43,X,-X,-7,-8,t(9;22)(q34;q11)	Oshimura and Sandberg 1977
45,XY,t(9;22),-17,-20,+mar	Rausen et al 1977
46,X?,t(9;22)	Bloomfield et al 1977
46,X?,t(9;22)	Bloomfield et al 1977
46,X?,t(9;22)(q34;q11)	Philip et al 1976
46,XX,-9,13q-,14q-,+t(9;13)(p?;q?)	Prigogina et al 1979
46,XX,t(9;22)	Chessells et al 1979
46,XX,t(9;22)	Forman et al 1977
46,XX,t(9;22)	Olah et al 1981a
46,XX,t(9;22)	Olah et al 1981a
46,XX,t(9;22)	Lessard and Le Prise 1982
46,XX,t(9;22)(q34;q11),t(2;9)(q21;p13),del(9)(p13)	Morse et al 1982a
46,XX,t(9;22)(q34;q11)/46,XX,t(9;22),del(7)(q11)/45,XX, t(9;22),-7	Shabtai and Halbrecht 1981b
46,XY,9q+	Mitelman 1983
46,XY,del(3)(p?),t(9;22)(q34;q11)	Mitelman 1983
46,XY,del(3)(p?),t(9;22)(q34;q11)	
46,XY,del(5)(q12q23),del(9)(p21)	Abe et al 1979
46,XY,t(4;11)(q21;q23)	Oshimura et al 1977a
46-49,XY,t(4;11)(q21;q23),+6,+8,+13,+17,t(1;13)(p22;q12), t(7;9)(q11;q34)	
46,XY,t(8;9)(q22;p24)/45,XY,t(8;9),-10,+12,-21	Morse et al 1982a
46,XY,t(9;22)	Gibbs et al 1977
46,XY,t(9;22)	Chessells et al 1979
46,XY,t(9;22)	Chessells et al 1979
46,XY,t(9;22)	Chessells et al 1979
46,XY,t(9;22)(q34;q11)	Secker Walker and Hardy 1976
46,XY,t(9;22)(q34;q11)	
46,XY,t(9;22)(q34;q11)	Cheson et al 1980

Karyotype	Reference
46,XY,t(9;22)(q34;q11)	Nordenson 1983
46,XY,t(9;22)(q34;q11)	Mitelman 1983
46,XY,t(9;22)	Schmidt et al 1975
45,X,-Y,t(9;22)	
46,XY,t(9;22)	
48,XY,t(4;12)(q?;q?),t(6;15)(q?;q?),t(9;22)(q34;q11),11q-, +20,+21	Nordenson 1983
48-53,XX,t(4;11)(q21;q23),+6,+7,+9,+13,+21	Prigogina et al 1979
49,XX,del(4)(p15),t(4;11)(q21;q23),+5,+8,del(9)(p13), del(9)(p13),t(12;?)(q24;?),+mar	Morse et al 1982b
54,XX,7p+,+10,+10,+14,+14,+18,+18,+21,+21/27,X,-X,-1,-2,-3, -4,-5,-6,-7,7p+,-8,-9,-11,-12,-13,-15,-16,-17,-19,-20, -22	Oshimura et al 1977a
54-55,XY,+5,+8,+9,+14,+21,+mar	Prigogina et al 1979
55,XX,+2,+4,+9,+14,+15,+19,+21,+21	Oshimura et al 1977a

ALL, FAB type L1

Karyotype	Reference
26,XX,-1,-2,-3,-4,-5,-6,-7,-8,-9,-10,-11,-12,-13,-14,-15, -16,-17,-19,-20,-22	Hoeltge et al 1982
28,XX,-1,-2,-3,-4,-5,-6,-7,-8,-9,-11,-12,-13,-15,-16,-17, -19,-20,-22/56,XX,+X,+X,+10,+10,+14,+14,+18,+18,+21,+21	Brodeur et al 1981a
34-36,XY,-2,-3,-4,-7,-9,-12,-13,-14,-16,-17,-20,+22	Sandberg et al 1982c
46,XX,del(7)(p15),del(9)(p22)	Kaneko et al 1982e
46,XX,del(9)(p21)	Kaneko et al 1982e
46,XX,t(12;17)(p13;q12)/47,XX,t(12;17),+mar	Kaneko et al 1982e
46,XX,del(9)(p21),t(12;17)(p13;q12)	
46,XX,t(9;22)	Sandberg et al 1980
46,XY,del(9)(p22),del(20)(p12)	Kaneko et al 1982e
46,XY,t(9;22)	Priest et al 1980
46,XY,t(9;22)	Priest et al 1980
46,XY,t(9;22)	Sandberg et al 1980
47,XY,t(22;?),-1,-1,-2,-3,-9,+6mar	Sandberg et al 1980
52-56,XY,t(9;22),+4,+6,+9,+10,+14,+15,+20,+21,+22q-	Sandberg et al 1980
90,XXYY,-9,-9,+2mar	Kaneko et al 1982e

ALL, FAB type L2

Karyotype	Reference
26,XX,-1,-2,-3,-4,-5,-6,-7,-8,-9,-10,-11,-12,-13,-15,-16, -17,-18,-19,-20,-22	Brodeur et al 1981a
34-37,XX,cx,t(9;22)	Priest et al 1980
44,XX,-8,t(13;14),t(9;22)"c"	Takeuchi et al 1981
46,XX,-9,-10,t(9;18)(p24;q12),+2mar	Kaneko et al 1981b
46,XX,t(9;22)	Borgström et al 1981
46,XX,t(9;22)(q34;q11)	Mitelman 1983
46,XX,t(9;22),-20,+mar	Borgström et al 1981
46,XY,del(6)(q21orq23),del(9)(p22),del(11)(q13)/46,XY, del(6),i(6p),del(7)(p15),del(9)(p22),del(11)(q13)	Kaneko et al 1982e
46,XY,t(9;22)	Sandberg et al 1980
46,XY,t(9;22)	Sandberg et al 1980
46,XY,t(9;22)(q34;q11)	Roozendaal et al 1981
46,XY,t(9;22)(q34;q11)	
47,XY,+18	Kaneko et al 1982e
46,XY,-2,-5,-6,-7,+8,-9,-10,-11,-12,-16,+18,+t(2;?)(q35;?), +t(11;?)(q21;?),+5mar	

47,XY,t(9;22)(q34;q11),+del(7)(q11)/48,XY,t(9;22),+7, +del(7)	Mitelman 1983
47,XY,t(9;22)(q34;q11),+del(7)(q11)	
48,XX,t(9;22),+8,+18,/46,XX,t(9;22)	Sandberg et al 1980
48,XX,t(9;22),i(17q),+18,+22q-	
49,XY,+7,+12,-13,t(6;18)(p25;q21),t(11;14)(q23;q32), +t(9;?)(p24;?),+t(1;13)(q12;p13)	Kaneko et al 1982e
52,XX,t(9;22),+22q-,+1-5mar	Borgström et al 1981
52-54,XX,+6,+7,+9,+12,+3mar	Mitelman 1983
53,XY,t(9;22),+4,+5,+6,+7,+8,+17,+21	Sandberg et al 1980
56,XY,t(9;22)(q34;q11),+4,+5,+8,+9,+11,+16,+17,+18,+19,+20	Mitelman 1983

ALL, FAB type L3

47,XY,dup(1)(q21q32),del(6)(q23),t(8;14)(q23;q32),+9	Roos et al 1982

Chronic lymphocytic leukemia

44-48,XY,t(Y;9)(q12;q13),del(1)(p22p32),t(4;6)(p16;q15), del(6)(q15),t(12;14)(q15;q32),+mar	Robert et al 1982
45,XX,-8,-14,-15,-22,+2,+9,+1p+,+mar	Finan et al 1978
45-46,XY,-8,9q+,-10,13p+,-14,-15,-19,-20,+mar	Vahdati et al 1983a
46,XX,t(9;13)(p22;q12)	Nowell et al 1981
46,XY,9p+/43,X,-Y,-14,-22	Nordenson 1983
46,XY,t(11;14)(q13;q32)/46,XY,t(3;4;9)(p12;q21q31;q31)	Robert et al 1982
46,XY,t(1;7;9)(p36;q21;q12),t(12;14)(q13;q24)	Miyoshi et al 1979a
46,XY,t(9;11;14)(p22;q13;q32)	Nowell et al 1981
47,XY,+2,t(9;13),t(9;19),i(17q)	Finan et al 1978
47,XY,+3,+12,-20,/47,XY,+9,+12,-20/45,XY,-20/46,XY,+12,-20	Morita et al 1981

Prolymphocytic leukemia

44,XX,-2,-9,-10,11p+,+mar	Pittman et al 1982
44,XX,-7,-8,t(9;17)(q22;q25),-11	Pittman et al 1982
44,XY,-5,-8,-9,+mar	Pittman et al 1982

Lymphocytic leukemia, special type

45,Y,-X,t(Y;14)(q12;q32),-13,-17,del(2)(q33), t(3;13)(q21;q34),del(6)(q16orq22),9p+,9q+,10q+,16q+,18q+	Miyoshi et al 1981a
46,X,-Y,-1,1p+,del(9)(q32),12p+,14q+,+2mar	Ueshima et al 1981
46,XY,t(1;7)(p36;q22),del(9)(q12),t(12;14)(q13;q32)	Miyoshi et al 1981a

MONOCLONAL GAMMOPATHIES

Multiple myeloma

46,XX,t(9;22)(q34;q11)	Van Den Berghe et al 1979f
46,XY,+t(1;?)(p1?;?),+t(1;12)(q21;q24),del(6)(p11),-8,+9, -12,-13,-16,-19,22q+,+mar	Philip et al 1980
47,X,-X,inv(1)(p22q12),t(7;?)(q11;?),+9,t(11;?)(p15;?),-13, -14,t(14;?)(q11;?),+15,17q+	Philip et al 1980

48,X,-X,+1,+7,-8,+9,+21,del(1)(p36),4p+,del(5)(q22),9p+, 14q+	Liang et al 1979
52,X,-X,+t(1;?)(q12;?),+3,+7,+8,+9,+11,-13,+18,-21,+2mar	Philip et al 1980
52,XY,-1,+t(1;15)(q12;p1?),+t(1;16)(p36;p13), t(3;5)(q2?;q13),+3,+5,+7,+8,+9,+11,-15,+18,-20	Philip et al 1980
53,XY,+3,+5,-7,-8,+9,+11,-14,+20,del(1)(p34),del(6)(q25), +5mar	Liang et al 1979
55,XY,+1,+5,+9,+11,-12,+13,+15,del(1)(q32),del(6)(q25), t(9;19)(q13;q13),14q+,16p+,+4mar	Liang et al 1979
45,XY,-2,-2,-3,+5,-8,-10,+14,-15,-16,del(6)(q25),+4mar	
55,XY,+3,+5,+7,+9,+11,+14,+15,+21,+21/53,XY,+3,+5,-8,+9, +11,+14,+15,-19,+21,14q+,+3mar	Liang et al 1979

Plasma cell leukemia / Plasmocytoma

43,X,-X,-1,+t(1;8)(q11;q11),+t(1;9)(q21;p24), +t(1;10)(p31;q26),+t(1;?)(q12;?),-8,-9,-10,-13,-22	Ueshima et al 1983
44,X,-Y,+t(1;9)(q12q44;q34),t(6;8)(q13;p11),-9,-13	Ueshima et al 1983
45,XY,-5,del(6)(q11),t(6;7)(q?;q?),+9,-12,-13,-16,-17, t(9;22)(q34;q11),+2mar	Karpas et al 1982
78,XY,cx,del(1)(q21),t(1;17)(q12;p13),3q+,3p-,9p+,11p+, 14q+,16p+	Ueshima et al 1983
86,XX,cx,t(1;?)(q11;?),2q+,dup(7)(q22q36),t(8;?)(p11;?), t(9;13)(p23;q21)	Ueshima et al 1983

SOLID TUMORS

UNCLASSIFIED NEOPLASMS

EPITHELIAL NEOPLASMS

Carcinoma, NOS

Karyotype	Reference
44,XX,ins(1;1)(q32;q12q31),i(2q),+3,t(9;11)(q13;p15),-12,-15,17p+,-18	Atkin and Baker 1982a
44,XY,+5,-8,-11,-13,-13,-14,-14,-15,+19,-22,-22,t(1;11),5q-,9q+,+5mar	Reichmann et al 1981
44-50,XX,+3,dup(1)(q21q32),t(10;11)(q26;q25),t(2;17),t(6;13),t(9;15),del(10)(q24)	Riet-Fox et al 1979
45,XY,-4,-7,+8,-9,-12,+13,-17,-18,-19,-20,+21,+3mar	Reichmann et al 1981
46,XX,del(2)(p24),+5,+8,t(9;17)(p?;p?),del(16)(q21),-20,-21	Mark 1975b
46,XX,t(3;17)(p24;p13)/46,XX,t(5;17)(q22;p13)/46,XX,t(9;18)(q13;q21)/46,XX,t(12;15)(q13;q22)	Stenman et al 1982
46,XY,-4,-5,-11,-13,-17,+20,+21,1p+,9p-,+3mar	Reichmann et al 1981
49,X,-Y,-3,-5,-14,+15,-16,-17,+19,+21,+22,1q+,5q-,6q-,8p-,9p-,11p-,18p+,20q-,22q-,t(Y;14),+3mar	Reichmann et al 1981
50,XY,+X,+7,+8,+9	Reichmann et al 1981
52,XY,+X,+Y,-2,+7,+8,+9,+10,+13,-17,+i(1q)	Reichmann et al 1981
54,XY,+1,+5,+9,+17,+19,+20,+21,+21	Sonta and Sandberg 1978a
58,XX,+X,+1,+2,+3,5p-,+6,+7,+9,11q+,-13,-14,-15,+16,+19,+20,+6mar	Kovacs 1978a
64-68,XX,cx,t(Y;13),i(9p),i(9q),dmin	Reichmann et al 1981
66,XX,t(Y;13),1q+,i(4q),i(9p),i(9q),cx	Martin et al 1979
80-83,XX,cx,t(2;5;16),t(2;5;18),t(11;13),11q+,t(8;10),del(2)(q24),del(6)(q21),i(14q),dup(9)(q11q12)	Riet-Fox et al 1979
82,XX,cx,2q+,5q-,9q+,i(17q)	Reichmann et al 1981

Carcinoma NOS, metastatic

Karyotype	Reference
??,X,-X,t(1;?),hsr(1),6q-,t(9;?),dic(13),dmin,cx	Bartnitzke and Bullerdiek 1983
??,XX,t(1;9),t(2;?),cx	Bartnitzke and Bullerdiek 1983
41-42,XX,t(1;18)(p13orp21;q11),3p-,t(3;?)(p25;?),t(8;?)(p2?;?),t(9;?)(q34;?),cx	Bartnitzke and Bullerdiek 1983
45,X,Xq-,t(1;?)(p11;?),t(1;?)(q23;?),t(3;?),t(5;?)(q2?;?),del(9)(p2?),cx	Bartnitzke and Bullerdiek 1983
46,XX,t(9;22)(q34;q11)"s"	Togawa et al 1981
52,XX,+3,+5,+7,+8,+9,+11,12q+,13q-,-14,+16,i(17q)"s"	
46,XY,+2,-6,-9,del(22)(q11),+mar	Hansson and Korsgaard 1974
65,XX,i(1q),t(1;9),der(5),t(6;10)(p?;q?),t(14;?),der(21),cx	Bartnitzke and Bullerdiek 1983
79-80,XX,t(1;?)(q12;?),t(1;2),t(3;?),t(7;?)(q36;?),t(8;?),t(9;?)(q34;?),cx	Bartnitzke and Bullerdiek 1983
85-90,XY,i(9q),del(1)(q2?),t(7;?)(q35;?),t(3;?)(p11;?),t(21;?)(q11;?)	Kakati et al 1975

Small cell carcinoma

44,XY,cx,del(1)(q31),t(2;3)(p?;q?),del(3)(p14p23), t(9;13)(p11;q11),del(11)(p15),12q+,del(X)(q22)	Whang-Peng et al 1982b
67,XY,cx,inv(1)(q32p36),dup(1)(q32q44),del(3)(p14q13), del(3)(p14p23),t(7;13)(p11;q11),del(9)(q11),del(11)(p11), t(12;19)(q24;p13)	Whang-Peng et al 1982b

Papillary carcinoma

54,XY,+del(2)(q11),t(2;14)(q11;q32),t(3;11)(p13;p15), +i(5p),+8,+9,+12,+16,+17,+21,+2mar	Pathak et al 1982

Adenoma

46,X,t(X;4)(q25;p16),del(2)(p13p21),del(5)(q15q31),-7, t(9;12)(p13;q13),del(11)(q11q14), t(7;9;12)(p11p22q11;p13p24;q13)/45,X,-X	Mark et al 1980
46,XX,t(1;3)(p21;p21),t(1;13)(p36;q14),t(3;8)(p21;q12), t(1;5;20)(p21;p14q12;p13),t(5;8)(p14;q12), t(X;9)(q27;q12)	Mark et al 1982a
46,XX,t(3;5)(p21;p15),del(8)(p12q12),t(8;9)(q12;q34)	Mark et al 1980
46,XX,t(9;12)(p13p22;q13q15)/47,XX,+8,t(9;12)	Mark et al 1981a

Adenocarcinoma

44,XX,+5,-9,-13,-13,-15,-17,-18,+22,+2mar	Martin et al 1979
49,XY,+7,-9,+11,+12,+mar	Couturier-Turpin et al 1982
54,XX,+5,+6,+7,+9,+11,+21,+21,+del(1)(q32)	Sonta et al 1977
71,XX,cx,1q-,4q+,6q-,9q+,10q-,12q+,17q+,14q+	Wake et al 1981

Adenocarcinoma, metastatic

??,X,del(X)(q21),del(1)(q12),i(7q),del(9)(q31), del(16)(q23)	Ayraud et al 1977
??,XX,t(1;16),2q-,del(3)(p14),del(4)(q22),del(6)(q22), i(7q),del(9)(q31),t(13;13)	Ayraud et al 1977
40,XX,t(4;5;8),t(3;?)(q21;?),t(1;11)(q21;p11), t(9;11)(q11;p11),del(1)(p36)	Kakati et al 1975
47-48,X,Xq+,t(3;9)(q12;p11),t(3;3)(q29;p25),del(3)(p25), t(4;?)(q35;?),+8	Kakati et al 1975

Adenomatous polyp

48,XX,+8,+14/48,XX,-3,+8,+9,+14	Mitelman et al 1974b

MESENCHYMAL NEOPLASMS

Leiomyosarcoma

42,XX,del(1)(p12orp13),del(11)(q13orq14),+7,-9,-13,-14,-15, -18,+19,-22 — Mark 1976

Mesothelioma, malignant

41,XY,del(1)(p13),t(1;3)(p13;q11),t(7;7)(q11;q36), t(9;12)(p24;q13),del(12)(q13),del(13)(q12q14),-14,-15, t(17;?)(p13;?),i(19p),-20,-21,-22 — Mark 1978

43,XY,t(1;1)(q42;q32p22),t(2;6)(p25;p21),inv(3)(p24q13), t(5;18)(p15;q11),-6,+9,del(13)(q24q14),-14,-18,-22 — Mark 1978

MELANOMAS

Malignant melanoma

44,X?,+2,+9,-6,-8,-15,-21,9p-,19q+ — Wake et al 1981

46,XY,5q+,+7,-8,-9,-10,-13,-20,+21,+3mar — McCulloch et al 1976

78-82,XY,t(4;9)(q21;q21),del(9)(p13),t(14;14)(p11;q12), del(2)(q24),t(14;?)(q11;?),t(21;?)(q11;?),del(17)(q11) — Kakati et al 1977

Malignant melanoma, metastatic

41,XY,dup(1)(q21q32),t(1;7)(q25;q36),t(7;9)(q11;q11), del(9)(q13),-3,-4,-5,-8,-10,-11,-16,+18,-20,-22 — Kakati et al 1977

41-43,XY,del(1)(q21q25),i(1q),t(1;9)(p22;q34), t(11;12)(q11;p11),t(10;12)(q26;q13),t(4;?)(q35;?), del(5)(p13),t(14;?)(p12;?),i(17q),t(18;?)(p11;?), del(6)(q21) — Kakati et al 1977

45,XY,t(1;9;13)(q?;q?;q?),r(2),t(1;9)(p?;q?),i(8q), t(11;?)(q?;?),-10,-13,+mar — Chen and Shaw 1973

80,XY,t(11;12)(q23;q13),del(5)(q13),del(5)(p11), del(9)(q13),t(5;9),t(14;?)(q11;?),t(7;12)(q22;q24), del(1)(p13),del(11)(q23) — Kakati et al 1977

NEUROGENIC NEOPLASMS

Oligodendroglioma

45,XY,t(21;22)(p11;q11),7q-,9q-,14q+ — Yamada et al 1980

Neuroblastoma

45,XY,-3,del(1)(p13),t(1;9)(q11;p11),t(9;15)(p11;p11), hsr(16p) — Brodeur et al 1981b

47,XX,+7,9q+,22q+ — Brodeur et al 1977

Retinoblastoma

Karyotype	Reference
44,XY,1p+,-3,t(3;12)(q?;p?),+6q-,der(9),13p+,-14,-16	Kusnetsova et al 1982
46,XX,inv(1),inv(2),inv(3),inv(4),t(6;7),inv(9),t(11;?), 11p-,inv(12),i(17q),-20	Gardner et al 1982
46,XY,+4,t(5;13)(q35;q21),t(7;17)(p22;q12),t(1;9)(q22;p24), -10,del(16)(q11)	Gardner et al 1982

Meningioma, benign

Karyotype	Reference
38,X,-Y,-22,-21,-17,-14,-13,-10,-9,-4,-1p-,+mar	Zankl et al 1975a
41-42,XY,-1,-5,-8,9q+,-14,-18,-22,+r	Mark 1973b
43,XX,-1,-6,-7,-11,-22,+2mar/40-45,XX,-1,-6,-7,-9,-11,-19, -22,+2mar	Mark et al 1972b
44,XX,-8,-22/40-44,XX,-8,-9,-19,-20,-21,-22	Mark et al 1972a
44,XY,-9,-22,+del(22q)	Mark 1973a
45,XX,-22/40-44,X,-X,-8,-9,-15,-21,-22	Mark et al 1972b
45,XX,-22/45-46,XX,-22,-9,+6,+14,+del(22q)	Mark 1973a
46,XX,-22,+mar/44-45,XX,-8,-9,-22,+mar	Mark et al 1972b
47,XX,-22,+7,+9	Zankl et al 1975a
50,XX,+9,+10,+12,+15,+20,-22	Mark 1973c
52,X,-Y,-22,+20,+19,+17,+15,+11,+9,+7,+5	Zankl et al 1975a

EMBRYONAL AND MISCELLANEOUS NEOPLASMS

Blastoma, NOS

Karyotype	Reference
46,XY,t(4;?)(p14;?),t(9;11)(q22;p14),del(11)(p13p15), del(11)(q21q23)	Slater and Kraker 1982

LYMPHOMAS

1. UNCLASSIFIED AND MISCELLANEOUS LYMPHOMAS

Non-Hodgkin's lymphoma, NOS

Karyotype	Reference
44-47,XY,3q+,del(4)(p13),del(6)(p23),+7,del(9)(p13)	Kristoffersson 1983
46,X,-X,del(1)(q22),t(2;11)(p21;q21),t(9;15)(q11;p12), del(9)(q11),-14,t(14;17)(q23;q23),-16,+mar	Kristoffersson 1983
46-48,XX,t(1;?)(q21orq22;?),t(2;?)(q37;?),-8,+9,+12, 16q+"s"	Mark et al 1976
48-54,XX,+X,3p+,-4,-6,-8,-9,-10,+12,-15,+7-10mar	Pearson et al 1982
52,XY,+2,+3,t(5;9)(q22;q32),+6,+12,+14,+20	Reeves and Pickup 1980

Non-Hodgkin's lymphoma, lymphocytic, NOS

Karyotype	Reference
42-45,XX,-4,-5,-9,-16,+mar	Fleischmann et al 1976
45-46,XY,-1,3q+,del(9)(q22),+14,+17	Reeves 1973
46,XY,del(1)(p22),del(9)(q22),del(10)(p13)	Reeves 1973
46-47,XX,t(1;1)(q25;p36),dup(1)(q25q44),t(14;?)(q32;?),+9	Slavutsky et al 1981

Karyotype	Reference
47,XY,+7,-9,-11,1p+,6q-,10p-,14q-, t(14;?;14)(q24orq32;?;q13),15q-,22q-	Fukuhara et al 1979b
47,XY,t(3;4)(q29;p12),del(6)(p22),+7,del(9)(p13)	Kristoffersson et al 1981
47-48,XY,+7,-11,-14,-15,4q+,9p-,14q+	Fleischman and Prigogina 1977
48-49,XY,1q-,+3,+del(3)(p?q?),-4,-6,+7,-9,-17,-18,-19, +5mar	Fleischmann et al 1976
51,XX,+5,+9,+11,+del(11)(q?),+del(12)(q?),14q+,i(17q)	Prieto et al 1978a

Non-Hodgkin's lymphoma, histiocytic, NOS

Karyotype	Reference
46,XX,del(1)(p31p35),+3,del(9)(p11),t(10;?)(q26;?), del(11)(q13),-13,t(11;14)(q13;q32),t(17;21), t(9;22)(q11;p13),+r	Mark et al 1977
46,XY,1p-,6q-,t(11;14)(q21;q32)	Kakati et al 1980
48,XY,1p-,+3,6q-,+7,9p-,t(11;14)(q21;q32)	
47,XY,-6,-14,1q-,+1p-,2q+,3q+,9p+,10p-,10p+,12p-,18q+,22p+, t(2;14)(q37?;q13?)	Fukuhara et al 1979b
48,XY,del(3)(p21),del(9)(q22),11q+,12q+,del(6)(q23orq25), 15q+,del(12)(q15)	Reeves 1973
48,XY,t(3;8)(q22;q11),t(5;10)(q33;q22),del(7)(q21), t(4;9)(q11;p24),t(10;13)(q22;q21),r(12),del(19)(p13), del(22)(q11),+7	Mark et al 1978
50,XY,t(2;?)(p25;?),del(2)(p13),+i(3p),del(6)(p21),+7,+8, +9,+9,-11,-13,+14,-15,+20,+21,-22	Mark et al 1978
44,XY,del(2)(p13),t(6;?)(p21;?),del(11)(q13), t(11;14)(q13;q31),i(15q),-20,-22	

Non-Hodgkin's lymphoma, T-cell, NOS

Karyotype	Reference
47,XX,t(2;?),inv(3),4p-,del(5)(q14q32),6p+,6q+,t(7;7),9q+, 10p-,13q-,t(14;14),-15,-15,16q-,17p+,+21,+2mar	Gaeke et al 1981

Angioimmunoblastic lymphadenopathy

Karyotype	Reference
47,XY,+3/47,XY,+3,-9,+21/48,XY,+3,+8,+9,-20	Hossfeld et al 1976
47,XY,del(6)(q21q23orq25),-9,+2mar	Kaneko et al 1982c

2. HODGKIN'S LYMPHOMAS

Hodgkin's disease, NOS

Karyotype	Reference
52,XX,+1,-9,-15,+7mar	Fleischmann and Krizsa 1977

Hodgkin's disease, mixed cellularity

Karyotype	Reference
48,XY,+9,+12,t(18;?)(p11;?)	Hossfeld and Schmidt 1978
81,XX,14q+,i(18q),i(18p),del(6)(q21),t(9;15)(q11;p12), del(16)(q13)	Reeves 1973

Hodgkin's disease, lymphocytic depletion

Karyotype	Reference
45,X,t(X;1)(q28;q11),t(3;5)(q29;q33),del(8)(q22),-9, del(11)(p13),-17	Hossfeld and Schmidt 1978
60,XY,+del(1)(p21),t(1;5)(q44;q33),dup(5)(q12), t(1;5)(p21q23;q13),del(6)(q22),t(9;?)(q12;?), t(3;12)(q11;q24),t(14;?)(p11;?),i(17q)	Hossfeld and Schmidt 1978

3A. NON-HODGKIN'S LYMPHOMAS - RAPPAPORT CLASSIFICATION

Lymphocytic, poorly differentiated, NOS

Karyotype	Reference
46,XY,del(6)(q21),-8,-9,t(6;14)(q21;q32),+16,+20	Mark et al 1979

Lymphocytic, poorly differentiated, diffuse

Karyotype	Reference
43,XX,-4,-5,-9,i(17q),-18,-21,+2mar	Goh et al 1980
44,XX,t(11;14)(q13;q32),-3,-9,-13,1p+,+2mar	Fukuhara et al 1979a
47,XY,-1,+7,-9,+9q+,+9p+,+10,14q+,-17,t(17;18)(p11;q11)	Kristoffersson et al 1981
47,XY,t(14;14)(q32;q?),+7,-9,-11,1p+,6q-,10p-,15q-,22q-, +2mar	Fukuhara and Rowley 1978
47,XY,t(14;18)(q32;q21),-9,-10,+12,-16,8p-,13q+,+3mar	Fukuhara and Rowley 1978
50,XX,+3,+4,+10,+18/51,XX,+3,+4,+9,+10,+18	Mark et al 1979
88,XX,del(1)(p22),del(3)(q21),t(3;9)(q?;q?),del(6)(q21), t(22;?)(q13;?)	Mark et al 1979

Lymphocytic, poorly differentiated, nodular

Karyotype	Reference
92,XY,t(2;9)(q11;p11),t(5;?)(p15;?),t(17;?)(q25;?),17q+, del(1)(p22),del(3)(p21),del(X)(q24),del(11)(q23)	Reeves 1973

Mixed cell, diffuse

Karyotype	Reference
46,XY,-4,-9,del(6)(p23),+2mar	Kaneko et al 1982a

Mixed cell, nodular

Karyotype	Reference
50,XX,t(3;6)(q29;p21),del(9)(p13q22),+11,+12,+21	Mark et al 1976

Histiocytic, diffuse

Karyotype	Reference
45,X,-Y,t(14;15)(q32;q21orq22),-15,+18,-22,1q+,3p+,5p-,7p-, 9p+,12p+,16q+,17p-,+2mar	Fukuhara and Rowley 1978
45,X,del(X)(q24),del(1)(q21),+del(1)(p11),del(2)(q22), del(3)(q21),del(6)(p11),del(6)(q15),t(7;9)(p22;q13), t(8;10)(p23;q21),del(9)(q13),del(10)(q21), t(2;11)(q22;q23),-13,del(14)(q24),del(15)(q22), del(16)(q22),+17,-19	Mark et al 1979
46,X,-X,-2,-10,+11,-13,-14,-17,del(1)(p22p32), del(3)(p13p21),del(6)(q21),del(9)(p13),+t(2;?)(p23;?), +t(14;?)(q32;?),t(18;?)(q23;?),+t(X;13)(q26;q12),+mar	Kaneko et al 1982a
47,X,-X,t(4;14)(q?;q32),-18,1q+,2p+,6q-,7q+,9q-,12q+,+3mar	Fukuhara and Rowley 1978

47,XY,-11,-22,del(1)(q42),del(2)(q33),del(5)(q22q31), del(7)(q32),del(9)(p22),t(21;22)(q22;q11), +t(11;?)(q23;?),+t(21;22)(q22;q11),+mar — Kaneko et al 1982a

50,XX,+3,+3,-4,-6,-9,+12,-17,+18,+i(6p),+3mar — Kaneko et al 1982a

73,XX,cx,del(7)(q32),del(9)(q22),del(10)(p14), t(7;?)(p22;?),t(10;?)(p15;?),t(11;?)(q25;?), t(17;?)(q25;?),t(19;?)(q13;?) — Kaneko et al 1982a

82,XX,del(X)(q13),i(1q),i(1p),del(7)(p11),i(9p), del(14)(q22),t(1;14)(q23;q32),i(15q),i(17q) — Mark et al 1979

3B. NON-HODGKIN'S LYMPHOMAS - KIEL CLASSIFICATION

Lymphocytic, T-zone

44-46,X,+t(Y;14)(q24;q12),-3,del(3)(q21),+9,14q+,+mar — Gödde-Salz et al 1981a

Immunocytoma

49,XX,+3,del(6)(q14),+7,+18/49,XX,+3,del(6)(q14), del(9)(q22),+12,+18 — Kristoffersson 1983

Centroblastic-centrocytic, diffuse

48-50,XY,-1,+7,9q+,+9p+,+10,+14q+,-17,-18, +t(17;18)(p11;q11),-19,+20,+22,+mar — Kristoffersson 1983

48-51,XY,del(1)(p33q23q25),-2,+del(3)(p22),+4,4p+,+5,+6p+, +7,+8,+9,+21,+22 — Kristoffersson 1983

45-51,XY,del(1)(p33q23q25),del(2)(q14),+del(3)(p22),+4,+5, +7,+8,-12

Centroblastic-centrocytic, follicular

46,XY,t(1;?)(p36;?),t(1;1)(p36;q1?),del(7)(q11), del(7)(q22),del(9)(p11),t(14;?)(q32;?) — Reeves and Pickup 1980

46-47,XX,t(8;9)(q13;p24),t(14;18)(q32;q21),+22 — Kristoffersson 1983

Centroblastic, diffuse

46-49,XY,1p+,+2,-6,+del(9)(p21),+12,+13,14q+?,i(17q),+18 — Kristoffersson 1983

47,XY,14q+,+22

46-51,XX,+3,del(6)(q21),+del(6)(q21),+7,-8,del(9)(q11),+13, -16,-17,+19,-20,-21,+1-5mar — Kristoffersson 1983

Lymphoblastic, Burkitt's type

43,XY,+1,+2,+3,-7,-8,-9,-12,-16,-22,del(1)(q25), del(2)(p11),del(10)(q22),14q+ — Biggar et al 1981

45,X,-Y,t(1;14)(q22;q32),t(8;22)(q24;q11),3q+,6q-,6q+,7q+, 9p-,14q-,18q- — Berger et al 1981c

45-47,XY,t(14;?)(q32;?),del(3)(p?),-9,-22 — Philip et al 1977b

46,XY,t(2;8;9)(p11;q23;q21q31) — Philip et al 1981

46-47,XY,t(14;?)(q32;?),dup(1)(q21q32),t(2;3)(q13;q29),+5, 6p-,-8,del(9)(q?),+12,+18,+mar — Miyoshi et al 1981c

48,XX,+5,+7,-9,+10,+11,-16,del(5)(p13q31),del(10)(p11q25), del(11)(q12) — Biggar et al 1981

48-53,X?,t(8;14)(q24;q32),+X,+2,+3,+9,+10,+11,+12,+20 — Zech et al 1976a

Lymphoblastic, convoluted type

43-46,XY,del(1)(p32),del(2)(p21),-6,t(7;9)(p11;q11),-9,-10, t(11;15)(p11;q11),-11,del(12)(q14),-13,-14,i(17q),-17, +18,del(19)(p11),-22,+2mar — Kristoffersson 1983

46,X,-Y,t(1;?)(q21orq23;?),+mar/46,X,-Y,-1,-2,+der(1), +t(2;?)(q33orq35;?),+mar/48,X,-Y,-5,-14,+18,+18,+20, del(9)(p13),t(1;14),+dup(5)(q13q35) — Kaneko et al 1982d

46,XY,t(2;9)(q33;q34) — Kaneko et al 1982d

47,XY,-7,-9,del(6)(q23?),t(14;18)(p11;q11),+t(9;?)(q34;?), +2mar/47,XY,-7,-9,del(6),del(1)(q32),t(14;18),t(9;?), +t(1;?)(q23orq25;?) — Kaneko et al 1982d

Immunoblastic

45,XY,del(1)(p22orp23),del(3)(q11),del(3)(q21),del(8)(p11), del(9)(p11),del(9)(q12),t(12;?)(q24;?) — Reeves and Stathopoulos 1976

78-83,XY,+del(1)(q11),+5,+7,+12,+13q+,+14q+,+14q+,+20,+mar — Kristoffersson 1983

80-86,XY,del(1)(q11),+3,+7,+9,+12,+13q+,+14q+,+14q+,+15q+, +18,+20,+mar

3C. NON-HODGKIN'S LYMPHOMAS - LUKES AND COLLINS CLASSIFICATION

Mycosis fungoides

48,XX,del(1)(p22),t(2;8;14)(q37;q24;q24),+5,del(7)(p13), del(9)(q22),t(9;18)(q11;q24),del(10)(p13), del(13)(q12q14),del(5)(q15q22),del(5)(q13),del(5)(q15) — Fukuhara et al 1978a

46,XX,t(1;14)(q32;q32)

Sezary's syndrome

44-46,XX,t(2;9),t(2;13),17p+,19p+,-21,-22 — Edelson et al 1979

46,XY,-1,-10,+t(1;9),+t(10;11),9q-,11q- — Nowell et al 1982

47,XY,-9,-13,-17,-19,+del(9),+t(9;13),+t(9;19),+i(17q) — Nowell et al 1982

50,XX,1q+,2p-,+3q-,+6q-,+9,-15,+21,+mar — Nowell et al 1982

Follicular center cell, diffuse, large cleaved

45,XY,-6,-9,-17,2q-,3p-,3q-,4q-,5q-,6p+,7p+,7q+,+7q-,9p-, 12q+,13p+,20p-,+mar — Fukuhara et al 1978b

47,XY,-6,-14,1q-,+1p-,2q+,3q+,9p+,10p-,10p+,12p-,18q+,19p+, 19q+,+22p+,+mar — Fukuhara et al 1978b

Follicular center cell, diffuse, small non-cleaved

44,X,-Y,t(1;14)(p21;q13),5p+,del(6)(q13),del(8)(p21),9p+, 9p-,t(11;21)(q23;q22),13q+,-10,+12,-17p+,der(19),22q+ — Fukuhara et al 1979a

Immunoblastic type (B-cell)

48,XY,t(4;12)(q?;q?),t(6;15)(q?;q?),t(9;22)(q34;q11),11q-, +20,+21 — Nordenson et al 1983

3D. NON-HODGKIN'S LYMPHOMAS - WORKING FORMULATION

Small lymphocytic

46,XX,t(1;8),t(8;9) — Yunis et al 1982

Follicular, mixed small cleaved and large cell

79,XY,cx,t(1;1),9p-,t(14;18)(q32;q21) — Yunis et al 1982

Diffuse, small cleaved cell

49,XY,dup(1)(q?),8p-,-9,+14,+17,+20,+mar — Yunis et al 1982

Diffuse, mixed small and large cell

46,XY,inv(9)(p13q21) — Yunis et al 1982
47,XX,+9,t(9;15)(p13;q11),t(12;14)/49,XX,+3,+5,+9,t(9;15), t(12;14) — Yunis et al 1982

Diffuse, large cell

54,XY,dup(1)(q?),inv(5),+3,+3,+9,+13,+18,+18,+20,+mar — Yunis et al 1982

Large cell, immunoblastic

45,XY,ins(2;21),t(8;14)(q24;q32),t(9;19),t(17;21),-21 — Yunis et al 1982

Chromosome 10

HEMATOLOGICAL DISORDERS

UNCLASSIFIED LEUKEMIAS

NONLYMPHOCYTIC LEUKEMIAS

Acute nonlymphocytic leukemia (ANLL)

Karyotype	Reference
??,X?,+t(1;10)(q21;q22)	Anglani et al 1981
??,XY,+6,+10,+11,+13	Philip et al 1978a
43,XX,-1,3p+,3q-,4q+,-5,-5,-7,+9,+9,10p+,+10,+10,10p-,10q-, +11q-,-13,-15,-16,-17,-18,-21"s"	Whang-Peng et al 1979
44,XX,-9,-10,-17,del(14)(q22),t(9;10)(q22orq34;q11orq22)	Oshimura et al 1976b
46,XY,-16,del(10)(p11or12),+r	Oshimura et al 1976b
46,XY,t(9;10)(p24;p12),t(3;16),t(4;15)(p13;q14)"c"	Van Den Berghe et al 1979a
47,XX,del(1)(p22),+10"s"	Geraedts et al 1980a
47,XY,-8,10q+,+2mar	Bernard et al 1982a
48,XX,+8,t(9;22),+10	Chrz et al 1983
49,X,t(X;10)(p11;p11),+del(1)(p22),+del(1)(p22), del(5)(q23),+7/49,X,t(X;10)(p11;p11),+del(1)(p22),+7,+8	Mitelman 1983
50,XX,t(9;22)(q34;q11),+6,+8,+10,+del(9)(p13), +del(22)(q11)	Abe and Sandberg 1979
52,XX,1p+,+10,+13,+19,+21,+21,+22"c"	Berger et al 1973

ANLL, FAB type M1

Karyotype	Reference
45,XX,-10,t(14;19)(p?;p?)	Takeuchi et al 1981
45,XY,t(2;5)(q22orq23;q14q15),6p-,-7,t(6;7)(p21;p22q21), t(7;10)(p21;q21),t(14;19)(q11;q13)"s"	Hagemeijer et al 1981a
46,XY,t(10;12)(q?;p?)	Lessard and Le Prise 198
46,XY,t(3;5;10)(p?;q?;p?)	Lessard and Le Prise 198
46,XY,t(9;22)(q34;q11)	Sasaki et al 1983
46,XY,t(9;20;22)(q34;q12;q11)/46,XY,t(9;22)(q34;q11),-6,-7, -7,-8,10q+,+t(7;8),i(8q)	
47-48,XY,+8,t(9;22)(q34;q11),-20,+t(1;20)(q21;q13), t(10;21)(p11;q22),+22q-	Sasaki et al 1983

ANLL, FAB type M2

Karyotype	Reference
46,XX,t(4;10)(q31;q22),t(6;9)(p23;q34)	Kaneko et al 1982b
46,XY,t(8;21)(q22;q22)/45,X,-Y,t(8;21)/48,XY,t(8;21),+der(21),t(10;15)(p15;q22),+r	Sakurai et al 1982b
47,XX,t(11;17)(q23;q25),+der(17)/48,XX,t(11;17)(q23;q25),+der(17),+10	Mitelman 1983
48,XX,t(11;17)(q23;q25),+der(17),+10	

ANLL, FAB type M1+M2

Karyotype	Reference
45,XY,+t(1;10)(q23;q26),t(8;21)(q22;q22),-10/45,X,-Y,+t(1;10),t(8;21),-10	Fitzgerald et al 1983
45,XY,-8,-10,+21/47,XY,+21	Alimena et al 1977
46,XY,t(9;22),del(10)(q22)	Wayne et al 1979
47,XY,t(9;22),+8,del(10)(q22)/48-50,XY,t(9;22),+8,+9,del(10)(q22),+10,+22q-	
46,XY,t(9;22),t(7;10)(q34;q22)	Wayne et al 1979
47,XX,+10,t(11;17)(q23;q25)	Mitelman et al 1981
47,XX,+8,+10,-19	Mitelman et al 1981
48,XX,t(1;3)(q44;p14),del(3)(q21p14),-5,+10,+11,+21	Mitelman et al 1981

ANLL, FAB type M3

Karyotype	Reference
46,XX,2q-,6p-,7q-,10q+,14q+,-16,+mar	Hagemeijer et al 1981a
46,XX,t(15;17)(q26;q22)/47,XX,t(15;17),-7,+8,-9,-10,+2mar	Van Den Berghe et al 1979c
47,XX,+10,t(15;17)(q26;q22)	Brodeur et al 1983

ANLL, FAB type M4

Karyotype	Reference
42,XY,-5,-9,-10,-15,-16,-17,-20,-21,-22,+5mar	Testa et al 1979
43,XY,+1,-5,+6,-9,-10,-15,-16,-17,-20,-21,-22,+6mar	
44,XY,-6,-10	Larson et al 1982
44,XY,t(3;5)(q29;q13),t(7;12)(q12;q14),t(7;10)(q12;q21),-8,-17,-18,-20	Mitelman et al 1981
46,XX,-1,-3,-5,+10,+11,-13,-14,+21,+del(3?)(p?q?),?t(1;3;5)(p?;p?q?;q?)	Mitelman 1983
46,XX,t(5;10)(q35;q23)	Kaneko et al 1982b
46,XY,t(10;11)(p14;q13)	Brodeur et al 1983
46-47,XY,+10,-12	Mitelman et al 1981
47-48,XY,+8,+10,+21,+22	Mitelman et al 1981

ANLL, FAB type M5

Karyotype	Reference
46,XX,ins(10;11)(p11;q23q24)/52,XX,+4,+8,ins(10;11),+12,+16,+19,+20	Kaneko et al 1982b
46,XY,-10,-12,16q+,+2mar	Brodeur et al 1983
46,XY,del(11)(q13q23),t(10;11)(p25;q23),t(1;7)(p21;q36)	Yunis et al 1981
46,XY,t(2;11)(q37;q12),t(2;12)(q24;q24),10p+	Mitelman 1983
46,XY,t(2;11)(q37;q12),t(2;12)(q24;q24),10p+	
46,XY,t(6;11)(q26;q22)/52-53,XY,t(6;11)(q26;q22),+der(6),+3,+10,+13,+18,+19,+21,dmin	Hagemeijer et al 1981a

Karyotype	Reference
46,XY,t(6;11)(q27;q23)/51-52,XY,t(6;11),+der(6),+3,+19,+19, +21,dmin/53,XY,t(6;11),+der(6),+4,+10,+13,+18,+19,+21	Löwenberg et al 1982
47,XX,del(11)(q14),+t(10;11)(p?;q14)	Bernstein et al 1982b
50,XX,+6,+8,10p-,+12,+13	Nordenson 1983
51-52,XX,+3,+6,+9,-10,+18,+19,del(1)(p22),del(1)(p22), dup(11)(q11q21),+t(10;?)(p13;?)	Kaneko et al 1982b
57-58,XX,+1,+3,-4,-5,+6,-8,+11,+12,-17,+19,-20, del(10)(q24),+8-9mar	Testa et al 1979
60-61,XX,Xp+,+1,+2,-4,+6,-7,+19,+20,del(2)(q31), del(10)(q24),+12-13mar	

<u>ANLL, FAB type M5a</u>

Karyotype	Reference
46,XY,del(10)(p12),del(11)(q23),der(8),ins(10)(p12)	Berger et al 1982a
46,XY,t(1;9)(q11;q34),t(10;11)(p14;q14)	Berger et al 1982a
46,XY,t(6;10;11)(p22;p14;q14)	Berger et al 1982a
47,XX,+8,t(10;11)(p13;q14)	Berger et al 1982a

<u>ANLL, FAB type M4+M5</u>

Karyotype	Reference
47,XY,+8/48,XY,+8,+10/48,XY,+21,+22	Alimena et al 1977
66,XY,+1,+1,+2,+4,+6,+6,+7,+8,+8,+10,+11,+11,+12,+12,+18, +19,+21,+21,+22,+22	Li et al 1983

<u>ANLL, FAB type M6</u>

Karyotype	Reference
40-45,XY,-4,-6,-7,-10,-11,-20,+1-3mar	Mitelman et al 1981
44-45,XY,-2,-4,-7,-10,-13,-16,-17,-22,+3mar	Mitelman et al 1981
45-49,XY,-4,-7,-8,-10,+14,+16,-21,-22,+2mar	Mitelman et al 1981
47,XY,+Y,3p-,5q-,10q-	Garson 1980

<u>Chronic myeloid leukemia, t(9;22)</u>

Karyotype	Reference
26,XX,t(9;22)(q34;q11),-1,-2,-3,-4,-5,-6,-7,-9,-10,-11,-12, -13,-14,-15,-16,-17,-18,-19,-20	Hartley and McBeath 1981
35,XX,t(9;22)(q34;q11),-3,-4,-5,+8,-9,-10,-11,-12,-13,-14, -15,-16,-17/66-72,XX,t(9;22)	Pedersen and Boesen 1983
43-46,XX,t(9;22),-5,-7,-10,-17,-19,5p+,14q+,+r(?),+mar	Sadamori et al 1981b
46,X?,t(9;22),19q+/46,X?,t(9;22),4p+,10q+,19q+	Fleischman et al 1981
46,XX,t(9;22)(q34;q11)/51,XX,t(9;22),+6,+8,+10,+19,+22q-	Mitelman 1983
46,XX,t(9;22),-10,+12,-19,+22q-	Stoll and Oberling 1978
46,XY,t(9;22)"s"	Hossfeld 1975b
46,XY,t(9;22)/51,XY,t(9;22),+10,+14,+19,+22q-,+mar"s"	
51,XY,t(9;22),+10,+14,+19,+22q-,+mar"s"	
51,XY,t(9;22),+10,+14,+19,+22q-,+mar"s"	
46,XY,t(9;22)(q34;q11)/50,XY,t(9;22),t(1;15)(p36;q22),+8, +10,+19,+22q-	Misawa 1978
46,XY,t(9;22),10q-	Lilleyman et al 1978
46,XY,t(9;22),t(7;10)(q34;q22)	Sharp et al 1976
47,XY,t(9;22)(q34;q11),+8,i(17)(q11)/50,XY, t(9;22)(q34;q11),i(17)(q11),+4,+6,+8,+8/47,XY, t(9;22)(q34;q11),t(3;17)(q21;p1?),del(10)(q24), i(17)(q11),+mar	Hayata et al 1975b
47,XY,t(9;22)(q34;q11),t(3;5),-8,-10,+17,+22q-,+mar	Alimena et al 1982b

47,XY,t(9;22),+22q-/56,XY,t(9;22),+1,+2,+5,+6,+7,+8,+10, +12,+17,+22q-	Stoll and Oberling 1978
47,XY,t(9;22),-10,+17,+19,-22,+22q-	Stoll and Oberling 1978
48,XY,t(9;22),+22q-,+22q-/47,XY,t(9;22),+17/46,XY,t(9;22), -4,-7,+6,+17	Stoll and Oberling 1978
45,XY,t(9;22),-21/46,XY,t(9;22),-10,+17	
48,XY,t(9;22),-22,+17,+10,+22q-	Stoll and Oberling 1978
48,XY,t(9;22),-22,+17,+20,+22q-	
49,XX,t(9;22)(q34;q11),8q-,+10,+14,+22q-	Alimena et al 1982b
49,XX,t(9;22),+10,+21,del(22q)	Castleman et al 1973
49,XY,t(9;22),+10,+14,+17	Hossfeld 1975b
49,XY,t(9;22),+10,-17,+19,+del(22q),+mar	Hossfeld 1975b
49,XY,t(9;22),+22q-,+8,+10	Rozynkowa et al 1977
49,XY,t(9;22),+8,+i(17q),+22q-/50,XY,t(9;22),+8,+10, +i(17q),+22q-	Olah and Rak 1981
49,XY,t(9;22),+9,+10,+12/45,XY,t(5;17)(p11;q21)	Hagemeijer et al 1981b
50,XY,t(9;22),+del(22q),+6,+10,+19	Prigogina et al 1978
50-58,XY,+X,t(9;22)(q34;q11),+7,+8,+9,+10,+11,+12, +del(15)(q22),+16,+18,+19,+22q-	Kwan et al 1977
46,XY,t(9;22)	
52,XY,t(9;22),+1,+8,+10,+11,+21,+22q-	Sadamori et al 1980
53,XX,t(9;22)(q34;q11),+8,+10,+19,+20,+21,+22q-	Greenberg et al 1978
53,XY,+Y,t(9;22)(q34;q11),+3,+6,+8,+9,+10,+12,-15,+22q-/56, XY,+Y,t(9;22),+3,+6,+8,+8,+9,+10,+16,+19,+22q-	Stoll and Oberling 1982
56,XY,+Y,t(9;22),+3,+6,+8,+8,+9,+10,+12,+19,+22q-	
53,XY,t(9;22),+6,+9q+,+10,+12,+19,+21,+22q-/54,XY,t(9;22), +6,+8,+9q+,+10,+12,+19,+21,+22q-	Sonta and Sandberg 1977b
53-56,XY,+Y,t(9;22),+del(22q),+6,+10,+19,+8,+11,+21	Prigogina et al 1978
54,XY,t(9;22)(q34;q11),+8,+10,+11,+13,+14,+19,+21,+21	Hayata et al 1975b
55,XY,t(9;22),+3,+6,+8,+8,+9,+10,+16,+19,+22q-/53,XY, t(9;22),-15,+3,+6,+8,+8,+9,+10,+12,+22q-	Stoll and Oberling 1978
55,XY,t(9;22),+6,+10,+10,+11,+13,+14,+14,+19,+21	Seabright 1983

Chronic myeloid leukemia, aberrant translocation

??,X?,t(9;10;22)	Geraedts et al 1977
46,X?,t(10;22)(q26;q11)	Hossfeld 1983
46,XX,t(10;14;22)(q26;q24;q11)	Shabtai et al 1980
46,XX,t(2;9;22)(q23;q34;q11)	Pasquali et al 1979b
46,XX,t(2;9;22)/50,XX,t(2;9;22),+6,+10,+19,+der(22)	
46,XX,t(9;15;22)(q34;q11;q11)	Hays et al 1981
54-57,XX,t(9;15;22),+5,+7,+8,+10,+11,+12,+17,+22q-	
46,XY,t(10;22)(p15?;q11)/45,X,-Y,t(10;22)	Testa et al 1982
46,XY,t(10;22)(q26;q11)	Misawa 1978
46,XY,t(10;22)(q26;q11)	Fleischman et al 1981
46,XY,t(3;9;10;22)(p13;q34;q22;q11),17q+	Seabright 1983
46,XY,t(9;10;15;19;22)(q34;q22;q21;q13;q12)	Tomiyasu et al 1982
46,XY,t(9;10;22)(q34;q11;q11)	Hayata et al 1975b
47,XY,t(10;22)(q26;q11),+8,i(17q)	Prigogina et al 1978

Eosinophilic leukemia

47,XX,+10 Goldman et al 1975
49,XY,+10,+15,+19,del(3)(q?) Huang et al 1979

MYELOPROLIFERATIVE DISORDERS

Polycythemia vera

??,X?,+2,-10 Vykoupil et al 1980
??,X?,+2,-10 Vykoupil et al 1980
??,X?,+2,-8,-10 Vykoupil et al 1980
??,X?,t(9;22),+8,+10,+12,+16 Vykoupil et al 1980
44-45,XY,-10,-13,-16,-17,-20 Wurster-Hill et al 1976

Myelosclerosis / Myelofibrosis

44,XY,5p-,-7,-10,11q-,i(17q)/45,XY,-7,18q-/45,XY,2p-,2q-, 5p-,6q-,9q-,10q-,-11,+i(17q) Whang-Peng et al 1978
46,XY,-16,+18 Whang-Peng et al 1978
46,XY,7q-,8q-,10q-,11p-,11q+,12q-,-16,+18

DYSMYELOPOIETIC SYNDROMES

Preleukemia, NOS

43,X,-Y,-3,5q-,-7,-10,-16,-22,+3mar"s" Pedersen-Bjergaard et al 1981
45,XY,-2,-4,-6,-8,-10,-10,+17,+18,+3mar Panani et al 1980

Refractory anemia, NOS

46,XX,del(5)(q?),del(10)(p13)/47,XX,del(5),del(10),+mar Mitelman 1983

Dysmyelopoietic syndrome, special type

45,XX,t(13;14)(p13;q11),del(5)(q14)/50,XX,t(13;14),+1,+6, -7,+8,+10,del(5),+mar"c""s" Rowley et al 1981a

SPECIAL LEUKEMIAS

Hairy cell leukemia

44,X,del(X)(q22),del(6)(q23),-7,-8,-10,del(11)(p15q21),-21, 14q+,+r,+mar Sadamori and Sandberg 1983a

LYMPHOCYTIC LEUKEMIAS

Acute lymphocytic leukemia (ALL)

Karyotype	Reference
46,XX,-8,-10,del(6)(q23),+2mar	Secker-Walker et al 1979
46,XY,t(8;9)(q22;p24)/45,XY,t(8;9),-10,+12,-21	Morse et al 1982a
49,XX,+10,-21,+3r"c"	Stern et al 1979b
54,XX,7p+,+10,+10,+14,+14,+18,+18,+21,+21/27,X,-X,-1,-2,-3, -4,-5,-6,-7,7p+,-8,-9,-11,-12,-13,-15,-16,-17,-19,-20, -22	Oshimura et al 1977a
55,XX,+4,+5,+6,+10,+17,+21,+3mar	Humbert et al 1978
55,XY,+X,+3,+6,+10,+13,+14,+16,+21,+21	Oshimura et al 1977a
57,XY,+X,+5,+6,i(7q),+10,+12,+13,+15,+18,+21,+21,+22	Oshimura et al 1977a

ALL, FAB type L1

Karyotype	Reference
26,XX,-1,-2,-3,-4,-5,-6,-7,-8,-9,-10,-11,-12,-13,-14,-15, -16,-17,-19,-20,-22	Hoeltge et al 1982
28,XX,-1,-2,-3,-4,-5,-6,-7,-8,-9,-11,-12,-13,-15,-16,-17, -19,-20,-22/56,XX,+X,+X,+10,+10,+14,+14,+18,+18,+21,+21	Brodeur et al 1981a
47,XY,-7,-10,-12,+17,del(11)(p11),del(16)(q22), +t(10;?)(q22;?),+t(12;?)(q15;?),+mar	Kaneko et al 1982e
52-56,XY,t(9;22),+4,+6,+9,+10,+14,+15,+20,+21,+22q-	Sandberg et al 1980
56,XY,+X,+4,+6,+10,+15,+17,+18,+21,+21,+mar	Kaneko et al 1982e
58,XX,+4,+6,+7,+10,+14,+14,+21,+del(18)(q21),+4mar	Kaneko et al 1982e

ALL, FAB type L2

Karyotype	Reference
26,XX,-1,-2,-3,-4,-5,-6,-7,-8,-9,-10,-11,-12,-13,-15,-16, -17,-18,-19,-20,-22	Brodeur et al 1981a
46,XX,-9,-10,t(9;18)(p24;q12),+2mar	Kaneko et al 1981b
46,XY,-4,+6,-8,-10,-11,-15,-17,+22,del(6)(q21),+i(17q), +3mar	Kaneko et al 1981b
47,XY,+10	Mitelman 1983
47,XY,+18	Kaneko et al 1982e
46,XY,-2,-5,-6,-7,+8,-9,-10,-11,-12,-16,+18,+t(2;?)(q35;?), +t(11;?)(q21;?),+5mar	
47,XY,-10,-18,+3mar	Takeuchi et al 1981
56,XX,+1,+6,+10,+13,+15,+16,+17,+18,+20,+21	Mitelman 1983

Chronic lymphocytic leukemia

Karyotype	Reference
??,XY,1q-,t(1;10)(q11;p15),t(1;12)(q11;p12)	Miyamoto et al 1981b
41,XX,t(14;14)(q11;q32),t(1;13)(q44;q11),t(15;18)(q11;q23), 10q+,-16,-20	Kaiser McCaw et al 1975
45-46,XY,-8,9q+,-10,13p+,-14,-15,-19,-20,+mar	Vahdati et al 1983a
46,XY,t(1;10)(q22;q26)	Fleischman and Prigogina 1977
48,XX,+3,+18,t(7;10)(p?;q?)	Hurley et al 1980

Prolymphocytic leukemia

Karyotype	Reference
44,XX,-2,-9,-10,11p+,+mar	Pittman et al 1982
47,XY,-10,-10,+3mar	Diamond et al 1980

Lymphocytic leukemia, special type

Karyotype	Reference
45,Y,-X,t(Y;14)(q12;q32),-13,-17,del(2)(q33), t(3;13)(q21;q34),del(6)(q16orq22),9p+,9q+,10q+,16q+,18q+	Miyoshi et al 1981a
46,XY,-10,-13,5p+,+t(10;13)(p13;q11q32),21p+	Ueshima et al 1981
48,XX,+3,-6,+7,del(10)(p13q24),10q+,+mar	Ueshima et al 1981
49,XX,+7,+8,+12,8q+,8p+,del(10)(q23),21q+/49,XX,+8,+12, +del(7)(p13),8q+,8p+,10q-,21q+	Ueshima et al 1981

MONOCLONAL GAMMOPATHIES

Multiple myeloma

Karyotype	Reference
45,X,del(X)(q25),t(3;8;22)(q13;q34;q11),-5,-7,-10,-11, del(13)(q13?),+16,19q+,+3mar	Van Den Berghe et al 1979f
45,XY,del(1),+del(1),-10,-11,-11,+t(11;?),-13,-14,-20, +3mar	Philip et al 1980
47,XY,dup(1)(q11q32),t(1;10)(p22;p15),inv(6)(p21q13), t(11;14)(q23;q32),21p+,+mar	Liang et al 1979
55,XY,+1,+5,+9,+11,-12,+13,+15,del(1)(q32),del(6)(q25), t(9;19)(q13;q13),14q+,16p+,+4mar	Liang et al 1979
45,XY,-2,-2,-3,+5,-8,-10,+14,-15,-16,del(6)(q25),+4mar	

Plasma cell leukemia / Plasmocytoma

Karyotype	Reference
43,X,-X,-1,+t(1;8)(q11;q11),+t(1;9)(q21;p24), +t(1;10)(p31;q26),+t(1;?)(q12;?),-8,-9,-10,-13,-22	Ueshima et al 1983
44-45,XY,+10,-17,-21	Wetter et al 1980

SOLID TUMORS

UNCLASSIFIED NEOPLASMS

Malignant neoplasm, NOS

Karyotype	Reference
47,XX,-2,-5,+6,-10,11q+,-13,+18,-21,t(1;?)(q21;?), t(1;13)(q21;p11),t(1;6)(q11;q23)	Kovacs 1978a

EPITHELIAL NEOPLASMS

Epithelial tumor NOS, uncertain benign or malignant

Karyotype	Reference
47,XX,+10	Knoerr-Gaertner et al 1977
47,XX,+10	Knoerr-Gaertner et al 1977

Carcinoma, NOS

Karyotype	Reference
32-35,XX,cx,del(1)(p12),dup(1)(q21q32),2q+,5q+,7p+,11p+, hsr(10),t(11;12)(q13;q24),12p+,13q+	Kusyk et al 1982
43-49,XX,-1,-8,+10,-11,t(1;11)(q11;q12)	Kusyk et al 1982
44-50,XX,+3,dup(1)(q21q32),t(10;11)(q26;q25),t(2;17), t(6;13),t(9;15),del(10)(q24)	Riet-Fox et al 1979
46,XX,+2,-7,-10,+13,-16,-17,+20,+5q-,15p+,15p+,18p+	Reichmann et al 1981
47,XY,-5,+7,-10,-13,-14,4q-,5q-,+4mar	Reichmann et al 1981
48,XX,+del(1)(p32),-10,-21,+mar	Atkin and Baker 1982a
48,XY,+8,-17,-18,+19,+20,t(10;18),t(15;17)	Reichmann et al 1981
52,XY,+X,+Y,-2,+7,+8,+9,+10,+13,-17,+i(1q)	Reichmann et al 1981
55,XY,t(1;17)(p36;q21),+2,+2,+4,+8,+8,-10,+15,+18,+20,+21, +2mar	Sonta and Sandberg 1978a
58,XY,del(3)(p14),+4,+7,+8,+10,+11,+12,+14,+15,+17,+18,+19, +21	Kovacs 1978a
60-63,XX,cx,t(5;18),t(3;6),t(3;22),del(10)(q24), del(4)(p1?)	Riet-Fox et al 1979
63,XX,-1,+3,+5,+6,+7,+8,+10,+11,+12,-13,-15,+20,+21,+22, +9mar	Sandberg 1977
69-70,XX,cx,t(13;13),t(1;10),1p+,del(1)(p22)	Riet-Fox et al 1979
80-83,XX,cx,t(2;5;16),t(2;5;18),t(11;13),11q+,t(8;10), del(2)(q24),del(6)(q21),i(14q),dup(9)(q11q12)	Riet-Fox et al 1979

Carcinoma NOS, metastatic

Karyotype	Reference
35,XX,-1,-3,-4,-5,-7,-8,-10,-11,-12,-13,-14,-15,-16,-17, -18,-19,-20,-21,-22,+8mar	Pathak et al 1979
60,XX,t(1;?),t(7;?)(q3?;?),i(10q),der(11),t(14;21),cx	Bartnitzke and Bullerdiek 1983
64-100,XX,t(6;?)(p11q27;?),t(4;10)(q?;q?),t(3;?)(q25;?), del(1)(q2?),t(6;11)(p21;q13)	Kakati et al 1975
65,XX,i(1q),t(1;9),der(5),t(6;10)(p?;q?),t(14;?),der(21), cx	Bartnitzke and Bullerdiek 1983

Large cell carcinoma

60,XY,del(5)(p11),t(6;10)(q11;p11),t(7;11)(p22;q13), t(14;18)(q11;q11),del(2)(q21),t(6;?)(p11;?)	Pickthall 1976

Small cell carcinoma

40,XY,cx,t(1;16)(q21;q24),t(1;19)(q23;q13),del(3)(p21p24), del(3)(p23q26),del(10)(q22),del(11)(q23), t(5;13)(q13;q32),del(17)(p12),dmin	Whang-Peng et al 1982b

Papillary carcinoma

62,XX,cx,1p-,1q-,t(6;14)(q21;q24),10q-	Wake et al 1980
68,XX,cx,t(6;14)(q21;q24),del(1)(p?q?),1p+,4q-,8q-,10q+	Wake et al 1980

Adenocarcinoma

38,X,-X,-2,-3,-10,-14,-22,del(1)(q25),del(6)(q15), del(14)(p11),dmin	Trent and Salmon 1981
43-46,XY,i(10q),15q+,-17,+19,-21,-22,dmin	Couturier-Turpin et al 1982
44,XX,-1,+5,-15,-18,2q+/44,XX,-1,+5,-6,-15,-18,2q+,10p-, +mar	Martin et al 1979
44-50,X,-X,-1,t(1;?)(q21;?),del(2)(q33),4p+,-5,-7,+8, del(10)(q12),t(10;13)(q12;q34),del(17)(q25),-20	Kusyk et al 1982
47,XY,-10,-12,+17,+18,+20	Couturier-Turpin et al 1982
48-48,XX,+3,+7,+8,-1,-5,-14,-17,5q+,6q+,del(1)(q42), del(5)(q3?),del(10)(q24),del(14)(q2?)	Riet-Fox et al 1979
50,XY,2q+,+5,+10,+11,-16,-18,+20,+20,+mar	Couturier-Turpin et al 1982
71,XX,cx,1q-,4q+,6q-,9q+,10q-,12q+,17q+,14q+	Wake et al 1981
72,XX,cx,del(2)(p23),hsr(3)(q13q28),del(6)(q21), t(10;11)(p12;p13)	Trent and Salmon 1981

Adenocarcinoma, metastatic

40-45,XY,-3,-8,-18,+5,+10,11p+,14q+,del(4)(q32)	Ayraud 1975

MESENCHYMAL NEOPLASMS

MELANOMAS

Malignant melanoma

Karyotype	Reference
46,XY,5q+,+7,-8,-9,-10,-13,-20,+21,+3mar	McCulloch et al 1976

Malignant melanoma, metastatic

Karyotype	Reference
41,XY,dup(1)(q21q32),t(1;7)(q25;q36),t(7;9)(q11;q11), del(9)(q13),-3,-4,-5,-8,-10,-11,-16,+18,-20,-22	Kakati et al 1977
41-43,XY,del(1)(q21q25),i(1q),t(1;9)(p22;q34), t(11;12)(q11;p11),t(10;12)(q26;q13),t(4;?)(q35;?), del(5)(p13),t(14;?)(p12;?),i(17q),t(18;?)(p11;?), del(6)(q21)	Kakati et al 1977
45,XY,t(1;9;13)(q?;q?;q?),r(2),t(1;9)(p?;q?),i(8q), t(11;?)(q?;?),-10,-13,+mar	Chen and Shaw 1973

NEUROGENIC NEOPLASMS

Neuroblastoma

Karyotype	Reference
47,XY,1p+,4p+,6q+,10q+,12q+,+19	Brodeur et al 1977
82,XXY,cx,del(1)(p22),del(10)(q22),del(17)(p11)	Brodeur et al 1981b

Retinoblastoma

Karyotype	Reference
46,X,-Y,1p+,+1p-,5q+,7p+,10q-,16q-,17q+,21p+	Kusnetsova et al 1982
46,XY,+4,t(5;13)(q35;q21),t(7;17)(p22;q12),t(1;9)(q22;p24), -10,del(16)(q11)	Gardner et al 1982
46,XY,dup(1)(q25q44),t(7;10)(q36;q11),-8,-10, t(12;13)(q11;p11),dup(13)(q22q34),+18,+22	Gardner et al 1982

Meningioma, benign

Karyotype	Reference
38,X,-Y,-22,-21,-17,-14,-13,-10,-9,-4,-1p-,+mar	Zankl et al 1975a
41,X,-Y,-22,-18,-15,-10,1p-	Zankl et al 1975a
41,XY,+del(1)(p?),-5,-10,-17,-20,-22	Zang and Zankl 1983
44,Y,-X,-22,-14,+mar/43,Y,-X,-22,-14,-15,-11,-10,+3mar	Zankl et al 1975a
50,XX,+9,+10,+12,+15,+20,-22	Mark 1973c
51,XY,+5,6q+,+12,+16,+17,+18,+20,-22/52,XY,+5,+8,+10,+12, +17,+18,+20,-22	Zang and Zankl 1983

EMBRYONAL AND MISCELLANEOUS NEOPLASMS

Blastoma, NOS

40-47,X,-X,cx,7q+,+del(10)(p13),+del(12)(p11), del(11)(p11p14)	Slater and Kraker 1982

Teratoma, malignant

48,X,-Y,+1,del(5)(q12q32),+8,+10	Tricot et al 1983
67,XY,+1,+3,+4,+5,+6,+7,+7,+8,+10,+11,+13,+13,+14,+16,+17, +18,+20,+20,+21,+21	Sonta et al 1977

LYMPHOMAS

1. UNCLASSIFIED AND MISCELLANEOUS LYMPHOMAS

Non-Hodgkin's lymphoma, NOS

48-54,XX,+X,3p+,-4,-6,-8,-9,-10,+12,-15,+7-10mar	Pearson et al 1982

Non-Hodgkin's lymphoma, lymphocytic, NOS

46,XY,del(1)(p22),del(9)(q22),del(10)(p13)	Reeves 1973
47,XY,+7,-9,-11,1p+,6q-,10p-,14q-, t(14;?;14)(q24orq32;?;q13),15q-,22q-	Fukuhara et al 1979b

Non-Hodgkin's lymphoma, histiocytic, NOS

45,X,del(Y)(q12),t(4;10)(q12;q24),t(6;7)(q21;q32), del(8)(p21)	Mark et al 1978
45-47,XX,t(10;14)(q21;q32),+8,-18,-21	Mark et al 1977
46,XX,del(1)(p31p35),+3,del(9)(p11),t(10;?)(q26;?), del(11)(q13),-13,t(11;14)(q13;q32),t(17;21), t(9;22)(q11;p13),+r	Mark et al 1977
47,XY,-6,-14,1q-,+1p-,2q+,3q+,9p+,10p-,10p+,12p-,18q+,22p+, t(2;14)(q37?;q13?)	Fukuhara et al 1979b
48,XY,t(3;8)(q22;q11),t(5;10)(q33;q22),del(7)(q21), t(4;9)(q11;p24),t(10;13)(q22;q21),r(12),del(19)(p13), del(22)(q11),+7	Mark et al 1978
82,XXX,t(3;12)(q13;q15),t(3;8)(q22;q11),del(3)(q11), del(5)(q31),i(7q),del(10)(q24),t(10;10)(q24;q26), inv(11)(p15q21),del(12)(q13),del(13)(q14)	Mark et al 1978
89,XX,i(1q),del(3)(p15p23),del(10)(q24),del(12)(p12), i(17q)	Mark 1977

Non-Hodgkin's lymphoma, T-cell, NOS

47,XX,t(2;?),inv(3),4p-,del(5)(q14q32),6p+,6q+,t(7;7),9q+, 10p-,13q-,t(14;14),-15,-15,16q-,17p+,+21,+2mar	Gaeke et al 1981

Angioimmunoblastic lymphadenopathy

56,XY,+5,+5,+6,+10,+15,+15,+19,+20,+21,+22	Kaneko et al 1982c

2. HODGKIN'S LYMPHOMAS

Hodgkin's disease, mixed cellularity

61-80,XX,del(1)(q32),i(1q),del(2)(p13),del(6)(q13or15), del(6)(q23or25),15q+,12q+,del(5)(q31),del(7)(q11), del(10)(q26),del(3)(p13)	Reeves 1973

Hodgkin's disease, lymphocytic depletion

46,XY,t(10;14)(q22;q32),t(15;?)(p11;?),-16,+mar	Hossfeld and Schmidt 1978
48,X,-Y,t(1;?)(q44;?),i(1q),del(1)(q21),del(3)(q21),-10, t(3;11)(q23;q21),+14,t(14;?)(q32;?),+15,-17,+19,+20,-22	Hossfeld and Schmidt 1978

3A. NON-HODGKIN'S LYMPHOMAS - RAPPAPORT CLASSIFICATION

Lymphocytic, well differentiated, diffuse

48,XY,1q+,3q+,+i(3p),-6,-10,t(14;?)(q12;?),+16,17p-,+18, +mar	Miyamoto et al 1981a

Lymphocytic, poorly differentiated, diffuse

46,XY,t(14;18)(q32;q21),t(10;14)(q24;q32),-1,-17,-22,+3mar	Fukuhara and Rowley 1978
47,XY,-1,+7,-9,+9q+,+9p+,+10,14q+,-17,t(17;18)(p11;q11)	Kristoffersson et al 1981
47,XY,t(14;14)(q32;q?),+7,-9,-11,1p+,6q-,10p-,15q-,22q-, +2mar	Fukuhara and Rowley 1978
47,XY,t(14;18)(q32;q21),-9,-10,+12,-16,8p-,13q+,+3mar	Fukuhara and Rowley 1978
47,XY,t(1;4)(q23;q33),+3,t(10;15),t(11;14)(q13;q32),r(13), r(13)+	San Roman et al 1982
48,XX,1q-,+3,+4,6q-,+7,10q-,-15	Kakati et al 1980
50,XX,+3,+4,+10,+18/51,XX,+3,+4,+9,+10,+18	Mark et al 1979

Mixed cell, diffuse

53,XY,t(8;14)(q24;q32),t(14;18)(q32;q21),+20,1q-,+1p-,2q-, 4q+,5q-,5p+,6q-,+7q+,10q+,11q+,11q-,13p+,+4mar	Fukuhara and Rowley 1978

Histiocytic, diffuse

45,X,del(X)(q24),del(1)(q21),+del(1)(p11),del(2)(q22), del(3)(q21),del(6)(p11),del(6)(q15),t(7;9)(p22;q13), t(8;10)(p23;q21),del(9)(q13),del(10)(q21), t(2;11)(q22;q23),-13,del(14)(q24),del(15)(q22), del(16)(q22),+17,-19 — Mark et al 1979

46,X,-X,-2,-10,+11,-13,-14,-17,del(1)(p22p32), del(3)(p13p21),del(6)(q21),del(9)(p13),+t(2;?)(p23;?), +t(14;?)(q32;?),t(18;?)(q23;?),+t(X;13)(q26;q12),+mar — Kaneko et al 1982a

47,XY,-10,-13,+15,-16,+19,+20,del(6)(q21),del(8)(q22), t(2;14)(q21;q32),t(4;4)(q31;q35),+t(10;?)(p11;?) — Kaneko et al 1982a

49,XY,+del(2)(q31),+3,del(10)(q22),+17,+t(10;17)(q22;p13), -19 — Mark et al 1979

73,XX,cx,del(7)(q32),del(9)(q22),del(10)(p14), t(7;?)(p22;?),t(10;?)(p15;?),t(11;?)(q25;?), t(17;?)(q25;?),t(19;?)(q13;?) — Kaneko et al 1982a

3B. NON-HODGKIN'S LYMPHOMAS - KIEL CLASSIFICATION

Immunocytoma

47,XX,+3,+del(3)(q11),del(10)(q11),-21 — Kristoffersson 1983

46,XY,t(10;10)(p15;q24),dup(12)(q13q22) — Gahrton et al 1982

Centroblastic-centrocytic, diffuse

48-50,XY,-1,+7,9q+,+9p+,+10,+14q+,-17,-18, +t(17;18)(p11;q11),-19,+20,+22,+mar — Kristoffersson 1983

Centroblastic-centrocytic, follicular

46-47,XY,del(4)(q28),-10,+12,-18,del(18)(q22) — Kristoffersson 1983

52,XX,+X,+7,+8,del(10)(q24),+12,t(13;?)(q22;?), t(14;?)(q32;?),del(18)(q21),+19,+mar — Reeves and Pickup 1980

Lymphoblastic, Burkitt's type

??,XX,dup(1)(q12orq21q31),t(8;14),+10q- — Douglass et al 1980a

43,XY,+1,+2,+3,-7,-8,-9,-12,-16,-22,del(1)(q25), del(2)(p11),del(10)(q22),14q+ — Biggar et al 1981

45,X?,t(10;11),t(10;13) — Zech et al 1976a

48,XX,+5,+7,-9,+10,+11,-16,del(5)(p13q31),del(10)(p11q25), del(11)(q12) — Biggar et al 1981

48-53,X?,t(8;14)(q24;q32),+X,+2,+3,+9,+10,+11,+12,+20 — Zech et al 1976a

Lymphoblastic, convoluted type

43-46,XY,del(1)(p32),del(2)(p21),-6,t(7;9)(p11;q11),-9,-10, t(11;15)(p11;q11),-11,del(12)(q14),-13,-14,i(17q),-17, +18,del(19)(p11),-22,+2mar — Kristoffersson 1983

3C. NON-HODGKIN'S LYMPHOMAS - LUKES AND COLLINS CLASSIFICATION

Mycosis fungoides

48,XX,del(1)(p22),t(2;8;14)(q37;q24;q24),+5,del(7)(p13), del(9)(q22),t(9;18)(q11;q24),del(10)(p13), del(13)(q12q14),del(5)(q15q22),del(5)(q13),del(5)(q15) — Fukuhara et al 1978a
46,XX,t(1;14)(q32;q32)

Sezary's syndrome

46,XX,inv(10)(p23q32) — Kristoffersson 1983
46,XY,-1,-10,+t(1;9),+t(10;11),9q-,11q- — Nowell et al 1982
50,XY,-10,-16,+6mar — Nowell et al 1982

Follicular center cell, diffuse, large cleaved

47,XY,-6,-14,1q-,+1p-,2q+,3q+,9p+,10p-,10p+,12p-,18q+,19p+, 19q+,+22p+,+mar — Fukuhara et al 1978b

Follicular center cell, diffuse, small non-cleaved

44,X,-Y,t(1;14)(p21;q13),5p+,del(6)(q13),del(8)(p21),9p+, 9p-,t(11;21)(q23;q22),13q+,-10,+12,-17p+,der(19),22q+ — Fukuhara et al 1979a

3D. NON-HODGKIN'S LYMPHOMAS - WORKING FORMULATION

Small lymphocytic

46,XY,t(10;18) — Yunis et al 1982

Follicular, small cleaved cell

46,XY,t(1;2),t(4;10),t(14;18)(q32;q21),t(1;22) — Yunis et al 1982

Follicular, mixed small cleaved and large cell

50,Y,inv(X),+3,+7,+10,t(12;12),+12,t(14;18)(q32;q21),+18, 18q-,-21 — Yunis et al 1982

Follicular, large cell

50,XY,1q-,t(2;3),+3,+10,t(1;12),13q-,+15,-19,t(3;19),+21 — Yunis et al 1982

Chromosome 11

HEMATOLOGICAL DISORDERS

UNCLASSIFIED LEUKEMIAS

Acute leukemia, NOS

Karyotype	Reference
41-44,XX,5q-,i(8q),11q+,+mar	Prigogina et al 1979
47,XX,+19/46,XX,-11,+mar	Hartley and Sainsbury 1981

NONLYMPHOCYTIC LEUKEMIAS

Acute nonlymphocytic leukemia (ANLL)

Karyotype	Reference
??,XY,+6,+10,+11,+13	Philip et al 1978a
43,X,-Y,t(1;8)(p22;q24),-1,-11,-11,-13,-16,-17,-22, t(1;22)(q11;p11),t(11;13)(q23q25;q12q14),t(17;?)(p11;?), +mar	Oshimura et al 1976b
43,XX,-1,3p+,3q-,4q+,-5,-5,-7,+9,+9,10p+,+10,+10,10p-,10q-, +11q-,-13,-15,-16,-17,-18,-21"s"	Whang-Peng et al 1979
43,XX,t(1;20),-5,-6,7q-,11q-,-12,17q+,-18,19q+,-21	Mitelman 1983
44,XX,t(8;11)(p23;p13 or p15),-20/46,XX,del(7)(q11), t(1;3)(q2?;q2?),-20,-21,-22,+3mar	Philip et al 1978a
44,XY,del(5)(q21),del(7)(q11),del(11)(q11), t(7;11;12)(q11;q11;p13),del(13)(q11),-20,-21/45,XY, t(3;12)(q11;p13),del(5)(q21),del(7)(q11),del(13)(q11), -20	Mitelman 1983
45,X,-Y,t(6;11)(p2?;q13),t(8;11)(q22;q13),del(8)(q22)	Philip et al 1978a
46,XX,del(1)(p32),del(5)(q13),+8,t(9;?)(p11;?), t(11;22)(q11;q13),-16,-17,-19,+3mar	Muir et al 1977
46,XX,del(11)(q23)	Hustinx and Rutten 1983
46,XX,del(11)(q23)	Bernard et al 1982a
46,XX,del(5)(q12q31),del(11)(q23)	Sadamori et al 1981a
46,XX,t(11;?)(q25;?)"s"	Mitelman 1983
46,XX,t(3;11)(q12;p13)"s"	Geraedts et al 1980a
46,XY,-11,+16"s"	Berger et al 1981a
46,XY,1p+,12p-,?t(1;12),?11p+,inv(12)(p?),-2,+mar	Mitelman 1983
46,XY,del(11)(q23)	Muir et al 1977
46,XY,dup(1)(q25q44),t(11;17)(p15;q21)/46,XY, t(11;17)(p15;q21)/47,XY,t(11;17)(p15;q21),+del(1)(p11)	Mitelman 1983

Karyotype	Reference
46,XY,dup(1)(q25q44),t(11;17)(p15;q21)/46,XY,t(11;17)(p15;q21)	
46,XY,dup(1),t(11;17)/46,XY,t(11;17)/47,XY,t(11;17),+del(1)	
46,XY,inv(9),t(11;15)(q11;q26),t(11;13)(q11;q34)"c"	Philip et al 1978a
46,XY,t(11;22)(q13;p11)"s"	Oshimura et al 1976b
46,Y,t(X;11)(q22.3;q23.3),ins(X;11)(q22.3;p13p15),dup(1)(q11q21),t(4;6)(p16;p21),t(7;12)(p13;q15)	Nacheva et al 1982
47,X,Xq-,t(3;5),-5,t(11;?),t(12;17),-18,-21,+6mar	Pedersen-Bjergaard et al 1980
50,XX,4q-,-11,+5mar	Bernard et al 1982a
57,XY,+6,+11,+17,+20,+21"s"	Berger et al 1981a

<u>ANLL, FAB type M1</u>

Karyotype	Reference
43,XX,del(5)(q12q21),-7,11p+,-15,-17	Kerkhofs et al 1982
44,XX,-5,-7,-20,-22,+2mar/44,XX,-3,-11,-11p+,-22	Prieto et al 1981
45,X,-X,-5,-7,-11,-17,+21,+3mar"s"	Zaccaria et al 1983b
46,XX,-1,inv(1),3q+,5q-,t(13;17),-18,+mar/51,XX,-1,inv(1),3q+,5q-,t(13;17),-18,+6,+9,+9,+11,+11,+mar	Hagemeijer et al 1981a
46,XX,t(9;11)(q34;q13)	Alimena 1983
46,XY,?del(11)(q?)"s"	Mitelman 1983
47,XX,+11	Yunis et al 1981
47,XX,+11"s"	Mitelman 1983
47,XX,+11"s"	
48,XY,+8,+8,t(11;19)(q23;p12orp13)	Hagemeijer et al 1981a
58,XY,+1,+4,+5,+6,+8,+11,+13,+13,+14,+14,+15,+21,+21,t(9;22)(q34;q11)	Alimena 1983

<u>ANLL, FAB type M2</u>

Karyotype	Reference
45,X,-Y,del(8)(q22)/45,X,-Y,del(8)(q22),del(11)(q21)/46,XY,del(8)(q22)	Sakurai et al 1982b
45,X,-Y,t(8;21)(q22;q22)/46,X,-Y,t(8;21)(q22;q22),+t(11;12)(p11;q13)	Sakurai et al 1982b
46,XX,1q+,2q+,-3,-11,+r,+mar	Berger et al 1981a
46,XX,t(11;19)(q23;p13orq13)	Kaneko et al 1982b
46,XX,t(9;11)(p22;q24)	Yunis et al 1981
47,XX,t(11;17)(q23;q25),+der(17)/48,XX,t(11;17)(q23;q25),+der(17),+10	Mitelman 1983
48,XX,t(11;17)(q23;q25),+der(17),+10	
47,XX,t(8;11;21),+8	Golomb et al 1978c
48,XX,-5,t(8;?)(p23;?),del(11)(q23),t(15;?)(p11;?),-21,+4mar"s"	Rowley et al 1977a
53,XX,+1,-4,-5,+6,-7,+8,+11,+12,+21,+21,+3mar"s"	Zaccaria et al 1983b

<u>ANLL, FAB type M1+M2</u>

Karyotype	Reference
37-45,XY,-7,-12,-13,-14,-16,-18,-19,-21,del(4)(q29),del(5)(q24),der(9),der(17),11q+,12p+,t(14;?)	Fitzgerald et al 1983
42-43,XX,cx,t(11;15)(q11;q11),t(9;?)(p?;?),-15,-17,-18,-20,-21	Fitzgerald et al 1983
45,X,-Y,+del(5)(q11),del(11)(q21),-14,-16,-18,-20,+2mar,+r	Mitelman et al 1981
45,XX,-5,-11,-13,-18,-21,+4mar,dmin"s"	Sandberg et al 1982a
45,XY,-7,t(11;21)(q25;q11)	Fitzgerald et al 1983

45,XY,-7/47,XY,+11 Li et al 1983
45,XY,t(5;11),t(5;17) Garson 1980
45-46,XX,4q+,t(5;11)(q35;p13) Mitelman et al 1981
45-46,XX,t(5;11),4q+,-21 Alimena et al 1977
45-46,XY,t(2;6)(q37;q23),-7,del(11)(q21),del(22)(q11) Mitelman et al 1981
47,XX,+19,t(1;11) Benedict et al 1979
46,XX,t(11;12)(p14;q13),del(12)(q13)"s" Prigogina et al 1979
46,XY,del(11)(q?) Benedict et al 1979
46,XY,del(11)(q23),del(12)(p11) Yamada and Furusawa 1976
46,XY,del(11)(q?) Garson 1980
46,XY,del(7)(q31),del(11)(q14),del(12)(p12) Mitelman et al 1981
46,XY,t(1;11)(q21;p15),del(1)(p22)"s" Sandberg et al 1982a
46,XY,t(1;6;11)(q12;q23;p15) Yamada and Furusawa 1976
47,X,-X,+del(1)(p11),del(5)(q13),t(5;12)(q21;q24), t(11;12)(q25;q13),t(13;14)(q11;p11),+16,+21 Mitelman et al 1981
47,XX,+10,t(11;17)(q23;q25) Mitelman et al 1981
47,XY,+11 Fitzgerald et al 1983
47,XY,+11 Li et al 1983
47,XY,+8,r(11),t(8;16)(q?;q?) Garson 1980
48,XX,+1,del(5)(q13orq15),+11"s" Rowley and Potter 1976
48,XX,-5,del(5)(q13q31),+8,+t(11;?)(p14;?),-17,-19,+21, +2mar,dmin"s" Testa et al 1981b
48,XX,t(1;3)(q44;p14),del(3)(q21p14),-5,+10,+11,+21 Mitelman et al 1981
48,XY,+11,+21 Fitzgerald et al 1983
57,XY,+6,+11,+13,+15,+21,+21,+21,+21,+21,+21,+22"c" Shiraishi et al 1982

ANLL, FAB type M3

46,XY,t(15;17)(q?;q?),t(3;11)(q?;q?),t(7;19)(q11;p13) Trujillo and Cork 1981
47,XY,del(11)(q23),t(15;17)(q24?;q22?),+mar Bernard et al 1982b

ANLL, FAB type M4

43,XY,-2,-3,-6,-7,-12,-15,-21,-22,del(2)(p11),del(11)(p14), t(14;?)(q32;?),t(17;?)(p13;?),del(22)(q11),+5mar"s" Rowley et al 1981a
45,X,-X,t(8;17;21)(q22;q23;q22) Testa et al 1979
45,X,-X,t(8;17;21)(q22;q23;q22)/45,X,-X, t(1;11)(p36q32;q13),t(8;17;21)
46,XX,-1,-3,-5,+10,+11,-13,-14,+21,+del(3?)(p?q?), ?t(1;3;5)(p?;p?q?;q?) Mitelman 1983
46,XX,inv(11)(p12q23),t(11;13)(p12q23;q21) Hagemeijer et al 1981a
46,XX,t(11;?)(q23;?) Rowley and Potter 1976
46,XX,t(2;11;12;17)(q31;p11;p13;q23) Fitzgerald et al 1983
46,XX,t(X;11)(q25;q23) Prigogina et al 1979
46,XY,del(11)(q23) Fitzgerald et al 1983
46,XY,t(10;11)(p14;q13) Brodeur et al 1983
46,XY,t(11;17)(q23;q27),del(11)(q13q23),t(15;19)(q21;p13) Yunis et al 1981
46,XY,t(11;19)(q23;q13) Prigogina et al 1979
46,XY,t(6;11)(q27;q23) Yunis et al 1981
47,XY,+11 Hagemeijer et al 1981a
47,XY,+8/48,XY,+8,+9/49,XY,+8,+9,+11 Hagemeijer et al 1981a

<u>ANLL, FAB type M5</u>

45,XY,-5,11p+ Brodeur et al 1983
46,XX,del(11)(q22) Alimena 1983
46,XX,ins(10;11)(p11;q23q24)/52,XX,+4,+8,ins(10;11),+12,
+16,+19,+20 Kaneko et al 1982b
46,XX,t(11;17)(q23;q23) Berger et al 1981f
46,XX,t(11;17)(q23;q21) Zaccaria et al 1982
46,XX,t(9;11)(p21;q23) Hagemeijer et al 1982b
46,XX,t(9;11)(p21;q23) Brodeur et al 1983
46,XY,del(11)(q13q23),t(10;11)(p25;q23),t(1;7)(p21;q36) Yunis et al 1981
46,XY,del(11)(q22) Brodeur et al 1983
46,XY,t(17;21)(q11;q11orq21),+i(8q),del(5)(q11q13),9p+,
11q+ Fitzgerald et al 1983
46,XY,t(2;11)(q37;q12),t(2;12)(q24;q24),10p+ Mitelman 1983
46,XY,t(2;11)(q37;q12),t(2;12)(q24;q24),10p+
46,XY,t(6;11)(q26;q22)/52-53,XY,t(6;11)(q26;q22),+der(6),
+3,+10,+13,+18,+19,+21,dmin Hagemeijer et al 1981a
46,XY,t(6;11)(q27;q23)/51-52,XY,t(6;11),+der(6),+3,+19,+19,
+21,dmin/53,XY,t(6;11),+der(6),+4,+10,+13,+18,+19,+21 Löwenberg et al 1982
46,XY,t(9;11)(p21;q23) Hagemeijer et al 1982b
46,XY,t(9;11)(p21;q23) Hagemeijer et al 1981a
46,XY,t(9;11)(p21;q23) Brodeur et al 1983
46,XY,t(9;11)(p22;q13) Michalski et al 1983
46,Y,-X,-1,-7,-8,-14,-15,-17,-20,inv(6)(p23q13),
del(11)(q23),+t(X;?)(p11;?),+t(1;?)(p24;?),
+t(12;?)(q22;?),+t(17;?)(p11;?),+4mar Kaneko et al 1982b
47,XX,del(11)(q14),+t(10;11)(p?;q14) Bernstein et al 1982b
47,XY,+21/48,XY,+11,+21 Brodeur et al 1983
47,XYY,del(11)(q23)"c" Alimena 1983
49,XX,+4,+8,+8,t(9;11)(p22;q23) Kaneko et al 1982b
49-50,XY,-21,11q+,19q+,+3-4mar"s" Prigogina et al 1979
50,XX,t(9;11)(p21;q23),+4mar Hagemeijer et al 1982b
51,XX,+6,+8,+14,+19,+21,t(11;?)(q21orq24;?) Soukup and Neely 1981
51-52,XX,+3,+6,+9,-10,+18,+19,del(1)(p22),del(1)(p22),
dup(11)(q11q21),+t(10;?)(p13;?) Kaneko et al 1982b
57-58,XX,+1,+3,-4,-5,+6,-8,+11,+12,-17,+19,-20,
del(10)(q24),+8-9mar Testa et al 1979
60-61,XX,Xp+,+1,+2,-4,+6,-7,+19,+20,del(2)(q31),
del(10)(q24),+12-13mar

<u>ANLL, FAB type M5a</u>

44,Y,-X,-5,del(11)(q22),17p+,+r Berger et al 1982a
46,XX,t(11;11)(q22;q25) Berger et al 1982a
46,XX,t(11;17)(q23;q23) Berger et al 1982a
46,XY,del(10)(p12),del(11)(q23),der(8),ins(10)(p12) Berger et al 1982a
46,XY,del(11)(q24) Berger et al 1982a
46,XY,t(1;9)(q11;q34),t(10;11)(p14;q14) Berger et al 1982a
46,XY,t(6;10;11)(p22;p14;q14) Berger et al 1982a
46,XY,t(8;13)(q24;q12),del(11)(q23),t(11;13)(p12;q12),13p+ Berger et al 1982a
47,XX,+8,t(10;11)(p13;q14) Berger et al 1982a
47,XX,+8,t(9;11)(p22;q24) Dewald et al 1983
47,XY,+6,t(11;19)(q24;p13) Berger et al 1982a

47,XY,+8,t(9;11)(p22;q24)	Dewald et al 1983
48,XX,t(7;8)(p11;q11),+i(8q),+i(8q),del(11)(q24)	Berger et al 1982a
48,XY,+4,+8,t(11;17)(q24;q21)"s"	Dewald et al 1983

ANLL, FAB type M5b

46,XX,t(9;11)(p22;q24)	Dewald et al 1983
47,XXX,t(1;11)(q21;q24)"c"	Berger et al 1982a

ANLL, FAB type M4+M5

45,XY,del(11)(p11),t(12;?)(p13;?),del(16)(q13), t(13;14)"c""s"	Nora et al 1979
66,XY,+1,+1,+2,+4,+6,+6,+7,+8,+8,+10,+11,+11,+12,+12,+18, +19,+21,+21,+22,+22	Li et al 1983

ANLL, FAB type M6

??,XY,t(4;?)(q23;?),t(5;17),t(8;9),+8,+11, t(12;16)(p12;p12)	Hustinx and Rutten 1983
40-45,XY,-4,-6,-7,-10,-11,-20,+1-3mar	Mitelman et al 1981
43,X,-X,del(5)(q14q34),-7,11q-,-12,12p+,t(12;17),t(20;21)	Kerkhofs et al 1982
43,XY,-2,4q-,-9,11q-,-15,-19,-20,+2mar/46,XY,-15,-20,+2mar	Li et al 1983
43-45,XX,der(1),inv(2)(p25q14),del(3)(q25),-4,-5,-6,6q-,-7, 7p+,11p+,-12,-17,-21,+r,+1-4mar"s"	Smadja et al 1983b
44,X,del(Y)(q?),+Xq-,3p-,3q-,4q-,-6,+7q-,11q+,-15,-18"s"	Bradley et al 1982
44,XX,del(5)(q13q31orq33),del(11)(q21),-17,-21"c""s"	Papa et al 1979
44,XY,t(3;5;11),17p+,-22	Garson 1980
46,XX,-7,-16,-21,+t(2;7)(p11;q11),del(2)(q33),del(11)(q22), +11p+,14p+,+t(16;21;21;21)(p13;q22;q11;q22)"s"	Rowley et al 1981a
46-47,XX,del(19)(p?q?),11q+,+21"s"	Whang-Peng et al 1977
48,XX,+8,+9,+9,-11,del(4)(q23)"s"	Weinfeld et al 1977a
72,XY,cx,t(2;15)(q31;q15),del(6)(p22q15),t(8;13)(q23?;q13), t(11;14)(p12;q13),i(16p),t(15;18)(q15;p11)	Douglass and Freeman 1983

Chronic myeloid leukemia, t(9;22)

44,XY,t(9;22),-17,-20	Stoll and Oberling 1978
45,XY,t(9;22),-8,-11,-18,+2mar	
47,XY,t(9;22),+7,i(17q)	
45,X,-Y,t(9;22)/45,X,-Y,t(9;22),t(1;11)(p?;p?)"s"	Hagemeijer et al 1980b
45,XX,t(9;22),-11	Chrz et al 1983
45,XY,t(9;22),+del(1)(q12),-3,-7,del(11)(q21),-19,+mar	Lyall and Garson 1978
45,XY,t(9;22),-20,-22,+22q-/46,XY,t(9;22),-20,-22,+8,+22q-	Stoll and Oberling 1978
44,XY,t(9;22),-17,-22,-22,+22q-/46,XY,t(9;22),-20,+22q-/47, XY,t(9;22),+11,-22,+22q-	
46,X,del(X)(q?),t(9;22),5q-,inv(11),16q+,20q+	Hagemeijer et al 1980b
46,XX,t(9;22),inv(11)	Lilleyman et al 1978
46,XX,t(9;22),t(11;17)(q?;p?)	Borgström et al 1982
46,XX,t(9;22),t(11;17)(q13;p11)	Seabright 1983
46,XX,t(9;22),t(2;11)(p25;p12)	Miyamoto 1980
46,XX,t(9;22)/47,XX,t(9;22),+11/48,XX,t(9;22),+11, +del(22)(q?)	Philip 1975a
46,XY,t(9;22)(q34;q11),11p-/47-48,XY,t(9;22),+del(22)(q11), 11p-,+8/46,XY,t(9;22),11p-,dic(22q-)	Prigogina et al 1978

Karyotype	Reference
46,XY,t(9;22),dic(22q-)	
46,XY,t(9;22),dic(22q-)	
46,XY,t(9;22)(q34;q11)/47,XY,t(9;22),+22q-/47,XY,t(9;22),-20,t(11;?),+22q-/48,XY,t(9;22),+8,-20,+t(11;?),+22q-	Misawa 1978
46,XY,t(9;22)(q34;q11),t(7;11)(p14;p15)/47,XY,t(9;22),t(7;11),+22q-	Tomiyasu et al 1982
46,XY,t(9;22)(q34;q11)	Ishihara et al 1982
27,X,-Y,t(9;22),-1,-2,-3,-4,-5,-6,-7,-9,-11,-13,-14,-15,-16,-17,-18,-19,-20,-22	
46,XY,t(9;22),r(11)	Sadamori et al 1980
47,XX,t(9;22)/47,XX,t(9;22),t(5;11)(p15;q22),del(5)(q21q31),+del(8)(q21q23)	Hagemeijer et al 1980b
48,XY,t(9;22),+8,-9,-11,-12,-14,-19,+del(22q),+5mar	Sharp et al 1976
52,XY,t(9;22),+1,+8,+10,+11,+21,+22q-	Sadamori et al 1980
53,XX,t(9;22),+8,+11,+12,+16,+17,+19,+del(22q)/52,XX,t(9;22),+8,+11,+17,+19,+del(22q)	Gahrton et al 1974b
53-56,XY,+Y,t(9;22),+del(22q),+6,+10,+19,+8,+11,+21	Prigogina et al 1978
55,XY,t(9;22),+6,+10,+10,+11,+13,+14,+14,+19,+21	Seabright 1983

Chronic myeloid leukemia, aberrant translocation

Karyotype	Reference
??,X?,t(11;22)(p15;q11)	Potter et al 1981
??,X?,t(3;4;9;11;22)	Potter et al 1981
46,X?,t(11;22)	Stoll 1983
46,X?,t(7;9;11;22)(p11;q34;q22;q11)	Ishihara 1983
46,X?,t(9;11;22)(q34;q13;q11)	Carbonell et al 1982a
46,X?,t(9;11;22)(q34;q13;q11)	Ishihara 1983
46,XX,t(11;14;22)(q13;q32;q11)	Kolitz et al 1981
46,XX,t(12;22)(q34;q11),t(9;11)(q34;q11)	Seabright 1983
46,XX,t(1;11;22)(q24;p11;q11)	Najfeld et al 1983
46,XX,t(6;9;11;22)(p21;q34;q13;q11)	Carbonell et al 1980
46,XX,t(9;11;22)(q34;q13;q11)	Sessarego et al 1982
46,XX,t(9;15;22)(q34;q11;q11)	Hays et al 1981
54-57,XX,t(9;15;22),+5,+7,+8,+10,+11,+12,+17,+22q-	
46,XY,t(11;22)(p15;q11)	Muldal et al 1975
46,XY,t(11;22)(p15;q11)	Seabright 1983
46,XY,t(11;22)(q23;q11)	Seabright 1983
46,XY,t(9;11;22)(q34;q13;q11)/52,XY,t(9;11;22),+6,+9q+,+11q-,+18,+19,+del(22q),1q+	Lawler et al 1976
46,XY,t(9;11;22)(q34;q13;q11)	Fleischman et al 1981
46,XY,t(9;9;11;22)(q34;p24;p13;q11)	Gahrton et al 1977
50,XY,del(22)(q11),+8,+11,+21,+21/51,XY,del(22)(q11),+8,+11,+13,+21,+21	Sonta and Sandberg 1978b
50,XY,del(22)(q11),+8,+11,+21,+21,/51,XY,del(22)(q11),+8,+11,+13,+21,+21	

Chronic myeloid leukemia, Ph1 negative

Karyotype	Reference
46,XX,t(3;11)(q21;p15)	Mitelman 1983
46,XX,t(3;11)(q21;p15)/49,XX,t(3;11)(q21;p15),+8,+19,+21	
46,XX,t(3;11)/49,XX,t(3;11),+8,+19,+21	
46,XX,t(9;11)(q34;q13)	Warburton and Shah 1976
47,XY,t(11;18)(q23;q12),+del(18)(p11q12)	Bagby Jr et al 1978

MYELOPROLIFERATIVE DISORDERS

Polycythemia vera

Karyotype	Reference
??,X?,+6,+11,+14,+22	Vykoupil et al 1980
??,X?,+8,+11,+22	Vykoupil et al 1980
45,X,-Y	Berger and Bernheim 1979
54-57,X,-Y,+6,+8,+11,+17,+20,+21,t(1;22)(p12;q13)	
45,XY,-16,-17,+21/48,XY,+Y,+11,+16,-17,+22	Wurster-Hill et al 1976
46,X?,5q-,11q-	Lessard and Le Prise 1983
46,XX,t(2;11)(p13;q21)/46,XX,t(1;15)(p1?;q1?), t(2;11)(p13;q21),del(16)(p12)/51,XX,+3,+8,+8,+19,t(1;15), t(2;11),del(16)	Testa et al 1981b
46,XY,del(11)(q13),del(13)(q?)	Shiraishi et al 1975
46,XY,del(5)(q15),del(20)(q11)/46,XY,del(11)(q22)/46,XY, del(13)(q14q32)	Testa et al 1981b
46,XY,del(6)(p22),del(11)(q26),del(12)(p12),del(13)(q13)	Kohno et al 1979b
46,XY,t(11;20)(p15;q11)	Berger 1975
47,XX,+9/49,XX,+9,+20,+20,t(4;5),-7,-11,-12,+3mar	Zech et al 1976b
47,XY,+9	Van Den Berghe et al 1979g
47,XY,del(1)(q12),t(1;18)(q12;q23),+9,del(20)(q12)/47,XY, +9,del(11)(q21),del(20)(q12)/47,XY,+9,del(20)(q12)	
47,XY,+9/47,XY,del(5)(q14q32),del(8)(q?),+9,del(11)(q21), del(13)(q21)/47,XY,del(5)(q14q32),del(6)(q?),+9, del(11)(q21),del(13)(q21),del(1)(q12),del(20)(q12)	
48,XX,+8,del(11)(q21?),+mar	Testa et al 1981b

Myelosclerosis / Myelofibrosis

Karyotype	Reference
44,XY,-5,-7,-11,-12,t(7;12)(p?;q?),t(17;18)(p11;p11), +mar"s"	Nowell and Finan 1978
44,XY,5p-,-7,-10,11q-,i(17q)/45,XY,-7,18q-/45,XY,2p-,2q-, 5p-,6q-,9q-,10q-,-11,+i(17q)	Whang-Peng et al 1978
46,XX,del(11)(q14),del(20)(q11)	Mitelman 1983
46,XX,del(11)(q14),del(20)(q11)	
46,XY,-16,+18	Whang-Peng et al 1978
46,XY,7q-,8q-,10q-,11p-,11q+,12q-,-16,+18	
46,XY,del(11)(q23)/46,XY,del(12)(p11)	Hustinx and Rutten 1983
51,XX,+1,del(2)(q33),+6,+9,+11,+17,-19,+20q+	Najfeld et al 1978b

Idiopathic thrombocythemia

Karyotype	Reference
46,XX,t(11;21)(q25;q21)	Zaccaria et al 1980
46,XX,t(11;21)(q25;q11)	Petit and Van Den Berghe 1979a

DYSMYELOPOIETIC SYNDROMES

Preleukemia, NOS

Karyotype	Reference
45,XX,-7/48,XX,+t(1;7),-7,+11,+13"s"	Pedersen-Bjergaard et al 1981
45,XY,-2,-5,-7,-8,-11,-12,-13,-14,+t(2;5),+t(11;12), +t(16;17),+17,+3mar	Watt et al 1982
46,XX,del(5)(q15q31),del(11)(q14)	Swolin et al 1981
46,XX,t(1;3;11)(p1?;q2?;q?)	Ruutu et al 1977b
46,XX,t(3;12)(q21;p13),del(7)(q11),del(11)(q11), t(7;11)(q11;q11),t(11;12)(q22;p13)	Mitelman 1983
46,XY,11q-,16p+"s"	Anderson and Bagby 1982
46,XY,5q-,-11,+18,18p+,18p+	Swansbury and Lawler 1980
46,XY,5q-,del(11)(q14)	Swansbury and Lawler 1980
46,XY,del(11)(q14)	Mitelman 1983
46,XY,del(11)(q?)	Panani et al 1977
47,X,Xp-,1p+,5q-,6q+,11q-,11p+,+11,12p-	Swansbury and Lawler 1980
50,XY,-5,-21,+1,+9,+11,17p+,+1-3mar	Borgström et al 1982

Refractory anemia, NOS

Karyotype	Reference
46,XX,del(5)(q13q34)/46,XX,t(9;11)/47,XX,+8	Swolin et al 1981
46,XX,del(5)(q14q32),t(11;21)(q25;q21)	Van Den Berghe et al 1979e

Acquired idiopathic sideroblastic anemia

Karyotype	Reference
46,X,t(X;11)(q13;p15)	Dewald et al 1982
46,XX,del(11)(q14)	Kardon et al 1982a
46,XY,del(11)(q14)	Mufti et al 1982
46,XY,del(11)(q23)	Hyder et al 1978
46,XY,t(2;11)(p21;q25),del(5)(q13q31)/46,XY, t(2;11)(p21;q25),del(11)(q14)	Kardon et al 1982a

Refractory anemia without excess of blasts

Karyotype	Reference
47,XY,5q-,-7,+2mar/48,XY,-8,-9,+11,+r,+2mar	Teerenhovi et al 1981

Refractory anemia with excess of blasts

Karyotype	Reference
44-47,XY,-3,-5,-7,-12,-16,-20,del(1)(q32),3p+,del(11)(q22), 3-6mar,+r	Streuli et al 1980
45,XY,-5,11q+,-11,-18,+mar	Berger et al 1981a
46,XX,del(5)(q15)/49-51,XX,+1,del(5)(q15),+11,-14,+22,+mar	Swolin et al 1981

Thrombocytopenia

Karyotype	Reference
46,XX,11q+	Tricot et al 1982
46,XX,t(11;?)(q14;?)	Tricot and Van Den Berghe 1982
46,XY,t(11;21)(q22;q21)	Tricot and Van Den Berghe 1982
46,XY,t(11;21)(q25;q?)	Tricot et al 1982

Pancytopenia

Karyotype	Reference
46,XX,-4,-5,-6,-7,-11,-13,-16,+19,+20,t(7;13)(q?;q?),+4mar/43,XX,-2,-3,-5,-7,-16,+19,15q+,+mar	Nowell and Finan 1978
46,XY,t(1;11)(q11orq12;q25)	Najfeld et al 1978a
47,XX,+del(11)(q?),21q-	Nowell and Finan 1978

SPECIAL LEUKEMIAS

Hairy cell leukemia

Karyotype	Reference
44,X,del(X)(q22),del(6)(q23),-7,-8,-10,del(11)(p15q21),-21,14q+,+r,+mar	Sadamori and Sandberg 1983a

LYMPHOCYTIC LEUKEMIAS

Acute lymphocytic leukemia (ALL)

Karyotype	Reference
??,X?,t(11;22)(p15;q12)	Pittman et al 1979
28,XX,-1,-2,-3,-4,-5,-7,-8,-9,-11,-12,-13,-14,-15,-16,-17,-19,-20,-22	Kaneko and Sakurai 1980
32,XY,-2,-3,-4,-5,-7,-8,-11,-12,-13,-14,-15,-16,-17,-20	Shabtai and Halbrecht 1981b
46,X?,t(4;11)(q21;q23)	Parkin et al 1982
46,X?,t(4;11)(q21;q23)	Parkin et al 1982
46,XX,-11,+t(1;11)(q21;q23)	Mamaeva et al 1983
46,XX,del(11)(q13q23),ins(19;11)(p13;q13q23)	Abe et al 1983
46,XX,t(3;11)(p11;p1?)	Inoue et al 1977
46,XX,t(4;11)(q13;q22)	Van Den Berghe et al 1979b
46,XX,t(4;11)(q13;q22)	Van Den Berghe et al 1979b
46,XX,t(4;11)(q21;q23)	Prigogina et al 1979
46,XX,t(4;11)(q21;q23)/49-50,XX,t(4;11),+8,+13,+19,+4q-	Prigogina et al 1979
46,XX,t(4;11)(q21;q23)	Morse et al 1982b
46,XX,t(4;11)(q21;q23)	Arthur et al 1982
46,XX,t(4;11)(q21;q23),del(17)(p11)	
46,XX,t(4;11)(q21;q23)	Arthur et al 1982
46,XX,t(4;11)(q21;q23)	Arthur et al 1982
46,XX,t(4;11)(q21;q23)	Arthur et al 1982
46,XX,t(4;11)(q21;q23),i(Xq)	Arthur et al 1982
46,XX,t(4;11)(q21;q23)	Esseltine et al 1982
46,XY,del(11)(q23)	Mitelman 1983
46,XY,del(6)(q15),-7,+11	Prigogina et al 1979
46,XY,t(4;11)(q13;q22)	Van Den Berghe et al 1979b
46,XY,t(4;11)(q21;q23)	Prigogina et al 1979
46,XY,t(4;11)(q21;q23)	Oshimura et al 1977a
46-49,XY,t(4;11)(q21;q23),+6,+8,+13,+17,t(1;13)(p22;q12),t(7;9)(q11;q34)	
46,XY,t(4;11)(q21;q23),i(7q)	Morse et al 1982b

Karyotype	Reference
46,XY,t(4;11)(q21;q23)	Arthur et al 1982
46,XY,t(4;11)(q22;q25)	Mitelman 1983
46,XY,t(8;22)(q24;q12),t(3;11)(p12p21;q23),inv(2)(p11q13), del(2)(p21)/46,XY,t(8;22)(q24;q12),dup(1)(q23q44)"s"	Fonatsch et al 1982
47,XX,+mar/50,X,-X,-11,+6mar	Morse et al 1978
47,XY,t(4;11)(q21;q23),1p+,del(2)(p16),del(6)(q21),i(7q), del(8)(q21),+mar	Arthur et al 1982
48,XY,t(4;12)(q?;q?),t(6;15)(q?;q?),t(9;22)(q34;q11),11q-, +20,+21	Nordenson 1983
48-53,XX,t(4;11)(q21;q23),+6,+7,+9,+13,+21	Prigogina et al 1979
49,XX,del(4)(p15),t(4;11)(q21;q23),+5,+8,del(9)(p13), del(9)(p13),t(12;?)(q24;?),+mar	Morse et al 1982b
52-56,XY,+5,+8,+11,+19,+20,+21,del(6)(q15),+mar	Prigogina et al 1979
54,XX,7p+,+10,+10,+14,+14,+18,+18,+21,+21/27,X,-X,-1,-2,-3, -4,-5,-6,-7,7p+,-8,-9,-11,-12,-13,-15,-16,-17,-19,-20, -22	Oshimura et al 1977a
57,XX,+3,+4,+8,11p+,+14,+16,+17,+18,+21,+22,+2mar/47,XX,+8	Morse et al 1978

ALL, FAB type L1

Karyotype	Reference
26,XX,-1,-2,-3,-4,-5,-6,-7,-8,-9,-10,-11,-12,-13,-14,-15, -16,-17,-19,-20,-22	Hoeltge et al 1982
28,XX,-1,-2,-3,-4,-5,-6,-7,-8,-9,-11,-12,-13,-15,-16,-17, -19,-20,-22/56,XX,+X,+X,+10,+10,+14,+14,+18,+18,+21,+21	Brodeur et al 1981a
46,XX,t(1;19)(q21;q13),del(3)(q23),del(11)(q21)	Kaneko et al 1982e
46,XY,-11,+t(1;11)(q21;q14)	Kaneko et al 1982e
46,XY,del(2)(q32),-11,-12,+2mar/46,X,-Y,del(2),-4,-5,-11, -12,-21,+6	Kaneko et al 1981b
47,XY,-7,-10,-12,+17,del(11)(p11),del(16)(q22), +t(10;?)(q22;?),+t(12;?)(q15;?),+mar	Kaneko et al 1982e

ALL, FAB type L2

Karyotype	Reference
26,XX,-1,-2,-3,-4,-5,-6,-7,-8,-9,-10,-11,-12,-13,-15,-16, -17,-18,-19,-20,-22	Brodeur et al 1981a
45,X,-X,-1,-14,-17,+21,t(11;12)(p15;p11),+i(17q), +t(1;?)(p34;?)	Kaneko et al 1981b
46,X,-X,-15,del(11)(q21),del(12)(p11),+2mar	Kaneko et al 1981b
46,XY,-4,+6,-8,-10,-11,-15,-17,+22,del(6)(q21),+i(17q), +3mar	Kaneko et al 1981b
46,XY,del(6)(q21orq23),del(9)(p22),del(11)(q13)/46,XY, del(6),i(6p),del(7)(p15),del(9)(p22),del(11)(q13)	Kaneko et al 1982e
46,XY,t(11;14)(q13;p13)	Kaneko et al 1982e
46,XY,t(4;11)(q21;q23)	Weh and Hossfeld 1982
47,XX,+X,t(4;11)(q21;q23)	Weh and Hossfeld 1982
47,XY,+18	Kaneko et al 1982e
46,XY,-2,-5,-6,-7,+8,-9,-10,-11,-12,-16,+18,+t(2;?)(q35;?), +t(11;?)(q21;?),+5mar	
47,XY,-11,+16,+22	Nordenson 1983
47,XY,-11,-17,+3mar	Mitelman 1983
49,XY,+7,+12,-13,t(6;18)(p25;q21),t(11;14)(q23;q32), +t(9;?)(p24;?),+t(1;13)(q12;p13)	Kaneko et al 1982e
56,XY,t(9;22)(q34;q11),+4,+5,+8,+9,+11,+16,+17,+18,+19,+20	Mitelman 1983

ALL, FAB type L3

46,XX,-16,-17,-22,+8,11q-,t(8;14),+1-2mar — Borgström et al 1981

Chronic lymphocytic leukemia

46,XX,t(11;14)(q13;q32) — Vahdati et al 1983a
46,XX,t(6;7)(q?;q?),t(7;13)(q?;q?),t(11;14)(q11;q32),17p+ — Schröder et al 1981
46,XX,t(7;11)(q21;p11q14) — Vahdati et al 1983a
46,XY,2p+,del(11)(q22) — Fleischman and Prigogina 1977
46,XY,del(11)(q22),+12,-13,-21,-22,+3mar/45,XY,del(11),+12, -13,-15,-21,-22,+2mar/46,XY,del(11),+12,+12,-13,-21,-22, +2mar/46,XY,t(1;8)(p22;q24) — Robert et al 1982
46,XY,i(17q)/45,XY,-11,i(17q) — Morita et al 1981
46,XY,t(11;14)(q13;q32)/46,XY,t(3;4;9)(p12;q21q31;q31) — Robert et al 1982
46,XY,t(11;14)(q13;q32) — Vahdati et al 1983a
46,XY,t(11;17)(q21;q25) — Fleischman and Prigogina 1977
46,XY,t(3;6)(q12;q26),t(11;14)(q13;q32) — San Roman et al 1982
46,XY,t(9;11;14)(p22;q13;q32) — Nowell et al 1981
47,XX,+12/47,XX,inv(11)(p15q22),+del(12)(q12orq21), t(12;15)(q15orq21;p13) — Najfeld et al 1980
47,XXY,t(3;8),t(11;14)(q11;q32),+i(18q)"c" — Finan et al 1978
48,XX,+3,+12/47,XX,+12/46,XX,-11,+12/45,XX,-11 — Morita et al 1981
48,XY,+12,+19/48,XY,t(1;11)(q31;q22),+12,+19 — Vahdati et al 1983a

Prolymphocytic leukemia

44,X,-Y,-11,-12,-13,-13,-14,-15,-15,-18,-18,+8mar — Diamond et al 1980
44,XX,-2,-9,-10,11p+,+mar — Pittman et al 1982
44,XX,-7,-8,t(9;17)(q22;q25),-11 — Pittman et al 1982

Lymphocytic leukemia, special type

48,XX,-6,-11,14q+,+18,+20,+22 — Imamura et al 1981

MONOCLONAL GAMMOPATHIES

Multiple myeloma

44,XY,del(3)(q25),t(5;12)(q15;q14),del(7)(q22), del(11)(q24),-14,-18,+mar — Mitelman 1983
45,X,-Y,-1,+t(1;3)(q3?;q2?),-3,+11,14q+/45,X,-Y,-1,-1, +t(1;1)(q2?;q3?),14q+ — Philip et al 1980
45,X,del(X)(q25),t(3;8;22)(q13;q34;q11),-5,-7,-10,-11, del(13)(q13?),+16,19q+,+3mar — Van Den Berghe et al 1979f
45,XY,del(1),+del(1),-10,-11,-11,+t(11;?),-13,-14,-20, +3mar — Philip et al 1980
46,XY,ins(1)(p36q21),t(11;14)(q13;q32) — Liang et al 1979
47,X,-X,inv(1)(p22q12),t(7;?)(q11;?),+9,t(11;?)(p15;?),-13, -14,t(14;?)(q11;?),+15,17q+ — Philip et al 1980

47,X,-Y,+t(1;15)(q12;p11),+3,+11,-15,+19,-22/47,X,-Y,+3,+11,+19,-22	Philip et al 1980
47,XY,dup(1)(q11q32),t(1;10)(p22;p15),inv(6)(p21q13),t(11;14)(q23;q32),21p+,+mar	Liang et al 1979
52,X,-X,+t(1;?)(q12;?),+3,+7,+8,+9,+11,-13,+18,-21,+2mar	Philip et al 1980
52,XY,-1,+t(1;15)(q12;p1?),+t(1;16)(p36;p13),t(3;5)(q2?;q13),+3,+5,+7,+8,+9,+11,-15,+18,-20	Philip et al 1980
53,XY,+3,+5,-7,-8,+9,+11,-14,+20,del(1)(p34),del(6)(q25),+5mar	Liang et al 1979
55,XY,+1,+5,+9,+11,-12,+13,+15,del(1)(q32),del(6)(q25),t(9;19)(q13;q13),14q+,16p+,+4mar	Liang et al 1979
45,XY,-2,-2,-3,+5,-8,-10,+14,-15,-16,del(6)(q25),+4mar	
55,XY,+3,+5,+7,+9,+11,+14,+15,+21,+21/53,XY,+3,+5,-8,+9,+11,+14,+15,-19,+21,14q+,+3mar	Liang et al 1979

Plasma cell leukemia / Plasmocytoma

44,X,-Y,del(1)(p11),+del(1)(q11),t(11;14)(q11;q32),14q+,-21,-22	Ueshima et al 1983
46,XX,t(1;6)(p22p34;p25),t(11;14)(q13;q32)	Gahrton et al 1980b
78,XY,cx,del(1)(q21),t(1;17)(q12;p13),3q+,3p-,9p+,11p+,14q+,16p+	Ueshima et al 1983

SOLID TUMORS

UNCLASSIFIED NEOPLASMS

Malignant neoplasm, NOS

Karyotype	Reference
47,XX,-2,-5,+6,-10,11q+,-13,+18,-21,t(1;?)(q21;?), t(1;13)(q21;p11),t(1;6)(q11;q23)	Kovacs 1978a

EPITHELIAL NEOPLASMS

Carcinoma, NOS

Karyotype	Reference
32-35,XX,cx,del(1)(p12),dup(1)(q21q32),2q+,5q+,7p+,11p+, hsr(10),t(11;12)(q13;q24),12p+,13q+	Kusyk et al 1982
43,X,-X,t(1;11)(q21;q23),3q+,4q+,5q+,-11,-13,-16,17p+	Atkin and Baker 1982a
43-49,XX,-1,-8,+10,-11,t(1;11)(q11;q12)	Kusyk et al 1982
44,XX,ins(1;1)(q32;q12q31),i(2q),+3,t(9;11)(q13;p15),-12, -15,17p+,-18	Atkin and Baker 1982a
44,XY,+5,-8,-11,-13,-13,-14,-14,-15,+19,-22,-22,t(1;11), 5q-,9q+,+5mar	Reichmann et al 1981
44-50,XX,+3,dup(1)(q21q32),t(10;11)(q26;q25),t(2;17), t(6;13),t(9;15),del(10)(q24)	Riet-Fox et al 1979
46,XX,t(3;11)(q12;q14),-6,+7,-8,-15,t(16;?)(q23;?),+2r	Mark 1975b
46,XY,-4,-5,-11,-13,-17,+20,+21,1p+,9p-,+3mar	Reichmann et al 1981
47,XX,+1,+3,t(11;14)(q?;q?),-12,-21,+r	Atkin and Baker 1982a
48-52,XX,+2,+11,+12,i(13q)	Riet-Fox et al 1979
49,X,-Y,-3,-5,-14,+15,-16,-17,+19,+21,+22,1q+,5q-,6q-,8p-, 9p-,11p-,18p+,20q-,22q-,t(Y;14),+3mar	Reichmann et al 1981
49,XX,+3,+11,-7,i(8q)	Riet-Fox et al 1979
49,XX,t(3;11)(q?;q?),+5q-,+6q-,+i(11q)	Atkin and Baker 1982a
50,XX,inv(3)(p13q25),+8,+14,+21/63,XX,inv(3),del(11)(p13), cx	Sonta and Sandberg 1978a
54,XX,+1,+2,+8,+11,+13,+14,+21,+21	Sonta et al 1977
57-61,XX,cx,t(11;16)(q?;q?),inv(3)(p14q29)	Riet-Fox et al 1979
58,XX,+X,+1,+2,+3,5p-,+6,+7,+9,11q+,-13,-14,-15,+16,+19, +20,+6mar	Kovacs 1978a
58,XY,del(3)(p14),+4,+7,+8,+10,+11,+12,+14,+15,+17,+18,+19, +21	Kovacs 1978a
63,XX,-1,+3,+5,+6,+7,+8,+10,+11,+12,-13,-15,+20,+21,+22, +9mar	Sandberg 1977
71-75,XX,cx,11p+,17q+,del(1)(p31),del(3)(p13),del(4)(p1?)	Riet-Fox et al 1979
73,XY,+1,+7,+11,+13,-16,+17,+18,+19	Sonta and Sandberg 1978a
80-83,XX,cx,t(2;5;16),t(2;5;18),t(11;13),11q+,t(8;10), del(2)(q24),del(6)(q21),i(14q),dup(9)(q11q12)	Riet-Fox et al 1979

Carcinoma NOS, metastatic

Karyotype	Reference
35,XX,-1,-3,-4,-5,-7,-8,-10,-11,-12,-13,-14,-15,-16,-17, -18,-19,-20,-21,-22,+8mar	Pathak et al 1979
40,XX,cx,t(1;11)(q?;q?)	Jones Cruciger et al 1976
43,XX,cx,t(1;11)(q?;q?)	Jones Cruciger et al 1976
46,XX,t(9;22)(q34;q11)"s"	Togawa et al 1981
52,XX,+3,+5,+7,+8,+9,+11,12q+,13q-,-14,+16,i(17q)"s"	

Karyotype	Reference
60,XX,t(1;?),t(7;?)(q3?;?),i(10q),der(11),t(14;21),cx	Bartnitzke and Bullerdiek 1983
64-100,XX,t(6;?)(p11q27;?),t(4;10)(q?;q?),t(3;?)(q25;?), del(1)(q2?),t(6;11)(p21;q13)	Kakati et al 1975

Large cell carcinoma

Karyotype	Reference
55-70,XY,t(4;5;14),del(1)(q11q21),t(7;11)(p22;q13), t(7;21)(q11;q22),t(3;8)(p?;q?),t(6;11)(p11;p11),i(16q), t(3;?)(q11;?),t(21;?)(q11;?),t(13;?)	Kakati et al 1975
60,XY,del(5)(p11),t(6;10)(q11;p11),t(7;11)(p22;q13), t(14;18)(q11;q11),del(2)(q21),t(6;?)(p11;?)	Pickthall 1976

Small cell carcinoma

Karyotype	Reference
40,XY,cx,t(1;16)(q21;q24),t(1;19)(q23;q13),del(3)(p21p24), del(3)(p23q26),del(10)(q22),del(11)(q23), t(5;13)(q13;q32),del(17)(p12),dmin	Whang-Peng et al 1982b
44,XY,cx,del(1)(q31),t(2;3)(p?;q?),del(3)(p14p23), t(9;13)(p11;q11),del(11)(p15),12q+,del(X)(q22)	Whang-Peng et al 1982b
67,XY,cx,inv(1)(q32p36),dup(1)(q32q44),del(3)(p14q13), del(3)(p14p23),t(7;13)(p11;q11),del(9)(q11),del(11)(p11), t(12;19)(q24;p13)	Whang-Peng et al 1982b
69,XY,cx,del(3)(p14q24),t(11;14)(p11;p11), t(11;13)(p11;p11),del(X)(q23)	Whang-Peng et al 1982b

Papillary carcinoma

Karyotype	Reference
41,XX,cx,1p-,1q-,3p+,3q+,6q-,11q-	Wake et al 1980
54,XY,+del(2)(q11),t(2;14)(q11;q32),t(3;11)(p13;p15), +i(5p),+8,+9,+12,+16,+17,+21,+2mar	Pathak et al 1982

Adenoma

Karyotype	Reference
46,X,t(X;4)(q25;p16),del(2)(p13p21),del(5)(q15q31),-7, t(9;12)(p13;q13),del(11)(q11q14), t(7;9;12)(p11p22q11;p13p24;q13)/45,X,-X	Mark et al 1980

Adenocarcinoma

Karyotype	Reference
37,X,-X,-3,-8,-15,del(2)(p23),del(6)(q15),t(11;?)(p12;?)	Trent and Salmon 1981
40-42,X,-X,t(1;?)(p13;?),del(1)(p12),2q+,del(3)(q21), del(6)(q13),del(7)(q11),t(11;11)(p15;q13), t(12;12)(q24;?),13p+,15p+,i(22q)	Kusyk et al 1981
48,XX,+11,+mar	Couturier-Turpin et al 1982
49,XY,+7,-9,+11,+12,+mar	Couturier-Turpin et al 1982
50,XY,2q+,+5,+10,+11,-16,-18,+20,+20,+mar	Couturier-Turpin et al 1982
54,XX,+5,+6,+7,+9,+11,+21,+21,+del(1)(q32)	Sonta et al 1977
72,XX,cx,del(2)(p23),hsr(3)(q13q28),del(6)(q21), t(10;11)(p12;p13)	Trent and Salmon 1981

Adenocarcinoma, metastatic

40,XX,t(4;5;8),t(3;?)(q21;?),t(1;11)(q21;p11), t(9;11)(q11;p11),del(1)(p36) — Kakati et al 1975
40-45,XY,-3,-8,-18,+5,+10,11p+,14q+,del(4)(q32) — Ayraud 1975

MESENCHYMAL NEOPLASMS

Liposarcoma

50,XY,+3,+6,+8,+11,+14,-22,17p+ — Sonta et al 1977

Leiomyosarcoma

42,XX,del(1)(p12orp13),del(11)(q13orq14),+7,-9,-13,-14,-15, -18,+19,-22 — Mark 1976

MELANOMAS

Malignant melanoma, metastatic

41,XY,dup(1)(q21q32),t(1;7)(q25;q36),t(7;9)(q11;q11), del(9)(q13),-3,-4,-5,-8,-10,-11,-16,+18,-20,-22 — Kakati et al 1977
41-43,XY,del(1)(q21q25),i(1q),t(1;9)(p22;q34), t(11;12)(q11;p11),t(10;12)(q26;q13),t(4;?)(q35;?), del(5)(p13),t(14;?)(p12;?),i(17q),t(18;?)(p11;?), del(6)(q21) — Kakati et al 1977
45,XY,t(1;9;13)(q?;q?;q?),r(2),t(1;9)(p?;q?),i(8q), t(11;?)(q?;?),-10,-13,+mar — Chen and Shaw 1973
80,XY,t(11;12)(q23;q13),del(5)(q13),del(5)(p11), del(9)(q13),t(5;9),t(14;?)(q11;?),t(7;12)(q22;q24), del(1)(p13),del(11)(q23) — Kakati et al 1977

NEUROGENIC NEOPLASMS

Astrocytoma, grade III, IV

44,XY,-4,-17,-22,1q-,5p-,-11q+,+mar — Yamada et al 1980

Neuroblastoma

45,XX,+14,-15,-18,del(1)(p32),dup(1)(p13p31),del(6)(q23), del(11)(q23) — Brodeur et al 1981b
46,XX,t(1;?)(p32;?),11p+,17q+,dmin — Gilbert et al 1982
46,XY,t(1;11)(p?;p?),dmin — Gilbert et al 1982
47,XY,+1,11q-,16q+,19q+ — Gilbert et al 1982
53,XX,+1,+4,+15,+20,dup(1)(p13p32),t(3;15)(q13;q26), del(4)(q31),t(7;11)(q36;q13) — Brodeur et al 1981b

Retinoblastoma

46,XX,inv(1),inv(2),inv(3),inv(4),t(6;7),inv(9),t(11;?), 11p-,inv(12),i(17q),-20 — Gardner et al 1982
51,XY,del(6)(q21),+8,+11,-12,t(13;18)(p11;q11),14q+,-16, +19,+22,+2mar — Hossfeld 1978

Meningioma, benign

43,XX,-1,-6,-7,-11,-22,+2mar/40-45,XX,-1,-6,-7,-9,-11,-19, -22,+2mar — Mark et al 1972b
44,Y,-X,-22,-14,+mar/43,Y,-X,-22,-14,-15,-11,-10,+3mar — Zankl et al 1975a
52,X,-Y,-22,+20,+19,+17,+15,+11,+9,+7,+5 — Zankl et al 1975a

EMBRYONAL AND MISCELLANEOUS NEOPLASMS

Blastoma, NOS

40-47,X,-X,cx,7q+,+del(10)(p13),+del(12)(p11), del(11)(p11p14) — Slater and Kraker 1982
46,XY,t(4;?)(p14;?),t(9;11)(q22;p14),del(11)(p13p15), del(11)(q21q23) — Slater and Kraker 1982

Teratoma, malignant

67,XY,+1,+3,+4,+5,+6,+7,+7,+8,+10,+11,+13,+13,+14,+16,+17, +18,+20,+20,+21,+21 — Sonta et al 1977

Seminoma

53,XY,+3,+4,+6,+7,+11,+13,+19 — Sonta et al 1977

LYMPHOMAS

1. UNCLASSIFIED AND MISCELLANEOUS LYMPHOMAS

Malignant lymphoma, NOS

47,XX,t(1;14)(q23;q32),+3,t(6;21)(p25;q11),+8, t(7;11)(q11;q25),-21 — Yamada et al 1980

Non-Hodgkin's lymphoma, NOS

46,X,-X,del(1)(q22),t(2;11)(p21;q21),t(9;15)(q11;p12), del(9)(q11),-14,t(14;17)(q23;q23),-16,+mar — Kristoffersson 1983
50,XX,+X,t(1;11)(q25;q23),+3,+7,-13,t(14;18)(p?;p?),-15, t(19;?)(q13;?),+3mar — Reeves and Pickup 1980

<u>Non-Hodgkin's lymphoma, lymphocytic, NOS</u>

Karyotype	Reference
42-43,XX,t(11;14)(q13;q32)	Fleischman and Prigogina 1977
45,X,-Y,t(11;14)(q13;q32),t(11;19)(q13;q13)	Fleischman and Prigogina 1977
46,XY,t(11;14)(q13;q32)	Fleischman and Prigogina 1977
47,XY,+7,-9,-11,1p+,6q-,10p-,14q-, t(14;?;14)(q24orq32;?;q13),15q-,22q-	Fukuhara et al 1979b
47-48,XY,+7,-11,-14,-15,4q+,9p-,14q+	Fleischman and Prigogina 1977
51,XX,+5,+9,+11,+del(11)(q?),+del(12)(q?),14q+,i(17q)	Prieto et al 1978a

<u>Non-Hodgkin's lymphoma, histiocytic, NOS</u>

Karyotype	Reference
46,XX,del(1)(p31p35),+3,del(9)(p11),t(10;?)(q26;?), del(11)(q13),-13,t(11;14)(q13;q32),t(17;21), t(9;22)(q11;p13),+r	Mark et al 1977
46,XY,1p-,6q-,t(11;14)(q21;q32)	Kakati et al 1980
48,XY,1p-,+3,6q-,+7,9p-,t(11;14)(q21;q32)	
47,XY,t(11;11)(p23;q13),+18,del(19)	Fleischman and Prigogina 1977
47-48,XY,+X,-13,-15,t(6;11)(p?;q?),t(13;14;?)	Fleischman and Prigogina 1977
47-50,XY,t(1;11)(q?;p?),1q-,1p-,3q+,11p+,t(11;17)	Fleischman and Prigogina 1977
48,XY,del(3)(p21),del(9)(q22),11q+,12q+,del(6)(q23orq25), 15q+,del(12)(q15)	Reeves 1973
50,XY,t(2;?)(p25;?),del(2)(p13),+i(3p),del(6)(p21),+7,+8, +9,+9,-11,-13,+14,-15,+20,+21,-22	Mark et al 1978
44,XY,del(2)(p13),t(6;?)(p21;?),del(11)(q13), t(11;14)(q13;q31),i(15q),-20,-22	
82,XXX,t(3;12)(q13;q15),t(3;8)(q22;q11),del(3)(q11), del(5)(q31),i(7q),del(10)(q24),t(10;10)(q24;q26), inv(11)(p15q21),del(12)(q13),del(13)(q14)	Mark et al 1978

2. HODGKIN'S LYMPHOMAS

<u>Hodgkin's disease, mixed cellularity</u>

Karyotype	Reference
90-102,XY,t(3;11)(p14;q21)	Reeves 1973

<u>Hodgkin's disease, lymphocytic depletion</u>

Karyotype	Reference
45,X,t(X;1)(q28;q11),t(3;5)(q29;q33),del(8)(q22),-9, del(11)(p13),-17	Hossfeld and Schmidt 197
46,XY,+2,t(2;?)(q25;?),t(3;11)(q29;p13),t(4;11)(q33;q13), -13	Hossfeld and Schmidt 197
48,X,-Y,t(1;?)(q44;?),i(1q),del(1)(q21),del(3)(q21),-10, t(3;11)(q23;q21),+14,t(14;?)(q32;?),+15,-17,+19,+20,-22	Hossfeld and Schmidt 197

3A. NON-HODGKIN'S LYMPHOMAS - RAPPAPORT CLASSIFICATION

Lymphocytic, poorly differentiated, diffuse

44,XX,t(11;14)(q13;q32),-3,-9,-13,1p+,+2mar	Fukuhara et al 1979a
45,XY,t(14;18)(q32;q21),+2,-5,-6,-11	Fukuhara and Rowley 1978
47,XY,t(14;14)(q32;q?),+7,-9,-11,1p+,6q-,10p-,15q-,22q-, +2mar	Fukuhara and Rowley 1978
47,XY,t(1;4)(q23;q33),+3,t(10;15),t(11;14)(q13;q32),r(13), r(13)	San Roman et al 1982
87,XY,del(1)(q23),del(3)(p14),del(6)(q21),del(7)(p11), del(11)(q21),t(17;?)(p13;?)	Mark et al 1979

Lymphocytic, poorly differentiated, nodular

48,XY,t(14;18)(q32;q21),-6,-11,+12,+del(1)(p?),3q+,+2mar	Fukuhara et al 1979a
92,XY,t(2;9)(q11;p11),t(5;?)(p15;?),t(17;?)(q25;?),17q+, del(1)(p22),del(3)(p21),del(X)(q24),del(11)(q23)	Reeves 1973

Mixed cell, diffuse

50,XY,+11,+12,-15,+del(3)(q11),+del(3)(q11), +t(X;1)(p22;q11)	Kaneko et al 1982a
53,XY,t(8;14)(q24;q32),t(14;18)(q32;q21),+20,1q-,+1p-,2q-, 4q+,5q-,5p+,6q-,+7q+,10q+,11q+,11q-,13p+,+4mar	Fukuhara and Rowley 1978

Mixed cell, nodular

50,XX,t(3;6)(q29;p21),del(9)(p13q22),+11,+12,+21	Mark et al 1976

Histiocytic, diffuse

45,X,del(X)(q24),del(1)(q21),+del(1)(p11),del(2)(q22), del(3)(q21),del(6)(p11),del(6)(q15),t(7;9)(p22;q13), t(8;10)(p23;q21),del(9)(q13),del(10)(q21), t(2;11)(q22;q23),-13,del(14)(q24),del(15)(q22), del(16)(q22),+17,-19	Mark et al 1979
46,X,-X,-11,-18,+t(11;?)(q23;?),t(18;?)(q23;?),+mar	Kaneko et al 1982a
46,X,-X,-2,-10,+11,-13,-14,-17,del(1)(p22p32), del(3)(p13p21),del(6)(q21),del(9)(p13),+t(2;?)(p23;?), +t(14;?)(q32;?),t(18;?)(q23;?),+t(X;13)(q26;q12),+mar	Kaneko et al 1982a
46,X,-X,-4,-11,-17,+18,-19,del(7)(p15),del(8)(q22),+i(17q), +t(4;19)(p11;q13),t(11;?)(q23;?)	Kaneko et al 1982a
47,X,+X,-Y,+12,-14,del(11)(q21q23),del(16)(q22), t(14;?)(q32;?)	Kaneko et al 1982a
47,XX,-1,-2,-6,+8,-14,+16,-19,del(11)(q23), t(8;14)(q22;q32),+t(2;6)(q21;p23),del(6)(q21), +t(1;14)(p11;q11),+t(19;?)(q13;?),+t(14;?)(q11;?)	Kaneko et al 1982a
47,XY,-11,-22,del(1)(q42),del(2)(q33),del(5)(q22q31), del(7)(q32),del(9)(p22),t(21;22)(q22;q11), +t(11;?)(q23;?),+t(21;22)(q22;q11),+mar	Kaneko et al 1982a
51,XY,+X,-2,+5,+11,-12,+21,del(6)(q21),+t(2;?)(q33orq35;?), +t(12;?)(p11;?),+mar	Kaneko et al 1982a

73,XX,cx,del(7)(q32),del(9)(q22),del(10)(p14), t(7;?)(p22;?),t(10;?)(p15;?),t(11;?)(q25;?), t(17;?)(q25;?),t(19;?)(q13;?) — Kaneko et al 1982a

3B. NON-HODGKIN'S LYMPHOMAS - KIEL CLASSIFICATION

Centrocytic

44,XY,del(4)(p13),-11,-13,t(13;14)(q11;q32) — Kristoffersson 1983

Centroblastic-centrocytic, diffuse

46,XX,del(11)(q23) — Kristoffersson 1983
46,XY,del(6)(q12),t(7;11)(p11;q11) — Kristoffersson 1983

Centroblastic-centrocytic, follicular

46-47,XX,del(1)(q11q22),+11 — Kristoffersson 1983

Lymphoblastic, Burkitt's type

45,X?,t(10;11),t(10;13) — Zech et al 1976a
45,XY,-2,5q+,11q+ — Biggar et al 1981
47,XX,t(8;22)(q24;q11),+8,t(11;21)(q14;q22) — Berger and Bernheim 198[illegible]
47,XY,+11,-14,+21,inv(1)(p11p31),11q+,21p+ — Biggar et al 1981
48,XX,+5,+7,-9,+10,+11,-16,del(5)(p13q31),del(10)(p11q25), del(11)(q12) — Biggar et al 1981
48-53,X?,t(8;14)(q24;q32),+X,+2,+3,+9,+10,+11,+12,+20 — Zech et al 1976a
50,XX,+1,+2,+3,+11,del(2)(p14),11p+,t(8;14)(q24;q32) — Biggar et al 1981

Lymphoblastic, convoluted type

43-46,XY,del(1)(p32),del(2)(p21),-6,t(7;9)(p11;q11),-9,-10, t(11;15)(p11;q11),-11,del(12)(q14),-13,-14,i(17q),-17, +18,del(19)(p11),-22,+2mar — Kristoffersson 1983

Immunoblastic

48-49,XY,+3,+7,+11,t(12;?)(p13;?),t(13;?)(p13;?),-16,-20, +21,+22 — Reeves and Pickup 1980
51,XY,t(1;6)(p22;q13orq15),t(2;13)(q21;q14), t(8;16)(q24;p11),+11,+12,del(15)(q22),+16,+20,+mar — Reeves and Pickup 1980

3C. NON-HODGKIN'S LYMPHOMAS - LUKES AND COLLINS CLASSIFICATION

Mycosis fungoides

46-48,XX,del(1)(q42),+3,-11,-12,+2-6mar — Edelson et al 1979

Sezary's syndrome

46,XY,-1,-10,+t(1;9),+t(10;11),9q-,11q- — Nowell et al 1982

Follicular center cell, diffuse, small non-cleaved

44,X,-Y,t(1;14)(p21;q13),5p+,del(6)(q13),del(8)(p21),9p+, 9p-,t(11;21)(q23;q22),13q+,-10,+12,-17p+,der(19),22q+ — Fukuhara et al 1979a

Immunoblastic type (B-cell)

48,XY,t(4;12)(q?;q?),t(6;15)(q?;q?),t(9;22)(q34;q11),11q-, +20,+21 — Nordenson et al 1983

3D. NON-HODGKIN'S LYMPHOMAS - WORKING FORMULATION

Small lymphocytic

46,XY,del(11)(q11q23) — Yunis et al 1982
46,XY,del(11)(q11q23) — Yunis et al 1982

Follicular, small cleaved cell

46,XY,6q-,11q- — Yunis et al 1982
49,XY,+X,+3,ins(11;3),t(14;18)(q32;q21),+18 — Yunis et al 1982

Diffuse, large cell

46,XY,t(11;14)/47,XY,t(11;14),+3 — Yunis et al 1982

Chromosome 12

HEMATOLOGICAL DISORDERS

UNCLASSIFIED LEUKEMIAS

Acute leukemia, NOS

Karyotype	Reference
45,X,-Y/46,X,-Y,+12	Najfeld et al 1981
46,XX,dup(1)(p31p36),dup(7)(p15p21),dup(3)(q12q29), ins(12)(q12)"s"	Berger et al 1977

NONLYMPHOCYTIC LEUKEMIAS

Acute nonlymphocytic leukemia (ANLL)

Karyotype	Reference
39,X,-X,-3,-5,-13,-14,-15,-16,-17,-21,-22,del(7)(p11), t(12;?)(p13;?),t(16;?)(p13;?),+3mar	Mitelman 1983
39,X,-X,-3,-5,-13,-14,-15,-16,-17,-21,-22,del(7)(p11), t(12;?)(p13;?),t(16;?)(p13;?),+3mar	
42,XX,-5,-12,-14,-16,-18,t(12;16)(q11;q22q24)"s"	Oshimura et al 1976b
42,XY,-5,t(12;?)(p11;?),-18,-22,-22	Hustinx and Rutten 1983
43,XX,t(1;20),-5,-6,7q-,11q-,-12,17q+,-18,19q+,-21	Mitelman 1983
44,X,-Y,-3,+4,-12,+15,-20	Bernard et al 1982a
44,XY,5q-,-7,12p-,-22"s"	Whang-Peng et al 1979
44,XY,del(5)(q21),del(7)(q11),del(11)(q11), t(7;11;12)(q11;q11;p13),del(13)(q11),-20,-21/45,XY, t(3;12)(q11;p13),del(5)(q21),del(7)(q11),del(13)(q11), -20	Mitelman 1983
45,XX,-7,12q+"s"	Mitelman 1983
45,XX,del(3)(p11),t(5;7;12)(q31;q22;q13-24),t(5;?)(q13;?)	Philip et al 1978a
46,XX,12q+,-22,+mar	Chrz et al 1983
46,XX,12q+	Bernard et al 1982a
46,XY,+del(7)(p11),-8,del(12)(p11),t(21;?)(q22;?)/50,XY,+Y, +del(7)(p11),del(12)(p11),+17,t(17;?)(q25;?), t(21;?)(q22;?),+22	Mitelman 1983
46,XY,+del(7)(p11),-8,del(12)(p11),t(21;?)(q22;?)/47,XY,+9	
46,XY,+t(1;12)(q21;p13),-12	Trent et al 1983a
46,XY,1p+,12p-,?t(1;12),?11p+,inv(12)(p?),-2,+mar	Mitelman 1983
46,XY,del(12)(q14q23)	Mitelman 1983
46,Y,t(X;11)(q22.3;q23.3),ins(X;11)(q22.3;p13p15), dup(1)(q11q21),t(4;6)(p16;p21),t(7;12)(p13;q15)	Nacheva et al 1982

Karyotype	Reference
47,X,Xq-,t(3;5),-5,t(11;?),t(12;17),-18,-21,+6mar	Pedersen-Bjergaard et al 1980
47,XY,+12	Sasaki et al 1976
48,XX,+12,+r	Bernard et al 1982a
48-50,XX,+8,+9,+del(12)(p11)"s"	Weinfeld et al 1977a

ANLL, FAB type M1

Karyotype	Reference
39,XY,-5,-7,-12,-16,-17,-20,-21	Berger et al 1981a
41,XY,-1,-2,-4,-5,-12,-15,-17,-17,-18,-22,+5mar	Sessarego et al 1981a
45,XX,del(5)(q12q31),-7,12p-"s"	Kerkhofs et al 1982
46,XY,+del(12)(p11q23),-13	Brodeur et al 1983
46,XY,t(10;12)(q?;p?)	Lessard and Le Prise 1983
46,XY,t(12;14)(q24;q32)	Brodeur et al 1983
46,XY,t(12;22)(p12;q11)	Hagemeijer et al 1981a

ANLL, FAB type M2

Karyotype	Reference
42,XY,5q-,7q-,-8,-12,-17,-21,-22,+i(17q),+2mar	Berger et al 1981a
43,XY,-5,-7,-12,-16,4p+,17p+,20q-,+mar"s"	Rowley et al 1981a
45,X,-Y,t(8;21)(q22;q22)/46,X,-Y,t(8;21)(q22;q22),+t(11;12)(p11;q13)	Sakurai et al 1982b
45,XY,-7,del(5)(q13),t(12;17)(q13?;p12?)"s"	Rowley et al 1981a
46,XX,del(4)(q31q34),t(7;12)(q21;q13)	Yunis et al 1981
46,XY,-7,-12,-17,del(5)(q14),8q+,del(17)(p11),+mar	Testa et al 1979
45,XY,-7,+8,-12,-17,del(5)(q14),8q+,t(12;17)	
47,XX,+8,t(12;22)(p13;q12)	Mitelman 1983
47,XX,+8,t(12;22)(p13;q12)	
47,XX,+8,t(12;22)(p13;q12)	
47,XX,1p+,der(5),-6,12p+,-13,-14,-16,-17,-17,-21,+22,+7mar,dmin	Cooperman and Klinger 1981
53,XX,+1,-4,-5,+6,-7,+8,+11,+12,+21,+21,+3mar"s"	Zaccaria et al 1983b

ANLL, FAB type M1+M2

Karyotype	Reference
37-45,XY,-7,-12,-13,-14,-16,-18,-19,-21,del(4)(q29),del(5)(q24),der(9),der(17),11q+,12p+,t(14;?)	Fitzgerald et al 1983
42-44,XX,+8,-12,-13,-14,t(2;5)(q12;q35)	Mitelman et al 1981
43,XX,del(5)(q12q32),-7,-12,-18,t(7;12;21)(q11;q12;q22)	Petit and Van Den Berghe 1979b
44,XY,-5,-12,-18,+mar,+dmin	Prigogina et al 1979
44,XY,-5,t(5;12)(q12;q24),i(18)(q11),-19	Mitelman et al 1981
45,XX,t(3;6)(p23;p25),del(5)(q13),-12,i(17q)	Mitelman et al 1981
45,XY,-5,-19,-20,-22,+t(7;12)(q11;q13),+r,+3mar	Fitzgerald et al 1983
46,XX,del(5)(q12q32),del(12)(p11)	Petit and Van Den Berghe 1979b
46,XX,t(11;12)(p14;q13),del(12)(q13)"s"	Prigogina et al 1979
46,XY,del(11)(q23),del(12)(p11)	Yamada and Furusawa 1976
46,XY,del(7)(q31),del(11)(q14),del(12)(p12)	Mitelman et al 1981
46,XY,t(12;16)(p11;p13)	Benedict et al 1979
46,XY,t(X;12)(p?;q?)	Garson 1980
46-50,XY,+Y,+del(7)(p11),-8,+9,del(12)(p11),+17,21q+,+22	Mitelman et al 1981

47,X,-X,+del(1)(p11),del(5)(q13),t(5;12)(q21;q24), t(11;12)(q25;q13),t(13;14)(q11;p11),+16,+21	Mitelman et al 1981
47,XY,+19,t(7;12)(q21;p13),del(4)(q26),del(5)(p12)	Mitelman et al 1981
47,XY,+i(12p)"s"	Sandberg et al 1982a

ANLL, FAB type M3

45,X,-Y/46,XY,del(17)(q22)	Van Den Berghe et al 1979c
46,XY,del(12)(q11)	
45-51,XY,t(3;12)(p14;q24),t(13;18)(q11;p11),+1,+6,+16,+19, +21,+22	Mitelman et al 1981
46,XY,t(3;18),t(3;12),3p-	Teerenhovi et al 1978
47,XY,del(12)(p11),del(20)(q11),+9	Mitelman et al 1981

ANLL, FAB type M4

43,XY,-2,-3,-6,-7,-12,-15,-21,-22,del(2)(p11),del(11)(p14), t(14;?)(q32;?),t(17;?)(p13;?),del(22)(q11),+5mar"s"	Rowley et al 1981a
44,XY,del(3)(p22),-9,-12,t(17;?)(p13;?)"s"	Mitelman 1983
44,XY,del(5)(q15q23),del(7)(q22),-9,-12,-13,del(15)(q22), +mar/46,X,-Y,del(5),del(7),+8,-12,-13,del(15),+mar,dmin	Alimena 1983
44,XY,t(3;5)(q29;q13),t(7;12)(q12;q14),t(7;10)(q12;q21),-8, -17,-18,-20	Mitelman et al 1981
46,XX,-12,+21"s"	Berger et al 1981a
46,XX,t(2;11;12;17)(q31;p11;p13;q23)	Fitzgerald et al 1983
46,XY,t(3;12)(p14;q24)	Sandberg et al 1982b
46,XY,t(3;12)(p21;q34)	Prieto et al 1981
46,XY,t(3;12)(q21;q13),del(5)(q14),ins(8;5)(q22;q14q22), -17,+2mar	Mitelman 1983
46-47,XY,+10,-12	Mitelman et al 1981
47,XX,+19,20p+/50,XX,+5,+8,+12,+mar	Morse et al 1978
47,XY,t(7;12)(q36;p13),+8	Hagemeijer et al 1981a

ANLL, FAB type M5

45,XY,-12	Mitelman et al 1981
46,XX,ins(10;11)(p11;q23q24)/52,XX,+4,+8,ins(10;11),+12, +16,+19,+20	Kaneko et al 1982b
45,XY,-12	Mitelman 1983
46,XY,-10,-12,16q+,+2mar	Brodeur et al 1983
46,XY,t(2;11)(q37;q12),t(2;12)(q24;q24),10p+	Mitelman 1983
46,XY,t(2;11)(q37;q12),t(2;12)(q24;q24),10p+	
46,Y,-X,-1,-7,-8,-14,-15,-17,-20,inv(6)(p23q13), del(11)(q23),+t(X;?)(p11;?),+t(1;?)(p24;?), +t(12;?)(q22;?),+t(17;?)(p11;?),+4mar	Kaneko et al 1982b
50,XX,+6,+8,10p-,+12,+13	Nordenson 1983
57-58,XX,+1,+3,-4,-5,+6,-8,+11,+12,-17,+19,-20, del(10)(q24),+8-9mar	Testa et al 1979
60-61,XX,Xp+,+1,+2,-4,+6,-7,+19,+20,del(2)(q31), del(10)(q24),+12-13mar	

ANLL, FAB type M5a

Karyotype	Reference
46,XX,t(1;12)"c"	Berger et al 1982a
47,XX,+12	Berger et al 1982a

ANLL, FAB type M4+M5

Karyotype	Reference
42,XX,-5,-7,-12,-15,-16,-18,+2mar	Li et al 1983
45,XY,del(11)(p11),t(12;?)(p13;?),del(16)(q13), t(13;14)"c""s"	Nora et al 1979
48,XY,+i(1q),+3,+12,-14	Shiraishi et al 1982
66,XY,+1,+1,+2,+4,+6,+6,+7,+8,+8,+10,+11,+11,+12,+12,+18, +19,+21,+21,+22,+22	Li et al 1983

ANLL, FAB type M6

Karyotype	Reference
??,XY,t(4;?)(q23;?),t(5;17),t(8;9),+8,+11, t(12;16)(p12;p12)	Hustinx and Rutten 1983
43,X,-X,del(5)(q14q34),-7,11q-,-12,12p+,t(12;17),t(20;21)	Kerkhofs et al 1982
43,XX,del(6)(q15orq21),-7,-12"s"	Rowley et al 1977a
43,XY,-4,5q-,t(4;6),-7,dic(12),-15"s"	Hagemeijer et al 1981a
43-45,XX,der(1),inv(2)(p25q14),del(3)(q25),-4,-5,-6,6q-,-7, 7p+,11p+,-12,-17,-21,+r,+1-4mar"s"	Smadja et al 1983b
46,XX,t(4;9)(q21;q34),t(4;13)(q12;q12),del(13)(q12orq13), del(12)(p11)	Fitzgerald et al 1983
46,XY,5p-,-9,-12,13q-,+2mar	Nordenson 1983
48,XX,+t(1;12)(p22;p13),+21,del(22)(q11)	Brodeur et al 1983
49,XY,+8,+12,+17,t(17;21)(p11;q22)"s"	Yamada and Furusawa 1976

Chronic myeloid leukemia, t(9;22)

Karyotype	Reference
26,XX,t(9;22)(q34;q11),-1,-2,-3,-4,-5,-6,-7,-9,-10,-11,-12, -13,-14,-15,-16,-17,-18,-19,-20	Hartley and McBeath 1981
35,XX,t(9;22)(q34;q11),-3,-4,-5,+8,-9,-10,-11,-12,-13,-14, -15,-16,-17/66-72,XX,t(9;22)	Pedersen and Boesen 1983
45,X,-Y,t(9;22)/49,X,-Y,t(9;22),+t(9;22),+8,+12	Miyamoto 1980
46,XX,t(9;22)(q34;q11),t(9;12)(q34;q13)	Dor et al 1977
46,XX,t(9;22),-10,+12,-19,+22q-	Stoll and Oberling 1978
46,XX,t(9;22),12p-,i(17q)	Lilleyman et al 1978
46,XX,t(9;22),t(12;21)(q13;q22)	Fleischman et al 1981
46,XX,t(9;22)/48,XX,t(9;22),+19,+22q-/50,XX,t(9;22),+8,+12, +19,+22q-/50,XX,t(9;22),+8,+14,+17,+19,+21	Hagemeijer et al 1980b
46,XY,t(9;22)	Mitelman 1983
50,XY,t(9;22)(q34;q11),del(3)(q11),+4,+5,+12,+22q-	
46,XY,t(9;22)(q34;q11)	Como and Graze 1979
35,XY,t(9;22)(q34;q11),-3,-4,-5,-7,-9,-11,-12,-13,-15,-16, +3mar	
46,XY,t(9;22)(q34;q11),21q+/50,XY,t(9;22),+7,+8,+12,+15, 21q+	Tomiyasu et al 1982
46,XY,t(9;22),-8,12q+,i(17q),+21/46,XY,t(9;22),-8,-8,12q+, +i(17q),+21	Olah and Rak 1981
46,XY,t(9;22),t(12;19)(q13;q13)/46,XY,t(9;22), t(1;6)(p36;p21)	Seabright 1983
46,XY,t(9;22),t(4;12;15)(q21;q24;q21)	Hartley and McBeath 1981

Karyotype	Reference
47,X?,t(9;22),+21/48,X?,t(9;22),+21,+21/50-52,X?,t(9;22), +8,+9,+12,+19,+21,+21	Carbonell et al 1982a
47,XY,t(9;22),+22q-,+8,t(12;13)(q?;q?),-13,i(17q)	Rozynkowa et al 1977
47,XY,t(9;22),+22q-/56,XY,t(9;22),+1,+2,+5,+6,+7,+8,+10, +12,+17,+22q-	Stoll and Oberling 1978
47,XY,t(9;22),+8,t(12;13)(q11;q34)	Fleischman et al 1981
48,XY,t(9;22),+12,+21	Lilleyman et al 1978
48,XY,t(9;22),+8,-9,-11,-12,-14,-19,+del(22q),+5mar	Sharp et al 1976
48-51,XY,t(9;22)(q34;q?),+7,+12,+C,+G	Kaffe et al 1974
49,XX,t(9;22),+12,+13,i(17q),+19	Seabright 1983
49,XY,t(9;22),+9,+10,+12/45,XY,t(5;17)(p11;q21)	Hagemeijer et al 1981b
50,XX,t(9;22),t(2;12)(p13;q24),+8,+8,+12q+,+12q+,+del(22q)	Sharp et al 1976
50-58,XY,+X,t(9;22)(q34;q11),+7,+8,+9,+10,+11,+12, +del(15)(q22),+16,+18,+19,+22q-	Kwan et al 1977
46,XY,t(9;22)	
51-52,XY,t(9;22),+del(22q),+7,+8,+12,+19,+21	Prigogina et al 1978
52,XY,t(9;22),+8,+9,+12,+19,+21,+22q-	Nordenson 1983
53,XX,t(9;22),+8,+11,+12,+16,+17,+19,+del(22q)/52,XX, t(9;22),+8,+11,+17,+19,+del(22q)	Gahrton et al 1974b
53,XY,+Y,t(9;22)(q34;q11),+3,+6,+8,+9,+10,+12,-15,+22q-/56, XY,+Y,t(9;22),+3,+6,+8,+8,+9,+10,+16,+19,+22q-	Stoll and Oberling 1982
56,XY,+Y,t(9;22),+3,+6,+8,+8,+9,+10,+12,+19,+22q-	
53,XY,t(9;22),+6,+9q+,+10,+12,+19,+21,+22q-/54,XY,t(9;22), +6,+8,+9q+,+10,+12,+19,+21,+22q-	Sonta and Sandberg 1977b
55,XX,t(9;22),+3,+8,+12,+14,+15,+19,+20,+22,+22q-	Lilleyman et al 1978
55,XY,t(9;22),+3,+6,+8,+8,+9,+10,+16,+19,+22q-/53,XY, t(9;22),-15,+3,+6,+8,+8,+9,+10,+12,+22q-	Stoll and Oberling 1978
59,X,-X,t(9;22),+4,+6,+6,t(7;?)(p15;?),+der(7),+8,+8,+12q+, +17,+19,+19,+21,+22,+22q-,+22q-	Miyamoto 1980

Chronic myeloid leukemia, aberrant translocation

Karyotype	Reference
??,X?,t(12;22)	Geraedts et al 1977
??,X?,t(12;22)(p13;q11)	Potter et al 1981
??,X?,t(12;22)(q24;q11)	Potter et al 1981
46,XX,t(12;22)(p13;q13)	Verma and Dosik 1979
46,XX,t(12;22)(p13;q11)	Van Der Blij-Philipsen et al 1977
46,XX,t(12;22)(p13;q11),t(8;17)(q13;q23)	Swolin et al 1983
46,XX,t(12;22)(q12;q11)	Marinello et al 1981
46,XX,t(12;22)(q34;q11),t(9;11)(q34;q11)	Seabright 1983
46,XX,t(9;15;22)(q34;q11;q11)	Hays et al 1981
54-57,XX,t(9;15;22),+5,+7,+8,+10,+11,+12,+17,+22q-	
46,XY,t(12;22)(p13;q11)	Engel et al 1977
46,XY,t(9;12;22)(q34;q15;q11)	Borgström 1981

Chronic myeloid leukemia, Ph1 negative

Karyotype	Reference
46,XY,t(12;13)(q11;p11)/46,XY,+8,-13,t(12;13)(q11;p11)	Labal de Vinuesa et al 1981

Eosinophilic leukemia

48,XY,t(9;22),r(12),+22,+22q-	Hartley 1983

MYELOPROLIFERATIVE DISORDERS

Polycythemia vera

??,X?,t(9;22),+8,+10,+12,+16	Vykoupil et al 1980
46,XX,del(12)(p11)/46,XX,del(12)(p11),del(20)(q11)	Westin 1976
44,X,-X,del(5)(q14q32),-7	Van Den Berghe et al 1979g
44,XX,del(5)(q14q32),-6,-7,-8,t(12;?)(p12;?), t(16;18)(q11;q12)	
46,XY,del(6)(p22),del(11)(q26),del(12)(p12),del(13)(q13)	Kohno et al 1979b
46,XY,t(12;17)(q13;p11)/47,X,-Y,+t(Y;1)(q12;q21),+9	Testa et al 1981b
47,XX,+9/49,XX,+9,+20,+20,t(4;5),-7,-11,-12,+3mar	Zech et al 1976b
47,XY,+12	Wurster-Hill et al 1976

Myelosclerosis / Myelofibrosis

??,XY,t(1;12)(q11;q24)	Miyamoto et al 1981b
44,XY,-5,-7,-11,-12,t(7;12)(p?;q?),t(17;18)(p11;p11), +mar"s"	Nowell and Finan 1978
46,XX,?der(12)	Mitelman 1983
46,XX,t(5;12)(q14;q13)/45,XX,-16	Whang-Peng et al 1978
46,XY,-16,+18	Whang-Peng et al 1978
46,XY,7q-,8q-,10q-,11p-,11q+,12q-,-16,+18	
46,XY,del(11)(q23)/46,XY,del(12)(p11)	Hustinx and Rutten 1983

Chronic myeloproliferative disease, NOS

46,XY,t(4;12)(q35;q11),i(17q),del(20)(q11)	Berger et al 1975

Myeloproliferative disorder, special type

45,XX,t(1;?)(p36;?),-5,-7,-12,t(13;?)(q34;?),-17,-22, +4mar/45,XX,t(1;?),-2,-3,-5,-7,-12,-17,-22,+8mar"s"	Rowley et al 1977a
46,XX,t(3;12)(q29;q24)	Sandberg et al 1982b
46,XY,-12,+mar	Fleischmann and Krizsa 1975

DYSMYELOPOIETIC SYNDROMES

Preleukemia, NOS

43,X,-X,-3,-13,4q-,8q-,12q+	Geraedts et al 1980a
45,XX,-1,-4,5q-,-7,-12,-18,14q+,+4mar"s"	Pedersen-Bjergaard et al 1981
45,XY,-2,-5,-7,-8,-11,-12,-13,-14,+t(2;5),+t(11;12), +t(16;17),+17,+3mar	Watt et al 1982
46,XX,t(3;12)(q21;p13),del(7)(q11),del(11)(q11), t(7;11)(q11;q11),t(11;12)(q22;p13)	Mitelman 1983
46,XX,t(5;12)(?q15;q13),-17,+mar"s"	Anderson and Bagby 1982
46,XY,del(12)(p11)/46,XY,del(12)(p11),del(20)(q11)	Mitelman 1983
46,XY,del(12)(p11)/46,XY,del(12)(p11),del(20)(q11)	
46,XY,del(12)(p11),del(20)(q11)	
46,XY,t(3;12)(q?;p?)	Anderson and Bagby 1982
47,X,Xp-,1p+,5q-,6q+,11q-,11p+,+11,12p-	Swansbury and Lawler 1980
47-48,XY,+3,+9,+12	Panani et al 1977

Chronic myelomonocytic leukemia

46,XY,del(12)(p12)	Kardon et al 1982a

Refractory anemia, NOS

46,XX,del(5)(q15),inv(12)(p13q21)	Mitelman 1983

Refractory anemia without excess of blasts

44,XX,5q-,-7,+8,-12,-13,-17,-20,+2mar	Teerenhovi et al 1981

Refractory anemia with excess of blasts

43,XX,3q-,del(5)(q13q33),7q-,-12,14p+,16q+,-17,-18,18q+	Kerkhofs et al 1982
44-47,XY,-3,-5,-7,-12,-16,-20,del(1)(q32),3p+,del(11)(q22), 3-6mar,+r	Streuli et al 1980

Pancytopenia

46,XX,-7,+8,-12,-16,-17,t(6;12)(q?;q?),5q-,6q-,+mar	Nowell and Finan 1978

Dysmyelopoietic syndrome, special type

45,XY,del(2)(q33),-5,del(7)(q22q32),-12,-15, t(12;15)(q11;q26),del(19)(q12),+r"s"	Albain et al 1983

SPECIAL LEUKEMIAS

Hairy cell leukemia

46,X,-Y,+12	Golomb et al 1978b
46,X,-Y,t(12;?)(q14;?)	Golomb et al 1978b

LYMPHOCYTIC LEUKEMIAS

Acute lymphocytic leukemia (ALL)

28,XX,-1,-2,-3,-4,-5,-7,-8,-9,-11,-12,-13,-14,-15,-16,-17, -19,-20,-22	Kaneko and Sakurai 1980
32,XY,-2,-3,-4,-5,-7,-8,-11,-12,-13,-14,-15,-16,-17,-20	Shabtai and Halbrecht 1981b
46,XX,2q+,4q-,12p+,13q+,14q+,-20,+21	Whang-Peng et al 1976c
46,XX,t(4;12)(q12;p13)	Mitelman 1983
46,XY,+2,t(6;?)(p21;?),+7,-12,-13,14q+,17p+	Oshimura et al 1977a
46,XY,t(12;17),14q+	Minowada et al 1977
46,XY,t(8;9)(q22;p24)/45,XY,t(8;9),-10,+12,-21	Morse et al 1982a
46-47,XY,del(5)(q15),del(12)(p12),del(17)(p11),+8	Prigogina et al 1979
47,X,del(Y)(q11),del(3)(q11),t(Y;12)/50,X,del(Y)(q11), del(3)(q11),del(1)(p31),t(Y;12),+4mar	Zuelzer et al 1976
47,XX,+12	Oshimura et al 1977a
47,XY,t(7;12)(q22;p13),+19	Morse et al 1979b
48,XY,t(4;12)(q?;q?),t(6;15)(q?;q?),t(9;22)(q34;q11),11q-, +20,+21	Nordenson 1983
49,XX,del(4)(p15),t(4;11)(q21;q23),+5,+8,del(9)(p13), del(9)(p13),t(12;?)(q24;?),+mar	Morse et al 1982b
54,XX,7p+,+10,+10,+14,+14,+18,+18,+21,+21/27,X,-X,-1,-2,-3, -4,-5,-6,-7,7p+,-8,-9,-11,-12,-13,-15,-16,-17,-19,-20, -22	Oshimura et al 1977a
54-55,XX,+5,+12,+14,+16,+17,+18,+21	Prigogina et al 1979
54-57,XY,+3,+5,+6,+7,+12,+13,+14,+21,+2mar	Prigogina et al 1979
56,XX,+3,+6,+12,+13,+14,+15,+18,+21,+r	Prigogina et al 1979
57,XY,+X,+5,+6,i(7q),+10,+12,+13,+15,+18,+21,+21,+22	Oshimura et al 1977a

ALL, FAB type L1

26,XX,-1,-2,-3,-4,-5,-6,-7,-8,-9,-10,-11,-12,-13,-14,-15, -16,-17,-19,-20,-22	Hoeltge et al 1982
28,XX,-1,-2,-3,-4,-5,-6,-7,-8,-9,-11,-12,-13,-15,-16,-17, -19,-20,-22/56,XX,+X,+X,+10,+10,+14,+14,+18,+18,+21,+21	Brodeur et al 1981a
34-36,XY,-2,-3,-4,-7,-9,-12,-13,-14,-16,-17,-20,+22	Sandberg et al 1982c
46,XX,t(12;17)(p13;q12)/47,XX,t(12;17),+mar	Kaneko et al 1982e
46,XX,del(9)(p21),t(12;17)(p13;q12)	
46,XY,del(2)(q32),-11,-12,+2mar/46,X,-Y,del(2),-4,-5,-11, -12,-21,+6	Kaneko et al 1981b
47,XY,-7,-10,-12,+17,del(11)(p11),del(16)(q22), +t(10;?)(q22;?),+t(12;?)(q15;?),+mar	Kaneko et al 1982e

ALL, FAB type L2

Karyotype	Reference
26,XX,-1,-2,-3,-4,-5,-6,-7,-8,-9,-10,-11,-12,-13,-15,-16,-17,-18,-19,-20,-22	Brodeur et al 1981a
45,X,-X,-1,-14,-17,+21,t(11;12)(p15;p11),+i(17q),+t(1;?)(p34;?)	Kaneko et al 1981b
46,X,-X,-15,del(11)(q21),del(12)(p11),+2mar	Kaneko et al 1981b
46,XX,t(5;8;14)(q11;q24;q32),t(12;22)(p13;q11),-18,+t(18;?)(q23;?)	Kaneko et al 1982e
47,XY,+18	Kaneko et al 1982e
46,XY,-2,-5,-6,-7,+8,-9,-10,-11,-12,-16,+18,+t(2;?)(q35;?),+t(11;?)(q21;?),+5mar	
49,XY,+7,+12,-13,t(6;18)(p25;q21),t(11;14)(q23;q32),+t(9;?)(p24;?),+t(1;13)(q12;p13)	Kaneko et al 1982e
52-54,XX,+6,+7,+9,+12,+3mar	Mitelman 1983

Chronic lymphocytic leukemia

Karyotype	Reference
??,XY,1q-,t(1;10)(q11;p15),t(1;12)(q11;p12)	Miyamoto et al 1981b
44,XX,6q-,i(8q),12p-,-14,t(14;14)(q31;q12),-20,20p+	Sparkes et al 1980
44-48,XY,t(Y;9)(q12;q13),del(1)(p22p32),t(4;6)(p16;q15),del(6)(q15),t(12;14)(q15;q32),+mar	Robert et al 1982
45,XX,-2,t(8;12)(q?;q?)	Pittman et al 1982
46,XY,+12,-17/47,XY,+12,-17,+mar	Robert et al 1982
46,XY,+12,-20/45,XY,-20	Robert et al 1982
46,XY,del(11)(q22),+12,-13,-21,-22,+3mar/45,XY,del(11),+12,-13,-15,-21,-22,+2mar/46,XY,del(11),+12,+12,-13,-21,-22,+2mar/46,XY,t(1;8)(p22;q24)	Robert et al 1982
46,XY,t(1;7;9)(p36;q21;q12),t(12;14)(q13;q24)	Miyoshi et al 1979a
47,XX,+12	Hurley et al 1980
47,XX,+12	Vahdati et al 1983a
47,XX,+12,del(14)(q24)	Robert et al 1982
47,XX,+12,del(14)(q?)	Schröder et al 1981
47,XX,+12,del(2)(q31)	Gahrton et al 1980c
47,XX,+12,del(2)(q31)	Robert et al 1982
47,XX,+12	Robert et al 1982
47,XX,+12	Robert et al 1982
47,XX,+12/47,XX,inv(11)(p15q22),+del(12)(q12orq21),t(12;15)(q15orq21;p13)	Najfeld et al 1980
47,XY,+12	Schröder et al 1981
47,XY,+12	Morita et al 1981
47,XY,+12	Morita et al 1981
47,XY,+12	Shabtai 1983
47,XY,+12	Vahdati et al 1983a
47,XY,+12,del(22)(q13)	Morita et al 1981
47,XY,+3,+12,-20,/47,XY,+9,+12,-20/45,XY,-20/46,XY,+12,-20	Morita et al 1981
48,XX,+3,+12/47,XX,+12/46,XX,-11,+12/45,XX,-11	Morita et al 1981
48,XY,+12,+19/48,XY,t(1;11)(q31;q22),+12,+19	Vahdati et al 1983a
49,XY,+12,+18,+19	Finan et al 1978

Prolymphocytic leukemia

44,X,-Y,-11,-12,-13,-13,-14,-15,-15,-18,-18,+8mar — Diamond et al 1980

Lymphocytic leukemia, special type

46,X,-Y,-1,1p+,del(9)(q32),12p+,14q+,+2mar — Ueshima et al 1981
46,XY,t(1;7)(p36;q22),del(9)(q12),t(12;14)(q13;q32) — Miyoshi et al 1981a
49,XX,+7,+8,+12,8q+,8p+,del(10)(q23),21q+/49,XX,+8,+12, +del(7)(p13),8q+,8p+,10q-,21q+ — Ueshima et al 1981
73,XX,cx,12p+ — Ueshima et al 1981

MONOCLONAL GAMMOPATHIES

Multiple myeloma

44,XY,del(3)(q25),t(5;12)(q15;q14),del(7)(q22), del(11)(q24),-14,-18,+mar — Mitelman 1983
46,XY,+t(1;?)(p1?;?),+t(1;12)(q21;q24),del(6)(p11),-8,+9, -12,-13,-16,-19,22q+,+mar — Philip et al 1980
46,XY,-7,-12,+2mar — Chrz et al 1983
55,XY,+1,+5,+9,+11,-12,+13,+15,del(1)(q32),del(6)(q25), t(9;19)(q13;q13),14q+,16p+,+4mar — Liang et al 1979
45,XY,-2,-2,-3,+5,-8,-10,+14,-15,-16,del(6)(q25),+4mar

Plasma cell leukemia / Plasmocytoma

45,XY,-5,del(6)(q11),t(6;7)(q?;q?),+9,-12,-13,-16,-17, t(9;22)(q34;q11),+2mar — Karpas et al 1982

SOLID TUMORS

UNCLASSIFIED NEOPLASMS

EPITHELIAL NEOPLASMS

Carcinoma, NOS

32-35,XX,cx,del(1)(p12),dup(1)(q21q32),2q+,5q+,7p+,11p+, hsr(10),t(11;12)(q13;q24),12p+,13q+	Kusyk et al 1982
41,XX,t(1;1;?)(p36q21;q12;?),t(1;?)(q12;?),del(6)(q21), t(5;12)(q?;q?),del(5)(q12q32)	Atkin and Baker 1982a
44,XX,ins(1;1)(q32;q12q31),i(2q),+3,t(9;11)(q13;p15),-12, -15,17p+,-18	Atkin and Baker 1982a
45,XY,-4,-7,+8,-9,-12,+13,-17,-18,-19,-20,+21,+3mar	Reichmann et al 1981
46,XX,+8,+12,+21,-2,-15,-16	Wake et al 1981
46,XX,t(3;17)(p24;p13)/46,XX,t(5;17)(q22;p13)/46,XX, t(9;18)(q13;q21)/46,XX,t(12;15)(q13;q22)	Stenman et al 1982
47,XX,+1,+3,t(11;14)(q?;q?),-12,-21,+r	Atkin and Baker 1982a
48-52,XX,+2,+11,+12,i(13q)	Riet-Fox et al 1979
50,XX,+X,+8,+12,+21	Sonta et al 1977
58,XY,del(3)(p14),+4,+7,+8,+10,+11,+12,+14,+15,+17,+18,+19, +21	Kovacs 1978a
63,XX,-1,+3,+5,+6,+7,+8,+10,+11,+12,-13,-15,+20,+21,+22, +9mar	Sandberg 1977
79,XX,cx,i(1q),t(1;3)(p11;q29),12q+,15p+,i(5p)	Kovacs 1978a
79,XX,cx,i(1q),t(1;3)(p13;q29),t(1;14)(p13;p13), t(12;?)(q24;?),t(15;21)(p11;q11),i(5p),r	Kovacs 1981

Carcinoma NOS, metastatic

35,XX,-1,-3,-4,-5,-7,-8,-10,-11,-12,-13,-14,-15,-16,-17, -18,-19,-20,-21,-22,+8mar	Pathak et al 1979
45-50,XX,-2,+7,-22,t(1;12),t(13;14),inv(7)(q?)	Ayraud 1975
46,XX,t(9;22)(q34;q11)"s"	Togawa et al 1981
52,XX,+3,+5,+7,+8,+9,+11,12q+,13q-,-14,+16,i(17q)"s"	
59,XX,cx,t(1;12)(q?;q?)	Jones Cruciger et al 197[illegible]
66,XX,cx,t(1;12)(q?;q?)	Jones Cruciger et al 197[illegible]
100,XY,t(12;16),inv(3)(q?),del(1)(q?),del(1)(p?),cx	Ayraud 1975

Small cell carcinoma

44,XY,cx,del(1)(q31),t(2;3)(p?;q?),del(3)(p14p23), t(9;13)(p11;q11),del(11)(p15),12q+,del(X)(q22)	Whang-Peng et al 1982b
60,XY,cx,del(3)(p14q23),del(3)(p14),12q+	Whang-Peng et al 1982b
67,XY,cx,inv(1)(q32p36),dup(1)(q32q44),del(3)(p14q13), del(3)(p14p23),t(7;13)(p11;q11),del(9)(q11),del(11)(p11), t(12;19)(q24;p13)	Whang-Peng et al 1982b

Papillary carcinoma

Karyotype	Reference
42-54,X,-X,-14,-16,-16,+1,+3,+12,+20,1p-,6q-	Wake et al 1981
50,X,-X,+1,+3,+12,+20,-14,-16,1p-,6q-	Wake et al 1980
54,XY,+del(2)(q11),t(2;14)(q11;q32),t(3;11)(p13;p15), +i(5p),+8,+9,+12,+16,+17,+21,+2mar	Pathak et al 1982
69,XX,cx,1p+,1q-,1p-,1q-,2q+,3q+,4q+,6p+,12q+,14q+	Wake et al 1980

Adenoma

Karyotype	Reference
46,X,t(X;4)(q25;p16),del(2)(p13p21),del(5)(q15q31),-7, t(9;12)(p13;q13),del(11)(q11q14), t(7;9;12)(p11p22q11;p13p24;q13)/45,X,-X	Mark et al 1980
46,XX,del(12)(q13q15)/46,XX,del(12)(q13q15),dmin	Mark et al 1982b
46,XX,t(9;12)(p13p22;q13q15)/47,XX,+8,t(9;12)	Mark et al 1981a

Adenocarcinoma

Karyotype	Reference
40-42,X,-X,t(1;?)(p13;?),del(1)(p12),2q+,del(3)(q21), del(6)(q13),del(7)(q11),t(11;11)(p15;q13), t(12;12)(q24;?),13p+,15p+,i(22q)	Kusyk et al 1981
47,XY,-10,-12,+17,+18,+20	Couturier-Turpin et al 1982
49,XY,+7,-9,+11,+12,+mar	Couturier-Turpin et al 1982
68-74,XX,cx,t(13;12),17q+,t(16;16),1q-,i(12p)	Riet-Fox et al 1979
71,XX,cx,1q-,4q+,6q-,9q+,10q-,12q+,17q+,14q+	Wake et al 1981

Adenocarcinoma, metastatic

Karyotype	Reference
??,X,del(X)(q21),t(1;16)(q?;p?),del(12)(p12),del(16)(q21)	Ayraud et al 1977
39-41,XX,-3,-7,-8,der(1),der(2),der(4),der(12),cx	Tiepolo and Zuffardi 1973
49,XX,1q+,+3,4q+,+7,+8,12q+,17q+	Granberg et al 1973

Adenomatous polyp

Karyotype	Reference
48-52,X,-Y,+7,+8,+13,+14,+16,+18,-20,+12q-,1q+	Reichmann et al 1982b

MESENCHYMAL NEOPLASMS

Mesothelioma, malignant

Karyotype	Reference
41,XY,del(1)(p13),t(1;3)(p13;q11),t(7;7)(q11;q36), t(9;12)(p24;q13),del(12)(q13),del(13)(q12q14),-14,-15, t(17;?)(p13;?),i(19p),-20,-21,-22	Mark 1978

MELANOMAS

Malignant melanoma, metastatic

Karyotype	Reference
41-43,XY,del(1)(q21q25),i(1q),t(1;9)(p22;q34), t(11;12)(q11;p11),t(10;12)(q26;q13),t(4;?)(q35;?), del(5)(p13),t(14;?)(p12;?),i(17q),t(18;?)(p11;?), del(6)(q21)	Kakati et al 1977
80,XY,t(11;12)(q23;q13),del(5)(q13),del(5)(p11), del(9)(q13),t(5;9),t(14;?)(q11;?),t(7;12)(q22;q24), del(1)(p13),del(11)(q23)	Kakati et al 1977

NEUROGENIC NEOPLASMS

Neuroblastoma

Karyotype	Reference
47,XY,1p+,4p+,6q+,10q+,12q+,+19	Brodeur et al 1977
89,XX,12q+,13p+	Brodeur et al 1977

Retinoblastoma

Karyotype	Reference
44,XY,1p+,-3,t(3;12)(q?;p?),+6q-,der(9),13p+,-14,-16	Kusnetsova et al 1982
46,XX,+i(6p),12q+,19p+,-22	Kusnetsova et al 1982
46,XX,inv(1),inv(2),inv(3),inv(4),t(6;7),inv(9),t(11;?), 11p-,inv(12),i(17q),-20	Gardner et al 1982
46,XY,dup(1)(q25q44),t(7;10)(q36;q11),-8,-10, t(12;13)(q11;p11),dup(13)(q22q34),+18,+22	Gardner et al 1982
47,XX,12p+,12q+,+i(17q),t(4;16)(q35;q24),-16	Gardner et al 1982
47,XY,5q+,+i(7q)/47,XY,+i(6p),12q+	Kusnetsova et al 1982
51,XY,del(6)(q21),+8,+11,-12,t(13;18)(p11;q11),14q+,-16, +19,+22,+2mar	Hossfeld 1978

Meningioma, benign

Karyotype	Reference
45,XX,-22/44-45,XX,-8,-12,-21,-22	Mark et al 1972b
45,XY,-22/44,XY,-12,-22	Mark 1973a
47,XX,+12,del(22)(q11)/46,XX,-8,+12,-17,del(22)(q11)	Mark et al 1972a
50,XX,+9,+10,+12,+15,+20,-22	Mark 1973c
51,XY,+5,6q+,+12,+16,+17,+18,+20,-22/52,XY,+5,+8,+10,+12, +17,+18,+20,-22	Zang and Zankl 1983

EMBRYONAL AND MISCELLANEOUS NEOPLASMS

Blastoma, NOS

Karyotype	Reference
40-47,X,-X,cx,7q+,+del(10)(p13),+del(12)(p11), del(11)(p11p14)	Slater and Kraker 1982

Teratoma, malignant

Karyotype	Reference
47,XX,+12	Kusyk et al 1982

Seminoma

Karyotype	Reference
??,XY,cx,i(12p)	Atkin and Baker 1982b
??,XY,cx,i(12p)	Atkin and Baker 1982b
??,XY,cx,i(12p)	Atkin and Baker 1982b
??,XY,cx,i(12p)	Atkin and Baker 1982b

LYMPHOMAS

1. UNCLASSIFIED AND MISCELLANEOUS LYMPHOMAS

Non-Hodgkin's lymphoma, NOS

Karyotype	Reference
46-48,XX,t(1;?)(q21orq22;?),t(2;?)(q37;?),-8,+9,+12, 16q+"s"	Mark et al 1976
48-54,XX,+X,3p+,-4,-6,-8,-9,-10,+12,-15,+7-10mar	Pearson et al 1982
52,XY,+2,+3,t(5;9)(q22;q32),+6,+12,+14,+20	Reeves and Pickup 1980

Non-Hodgkin's lymphoma, lymphocytic, NOS

Karyotype	Reference
48,XX,del(1)(q22q25),i(mar1),del(12)(p12),dup(14)(q24q32)	Mark 1975a
50,XY,+3,+3,+12q-,+18,+14q+	Fleischman and Prigogina 1977
51,XX,+5,+9,+11,+del(11)(q?),+del(12)(q?),14q+,i(17q)	Prieto et al 1978a

Non-Hodgkin's lymphoma, histiocytic, NOS

Karyotype	Reference
47,XX,t(1;18)(p36;q21),+7,t(8;12)(q11;q15), t(8;12;13)(q11q22;q15;q14),del(13)(q14),t(8;14)(q22;q32)	Mark et al 1978
47,XY,-6,-14,1q-,+1p-,2q+,3q+,9p+,10p-,10p+,12p-,18q+,22p+, t(2;14)(q37?;q13?)	Fukuhara et al 1979b
48,XY,del(3)(p21),del(9)(q22),11q+,12q+,del(6)(q23orq25), 15q+,del(12)(q15)	Reeves 1973
48,XY,t(3;8)(q22;q11),t(5;10)(q33;q22),del(7)(q21), t(4;9)(q11;p24),t(10;13)(q22;q21),r(12),del(19)(p13), del(22)(q11),+7	Mark et al 1978
82,XXX,t(3;12)(q13;q15),t(3;8)(q22;q11),del(3)(q11), del(5)(q31),i(7q),del(10)(q24),t(10;10)(q24;q26), inv(11)(p15q21),del(12)(q13),del(13)(q14)	Mark et al 1978
89,XX,i(1q),del(3)(p15p23),del(10)(q24),del(12)(p12), i(17q)	Mark 1977

Malignant histiocytosis

Karyotype	Reference
50,XY,+5,+12,+13,+21	Sonta et al 1977

2. HODGKIN'S LYMPHOMAS

Hodgkin's disease, mixed cellularity

Karyotype	Reference
48,XY,+9,+12,t(18;?)(p11;?)	Hossfeld and Schmidt 197
61-80,XX,del(1)(q32),i(1q),del(2)(p13),del(6)(q13or15), del(6)(q23or25),15q+,12q+,del(5)(q31),del(7)(q11), del(10)(q26),del(3)(p13)	Reeves 1973

Hodgkin's disease, lymphocytic depletion

Karyotype	Reference
60,XY,+del(1)(p21),t(1;5)(q44;q33),dup(5)(q12), t(1;5)(p21q23;q13),del(6)(q22),t(9;?)(q12;?), t(3;12)(q11;q24),t(14;?)(p11;?),i(17q)	Hossfeld and Schmidt 197

Hodgkin's disease, nodular sclerosis

Karyotype	Reference
47,XY,t(1;?)(p36;?),+2,t(2;6)(q31;q27),+8,del(12)(p11), t(17;?)(p13;?),-14,-20,+mar	Reeves and Pickup 1980

3A. NON-HODGKIN'S LYMPHOMAS - RAPPAPORT CLASSIFICATION

Lymphocytic, poorly differentiated, diffuse

Karyotype	Reference
47,XY,t(14;18)(q32;q21),-9,-10,+12,-16,8p-,13q+,+3mar	Fukuhara and Rowley 1978
48,XX,t(14;18)(q32;q21),1p+,12q+,+mar	Fukuhara and Rowley 1978
48-49,XX,2p-,+7,+i(8q),+12,14q+,15q-,18q-	Kakati et al 1980

Lymphocytic, poorly differentiated, nodular

Karyotype	Reference
46,XY,12q+/46,XY,t(14;18)(q32;q21)	Kristoffersson et al 198
48,XY,t(14;18)(q32;q21),-6,-11,+12,+del(1)(p?),3q+,+2mar	Fukuhara et al 1979a
52,X,-X,+2,-4,+5,+7,+8,-14,-15,-16,+20,+21, +t(1;15)(p11;q11),+t(12;17)(q24;q11),+t(14;?)(q32;?), +mar	Kaneko et al 1982a

Mixed cell, diffuse

Karyotype	Reference
50,XY,+11,+12,-15,+del(3)(q11),+del(3)(q11), +t(X;1)(p22;q11)	Kaneko et al 1982a
50,XY,+X,-1,+5,+7,+12,-14,-17,del(13)(q12q14), ins(1;1)(q21;q21q32),+t(14;?)(q32;?),+t(1;17)(q21;q25)	Kaneko et al 1982a

Mixed cell, nodular

Karyotype	Reference
50,XX,t(3;6)(q29;p21),del(9)(p13q22),+11,+12,+21	Mark et al 1976

Histiocytic, diffuse

Karyotype	Reference
45,X,-Y,t(14;15)(q32;q21orq22),-15,+18,-22,1q+,3p+,5p-,7p-, 9p+,12p+,16q+,17p-,+2mar	Fukuhara and Rowley 1978
47,X,+X,-Y,+12,-14,del(11)(q21q23),del(16)(q22), t(14;?)(q32;?)	Kaneko et al 1982a
47,X,-X,t(4;14)(q?;q32),-18,1q+,2p+,6q-,7q+,9q-,12q+,+3mar	Fukuhara and Rowley 1978
50,XX,+3,+3,-4,-6,-9,+12,-17,+18,+i(6p),+3mar	Kaneko et al 1982a
50,XX,del(1)(p22),+3,+7,+12,t(1;14)(p22;q32),+18	Mark et al 1979
50,XY,del(3)(p24),+7,+12,+14,+20	Kristoffersson et al 1981
51,XY,+X,-2,+5,+11,-12,+21,del(6)(q21),+t(2;?)(q33orq35;?), +t(12;?)(p11;?),+mar	Kaneko et al 1982a

3B. NON-HODGKIN'S LYMPHOMAS - KIEL CLASSIFICATION

Immunocytoma

Karyotype	Reference
46,XX,1p+,+2,del(17)(p12),17p+	Kristoffersson 1983
46,XX,1p+	
47,XX,1p+,+12,del(17)(p12)	
46,XY,t(10;10)(p15;q24),dup(12)(q13q22)	Gahrton et al 1982
49,XX,+3,del(6)(q14),+7,+18/49,XX,+3,del(6)(q14), del(9)(q22),+12,+18	Kristoffersson 1983

Centroblastic-centrocytic, diffuse

Karyotype	Reference
48-51,XY,del(1)(p33q23q25),-2,+del(3)(p22),+4,4p+,+5,+6p+, +7,+8,+9,+21,+22	Kristoffersson 1983
45-51,XY,del(1)(p33q23q25),del(2)(q14),+del(3)(p22),+4,+5, +7,+8,-12	

Centroblastic-centrocytic, follicular

Karyotype	Reference
46,XX,t(12;?)(q15;?)	Reeves and Pickup 1980
46,XY,12q+	Kristoffersson 1983
46,XY,t(14;18)(q32;q21)	
46-47,XY,del(4)(q28),-10,+12,-18,del(18)(q22)	Kristoffersson 1983
47,XY,t(1;14)(q21;q32),+12,-16,+19,t(14;?)(q32;?)	Reeves and Pickup 1980
52,XX,+X,+7,+8,del(10)(q24),+12,t(13;?)(q22;?), t(14;?)(q32;?),del(18)(q21),+19,+mar	Reeves and Pickup 1980

Centroblastic, diffuse

Karyotype	Reference
46-49,XY,1p+,+2,-6,+del(9)(p21),+12,+13,14q+?,i(17q),+18	Kristoffersson 1983
47,XY,14q+,+22	

Lymphoblastic, Burkitt's type

??,XY,t(8;14),12q- Douglass et al 1980a
43,XY,+1,+2,+3,-7,-8,-9,-12,-16,-22,del(1)(q25), del(2)(p11),del(10)(q22),14q+ Biggar et al 1981
46-47,XY,t(14;?)(q32;?),dup(1)(q21q32),t(2;3)(q13;q29),+5, 6p-,-8,del(9)(q?),+12,+18,+mar Miyoshi et al 1981c
47,XX,+12,t(8;14)(q24;q32),del(12)(p11q22),del(17)(q21) Biggar et al 1981
48-53,X?,t(8;14)(q24;q32),+X,+2,+3,+9,+10,+11,+12,+20 Zech et al 1976a
49,XY,+3,+7,+12,14q+ Biggar et al 1981

Lymphoblastic, convoluted type

43-46,XY,del(1)(p32),del(2)(p21),-6,t(7;9)(p11;q11),-9,-10, t(11;15)(p11;q11),-11,del(12)(q14),-13,-14,i(17q),-17, +18,del(19)(p11),-22,+2mar Kristoffersson 1983
45,XY,t(3;?)(q29;?),t(8;?)(q2?;?),t(12;17)(p13;q11) Kaneko et al 1982d

Immunoblastic

45,XY,del(1)(p22orp23),del(3)(q11),del(3)(q21),del(8)(p11), del(9)(p11),del(9)(q12),t(12;?)(q24;?) Reeves and Stathopoulos 1976
48-49,XY,+3,+7,+11,t(12;?)(p13;?),t(13;?)(p13;?),-16,-20, +21,+22 Reeves and Pickup 1980
48-49,XY,del(3)(p23),+7,+12,+14,+20 Kristoffersson 1983
51,XY,+X,t(3;3)(q11;q27),t(8;?)(q2?;?),+7,+12,+21,+21 Reeves and Pickup 1980
51,XY,t(1;6)(p22;q13orq15),t(2;13)(q21;q14), t(8;16)(q24;p11),+11,+12,del(15)(q22),+16,+20,+mar Reeves and Pickup 1980
78-83,XY,+del(1)(q11),+5,+7,+12,+13q+,+14q+,+14q+,+20,+mar Kristoffersson 1983
80-86,XY,del(1)(q11),+3,+7,+9,+12,+13q+,+14q+,+14q+,+15q+, +18,+20,+mar

3C. NON-HODGKIN'S LYMPHOMAS - LUKES AND COLLINS CLASSIFICATION

Mycosis fungoides

46,XY,6q-,der(12) Nowell et al 1982
46-48,XX,del(1)(q42),+3,-11,-12,+2-6mar Edelson et al 1979

Sezary's syndrome

46,XY,2q+,der(12),15q+ Nowell et al 1982

Follicular center cell, diffuse, large cleaved

45,XY,-6,-9,-17,2q-,3p-,3q-,4q-,5q-,6p+,7p+,7q+,+7q-,9p-, 12q+,13p+,20p-,+mar Fukuhara et al 1978b
47,XY,-6,-14,1q-,+1p-,2q+,3q+,9p+,10p-,10p+,12p-,18q+,19p+, 19q+,+22p+,+mar Fukuhara et al 1978b

Follicular center cell, diffuse, small non-cleaved

44,X,-Y,t(1;14)(p21;q13),5p+,del(6)(q13),del(8)(p21),9p+, 9p-,t(11;21)(q23;q22),13q+,-10,+12,-17p+,der(19),22q+	Fukuhara et al 1979a

Immunoblastic type (B-cell)

48,XY,t(4;12)(q?;q?),t(6;15)(q?;q?),t(9;22)(q34;q11),11q-, +20,+21	Nordenson et al 1983

3D. NON-HODGKIN'S LYMPHOMAS - WORKING FORMULATION

Small lymphocytic

47,XX,+12	Yunis et al 1982
47,XY,+12	Yunis et al 1982
47,XY,+12	Yunis et al 1982
47,XY,+12	Yunis et al 1982

Follicular, small cleaved cell

44,XY,+der(1),+2,t(2;13),-3,5q-,-12,t(14;18)(q32;q21), i(17q)	Yunis et al 1982

Follicular, mixed small cleaved and large cell

50,Y,inv(X),+3,+7,+10,t(12;12),+12,t(14;18)(q32;q21),+18, 18q-,-21	Yunis et al 1982

Follicular, large cell

50,XY,1q-,t(2;3),+3,+10,t(1;12),13q-,+15,-19,t(3;19),+21	Yunis et al 1982

Diffuse, mixed small and large cell

47,XX,+9,t(9;15)(p13;q11),t(12;14)/49,XX,+3,+5,+9,t(9;15), t(12;14)	Yunis et al 1982

Large cell, immunoblastic

99,XY,cx,5p-,5q-,6q-,7p-,i(12q)	Yunis et al 1982

Small non-cleaved cell

46,X,-X,t(8;14)(q24;q32),dup(13),+r/45,X,-X,t(8;14),-12, t(1;15),der(22),+r	Yunis et al 1982

Chromosome 13

HEMATOLOGICAL DISORDERS

UNCLASSIFIED LEUKEMIAS

Acute leukemia, NOS

54,XX,+X,+3,+6,+8,+8,+13,+19,+22 — Muir et al 1977

NONLYMPHOCYTIC LEUKEMIAS

Acute nonlymphocytic leukemia (ANLL)

Karyotype	Reference
??,XY,+6,+10,+11,+13	Philip et al 1978a
39,X,-X,-3,-5,-13,-14,-15,-16,-17,-21,-22,del(7)(p11), t(12;?)(p13;?),t(16;?)(p13;?),+3mar	Mitelman 1983
39,X,-X,-3,-5,-13,-14,-15,-16,-17,-21,-22,del(7)(p11), t(12;?)(p13;?),t(16;?)(p13;?),+3mar	
43,X,-Y,t(1;8)(p22;q24),-1,-11,-11,-13,-16,-17,-22, t(1;22)(q11;p11),t(11;13)(q23q25;q12q14),t(17;?)(p11;?), +mar	Oshimura et al 1976b
43,XX,-1,3p+,3q-,4q+,-5,-5,-7,+9,+9,10p+,+10,+10,10p-,10q-, +11q-,-13,-15,-16,-17,-18,-21"s"	Whang-Peng et al 1979
44,XY,del(5)(q21),del(7)(q11),del(11)(q11), t(7;11;12)(q11;q11;p13),del(13)(q11),-20,-21/45,XY, t(3;12)(q11;p13),del(5)(q21),del(7)(q11),del(13)(q11), -20	Mitelman 1983
45,XX,t(13;14),5q-"c""s"	Rowley 1976
45,XY,-9,13q-"s"	Geraedts et al 1980a
46,X?,t(1;13)(p?;q?)	Wurster-Hill 1983
46,XX,t(13;14)(q?;q?)	Bernard et al 1982a
46,XY,13q+	Bernard et al 1982a
46,XY,inv(9),t(11;15)(q11;q26),t(11;13)(q11;q34)"c"	Philip et al 1978a
47,XX,+13	Hsu et al 1979
47,XX,+t(1;16),-16,-16,+13,+20	Prieto et al 1978b
47,XX,-4,+13,t(1;4)(q22q25;p14p16)	Oshimura et al 1976b
47,XX,t(3;5)(q21;q31),t(1;13)(p36;q14),+8,+22	Oshimura et al 1976b
47,XX,t(9;22)(q34;q11),+22q-/48,XX,t(9;22),+8,+22q-	Mitelman 1983
49,XX,t(9;22)(q34;q11),+8,+13,+22q-	
49,XX,t(9;22)(q34;q11),+8,+22q-,+mar	

Karyotype	Reference
50,XX,t(9;22)(q34;q11),dup(1)(q25q44),+del(3)(p11),+8, +22q-,+mar	
47,XY,+13	Bernard et al 1982a
47,XY,+13/48,XY,+13,+13	Mitelman 1983
49,XY,t(5;?)(q35;?),+6,+13,+14	Philip et al 1978a
50,XY,+6,+13,+16,+mar	Chrz et al 1983
51,XX,del(5)(q?),+7,+8,+13,+17,+22	Van Den Berghe et al 1976
51,XY,+8,+9,+13,+14,+21	Oshimura et al 1976b
52,XX,1p+,+10,+13,+19,+21,+21,+22"c"	Berger et al 1973

ANLL, FAB type M1

Karyotype	Reference
46,XX,-1,inv(1),3q+,5q-,t(13;17),-18,+mar/51,XX,-1,inv(1), 3q+,5q-,t(13;17),-18,+6,+9,+9,+11,+11,+mar	Hagemeijer et al 1981a
46,XY,+del(12)(p11q23),-13	Brodeur et al 1983
46,XY,-2,-5,-7,-13,+16,-21,-21,+5mar	Sessarego et al 1981a
48,XY,+8,+13,inv(9)"c"	Yunis et al 1981
58,XY,+1,+4,+5,+6,+8,+11,+13,+13,+14,+14,+15,+21,+21, t(9;22)(q34;q11)	Alimena 1983

ANLL, FAB type M2

Karyotype	Reference
46,XX,+1p-,+9,-16,-16/46,XX,+1p-,+13,-16,-16	Prieto et al 1981
46,XX,t(3;13)(p14;q34)	Lessard and Le Prise 1983
46,XY,t(13;14)(q22;q24)	Yunis et al 1981
47,XX,+13/47,XX,+21/48,XX,+13,+21	Prieto et al 1981
47,XX,1p+,der(5),-6,12p+,-13,-14,-16,-17,-17,-21,+22,+7mar, dmin	Cooperman and Klinger 1981
47,XY,+13	Golomb et al 1978c

ANLL, FAB type M1+M2

Karyotype	Reference
37-45,XY,-7,-12,-13,-14,-16,-18,-19,-21,del(4)(q29), del(5)(q24),der(9),der(17),11q+,12p+,t(14;?)	Fitzgerald et al 1983
42-44,XX,+8,-12,-13,-14,t(2;5)(q12;q35)	Mitelman et al 1981
43,XX,1p+,-4,-5,-6,-7,-13,-17,del(18)(q21),+3mar	Rowley and Potter 1976
45,XX,-5,-11,-13,-18,-21,+4mar,dmin"s"	Sandberg et al 1982a
46,XX,del(1)(q32)/46,XX,del(1)(q32),del(13)(q22)	Fitzgerald et al 1983
46,XX,del(13)(q21)	Prigogina et al 1979
46,XX,t(6;9)(p23;q34)	Testa et al 1979
48,XX,t(6;9)(p23;q34),+8,+13	
47,X,-X,+del(1)(p11),del(5)(q13),t(5;12)(q21;q24), t(11;12)(q25;q13),t(13;14)(q11;p11),+16,+21	Mitelman et al 1981
47,X,-Y,t(1;13)(q32p32q11;q11),del(3)(q21), t(8;21)(q22;q22),del(9)(q22),+8,+18	Sakurai et al 1982b
47,XY,del(3)(q12),del(5)(q12),t(6;?)(q11;?),-13,+2mar"s"	Testa et al 1981b
47-49,XY,-5,del(5)(q15),-7,i(8q),dic(9),-13,-16,-22, +6-9mar"s"	Testa et al 1981b
48,XY,+13,+mar	Benedict et al 1979
48,XY,+8,+13	Fitzgerald et al 1983
57,XY,+6,+11,+13,+15,+21,+21,+21,+21,+21,+21,+22"c"	Shiraishi et al 1982

ANLL, FAB type M3

45-51,XY,t(3;12)(p14;q24),t(13;18)(q11;p11),+1,+6,+16,+19, +21,+22	Mitelman et al 1981

ANLL, FAB type M4

44,XY,del(5)(q15q23),del(7)(q22),-9,-12,-13,del(15)(q22), +mar/46,X,-Y,del(5),del(7),+8,-12,-13,del(15),+mar,dmin	Alimena 1983
45,XX,del(2)(q31),del(7)(p13),-13	Rowley and Potter 1976
46,XX,-1,-3,-5,+10,+11,-13,-14,+21,+del(3?)(p?q?), ?t(1;3;5)(p?;p?q?;q?)	Mitelman 1983
46,XX,13q+	Mitelman et al 1981
46,XX,inv(11)(p12q23),t(11;13)(p12q23;q21)	Hagemeijer et al 1981a
46,XX,t(3;13)(q29;q12),del(5)(q13q31)/46,XX, t(1;18)(q44;q21),t(3;13),del(5)	Alimena 1983
46,XY,-13,t(13;13)"s"	Berger et al 1981a
46,XY,13q+	Brodeur et al 1983
47,XY,+13/49,XY,+8,+13,+mar	Muir et al 1977
49,XX,+8,+13,+21	Prigogina et al 1979

ANLL, FAB type M5

46,XY,t(3;9;13)(p21;q22;q14)	Benedict et al 1979
46,XY,t(6;11)(q26;q22)/52-53,XY,t(6;11)(q26;q22),+der(6), +3,+10,+13,+18,+19,+21,dmin	Hagemeijer et al 1981a
46,XY,t(6;11)(q27;q23)/51-52,XY,t(6;11),+der(6),+3,+19,+19, +21,dmin/53,XY,t(6;11),+der(6),+4,+10,+13,+18,+19,+21	Löwenberg et al 1982
47,XY,+8,t(1;13)(p31;q11)/48,XY,+8,+18q-	Hagemeijer et al 1981a
47,XY,1p+,1q+,2p+,-3,4q+,+13,+mar	Kaneko et al 1982b
48,XY,+9,+13	Yunis et al 1981
50,XX,+6,+8,10p-,+12,+13	Nordenson 1983

ANLL, FAB type M5a

46,XY,+8,+8,-15,t(13;15)(p11;q11)	Berger et al 1982a
46,XY,t(8;13)(q24;q12),del(11)(q23),t(11;13)(p12;q12),13p+	Berger et al 1982a

ANLL, FAB type M4+M5

45,XY,del(11)(p11),t(12;?)(p13;?),del(16)(q13), t(13;14)"c""s"	Nora et al 1979
53,XY,+13,-14,+16,+16,+19,+20,+3mar	Li et al 1983

ANLL, FAB type M6

43,X,-Y,-4,-13,-14,-15,+2mar/46,X,-Y,-3,+9,+mar	Li et al 1983
44,XY,t(2;13)(q37;q12?),t(13;?)(p11;?),-15"s"	Weinfeld et al 1977a
44-45,XY,-2,-4,-7,-10,-13,-16,-17,-22,+3mar	Mitelman et al 1981
45,XY,-13	Li et al 1983
46,XX,t(4;9)(q21;q34),t(4;13)(q12;q12),del(13)(q12orq13), del(12)(p11)	Fitzgerald et al 1983
46,XY,5p-,-9,-12,13q-,+2mar	Nordenson 1983

Karyotype	Reference
49,XX,+6,+6,+21,der(13)(q12q21)	Bernstein et al 1982b
72,XY,cx,t(2;15)(q31;q15),del(6)(p22q15),t(8;13)(q23?;q13), t(11;14)(p12;q13),i(16p),t(15;18)(q15;p11)	Douglass and Freeman 1983

Chronic myeloid leukemia, t(9;22)

Karyotype	Reference
26,XX,t(9;22)(q34;q11),-1,-2,-3,-4,-5,-6,-7,-9,-10,-11,-12, -13,-14,-15,-16,-17,-18,-19,-20	Hartley and McBeath 1981
35,XX,t(9;22)(q34;q11),-3,-4,-5,+8,-9,-10,-11,-12,-13,-14, -15,-16,-17/66-72,XX,t(9;22)	Pedersen and Boesen 1983
45,X,-Y,t(9;22),del(13)(q11orq13q21)	Kohno et al 1979b
45,X?,t(9;22),-13/45,X?,t(9;22),-13,-20,+mar	Carbonell et al 1982a
45,XY,t(9;22)(q34;q11),-13,-17,+t(13;17)(p11;q11)	Tomiyasu et al 1982
46,X,-Y,t(9;22)(q34;q12),+13/47-48,X,-Y,t(9;22),+8,+13, +22q-/47,XY,t(9;22),+22q-	Izakovic et al 1982
46,XX,t(9;22)(q34;q11)/46,XX,t(9;22),-7,-17,+13,+14	Sonta and Sandberg 1978b
46,XX,t(9;22)(q34;q11)	Rajasekariah et al 1982
50,XX,t(9;22)(q34;q11),t(9;22)(q34;q11),+8,+11,+13, t(16;17)(p11;p12),+22	
46,XX,t(9;22),13q+	Prigogina et al 1978
46,XX,t(9;22),t(8;13)(p12;q34),+der(8),-20	Lyall and Garson 1978
46,XY,t(9;22)(q34;q11)	Como and Graze 1979
35,XY,t(9;22)(q34;q11),-3,-4,-5,-7,-9,-11,-12,-13,-15,-16, +3mar	
46,XY,t(9;22)(q34;q11),t(6;13)"c"	Tomiyasu et al 1982
46,XY,t(9;22)(q34;q11)	Ishihara et al 1982
27,X,-Y,t(9;22),-1,-2,-3,-4,-5,-6,-7,-9,-11,-13,-14,-15, -16,-17,-18,-19,-20,-22	
46,XY,t(9;22),del(13)(q21)	Kohno et al 1979b
46,XY,t(9;22)/46,XY,t(9;22),del(5)(q12),del(13)(q21), t(17;21)(p11;q11)	Bernstein et al 1980b
46,XY,t(9;22)/48,XY,t(9;22),+8,+22q-/49,XY,t(9;22),+8,13q-, +19,+22q-	Hagemeijer et al 1980b
47,XX,t(9;22)(q34;q12)/62,XX,cx,t(9;22)(q34;q12), del(1)(q25),del(1)(q11),del(1)(q21),del(1)(q11q25), del(2)(p12),del(3)(q21),del(6)(q23),i(13q),+22q-,+22q-, +22q-,+22q-	Hjorth et al 1980
47,XY,t(9;22)(q34;q11),-7,13q+,20p+,+21,+mar	Alimena et al 1982b
47,XY,t(9;22),+22q-,+8,t(12;13)(q?;q?),-13,i(17q)	Rozynkowa et al 1977
47,XY,t(9;22),+22q-/46,XY,t(9;22),-9,+22q-/46,XY,t(9;22), t(3;5)(p?;q?),t(6;13)(p?;q?)	Kohno and Sandberg 1980
47,XY,t(9;22),+8,t(12;13)(q11;q34)	Fleischman et al 1981
47,XY,t(9;22),+8/50,XY,t(9;22),+8,+13,+i(17q),+22q-	Miyamoto 1980
48,XX,t(9;22)(q34;q11),t(13;14)(q14;q24orq32),+19,+22q-	Carbone et al 1982a
48,XX,t(9;22)(q34;q11),del(13)(q12q32),+19,+22q-	Tomiyasu et al 1982
48,XY,t(9;22),+8,+13,-15,t(9;15),+i(22q)	Stoll and Oberling 1978
50,XY,t(9;22),+5,+8,+13,-15,t(9;15),+i(22q),+22q-	
48-53,X?,t(9;22),+8,+8,+13,+14,+19,+21,+22q-	Carbonell et al 1982a
49,XX,t(9;22),+12,+13,i(17q),+19	Seabright 1983
49,XY,t(9;22),+13,+22q-,+22q-/52,XY,t(9;22),+13,+13,+17, +17,+22q-,+22q-	Stoll and Oberling 1978
54,XY,t(9;22)(q34;q11),+8,+10,+11,+13,+14,+19,+21,+21	Hayata et al 1975b
55,XY,t(9;22),+6,+10,+10,+11,+13,+14,+14,+19,+21	Seabright 1983

Chronic myeloid leukemia, aberrant translocation

??,X?,t(9;13;22)(q34;q22;q11)	Potter et al 1981
46,X?,t(13;22)(q34;q11),t(8;9)(q22;q34)	Hossfeld 1983
46,XX,inv(5)(p14q12),t(9;13;17;22)(q31;q11;q12;q12)	Fraisse et al 1980
46,XX,t(17;22),inv(5),t(9;13)	Freycon et al 1982
46,XY,22q-"c"	Mitelman 1983
46,XY,22q-"c"	
46,XY,22q-"c"	
46,XY,t(13;17;22)(p13;q12;q11)"c"	
46,XY,t(13;17;22)(p13;q12;q11)"c"	
46,XY,t(13;17;22)(p13;q12;q11)"c"	
46,XY,t(13;17;22)(p13;q12;q11)"c"	
46,XY,t(13;17;22)(p13;q12;q11)"c"	
46,XY,t(13;22)(p13;q11)	Hayata et al 1975b
46,XY,t(13;22)(q34;q11)	Seabright 1983
46,XY,t(5;13)(q33;q14),del(22)(q11)	Rolovic et al 1983
46,XY,t(9;13;22)(q13-21;q13;q11)	Hayata et al 1975b
46,XY,t(9;13;22)(q34;q14;q11)	Seabright 1983
47,XY,+Y,t(9;13;15;22)(q34;q12;q22;q11)	Potter et al 1975
50,XY,del(22)(q11),+8,+11,+21,+21/51,XY,del(22)(q11),+8,+11,+13,+21,+21	Sonta and Sandberg 1978b
50,XY,del(22)(q11),+8,+11,+21,+21,/51,XY,del(22)(q11),+8,+11,+13,+21,+21	

Chronic myeloid leukemia, Ph1 negative

46,XY,del(13)(q13)	McGlave et al 1982
46,XY,t(12;13)(q11;p11)/46,XY,+8,-13,t(12;13)(q11;p11)	Labal de Vinuesa et al 1981
47,XX,+13	Hsu et al 1974a

MYELOPROLIFERATIVE DISORDERS

Polycythemia vera

??,X?,+2,+del(13)	Vykoupil et al 1980
??,X?,4q+,+13,15q-,20q-	Vykoupil et al 1980
44-45,XY,-10,-13,-16,-17,-20	Wurster-Hill et al 1976
46,XX,13q-	Wurster-Hill et al 1976
46,XX,del(13)(q11q22)	Kohno et al 1979b
46,XY,del(11)(q13),del(13)(q?)	Shiraishi et al 1975
46,XY,del(13)(q11orq13q21)	Kohno et al 1979b
46,XY,del(13)(q21)	Kohno et al 1979b
46,XY,del(5)(q15),del(20)(q11)/46,XY,del(11)(q22)/46,XY,del(13)(q14q32)	Testa et al 1981b
46,XY,del(6)(p22),del(11)(q26),del(12)(p12),del(13)(q13)	Kohno et al 1979b
46,XY,t(1;13)(q12;p12)	Hsu et al 1977
47,XY,+9	Van Den Berghe et al 1979g
47,XY,del(1)(q12),t(1;18)(q12;q23),+9,del(20)(q12)/47,XY,+9,del(11)(q21),del(20)(q12)/47,XY,+9,del(20)(q12)	

47,XY,+9/47,XY,del(5)(q14q32),del(8)(q?),+9,del(11)(q21), del(13)(q21)/47,XY,del(5)(q14q32),del(6)(q?),+9, del(11)(q21),del(13)(q21),del(1)(q12),del(20)(q12)	
48,XY,+3,-5,t(13;14)(q?;q?),+2mar	Nowell and Finan 1978

Myelosclerosis / Myelofibrosis

46,X,del(X)(p?q?),del(5)(q?),13q+,15q-"s"	Whang-Peng et al 1978
46,XY,9q+,13q-,20q-/47,XY,-7,+2mar	Chrz et al 1983
46,XY,t(1;2)(q?;q?),-8,inv(13)(q12q31),i(17q)	Whang-Peng et al 1978
47,XX,+13	Hsu et al 1979
47,XX,t(13;14),+8,+9"c"	Greef et al 1982

Idiopathic thrombocythemia

46,XY,del(13)(q21)	Mitelman 1983
46,XY,del(13)(q21)	

Chronic myeloproliferative disease, NOS

46,X?,t(8;13)(p?;q?)	Lessard and Le Prise 1983

Myeloproliferative disorder, special type

43,XX,-5,-13,-18/44,XX,-5,t(13;?)(p11;?),-18"s"	Rowley et al 1977a
45,XX,t(1;?)(p36;?),-5,-7,-12,t(13;?)(q34;?),-17,-22, +4mar/45,XX,t(1;?),-2,-3,-5,-7,-12,-17,-22,+8mar"s"	Rowley et al 1977a
46,XY,t(1;13)(p36;q13),del(3)(p21),-5,inv(7)(p11q22), t(16;16)(q22;q24),+mar	Clare et al 1982

DYSMYELOPOIETIC SYNDROMES

Preleukemia, NOS

43,X,-X,-3,-13,4q-,8q-,12q+	Geraedts et al 1980a
45,XX,-7/48,XX,+t(1;7),-7,+11,+13"s"	Pedersen-Bjergaard et al 1981
45,XY,-2,-5,-7,-8,-11,-12,-13,-14,+t(2;5),+t(11;12), +t(16;17),+17,+3mar	Watt et al 1982

Refractory anemia, NOS

46,XX,t(2;4;13)(p11;q12;q34),del(5)(q15q31)	Kerkhofs et al 1982
47,XY,+13,i(17q)	Prieto et al 1976

Acquired idiopathic sideroblastic anemia

46,XX,del(5)(q13q34)/46,X,Xq-,5q-,13q-"s"	Kerkhofs et al 1982

Refractory anemia without excess of blasts

Karyotype	Reference
44,XX,5q-,-7,+8,-12,-13,-17,-20,+2mar	Teerenhovi et al 1981

Refractory anemia with excess of blasts

Karyotype	Reference
44,XX,5q-,-13,-17,-20,-22,+2mar	Teerenhovi et al 1981
56-64,XY,+1,-2,+6,+9,+13,+15,+19,+21,10-13mar	Streuli et al 1980

Pancytopenia

Karyotype	Reference
46,XX,-4,-5,-6,-7,-11,-13,-16,+19,+20,t(7;13)(q?;q?), +4mar/43,XX,-2,-3,-5,-7,-16,+19,15q+,+mar	Nowell and Finan 1978

Aplastic anemia

Karyotype	Reference
46,XX,del(13)(q13)	Kohno et al 1979b

Dysmyelopoietic syndrome, special type

Karyotype	Reference
45,XX,t(13;14)(p13;q11),del(5)(q14)/50,XX,t(13;14),+1,+6, -7,+8,+10,del(5),+mar"c""s"	Rowley et al 1981a

SPECIAL LEUKEMIAS

LYMPHOCYTIC LEUKEMIAS

Acute lymphocytic leukemia (ALL)

Karyotype	Reference
28,XX,-1,-2,-3,-4,-5,-7,-8,-9,-11,-12,-13,-14,-15,-16,-17, -19,-20,-22	Kaneko and Sakurai 1980
32,XY,-2,-3,-4,-5,-7,-8,-11,-12,-13,-14,-15,-16,-17,-20	Shabtai and Halbrecht 1981b
45,XX,-8,t(13;?)(q?;?),19q+	Prigogina et al 1979
45,XX,t(14;14)(q34;q11),t(13;15)(q11;p11), t(13;17)(p11;q11),-16,18q+,+3mar	Levitt et al 1978
46,XX,-9,13q-,14q-,+t(9;13)(p?;q?)	Prigogina et al 1979
46,XX,2q+,4q-,12p+,13q+,14q+,-20,+21	Whang-Peng et al 1976c
46,XX,t(4;11)(q21;q23)/49-50,XX,t(4;11),+8,+13,+19,+4q-	Prigogina et al 1979
46,XY,+2,t(6;?)(p21;?),+7,-12,-13,14q+,17p+	Oshimura et al 1977a
46,XY,13q+,21q-	Whang-Peng et al 1976c
46,XY,t(4;11)(q21;q23)	Oshimura et al 1977a
46-49,XY,t(4;11)(q21;q23),+6,+8,+13,+17,t(1;13)(p22;q12), t(7;9)(q11;q34)	
47,XX,t(1;?)(q25;?),+13,del(1)(q32)	Oshimura et al 1977a
48,XX,+13,+22	Nordenson 1983
48-53,XX,t(4;11)(q21;q23),+6,+7,+9,+13,+21	Prigogina et al 1979
53,XX,+2,+5,i(7q),+8,+13,+21,+21,+22,+17q+	Oshimura et al 1977a
54,XX,7p+,+10,+10,+14,+14,+18,+18,+21,+21/27,X,-X,-1,-2,-3, -4,-5,-6,-7,7p+,-8,-9,-11,-12,-13,-15,-16,-17,-19,-20, -22	Oshimura et al 1977a
54-57,XY,+3,+5,+6,+7,+12,+13,+14,+21,+2mar	Prigogina et al 1979

55,XY,+X,+3,+6,+10,+13,+14,+16,+21,+21 Oshimura et al 1977a
56,XX,+3,+6,+12,+13,+14,+15,+18,+21,+r Prigogina et al 1979
57,XY,+X,+5,+6,i(7q),+10,+12,+13,+15,+18,+21,+21,+22 Oshimura et al 1977a
61,XX,+X,+1,+2,+3,+4,+5,+6,+8,del(6)(q21),+13,+14,+15,+16, +18,+21,+22 Oshimura et al 1977a

ALL, FAB type L1

26,XX,-1,-2,-3,-4,-5,-6,-7,-8,-9,-10,-11,-12,-13,-14,-15, -16,-17,-19,-20,-22 Hoeltge et al 1982
28,XX,-1,-2,-3,-4,-5,-6,-7,-8,-9,-11,-12,-13,-15,-16,-17, -19,-20,-22/56,XX,+X,+X,+10,+10,+14,+14,+18,+18,+21,+21 Brodeur et al 1981a
34-36,XY,-2,-3,-4,-7,-9,-12,-13,-14,-16,-17,-20,+22 Sandberg et al 1982c
47,XY,del(17)(q11q21),t(1;13)(p34;q14),+mar Kaneko et al 1981b
48,XX,+13,+19,/47,XX,+19 Kaneko et al 1982e

ALL, FAB type L2

26,XX,-1,-2,-3,-4,-5,-6,-7,-8,-9,-10,-11,-12,-13,-15,-16, -17,-18,-19,-20,-22 Brodeur et al 1981a
44,XX,-8,t(13;14),t(9;22)"c" Takeuchi et al 1981
49,XY,+7,+12,-13,t(6;18)(p25;q21),t(11;14)(q23;q32), +t(9;?)(p24;?),+t(1;13)(q12;p13) Kaneko et al 1982e
56,XX,+1,+6,+10,+13,+15,+16,+17,+18,+20,+21 Mitelman 1983

ALL, FAB type L3

46,XY,t(8;14)(q24;q32),13q+ Berger and Bernheim 1982
46,XY,t(8;14)(q24;q32),13q+ Berger and Bernheim 1982
46-47,XX,t(14;?)(q32),t(13;?)(q31;?),1q+,+7 Slater et al 1979

Chronic lymphocytic leukemia

41,XX,t(14;14)(q11;q32),t(1;13)(q44;q11),t(15;18)(q11;q23), 10q+,-16,-20 Kaiser McCaw et al 1975
45-46,XY,-8,9q+,-10,13p+,-14,-15,-19,-20,+mar Vahdati et al 1983a
46,XX,t(6;7)(q?;q?),t(7;13)(q?;q?),t(11;14)(q11;q32),17p+ Schröder et al 1981
46,XX,t(9;13)(p22;q12) Nowell et al 1981
46,XY,del(11)(q22),+12,-13,-21,-22,+3mar/45,XY,del(11),+12, -13,-15,-21,-22,+2mar/46,XY,del(11),+12,+12,-13,-21,-22, +2mar/46,XY,t(1;8)(p22;q24) Robert et al 1982
46,XY,del(13)(q21) Kohno et al 1979b
47,XY,+2,t(9;13),t(9;19),i(17q) Finan et al 1978
50,XX,+5,+5,t(13;?)(q34;?),+22 Nordenson 1983

Prolymphocytic leukemia

44,X,-Y,-11,-12,-13,-13,-14,-15,-15,-18,-18,+8mar Diamond et al 1980

Lymphocytic leukemia, special type

45,Y,-X,t(Y;14)(q12;q32),-13,-17,del(2)(q33), t(3;13)(q21;q34),del(6)(q16orq22),9p+,9q+,10q+,16q+,18q+	Miyoshi et al 1981a
46,XY,-10,-13,5p+,+t(10;13)(p13;q11q32),21p+	Ueshima et al 1981

MONOCLONAL GAMMOPATHIES

Multiple myeloma

45,X,del(X)(q25),t(3;8;22)(q13;q34;q11),-5,-7,-10,-11, del(13)(q13?),+16,19q+,+3mar	Van Den Berghe et al 1979f
45,XY,del(1),+del(1),-10,-11,-11,+t(11;?),-13,-14,-20, +3mar	Philip et al 1980
46,XY,+t(1;?)(p1?;?),+t(1;12)(q21;q24),del(6)(p11),-8,+9, -12,-13,-16,-19,22q+,+mar	Philip et al 1980
47,X,-X,inv(1)(p22q12),t(7;?)(q11;?),+9,t(11;?)(p15;?),-13, -14,t(14;?)(q11;?),+15,17q+	Philip et al 1980
52,X,-X,+t(1;?)(q12;?),+3,+7,+8,+9,+11,-13,+18,-21,+2mar	Philip et al 1980
55,XY,+1,+5,+9,+11,-12,+13,+15,del(1)(q32),del(6)(q25), t(9;19)(q13;q13),14q+,16p+,+4mar	Liang et al 1979
45,XY,-2,-2,-3,+5,-8,-10,+14,-15,-16,del(6)(q25),+4mar	

Plasma cell leukemia / Plasmocytoma

43,X,-X,-1,+t(1;8)(q11;q11),+t(1;9)(q21;p24), +t(1;10)(p31;q26),+t(1;?)(q12;?),-8,-9,-10,-13,-22	Ueshima et al 1983
44,X,-Y,+t(1;9)(q12q44;q34),t(6;8)(q13;p11),-9,-13	Ueshima et al 1983
45,XY,-5,del(6)(q11),t(6;7)(q?;q?),+9,-12,-13,-16,-17, t(9;22)(q34;q11),+2mar	Karpas et al 1982
86,XX,cx,t(1;?)(q11;?),2q+,dup(7)(q22q36),t(8;?)(p11;?), t(9;13)(p23;q21)	Ueshima et al 1983

SOLID TUMORS

UNCLASSIFIED NEOPLASMS

Malignant neoplasm, NOS

Karyotype	Reference
47,XX,-2,-5,+6,-10,11q+,-13,+18,-21,t(1;?)(q21;?), t(1;13)(q21;p11),t(1;6)(q11;q23)	Kovacs 1978a

EPITHELIAL NEOPLASMS

Carcinoma, NOS

Karyotype	Reference
32-35,XX,cx,del(1)(p12),dup(1)(q21q32),2q+,5q+,7p+,11p+, hsr(10),t(11;12)(q13;q24),12p+,13q+	Kusyk et al 1982
43,X,-X,t(1;11)(q21;q23),3q+,4q+,5q+,-11,-13,-16,17p+	Atkin and Baker 1982a
44,XY,+5,-8,-11,-13,-13,-14,-14,-15,+19,-22,-22,t(1;11), 5q-,9q+,+5mar	Reichmann et al 1981
44-50,XX,+3,dup(1)(q21q32),t(10;11)(q26;q25),t(2;17), t(6;13),t(9;15),del(10)(q24)	Riet-Fox et al 1979
45,XY,-4,-7,+8,-9,-12,+13,-17,-18,-19,-20,+21,+3mar	Reichmann et al 1981
46,XX,+2,-7,-10,+13,-16,-17,+20,+5q-,15p+,15p+,18p+	Reichmann et al 1981
46,XY,-4,-5,-11,-13,-17,+20,+21,1p+,9p-,+3mar	Reichmann et al 1981
47,XX,+8,+13,-14	Sonta et al 1977
47,XY,-5,+7,-10,-13,-14,4q-,5q-,+4mar	Reichmann et al 1981
47-48,Y,-X,-5,-8,+13,-14,-16,-18,-22,+7mar	Reichmann et al 1981
48-52,XX,+2,+11,+12,i(13q)	Riet-Fox et al 1979
49,XX,+7,+8,+13	Reichmann et al 1981
50,XX,+X,+8,+8,+13	Sonta and Sandberg 1978a
52,XY,+X,+Y,-2,+7,+8,+9,+10,+13,-17,+i(1q)	Reichmann et al 1981
54,XX,+1,+2,+8,+11,+13,+14,+21,+21	Sonta et al 1977
58,XX,+X,+1,+2,+3,5p-,+6,+7,+9,11q+,-13,-14,-15,+16,+19, +20,+6mar	Kovacs 1978a
63,XX,-1,+3,+5,+6,+7,+8,+10,+11,+12,-13,-15,+20,+21,+22, +9mar	Sandberg 1977
64-68,XX,cx,t(Y;13),i(9p),i(9q),dmin	Reichmann et al 1981
66,XX,t(Y;13),1q+,i(4q),i(9p),i(9q),cx	Martin et al 1979
69-70,XX,cx,t(13;13),t(1;10),1p+,del(1)(p22)	Riet-Fox et al 1979
73,XY,+1,+7,+11,+13,-16,+17,+18,+19	Sonta and Sandberg 1978a
73-81,XX,cx,2q+,i(13q)	Reichmann et al 1981
80-83,XX,cx,t(2;5;16),t(2;5;18),t(11;13),11q+,t(8;10), del(2)(q24),del(6)(q21),i(14q),dup(9)(q11q12)	Riet-Fox et al 1979

Carcinoma NOS, metastatic

Karyotype	Reference
??,X,-X,t(1;?),hsr(1),6q-,t(9;?),dic(13),dmin,cx	Bartnitzke and Bullerdiek 1983
35,XX,-1,-3,-4,-5,-7,-8,-10,-11,-12,-13,-14,-15,-16,-17, -18,-19,-20,-21,-22,+8mar	Pathak et al 1979
45-50,XX,-2,+7,-22,t(1;12),t(13;14),inv(7)(q?)	Ayraud 1975
46,XX,t(9;22)(q34;q11)"s"	Togawa et al 1981
52,XX,+3,+5,+7,+8,+9,+11,12q+,13q-,-14,+16,i(17q)"s"	
81-83,XX,t(1;3)(p22;p25),del(1)(p13),t(5;13)(q13;q14), del(4)(p12),i(2)(p11),i(5)(p11),t(4;5)(q12;q11q13)	Kakati et al 1976a

Large cell carcinoma

55-70,XY,t(4;5;14),del(1)(q11q21),t(7;11)(p22;q13), t(7;21)(q11;q22),t(3;8)(p?;q?),t(6;11)(p11;p11),i(16q), t(3;?)(q11;?),t(21;?)(q11;?),t(13;?) — Kakati et al 1975

Small cell carcinoma

40,XY,cx,t(1;16)(q21;q24),t(1;19)(q23;q13),del(3)(p21p24), del(3)(p23q26),del(10)(q22),del(11)(q23), t(5;13)(q13;q32),del(17)(p12),dmin — Whang-Peng et al 1982b
44,XY,cx,del(1)(q31),t(2;3)(p?;q?),del(3)(p14p23), t(9;13)(p11;q11),del(11)(p15),12q+,del(X)(q22) — Whang-Peng et al 1982b
67,XY,cx,inv(1)(q32p36),dup(1)(q32q44),del(3)(p14q13), del(3)(p14p23),t(7;13)(p11;q11),del(9)(q11),del(11)(p11), t(12;19)(q24;p13) — Whang-Peng et al 1982b
69,XY,cx,del(3)(p14q24),t(11;14)(p11;p11), t(11;13)(p11;p11),del(X)(q23) — Whang-Peng et al 1982b

Papillary carcinoma

41,XX,cx,t(6;14)(q21;q24),del(1)(p?q?),13q+ — Wake et al 1980
59,XX,cx,13q+,14q+ — Wake et al 1980

Adenoma

46,XX,t(1;3)(p21;p21),t(1;13)(p36;q14),t(3;8)(p21;q12), t(1;5;20)(p21;p14q12;p13),t(5;8)(p14;q12), t(X;9)(q27;q12) — Mark et al 1982a

Adenocarcinoma

??,XX,t(13;15)(q33;q15) — Trent and Davis 1979
40-42,X,-X,t(1;?)(p13;?),del(1)(p12),2q+,del(3)(q21), del(6)(q13),del(7)(q11),t(11;11)(p15;q13), t(12;12)(q24;?),13p+,15p+,i(22q) — Kusyk et al 1981
44,XX,+5,-9,-13,-13,-15,-17,-18,+22,+2mar — Martin et al 1979
44-50,X,-X,-1,t(1;?)(q21;?),del(2)(q33),4p+,-5,-7,+8, del(10)(q12),t(10;13)(q12;q34),del(17)(q25),-20 — Kusyk et al 1982
62-74,XX,cx,3q-,6q-,t(13;14),dmin — Hecht et al 1983
68-74,XX,cx,t(13;12),17q+,t(16;16),1q-,i(12p) — Riet-Fox et al 1979

Adenocarcinoma, metastatic

??,X,del(X)(q21),t(13;13)(q?;q?),del(16)(q21) — Ayraud et al 1977
??,XX,t(1;16),2q-,del(3)(p14),del(4)(q22),del(6)(q22), i(7q),del(9)(q31),t(13;13) — Ayraud et al 1977

Adenomatous polyp

48-52,X,-Y,+7,+8,+13,+14,+16,+18,-20,+12q-,1q+ Reichmann et al 1982b

MESENCHYMAL NEOPLASMS

Sarcoma, NOS

49,XX,+3,+13,+14 Sonta and Sandberg 1978a

Leiomyosarcoma

42,XX,del(1)(p12orp13),del(11)(q13orq14),+7,-9,-13,-14,-15, -18,+19,-22 Mark 1976

Rhabdomyosarcoma

83-87,XY,t(2;13)(q37;q14),cx Seidal et al 1982

Mesothelioma, malignant

41,XY,del(1)(p13),t(1;3)(p13;q11),t(7;7)(q11;q36), t(9;12)(p24;q13),del(12)(q13),del(13)(q12q14),-14,-15, t(17;?)(p13;?),i(19p),-20,-21,-22 Mark 1978
43,XY,t(1;1)(q42;q32p22),t(2;6)(p25;p21),inv(3)(p24q13), t(5;18)(p15;q11),-6,+9,del(13)(q24q14),-14,-18,-22 Mark 1978

MELANOMAS

Malignant melanoma

46,XY,5q+,+7,-8,-9,-10,-13,-20,+21,+3mar McCulloch et al 1976

Malignant melanoma, metastatic

24,X,cx,t(1;15),t(1;14)(q12;p11),del(7)(p13), t(6;7)(p11;q11),t(13;21)(p11;q11),+r Atkin and Baker 1981
45,XY,t(1;9;13)(q?;q?;q?),r(2),t(1;9)(p?;q?),i(8q), t(11;?)(q?;?),-10,-13,+mar Chen and Shaw 1973

NEUROGENIC NEOPLASMS

Neuroblastoma

46,XX,hsr(13p) Gilbert et al 1982
46,XX,t(1;?)(p32;?),2q+,7p+,13q+,hsr(5q),dmin Gilbert et al 1982
46,XY,del(1)(p32),6q+,13q+,16q+,dmin Gilbert et al 1982
47,XY,hsr(13) Gilbert et al 1982
83,XX,cx,hsr(1)(p34),hsr(13p) Gilbert et al 1982
89,XX,12q+,13p+ Brodeur et al 1977

Retinoblastoma

44,XY,1p+,-3,t(3;12)(q?;p?),+6q-,der(9),13p+,-14,-16	Kusnetsova et al 1982
46,XX,+del(1)(p?),-13	Balaban et al 1982
46,XX,3q+,del(13)(q12q14)	Balaban et al 1982
46,XY,+4,t(5;13)(q35;q21),t(7;17)(p22;q12),t(1;9)(q22;p24), -10,del(16)(q11)	Gardner et al 1982
46,XY,dup(1)(q25q44),t(7;10)(q36;q11),-8,-10, t(12;13)(q11;p11),dup(13)(q22q34),+18,+22	Gardner et al 1982
47,XX,+6,del(13)(q12q14),dmin	Balaban et al 1982
47,XX,del(13)(q14),+mar	Balaban et al 1982
47,XY,del(13)(q14),t(20;?)(q13;?),2q+,i(17q),dmin	Balaban et al 1982
51,XY,del(6)(q21),+8,+11,-12,t(13;18)(p11;q11),14q+,-16, +19,+22,+2mar	Hossfeld 1978

Meningioma, benign

38,X,-Y,-22,-21,-17,-14,-13,-10,-9,-4,-1p-,+mar	Zankl et al 1975a
45,XX,-22,+13,-14/44,XX,-22,+13,-14,-1	Mark 1973a

EMBRYONAL AND MISCELLANEOUS NEOPLASMS

Teratoma, malignant

67,XY,+1,+3,+4,+5,+6,+7,+7,+8,+10,+11,+13,+13,+14,+16,+17, +18,+20,+20,+21,+21	Sonta et al 1977

Seminoma

53,XY,+3,+4,+6,+7,+11,+13,+19	Sonta et al 1977

LYMPHOMAS

1. UNCLASSIFIED AND MISCELLANEOUS LYMPHOMAS

Non-Hodgkin's lymphoma, NOS

47-48,XY,t(14;?)(q32;?),16q-,-17,18q-,13q-,+7,+8,8q-	Catovsky et al 1977
50,XX,+X,t(1;11)(q25;q23),+3,+7,-13,t(14;18)(p?;p?),-15, t(19;?)(q13;?),+3mar	Reeves and Pickup 1980

Non-Hodgkin's lymphoma, lymphocytic, NOS

46,XX,+3,t(13;22)(p11;p11)	Fleischman and Prigogina 1977
47,XX,del(13)(q22),del(13)(q14),dup(14)(q24q32),+mar	Mark 1975a

Non-Hodgkin's lymphoma, histiocytic, NOS

46,XX,del(1)(p31p35),+3,del(9)(p11),t(10;?)(q26;?), del(11)(q13),-13,t(11;14)(q13;q32),t(17;21), t(9;22)(q11;p13),+r	Mark et al 1977
47,XX,t(1;18)(p36;q21),+7,t(8;12)(q11;q15), t(8;12;13)(q11q22;q15;q14),del(13)(q14),t(8;14)(q22;q32)	Mark et al 1978
47-48,XY,+X,-13,-15,t(6;11)(p?;q?),t(13;14;?)	Fleischman and Prigogina 1977
48,XY,t(3;8)(q22;q11),t(5;10)(q33;q22),del(7)(q21), t(4;9)(q11;p24),t(10;13)(q22;q21),r(12),del(19)(p13), del(22)(q11),+7	Mark et al 1978
50,XY,t(2;?)(p25;?),del(2)(p13),+i(3p),del(6)(p21),+7,+8, +9,+9,-11,-13,+14,-15,+20,+21,-22	Mark et al 1978
44,XY,del(2)(p13),t(6;?)(p21;?),del(11)(q13), t(11;14)(q13;q31),i(15q),-20,-22	
82,XXX,t(3;12)(q13;q15),t(3;8)(q22;q11),del(3)(q11), del(5)(q31),i(7q),del(10)(q24),t(10;10)(q24;q26), inv(11)(p15q21),del(12)(q13),del(13)(q14)	Mark et al 1978

Non-Hodgkin's lymphoma, T-cell, NOS

47,XX,t(2;?),inv(3),4p-,del(5)(q14q32),6p+,6q+,t(7;7),9q+, 10p-,13q-,t(14;14),-15,-15,16q-,17p+,+21,+2mar	Gaeke et al 1981

Malignant histiocytosis

50,XY,+5,+12,+13,+21	Sonta et al 1977

2. HODGKIN'S LYMPHOMAS

Hodgkin's disease, lymphocytic depletion

46,XY,+2,t(2;?)(q25;?),t(3;11)(q29;p13),t(4;11)(q33;q13), -13	Hossfeld and Schmidt 1978

3A. NON-HODGKIN'S LYMPHOMAS - RAPPAPORT CLASSIFICATION

Lymphocytic, poorly differentiated, diffuse

44,XX,t(11;14)(q13;q32),-3,-9,-13,1p+,+2mar	Fukuhara et al 1979a
47,XY,t(14;18)(q32;q21),-9,-10,+12,-16,8p-,13q+,+3mar	Fukuhara and Rowley 1978
47,XY,t(1;4)(q23;q33),+3,t(10;15),t(11;14)(q13;q32),r(13), r(13)	San Roman et al 1982

Mixed cell, diffuse

50,XY,+X,-1,+5,+7,+12,-14,-17,del(13)(q12q14), ins(1;1)(q21;q21q32),+t(14;?)(q32;?),+t(1;17)(q21;q25)	Kaneko et al 1982a
53,XY,t(8;14)(q24;q32),t(14;18)(q32;q21),+20,1q-,+1p-,2q-, 4q+,5q-,5p+,6q-,+7q+,10q+,11q+,11q-,13p+,+4mar	Fukuhara and Rowley 1978

Histiocytic, diffuse

45,X,del(X)(q24),del(1)(q21),+del(1)(p11),del(2)(q22), del(3)(q21),del(6)(p11),del(6)(q15),t(7;9)(p22;q13), t(8;10)(p23;q21),del(9)(q13),del(10)(q21), t(2;11)(q22;q23),-13,del(14)(q24),del(15)(q22), del(16)(q22),+17,-19 — Mark et al 1979
46,X,-X,-2,-10,+11,-13,-14,-17,del(1)(p22p32), del(3)(p13p21),del(6)(q21),del(9)(p13),+t(2;?)(p23;?), +t(14;?)(q32;?),t(18;?)(q23;?),+t(X;13)(q26;q12),+mar — Kaneko et al 1982a
47,XY,-10,-13,+15,-16,+19,+20,del(6)(q21),del(8)(q22), t(2;14)(q21;q32),t(4;4)(q31;q35),+t(10;?)(p11;?) — Kaneko et al 1982a
61,XY,cx,t(1;16)(q23;p11),dup(1)(q23q32),t(1;6)(q23;p21), t(3;6)(q21;p21),t(7;7)(q32;p22),13p+ — Fitzgerald et al 1980
89,XY,del(1)(p22),t(5;13)(q15;q34),t(14;20)(q32;q12), t(1;17)(p22;q25),t(1;19)(q21;p13),del(20)(q12) — Mark et al 1979

3B. NON-HODGKIN'S LYMPHOMAS - KIEL CLASSIFICATION

Lymphocytic, T-zone

46,XY,+t(Y;14)(q12;q24),3q-,-5,-13,+mar — Gödde-Salz et al 1981b

Centrocytic

44,XY,del(4)(p13),-11,-13,t(13;14)(q11;q32) — Kristoffersson 1983

Centroblastic-centrocytic, follicular

46-49,XX,1p+,3q-,4q+,+6,+13,-14,14q+,17p+ — Kristoffersson 1983
52,XX,+X,+7,+8,del(10)(q24),+12,t(13;?)(q22;?), t(14;?)(q32;?),del(18)(q21),+19,+mar — Reeves and Pickup 1980

Centroblastic, diffuse

46-49,XY,1p+,+2,-6,+del(9)(p21),+12,+13,14q+?,i(17q),+18 — Kristoffersson 1983
47,XY,14q+,+22
46-51,XX,+3,del(6)(q21),+del(6)(q21),+7,-8,del(9)(q11),+13, -16,-17,+19,-20,-21,+1-5mar — Kristoffersson 1983

Lymphoblastic, Burkitt's type

45,X?,t(10;11),t(10;13) — Zech et al 1976a
46,XY,t(8;14)(q24;q32),-13,16p+,+mar — Slater et al 1982
46-48,X?,t(8;14)(q24;q32),+7,15q-,+6,+13 — Zech et al 1976a

Lymphoblastic, convoluted type

43-46,XY,del(1)(p32),del(2)(p21),-6,t(7;9)(p11;q11),-9,-10, t(11;15)(p11;q11),-11,del(12)(q14),-13,-14,i(17q),-17, +18,del(19)(p11),-22,+2mar — Kristoffersson 1983
47,XY,t(1;2)(p34;p21),+4,t(13;18)(p13;p11)/47,XY,+3 — Kaneko et al 1982d

48,XX,+13,+t(1;5)(q21;p15)/48,XX,-5,+8,+13,del(1)(q32), +t(1;5)(q21;p15)	Kaneko et al 1982d

Immunoblastic

48-49,XY,+3,+7,+11,t(12;?)(p13;?),t(13;?)(p13;?),-16,-20, +21,+22	Reeves and Pickup 1980
51,XY,t(1;6)(p22;q13orq15),t(2;13)(q21;q14), t(8;16)(q24;p11),+11,+12,del(15)(q22),+16,+20,+mar	Reeves and Pickup 1980
78-83,XY,+del(1)(q11),+5,+7,+12,+13q+,+14q+,+14q+,+20,+mar	Kristoffersson 1983
80-86,XY,del(1)(q11),+3,+7,+9,+12,+13q+,+14q+,+14q+,+15q+, +18,+20,+mar	

3C. NON-HODGKIN'S LYMPHOMAS - LUKES AND COLLINS CLASSIFICATION

Mycosis fungoides

48,XX,del(1)(p22),t(2;8;14)(q37;q24;q24),+5,del(7)(p13), del(9)(q22),t(9;18)(q11;q24),del(10)(p13), del(13)(q12q14),del(5)(q15q22),del(5)(q13),del(5)(q15)	Fukuhara et al 1978a
46,XX,t(1;14)(q32;q32)	

Sezary's syndrome

42,XY,-13,-21,-22,1p+,17q+	Nowell et al 1982
44-46,XX,t(2;9),t(2;13),17p+,19p+,-21,-22	Edelson et al 1979
47,XY,-9,-13,-17,-19,+del(9),+t(9;13),+t(9;19),+i(17q)	Nowell et al 1982

Follicular center cell, diffuse, large cleaved

45,XY,-6,-9,-17,2q-,3p-,3q-,4q-,5q-,6p+,7p+,7q+,+7q-,9p-, 12q+,13p+,20p-,+mar	Fukuhara et al 1978b

Follicular center cell, diffuse, small non-cleaved

44,X,-Y,t(1;14)(p21;q13),5p+,del(6)(q13),del(8)(p21),9p+, 9p-,t(11;21)(q23;q22),13q+,-10,+12,-17p+,der(19),22q+	Fukuhara et al 1979a

Immunoblastic type (B-cell)

49,XX,+i(1q),+5,del(8)(q24),t(3;22)(q29;q11),+13	Berger et al 1983a

3D. NON-HODGKIN'S LYMPHOMAS - WORKING FORMULATION

Follicular, small cleaved cell

44,XY,+der(1),+2,t(2;13),-3,5q-,-12,t(14;18)(q32;q21), i(17q)	Yunis et al 1982

Follicular, large cell

47,XX,t(2;13),t(14;18)(q32;q21),+21	Yunis et al 1982
50,XY,1q-,t(2;3),+3,+10,t(1;12),13q-,+15,-19,t(3;19),+21	Yunis et al 1982

Diffuse, large cell

54,XY,dup(1)(q?),inv(5),+3,+3,+9,+13,+18,+18,+20,+mar	Yunis et al 1982

Small non-cleaved cell

46,X,-X,t(8;14)(q24;q32),dup(13),+r/45,X,-X,t(8;14),-12, t(1;15),der(22),+r	Yunis et al 1982

Chromosome 14

HEMATOLOGICAL DISORDERS

UNCLASSIFIED LEUKEMIAS

NONLYMPHOCYTIC LEUKEMIAS

Acute nonlymphocytic leukemia (ANLL)

39,X,-X,-3,-5,-13,-14,-15,-16,-17,-21,-22,del(7)(p11), t(12;?)(p13;?),t(16;?)(p13;?),+3mar	Mitelman 1983
39,X,-X,-3,-5,-13,-14,-15,-16,-17,-21,-22,del(7)(p11), t(12;?)(p13;?),t(16;?)(p13;?),+3mar	
42,XX,-5,-12,-14,-16,-18,t(12;16)(q11;q22q24)"s"	Oshimura et al 1976b
43-44,XX,-5,-7,-14,+mar,dmin	Marinello and Levan 1982
43-44,XY,-5,-5,-14,-15,-16,-17,-18,-20,+5mar	Mitelman 1983
44,XX,-9,-10,-17,del(14)(q22),t(9;10)(q22orq34;q11orq22)	Oshimura et al 1976b
45,XX,t(13;14),5q-"c""s"	Rowley 1976
45,XY,-14	Trent et al 1983a
45,XY,t(1;14)(p22;p11),t(1;15)(p22;p11),-2,del(5)(p13),-17, +mar	Philip et al 1978a
46,XX,t(13;14)(q?;q?)	Bernard et al 1982a
46-49,XY,+8,+14,-17,+18,-21"s"	Cavallin-Ståhl et al 1977
47,XY,+8/48,XY,+8,+14	Bernard et al 1982a
49,XY,t(5;?)(q35;?),+6,+13,+14	Philip et al 1978a
51,XY,+3,+5,-9,+19,+20,+21,t(6;14)(p25;q11),t(9;?)(q22;?)	Philip et al 1978a
51,XY,+8,+9,+13,+14,+21	Oshimura et al 1976b
56,XY,+Y,+6,+14,+21,+t(5;6)(q?;p?),+del(1)(p22q25), +del(2)(q12),+del(3)(p12),+del(4)(q13),+del(5)(p14q14), +mar	Zuelzer et al 1976

ANLL, FAB type M1

45,XX,-10,t(14;19)(p?;p?)	Takeuchi et al 1981
45,XY,t(2;5)(q22orq23;q14q15),6p-,-7,t(6;7)(p21;p22q21), t(7;10)(p21;q21),t(14;19)(q11;q13)"s"	Hagemeijer et al 1981a
46,XY,t(12;14)(q24;q32)	Brodeur et al 1983
58,XY,+1,+4,+5,+6,+8,+11,+13,+13,+14,+14,+15,+21,+21, t(9;22)(q34;q11)	Alimena 1983

ANLL, FAB type M2

46,XX,t(14;21)(q24;q21)	Alimena 1983
46,XX,t(8;21),t(14;21)"c"	Tricot et al 1981
46,XY,del(5)(q14q33),t(6;14)(q23;q22)	Kerkhofs et al 1982
46,XY,t(13;14)(q22;q24)	Yunis et al 1981
47,X,-X,+8,t(8;21)(q22;q22),+14	Brodeur et al 1983
47,XX,1p+,der(5),-6,12p+,-13,-14,-16,-17,-17,-21,+22,+7mar, dmin	Cooperman and Klinger 1981

ANLL, FAB type M1+M2

37-45,XY,-7,-12,-13,-14,-16,-18,-19,-21,del(4)(q29), del(5)(q24),der(9),der(17),11q+,12p+,t(14;?)	Fitzgerald et al 1983
42-44,XX,+8,-12,-13,-14,t(2;5)(q12;q35)	Mitelman et al 1981
45,X,-Y,+del(5)(q11),del(11)(q21),-14,-16,-18,-20,+2mar,+r	Mitelman et al 1981
45,XY,-14	Li et al 1983
45-46,XX,+14,-7,-22,+16,-20	Mitelman et al 1981
45-46,XY,9p+,-14	Mitelman et al 1981
46,XX,+14,-22/46,XX,-7,+14,+16,-20,-22	Alimena et al 1977
47,X,-X,+del(1)(p11),del(5)(q13),t(5;12)(q21;q24), t(11;12)(q25;q13),t(13;14)(q11;p11),+16,+21	Mitelman et al 1981
50,XY,+8,+14,+19,+mar	Fitzgerald et al 1983
50-52,XY,+9,+14,+17,+19,+20,+21	Alimena et al 1977
50-52,XY,-5,-7,+9,+14,+17,+19,+20,+21,+mar	Mitelman et al 1981

ANLL, FAB type M3

45,XX,-4,-5,-14,t(21;?)(q22;?),+2mar"s"	Rowley et al 1977a
46,XX,2q-,6p-,7q-,10q+,14q+,-16,+mar	Hagemeijer et al 1981a
46,XY,t(14;21)"c"	Alimena 1983
46,XY,t(15;17),t(14;21)"c"	

ANLL, FAB type M4

43,XY,-2,-3,-6,-7,-12,-15,-21,-22,del(2)(p11),del(11)(p14), t(14;?)(q32;?),t(17;?)(p13;?),del(22)(q11),+5mar"s"	Rowley et al 1981a
46,XX,-1,-3,-5,+10,+11,-13,-14,+21,+del(3?)(p?q?), ?t(1;3;5)(p?;p?q?;q?)	Mitelman 1983
47,XY,+14	Li et al 1983

ANLL, FAB type M5

46,XX,t(14;17)(q13;q24)	Prigogina et al 1979
46,Y,-X,-1,-7,-8,-14,-15,-17,-20,inv(6)(p23q13), del(11)(q23),+t(X;?)(p11;?),+t(1;?)(p24;?), +t(12;?)(q22;?),+t(17;?)(p11;?),+4mar	Kaneko et al 1982b
51,XX,+6,+8,+14,+19,+21,t(11;?)(q21orq24;?)	Soukup and Neely 1981

ANLL, FAB type M5a

47,XY,+8,9q+,del(14)(q23),17p+	Berger et al 1982a

ANLL, FAB type M4+M5

45,XY,del(11)(p11),t(12;?)(p13;?),del(16)(q13), t(13;14)"c""s"	Nora et al 1979
48,XY,+i(1q),+3,+12,-14	Shiraishi et al 1982
53,XY,+13,-14,+16,+16,+19,+20,+3mar	Li et al 1983

ANLL, FAB type M6

43,X,-Y,-4,-13,-14,-15,+2mar/46,X,-Y,-3,+9,+mar	Li et al 1983
44-47,XY,+2,-3,-5,-7,+8,-14,-19,+2mar	Mitelman et al 1981
45-49,XY,-4,-7,-8,-10,+14,+16,-21,-22,+2mar	Mitelman et al 1981
46,XX,-7,-16,-21,+t(2;7)(p11;q11),del(2)(q33),del(11)(q22), +11p+,14p+,+t(16;21;21;21)(p13;q22;q11;q22)"s"	Rowley et al 1981a
51,XY,+1,+2,+6,-7,+8,-14,+15,+21,+mar"s"	Rowley et al 1981a
72,XY,cx,t(2;15)(q31;q15),del(6)(p22q15),t(8;13)(q23?;q13), t(11;14)(p12;q13),i(16p),t(15;18)(q15;p11)	Douglass and Freeman 1983

Chronic myeloid leukemia, t(9;22)

26,XX,t(9;22)(q34;q11),-1,-2,-3,-4,-5,-6,-7,-9,-10,-11,-12, -13,-14,-15,-16,-17,-18,-19,-20	Hartley and McBeath 1981
35,XX,t(9;22)(q34;q11),-3,-4,-5,+8,-9,-10,-11,-12,-13,-14, -15,-16,-17/66-72,XX,t(9;22)	Pedersen and Boesen 1983
43-46,XX,t(9;22),-5,-7,-10,-17,-19,5p+,14q+,+r(?),+mar	Sadamori et al 1981b
45,XX,t(9;22),-2,t(2;14)(q21;q32)	Sessarego et al 1979
46,XX,t(9;22)(q34;q11)/45,XX,t(9;22),t(8;14)(q11;q32), -der(8)	Hayata et al 1975b
46,XX,t(9;22)(q34;q11)/46,XX,t(9;22),-7,-17,+13,+14	Sonta and Sandberg 1978b
46,XX,t(9;22)(q34;q11),14q+	Alimena et al 1982b
46,XX,t(9;22)/48,XX,t(9;22),+19,+22q-/50,XX,t(9;22),+8,+12, +19,+22q-/50,XX,t(9;22),+8,+14,+17,+19,+21	Hagemeijer et al 1980b
46,XY,t(9;22)"s"	Hossfeld 1975b
46,XY,t(9;22)/51,XY,t(9;22),+10,+14,+19,+22q-,+mar"s"	
51,XY,t(9;22),+10,+14,+19,+22q-,+mar"s"	
51,XY,t(9;22),+10,+14,+19,+22q-,+mar"s"	
46,XY,t(9;22)(q34;q11)	Ishihara et al 1982
27,X,-Y,t(9;22),-1,-2,-3,-4,-5,-6,-7,-9,-11,-13,-14,-15, -16,-17,-18,-19,-20,-22	
47,X?,t(9;22),+22q-/48,X?,t(9;22),+8,21q+,+22q-/49,X?, t(9;22),+8,t(14;?),21q+,+22q-	Carbonell et al 1982a
47,XY,t(9;22)(q34;q11),del(1)(q21),dup(1)(q?),+8,+9,-14, -16,-21,+22q-,+mar	Alimena et al 1982b
48,XX,t(9;22)(q34;q11),t(13;14)(q14;q24orq32),+19,+22q-	Carbone et al 1982a
48,XY,t(9;22),+8,-9,-11,-12,-14,-19,+del(22q),+5mar	Sharp et al 1976
48-53,X?,t(9;22),+8,+8,+13,+14,+19,+21,+22q-	Carbonell et al 1982a
49,XX,t(9;22)(q34;q11),8q-,+10,+14,+22q-	Alimena et al 1982b
49,XY,t(9;22),+10,+14,+17	Hossfeld 1975b
49,XY,t(9;22),+6,+8,1p+,14q+,+22q-	Sadamori et al 1980
49,XY,t(9;22),+9,+14,+22q-	Lyall and Garson 1978

51,XX,t(9;22),+8,+9,+14,+19,+22q-	Miyamoto 1980
51,XY,t(9;22),+8,+14,+17,+19,+19	Borgström et al 1982
52,XY,t(9;22),+8,+14,+15,+19,+21,+22q-	Sadamori et al 1980
54,XY,t(9;22)(q34;q11),+8,+10,+11,+13,+14,+19,+21,+21	Hayata et al 1975b
55,XX,t(9;22),+3,+8,+12,+14,+15,+19,+20,+22,+22q-	Lilleyman et al 1978
55,XY,t(9;22),+6,+10,+10,+11,+13,+14,+14,+19,+21	Seabright 1983

Chronic myeloid leukemia, aberrant translocation

46,X,t(X;9;22)(q27;q34;q12)	Dallapiccola and Alimena 1979
46,X,t(X;9;22)/47-51,X,t(X;9;22),+1,+4,+8,+9,-14,i(17q), -19,+22q-	
46,X?,t(14;22)	Rowley 1983
46,X?,t(14;22)(q32;q11)	Nowell 1983
46,X?,t(14;22)(q32;q11)	Ishihara 1983
46,X?,t(9;14;22)(q34;q2?;q11)	Nowell 1983
46,XX,t(10;14;22)(q26;q24;q11)	Shabtai et al 1980
46,XX,t(11;14;22)(q13;q32;q11)	Kolitz et al 1981
46,XX,t(14;22)(p13;q11)	Borgström 1981
46,XX,t(9;14;22)(q34;q24;q11)	Potter et al 1975
46,XX,t(9;14;22)(q34;q12orq13;q11)	Tanzer et al 1980a
46,XX,t(9;14;22)(q34;q11;q11)	Borgström 1981
46,XX,t(9;14;22)(q34;q13;q11)	Seabright 1983
46,XX,t(9;14;22)(q34;q21;q11)	Pasquali 1983
46,XY,t(14;22)(q32;q11)	Seabright 1983
46,XY,t(9;14;22)(q34;q24;q11)	Borgström 1981
46,XY,t(9;14;22)(q34;q21;q11)	Pasquali 1983
46,XY,t(9;14;22)(q34;q24;q11)	Tomiyasu et al 1982

Chronic myeloid leukemia, Ph1 negative

46,XY,t(6;14)(p21;q32)	Mintz et al 1979
47,XY,+14	Shashaty and Baumiller 1980

Eosinophilic leukemia

47,XX,+8,14q+	Chilcote et al 1982

MYELOPROLIFERATIVE DISORDERS

Polycythemia vera

??,X?,+6,+11,+14,+22	Vykoupil et al 1980
46,XX,20q-,t(14;20)	Wurster-Hill et al 1976
46,XY,del(7)(q22)/46,XY,del(7)(q22),inv(14)(p11q24)	Tsuchimoto et al 1974
48,XY,+3,-5,t(13;14)(q?;q?),+2mar	Nowell and Finan 1978

Myelosclerosis / Myelofibrosis

47,XX,t(13;14),+8,+9"c"	Greef et al 1982

DYSMYELOPOIETIC SYNDROMES

Preleukemia, NOS

45,XX,-1,-4,5q-,-7,-12,-18,14q+,+4mar"s"	Pedersen-Bjergaard et al 1981
45,XY,-2,-5,-7,-8,-11,-12,-13,-14,+t(2;5),+t(11;12),+t(16;17),+17,+3mar	Watt et al 1982

Refractory anemia, NOS

46,X?,t(7;14)(p?;?)	Lessard and Le Prise 1983

Refractory anemia without excess of blasts

46,XX,5q-/47,XX,+8/47,XX,+14/46,X,-X,+8	Teerenhovi et al 1981

Refractory anemia with excess of blasts

43,XX,3q-,del(5)(q13q33),7q-,-12,14p+,16q+,-17,-18,18q+	Kerkhofs et al 1982
46,XX,del(5)(q15)/49-51,XX,+1,del(5)(q15),+11,-14,+22,+mar	Swolin et al 1981

Pancytopenia

47,XX,+14	Nowell and Finan 1978

Dysmyelopoietic syndrome, special type

45,XX,t(13;14)(p13;q11),del(5)(q14)/50,XX,t(13;14),+1,+6,-7,+8,+10,del(5),+mar"c""s"	Rowley et al 1981a

SPECIAL LEUKEMIAS

Hairy cell leukemia

44,X,del(X)(q22),del(6)(q23),-7,-8,-10,del(11)(p15q21),-21,14q+,+r,+mar	Sadamori and Sandberg 1983a
47,XY,-2,3q+,14q-	Khalid et al 1981

LYMPHOCYTIC LEUKEMIAS

Acute lymphocytic leukemia (ALL)

28,XX,-1,-2,-3,-4,-5,-7,-8,-9,-11,-12,-13,-14,-15,-16,-17, -19,-20,-22	Kaneko and Sakurai 1980
32,XY,-2,-3,-4,-5,-7,-8,-11,-12,-13,-14,-15,-16,-17,-20	Shabtai and Halbrecht 1981b
44,XY,-14,-21	Alimena et al 1977
44-45,XY,-14,-17,-21	Alimena et al 1977
45,XX,5q+,14q+,-15	Chrz et al 1983
45,XX,t(14;14)(q34;q11),t(13;15)(q11;p11), t(13;17)(p11;q11),-16,18q+,+3mar	Levitt et al 1978
46,X?,t(14;18)(q32;q22)	Shabtai 1983
46,XX,-9,13q-,14q-,+t(9;13)(p?;q?)	Prigogina et al 1979
46,XX,2q+,4q-,12p+,13q+,14q+,-20,+21	Whang-Peng et al 1976c
46,XX,t(4;14)(q35;q12)	Prigogina et al 1979
46,XY,+2,t(6;?)(p21;?),+7,-12,-13,14q+,17p+	Oshimura et al 1977a
46,XY,-8,del(3)(q12q25),t(8;14)(q24;q32),+t(1;8)(p11;q11)	Kaneko et al 1980
46,XY,del(14)(q2?)	Secker-Walker et al 1979
46,XY,t(12;17),14q+	Minowada et al 1977
46,XY,t(14;22)(q32;q11)/46,XY,t(14;22),+14,-15	Ayraud et al 1975
49,XX,+14,+17,+21,7q-	Shabtai and Halbrecht 1981b
53,XY,+X,+4,+8,+14,+15,+17,+21	Oshimura et al 1977a
54,XX,+6,+7,+8,+14,+17,+18,+21,+22	Prigogina et al 1979
54,XX,7p+,+10,+10,+14,+14,+18,+18,+21,+21/27,X,-X,-1,-2,-3, -4,-5,-6,-7,7p+,-8,-9,-11,-12,-13,-15,-16,-17,-19,-20, -22	Oshimura et al 1977a
54-55,XX,+5,+12,+14,+16,+17,+18,+21	Prigogina et al 1979
54-55,XY,+5,+8,+9,+14,+21,+mar	Prigogina et al 1979
54-57,XY,+3,+5,+6,+7,+12,+13,+14,+21,+2mar	Prigogina et al 1979
55,XX,+2,+4,+9,+14,+15,+19,+21,+21	Oshimura et al 1977a
55,XY,+X,+3,+6,+10,+13,+14,+16,+21,+21	Oshimura et al 1977a
56,XX,+3,+6,+12,+13,+14,+15,+18,+21,+r	Prigogina et al 1979
57,XX,+3,+4,+8,11p+,+14,+16,+17,+18,+21,+22,+2mar/47,XX,+8	Morse et al 1978
61,XX,+X,+1,+2,+3,+4,+5,+6,+8,del(6)(q21),+13,+14,+15,+16, +18,+21,+22	Oshimura et al 1977a

ALL, FAB type L1

26,XX,-1,-2,-3,-4,-5,-6,-7,-8,-9,-10,-11,-12,-13,-14,-15, -16,-17,-19,-20,-22	Hoeltge et al 1982
28,XX,-1,-2,-3,-4,-5,-6,-7,-8,-9,-11,-12,-13,-15,-16,-17, -19,-20,-22/56,XX,+X,+X,+10,+10,+14,+14,+18,+18,+21,+21	Brodeur et al 1981a
34-36,XY,-2,-3,-4,-7,-9,-12,-13,-14,-16,-17,-20,+22	Sandberg et al 1982c
46,XY,del(3)(q12q25),t(8;14)(q24;q32),-8,+t(1;8)(p11;q11)	Kaneko et al 1982e
46,XY,t(8;14)(q11;q32)	Kardon et al 1982b
52-56,XY,t(9;22),+4,+6,+9,+10,+14,+15,+20,+21,+22q-	Sandberg et al 1980
53,XX,+1,+5,+14,+15,+19,+20,+22	Kaneko et al 1981b
58,XX,+4,+6,+7,+10,+14,+14,+21,+del(18)(q21),+4mar	Kaneko et al 1982e

ALL, FAB type L2

44,XX,-8,t(13;14),t(9;22)"c"	Takeuchi et al 1981
45,X,-X,-1,-14,-17,+21,t(11;12)(p15;p11),+i(17q), +t(1;?)(p34;?)	Kaneko et al 1981b
46,XX,t(5;8;14)(q11;q24;q32),t(12;22)(p13;q11),-18, +t(18;?)(q23;?)	Kaneko et al 1982e
46,XY,+1,-8,t(14;?)(q32;?)	Mitelman 1983
46,XY,1q+/46,XY,1q+,14q+	Borgström et al 1981
46,XY,t(11;14)(q13;p13)	Kaneko et al 1982e
46,XY,t(2;14)(p?;q?)	Prieto et al 1981
49,XY,+7,+12,-13,t(6;18)(p25;q21),t(11;14)(q23;q32), +t(9;?)(p24;?),+t(1;13)(q12;p13)	Kaneko et al 1982e

ALL, FAB type L3

??,XY,t(8;14)(q?;q?),t(Y;1)	Prieto et al 1981
46,XX,-16,-17,-22,+8,11q-,t(8;14),+1-2mar	Borgström et al 1981
46,XX,dup(1)(q21q32),t(8;14)(q24;q32)	Rossi et al 1982
46,XX,t(2;8;14)(p11orp12;q24;q32),t(21;21)(p11;q11), +t(21;21)	Ekblom et al 1982
46,XX,t(8;14)(q24;q32)	Berger and Bernheim 1982
46,XY,dup(1)(q23q32),t(1;6)(q21;q13),t(14;?)(q32;?), del(6)(q13),del(8)(p21)	Slater et al 1979
46,XY,t(8;14)	Borgström et al 1981
46,XY,t(8;14),1q+	
46,XY,t(8;14)	Borgström et al 1981
46,XY,t(8;14)	Borgström et al 1981
46,XY,t(8;14)(q23;q32),del(3)(p21)	Lessard and Le Prise 1983
46,XY,t(8;14)(q24;q32)"s"	Mitelman et al 1979b
46,XY,t(8;14)(q24;q32),13q+	Berger and Bernheim 1982
46,XY,t(8;14)(q24;q32)	Berger and Bernheim 1982
46,XY,t(8;14)(q24;q32),-15,+16	Berger and Bernheim 1982
46,XY,t(8;14)(q24;q32)/46,XY,t(8;14)(q24;q32), dup(1)(q21q24)	Berger and Bernheim 1982
46,XY,t(8;14)(q24;q32),13q+	Berger and Bernheim 1982
46,XY,t(8;14)(q24;q32)	Berger and Bernheim 1982
46,XY,t(8;14)(q24;q32),+8,-22	Berger and Bernheim 1982
46,XY,t(8;14)(q24;q32)/46,XY,t(8;14)(q24;q32), dup(1)(q21q24)	Berger and Bernheim 1982
46,XY,t(8;14)(q24;q32)	Berger and Bernheim 1982
46,XY,t(8;14)(q24;q32),del(3)(p24),18p+	Berger and Bernheim 1982
46,XY,t(8;14)(q24;q32)/46,XY,t(3;8;14)(q14;q24;q32)	Berger and Bernheim 1982
46,XY,t(8;14)(q24;q32)/46,XY,t(1;8;14)(q23;q24;q32)	Berger and Bernheim 1982
46,XY,t(8;14)(q?;q?)	Prieto et al 1981
46-47,XX,t(14;?)(q32),t(13;?)(q31;?),1q+,+7	Slater et al 1979
47,X?,t(8;14)(q24;q32),+8	Shabtai 1983
47,XX,t(8;14)(q22;q32),del(21)(q21),+mar	Alimena et al 1981
47,XX,t(8;14)(q24;q32),+mar/47,XX,t(8;14),del(21)(q21), +mar	Rossi et al 1982
47,XY,+8/47,XY,+8,14q+	Borgström et al 1981
47,XY,dup(1)(q21q32),del(6)(q23),t(8;14)(q23;q32),+9	Roos et al 1982

Chronic lymphocytic leukemia

41,XX,t(14;14)(q11;q32),t(1;13)(q44;q11),t(15;18)(q11;q23), 10q+,-16,-20	Kaiser McCaw et al 1975
44,XX,-3,14q+,-17	Najfeld et al 1980
44,XX,6q-,i(8q),12p-,-14,t(14;14)(q31;q12),-20,20p+	Sparkes et al 1980
44-48,XY,t(Y;9)(q12;q13),del(1)(p22p32),t(4;6)(p16;q15), del(6)(q15),t(12;14)(q15;q32),+mar	Robert et al 1982
45,XX,-8,-14,-15,-22,+2,+9,+1p+,+mar	Finan et al 1978
45,XX,t(14;14)(q32;?)	Saxon et al 1979
45-46,XY,-8,9q+,-10,13p+,-14,-15,-19,-20,+mar	Vahdati et al 1983a
46,X?,del(14)(q24)	Shabtai 1983
46,XX,-3,t(2;3),t(14;18)(q32;q?),2q-,17q+	Finan et al 1978
46,XX,14q+	Morita et al 1981
46,XX,t(11;14)(q13;q32)	Vahdati et al 1983a
46,XX,t(14;22)(q24q32;q11)	Nowell et al 1981
46,XX,t(6;7)(q?;q?),t(7;13)(q?;q?),t(11;14)(q11;q32),17p+	Schröder et al 1981
46,XY,14q+	Morita et al 1981
46,XY,9p+/43,X,-Y,-14,-22	Nordenson 1983
46,XY,del(6)(p12)/46,XY,t(14;18)(q32;q21)	Robert et al 1982
46,XY,t(11;14)(q13;q32)/46,XY,t(3;4;9)(p12;q21q31;q31)	Robert et al 1982
46,XY,t(11;14)(q13;q32)	Vahdati et al 1983a
46,XY,t(1;7;9)(p36;q21;q12),t(12;14)(q13;q24)	Miyoshi et al 1979a
46,XY,t(3;6)(q12;q26),t(11;14)(q13;q32)	San Roman et al 1982
46,XY,t(8;14)(q24;q32),19p-	Fleischman and Prigogina 1977
46,XY,t(9;11;14)(p22;q13;q32)	Nowell et al 1981
47,XX,+12,del(14)(q24)	Robert et al 1982
47,XX,+12,del(14)(q?)	Schröder et al 1981
47,XX,+7/49,XX,+14,+17,+21	Shabtai 1983
47,XXY,t(3;8),t(11;14)(q11;q32),+i(18q)"c"	Finan et al 1978

Prolymphocytic leukemia

44,X,-Y,-11,-12,-13,-13,-14,-15,-15,-18,-18,+8mar	Diamond et al 1980

Lymphocytic leukemia, special type

45,XX,-1,2q+,14q+	Ueshima et al 1981
45,Y,-X,t(Y;14)(q12;q32),-13,-17,del(2)(q33), t(3;13)(q21;q34),del(6)(q16orq22),9p+,9q+,10q+,16q+,18q+	Miyoshi et al 1981a
46,X,-Y,-1,1p+,del(9)(q32),12p+,14q+,+2mar	Ueshima et al 1981
46,XY,t(1;7)(p36;q22),del(9)(q12),t(12;14)(q13;q32)	Miyoshi et al 1981a
48,XX,-6,-11,14q+,+18,+20,+22	Imamura et al 1981

MONOCLONAL GAMMOPATHIES

Multiple myeloma

42,XX,14q+,-?,+mar	Wurster-Hill et al 1973
44,XY,del(3)(q25),t(5;12)(q15;q14),del(7)(q22), del(11)(q24),-14,-18,+mar	Mitelman 1983
45,X,-Y,-1,+t(1;3)(q3?;q2?),-3,+11,14q+/45,X,-Y,-1,-1, +t(1;1)(q2?;q3?),14q+	Philip et al 1980
45,XY,del(1),+del(1),-10,-11,-11,+t(11;?),-13,-14,-20, +3mar	Philip et al 1980
46,XY,14q+	Nordenson 1983
46,XY,ins(1)(p36q21),t(11;14)(q13;q32)	Liang et al 1979
47,X,-X,inv(1)(p22q12),t(7;?)(q11;?),+9,t(11;?)(p15;?),-13, -14,t(14;?)(q11;?),+15,17q+	Philip et al 1980
47,XX,14q+,+mar	Philip et al 1980
47,XY,dup(1)(q11q32),t(1;10)(p22;p15),inv(6)(p21q13), t(11;14)(q23;q32),21p+,+mar	Liang et al 1979
48,X,-X,+1,+7,-8,+9,+21,del(1)(p36),4p+,del(5)(q22),9p+, 14q+	Liang et al 1979
53,XY,+3,+5,-7,-8,+9,+11,-14,+20,del(1)(p34),del(6)(q25), +5mar	Liang et al 1979
55,XY,+1,+5,+9,+11,-12,+13,+15,del(1)(q32),del(6)(q25), t(9;19)(q13;q13),14q+,16p+,+4mar	Liang et al 1979
45,XY,-2,-2,-3,+5,-8,-10,+14,-15,-16,del(6)(q25),+4mar	
55,XY,+3,+5,+7,+9,+11,+14,+15,+21,+21/53,XY,+3,+5,-8,+9, +11,+14,+15,-19,+21,14q+,+3mar	Liang et al 1979

Plasma cell leukemia / Plasmocytoma

44,X,-Y,del(1)(p11),+del(1)(q11),t(11;14)(q11;q32),14q+, -21,-22	Ueshima et al 1983
46,X?,14q+	Wurster-Hill et al 1973
46,XX,t(1;6)(p22p34;p25),t(11;14)(q13;q32)	Gahrton et al 1980b
46,XY,+t(1;16)(q21;p13),del(6)(q21),-16,-18,+mar	Ueshima et al 1983
47,XY,+t(1;16),del(6)(q21),+t(8;15)(p23;q12),-8,14q+,-15, -16,-18,+3mar	
46,XY,t(8;14)(q24;q32)/47,XY,t(8;14),+7/48,XY,t(8;14),+7, +18	Yamada et al 1983b
78,XY,cx,del(1)(q21),t(1;17)(q12;p13),3q+,3p-,9p+,11p+, 14q+,16p+	Ueshima et al 1983

SOLID TUMORS

UNCLASSIFIED NEOPLASMS

EPITHELIAL NEOPLASMS

Carcinoma, NOS

39-41,XX,del(1)(q32),1p+,t(1;14)(q21;p11),cx	Kovacs 1978a
44,XY,+5,-8,-11,-13,-13,-14,-14,-15,+19,-22,-22,t(1;11), 5q-,9q+,+5mar	Reichmann et al 1981
44-45,XX,t(1;4)(q11;q35),+3,5q-,-14,i(17q),-21	Atkin and Baker 1982a
45,XY,-1,-6,-14,+21,+2mar/45,XY,t(1;14)(q11?;q24?),-6,-16, +21	Sonta and Sandberg 1978a
45,XY,-7,+8,-14	Sandberg 1977
47,XX,+1,+3,t(11;14)(q?;q?),-12,-21,+r	Atkin and Baker 1982a
47,XX,+8,+13,-14	Sonta et al 1977
47,XY,-5,+7,-10,-13,-14,4q-,5q-,+4mar	Reichmann et al 1981
47-48,Y,-X,-5,-8,+13,-14,-16,-18,-22,+7mar	Reichmann et al 1981
49,X,-Y,-3,-5,-14,+15,-16,-17,+19,+21,+22,1q+,5q-,6q-,8p-, 9p-,11p-,18p+,20q-,22q-,t(Y;14),+3mar	Reichmann et al 1981
50,XX,inv(3)(p13q25),+8,+14,+21/63,XX,inv(3),del(11)(p13), cx	Sonta and Sandberg 1978a
54,XX,+1,+2,+8,+11,+13,+14,+21,+21	Sonta et al 1977
58,XX,+X,+1,+2,+3,5p-,+6,+7,+9,11q+,-13,-14,-15,+16,+19, +20,+6mar	Kovacs 1978a
58,XY,del(3)(p14),+4,+7,+8,+10,+11,+12,+14,+15,+17,+18,+19, +21	Kovacs 1978a
79,XX,cx,i(1q),t(1;3)(p13;q29),t(1;14)(p13;p13), t(12;?)(q24;?),t(15;21)(p11;q11),i(5p),r	Kovacs 1981
80-83,XX,cx,t(2;5;16),t(2;5;18),t(11;13),11q+,t(8;10), del(2)(q24),del(6)(q21),i(14q),dup(9)(q11q12)	Riet-Fox et al 1979

Carcinoma NOS, metastatic

35,XX,-1,-3,-4,-5,-7,-8,-10,-11,-12,-13,-14,-15,-16,-17, -18,-19,-20,-21,-22,+8mar	Pathak et al 1979
45-50,XX,-2,+7,-22,t(1;12),t(13;14),inv(7)(q?)	Ayraud 1975
46,XX,t(9;22)(q34;q11)"s"	Togawa et al 1981
52,XX,+3,+5,+7,+8,+9,+11,12q+,13q-,-14,+16,i(17q)"s"	
48,XX,t(3;?)(q2?;?),t(7;14)(p21;q1?),cx	Bartnitzke and Bullerdiek 1983
60,XX,t(1;?),t(7;?)(q3?;?),i(10q),der(11),t(14;21),cx	Bartnitzke and Bullerdiek 1983
65,XX,i(1q),t(1;9),der(5),t(6;10)(p?;q?),t(14;?),der(21), cx	Bartnitzke and Bullerdiek 1983

Large cell carcinoma

55-70,XY,t(4;5;14),del(1)(q11q21),t(7;11)(p22;q13), t(7;21)(q11;q22),t(3;8)(p?;q?),t(6;11)(p11;p11),i(16q), t(3;?)(q11;?),t(21;?)(q11;?),t(13;?)	Kakati et al 1975
60,XY,del(5)(p11),t(6;10)(q11;p11),t(7;11)(p22;q13), t(14;18)(q11;q11),del(2)(q21),t(6;?)(p11;?)	Pickthall 1976

Small cell carcinoma

44,XX,cx,del(3)(p14p23),t(14;14)(p11;q11)	Whang-Peng et al 1982b
69,XY,cx,del(3)(p14q24),t(11;14)(p11;p11), t(11;13)(p11;p11),del(X)(q23)	Whang-Peng et al 1982b

Papillary carcinoma

34-36,XX,cx,1p-,1q-,6q-,14q+	Wake et al 1980
41,XX,cx,t(6;14)(q21;q24),del(1)(p?q?),13q+	Wake et al 1980
42-54,X,-X,-14,-16,-16,+1,+3,+12,+20,1p-,6q-	Wake et al 1981
46,XX,t(6;14)(q21;q24),1p-,1q-,3p+,3q+	Wake et al 1980
50,X,-X,+1,+3,+12,+20,-14,-16,1p-,6q-	Wake et al 1980
54,XY,+del(2)(q11),t(2;14)(q11;q32),t(3;11)(p13;p15), +i(5p),+8,+9,+12,+16,+17,+21,+2mar	Pathak et al 1982
56-63,XX,cx,t(6;14)(q21;q24),6q-,6q+	Wake et al 1980
59,XX,cx,13q+,14q+	Wake et al 1980
62,XX,cx,1p-,1q-,t(6;14)(q21;q24),10q-	Wake et al 1980
62,XX,cx,6q-,14q+	Wake et al 1980
68,XX,cx,t(6;14)(q21;q24)	Wake et al 1980
68,XX,cx,t(6;14)(q21;q24),del(1)(p?q?),1p+,4q-,8q-,10q+	Wake et al 1980
69,XX,cx,1p+,1q-,1p-,1q-,2q+,3q+,4q+,6p+,12q+,14q+	Wake et al 1980

Adenoma

46,XY,t(6;21)(q13;q22)/46,XY,t(1;14)(q42;p12)/45,X,-Y	Mark et al 1981b
45,X,-Y	
46,Y,t(X;15)(q24;q15)/45,X,-Y	
46,Y,-X,ins(X;3)(q22;q25q27),del(3)(q25),t(8;14)(q12;q22)	Mark et al 1981a

Adenocarcinoma

38,X,-X,-2,-3,-10,-14,-22,del(1)(q25),del(6)(q15), del(14)(p11),dmin	Trent and Salmon 1981
38,X,-X,-8,-14,-20,+3-5mar	Trent and Salmon 1981
46-48,XX,+3,+7,+8,-1,-5,-14,-17,5q+,6q+,del(1)(q42), del(5)(q3?),del(10)(q24),del(14)(q2?)	Riet-Fox et al 1979
62-74,XX,cx,3q-,6q-,t(13;14),dmin	Hecht et al 1983
71,XX,cx,1q-,4q+,6q-,9q+,10q-,12q+,17q+,14q+	Wake et al 1981

Adenocarcinoma, metastatic

40-45,XY,-3,-8,-18,+5,+10,11p+,14q+,del(4)(q32)	Ayraud 1975
56-64,XX,del(X)(q22),del(1)(p12),del(5)(p21),del(7)(q31), t(14;18)(q11;q11),18q+,i(22q)	Berger and Lacour 1974

Adenomatous polyp

45-47,XY,-7,+14	Mitelman et al 1974b
47,XX,+14	Mitelman et al 1974b
47,XY,+14	Mitelman et al 1974b
48,XX,+8,+14	Mitelman et al 1974b
48,XX,+8,+14/48,XX,-3,+8,+9,+14	Mitelman et al 1974b
48-52,X,-Y,+7,+8,+13,+14,+16,+18,-20,+12q-,1q+	Reichmann et al 1982b

MESENCHYMAL NEOPLASMS

Sarcoma, NOS

Karyotype	Reference
49,XX,+3,+13,+14	Sonta and Sandberg 1978a

Liposarcoma

Karyotype	Reference
50,XY,+3,+6,+8,+11,+14,-22,17p+	Sonta et al 1977

Leiomyosarcoma

Karyotype	Reference
42,XX,del(1)(p12orp13),del(11)(q13orq14),+7,-9,-13,-14,-15, -18,+19,-22	Mark 1976

Mesothelioma, malignant

Karyotype	Reference
40-46,XY,+16,+20,-22,t(14;15),del(2)(p?),i(5p)	Ayraud 1975
41,XY,del(1)(p13),t(1;3)(p13;q11),t(7;7)(q11;q36), t(9;12)(p24;q13),del(12)(q13),del(13)(q12q14),-14,-15, t(17;?)(p13;?),i(19p),-20,-21,-22	Mark 1978
43,XY,t(1;1)(q42;q32p22),t(2;6)(p25;p21),inv(3)(p24q13), t(5;18)(p15;q11),-6,+9,del(13)(q24q14),-14,-18,-22	Mark 1978
46,XX,t(14;14)(q12;q32)	Shah-Reddy et al 1982
86,X?,cx,del(1)(p?q?),2p-,2p+,6q-,14q+	Wake et al 1981

MELANOMAS

Malignant melanoma

Karyotype	Reference
78-82,XY,t(4;9)(q21;q21),del(9)(p13),t(14;14)(p11;q12), del(2)(q24),t(14;?)(q11;?),t(21;?)(q11;?),del(17)(q11)	Kakati et al 1977

Malignant melanoma, metastatic

Karyotype	Reference
24,X,cx,t(1;15),t(1;14)(q12;p11),del(7)(p13), t(6;7)(p11;q11),t(13;21)(p11;q11),+r	Atkin and Baker 1981
41-43,XY,del(1)(q21q25),i(1q),t(1;9)(p22;q34), t(11;12)(q11;p11),t(10;12)(q26;q13),t(4;?)(q35;?), del(5)(p13),t(14;?)(p12;?),i(17q),t(18;?)(p11;?), del(6)(q21)	Kakati et al 1977
80,XY,t(11;12)(q23;q13),del(5)(q13),del(5)(p11), del(9)(q13),t(5;9),t(14;?)(q11;?),t(7;12)(q22;q24), del(1)(p13),del(11)(q23)	Kakati et al 1977

NEUROGENIC NEOPLASMS

Oligodendroglioma

Karyotype	Reference
45,XY,t(21;22)(p11;q11),7q-,9q-,14q+	Yamada et al 1980

Neuroblastoma

Karyotype	Reference
45,XX,+14,-15,-18,del(1)(p32),dup(1)(p13p31),del(6)(q23), del(11)(q23)	Brodeur et al 1981b
45,XY,-21,t(1;14)(p32;q31),t(8;?)(q24;?),dup(14)(q11q24)	Brodeur et al 1981b

Retinoblastoma

Karyotype	Reference
44,XY,1p+,-3,t(3;12)(q?;p?),+6q-,der(9),13p+,-14,-16	Kusnetsova et al 1982
47,XY,dup(1)(q12q44),del(2)(q13),+del(6)(q16), t(14;?)(q32;?),i(17q),+20,-22	Gardner et al 1982
51,XY,del(6)(q21),+8,+11,-12,t(13;18)(p11;q11),14q+,-16, +19,+22,+2mar	Hossfeld 1978

Meningioma, benign

Karyotype	Reference
38,X,-Y,-22,-21,-17,-14,-13,-10,-9,-4,-1p-,+mar	Zankl et al 1975a
41-42,XY,-1,-5,-8,9q+,-14,-18,-22,+r	Mark 1973b
44,Y,-X,-22,-14,+mar/43,Y,-X,-22,-14,-15,-11,-10,+3mar	Zankl et al 1975a
45,XX,-22,+13,-14/44,XX,-22,+13,-14,-1	Mark 1973a
45,XX,-22/45-46,XX,-22,-9,+6,+14,+del(22q)	Mark 1973a

EMBRYONAL AND MISCELLANEOUS NEOPLASMS

Teratoma, malignant

Karyotype	Reference
67,XY,+1,+3,+4,+5,+6,+7,+7,+8,+10,+11,+13,+13,+14,+16,+17, +18,+20,+20,+21,+21	Sonta et al 1977

LYMPHOMAS

1. UNCLASSIFIED AND MISCELLANEOUS LYMPHOMAS

Benign lymphomatous tumor, NOS

Karyotype	Reference
48,XY,t(8;14)(q24;q32),+17,+17q+"s"	Mitelman et al 1979a

Malignant lymphoma, NOS

Karyotype	Reference
46-47,XY,-5,-14,+1-3mar	Knuutila et al 1981
47,XX,t(1;14)(q23;q32),+3,t(6;21)(p25;q11),+8, t(7;11)(q11;q25),-21	Yamada et al 1980

Non-Hodgkin's lymphoma, NOS

Karyotype	Reference
46,X,-X,del(1)(q22),t(2;11)(p21;q21),t(9;15)(q11;p12), del(9)(q11),-14,t(14;17)(q23;q23),-16,+mar	Kristoffersson 1983
46-48,XX,del(1)(q32),+1,+3,r(7),14q+,-20,+mar	Kristoffersson 1983
47,X?,t(8;14)(q24;q32),+7	Zech et al 1976a
47-48,XY,t(14;?)(q32;?),16q-,-17,18q-,13q-,+7,+8,8q-	Catovsky et al 1977
50,XX,+X,t(1;11)(q25;q23),+3,+7,-13,t(14;18)(p?;p?),-15, t(19;?)(q13;?),+3mar	Reeves and Pickup 1980
52,XY,+2,+3,t(5;9)(q22;q32),+6,+12,+14,+20	Reeves and Pickup 1980

Non-Hodgkin's lymphoma, lymphocytic, NOS

Karyotype	Reference
42-43,XX,t(11;14)(q13;q32)	Fleischman and Prigogina 1977
45,X,-Y,t(11;14)(q13;q32),t(11;19)(q13;q13)	Fleischman and Prigogina 1977
45-46,XY,-1,3q+,del(9)(q22),+14,+17	Reeves 1973
46,XY,t(11;14)(q13;q32)	Fleischman and Prigogina 1977
46-47,XX,t(1;1)(q25;p36),dup(1)(q25q44),t(14;?)(q32;?),+9	Slavutsky et al 1981
47,XX,14q+,+17p+	Prieto et al 1978a
47,XX,del(13)(q22),del(13)(q14),dup(14)(q24q32),+mar	Mark 1975a
47,XY,+21,14q+"c"	Oshimura et al 1981b
47,XY,+7,-9,-11,1p+,6q-,10p-,14q-, t(14;?;14)(q24orq32;?;q13),15q-,22q-	Fukuhara et al 1979b
47-48,XY,+7,-11,-14,-15,4q+,9p-,14q+	Fleischman and Prigogina 1977
47-49,XX,+X,-14,i(17q),t(1;17),1p+	Fleischman and Prigogina 1977
48,XX,del(1)(q22q25),i(mar1),del(12)(p12),dup(14)(q24q32)	Mark 1975a
50,XY,+3,+3,+12q-,+18,+14q+	Fleischman and Prigogina 1977
51,XX,+5,+9,+11,+del(11)(q?),+del(12)(q?),14q+,i(17q)	Prieto et al 1978a

Non-Hodgkin's lymphoma, histiocytic, NOS

Karyotype	Reference
45-47,XX,t(10;14)(q21;q32),+8,-18,-21	Mark et al 1977
46,XX,del(1)(p31p35),+3,del(9)(p11),t(10;?)(q26;?), del(11)(q13),-13,t(11;14)(q13;q32),t(17;21), t(9;22)(q11;p13),+r	Mark et al 1977
46,XY,1p-,6q-,t(11;14)(q21;q32)	Kakati et al 1980
48,XY,1p-,+3,6q-,+7,9p-,t(11;14)(q21;q32)	
46,XY,1q+,t(14;?)(q32;?),18p-	Kakati et al 1980
47,XX,del(3)(q11),14q+	Nordenson 1983
47,XX,t(1;18)(p36;q21),+7,t(8;12)(q11;q15), t(8;12;13)(q11q22;q15;q14),del(13)(q14),t(8;14)(q22;q32)	Mark et al 1978
47,XY,-6,-14,1q-,+1p-,2q+,3q+,9p+,10p-,10p+,12p-,18q+,22p+, t(2;14)(q37?;q13?)	Fukuhara et al 1979b
47-48,XY,+X,-13,-15,t(6;11)(p?;q?),t(13;14;?)	Fleischman and Prigogina 1977
50,XY,dup(1)(q12q32),+2,4p+,del(4)(q21),t(8;14)(q24;q32), +17p+,18q+,+20	Brynes et al 1978

50,XY,t(2;?)(p25;?),del(2)(p13),+i(3p),del(6)(p21),+7,+8, +9,+9,-11,-13,+14,-15,+20,+21,-22 — Mark et al 1978
44,XY,del(2)(p13),t(6;?)(p21;?),del(11)(q13), t(11;14)(q13;q31),i(15q),-20,-22

Non-Hodgkin's lymphoma, T-cell, NOS

46-47,XY,del(6)(q14),14q+,+mar — Kristoffersson 1983
47,XX,t(2;?),inv(3),4p-,del(5)(q14q32),6p+,6q+,t(7;7),9q+, 10p-,13q-,t(14;14),-15,-15,16q-,17p+,+21,+2mar — Gaeke et al 1981

Angioimmunoblastic lymphadenopathy

46,XX,t(8;14)(q?;q32) — Mazur et al 1979
46-47,XY,14q+,+16 — Kristoffersson 1983

2. HODGKIN'S LYMPHOMAS

Hodgkin's disease, mixed cellularity

45,XX,-14,del(6)(q21) — Bernstein et al 1981
47,XX,+X,t(6;14)(q15;q32) — Reeves and Pickup 1980
81,XX,14q+,i(18q),i(18p),del(6)(q21),t(9;15)(q11;p12), del(16)(q13) — Reeves 1973

Hodgkin's disease, lymphocytic depletion

46,XY,t(10;14)(q22;q32),t(15;?)(p11;?),-16,+mar — Hossfeld and Schmidt 1978
48,X,-Y,t(1;?)(q44;?),i(1q),del(1)(q21),del(3)(q21),-10, t(3;11)(q23;q21),+14,t(14;?)(q32;?),+15,-17,+19,+20,-22 — Hossfeld and Schmidt 1978
60,XY,+del(1)(p21),t(1;5)(q44;q33),dup(5)(q12), t(1;5)(p21q23;q13),del(6)(q22),t(9;?)(q12;?), t(3;12)(q11;q24),t(14;?)(p11;?),i(17q) — Hossfeld and Schmidt 1978

Hodgkin's disease, nodular sclerosis

47,XY,t(1;?)(p36;?),+2,t(2;6)(q31;q27),+8,del(12)(p11), t(17;?)(p13;?),-14,-20,+mar — Reeves and Pickup 1980

3A. NON-HODGKIN'S LYMPHOMAS - RAPPAPORT CLASSIFICATION

Lymphocytic, well differentiated, diffuse

48,XY,1q+,3q+,+i(3p),-6,-10,t(14;?)(q12;?),+16,17p-,+18, +mar — Miyamoto et al 1981a

Lymphocytic, poorly differentiated, NOS

46,XY,del(6)(q21),-8,-9,t(6;14)(q21;q32),+16,+20 — Mark et al 1979

Lymphocytic, poorly differentiated, diffuse

Karyotype	Reference
44,XX,t(11;14)(q13;q32),-3,-9,-13,1p+,+2mar	Fukuhara et al 1979a
45,XY,t(14;18)(q32;q21),+2,-5,-6,-11	Fukuhara and Rowley 1978
46,XY,t(14;18)(q32;q21),t(10;14)(q24;q32),-1,-17,-22,+3mar	Fukuhara and Rowley 1978
47,XY,-1,+7,-9,+9q+,+9p+,+10,14q+,-17,t(17;18)(p11;q11)	Kristoffersson et al 1981
47,XY,t(14;14)(q32;q?),+7,-9,-11,1p+,6q-,10p-,15q-,22q-, +2mar	Fukuhara and Rowley 1978
47,XY,t(14;18)(q32;q21),-9,-10,+12,-16,8p-,13q+,+3mar	Fukuhara and Rowley 1978
47,XY,t(1;4)(q23;q33),+3,t(10;15),t(11;14)(q13;q32),r(13), r(13)	San Roman et al 1982
48,XX,t(14;18)(q32;q21),1p+,12q+,+mar	Fukuhara and Rowley 1978
48,XY,t(14;?)(q32;?),21q+,+mar	Fukuhara et al 1979a
48-49,XX,2p-,+7,+i(8q),+12,14q+,15q-,18q-	Kakati et al 1980

Lymphocytic, poorly differentiated, nodular

Karyotype	Reference
45-47,XX,14q+,del(18)(q21),+del(1)(p34q32)	Reeves 1973
46,XX,del(20)(q11),t(14;18)(q32;q21)	Kaneko et al 1982a
46,XY,12q+/46,XY,t(14;18)(q32;q21)	Kristoffersson et al 1981
48,XY,t(14;18)(q32;q21),-6,-11,+12,+del(1)(p?),3q+,+2mar	Fukuhara et al 1979a
52,X,-X,+2,-4,+5,+7,+8,-14,-15,-16,+20,+21, +t(1;15)(p11;q11),+t(12;17)(q24;q11),+t(14;?)(q32;?), +mar	Kaneko et al 1982a

Mixed cell, diffuse

Karyotype	Reference
50,XY,+X,-1,+5,+7,+12,-14,-17,del(13)(q12q14), ins(1;1)(q21;q21q32),+t(14;?)(q32;?),+t(1;17)(q21;q25)	Kaneko et al 1982a
53,XY,t(8;14)(q24;q32),t(14;18)(q32;q21),+20,1q-,+1p-,2q-, 4q+,5q-,5p+,6q-,+7q+,10q+,11q+,11q-,13p+,+4mar	Fukuhara and Rowley 1978

Histiocytic, diffuse

Karyotype	Reference
45,X,-Y,t(14;15)(q32;q21orq22),-15,+18,-22,1q+,3p+,5p-,7p-, 9p+,12p+,16q+,17p-,+2mar	Fukuhara and Rowley 1978
45,X,del(X)(q24),del(1)(q21),+del(1)(p11),del(2)(q22), del(3)(q21),del(6)(p11),del(6)(q15),t(7;9)(p22;q13), t(8;10)(p23;q21),del(9)(q13),del(10)(q21), t(2;11)(q22;q23),-13,del(14)(q24),del(15)(q22), del(16)(q22),+17,-19	Mark et al 1979
46,X,-X,-2,-10,+11,-13,-14,-17,del(1)(p22p32), del(3)(p13p21),del(6)(q21),del(9)(p13),+t(2;?)(p23;?), +t(14;?)(q32;?),t(18;?)(q23;?),+t(X;13)(q26;q12),+mar	Kaneko et al 1982a
47,X,+X,-Y,+12,-14,del(11)(q21q23),del(16)(q22), t(14;?)(q32;?)	Kaneko et al 1982a
47,X,-X,t(4;14)(q?;q32),-18,1q+,2p+,6q-,7q+,9q-,12q+,+3mar	Fukuhara and Rowley 1978
47,XX,-1,-2,-6,+8,-14,+16,-19,del(11)(q23), t(8;14)(q22;q32),+t(2;6)(q21;p23),del(6)(q21), +t(1;14)(p11;q11),+t(19;?)(q13;?),+t(14;?)(q11;?)	Kaneko et al 1982a
47,XY,-10,-13,+15,-16,+19,+20,del(6)(q21),del(8)(q22), t(2;14)(q21;q32),t(4;4)(q31;q35),+t(10;?)(p11;?)	Kaneko et al 1982a
47,XY,cx,-14	Kakati et al 1980
50,XX,del(1)(p22),+3,+7,+12,t(1;14)(p22;q32),+18	Mark et al 1979

50,XY,del(3)(p24),+7,+12,+14,+20 — Kristoffersson et al 1981
50,XY,t(8;14)(q24;q32),+2,+20,1q+,4p+,+4q+,17p+,18q+ — Fukuhara and Rowley 1978
82,XX,del(X)(q13),i(1q),i(1p),del(7)(p11),i(9p), del(14)(q22),t(1;14)(q23;q32),i(15q),i(17q) — Mark et al 1979
84,XY,t(1;14)(q2?;q24orq32),cx — Fukuhara and Rowley 1978
89,XY,del(1)(p22),t(5;13)(q15;q34),t(14;20)(q32;q12), t(1;17)(p22;q25),t(1;19)(q21;p13),del(20)(q12) — Mark et al 1979

3B. NON-HODGKIN'S LYMPHOMAS - KIEL CLASSIFICATION

Lymphocytic, T-zone

44-46,X,+t(Y;14)(q24;q12),-3,del(3)(q21),+9,14q+,+mar — Gödde-Salz et al 1981a
46,XY,+t(Y;14)(q12;q24),3q-,-5,-13,+mar — Gödde-Salz et al 1981b

Centrocytic

44,XY,del(4)(p13),-11,-13,t(13;14)(q11;q32) — Kristoffersson 1983

Centroblastic-centrocytic, diffuse

48-50,XY,-1,+7,9q+,+9p+,+10,+14q+,-17,-18, +t(17;18)(p11;q11),-19,+20,+22,+mar — Kristoffersson 1983

Centroblastic-centrocytic, follicular

45-46,XY,-7,t(14;18)(q32;q21) — Kristoffersson 1983
46,XY,12q+ — Kristoffersson 1983
46,XY,t(14;18)(q32;q21)
46,XY,t(1;?)(p36;?),t(1;1)(p36;q1?),del(7)(q11), del(7)(q22),del(9)(p11),t(14;?)(q32;?) — Reeves and Pickup 1980
46-47,XX,t(8;9)(q13;p24),t(14;18)(q32;q21),+22 — Kristoffersson 1983
46-49,XX,1p+,3q-,4q+,+6,+13,-14,14q+,17p+ — Kristoffersson 1983
47,XY,t(1;14)(q21;q32),+12,-16,+19,t(14;?)(q32;?) — Reeves and Pickup 1980
48,XX,+X,t(1;?)(p36;?),del(2)(p21),t(3;?)(q27orq29;?), t(14;?)(q32;?),t(17;18) — Reeves and Pickup 1980
52,XX,+X,+7,+8,del(10)(q24),+12,t(13;?)(q22;?), t(14;?)(q32;?),del(18)(q21),+19,+mar — Reeves and Pickup 1980

Centroblastic, diffuse

42-45,XY,1p+,-6,-8,14q+,14q-,-16,i(17q),+18,-20,+mar — Kristoffersson 1983
46-49,XY,1p+,+2,-6,+del(9)(p21),+12,+13,14q+?,i(17q),+18 — Kristoffersson 1983
47,XY,14q+,+22

Centroblastic, follicular

47,XY,del(6)(q23),+8,14q+?,i(17q) — Kristoffersson 1983

Lymphoblastic, unclassified

Karyotype	Reference
45,XY,3p-,14q+,-C	Kristoffersson 1983

Lymphoblastic, Burkitt's type

Karyotype	Reference
??,X?,t(8;14)(q24.1;q32.5)	Manolova et al 1979b
??,X?,t(8;14)(q24.1;q32.5)	Manolova et al 1979b
??,X?,t(8;14)(q24.1;q32.5)	Manolova et al 1979b
??,X?,t(8;14)(q24.1;q32.5)	Manolova et al 1979b
??,X?,t(8;14)(q24.1;q32.5)	Manolova et al 1979b
??,XX,14q+	Manolov and Manolova 1972
??,XX,14q+	Manolov and Manolova 1972
??,XX,14q+	Manolov and Manolova 1972
??,XX,14q+	
??,XX,6p-,t(8;14)	Douglass et al 1980a
??,XX,dup(1)(q12orq21q31),t(8;14),+10q-	Douglass et al 1980a
??,XY,-5,+6	Douglass et al 1980a
??,XY,1p+,t(8;14)	
??,XY,14q+	Manolov and Manolova 1972
??,XY,14q+	Manolov and Manolova 1972
??,XY,14q+	Manolov and Manolova 1972
??,XY,14q+	Manolov and Manolova 1972
??,XY,14q+	Manolov and Manolova 1972
??,XY,14q+	Manolov and Manolova 1972
??,XY,14q+	Manolov and Manolova 1972
??,XY,14q+	
??,XY,dup(1)(q12orq21q31),3p-,3q-,+6p-,+8,+14,+15,+18,-21	Douglass et al 1980a
??,XY,dup(1)(q?q31),t(8;14)	Douglass et al 1980a
??,XY,t(8;14),+15,+21	Douglass et al 1980a
??,XY,t(8;14),12q-	Douglass et al 1980a
43,XY,+1,+2,+3,-7,-8,-9,-12,-16,-22,del(1)(q25), del(2)(p11),del(10)(q22),14q+	Biggar et al 1981
45,X,-Y,t(1;14)(q22;q32),t(8;22)(q24;q11),3q+,6q-,6q+,7q+, 9p-,14q-,18q-	Berger et al 1981c
45,X?,t(8;14)(q24;q32),t(7;16),t(6;17)	Zech et al 1976a
45-47,XY,t(14;?)(q32;?),del(3)(p?),-9,-22	Philip et al 1977b
46,X?,14q+	Zech et al 1976a
46,X?,14q+	Zech et al 1976a
46,X?,14q+	Zech et al 1976a
46,X?,t(8;14)(q23;q32)	Kaiser McCaw et al 1977
46,X?,t(8;14)(q24;q32)	Zech et al 1976a
46,X?,t(8;14)(q24;q32)	Zech et al 1976a
46,X?,t(8;14)(q24;q32)	Zech et al 1976a
46,XX,1p+,14q+,+mar,+r	Kristoffersson 1983
46,XX,dup(1)(q12orq21q31),t(8;14)	Douglass et al 1980a
46,XX,dup(1)(q12orq21q31),t(8;14)	Douglass et al 1980a
46,XX,dup(1)(q12q32),t(8;14)(q24;q32)	Miyamoto et al 1982a
46,XX,t(8;14)(q24;q32),dup(1)(q23q32),del(3)(p25)	Slater et al 1982
46,XX,t(8;14)(q24;q32)	Biggar et al 1981
46,XX,t(8;14)(q24;q32)	Berger and Bernheim 1982
46,XX,t(8;14),16q-	Douglass et al 1980a
46,XY,del(2)(p12),t(8;14)(q24;q32)	Biggar et al 1981
46,XY,t(4;5;7)(q13;p13;p22),t(8;14)(q23;q32)	Kaiser McCaw et al 1977

46,XY,t(8;14)	Miyoshi et al 1981c
46,XY,t(8;14)	Douglass et al 1980a
46,XY,t(8;14)	Douglass et al 1980a
46,XY,t(8;14)	Sakurai et al 1983
46,XY,t(8;14)	Sakurai et al 1983
46,XY,t(8;14)	Sakurai et al 1983
46,XY,t(8;14)(q23;q32)	Kaiser McCaw et al 1977
46,XY,t(8;14)(q24;q32),17p-,17q+	Miyoshi et al 1981c
46,XY,t(8;14)(q24;q32),-13,16p+,+mar	Slater et al 1982
46,XY,t(8;14)(q24;q32)	Biggar et al 1981
46,XY,t(8;14)(q24;q32)	Berger and Bernheim 1982
46,XY,t(8;14)(q24;q32)	Berger and Bernheim 1982
46,XY,t(8;14)(q24;q32)	Berger and Bernheim 1982
46,XY,t(8;14)(q24;q32)	Berger and Bernheim 1982
46,XY,t(8;14)(q24;q32)	Berger and Bernheim 1982
46-47,XY,t(14;?)(q32;?),dup(1)(q21q32),t(2;3)(q13;q29),+5, 6p-,-8,del(9)(q?),+12,+18,+mar	Miyoshi et al 1981c
46-48,X?,t(8;14)(q24;q32),+7,15q-,+6,+13	Zech et al 1976a
46-52,X?,t(8;14)(q24;q32),+1,+3,+5,+7,i(17q),+X	Zech et al 1976a
47,X?,t(8;14)(q23;q32),+mar	Kaiser McCaw et al 1977
47,X?,t(8;14)(q24;q32)	Zech et al 1976a
47,X?,t(8;14)(q24;q32)	Zech et al 1976a
47,X?,t(8;14)(q24;q32),+7	Zech et al 1976a
47,XX,+12,t(8;14)(q24;q32),del(12)(p11q22),del(17)(q21)	Biggar et al 1981
47,XX,+14,t(8;14)(q24;q32)	Biggar et al 1981
47,XX,t(8;14)(q24;q32),dup(1)(q24q32),del(2)(q34), del(5)(q31q34),+8"s"	Berger and Bernheim 1982
47,XY,+11,-14,+21,inv(1)(p11p31),11q+,21p+	Biggar et al 1981
47-49,XY,t(8;14)(q23;q32),+7,+t(X;1)(p21;q21)	Kakati et al 1979
48,X?,t(8;14)(q23;q32),+2mar	Kaiser McCaw et al 1977
48-53,X?,t(8;14)(q24;q32),+X,+2,+3,+9,+10,+11,+12,+20	Zech et al 1976a
49,XY,+3,+7,+12,14q+	Biggar et al 1981
50,XX,+1,+2,+3,+11,del(2)(p14),11p+,t(8;14)(q24;q32)	Biggar et al 1981

Lymphoblastic, convoluted type

43-46,XY,del(1)(p32),del(2)(p21),-6,t(7;9)(p11;q11),-9,-10, t(11;15)(p11;q11),-11,del(12)(q14),-13,-14,i(17q),-17, +18,del(19)(p11),-22,+2mar	Kristoffersson 1983
46,X,-Y,t(1;?)(q21orq23;?),+mar/46,X,-Y,-1,-2,+der(1), +t(2;?)(q33orq35;?),+mar/48,X,-Y,-5,-14,+18,+18,+20, del(9)(p13),t(1;14),+dup(5)(q13q35)	Kaneko et al 1982d
47,XY,-7,-9,del(6)(q23?),t(14;18)(p11;q11),+t(9;?)(q34;?), +2mar/47,XY,-7,-9,del(6),del(1)(q32),t(14;18),t(9;?), +t(1;?)(q23orq25;?)	Kaneko et al 1982d

Immunoblastic

47-48,XX,+del(3)(q11),+17/49,XX,+del(3)(q11),+16,+17	Kristoffersson 1983
45-47,XX,1p+,-14,-15,+16,+17,-19,+mar	
48-49,XY,del(3)(p23),+7,+12,+14,+20	Kristoffersson 1983
78-83,XY,+del(1)(q11),+5,+7,+12,+13q+,+14q+,+14q+,+20,+mar	Kristoffersson 1983
80-86,XY,del(1)(q11),+3,+7,+9,+12,+13q+,+14q+,+14q+,+15q+, +18,+20,+mar	

3C. NON-HODGKIN'S LYMPHOMAS - LUKES AND COLLINS CLASSIFICATION

Mycosis fungoides

47,XY,+Xp-,t(4;14),+14	Edelson et al 1979
48,XX,del(1)(p22),t(2;8;14)(q37;q24;q24),+5,del(7)(p13), del(9)(q22),t(9;18)(q11;q24),del(10)(p13), del(13)(q12q14),del(5)(q15q22),del(5)(q13),del(5)(q15)	Fukuhara et al 1978a
46,XX,t(1;14)(q32;q32)	

Follicular center cell, follicular, small cleaved

45,X,-X,t(14;?)(q?;?)	Fukuhara et al 1979a

Follicular center cell, diffuse, large cleaved

47,XY,-6,-14,1q-,+1p-,2q+,3q+,9p+,10p-,10p+,12p-,18q+,19p+, 19q+,+22p+,+mar	Fukuhara et al 1978b

Follicular center cell, diffuse, small non-cleaved

44,X,-Y,t(1;14)(p21;q13),5p+,del(6)(q13),del(8)(p21),9p+, 9p-,t(11;21)(q23;q22),13q+,-10,+12,-17p+,der(19),22q+	Fukuhara et al 1979a

3D. NON-HODGKIN'S LYMPHOMAS - WORKING FORMULATION

Small lymphocytic

63,XY,cx,t(14;17)	Yunis et al 1982

Follicular, small cleaved cell

44,XY,+der(1),+2,t(2;13),-3,5q-,-12,t(14;18)(q32;q21), i(17q)	Yunis et al 1982
46,XX,t(2;3),i(6p),t(14;18)(q32;q21)	Yunis et al 1982
46,XY,t(14;18)(q32;q21)	Yunis et al 1982
46,XY,t(14;18)(q32;q21)	Yunis et al 1982
46,XY,t(14;18)(q32;q21)	Yunis et al 1982
46,XY,t(14;18)(q32;q21)	Yunis et al 1982
46,XY,t(1;2),t(4;10),t(14;18)(q32;q21),t(1;22)	Yunis et al 1982
47,XX,t(X;1),ins(1;8),t(2;8),t(14;18)(q32;q21),+17	Yunis et al 1982
47,XY,+7,t(14;18)(q32;q21)	Yunis et al 1982
47,XY,+X,t(2;3;17),6q-,t(14;18)(q32;q21)	Yunis et al 1982
48,X,r(Y),+2,+3,t(14;18)(q32;q21),t(1;22)	Yunis et al 1982
49,XY,+X,+3,ins(11;3),t(14;18)(q32;q21),+18	Yunis et al 1982

Follicular, mixed small cleaved and large cell

50,Y,inv(X),+3,+7,+10,t(12;12),+12,t(14;18)(q32;q21),+18, 18q-,-21 — Yunis et al 1982
79,XY,cx,t(1;1),9p-,t(14;18)(q32;q21) — Yunis et al 1982

Follicular, large cell

47,XX,t(2;13),t(14;18)(q32;q21),+21 — Yunis et al 1982
75,XX,cx,i(6p),t(14;18)(q32;q21),i(17q) — Yunis et al 1982

Diffuse, small cleaved cell

49,XY,dup(1)(q?),8p-,-9,+14,+17,+20,+mar — Yunis et al 1982

Diffuse, mixed small and large cell

47,XX,+9,t(9;15)(p13;q11),t(12;14)/49,XX,+3,+5,+9,t(9;15), t(12;14) — Yunis et al 1982

Diffuse, large cell

46,XY,t(11;14)/47,XY,t(11;14),+3 — Yunis et al 1982
47,XX,+X,6q-,t(14;17) — Yunis et al 1982

Large cell, immunoblastic

45,XY,ins(2;21),t(8;14)(q24;q32),t(9;19),t(17;21),-21 — Yunis et al 1982
48,Y,t(X;?),t(1;?),t(3;6),+5,-6,t(8;14)(q24;q32),i(17q), +mar — Yunis et al 1982

Small non-cleaved cell

46,X,-X,t(8;14)(q24;q32),dup(13),+r/45,X,-X,t(8;14),-12, t(1;15),der(22),+r — Yunis et al 1982
46,XX,dup(1)(q?),t(8;14)(q24;q32) — Yunis et al 1982
47,XY,t(8;14)(q24;q32),+mar — Yunis et al 1982

Chromosome 15

HEMATOLOGICAL DISORDERS

UNCLASSIFIED LEUKEMIAS

NONLYMPHOCYTIC LEUKEMIAS

Acute nonlymphocytic leukemia (ANLL)

Karyotype	Reference
39,X,-X,-3,-5,-13,-14,-15,-16,-17,-21,-22,del(7)(p11), t(12;?)(p13;?),t(16;?)(p13;?),+3mar	Mitelman 1983
39,X,-X,-3,-5,-13,-14,-15,-16,-17,-21,-22,del(7)(p11), t(12;?)(p13;?),t(16;?)(p13;?),+3mar	
43,XX,-1,3p+,3q-,4q+,-5,-5,-7,+9,+9,10p+,+10,+10,10p-,10q-, +11q-,-13,-15,-16,-17,-18,-21"s"	Whang-Peng et al 1979
43-44,XY,-5,-5,-14,-15,-16,-17,-18,-20,+5mar	Mitelman 1983
44,X,-Y,-3,+4,-12,+15,-20	Bernard et al 1982a
45,XY,-5,-15,+r"s"	Berger et al 1981a
45,XY,-7/50,XY,+8,+15,+19,+20	Berger et al 1981a
45,XY,t(1;14)(p22;p11),t(1;15)(p22;p11),-2,del(5)(p13),-17, +mar	Philip et al 1978a
45-47,XX,-2,-7,7q-,-8,+15,+22,+2mar"s"	Whang-Peng et al 1979
46,XX,del(5)(q12q31),-9,15p+,18p-,-21,+2mar	Kerkhofs et al 1982
46,XX,t(15;17)(q24;q21)	Yunis 1982
46,XX,t(15;17)(q?;q?)	Bernard et al 1982a
46,XY,inv(9),t(11;15)(q11;q26),t(11;13)(q11;q34)"c"	Philip et al 1978a
46,XY,t(15;17)(q?;q?)	Bernard et al 1982a
46,XY,t(9;10)(p24;p12),t(3;16),t(4;15)(p13;q14)"c"	Van Den Berghe et al 1979a
47,XY,+15	Philip et al 1978a
48,XY,+15,+21	Bernard et al 1982a
48,XY,+8,+8/47,XY,+8,+8,-15	Bernard et al 1982a
48-49,XX,-5,del(15)(q?),+19,t(5;?)(q13;?)	Philip et al 1978a

ANLL, FAB type M1

Karyotype	Reference
41,XY,-1,-2,-4,-5,-12,-15,-17,-17,-18,-22,+5mar	Sessarego et al 1981a
43,XX,del(5)(q12q21),-7,11p+,-15,-17	Kerkhofs et al 1982
58,XY,+1,+4,+5,+6,+8,+11,+13,+13,+14,+14,+15,+21,+21, t(9;22)(q34;q11)	Alimena 1983

ANLL, FAB type M2

Karyotype	Reference
46,XY,t(8;21)(q22;q22)/45,X,-Y,t(8;21)/48,XY,t(8;21), +der(21),t(10;15)(p15;q22),+r	Sakurai et al 1982b
48,XX,-5,t(8;?)(p23;?),del(11)(q23),t(15;?)(p11;?),-21, +4mar"s"	Rowley et al 1977a
48,XY,+1,+15	Lessard and Le Prise 1983

ANLL, FAB type M1+M2

Karyotype	Reference
42-43,XX,cx,t(11;15)(q11;q11),t(9;?)(p?;?),-15,-17,-18,-20, -21	Fitzgerald et al 1983
44,XX,-15,-17,dmin/43,XX,-15,-17,-20,dmin"s"	Sandberg et al 1982a
45,XY,-15	Nordenson 1983
45,XY,t(19;22)(p13;q11),t(1;7)(q21;q22),t(7;9),+del(8)(p?), -15	Oshimura and Sandberg 1977
46,XY,t(9;22)(q34;q11),t(2;15)(q37;q12)	Mitelman et al 1981
46-47,XX,del(1)(q32),+8,t(15;17)(q25or26;q22or23),+18	Mitelman et al 1981
48,X,-Y,1p-,+3p-,+15,+mar"s"	Sandberg et al 1982a
57,X,-Y,-6,+8,+15,+15,+19,+19,+20,+20,+21,+21,+22,+3mar	Fitzgerald et al 1983
57,XY,+6,+11,+13,+15,+21,+21,+21,+21,+21,+21,+22"c"	Shiraishi et al 1982

ANLL, FAB type M3

Karyotype	Reference
45,X,-Y,t(15;17)(q22;q12)	Kaneko et al 1982b
45,XY,t(15;17),-19	Ferro et al 1982
46,X?,t(15;17)	Sakurai et al 1982a
46,X?,t(15;17)	Sakurai et al 1982a
46,X?,t(15;17)	Sakurai et al 1982a
46,X?,t(15;17)	Sakurai et al 1982a
46,X?,t(15;17)	Sakurai et al 1982a
46,X?,t(15;17)	Sakurai et al 1982a
46,X?,t(15;17)(q?;q?)	Hartley 1983
46,XX,t(15;17)	Alimena 1983
46,XX,4q+,t(15;17)	Rochon and Vaillancourt 1981
46,XX,del(2)(q32q34),t(15;17)(q25;q22)	Webber and Garson 1983
46,XX,der(15)(q22),i(17)(q12)	Hagemeijer et al 1982a
46,XX,t(15;17)	Alimena 1983
46,XX,t(15;17)	Alimena 1983
46,XX,t(15;17)	Alimena 1983
46,XX,t(15;17)	Alimena 1983
46,XX,t(15;17)	Alimena 1983
46,XX,t(15;17)	Alimena 1983
46,XX,t(15;17)(q22;q12)/46,XX,15q+,i(17q-)/47,XX,t(15;17), +8	Kondo and Sasaki 1982
46,XX,t(15;17)(q22;q12)	Kondo and Sasaki 1982
46,XX,t(15;17)(q22;q12)	Kondo and Sasaki 1982
46,XX,t(15;17)(q22;q21)	Berger et al 1981e
46,XX,t(15;17)(q22;q21)	Berger et al 1981e
46,XX,t(15;17)(q22;q21)	Berger et al 1981e
46,XX,t(15;17)(q22;q21)	Berger et al 1981e
46,XX,t(15;17)(q22;q21)	Berger et al 1981e
46,XX,t(15;17)(q22;q21)	Berger et al 1981e

46,XX,t(15;17)(q22;q21)/47,XX,t(15;17),+del(16)(q11)	Berger et al 1981e
46,XX,t(15;17)(q22;q21)/46,XX,t(15;17),t(1;4)(p35;q21), del(X)(q25)	Berger et al 1981e
46,XX,t(15;17)(q22;q12)	Hagemeijer et al 1981a
46,XX,t(15;17)(q22;q21)	Hagemeijer et al 1981a
46,XX,t(15;17)(q22;q21)	Mitelman 1983
46,XX,t(15;17)(q22;q21)	
46,XX,t(15;17)(q22;q21)	Okada et al 1977
46,XX,t(15;17)(q22;q21)	Mitelman 1983
46,XX,t(15;17)(q22;q21)	Alimena 1983
46,XX,t(15;17)(q22;q21)	Pasquali 1983
46,XX,t(15;17)(q22;q21)	Pasquali 1983
46,XX,t(15;17)(q22;q21)	Pasquali 1983
46,XX,t(15;17)(q22;q21)	Pasquali 1983
46,XX,t(15;17)(q24;q21)	Hurd et al 1982
46,XX,t(15;17)(q24;q21)	Hurd et al 1982
46,XX,t(15;17)(q24;q21)	Yunis et al 1981
46,XX,t(15;17)(q24;q22)	Alimena 1983
46,XX,t(15;17)(q25;q22)	Golomb et al 1979
46,XX,t(15;17)(q25;q22)	Golomb et al 1979
46,XX,t(15;17)(q25;q22)	Golomb et al 1980
46,XX,t(15;17)(q25;q22)	Bernstein et al 1982b
46,XX,t(15;17)(q25;q22)	Ferro et al 1982
46,XX,t(15;17)(q25;q22)"s"	Sheer et al 1982
46,XX,t(15;17)(q25;q22)	Fraisse et al 1981a
46,XX,t(15;17)(q25;q22)	Fraisse et al 1981a
46,XX,t(15;17)(q25;q22)	Alimena 1983
46,XX,t(15;17)(q26;q22)	Van Den Berghe et al 1979c
46,XX,t(15;17)(q26;q22)	Van Den Berghe et al 1979c
46,XX,t(15;17)(q26;q22),del(7)(q33orq35)	Van Den Berghe et al 1979c
46,XX,t(15;17)(q26;q22)	Van Den Berghe et al 1979c
46,XX,t(15;17)(q26;q22)	Van Den Berghe et al 1979c
46,XX,t(15;17)(q26;q22)/47,XX,t(15;17),-7,+8,-9,-10,+2mar	Van Den Berghe et al 1979c
46,XX,t(15;17)(q?;q?)	Lessard and Le Prise 198
46,XX,t(2;15;17)(q21?;q25?;q21?;),9p+	Bernstein et al 1982b
46,XY,del(15)(q22q26)/45,XY,-7	Zahavi et al 1982
46,XY,t(14;21)"c"	Alimena 1983
46,XY,t(15;17),t(14;21)"c"	
46,XY,t(15;17)	Takeuchi et al 1981
46,XY,t(15;17)	Alimena 1983
46,XY,t(15;17)(q22;q21)	Kaneko and Sakurai 1977
46,XY,t(15;17)(q22;q12)	Kondo and Sasaki 1982
46,XY,t(15;17)(q22;q12)	Kondo and Sasaki 1982
46,XY,t(15;17)(q22;q12)	Kondo and Sasaki 1982
46,XY,t(15;17)(q22;q12)	Kondo and Sasaki 1982
46,XY,t(15;17)(q22;q12)	Kondo and Sasaki 1982
46,XY,t(15;17)(q22;q12orq21),i(17q-)	Chapelle et al 1981
46,XY,t(15;17)(q22;q21)	Berger et al 1981e
46,XY,t(15;17)(q22;q21),t(X;9)(q28;q22)	Berger et al 1981e

46,XY,t(15;17)(q22;q21)	Berger et al 1981e
46,XY,t(15;17)(q22;q21)	Berger et al 1981e
46,XY,t(15;17)(q22;q21)	Berger et al 1981e
46,XY,t(15;17)(q22;q21),8q+	Berger et al 1981e
46,XY,t(15;17)(q22;q21)	Berger et al 1981e
46,XY,t(15;17)(q22;q12)	Hagemeijer et al 1982a
46,XY,t(15;17)(q22;q12)/46,XY,t(15;17),t(21;?)(p11;?)/46, XY,del(4)(q22),t(15;17)	Kaneko et al 1982b
46,XY,t(15;17)(q22;q12)	Kaneko et al 1982b
46,XY,t(15;17)(q22;q21)	Mitelman 1983
46,XY,t(15;17)(q22;q12)	Alimena 1983
46,XY,t(15;17)(q22;q21)	Pasquali 1983
46,XY,t(15;17)(q22;q21)	Pasquali 1983
46,XY,t(15;17)(q24;q21)	Hurd et al 1982
46,XY,t(15;17)(q24;q21)	Hurd et al 1982
46,XY,t(15;17)(q24;q21)	Hurd et al 1982
46,XY,t(15;17)(q24;q21)	Hurd et al 1982
46,XY,t(15;17)(q24;q21)	Hurd et al 1982
46,XY,t(15;17)(q24;q21)	Hurd et al 1982
46,XY,t(15;17)(q24;q21)	Fitzgerald et al 1982b
46,XY,t(15;17)(q24;q21)	Fitzgerald et al 1982b
46,XY,t(15;17)(q24;q21)	Fitzgerald et al 1982b
46,XY,t(15;17)(q24;q21)	Fitzgerald et al 1982b
46,XY,t(15;17)(q24;q21)	Fitzgerald et al 1982b
46,XY,t(15;17)(q24?;q21?)	Yunis et al 1981
46,XY,t(15;17)(q25;q22)	Golomb et al 1979
46,XY,t(15;17)(q25;q22)/46,XY,t(15;17),del(7)(q22), del(9)(q22)	Golomb et al 1979
46,XY,t(15;17)(q25;q22)	Golomb et al 1979
46,XY,t(15;17)(q25;q22)	Golomb et al 1979
46,XY,t(15;17)(q25;q22)	Fraisse et al 1981a
46,XY,t(15;17)(q25?;q22)	Bernstein et al 1982b
46,XY,t(15;17)(q26;q22)/47,XY,t(15;17),+8	Van Den Berghe et al 1979c
46,XY,t(15;17)/46,XY,del(1)(q12)/47,XY,+8/47,XY, del(1)(q12),+8/47,XY,t(15;17),+8	
46,XY,t(15;17)(q26;q22)	Van Den Berghe et al 1979c
46,XY,t(15;17)(q26;q22)	Van Den Berghe et al 1979c
46,XY,t(15;17)(q26;q22)	Van Den Berghe et al 1979c
46,XY,t(15;17)(q?;q?),t(3;11)(q?;q?),t(7;19)(q11;p13)	Trujillo and Cork 1981
46,XY,t(15;17)/46,XY,t(15;17),t(1;9)(p36;q22)	Hurd et al 1982
47,X?,t(15;17),+8	Sakurai et al 1982a
47,XX,+10,t(15;17)(q26;q22)	Brodeur et al 1983
47,XX,t(15;17)(q22;q12),+8	Kondo and Sasaki 1982
47,XX,t(15;17)(q26;q22),+8,/47,XX,+8/46,XX, t(15;17)(q26;q22),-7,+8	Van Den Berghe et al 1979c
47,XX,t(15;17),+8/47,XX,+8	Sakurai et al 1982a
47,XY,+8,t(15;17)(q24;q21)	Fitzgerald et al 1982b
47,XY,del(11)(q23),t(15;17)(q24?;q22?),+mar	Bernard et al 1982b

47,XY,t(15;17)(q21?;q25?),+8/47,XY,t(15;17)(q21?;q25?), +9/48,XY,t(15;17),+8,+9	Van Den Berghe et al 1979c
47,XY,t(15;17)(q22;q21),+8	Berger et al 1981e
47,XY,t(15;17)(q24;q21),+21	Scheres et al 1978
47,XY,t(15;17),del(1)(q42),+del(1)(p22q42)	Sakurai et al 1982a

ANLL, FAB type M4

42,XY,-4,-15,-16,-22	Prieto et al 1981
42,XY,-5,-9,-10,-15,-16,-17,-20,-21,-22,+5mar	Testa et al 1979
43,XY,+1,-5,+6,-9,-10,-15,-16,-17,-20,-21,-22,+6mar	
43,XY,-2,-3,-6,-7,-12,-15,-21,-22,del(2)(p11),del(11)(p14), t(14;?)(q32;?),t(17;?)(p13;?),del(22)(q11),+5mar"s"	Rowley et al 1981a
44,XY,del(5)(q15q23),del(7)(q22),-9,-12,-13,del(15)(q22), +mar/46,X,-Y,del(5),del(7),+8,-12,-13,del(15),+mar,dmin	Alimena 1983
44-46,XY,+1,-5,-17,-18,-20,del(6)(q21),-15,t(1;6)(q11;q24), +3mar	Mitelman et al 1981
44-46,XY,+1,t(1;6)(q11;q27),-5,del(6)(q21),-15,-17,-18,-20, +mar	Alimena 1983
45,XY,del(5)(q13q31),del(7)(q22),del(15)(q22q24)	Alimena et al 1982a
46,XY,t(11;17)(q23;q27),del(11)(q13q23),t(15;19)(q21;p13)	Yunis et al 1981
49,XX,+5,+8,+15	Fitzgerald et al 1983
53,XX,+15,+18,+21,+21,+mar"c"	Benedict et al 1979

ANLL, FAB type M5

44,X,-Y,del(2)(p11),-15,-21,+mar	Weber et al 1979
46,Y,-X,-1,-7,-8,-14,-15,-17,-20,inv(6)(p23q13), del(11)(q23),+t(X;?)(p11;?),+t(1;?)(p24;?), +t(12;?)(q22;?),+t(17;?)(p11;?),+4mar	Kaneko et al 1982b

ANLL, FAB type M5a

46,XY,+8,+8,-15,t(13;15)(p11;q11)	Berger et al 1982a

ANLL, FAB type M4+M5

42,XX,-5,-7,-12,-15,-16,-18,+2mar	Li et al 1983
45,XX,-20/45,XX,-18,-20,+mar/46,XX,-15,+mar	Li et al 1983
45,XY,+9,-15,-22	Li et al 1983

ANLL, FAB type M6

43,X,-Y,-4,-13,-14,-15,+2mar/46,X,-Y,-3,+9,+mar	Li et al 1983
43,XY,-2,4q-,-9,11q-,-15,-19,-20,+2mar/46,XY,-15,-20,+2mar	Li et al 1983
43,XY,-3,-15,-15"s"	Sandberg et al 1982a
43,XY,-4,5q-,t(4;6),-7,dic(12),-15"s"	Hagemeijer et al 1981a
44,X,del(Y)(q?),+Xq-,3p-,3q-,4q-,-6,+7q-,11q+,-15,-18"s"	Bradley et al 1982
44,XY,t(2;13)(q37;q12?),t(13;?)(p11;?),-15"s"	Weinfeld et al 1977a
47,XX,+8/48,XX,+15,+18	Testa et al 1979
51,XY,+1,+2,+6,-7,+8,-14,+15,+21,+mar"s"	Rowley et al 1981a
51,XY,+Y,+8,+15,+2mar	Hagemeijer et al 1981a
72,XY,cx,t(2;15)(q31;q15),del(6)(p22q15),t(8;13)(q23?;q13), t(11;14)(p12;q13),i(16p),t(15;18)(q15;p11)	Douglass and Freeman 198[illegible]

Chronic myeloid leukemia, t(9;22)

26,XX,t(9;22)(q34;q11),-1,-2,-3,-4,-5,-6,-7,-9,-10,-11,-12, -13,-14,-15,-16,-17,-18,-19,-20	Hartley and McBeath 1981
35,XX,t(9;22)(q34;q11),-3,-4,-5,+8,-9,-10,-11,-12,-13,-14, -15,-16,-17/66-72,XX,t(9;22)	Pedersen and Boesen 1983
46,X?,t(9;22)(q34;q11),15q-	Lawler et al 1976
46,XX,t(9;22),-7,-15,+20,+i(17q)	Stoll and Oberling 1978
46,XX,t(9;22),der(3),del(15)(q22)	Fleischman et al 1981
46,XY,t(9;22)	Berger et al 1983b
46,XY,t(9;22)/46,XY,t(9;22),t(15;17)	
46,XY,t(9;22)(q34;q11)/50,XY,t(9;22),t(1;15)(p36;q22),+8, +10,+19,+22q-	Misawa 1978
46,XY,t(9;22)(q34;q11)	Como and Graze 1979
35,XY,t(9;22)(q34;q11),-3,-4,-5,-7,-9,-11,-12,-13,-15,-16, +3mar	
46,XY,t(9;22)(q34;q11),21q+/50,XY,t(9;22),+7,+8,+12,+15, 21q+	Tomiyasu et al 1982
46,XY,t(9;22)(q34;q11)	Ishihara et al 1982
27,X,-Y,t(9;22),-1,-2,-3,-4,-5,-6,-7,-9,-11,-13,-14,-15, -16,-17,-18,-19,-20,-22	
46,XY,t(9;22),del(15)(q15q22)	Kohno et al 1979b
46,XY,t(9;22),del(7)(q33?),del(15)(q?)	Kohno et al 1979b
46,XY,t(9;22),t(1;6)(q25;q25)/46,XY,t(9;22), t(1;15)(q12;p11)	Miyamoto 1980
46,XY,t(9;22),t(4;12;15)(q21;q24;q21)	Hartley and McBeath 1981
47,XX,t(9;22),+15	Sadamori et al 1980
47,XY,t(9;22),+15/48,XY,t(9;22),+17,+22q-	Stoll and Oberling 1978
47,XY,t(9;22),+8,15q-/48,XY,t(9;22),+8,15q-,+19	Olah and Rak 1981
48,XY,t(9;22),+8,+13,-15,t(9;15),+i(22q)	Stoll and Oberling 1978
50,XY,t(9;22),+5,+8,+13,-15,t(9;15),+i(22q),+22q-	
48,XY,t(9;22),+8,+22q-/61,XY,+Y,t(9;22),+4,+5,+6,+8,+8,+8, +8,+15,+17,+18,+19,+20,+22q-,+22q-	Wurster-Hill 1983
49,XY,t(9;22),+17,+19,+22q-/50,XY,t(9;22),+8,+15,+19, +22q-/50,XY,t(9;22),+8,+9,+19,+22q-	Hartley and McBeath 1981
50,XY,t(9;22),+8,+15,+17,+del(22q)	Sharp et al 1976
50-58,XY,+X,t(9;22)(q34;q11),+7,+8,+9,+10,+11,+12, +del(15)(q22),+16,+18,+19,+22q-	Kwan et al 1977
46,XY,t(9;22)	
52,XY,t(9;22),+8,+14,+15,+19,+21,+22q-	Sadamori et al 1980
53,XY,+Y,t(9;22)(q34;q11),+3,+6,+8,+9,+10,+12,-15,+22q-/56, XY,+Y,t(9;22),+3,+6,+8,+8,+9,+10,+16,+19,+22q-	Stoll and Oberling 1982
56,XY,+Y,t(9;22),+3,+6,+8,+8,+9,+10,+12,+19,+22q-	
55,XX,t(9;22),+3,+8,+12,+14,+15,+19,+20,+22,+22q-	Lilleyman et al 1978
55,XY,t(9;22),+3,+6,+8,+8,+9,+10,+16,+19,+22q-/53,XY, t(9;22),-15,+3,+6,+8,+8,+9,+10,+12,+22q-	Stoll and Oberling 1978

Chronic myeloid leukemia, aberrant translocation

??,X?,t(15;22)(p11;q11)	Potter et al 1981
??,X?,t(9;15;22)(q34;q15;q11)	Potter et al 1981
??,X?,t(9;15;22)(q34;q15;q11)	Potter et al 1981
46,XX,t(15;22)(p13;q11)	Jotterand-Bellomo 1978
46,XX,t(15;22)(q26;q11)	Hossfeld and Köhler 1979

46,XX,t(9;15;22)(q34;q11;q11)	Hays et al 1981
54-57,XX,t(9;15;22),+5,+7,+8,+10,+11,+12,+17,+22q-	
46,XY,t(4;22)(q35;q21),t(3;5)(q27;q22)	Sessarego et al 1981b
46,XY,t(4;22)(q35;q21),t(3;5)(q27;q22),i(17q)/50,XY, t(4;22),t(3;5),i(17q),+8,+15,+21,+22q-	
46,XY,t(9;10;15;19;22)(q34;q22;q21;q13;q12)	Tomiyasu et al 1982
46,XY,t(9;15;22)(q34;q15;q11)	Seabright 1983
47,XY,+Y,t(9;13;15;22)(q34;q12;q22;q11)	Potter et al 1975

Eosinophilic leukemia

45,X,-Y,del(15)(q22)"s"	Goffman et al 1983
49,XY,+10,+15,+19,del(3)(q?)	Huang et al 1979

MYELOPROLIFERATIVE DISORDERS

Polycythemia vera

??,X?,+15,+21	Vykoupil et al 1980
??,X?,4q+,+13,15q-,20q-	Vykoupil et al 1980
46,XX,-15,t(1;15)(q11;q11)	Wurster-Hill et al 1976
46,XX,t(2;11)(p13;q21)/46,XX,t(1;15)(p1?;q1?), t(2;11)(p13;q21),del(16)(p12)/51,XX,+3,+8,+8,+19,t(1;15), t(2;11),del(16)	Testa et al 1981b

Myelosclerosis / Myelofibrosis

46,X,del(X)(p?q?),del(5)(q?),13q+,15q-"s"	Whang-Peng et al 1978
46,XX,del(15)(q14q23?)	Kohno et al 1979b

DYSMYELOPOIETIC SYNDROMES

Refractory anemia with excess of blasts

56-64,XY,+1,-2,+6,+9,+13,+15,+19,+21,10-13mar	Streuli et al 1980

Erythrocytopenia

46,XX,del(15)(q21)	Kohno et al 1979b

Pancytopenia

46,XX,-4,-5,-6,-7,-11,-13,-16,+19,+20,t(7;13)(q?;q?), +4mar/43,XX,-2,-3,-5,-7,-16,+19,15q+,+mar	Nowell and Finan 1978

Dysmyelopoietic syndrome, special type

45,XY,del(2)(q33),-5,del(7)(q22q32),-12,-15, t(12;15)(q11;q26),del(19)(q12),+r"s"	Albain et al 1983

SPECIAL LEUKEMIAS

LYMPHOCYTIC LEUKEMIAS

Acute lymphocytic leukemia (ALL)

28,XX,-1,-2,-3,-4,-5,-7,-8,-9,-11,-12,-13,-14,-15,-16,-17,-19,-20,-22	Kaneko and Sakurai 1980
32,XY,-2,-3,-4,-5,-7,-8,-11,-12,-13,-14,-15,-16,-17,-20	Shabtai and Halbrecht 1981b
43,X,-X,-7,-15,t(9;22)	Gibbs et al 1977
45,XX,5q+,14q+,-15	Chrz et al 1983
45,XX,t(14;14)(q34;q11),t(13;15)(q11;p11),t(13;17)(p11;q11),-16,18q+,+3mar	Levitt et al 1978
46,XY,del(5)(q13),del(15)(q15)/46,XY,del(7)(q32)/46,XY,del(6)(q21)	Secker-Walker et al 1979
46,XY,t(14;22)(q32;q11)/46,XY,t(14;22),+14,-15	Ayraud et al 1975
48,XY,t(4;12)(q?;q?),t(6;15)(q?;q?),t(9;22)(q34;q11),11q-,+20,+21	Nordenson 1983
53,XY,+X,+4,+8,+14,+15,+17,+21	Oshimura et al 1977a
54,XX,7p+,+10,+10,+14,+14,+18,+18,+21,+21/27,X,-X,-1,-2,-3,-4,-5,-6,-7,7p+,-8,-9,-11,-12,-13,-15,-16,-17,-19,-20,-22	Oshimura et al 1977a
55,XX,+2,+4,+9,+14,+15,+19,+21,+21	Oshimura et al 1977a
56,XX,+3,+6,+12,+13,+14,+15,+18,+21,+r	Prigogina et al 1979
57,XY,+X,+5,+6,i(7q),+10,+12,+13,+15,+18,+21,+21,+22	Oshimura et al 1977a
61,XX,+X,+1,+2,+3,+4,+5,+6,+8,del(6)(q21),+13,+14,+15,+16,+18,+21,+22	Oshimura et al 1977a

ALL, FAB type L1

26,XX,-1,-2,-3,-4,-5,-6,-7,-8,-9,-10,-11,-12,-13,-14,-15,-16,-17,-19,-20,-22	Hoeltge et al 1982
28,XX,-1,-2,-3,-4,-5,-6,-7,-8,-9,-11,-12,-13,-15,-16,-17,-19,-20,-22/56,XX,+X,+X,+10,+10,+14,+14,+18,+18,+21,+21	Brodeur et al 1981a
52-56,XY,t(9;22),+4,+6,+9,+10,+14,+15,+20,+21,+22q-	Sandberg et al 1980
53,XX,+1,+5,+14,+15,+19,+20,+22	Kaneko et al 1981b
56,XY,+X,+4,+6,+10,+15,+17,+18,+21,+21,+mar	Kaneko et al 1982e

ALL, FAB type L2

26,XX,-1,-2,-3,-4,-5,-6,-7,-8,-9,-10,-11,-12,-13,-15,-16,-17,-18,-19,-20,-22	Brodeur et al 1981a
46,X,-X,-15,del(11)(q21),del(12)(p11),+2mar	Kaneko et al 1981b
46,XY,-4,+6,-8,-10,-11,-15,-17,+22,del(6)(q21),+i(17q),+3mar	Kaneko et al 1981b
56,XX,+1,+6,+10,+13,+15,+16,+17,+18,+20,+21	Mitelman 1983

ALL, FAB type L3

46,XY,t(8;14)(q24;q32),-15,+16	Berger and Bernheim 1982

Chronic lymphocytic leukemia

41,XX,t(14;14)(q11;q32),t(1;13)(q44;q11),t(15;18)(q11;q23), 10q+,-16,-20	Kaiser McCaw et al 1975
45,XX,-8,-14,-15,-22,+2,+9,+1p+,+mar	Finan et al 1978
45-46,XY,-8,9q+,-10,13p+,-14,-15,-19,-20,+mar	Vahdati et al 1983a
46,XY,del(11)(q22),+12,-13,-21,-22,+3mar/45,XY,del(11),+12, -13,-15,-21,-22,+2mar/46,XY,del(11),+12,+12,-13,-21,-22, +2mar/46,XY,t(1;8)(p22;q24)	Robert et al 1982
47,XX,+12/47,XX,inv(11)(p15q22),+del(12)(q12orq21), t(12;15)(q15orq21;p13)	Najfeld et al 1980

Prolymphocytic leukemia

44,X,-Y,-11,-12,-13,-13,-14,-15,-15,-18,-18,+8mar	Diamond et al 1980

Lymphocytic leukemia, special type

45,X,-X,-4,+t(4;7)(p16;q11),15q+	Ueshima et al 1981

MONOCLONAL GAMMOPATHIES

Multiple myeloma

47,X,-X,inv(1)(p22q12),t(7;?)(q11;?),+9,t(11;?)(p15;?),-13, -14,t(14;?)(q11;?),+15,17q+	Philip et al 1980
47,X,-Y,+t(1;15)(q12;p11),+3,+11,-15,+19,-22/47,X,-Y,+3, +11,+19,-22	Philip et al 1980
52,XY,-1,+t(1;15)(q12;p1?),+t(1;16)(p36;p13), t(3;5)(q2?;q13),+3,+5,+7,+8,+9,+11,-15,+18,-20	Philip et al 1980
55,XY,+1,+5,+9,+11,-12,+13,+15,del(1)(q32),del(6)(q25), t(9;19)(q13;q13),14q+,16p+,+4mar	Liang et al 1979
45,XY,-2,-2,-3,+5,-8,-10,+14,-15,-16,del(6)(q25),+4mar	
55,XY,+3,+5,+7,+9,+11,+14,+15,+21,+21/53,XY,+3,+5,-8,+9, +11,+14,+15,-19,+21,14q+,+3mar	Liang et al 1979

Plasma cell leukemia / Plasmocytoma

46,XY,+t(1;16)(q21;p13),del(6)(q21),-16,-18,+mar	Ueshima et al 1983
47,XY,+t(1;16),del(6)(q21),+t(8;15)(p23;q12),-8,14q+,-15, -16,-18,+3mar	

SOLID TUMORS

UNCLASSIFIED NEOPLASMS

EPITHELIAL NEOPLASMS

Carcinoma, NOS

44,XX,ins(1;1)(q32;q12q31),i(2q),+3,t(9;11)(q13;p15),-12,-15,17p+,-18	Atkin and Baker 1982a
44,XY,+5,-8,-11,-13,-13,-14,-14,-15,+19,-22,-22,t(1;11),5q-,9q+,+5mar	Reichmann et al 1981
44-50,XX,+3,dup(1)(q21q32),t(10;11)(q26;q25),t(2;17),t(6;13),t(9;15),del(10)(q24)	Riet-Fox et al 1979
46,XX,+2,-7,-10,+13,-16,-17,+20,+5q-,15p+,15p+,18p+	Reichmann et al 1981
46,XX,+8,+12,+21,-2,-15,-16	Wake et al 1981
46,XX,t(3;11)(q12;q14),-6,+7,-8,-15,t(16;?)(q23;?),+2r	Mark 1975b
46,XX,t(3;17)(p24;p13)/46,XX,t(5;17)(q22;p13)/46,XX,t(9;18)(q13;q21)/46,XX,t(12;15)(q13;q22)	Stenman et al 1982
47,XX,-7,+15,+17,+mar	Sonta and Sandberg 1978a
48,XX,-7,+15,+17,+mar	Sandberg 1977
48,XY,+8,-17,-18,+19,+20,t(10;18),t(15;17)	Reichmann et al 1981
49,X,-Y,-3,-5,-14,+15,-16,-17,+19,+21,+22,1q+,5q-,6q-,8p-,9p-,11p-,18p+,20q-,22q-,t(Y;14),+3mar	Reichmann et al 1981
55,XY,t(1;17)(p36;q21),+2,+2,+4,+8,+8,-10,+15,+18,+20,+21,+2mar	Sonta and Sandberg 1978a
58,XX,+X,+1,+2,+3,5p-,+6,+7,+9,11q+,-13,-14,-15,+16,+19,+20,+6mar	Kovacs 1978a
58,XY,del(3)(p14),+4,+7,+8,+10,+11,+12,+14,+15,+17,+18,+19,+21	Kovacs 1978a
58-61,XY,cx,t(Y;15)	Reichmann et al 1981
63,XX,-1,+3,+5,+6,+7,+8,+10,+11,+12,-13,-15,+20,+21,+22,+9mar	Sandberg 1977
79,XX,cx,i(1q),t(1;3)(p11;q29),12q+,15p+,i(5p)	Kovacs 1978a
79,XX,cx,i(1q),t(1;3)(p13;q29),t(1;14)(p13;p13),t(12;?)(q24;?),t(15;21)(p11;q11),i(5p),r	Kovacs 1981

Carcinoma NOS, metastatic

35,XX,-1,-3,-4,-5,-7,-8,-10,-11,-12,-13,-14,-15,-16,-17,-18,-19,-20,-21,-22,+8mar	Pathak et al 1979

Small cell carcinoma

47,XY,+del(1)(p?),+del(3)(q?),+del(3)(p?),-4,+8,-15,+18,-22	Wurster-Hill and Maurer 1978
58,XY,cx,del(1)(p13),del(1)(q21),del(3)(p21),inv(3)(p14p23),hsr(15)(p?)	Whang-Peng et al 1982b

Adenoma

46,XY,del(5)(q14q31),ins(15;5)(q22;q14q31)	Mark et al 1982a
46,XY,t(6;21)(q13;q22)/46,XY,t(1;14)(q42;p12)/45,X,-Y	Mark et al 1981b
45,X,-Y	
46,Y,t(X;15)(q24;q15)/45,X,-Y	

Adenocarcinoma

??,XX,t(13;15)(q33;q15)	Trent and Davis 1979
??,XX,t(1;15)(q32;q25)	Trent and Davis 1979
37,X,-X,-3,-8,-15,del(2)(p23),del(6)(q15),t(11;?)(p12;?)	Trent and Salmon 1981
40-42,X,-X,t(1;?)(p13;?),del(1)(p12),2q+,del(3)(q21), del(6)(q13),del(7)(q11),t(11;11)(p15;q13), t(12;12)(q24;?),13p+,15p+,i(22q)	Kusyk et al 1981
43-46,XY,i(10q),15q+,-17,+19,-21,-22,dmin	Couturier-Turpin et al 1982
44,XX,+5,-9,-13,-13,-15,-17,-18,+22,+2mar	Martin et al 1979
44,XX,-1,+5,-15,-18,2q+/44,XX,-1,+5,-6,-15,-18,2q+,10p-, +mar	Martin et al 1979

MESENCHYMAL NEOPLASMS

Leiomyosarcoma

42,XX,del(1)(p12orp13),del(11)(q13orq14),+7,-9,-13,-14,-15, -18,+19,-22	Mark 1976

Mesothelioma, malignant

40-46,XY,+16,+20,-22,t(14;15),del(2)(p?),i(5p)	Ayraud 1975
41,XY,del(1)(p13),t(1;3)(p13;q11),t(7;7)(q11;q36), t(9;12)(p24;q13),del(12)(q13),del(13)(q12q14),-14,-15, t(17;?)(p13;?),i(19p),-20,-21,-22	Mark 1978

MELANOMAS

Malignant melanoma

44,X?,+2,+9,-6,-8,-15,-21,9p-,19q+	Wake et al 1981

Malignant melanoma, metastatic

24,X,cx,t(1;15),t(1;14)(q12;p11),del(7)(p13), t(6;7)(p11;q11),t(13;21)(p11;q11),+r	Atkin and Baker 1981

NEUROGENIC NEOPLASMS

Neuroblastoma

Karyotype	Reference
45,XX,+14,-15,-18,del(1)(p32),dup(1)(p13p31),del(6)(q23), del(11)(q23)	Brodeur et al 1981b
45,XY,-3,del(1)(p13),t(1;9)(q11;p11),t(9;15)(p11;p11), hsr(16p)	Brodeur et al 1981b
53,XX,+1,+4,+15,+20,dup(1)(p13p32),t(3;15)(q13;q26), del(4)(q31),t(7;11)(q36;q13)	Brodeur et al 1981b

Retinoblastoma

Karyotype	Reference
46,XX,t(1;15)(q11q32;q12),t(1;16)(q21;q11),dup(17)(p11q25)	Gardner et al 1982
47,XY,t(X;1)(q28;p11),t(2;?)(q11;?),t(2;15)(p25;q11), t(3;?)(q27;?),4p+,t(5;?)(q35;?),i(17q),+mar	Gardner et al 1982

Meningioma, benign

Karyotype	Reference
41,X,-Y,-22,-18,-15,-10,1p-	Zankl et al 1975a
43,X,-Y,-22,-15	Zankl et al 1975a
44,XY,-22,-15	Zankl et al 1975a
44,Y,-X,-22,-14,+mar/43,Y,-X,-22,-14,-15,-11,-10,+3mar	Zankl et al 1975a
45,XX,-15,-22,1p-,2p-,4p-,+mar	Zankl et al 1979
45,XX,-22/40-44,X,-X,-8,-9,-15,-21,-22	Mark et al 1972b
45,XX,t(6;15)(p12;q26),-22	Zang and Zankl 1983
45,XY,-22/42-45,XY,-8,-15,-17,-20,-22	Mark et al 1972b
46,XX,22q-/43-45,XX,22q-,-15,-17	Mark et al 1972b
50,XX,+9,+10,+12,+15,+20,-22	Mark 1973c
52,X,-Y,-22,+20,+19,+17,+15,+11,+9,+7,+5	Zankl et al 1975a

EMBRYONAL AND MISCELLANEOUS NEOPLASMS

LYMPHOMAS

1. UNCLASSIFIED AND MISCELLANEOUS LYMPHOMAS

Non-Hodgkin's lymphoma, NOS

Karyotype	Reference
46,X,-X,del(1)(q22),t(2;11)(p21;q21),t(9;15)(q11;p12), del(9)(q11),-14,t(14;17)(q23;q23),-16,+mar	Kristoffersson 1983
48-54,XX,+X,3p+,-4,-6,-8,-9,-10,+12,-15,+7-10mar	Pearson et al 1982
50,XX,+X,t(1;11)(q25;q23),+3,+7,-13,t(14;18)(p?;p?),-15, t(19;?)(q13;?),+3mar	Reeves and Pickup 1980

Non-Hodgkin's lymphoma, lymphocytic, NOS

46-47,XY,-4,-15,-16,+3mar	Fleischmann et al 1976
47,XY,+7,-9,-11,1p+,6q-,10p-,14q-, t(14;?;14)(q24orq32;?;q13),15q-,22q-	Fukuhara et al 1979b
47-48,XY,+7,-11,-14,-15,4q+,9p-,14q+	Fleischman and Prigogina 1977
48,XY,+3,+3,-15,+mar	Reeves and Pickup 1980

Non-Hodgkin's lymphoma, histiocytic, NOS

45,XY,del(6)(q13),del(6)(q21),15q+,i(1q)	Reeves 1973
46,XX,2p+,-15,del(16)(q13)	Reeves 1973
47-48,XY,+X,-13,-15,t(6;11)(p?;q?),t(13;14;?)	Fleischman and Prigogina 1977
48,XY,del(3)(p21),del(9)(q22),11q+,12q+,del(6)(q23orq25), 15q+,del(12)(q15)	Reeves 1973
50,XY,t(2;?)(p25;?),del(2)(p13),+i(3p),del(6)(p21),+7,+8, +9,+9,-11,-13,+14,-15,+20,+21,-22	Mark et al 1978
44,XY,del(2)(p13),t(6;?)(p21;?),del(11)(q13), t(11;14)(q13;q31),i(15q),-20,-22	

Non-Hodgkin's lymphoma, T-cell, NOS

47,XX,t(2;?),inv(3),4p-,del(5)(q14q32),6p+,6q+,t(7;7),9q+, 10p-,13q-,t(14;14),-15,-15,16q-,17p+,+21,+2mar	Gaeke et al 1981

Angioimmunoblastic lymphadenopathy

48,XY,+3,+18/51,XY,+5,+15,+19,+21,+22	Kaneko et al 1982c
56,XY,+5,+5,+6,+10,+15,+15,+19,+20,+21,+22	Kaneko et al 1982c

2. HODGKIN'S LYMPHOMAS

Hodgkin's disease, NOS

46,XY,del(15)(q14q21)	Kohno et al 1979b
52,XX,+1,-9,-15,+7mar	Fleischmann and Krizsa 1977

Hodgkin's disease, mixed cellularity

61-80,XX,del(1)(q32),i(1q),del(2)(p13),del(6)(q13or15), del(6)(q23or25),15q+,12q+,del(5)(q31),del(7)(q11), del(10)(q26),del(3)(p13)	Reeves 1973
81,XX,14q+,i(18q),i(18p),del(6)(q21),t(9;15)(q11;p12), del(16)(q13)	Reeves 1973

Hodgkin's disease, lymphocytic depletion

Karyotype	Reference
46,XY,t(10;14)(q22;q32),t(15;?)(p11;?),-16,+mar	Hossfeld and Schmidt 1978
48,X,-Y,t(1;?)(q44;?),i(1q),del(1)(q21),del(3)(q21),-10, t(3;11)(q23;q21),+14,t(14;?)(q32;?),+15,-17,+19,+20,-22	Hossfeld and Schmidt 1978

3A. NON-HODGKIN'S LYMPHOMAS - RAPPAPORT CLASSIFICATION

Lymphocytic, poorly differentiated, diffuse

Karyotype	Reference
47,XY,t(14;14)(q32;q?),+7,-9,-11,1p+,6q-,10p-,15q-,22q-, +2mar	Fukuhara and Rowley 1978
47,XY,t(1;4)(q23;q33),+3,t(10;15),t(11;14)(q13;q32),r(13), r(13)	San Roman et al 1982
48,XX,1q-,+3,+4,6q-,+7,10q-,-15	Kakati et al 1980
48-49,XX,2p-,+7,+i(8q),+12,14q+,15q-,18q-	Kakati et al 1980

Lymphocytic, poorly differentiated, nodular

Karyotype	Reference
52,X,-X,+2,-4,+5,+7,+8,-14,-15,-16,+20,+21, +t(1;15)(p11;q11),+t(12;17)(q24;q11),+t(14;?)(q32;?), +mar	Kaneko et al 1982a

Mixed cell, diffuse

Karyotype	Reference
50,XY,+11,+12,-15,+del(3)(q11),+del(3)(q11), +t(X;1)(p22;q11)	Kaneko et al 1982a

Histiocytic, diffuse

Karyotype	Reference
45,X,-Y,t(14;15)(q32;q21orq22),-15,+18,-22,1q+,3p+,5p-,7p-, 9p+,12p+,16q+,17p-,+2mar	Fukuhara and Rowley 1978
45,X,del(X)(q24),del(1)(q21),+del(1)(p11),del(2)(q22), del(3)(q21),del(6)(p11),del(6)(q15),t(7;9)(p22;q13), t(8;10)(p23;q21),del(9)(q13),del(10)(q21), t(2;11)(q22;q23),-13,del(14)(q24),del(15)(q22), del(16)(q22),+17,-19	Mark et al 1979
47,XY,-10,-13,+15,-16,+19,+20,del(6)(q21),del(8)(q22), t(2;14)(q21;q32),t(4;4)(q31;q35),+t(10;?)(p11;?)	Kaneko et al 1982a
82,XX,del(X)(q13),i(1q),i(1p),del(7)(p11),i(9p), del(14)(q22),t(1;14)(q23;q32),i(15q),i(17q)	Mark et al 1979

3B. NON-HODGKIN'S LYMPHOMAS - KIEL CLASSIFICATION

Immunocytoma

Karyotype	Reference
45-46,XY,-8	Kristoffersson 1983
47,XY,15p+,+20/46,XY,del(5)(q23),del(6)(p22),+del(7)(p11),-21	

Lymphoblastic, Burkitt's type

Karyotype	Reference
??,XY,dup(1)(q12orq21q31),3p-,3q-,+6p-,+8,+14,+15,+18,-21	Douglass et al 1980a
??,XY,t(8;14),+15,+21	Douglass et al 1980a
46-48,X?,t(8;14)(q24;q32),+7,15q-,+6,+13	Zech et al 1976a

Lymphoblastic, convoluted type

Karyotype	Reference
43-46,XY,del(1)(p32),del(2)(p21),-6,t(7;9)(p11;q11),-9,-10,t(11;15)(p11;q11),-11,del(12)(q14),-13,-14,i(17q),-17,+18,del(19)(p11),-22,+2mar	Kristoffersson 1983

Immunoblastic

Karyotype	Reference
47-48,XX,+del(3)(q11),+17/49,XX,+del(3)(q11),+16,+17	Kristoffersson 1983
45-47,XX,1p+,-14,-15,+16,+17,-19,+mar	
47-48,XY,+4,del(4)(q21),+5,t(15;?)(q25;?),+19,-21	Reeves and Stathopoulos 1976
51,XY,t(1;6)(p22;q13orq15),t(2;13)(q21;q14),t(8;16)(q24;p11),+11,+12,del(15)(q22),+16,+20,+mar	Reeves and Pickup 1980
78-83,XY,+del(1)(q11),+5,+7,+12,+13q+,+14q+,+14q+,+20,+mar	Kristoffersson 1983
80-86,XY,del(1)(q11),+3,+7,+9,+12,+13q+,+14q+,+14q+,+15q+,+18,+20,+mar	

3C. NON-HODGKIN'S LYMPHOMAS - LUKES AND COLLINS CLASSIFICATION

Sezary's syndrome

Karyotype	Reference
46,XY,2q+,der(12),15q+	Nowell et al 1982
50,XX,1q+,2p-,+3q-,+6q-,+9,-15,+21,+mar	Nowell et al 1982

Immunoblastic type (B-cell)

Karyotype	Reference
48,XY,t(4;12)(q?;q?),t(6;15)(q?;q?),t(9;22)(q34;q11),11q-,+20,+21	Nordenson et al 1983

3D. NON-HODGKIN'S LYMPHOMAS - WORKING FORMULATION

Follicular, large cell

50,XY,1q-,t(2;3),+3,+10,t(1;12),13q-,+15,-19,t(3;19),+21 Yunis et al 1982

Diffuse, mixed small and large cell

47,XX,+9,t(9;15)(p13;q11),t(12;14)/49,XX,+3,+5,+9,t(9;15), t(12;14) Yunis et al 1982

Small non-cleaved cell

46,X,-X,t(8;14)(q24;q32),dup(13),+r/45,X,-X,t(8;14),-12, t(1;15),der(22),+r Yunis et al 1982

Chromosome 16

HEMATOLOGICAL DISORDERS

UNCLASSIFIED LEUKEMIAS

NONLYMPHOCYTIC LEUKEMIAS

Acute nonlymphocytic leukemia (ANLL)

Karyotype	Reference
39,X,-X,-3,-5,-13,-14,-15,-16,-17,-21,-22,del(7)(p11), t(12;?)(p13;?),t(16;?)(p13;?),+3mar	Mitelman 1983
39,X,-X,-3,-5,-13,-14,-15,-16,-17,-21,-22,del(7)(p11), t(12;?)(p13;?),t(16;?)(p13;?),+3mar	
42,XX,-5,-12,-14,-16,-18,t(12;16)(q11;q22q24)"s"	Oshimura et al 1976b
43,X,-Y,t(1;8)(p22;q24),-1,-11,-11,-13,-16,-17,-22, t(1;22)(q11;p11),t(11;13)(q23q25;q12q14),t(17;?)(p11;?), +mar	Oshimura et al 1976b
43,XX,-1,3p+,3q-,4q+,-5,-5,-7,+9,+9,10p+,+10,+10,10p-,10q-, +11q-,-13,-15,-16,-17,-18,-21"s"	Whang-Peng et al 1979
43-44,XY,-5,-5,-14,-15,-16,-17,-18,-20,+5mar	Mitelman 1983
45,X,-X,t(8;16)(q22;q24),+20,-21	Trent et al 1983a
46,XX,4q-,5q-,-16,-17,+2mar	Rowley 1976
46,XX,del(1)(p32),del(5)(q13),+8,t(9;?)(p11;?), t(11;22)(q11;q13),-16,-17,-19,+3mar	Muir et al 1977
46,XY,-11,+16"s"	Berger et al 1981a
46,XY,-16,del(10)(p11or12),+r	Oshimura et al 1976b
46,XY,t(9;10)(p24;p12),t(3;16),t(4;15)(p13;q14)"c"	Van Den Berghe et al 1979a
47,XX,+t(1;16),-16,-16,+13,+20	Prieto et al 1978b
48,XX,3q+,-5,+9,+16,+19	Nordenson 1983
50,XY,+6,+13,+16,+mar	Chrz et al 1983

ANLL, FAB type M1

Karyotype	Reference
39,XY,-5,-7,-12,-16,-17,-20,-21	Berger et al 1981a
46,XY,+del(2)(p13),i(9q),-16,del(17)(p11)	Brodeur et al 1983
46,XY,-2,-5,-7,-13,+16,-21,-21,+5mar	Sessarego et al 1981a
46,XY,del(16)(q21)	Bernstein et al 1982b

ANLL, FAB type M2

43,XY,-5,-7,-12,-16,4p+,17p+,20q-,+mar"s"	Rowley et al 1981a
45,XY,-5,-7,-8,+16,-17,+r,+mar	Berger et al 1981a
45,XY,-7,del(3)(p13),del(8)(q22),del(16)(p12)"s"	Rowley et al 1981a
46,XX,+1p-,+9,-16,-16/46,XX,+1p-,+13,-16,-16	Prieto et al 1981
47,XX,1p+,der(5),-6,12p+,-13,-14,-16,-17,-17,-21,+22,+7mar, dmin	Cooperman and Klinger 1981
55,XX,+16,+17,+18,+20,+21,+22,+3mar	Prieto et al 1981

ANLL, FAB type M1+M2

37-45,XY,-7,-12,-13,-14,-16,-18,-19,-21,del(4)(q29), del(5)(q24),der(9),der(17),11q+,12p+,t(14;?)	Fitzgerald et al 1983
42,X,-Y,4q+,-8,-16,17p+,-21,-22,+Yp+"s"	Sandberg et al 1982a
44,XY,-7,-16,-19,-21,-22,+3mar"s"	Sandberg et al 1982a
45,X,-Y,+del(5)(q11),del(11)(q21),-14,-16,-18,-20,+2mar,+r	Mitelman et al 1981
45-46,XX,+14,-7,-22,+16,-20	Mitelman et al 1981
46,XX,+14,-22/46,XX,-7,+14,+16,-20,-22	Alimena et al 1977
46,XX,t(16;16)(p13;q22)	Mitelman et al 1981
46,XX,t(16;16)(q22;p13)	Mitelman 1983
46,XX,t(8;16)(p11;q13)	Mitelman et al 1981
46,XY,t(12;16)(p11;p13)	Benedict et al 1979
46,XY,t(9;16)(p?;q?)	Garson 1980
47,X,-X,+del(1)(p11),del(5)(q13),t(5;12)(q21;q24), t(11;12)(q25;q13),t(13;14)(q11;p11),+16,+21	Mitelman et al 1981
47,XY,+16	Li et al 1983
47,XY,+8,r(11),t(8;16)(q?;q?)	Garson 1980
47-49,XY,-5,del(5)(q15),-7,i(8q),dic(9),-13,-16,-22, +6-9mar"s"	Testa et al 1981b

ANLL, FAB type M3

45-51,XY,t(3;12)(p14;q24),t(13;18)(q11;p11),+1,+6,+16,+19, +21,+22	Mitelman et al 1981
46,XX,2q-,6p-,7q-,10q+,14q+,-16,+mar	Hagemeijer et al 1981a
46,XX,t(15;17)(q22;q21)/47,XX,t(15;17),+del(16)(q11)	Berger et al 1981e

ANLL, FAB type M4

42,XY,-4,-15,-16,-22	Prieto et al 1981
42,XY,-5,-9,-10,-15,-16,-17,-20,-21,-22,+5mar	Testa et al 1979
43,XY,+1,-5,+6,-9,-10,-15,-16,-17,-20,-21,-22,+6mar	
42,XY,-7,i(8)(q11),-16,-17,-18,t(21;21)(q12;q11),+mar	Rowley and Potter 1976
44,XX,-16,-21	Garson 1980
45,X,-Y,del(16)(q21)	Hagemeijer et al 1981a
46,XX,16p+	Golomb et al 1978c
46,XX,inv(16)(p13q22)	Yunis et al 1981
46,XY,t(5;19),16p+,19p+	Golomb et al 1978c
47,XY,t(9;22)(q34;q11),t(16;?)(p13;?),-18,del(20)(p12q12), +22,+mar	Li et al 1981b

ANLL, FAB type M5

Karyotype	Reference
46,XX,ins(10;11)(p11;q23q24)/52,XX,+4,+8,ins(10;11),+12, +16,+19,+20	Kaneko et al 1982b
46,XY,-10,-12,16q+,+2mar	Brodeur et al 1983

ANLL, FAB type M4+M5

Karyotype	Reference
42,XX,-5,-7,-12,-15,-16,-18,+2mar	Li et al 1983
45,XY,del(11)(p11),t(12;?)(p13;?),del(16)(q13), t(13;14)"c""s"	Nora et al 1979
53,XY,+13,-14,+16,+16,+19,+20,+3mar	Li et al 1983

ANLL, FAB type M6

Karyotype	Reference
??,XY,t(4;?)(q23;?),t(5;17),t(8;9),+8,+11, t(12;16)(p12;p12)	Hustinx and Rutten 1983
44-45,XY,-2,-4,-7,-10,-13,-16,-17,-22,+3mar	Mitelman et al 1981
45,XY,del(4)(q23),del(5)(q13),-6,del(6)(q21),del(7)(p14), del(8)(p21),del(9)(q31),16q+	Rowley and Potter 1976
45-49,XY,-4,-7,-8,-10,+14,+16,-21,-22,+2mar	Mitelman et al 1981
46,XX,-7,-16,-21,+t(2;7)(p11;q11),del(2)(q33),del(11)(q22), +11p+,14p+,+t(16;21;21;21)(p13;q22;q11;q22)"s"	Rowley et al 1981a
46,XX,t(4;16)(q11;q23)	Hagemeijer et al 1981a
72,XY,cx,t(2;15)(q31;q15),del(6)(p22q15),t(8;13)(q23?;q13), t(11;14)(p12;q13),i(16p),t(15;18)(q15;p11)	Douglass and Freeman 1983

Chronic myeloid leukemia, t(9;22)

Karyotype	Reference
26,XX,t(9;22)(q34;q11),-1,-2,-3,-4,-5,-6,-7,-9,-10,-11,-12, -13,-14,-15,-16,-17,-18,-19,-20	Hartley and McBeath 1981
35,XX,t(9;22)(q34;q11),-3,-4,-5,+8,-9,-10,-11,-12,-13,-14, -15,-16,-17/66-72,XX,t(9;22)	Pedersen and Boesen 1983
44,XX,t(9;22),-16,-17/46,XX,t(9;22),-22,+22q-	Stoll and Oberling 1978
44,XX,t(9;22),-16,-17/46,XX,t(9;22),-22,+22q-	
44,XX,t(9;22),-16,-17/46,XX,t(9;22),-22,+22q-	
45,X,-Y,t(9;22)	Sonta and Sandberg 1977b
42,X,-Y,t(9;22),-7,-16,-17	
46,X,del(X)(q?),t(9;22),5q-,inv(11),16q+,20q+	Hagemeijer et al 1980b
46,X?,t(9;22)(q34;q11),t(16;17)	Lawler et al 1976
46,X?,t(9;22)(q34;q11),i(17q)	Lawler et al 1976
46,X?,t(9;22),-16,i(17q),+del(22)(q11)	
46,XX,t(9;22)(q34;q11),ins(11)(q23p12p15),del(16)(p11)	Beltran and Varela 1980
46,XX,t(9;22)(q34;q11)	Rajasekariah et al 1982
50,XX,t(9;22)(q34;q11),t(9;22)(q34;q11),+8,+11,+13, t(16;17)(p11;p12),+22	
46,XX,t(9;22),+C,-16,i(17q)	Lobb et al 1972
46,XX,t(9;22),16q-	Chrz et al 1983
46,XX,t(9;22),del(16)(q2?)	Bernstein et al 1980b
46,XY,t(9;22)(q34;q11)	Como and Graze 1979
35,XY,t(9;22)(q34;q11),-3,-4,-5,-7,-9,-11,-12,-13,-15,-16, +3mar	
46,XY,t(9;22)(q34;q11)	Ishihara et al 1982

Karyotype	Reference
27,X,-Y,t(9;22),-1,-2,-3,-4,-5,-6,-7,-9,-11,-13,-14,-15,-16,-17,-18,-19,-20,-22	
47,XX,t(9;22),+8,16q+,i(17q)	Sadamori et al 1980
47,XY,t(9;22)(q34;q11),del(1)(q21),dup(1)(q?),+8,+9,-14,-16,-21,+22q-,+mar	Alimena et al 1982b
47,XY,t(9;22),+8,i(17q)/47,XY,t(9;22),+16,i(17q)	Stoll and Oberling 1978
50-54,XY,t(9;22),+1,+5,+8,+9,-16,+17,+18,+19,+21,+22q-	Alimena et al 1980
50-55,XY,t(9;22)(q34;q11),+1,+5,+8,+9,-16,+17,+18,+19,+21,+22q-,+mar	Alimena et al 1982b
50-58,XY,+X,t(9;22)(q34;q11),+7,+8,+9,+10,+11,+12,+del(15)(q22),+16,+18,+19,+22q-	Kwan et al 1977
46,XY,t(9;22)	
53,XX,t(9;22),+8,+11,+12,+16,+17,+19,+del(22q)/52,XX,t(9;22),+8,+11,+17,+19,+del(22q)	Gahrton et al 1974b
53,XY,+Y,t(9;22)(q34;q11),+3,+6,+8,+9,+10,+12,-15,+22q-/56,XY,+Y,t(9;22),+3,+6,+8,+8,+9,+10,+16,+19,+22q-	Stoll and Oberling 1982
56,XY,+Y,t(9;22),+3,+6,+8,+8,+9,+10,+12,+19,+22q-	
55,XY,t(9;22),+3,+6,+8,+8,+9,+10,+16,+19,+22q-/53,XY,t(9;22),-15,+3,+6,+8,+8,+9,+10,+12,+22q-	Stoll and Oberling 1978

Chronic myeloid leukemia, aberrant translocation

Karyotype	Reference
46,XX,t(16;22)(p13;q11)	Lyall et al 1978
46,XY,t(16;22)(p13;q11)	Engel et al 1977
46,XY,t(16;22)(q24;q11)	Panani et al 1979
46,XY,t(16;22)(q24;q11),t(9;17)(q34;q21)/46,XY,t(16;22),t(9;17),del(7)(q22)	Mitelman 1983
46,XY,t(16;22)(q24;q11),t(9;17)(q34;q21)	

MYELOPROLIFERATIVE DISORDERS

Polycythemia vera

Karyotype	Reference
??,X?,t(9;22),+8,+10,+12,+16	Vykoupil et al 1980
44-45,XY,-10,-13,-16,-17,-20	Wurster-Hill et al 1976
45,XY,-16,-17,+21/48,XY,+Y,+11,+16,-17,+22	Wurster-Hill et al 1976
45,XY,-16	Westin et al 1976
44,X,-X,del(5)(q14q32),-7	Van Den Berghe et al 1979g
44,XX,del(5)(q14q32),-6,-7,-8,t(12;?)(p12;?),t(16;18)(q11;q12)	
46,XX,t(2;11)(p13;q21)/46,XX,t(1;15)(p1?;q1?),t(2;11)(p13;q21),del(16)(p12)/51,XX,+3,+8,+8,+19,t(1;15),t(2;11),del(16)	Testa et al 1981b

Myelosclerosis / Myelofibrosis

Karyotype	Reference
46,XX,t(5;12)(q14;q13)/45,XX,-16	Whang-Peng et al 1978
46,XY,-16,+18	Whang-Peng et al 1978
46,XY,7q-,8q-,10q-,11p-,11q+,12q-,-16,+18	
47,XX,+16	Whang-Peng et al 1978

<u>Idiopathic thrombocythemia</u>

Karyotype	Reference
46,XX,16q-	Köpf et al 1982

<u>Chronic myeloproliferative disease, NOS</u>

Karyotype	Reference
46,XY,-7,+19/47,XY,+19/52,XY,+16,+19,+19,+21,+22,+22	Chrz et al 1983

<u>Myeloproliferative disorder, special type</u>

Karyotype	Reference
46,XY,t(1;13)(p36;q13),del(3)(p21),-5,inv(7)(p11q22), t(16;16)(q22;q24),+mar	Clare et al 1982

DYSMYELOPOIETIC SYNDROMES

<u>Preleukemia, NOS</u>

Karyotype	Reference
43,X,-Y,-3,5q-,-7,-10,-16,-22,+3mar"s"	Pedersen-Bjergaard et al 1981
45,XX,-2,-3,5q-,-6,-9,16p-,17p+,+3mar	Swansbury and Lawler 1980
45,XY,-2,-5,-7,-8,-11,-12,-13,-14,+t(2;5),+t(11;12), +t(16;17),+17,+3mar	Watt et al 1982
46,XX,t(1;16)(q11;p11),t(1;20)(p?;p?)	Anderson and Bagby 1982
46,XY,11q-,16p+"s"	Anderson and Bagby 1982
46,XY,t(1;16)	Knuutila et al 1981

<u>Chronic myelomonocytic leukemia</u>

Karyotype	Reference
46,X?,t(3;16)(p13;q22)	Geary et al 1975

<u>Sideroblastic anemia, NOS</u>

Karyotype	Reference
46,XX,-5,+16	Seabright 1983

<u>Refractory anemia with excess of blasts</u>

Karyotype	Reference
43,XX,3q-,del(5)(q13q33),7q-,-12,14p+,16q+,-17,-18,18q+	Kerkhofs et al 1982
44-45,XX,del(3)(q?),-7,-16,-18,+1-2mar	Mitelman 1983
44-47,XY,-3,-5,-7,-12,-16,-20,del(1)(q32),3p+,del(11)(q22), 3-6mar,+r	Streuli et al 1980
45,XY,-16	Kardon et al 1982a

<u>Pancytopenia</u>

Karyotype	Reference
46,XX,-4,-5,-6,-7,-11,-13,-16,+19,+20,t(7;13)(q?;q?), +4mar/43,XX,-2,-3,-5,-7,-16,+19,15q+,+mar	Nowell and Finan 1978
46,XX,-7,+8,-12,-16,-17,t(6;12)(q?;q?),5q-,6q-,+mar	Nowell and Finan 1978
46,XX,t(6;16;16)(p?;p?;p?)	Nowell and Finan 1978

SPECIAL LEUKEMIAS

LYMPHOCYTIC LEUKEMIAS

Acute lymphocytic leukemia (ALL)

28,XX,-1,-2,-3,-4,-5,-7,-8,-9,-11,-12,-13,-14,-15,-16,-17, -19,-20,-22	Kaneko and Sakurai 1980
32,XY,-2,-3,-4,-5,-7,-8,-11,-12,-13,-14,-15,-16,-17,-20	Shabtai and Halbrecht 1981b
45,XX,t(14;14)(q34;q11),t(13;15)(q11;p11), t(13;17)(p11;q11),-16,18q+,+3mar	Levitt et al 1978
47,XX,+16	Yamada and Furusawa 1976
54,XX,7p+,+10,+10,+14,+14,+18,+18,+21,+21/27,X,-X,-1,-2,-3, -4,-5,-6,-7,7p+,-8,-9,-11,-12,-13,-15,-16,-17,-19,-20, -22	Oshimura et al 1977a
54-55,XX,+5,+12,+14,+16,+17,+18,+21	Prigogina et al 1979
55,XY,+X,+3,+6,+10,+13,+14,+16,+21,+21	Oshimura et al 1977a
57,XX,+3,+4,+8,11p+,+14,+16,+17,+18,+21,+22,+2mar/47,XX,+8	Morse et al 1978
61,XX,+X,+1,+2,+3,+4,+5,+6,+8,del(6)(q21),+13,+14,+15,+16, +18,+21,+22	Oshimura et al 1977a

ALL, FAB type L1

26,XX,-1,-2,-3,-4,-5,-6,-7,-8,-9,-10,-11,-12,-13,-14,-15, -16,-17,-19,-20,-22	Hoeltge et al 1982
28,XX,-1,-2,-3,-4,-5,-6,-7,-8,-9,-11,-12,-13,-15,-16,-17, -19,-20,-22/56,XX,+X,+X,+10,+10,+14,+14,+18,+18,+21,+21	Brodeur et al 1981a
34-36,XY,-2,-3,-4,-7,-9,-12,-13,-14,-16,-17,-20,+22	Sandberg et al 1982c
47,XY,-16,-17,-18,-20,+5mar	Kaneko et al 1982e
47,XY,-7,-10,-12,+17,del(11)(p11),del(16)(q22), +t(10;?)(q22;?),+t(12;?)(q15;?),+mar	Kaneko et al 1982e

ALL, FAB type L2

26,XX,-1,-2,-3,-4,-5,-6,-7,-8,-9,-10,-11,-12,-13,-15,-16, -17,-18,-19,-20,-22	Brodeur et al 1981a
46,XY,t(16;17)(p13;q23)	Kaneko et al 1982e
47,XY,+18	Kaneko et al 1982e
46,XY,-2,-5,-6,-7,+8,-9,-10,-11,-12,-16,+18,+t(2;?)(q35;?), +t(11;?)(q21;?),+5mar	
47,XY,-11,+16,+22	Nordenson 1983
56,XX,+1,+6,+10,+13,+15,+16,+17,+18,+20,+21	Mitelman 1983
56,XY,t(9;22)(q34;q11),+4,+5,+8,+9,+11,+16,+17,+18,+19,+20	Mitelman 1983

ALL, FAB type L3

46,XX,-16,-17,-22,+8,11q-,t(8;14),+1-2mar	Borgström et al 1981
46,XY,t(8;14)(q24;q32),-15,+16	Berger and Bernheim 1982

Chronic lymphocytic leukemia

Karyotype	Reference
41,XX,t(14;14)(q11;q32),t(1;13)(q44;q11),t(15;18)(q11;q23), 10q+,-16,-20	Kaiser McCaw et al 1975
44,XY,-8,del(16)(q21),-18	Pittman et al 1982

Prolymphocytic leukemia

Karyotype	Reference
47,XY,16q-,17p-,+mar	Pittman et al 1982

Lymphocytic leukemia, special type

Karyotype	Reference
45,Y,-X,t(Y;14)(q12;q32),-13,-17,del(2)(q33), t(3;13)(q21;q34),del(6)(q16orq22),9p+,9q+,10q+,16q+,18q+	Miyoshi et al 1981a

MONOCLONAL GAMMOPATHIES

Multiple myeloma

Karyotype	Reference
45,X,del(X)(q25),t(3;8;22)(q13;q34;q11),-5,-7,-10,-11, del(13)(q13?),+16,19q+,+3mar	Van Den Berghe et al 1979f
46,XY,+t(1;?)(p1?;?),+t(1;12)(q21;q24),del(6)(p11),-8,+9, -12,-13,-16,-19,22q+,+mar	Philip et al 1980
52,XY,-1,+t(1;15)(q12;p1?),+t(1;16)(p36;p13), t(3;5)(q2?;q13),+3,+5,+7,+8,+9,+11,-15,+18,-20	Philip et al 1980
55,XY,+1,+5,+9,+11,-12,+13,+15,del(1)(q32),del(6)(q25), t(9;19)(q13;q13),14q+,16p+,+4mar	Liang et al 1979
45,XY,-2,-2,-3,+5,-8,-10,+14,-15,-16,del(6)(q25),+4mar	

Macroglobulinemia

Karyotype	Reference
46,Y,-X,del(1)(q21),del(5)(q23),+del(5)(q23), +t(7;19)(p22;q13),del(8)(p12),+t(1;16)(q12;p13), +t(1;18)(q12;p11),-7,-16,-18,-19,+mar	Ueshima et al 1983

Plasma cell leukemia / Plasmocytoma

Karyotype	Reference
45,XY,-5,del(6)(q11),t(6;7)(q?;q?),+9,-12,-13,-16,-17, t(9;22)(q34;q11),+2mar	Karpas et al 1982
46,XY,+t(1;16)(q21;p13),del(6)(q21),-16,-18,+mar	Ueshima et al 1983
47,XY,+t(1;16),del(6)(q21),+t(8;15)(p23;q12),-8,14q+,-15, -16,-18,+3mar	
78,XY,cx,del(1)(q21),t(1;17)(q12;p13),3q+,3p-,9p+,11p+, 14q+,16p+	Ueshima et al 1983

SOLID TUMORS

UNCLASSIFIED NEOPLASMS

EPITHELIAL NEOPLASMS

Carcinoma, NOS

Karyotype	Reference
43,X,-X,t(1;11)(q21;q23),3q+,4q+,5q+,-11,-13,-16,17p+	Atkin and Baker 1982a
45,XY,-1,-6,-14,+21,+2mar/45,XY,t(1;14)(q11?;q24?),-6,-16,+21	Sonta and Sandberg 1978a
46,XX,+2,-7,-10,+13,-16,-17,+20,+5q-,15p+,15p+,18p+	Reichmann et al 1981
46,XX,+8,+12,+21,-2,-15,-16	Wake et al 1981
46,XX,del(2)(p24),+5,+8,t(9;17)(p?;p?),del(16)(q21),-20,-21	Mark 1975b
46,XX,t(3;11)(q12;q14),-6,+7,-8,-15,t(16;?)(q23;?),+2r	Mark 1975b
47-48,Y,-X,-5,-8,+13,-14,-16,-18,-22,+7mar	Reichmann et al 1981
49,X,-Y,-3,-5,-14,+15,-16,-17,+19,+21,+22,1q+,5q-,6q-,8p-,9p-,11p-,18p+,20q-,22q-,t(Y;14),+3mar	Reichmann et al 1981
57-61,XX,cx,t(11;16)(q?;q?),inv(3)(p14q29)	Riet-Fox et al 1979
58,XX,+X,+1,+2,+3,5p-,+6,+7,+9,11q+,-13,-14,-15,+16,+19,+20,+6mar	Kovacs 1978a
63-66,XX,cx,1p+,16q+,del(3)(p13)	Riet-Fox et al 1979
73,XY,+1,+7,+11,+13,-16,+17,+18,+19	Sonta and Sandberg 1978a
80-83,XX,cx,t(2;5;16),t(2;5;18),t(11;13),11q+,t(8;10),del(2)(q24),del(6)(q21),i(14q),dup(9)(q11q12)	Riet-Fox et al 1979

Carcinoma NOS, metastatic

Karyotype	Reference
35,XX,-1,-3,-4,-5,-7,-8,-10,-11,-12,-13,-14,-15,-16,-17,-18,-19,-20,-21,-22,+8mar	Pathak et al 1979
45,X?,+16,-18,1p-,3p-,16p-	Wake et al 1981
46,XX,t(9;22)(q34;q11)"s"	Togawa et al 1981
52,XX,+3,+5,+7,+8,+9,+11,12q+,13q-,-14,+16,i(17q)"s"	
47,XX,+4,+16,-21	Musilova et al 1981
100,XY,t(12;16),inv(3)(q?),del(1)(q?),del(1)(p?),cx	Ayraud 1975

Large cell carcinoma

Karyotype	Reference
55-70,XY,t(4;5;14),del(1)(q11q21),t(7;11)(p22;q13),t(7;21)(q11;q22),t(3;8)(p?;q?),t(6;11)(p11;p11),i(16q),t(3;?)(q11;?),t(21;?)(q11;?),t(13;?)	Kakati et al 1975

Small cell carcinoma

Karyotype	Reference
40,XY,cx,t(1;16)(q21;q24),t(1;19)(q23;q13),del(3)(p21p24),del(3)(p23q26),del(10)(q22),del(11)(q23),t(5;13)(q13;q32),del(17)(p12),dmin	Whang-Peng et al 1982b
110,XY,cx,del(1)(q41),del(3)(p13p23),del(3)(p13),t(3;16)(q24;p13),del(6)(q24)	Whang-Peng et al 1982b

Papillary carcinoma

42-54,X,-X,-14,-16,-16,+1,+3,+12,+20,1p-,6q-	Wake et al 1981
50,X,-X,+1,+3,+12,+20,-14,-16,1p-,6q-	Wake et al 1980
54,XY,+del(2)(q11),t(2;14)(q11;q32),t(3;11)(p13;p15), +i(5p),+8,+9,+12,+16,+17,+21,+2mar	Pathak et al 1982

Adenocarcinoma

47,XX,2q+,+16	Couturier-Turpin et al 1982
50,XY,2q+,+5,+10,+11,-16,-18,+20,+20,+mar	Couturier-Turpin et al 1982
68-74,XX,cx,t(13;12),17q+,t(16;16),1q-,i(12p)	Riet-Fox et al 1979

Adenocarcinoma, metastatic

??,X,del(X)(q21),del(1)(q12),i(7q),del(9)(q31), del(16)(q23)	Ayraud et al 1977
??,X,del(X)(q21),t(13;13)(q?;q?),del(16)(q21)	Ayraud et al 1977
??,X,del(X)(q21),t(1;16)(q?;p?),del(12)(p12),del(16)(q21)	Ayraud et al 1977
??,XX,del(3)(p14),del(4)(q22),del(4)(q13),del(16)(q21)	Ayraud et al 1977
??,XX,del(3)(p14),del(4)(q13),del(16)(q23)	Ayraud et al 1977
??,XX,t(1;16),2q-,del(3)(p14),del(4)(q22),del(6)(q22), i(7q),del(9)(q31),t(13;13)	Ayraud et al 1977
??,XX,t(1;16),del(1)(q12),del(4)(q22),del(16)(p11)	Ayraud et al 1977
43-46,XX,t(1;16),t(1;22),+r	Bertrand et al 1979
48,XX,+1,+3,+6,-16,1p-	Wake et al 1981

Adenomatous polyp

48-52,X,-Y,+7,+8,+13,+14,+16,+18,-20,+12q-,1q+	Reichmann et al 1982b

MESENCHYMAL NEOPLASMS

Mesothelioma, malignant

40-46,XY,+16,+20,-22,t(14;15),del(2)(p?),i(5p)	Ayraud 1975

MELANOMAS

Malignant melanoma, metastatic

41,XY,dup(1)(q21q32),t(1;7)(q25;q36),t(7;9)(q11;q11), del(9)(q13),-3,-4,-5,-8,-10,-11,-16,+18,-20,-22	Kakati et al 1977

NEUROGENIC NEOPLASMS

Neuroblastoma

Karyotype	Reference
44,X?,del(1)(p22),-4,-7,del(7)(q22),t(7;16)(q22q32;q24)	Trent 1983
45,XY,-3,del(1)(p13),t(1;9)(q11;p11),t(9;15)(p11;p11), hsr(16p)	Brodeur et al 1981b
46,XY,del(1)(p32),6q+,13q+,16q+,dmin	Gilbert et al 1982
47,X,-Y,del(1)(p?),+6,16q+	Brodeur et al 1977
47,XY,+1,11q-,16q+,19q+	Gilbert et al 1982

Retinoblastoma

Karyotype	Reference
44,XY,1p+,-3,t(3;12)(q?;p?),+6q-,der(9),13p+,-14,-16	Kusnetsova et al 1982
45,X,-X,+1p+,+i(6p),-16,17q+,-19,22p+	Kusnetsova et al 1982
46,X,-Y,1p+,+1p-,5q+,7p+,10q-,16q-,17q+,21p+	Kusnetsova et al 1982
46,XX,t(1;15)(q11q32;q12),t(1;16)(q21;q11),dup(17)(p11q25)	Gardner et al 1982
46,XY,+4,t(5;13)(q35;q21),t(7;17)(p22;q12),t(1;9)(q22;p24), -10,del(16)(q11)	Gardner et al 1982
46,XY,+del(1)(p13),del(16)(p11),-19	Gardner et al 1982
47,XX,12p+,12q+,+i(17q),t(4;16)(q35;q24),-16	Gardner et al 1982
47,XX,i(6p),+4q+,-16	Kusnetsova et al 1982
51,XY,del(6)(q21),+8,+11,-12,t(13;18)(p11;q11),14q+,-16, +19,+22,+2mar	Hossfeld 1978

Meningioma, benign

Karyotype	Reference
43-45,XX,-8,-16,-22	Mark et al 1972a
45,XX,-22/43-44,XX,-8,-16	Mark 1973a
45,XX,-22/43-44,XX,-2,-16,-22	Mark 1973a
45,XY,-22/42-44,XY,-8,-16,-19,-22	Mark 1973a
51,XY,+5,6q+,+12,+16,+17,+18,+20,-22/52,XY,+5,+8,+10,+12, +17,+18,+20,-22	Zang and Zankl 1983

EMBRYONAL AND MISCELLANEOUS NEOPLASMS

Teratoma, malignant

Karyotype	Reference
67,XY,+1,+3,+4,+5,+6,+7,+7,+8,+10,+11,+13,+13,+14,+16,+17, +18,+20,+20,+21,+21	Sonta et al 1977

LYMPHOMAS

1. UNCLASSIFIED AND MISCELLANEOUS LYMPHOMAS

Non-Hodgkin's lymphoma, NOS

Karyotype	Reference
46,X,-X,del(1)(q22),t(2;11)(p21;q21),t(9;15)(q11;p12), del(9)(q11),-14,t(14;17)(q23;q23),-16,+mar	Kristoffersson 1983
46-48,XX,t(1;?)(q21orq22;?),t(2;?)(q37;?),-8,+9,+12, 16q+"s"	Mark et al 1976
47-48,XY,t(14;?)(q32;?),16q-,-17,18q-,13q-,+7,+8,8q-	Catovsky et al 1977

Non-Hodgkin's lymphoma, lymphocytic, NOS

Karyotype	Reference
42-45,XX,-4,-5,-9,-16,+mar	Fleischmann et al 1976
46-47,XY,-4,-15,-16,+3mar	Fleischmann et al 1976

Non-Hodgkin's lymphoma, histiocytic, NOS

Karyotype	Reference
46,XX,2p+,-15,del(16)(q13)	Reeves 1973

Non-Hodgkin's lymphoma, T-cell, NOS

Karyotype	Reference
47,XX,t(2;?),inv(3),4p-,del(5)(q14q32),6p+,6q+,t(7;7),9q+, 10p-,13q-,t(14;14),-15,-15,16q-,17p+,+21,+2mar	Gaeke et al 1981

Angioimmunoblastic lymphadenopathy

Karyotype	Reference
46-47,XY,14q+,+16	Kristoffersson 1983

2. HODGKIN'S LYMPHOMAS

Hodgkin's disease, NOS

Karyotype	Reference
46,XY,-16,+mar	Fleischmann et al 1976
47,XY,-2,-5,+16,-17,+3mar	Fleischmann et al 1976

Hodgkin's disease, mixed cellularity

Karyotype	Reference
81,XX,14q+,i(18q),i(18p),del(6)(q21),t(9;15)(q11;p12), del(16)(q13)	Reeves 1973

Hodgkin's disease, lymphocytic depletion

Karyotype	Reference
46,XY,t(10;14)(q22;q32),t(15;?)(p11;?),-16,+mar	Hossfeld and Schmidt 1978

3A. NON-HODGKIN'S LYMPHOMAS - RAPPAPORT CLASSIFICATION

Lymphocytic, well differentiated, diffuse

Karyotype	Reference
48,XY,1q+,3q+,+i(3p),-6,-10,t(14;?)(q12;?),+16,17p-,+18,+mar	Miyamoto et al 1981a

Lymphocytic, poorly differentiated, NOS

Karyotype	Reference
46,XY,del(6)(q21),-8,-9,t(6;14)(q21;q32),+16,+20	Mark et al 1979

Lymphocytic, poorly differentiated, diffuse

Karyotype	Reference
47,XY,t(14;18)(q32;q21),-9,-10,+12,-16,8p-,13q+,+3mar	Fukuhara and Rowley 1978

Lymphocytic, poorly differentiated, nodular

Karyotype	Reference
52,X,-X,+2,-4,+5,+7,+8,-14,-15,-16,+20,+21,+t(1;15)(p11;q11),+t(12;17)(q24;q11),+t(14;?)(q32;?),+mar	Kaneko et al 1982a

Histiocytic, diffuse

Karyotype	Reference
45,X,-Y,t(14;15)(q32;q21orq22),-15,+18,-22,1q+,3p+,5p-,7p-,9p+,12p+,16q+,17p-,+2mar	Fukuhara and Rowley 1978
45,X,del(X)(q24),del(1)(q21),+del(1)(p11),del(2)(q22),del(3)(q21),del(6)(p11),del(6)(q15),t(7;9)(p22;q13),t(8;10)(p23;q21),del(9)(q13),del(10)(q21),t(2;11)(q22;q23),-13,del(14)(q24),del(15)(q22),del(16)(q22),+17,-19	Mark et al 1979
47,X,+X,-Y,+12,-14,del(11)(q21q23),del(16)(q22),t(14;?)(q32;?)	Kaneko et al 1982a
47,XX,-1,-2,-6,+8,-14,+16,-19,del(11)(q23),t(8;14)(q22;q32),+t(2;6)(q21;p23),del(6)(q21),+t(1;14)(p11;q11),+t(19;?)(q13;?),+t(14;?)(q11;?)	Kaneko et al 1982a
47,XY,-10,-13,+15,-16,+19,+20,del(6)(q21),del(8)(q22),t(2;14)(q21;q32),t(4;4)(q31;q35),+t(10;?)(p11;?)	Kaneko et al 1982a
61,XY,cx,t(1;16)(q23;p11),dup(1)(q23q32),t(1;6)(q23;p21),t(3;6)(q21;p21),t(7;7)(q32;p22),13p+	Fitzgerald et al 1980

Histiocytic, nodular

Karyotype	Reference
46,XY,-4,inv(16)(p11q24),+t(2;4)(p11;p11)	Kaneko et al 1982a

3B. NON-HODGKIN'S LYMPHOMAS - KIEL CLASSIFICATION

Centroblastic-centrocytic, follicular

47,XY,t(1;14)(q21;q32),+12,-16,+19,t(14;?)(q32;?) Reeves and Pickup 1980

Centroblastic, diffuse

42-45,XY,1p+,-6,-8,14q+,14q-,-16,i(17q),+18,-20,+mar Kristoffersson 1983
46-51,XX,+3,del(6)(q21),+del(6)(q21),+7,-8,del(9)(q11),+13, -16,-17,+19,-20,-21,+1-5mar Kristoffersson 1983

Lymphoblastic, Burkitt's type

43,XY,+1,+2,+3,-7,-8,-9,-12,-16,-22,del(1)(q25), del(2)(p11),del(10)(q22),14q+ Biggar et al 1981
45,X?,t(8;14)(q24;q32),t(7;16),t(6;17) Zech et al 1976a
46,XX,t(8;14),16q- Douglass et al 1980a
46,XY,t(8;14)(q24;q32),-13,16p+,+mar Slater et al 1982
48,XX,+5,+7,-9,+10,+11,-16,del(5)(p13q31),del(10)(p11q25), del(11)(q12) Biggar et al 1981

Immunoblastic

45-46,XX,+3,del(6)(q21),+4,+16,+21,-C,-C Kristoffersson 1983
47-48,XX,+del(3)(q11),+17/49,XX,+del(3)(q11),+16,+17 Kristoffersson 1983
45-47,XX,1p+,-14,-15,+16,+17,-19,+mar
48-49,XY,+3,+7,+11,t(12;?)(p13;?),t(13;?)(p13;?),-16,-20, +21,+22 Reeves and Pickup 1980
51,XY,t(1;6)(p22;q13orq15),t(2;13)(q21;q14), t(8;16)(q24;p11),+11,+12,del(15)(q22),+16,+20,+mar Reeves and Pickup 1980

3C. NON-HODGKIN'S LYMPHOMAS - LUKES AND COLLINS CLASSIFICATION

Sezary's syndrome

50,XY,-10,-16,+6mar Nowell et al 1982

3D. NON-HODGKIN'S LYMPHOMAS - WORKING FORMULATION

Chromosome 17

HEMATOLOGICAL DISORDERS

UNCLASSIFIED LEUKEMIAS

NONLYMPHOCYTIC LEUKEMIAS

Acute nonlymphocytic leukemia (ANLL)

Karyotype	Reference
39,X,-X,-3,-5,-13,-14,-15,-16,-17,-21,-22,del(7)(p11), t(12;?)(p13;?),t(16;?)(p13;?),+3mar	Mitelman 1983
39,X,-X,-3,-5,-13,-14,-15,-16,-17,-21,-22,del(7)(p11), t(12;?)(p13;?),t(16;?)(p13;?),+3mar	
43,X,-Y,t(1;8)(p22;q24),-1,-11,-11,-13,-16,-17,-22, t(1;22)(q11;p11),t(11;13)(q23q25;q12q14),t(17;?)(p11;?), +mar	Oshimura et al 1976b
43,XX,-1,3p+,3q-,4q+,-5,-5,-7,+9,+9,10p+,+10,+10,10p-,10q-, +11q-,-13,-15,-16,-17,-18,-21"s"	Whang-Peng et al 1979
43,XX,2q-,inv(3),-4,del(5)(q?),-7,17p+,-19	Rowley 1976
43,XX,t(1;20),-5,-6,7q-,11q-,-12,17q+,-18,19q+,-21	Mitelman 1983
43-44,XY,-5,-5,-14,-15,-16,-17,-18,-20,+5mar	Mitelman 1983
44,XX,-9,-10,-17,del(14)(q22),t(9;10)(q22orq34;q11orq22)	Oshimura et al 1976b
44,XY,-7,-17,-20,-22,+2mar"s"	Pedersen-Bjergaard et al 1982
45,X,-X,7q-,-8,+i(17q)	Nordenson 1983
45,XY,-17	Bernard et al 1982a
45,XY,t(1;14)(p22;p11),t(1;15)(p22;p11),-2,del(5)(p13),-17, +mar	Philip et al 1978a
46,XX,4q-,5q-,-16,-17,+2mar	Rowley 1976
46,XX,del(1)(p32),del(5)(q13),+8,t(9;?)(p11;?), t(11;22)(q11;q13),-16,-17,-19,+3mar	Muir et al 1977
46,XX,i(17q)	Bernard et al 1982a
46,XX,i(17q)	Nordenson 1983
46,XX,t(15;17)(q24;q21)	Yunis 1982
46,XX,t(15;17)(q?;q?)	Bernard et al 1982a
46,XX,t(1;17)(q2?;q21)	Philip et al 1978a
46,XY,+del(7)(p11),-8,del(12)(p11),t(21;?)(q22;?)/50,XY,+Y, +del(7)(p11),del(12)(p11),+17,t(17;?)(q25;?), t(21;?)(q22;?),+22	Mitelman 1983
46,XY,+del(7)(p11),-8,del(12)(p11),t(21;?)(q22;?)/47,XY,+9	
46,XY,-9,17p+,+mar	Borgström et al 1982

Karyotype	Reference
46,XY,del(2)(q31),-7,del(9)(p13),t(2;21)(q21;q22), t(2;?)(p21;?)/46,XY,t(1;?)(p22;?),t(17;?)(q25;?)	Philip et al 1978a
46,XY,dup(1)(q25q44),t(11;17)(p15;q21)/46,XY, t(11;17)(p15;q21)/47,XY,t(11;17)(p15;q21),+del(1)(p11)	Mitelman 1983
46,XY,dup(1)(q25q44),t(11;17)(p15;q21)/46,XY, t(11;17)(p15;q21)	
46,XY,dup(1),t(11;17)/46,XY,t(11;17)/47,XY,t(11;17), +del(1)	
46,XY,i(17q)	Engel et al 1975b
46,XY,t(15;17)(q?;q?)	Bernard et al 1982a
46-48,XX,t(17;?)(q25;?),+1-2mar"s"	Mitelman 1983
46-49,XX,t(17;?)(q25;?),+1-3mar"s"	
46-49,XY,+8,+14,-17,+18,-21"s"	Cavallin-Ståhl et al 1977
47,X,Xq-,t(3;5),-5,t(11;?),t(12;17),-18,-21,+6mar	Pedersen-Bjergaard et al 1980
47,XX,17q-,dmin	Bernard et al 1982a
47,XY,t(17;17),+8	Borgström et al 1982
51,XX,del(5)(q?),+7,+8,+13,+17,+22	Van Den Berghe et al 1976
57,XY,+6,+11,+17,+20,+21"s"	Berger et al 1981a

<u>ANLL, FAB type M1</u>

Karyotype	Reference
39,XY,-5,-7,-12,-16,-17,-20,-21	Berger et al 1981a
41,XY,-1,-2,-4,-5,-12,-15,-17,-17,-18,-22,+5mar	Sessarego et al 1981a
42,XX,-7,-17,-19,-21"s"	Papa et al 1979
43,XX,del(5)(q12q21),-7,11p+,-15,-17	Kerkhofs et al 1982
45,X,-X,-5,-7,-11,-17,+21,+3mar"s"	Zaccaria et al 1983b
46,XX,-1,inv(1),3q+,5q-,t(13;17),-18,+mar/51,XX,-1,inv(1), 3q+,5q-,t(13;17),-18,+6,+9,+9,+11,+11,+mar	Hagemeijer et al 1981a
46,XX,i(17q)	Bernstein et al 1982b
46,XY,+del(2)(p13),i(9q),-16,del(17)(p11)	Brodeur et al 1983
47,XY,-17,+2mar	Prieto et al 1981

<u>ANLL, FAB type M2</u>

Karyotype	Reference
42,XY,5q-,7q-,-8,-12,-17,-21,-22,+i(17q),+2mar	Berger et al 1981a
43,XX,-3,-4,-7,-19,del(5)(q14),17p+	Testa et al 1979
43,XX,-3,-4,-7,-19,del(2)(q31?),del(5)(q14)	
43,XY,-3,-4,-7,-7,-8,-8,-17,-20,+5mar	Berger et al 1981a
43,XY,-5,-7,-12,-16,4p+,17p+,20q-,+mar"s"	Rowley et al 1981a
44,XY,-7,17p+,-20,21q+"s"	Pedersen-Bjergaard et al 1981
45,X,-X,t(8;21)(q21;q21)	Hagemeijer et al 1981a
45,X,-X,t(8;21)(q21;q21),1p+,17p-	
45,XX,-17,-21,+mar	Oshimura et al 1982b
45,XY,-5,-7,-8,+16,-17,+r,+mar	Berger et al 1981a
45,XY,-5,-7,del(3)(p13),+t(5;17)(q11;q11)"s"	Rowley et al 1981a
45,XY,-7,del(5)(q13),t(12;17)(q13?;p12?)"s"	Rowley et al 1981a
45-47,XX,-9,-17,-20,del(4)(q21),del(5)(q12q31), t(9;17)(p13;p11),+mar	Testa et al 1979
46,XX,2q+,5q-,7q-,17q+"s"	Berger et al 1981a
46,XX,t(9;22)(q34;q11),i(17q)	Sasaki et al 1983
46,XY,-7,-12,-17,del(5)(q14),8q+,del(17)(p11),+mar	Testa et al 1979
45,XY,-7,+8,-12,-17,del(5)(q14),8q+,t(12;17)	
46,XY,t(1;17)(p36;q21)"s"	Rowley et al 1977a

47,XX,1p+,der(5),-6,12p+,-13,-14,-16,-17,-17,-21,+22,+7mar, dmin	Cooperman and Klinger 1981
47,XX,t(11;17)(q23;q25),+der(17)/48,XX,t(11;17)(q23;q25), +der(17),+10	Mitelman 1983
48,XX,t(11;17)(q23;q25),+der(17),+10	
55,XX,+16,+17,+18,+20,+21,+22,+3mar	Prieto et al 1981

ANLL, FAB type M1+M2

37-45,XY,-7,-12,-13,-14,-16,-18,-19,-21,del(4)(q29), del(5)(q24),der(9),der(17),11q+,12p+,t(14;?)	Fitzgerald et al 1983
42,X,-Y,4q+,-8,-16,17p+,-21,-22,+Yp+"s"	Sandberg et al 1982a
42-43,XX,cx,t(11;15)(q11;q11),t(9;?)(p?;?),-15,-17,-18,-20, -21	Fitzgerald et al 1983
43,XX,1p+,-4,-5,-6,-7,-13,-17,del(18)(q21),+3mar	Rowley and Potter 1976
44,XX,-15,-17,dmin/43,XX,-15,-17,-20,dmin"s"	Sandberg et al 1982a
44,XX,-3,-4,-5,+9,-17,-18,-21,-22,+5mar	Garson 1980
45,XX,t(3;6)(p23;p25),del(5)(q13),-12,i(17q)	Mitelman et al 1981
45,XY,t(5;11),t(5;17)	Garson 1980
46,XX,t(8;9)(q23or24;q31),-17,+20	Mitelman et al 1981
46,XY,i(17q)	Nordenson 1983
46,XY,t(9;22)/47,XY,t(9;22),+17/47,XY,t(9;22),+22q-	Nagao et al 1977
46-47,XX,del(1)(q32),+8,t(15;17)(q25or26;q22or23),+18	Mitelman et al 1981
46-50,XY,+Y,+del(7)(p11),-8,+9,del(12)(p11),+17,21q+,+22	Mitelman et al 1981
47,XX,+10,t(11;17)(q23;q25)	Mitelman et al 1981
48,XX,-5,del(5)(q13q31),+8,+t(11;?)(p14;?),-17,-19,+21, +2mar,dmin"s"	Testa et al 1981b
48,XY,+7,+8,+8,-17"s"	Sandberg et al 1982a
49,XX,+6,t(9;22)(q34;q11),+17,+21	Nordenson 1983
50-52,XY,+9,+14,+17,+19,+20,+21	Alimena et al 1977
50-52,XY,-5,-7,+9,+14,+17,+19,+20,+21,+mar	Mitelman et al 1981
92,XY,-17,+mar	Garson 1980

ANLL, FAB type M3

44,XX,-5,-8,-17,-18,+2mar	Gallagher et al 1979
45,X,-Y,t(15;17)(q22;q12)	Kaneko et al 1982b
45,X,-Y/46,XY,del(17)(q22)	Van Den Berghe et al 1979c
46,XY,del(12)(q11)	
45,XY,t(15;17),-19	Ferro et al 1982
46,X?,t(15;17)	Sakurai et al 1982a
46,X?,t(15;17)	Sakurai et al 1982a
46,X?,t(15;17)	Sakurai et al 1982a
46,X?,t(15;17)	Sakurai et al 1982a
46,X?,t(15;17)	Sakurai et al 1982a
46,X?,t(15;17)	Sakurai et al 1982a
46,X?,t(15;17)(q?;q?)	Hartley 1983
47,XX,+8,i(17q)	Chapelle et al 1981
46,XX,t(15;17)	Alimena 1983
46,XX,17q-	Brodeur et al 1983
46,XX,4q+,t(15;17)	Rochon and Vaillancourt 1981
46,XX,del(17)(q?)	Mitelman 1983

46,XX,del(2)(q32q34),t(15;17)(q25;q22)	Webber and Garson 1983
46,XX,der(15)(q22),i(17)(q12)	Hagemeijer et al 1982a
46,XX,t(15;17)	Alimena 1983
46,XX,t(15;17)	Alimena 1983
46,XX,t(15;17)	Alimena 1983
46,XX,t(15;17)	Alimena 1983
46,XX,t(15;17)	Alimena 1983
46,XX,t(15;17)	Alimena 1983
46,XX,t(15;17)(q22;q12)/46,XX,15q+,i(17q-)/47,XX,t(15;17),+8	Kondo and Sasaki 1982
46,XX,t(15;17)(q22;q12)	Kondo and Sasaki 1982
46,XX,t(15;17)(q22;q12)	Kondo and Sasaki 1982
46,XX,t(15;17)(q22;q21)	Berger et al 1981e
46,XX,t(15;17)(q22;q21)	Berger et al 1981e
46,XX,t(15;17)(q22;q21)	Berger et al 1981e
46,XX,t(15;17)(q22;q21)	Berger et al 1981e
46,XX,t(15;17)(q22;q21)	Berger et al 1981e
46,XX,t(15;17)(q22;q21)	Berger et al 1981e
46,XX,t(15;17)(q22;q21)/47,XX,t(15;17),+del(16)(q11)	Berger et al 1981e
46,XX,t(15;17)(q22;q21)/46,XX,t(15;17),t(1;4)(p35;q21),del(X)(q25)	Berger et al 1981e
46,XX,t(15;17)(q22;q12)	Hagemeijer et al 1981a
46,XX,t(15;17)(q22;q21)	Hagemeijer et al 1981a
46,XX,t(15;17)(q22;q21)	Mitelman 1983
46,XX,t(15;17)(q22;q21)	
46,XX,t(15;17)(q22;q21)	Okada et al 1977
46,XX,t(15;17)(q22;q21)	Mitelman 1983
46,XX,t(15;17)(q22;q21)	Alimena 1983
46,XX,t(15;17)(q22;q21)	Pasquali 1983
46,XX,t(15;17)(q22;q21)	Pasquali 1983
46,XX,t(15;17)(q22;q21)	Pasquali 1983
46,XX,t(15;17)(q22;q21)	Pasquali 1983
46,XX,t(15;17)(q24;q21)	Hurd et al 1982
46,XX,t(15;17)(q24;q21)	Hurd et al 1982
46,XX,t(15;17)(q24;q21)	Yunis et al 1981
46,XX,t(15;17)(q24;q22)	Alimena 1983
46,XX,t(15;17)(q25;q22)	Golomb et al 1979
46,XX,t(15;17)(q25;q22)	Golomb et al 1979
46,XX,t(15;17)(q25;q22)	Golomb et al 1980
46,XX,t(15;17)(q25;q22)	Bernstein et al 1982b
46,XX,t(15;17)(q25;q22)	Ferro et al 1982
46,XX,t(15;17)(q25;q22)"s"	Sheer et al 1982
46,XX,t(15;17)(q25;q22)	Fraisse et al 1981a
46,XX,t(15;17)(q25;q22)	Fraisse et al 1981a
46,XX,t(15;17)(q25;q22)	Alimena 1983
46,XX,t(15;17)(q26;q22)	Van Den Berghe et al 1979c
46,XX,t(15;17)(q26;q22)	Van Den Berghe et al 1979c
46,XX,t(15;17)(q26;q22),del(7)(q33orq35)	Van Den Berghe et al 1979c
46,XX,t(15;17)(q26;q22)	Van Den Berghe et al 1979c
46,XX,t(15;17)(q26;q22)	Van Den Berghe et al 1979c

Karyotype	Reference
46,XX,t(15;17)(q26;q22)/47,XX,t(15;17),-7,+8,-9,-10,+2mar	Van Den Berghe et al 1979c
46,XX,t(15;17)(q?;q?)	Lessard and Le Prise 1983
46,XX,t(2;15;17)(q21?;q25?;q21?;),9p+	Bernstein et al 1982b
46,XX,t(7;17)(q36;q22)	Yamada et al 1983a
46,XY,17q-	Prieto et al 1981
46,XY,i(17q)	Borgström et al 1982
46,XY,i(17q)	Nordenson 1983
46,XY,t(14;21)"c"	Alimena 1983
46,XY,t(15;17),t(14;21)"c"	
46,XY,t(15;17)	Takeuchi et al 1981
46,XY,t(15;17)	Alimena 1983
46,XY,t(15;17)(q22;q21)	Kaneko and Sakurai 1977
46,XY,t(15;17)(q22;q12)	Kondo and Sasaki 1982
46,XY,t(15;17)(q22;q12)	Kondo and Sasaki 1982
46,XY,t(15;17)(q22;q12)	Kondo and Sasaki 1982
46,XY,t(15;17)(q22;q12)	Kondo and Sasaki 1982
46,XY,t(15;17)(q22;q12)	Kondo and Sasaki 1982
46,XY,t(15;17)(q22;q12orq21),i(17q-)	Chapelle et al 1981
46,XY,t(15;17)(q22;q21)	Berger et al 1981e
46,XY,t(15;17)(q22;q21),t(X;9)(q28;q22)	Berger et al 1981e
46,XY,t(15;17)(q22;q21)	Berger et al 1981e
46,XY,t(15;17)(q22;q21)	Berger et al 1981e
46,XY,t(15;17)(q22;q21)	Berger et al 1981e
46,XY,t(15;17)(q22;q21),8q+	Berger et al 1981e
46,XY,t(15;17)(q22;q21)	Berger et al 1981e
46,XY,t(15;17)(q22;q12)	Hagemeijer et al 1982a
46,XY,t(15;17)(q22;q12)/46,XY,t(15;17),t(21;?)(p11;?)/46, XY,del(4)(q22),t(15;17)	Kaneko et al 1982b
46,XY,t(15;17)(q22;q12)	Kaneko et al 1982b
46,XY,t(15;17)(q22;q21)	Mitelman 1983
46,XY,t(15;17)(q22;q12)	Alimena 1983
46,XY,t(15;17)(q22;q21)	Pasquali 1983
46,XY,t(15;17)(q22;q21)	Pasquali 1983
46,XY,t(15;17)(q24;q21)	Hurd et al 1982
46,XY,t(15;17)(q24;q21)	Hurd et al 1982
46,XY,t(15;17)(q24;q21)	Hurd et al 1982
46,XY,t(15;17)(q24;q21)	Hurd et al 1982
46,XY,t(15;17)(q24;q21)	Hurd et al 1982
46,XY,t(15;17)(q24;q21)	Hurd et al 1982
46,XY,t(15;17)(q24;q21)	Fitzgerald et al 1982b
46,XY,t(15;17)(q24;q21)	Fitzgerald et al 1982b
46,XY,t(15;17)(q24;q21)	Fitzgerald et al 1982b
46,XY,t(15;17)(q24;q21)	Fitzgerald et al 1982b
46,XY,t(15;17)(q24;q21)	Fitzgerald et al 1982b
46,XY,t(15;17)(q24?;q21?)	Yunis et al 1981
46,XY,t(15;17)(q25;q22)	Golomb et al 1979
46,XY,t(15;17)(q25;q22)/46,XY,t(15;17),del(7)(q22), del(9)(q22)	Golomb et al 1979
46,XY,t(15;17)(q25;q22)	Golomb et al 1979
46,XY,t(15;17)(q25;q22)	Golomb et al 1979
46,XY,t(15;17)(q25;q22)	Fraisse et al 1981a
46,XY,t(15;17)(q25?;q22)	Bernstein et al 1982b
46,XY,t(15;17)(q26;q22)/47,XY,t(15;17),+8	Van Den Berghe et al 1979c

Karyotype	Reference
46,XY,t(15;17)/46,XY,del(1)(q12)/47,XY,+8/47,XY, del(1)(q12),+8/47,XY,t(15;17),+8	
46,XY,t(15;17)(q26;q22)	Van Den Berghe et al 1979c
46,XY,t(15;17)(q26;q22)	Van Den Berghe et al 1979c
46,XY,t(15;17)(q26;q22)	Van Den Berghe et al 1979c
46,XY,t(15;17)(q?;q?),t(3;11)(q?;q?),t(7;19)(q11;p13)	Trujillo and Cork 1981
46,XY,t(15;17)/46,XY,t(15;17),t(1;9)(p36;q22)	Hurd et al 1982
46,XY,t(1;17)(p36;q21)	Yamada et al 1983a
46,XY,t(3;5;17)(p21;q25?;q21?),4p+	Bernstein et al 1982b
47,X?,t(15;17),+8	Sakurai et al 1982a
47,XX,+10,t(15;17)(q26;q22)	Brodeur et al 1983
47,XX,t(15;17)(q22;q12),+8	Kondo and Sasaki 1982
47,XX,t(15;17)(q26;q22),+8,/47,XX,+8/46,XX, t(15;17)(q26;q22),-7,+8	Van Den Berghe et al 1979c
47,XX,t(15;17),+8/47,XX,+8	Sakurai et al 1982a
47,XY,+8,t(15;17)(q24;q21)	Fitzgerald et al 1982b
47,XY,del(11)(q23),t(15;17)(q24?;q22?),+mar	Bernard et al 1982b
47,XY,t(15;17)(q21?;q25?),+8/47,XY,t(15;17)(q21?;q25?), +9/48,XY,t(15;17),+8,+9	Van Den Berghe et al 1979c
47,XY,t(15;17)(q22;q21),+8	Berger et al 1981e
47,XY,t(15;17)(q24;q21),+21	Scheres et al 1978
47,XY,t(15;17),del(1)(q42),+del(1)(p22q42)	Sakurai et al 1982a

ANLL, FAB type M4

Karyotype	Reference
42,XY,-5,-9,-10,-15,-16,-17,-20,-21,-22,+5mar	Testa et al 1979
43,XY,+1,-5,+6,-9,-10,-15,-16,-17,-20,-21,-22,+6mar	
42,XY,-7,i(8)(q11),-16,-17,-18,t(21;21)(q12;q11),+mar	Rowley and Potter 1976
43,XY,-2,-3,-6,-7,-12,-15,-21,-22,del(2)(p11),del(11)(p14), t(14;?)(q32;?),t(17;?)(p13;?),del(22)(q11),+5mar"s"	Rowley et al 1981a
44,XY,del(3)(p22),-9,-12,t(17;?)(p13;?)"s"	Mitelman 1983
44,XY,t(3;5)(q29;q13),t(7;12)(q12;q14),t(7;10)(q12;q21),-8, -17,-18,-20	Mitelman et al 1981
44-46,XY,+1,-5,-17,-18,-20,del(6)(q21),-15,t(1;6)(q11;q24), +3mar	Mitelman et al 1981
44-46,XY,+1,t(1;6)(q11;q27),-5,del(6)(q21),-15,-17,-18,-20, +mar	Alimena 1983
45,X,-X,-9,-17,-18,-21,+5mar	Prieto et al 1981
45,X,-X,t(8;17;21)(q22;q23;q22)	Testa et al 1979
45,X,-X,t(8;17;21)(q22;q23;q22)/45,X,-X, t(1;11)(p36q32;q13),t(8;17;21)	
45,XX,2p-,-5,-9,-17,+r,+mar	Berger et al 1981a
45-46,XY,-7,+17	Mitelman et al 1981
47,XX,+6,-17,+t(1;17)(q21;p13)	Kaneko et al 1982b
46,XX,i(17)(q11)	Mitelman et al 1981
46,XX,t(2;11;12;17)(q31;p11;p13;q23)	Fitzgerald et al 1983
46,XY,-17,+21/47,XY,+21"s"	Papa et al 1979
46,XY,del(17)(q21)	Bernstein et al 1982b
46,XY,t(11;17)(q23;q27),del(11)(q13q23),t(15;19)(q21;p13)	Yunis et al 1981
46,XY,t(3;12)(q21;q13),del(5)(q14),ins(8;5)(q22;q14q22), -17,+2mar	Mitelman 1983
46-47,XY,+8,-17	Mitelman et al 1981

47,XX,+8/48,XX,+8,+17	Garson 1980
47,XY,+8/46,XY,i(17q)	Mitelman 1983
49,XX,del(3)(q25),del(7)(q22),+8,+17,+19	Alimena 1983

ANLL, FAB type M5

46,XX,17q+"s"	Berger et al 1981a
46,XX,t(11;17)(q23;q23)	Berger et al 1981f
46,XX,t(11;17)(q23;q21)	Zaccaria et al 1982
46,XX,t(14;17)(q13;q24)	Prigogina et al 1979
46,XY,-7,+17/45,XY,-7	Alimena 1983
46,XY,t(17;21)(q11;q11orq21),+i(8q),del(5)(q11q13),9p+, 11q+	Fitzgerald et al 1983
46,Y,-X,-1,-7,-8,-14,-15,-17,-20,inv(6)(p23q13), del(11)(q23),+t(X;?)(p11;?),+t(1;?)(p24;?), +t(12;?)(q22;?),+t(17;?)(p11;?),+4mar	Kaneko et al 1982b
47,XX,+8/48,XX,+8,+17/48,XX,+8,+20	Garson 1980
47,XY,+8/48,XY,+8,+17/48,XY,+8,+22	Garson 1980
48,XY,i(17q),+2mar	Seabright 1983
57-58,XX,+1,+3,-4,-5,+6,-8,+11,+12,-17,+19,-20, del(10)(q24),+8-9mar	Testa et al 1979
60-61,XX,Xp+,+1,+2,-4,+6,-7,+19,+20,del(2)(q31), del(10)(q24),+12-13mar	

ANLL, FAB type M5a

44,Y,-X,-5,del(11)(q22),17p+,+r	Berger et al 1982a
46,XX,t(11;17)(q23;q23)	Berger et al 1982a
47,XY,+8,9q+,del(14)(q23),17p+	Berger et al 1982a
48,XY,+4,+8,t(11;17)(q24;q21)"s"	Dewald et al 1983

ANLL, FAB type M5b

48,XY,17p+,+19,+mar	Berger et al 1982a

ANLL, FAB type M4+M5

46,XY,-17,+21/47,XY,+21	Alimena et al 1977
47,XY,+17	Alimena et al 1977
47,XY,+8/47,XY,+8,t(17;17)	Chapelle et al 1976

ANLL, FAB type M6

??,XY,t(4;?)(q23;?),t(5;17),t(8;9),+8,+11, t(12;16)(p12;p12)	Hustinx and Rutten 1983
43,X,-X,del(5)(q14q34),-7,11q-,-12,12p+,t(12;17),t(20;21)	Kerkhofs et al 1982
43,XX,5q-,-7,-17,-18	Garson 1980
43-45,XX,der(1),inv(2)(p25q14),del(3)(q25),-4,-5,-6,6q-,-7, 7p+,11p+,-12,-17,-21,+r,+1-4mar"s"	Smadja et al 1983b
44,XX,del(5)(q13q31orq33),del(11)(q21),-17,-21"c""s"	Papa et al 1979
44,XY,3p-,t(2;6),t(5;6),t(17;20)"s"	Sandberg et al 1982a
44,XY,t(3;5;11),17p+,-22	Garson 1980
44-45,XY,-2,-4,-7,-10,-13,-16,-17,-22,+3mar	Mitelman et al 1981
44-46,XY,-5,-17,-19,+mar	Mitelman et al 1981
45,XY,del(5)(q13q31),r(8),-17,-21,+mar	Kerkhofs et al 1982

Karyotype	Reference
47,XX,-5,i(17q),-19,+3mar	Najfeld 1976
49,XY,+8,+12,+17,t(17;21)(p11;q22)"s"	Yamada and Furusawa 1976

ANLL, special type

Karyotype	Reference
48,XY,-5,-17,+idic(17q),+idic(17q),+idic(17q),+mar	Atkin et al 1981

Chronic myeloid leukemia, t(9;22)

Karyotype	Reference
26,XX,t(9;22)(q34;q11),-1,-2,-3,-4,-5,-6,-7,-9,-10,-11,-12, -13,-14,-15,-16,-17,-18,-19,-20	Hartley and McBeath 1981
35,XX,t(9;22)(q34;q11),-3,-4,-5,+8,-9,-10,-11,-12,-13,-14, -15,-16,-17/66-72,XX,t(9;22)	Pedersen and Boesen 1983
43-46,XX,t(9;22),-5,-7,-10,-17,-19,5p+,14q+,+r(?),+mar	Sadamori et al 1981b
44,XX,t(9;22),-16,-17/46,XX,t(9;22),-22,+22q-	Stoll and Oberling 1978
44,XX,t(9;22),-16,-17/46,XX,t(9;22),-22,+22q-	
44,XX,t(9;22),-16,-17/46,XX,t(9;22),-22,+22q-	
44,XY,t(9;22),-17,-20	Stoll and Oberling 1978
45,XY,t(9;22),-8,-11,-18,+2mar	
47,XY,t(9;22),+7,i(17q)	
44,XY,t(9;22),17p+,-17,-18	Hagemeijer et al 1980b
45,X,-Y,t(9;22)	Sonta and Sandberg 1977b
42,X,-Y,t(9;22),-7,-16,-17	
45,X,-Y,t(9;22)(q34;q11),t(4;17)(q25;p13)	Lawler et al 1976
45,X?,t(9;22),-17	Rozynkowa et al 1977
45,X?,t(9;22),-7,i(17q)	Carbonell et al 1982a
45,XY,t(9;22)(q34;q11),-17	Sonta and Sandberg 1978b
45,XY,t(9;22)(q34;q11),-13,-17,+t(13;17)(p11;q11)	Tomiyasu et al 1982
45,XY,t(9;22),-17	Stoll and Oberling 1978
47,XY,t(9;22),+8,-17,+22q-	
45,XY,t(9;22),-20	Stoll and Oberling 1978
45,XY,t(9;22),-7,-8,+17	
45,XY,t(9;22),-20,-22,+22q-/46,XY,t(9;22),-20,-22,+8,+22q-	Stoll and Oberling 1978
44,XY,t(9;22),-17,-22,-22,+22q-/46,XY,t(9;22),-20,+22q-/47, XY,t(9;22),+11,-22,+22q-	
45,XY,t(9;22),-8	Stoll and Oberling 1978
45,X,-Y,t(9;22),i(17q)/45,XY,t(9;22),-8	
45,XY,t(9;22),t(17;20)(q12;q11),del(11)(p13)	Fleischman et al 1981
46,X,-Y,t(9;22),+22q-/47,X,-Y,t(9;22),+8,i(17q),+22q-	Bernstein et al 1980b
46,X?,t(9;22)(q34;q11)	Lawler et al 1976
46,X?,t(9;22),i(17q)	
46,X?,t(9;22)(q34;q11),t(16;17)	Lawler et al 1976
46,X?,t(9;22)(q34;q11),i(17q)	Lawler et al 1976
46,X?,t(9;22),-16,i(17q),+del(22)(q11)	
46,X?,t(9;22)(q34;q11)/46,X?,t(9;22),i(17q)	Lawler et al 1976
46,X?,t(9;22),6q-/46,X?,t(9;22),6q-,i(17q)	Carbonell et al 1982a
46,X?,t(9;22),i(17q)	Rozynkowa et al 1977
46,X?,t(9;22),i(17q)	Rozynkowa et al 1977
46,X?,t(9;22),i(17q)/47,X?,t(9;22),+8,i(17q)	Fleischman et al 1981
46,X?,t(9;22),i(17q)/47,X?,t(9;22),i(17q),+8	Fleischman et al 1981
46,X?,t(9;22),i(17q)	Carbonell et al 1982a
46,X?,t(9;22),i(17q)	Carbonell et al 1982a
46,X?,t(9;22),i(17q)	Carbonell et al 1982a
46,X?,t(9;22),i(17q)	Carbonell et al 1982a
46,X?,t(9;22),i(17q)/47,X?,t(9;22),+8,i(17q)	Carbonell et al 1982a

46,X?,t(9;22),i(17q)/47,X?,t(9;22),+8,i(17q)	Carbonell et al 1982a
46,X?,t(9;22),i(17q)/47,X?,t(9;22),+8,i(17q)	Carbonell et al 1982a
46,X?,t(9;22),inv(7)/47,X?,t(9;22),+8,i(17q)	Carbonell et al 1982a
46,X?,t(9;22)/47,X?,t(9;22),+8,i(17q)	Carbonell et al 1982a
46,XX,t(9;22)(q34;q11)/47,XX,t(9;22)(q34;q11),+8, i(17)(q11)	Rowley 1973a
46,XX,t(9;22)(q34;q11)/46,XX,t(9;22)(q34;q11),ins(1)(q32), 17p+	Hayata et al 1975b
46,XX,t(9;22)(q34;q11)/46,XX,t(9;22)(q34;q11),i(17)(q11)	Hayata et al 1975b
46,XX,t(9;22)(q34;q11)	Mitelman et al 1974a
46,XX,t(9;22)(q34;q11)	
46,XX,t(9;22)(q34;q11)	
46,XX,t(9;22)(q34;q11)	
46,XX,t(9;22)(q34;q11)	
46,XX,t(9;22)(q34;q11)/48,XX,t(9;22)(q34;q11),+17, del(22)(q11)	
46,XX,t(9;22)(q34;q11)/48,XX,t(9;22)(q34;q11),+17, del(22)(q11)	
46,XX,t(9;22)(q34;q11)/48,XX,t(9;22)(q34;q11),+17, +del(22)(q11)	
46,XX,t(9;22)(q34;q11),i(17q)/47,XX,t(9;22),i(17q),+i(17q)	Sonta and Sandberg 1978b
46,XX,t(9;22)(q34;q11)/46,XX,t(9;22),-7,-17,+13,+14	Sonta and Sandberg 1978b
46,XX,t(9;22)(q34;q11)/45,XX,t(9;22),-17	Sonta and Sandberg 1978b
46,XX,t(9;22)(q34;q11)/48,XX,t(9;22),+8,+17	Misawa 1978
46,XX,t(9;22)(q34;q11)/51,XX,t(9;22),+5,+8,+17,+19,+22q-	Misawa 1978
46,XX,t(9;22)(q34;q11),i(17q)	Watt et al 1977
46,XX,t(9;22)(q34;q11),17p+	Verma and Dosik 1981
46,XX,t(9;22)(q34;q11),i(17q)	McGlave et al 1982
46,XX,t(9;22)(q34;q11)	Rajasekariah et al 1982
50,XX,t(9;22)(q34;q11),t(9;22)(q34;q11),+8,+11,+13, t(16;17)(p11;p12),+22	
46,XX,t(9;22)(q34;q11),t(8;17)(p12;q21)	Petit and Van Den Berghe 1981
46,XX,t(9;22)(q34;q11),i(17q)	Alimena et al 1979
46,XX,t(9;22)(q34;q11),5q+,t(8;19),17p+	Alimena et al 1982b
46,XX,t(9;22)(q34;q11)/47,XX,t(9;22),+22q-/49,XX,t(9;22), +8,+17,+22q-/49,XX,t(9;22),+18,+19,+22q-/50,XX,t(9;22), +8,+17,+18,+22q-	Mitelman 1983
46,XX,t(9;22)(q34;q11),i(17q)/47,XX,t(9;22),+8,i(17q), +22q-	Tomiyasu et al 1982
46,XX,t(9;22)(q34;q11)	Mitelman 1983
46,XX,t(9;22)	
46,XX,t(9;22)	
46,XX,t(9;22)	
46,XX,t(9;22)	
46,XX,t(9;22)	
46,XX,t(9;22)	
46,XX,t(9;22)/48,XX,t(9;22),+17,+22q-	
46,XX,t(9;22)/48,XX,t(9;22),+17,+22q-	
46,XX,t(9;22)/48,XX,t(9;22),+17,+22q-	
46,XX,t(9;22),+C,-16,i(17q)	Lobb et al 1972
46,XX,t(9;22),-7,-15,+20,+i(17q)	Stoll and Oberling 1978
46,XX,t(9;22),12p-,i(17q)	Lilleyman et al 1978
46,XX,t(9;22),2q+,i(17q),19p-	Olah and Rak 1981
46,XX,t(9;22),dic(22q-),-17,+mar/46,XX,t(9;22)	Hossfeld 1975b

Karyotype	Reference
47,XX,t(9;22),dic(22q-),-17,+2mar	
47,XX,t(9;22),dic(22q-),-17,+2mar	
46,XX,t(9;22),i(17q)	Prigogina et al 1978
46,XX,t(9;22),i(17q)/47-49,XX,t(9;22),i(17q),+3,+8	Prigogina et al 1978
46,XX,t(9;22),i(17q)/47,XX,t(9;22),i(17q),+8	Prigogina et al 1978
46,XX,t(9;22),i(17q)	Lilleyman et al 1978
46,XX,t(9;22),i(17q)	Stoll and Oberling 1978
46,XX,t(9;22),i(17q)	Lyall and Garson 1978
46,XX,t(9;22),i(17q)	Kohno and Sandberg 1980
46,XX,t(9;22),i(17q)	Kohno and Sandberg 1980
46,XX,t(9;22),i(17q)	Miyamoto 1980
46,XX,t(9;22),i(17q)/47,XX,t(9;22),+8	Olah and Rak 1981
46,XX,t(9;22),i(17q)/46,XX,t(9;22)	Olah and Rak 1981
46,XX,t(9;22),i(17q)	Olah and Rak 1981
46,XX,t(9;22),i(17q)	Olah and Rak 1981
46,XX,t(9;22),i(17q)	Olah and Rak 1981
46,XX,t(9;22),i(17q)/47,XX,t(9;22),i(17q),+20/46,XX, t(9;22),-1,+20	Olah and Rak 1981
46,XX,t(9;22),i(17q)	Borgström et al 1982
46,XX,t(9;22),i(17q)	Borgström et al 1982
46,XX,t(9;22),i(17q)	Borgström et al 1982
46,XX,t(9;22),i(17q)	Borgström et al 1982
46,XX,t(9;22),i(17q)	Borgström et al 1982
46,XX,t(9;22),i(17q)	Seabright 1983
46,XX,t(9;22),i(17q)	Seabright 1983
46,XX,t(9;22),r(17)	Borgström et al 1982
46,XX,t(9;22),t(11;17)(q?;p?)	Borgström et al 1982
46,XX,t(9;22),t(11;17)(q13;p11)	Seabright 1983
46,XX,t(9;22),t(17;18)(p?;q?)	Sadamori et al 1980
46,XX,t(9;22),t(1;17)(p?;q?)	Borgström et al 1982
46,XX,t(9;22),t(X;17)(q?;q?),-17,+22q-	Borgström et al 1982
46,XX,t(9;22)/48,XX,t(9;22),+19,+22q-/50,XX,t(9;22),+8,+12, +19,+22q-/50,XX,t(9;22),+8,+14,+17,+19,+21	Hagemeijer et al 1980b
46,XY,t(9;22)	Hagemeijer et al 1981b
46,XY,t(9;22),r(18)	
44,XY,t(9;22),r(18),t(5;17)(q11;p11)	
46,XY,t(9;22)	Berger et al 1983b
46,XY,t(9;22)/46,XY,t(9;22),t(15;17)	
46,XY,t(9;22)(q34;q11)/46,XY,t(9;22)(q34;q11),i(17)(q11)	Hayata et al 1975b
46,XY,t(9;22)(q34;q11),i(17)(q11)	Hayata et al 1975b
46,XY,t(9;22)(q34;q11)/46,XY,t(9;22)(q34;q11),i(17)(q11)	Hayata et al 1975b
46,XY,t(9;22)(q34;q11),i(17q)	Sonta and Sandberg 1978b
46,XY,t(9;22)(q34;q11)/47,XY,t(9;22),i(17q),+del(22q)/49, XY,t(9;22),+8,+21,i(17q),+del(22q)	Sonta and Sandberg 1978b
46,XY,t(9;22)(q34;q11),i(17q)	Sonta and Sandberg 1978b
46,XY,t(9;22)(q34;q11),i(17q)	Sonta and Sandberg 1978b
46,XY,t(9;22)(q34;q11),-19,3q+,6q-,+17	Greenberg et al 1978
46,XY,t(9;22)(q34;q11),del(6)(q13)	Srivastava et al 1981
46,XY,t(9;22),del(6)(q13),i(17q)	
46,XY,t(9;22)(q34;q11),i(17q)	Alimena et al 1982b
46,XY,t(9;22)(q34;q11)	Mitelman 1983
46,XY,t(9;22)(q34;q11),i(17q)	
46,XY,t(9;22)(q34;q11),i(17q)	
46,XY,t(9;22)(q34;q11)/47,XY,t(9;22),+17	Mitelman 1983
46,XY,t(9;22)(q34;q11)/46,XY,t(9;22)(q34;q11),i(17q)	Mitelman 1983

46,XY,t(9;22)(q34;q11)/46,XY,t(9;22)(q34;q11),i(17q)	
46,XY,t(9;22)(q34;q11)/47,XY,t(9;22),+8,i(17q)	Mitelman 1983
47,XY,t(9;22)(q34;q11),+8,i(17q)/46,XY,t(9;22)	
46,XY,t(9;22)(q34;q11),+t(1;17)(q21;p11)/47,XY,t(9;22), +t(1;17),+18/48,XY,t(9;22),+t(1;17),+8,+19	Mamaeva et al 1983
46,XY,t(9;22)(q34;q11),i(17q)/47,XY,t(9;22),+8,i(17q)/47, XY,t(9;22),+22q-/48,XY,t(9;22),+8,i(17q),+19	Tomiyasu et al 1982
46,XY,t(9;22)(q34;q11)	Ishihara et al 1982
27,X,-Y,t(9;22),-1,-2,-3,-4,-5,-6,-7,-9,-11,-13,-14,-15, -16,-17,-18,-19,-20,-22	
46,XY,t(9;22),-5,-17,t(5;17),+22q-	Borgström et al 1982
46,XY,t(9;22),-8,12q+,i(17q),+21/46,XY,t(9;22),-8,-8,12q+, +i(17q),+21	Olah and Rak 1981
46,XY,t(9;22),17p-	Lilleyman et al 1978
46,XY,t(9;22),i(17q),20q-	Prigogina et al 1978
46,XY,t(9;22),i(17q)	Prigogina et al 1978
46,XY,t(9;22),i(17q)/47,XY,t(9;22),+8/48,XY,t(9;22),+8,+19	Hagemeijer et al 1980b
46,XY,t(9;22),i(17q)/47,XY,t(9;22),i(17q),+8/48,XY,t(9;22), i(17q),+8,+8/49,XY,t(9;22),i(17q),+8,+8,+22q-/50,XY, t(9;22),i(17q),+8,+8,+22q-,+19	Hagemeijer et al 1980b
46,XY,t(9;22),i(17q)	Miyoshi et al 1976
46,XY,t(9;22),i(17q)	Lilleyman et al 1978
46,XY,t(9;22),i(17q)	McDermott et al 1978
46,XY,t(9;22),i(17q)	Stoll and Oberling 1978
46,XY,t(9;22),i(17q)	Lyall and Garson 1978
46,XY,t(9;22),i(17q)/47,XY,t(9;22),+8,i(17q)/48,XY,t(9;22), +8,+22q-	Bernstein et al 1980b
46,XY,t(9;22),i(17q)/47,XY,t(9;22),+8,i(17q)	Bernstein et al 1980b
46,XY,t(9;22),i(17q)	Kohno and Sandberg 1980
46,XY,t(9;22),i(17q)	Carbonell et al 1981
46,XY,t(9;22),i(17q)/47,XY,t(9;22),+22q-/48,XY,t(9;22),+8, i(17q),+22q-	Miyamoto 1980
46,XY,t(9;22),i(17q)	Olah and Rak 1981
46,XY,t(9;22),i(17q)	Olah and Rak 1981
46,XY,t(9;22),i(17q)	Borgström et al 1982
46,XY,t(9;22),i(17q)	Borgström et al 1982
46,XY,t(9;22),i(17q)	Borgström et al 1982
46,XY,t(9;22),i(17q)	Borgström et al 1982
46,XY,t(9;22),i(17q)	Chrz et al 1983
46,XY,t(9;22),i(17q)	Chrz et al 1983
46,XY,t(9;22),i(17q)/47,XY,t(9;22),i(17q),+18	Chrz et al 1983
46,XY,t(9;22),r(18)/45,XY,t(9;22),-5,-17,t(5;17)(q11;p11), r(18)	Hagemeijer et al 1980b
46,XY,t(9;22),t(1;17)(q21;q25)	Prigogina et al 1978
46,XY,t(9;22),t(8;17)(q13;q25)	Fleischman et al 1981
46,XY,t(9;22),t(X;6)(q28?;q22?),t(5;17)(q13?;q25)	Bernstein et al 1980b
46,XY,t(9;22)/46,XY,t(9;22),i(17q)	Prigogina et al 1978
46-47,XY,t(9;22),+8,i(17q)	
46,XY,t(9;22)/46,XY,t(9;22),del(5)(q12),del(13)(q21), t(17;21)(p11;q11)	Bernstein et al 1980b
46,XY,t(9;22)/47,XY,t(9;22),i(17q),+8	Prigogina et al 1978
46,XY,t(9;22)/50,XY,t(9;22),+8,+17,+19,+22q-	Hagemeijer et al 1980b
46-47,XY,t(9;22),+8,i(17q)	Sharp et al 1976
47,X,-Y,t(9;22)(q34;q11),+8,i(17q),+22q-	Pasquali et al 1982b
47,X?,t(9;22)(q34;q11),+8,i(17)(q11)	Nowell et al 1975

Karyotype	Reference
47,X?,t(9;22),+22q-/51,X?,t(9;22),+8,+8,+17,+19,+22q-	Fleischman et al 1981
47,X?,t(9;22),+8,i(17q)/48,X?,t(9;22),+8,i(17q),+22q-	Carbonell et al 1982a
47,X?,t(9;22),+8,i(17q+),19p-	Carbonell et al 1982a
47,X?,t(9;22),+C,i(17q)	Lawler et al 1976
47,X?,t(9;22),del(6)(q14),+8,i(17q)	Fleischman et al 1981
47,X?,t(9;22),r(19),+22q-/48,X?,t(9;22),+17,+22q-/49,X?,	Carbonell et al 1982a
t(9;22),+8,+17,+22q-/50,X?,t(9;22),+8,+17,+r(19),	
+22q-/49,X?,t(9;22),1q+,+8,+17,+22q-	
47,XX,+C,+2,-16,t(9;22),i(17q)	Lobb et al 1972
47,XX,t(9;22)(q34;q11),+8,i(17q)	Alimena et al 1979
47,XX,t(9;22)(q34;q11),+8,i(17q)/48,XX,t(9;22),+8,i(17q),	Tomiyasu et al 1982
+22q-	
47,XX,t(9;22)(q34;q11),+t(1;17)(q21;p11),-17,+22q-	Tomiyasu et al 1982
47,XX,t(9;22),+17/48,XX,t(9;22),+17,+19	Stoll and Oberling 1978
47,XX,t(9;22),+8,16q+,i(17q)	Sadamori et al 1980
47,XX,t(9;22),+8,i(17q)	Lilleyman et al 1978
47,XX,t(9;22),+8,i(17q)	Lilleyman et al 1978
47,XX,t(9;22),+8,i(17q)	Stoll and Oberling 1978
47,XX,t(9;22),+8,i(17q)/48,XX,t(9;22),+i(17q),+22q-	
47,XX,t(9;22),+8,i(17q)	Lyall and Garson 1978
47,XX,t(9;22),+8,i(17q)	Borgström et al 1982
47,XX,t(9;22),+8,i(17q)	Borgström et al 1982
47,XX,t(9;22),+8,i(17q)	Borgström et al 1982
47,XX,t(9;22),+i(17q)	Borgström et al 1982
47,XX,t(9;22),i(17q),+8,6q-	Prigogina et al 1978
47,XX,t(9;22),i(17q),+8/48,XX,t(9;22),i(17q),+8,+19/49,XX,	Prigogina et al 1978
t(9;22),i(17q),+8,+19,+del(22)(q11)	
47,XX,t(9;22),i(17q),+22q-	Stoll and Oberling 1978
47,XX,t(9;22),i(17q),+22q-	Stoll and Oberling 1978
47,XX,t(9;22),i(17q),+8	Lessard and Le Prise 1982
47,XX,t(9;22),i(17q),+22q-	Seabright 1983
47,XXX,t(9;22),i(17q)"c"	Borgström et al 1982
47,XY,+C,t(9;22),i(17q)/46,XY,t(9;22),i(17q)	Lobb et al 1972
47,XY,t(9;22)(q34;q11),+8,i(17)(q11)/50,XY,	Hayata et al 1975b
t(9;22)(q34;q11),i(17)(q11),+4,+6,+8,+8/47,XY,	
t(9;22)(q34;q11),t(3;17)(q21;p1?),del(10)(q24),	
i(17)(q11),+mar	
47,XY,t(9;22)(q34;q11),-17,+t(1;7)(p11;q11),+22	Pasquali et al 1979a
47,XY,t(9;22)(q34;q12),idic(17q),+8	Whang-Peng et al 1981
47,XY,t(9;22)(q34;q11),t(8;17)(q12;q22),+22q-	Petit and Van Den Berghe 1981
47,XY,t(9;22)(q34;q11),+8,i(17q)	Pasquali et al 1982b
47,XY,t(9;22)(q34;q11),t(3;5),-8,-10,+17,+22q-,+mar	Alimena et al 1982b
47,XY,t(9;22)(q34;q11),del(6)(q11),+8,i(17q)	Mitelman 1983
47,XY,t(9;22)(q34;q11),del(6)(q11),+8,i(17q)	
47,XY,t(9;22)(q34;q11),+8,i(17q)	Tomiyasu et al 1982
47,XY,t(9;22)(q34;q11),+8,i(17q)/46,XY,t(9;22),i(17q)	Tomiyasu et al 1982
47,XY,t(9;22)(q34;q11),+8,i(17q)	Sadamori and Sandberg 1983b
47,XY,t(9;22),+15/48,XY,t(9;22),+17,+22q-	Stoll and Oberling 1978
47,XY,t(9;22),+17	Stoll and Oberling 1978
47,XY,t(9;22),+17	Stoll and Oberling 1978
47,XY,t(9;22),+17/47,XY,t(9;22),-22,+5,+22q-/49,XY,t(9;22),	Stoll and Oberling 1978
+5,+21,+22q-	
47,XY,t(9;22),+22q-,+8,t(12;13)(q?;q?),-13,i(17q)	Rozynkowa et al 1977

47,XY,t(9;22),+22q-/46,XY,t(9;22),i(17q)	Stoll and Oberling 1978
47,XY,t(9;22),+22q-/47,XY,t(9;22),+17	Stoll and Oberling 1978
47,XY,t(9;22),+22q-/56,XY,t(9;22),+1,+2,+5,+6,+7,+8,+10, +12,+17,+22q-	Stoll and Oberling 1978
47,XY,t(9;22),+22q-/48,XY,t(9;22),+8,+22q-/48,XY,t(9;22), +17,+22	Stoll and Oberling 1978
47,XY,t(9;22),+8	Sadamori et al 1980
47,XY,t(9;22),+8,i(17q)	
47,XY,t(9;22),+8,-17,-18,t(17;18)(p11;p11),+del(22q)	Sharp et al 1976
47,XY,t(9;22),+8,i(17q)/47,XY,t(9;22),+8	Chapelle et al 1976
47,XY,t(9;22),+8,i(17q)	Lilleyman et al 1978
47,XY,t(9;22),+8,i(17q)	McDermott et al 1978
47,XY,t(9;22),+8,i(17q)/47,XY,t(9;22),+16,i(17q)	Stoll and Oberling 1978
47,XY,t(9;22),+8,i(17q)	Lyall and Garson 1978
47,XY,t(9;22),+8,i(17q)	Lyall and Garson 1978
47,XY,t(9;22),+8,i(17q)/48,XY,t(9;22),+8,i(17q),+22q-	Lyall and Garson 1978
47,XY,t(9;22),+8,i(17q)	Miyamoto 1980
47,XY,t(9;22),+8,i(17q)	Borgström et al 1982
47,XY,t(9;22),+8,i(17q)	Borgström et al 1982
47,XY,t(9;22),+8/50,XY,t(9;22),+8,+13,+i(17q),+22q-	Miyamoto 1980
47,XY,t(9;22),+9q+,i(17q)	Kamada et al 1981
47,XY,t(9;22),-10,+17,+19,-22,+22q-	Stoll and Oberling 1978
47,XY,t(9;22),i(17q),+8	Prigogina et al 1978
47,XY,t(9;22),i(17q)/49,XY,t(9;22),i(1q),+8,+21,+22q-	Sonta and Sandberg 1977b
47,XY,t(9;22),i(17q),+mar	Chrz et al 1983
48,XX,t(9;22)(q34;q11),+17,+mar	Sadamori and Sandberg 1983b
48,XX,t(9;22),+17,+22q-	Stoll and Oberling 1978
48,XX,t(9;22),+8,+8,i(17q)	Borgström et al 1982
48,XX,t(9;22),+8,i(17q),+19/46,XX,t(9;22),i(17q)/47,XX, t(9;22),i(17q),+19	Hartley and McBeath 1981
48,XY,+X,t(9;22),+8,i(17q)	Borgström et al 1982
48,XY,t(9;22)(q34;q11),+8,i(17)(q11),+F	Rowley 1973a
48,XY,t(9;22)(q34;q11),+8,i(17q),+22q-	Tomiyasu et al 1982
48,XY,t(9;22),+17,+22q-	Stoll and Oberling 1978
48,XY,t(9;22),+17,+22q-	Stoll and Oberling 1978
48,XY,t(9;22),+22q-,+22q-/47,XY,t(9;22),+17/46,XY,t(9;22), -4,-7,+6,+17	Stoll and Oberling 1978
45,XY,t(9;22),-21/46,XY,t(9;22),-10,+17	
48,XY,t(9;22),+3,+17	Stoll and Oberling 1978
48,XY,t(9;22),+8,+22q-/61,XY,+Y,t(9;22),+4,+5,+6,+8,+8,+8, +8,+15,+17,+18,+19,+20,+22q-,+22q-	Wurster-Hill 1983
48,XY,t(9;22),+8,+8,i(17q)	Lilleyman et al 1978
48,XY,t(9;22),+8,i(17q),+19	Lilleyman et al 1978
48,XY,t(9;22),+8,i(17q)	Lilleyman et al 1978
48,XY,t(9;22),+8,i(17q),+22q-	Stoll and Oberling 1978
48,XY,t(9;22),+8,i(17q),+22	Stoll and Oberling 1978
48,XY,t(9;22),-22,+17,+10,+22q-	Stoll and Oberling 1978
48,XY,t(9;22),-22,+17,+20,+22q-	
49,X?,t(9;22),+17,+19,+22q-/50,X?,t(9;22),+8,+17,+19, +22q-/51,X?,t(9;22),+8,+9,+17,+19,+22q-	Carbonell et al 1982a
49,XX,t(9;22),+12,+13,i(17q),+19	Seabright 1983
49,XX,t(9;22),+17,+19,+22q-	Sadamori et al 1980
49,XX,t(9;22),+8,+17,+21/49,XX,t(9;22),+8,+17,+17,-19,+21	Olah and Rak 1981

49,XX,t(9;22),+8,17q-,+19,+22q-/49,XX,t(9;22),+8,i(17q),+19,+22q-	Olah and Rak 1981
49,XY,t(9;22)(q34;q11),+8,dup(9)(q13q22),+17,+19	Tomiyasu et al 1982
49,XY,t(9;22),+10,+14,+17	Hossfeld 1975b
49,XY,t(9;22),+10,-17,+19,+del(22q),+mar	Hossfeld 1975b
49,XY,t(9;22),+13,+22q-,+22q-/52,XY,t(9;22),+13,+13,+17,+17,+22q-,+22q-	Stoll and Oberling 1978
49,XY,t(9;22),+17,+19,+22q-/50,XY,t(9;22),+8,+15,+19,+22q-/50,XY,t(9;22),+8,+9,+19,+22q-	Hartley and McBeath 1981
49,XY,t(9;22),+8,+17,+mar	Borgström et al 1982
49,XY,t(9;22),+8,+i(17q),+22q-/50,XY,t(9;22),+8,+10,+i(17q),+22q-	Olah and Rak 1981
49,XY,t(9;22),+9,+10,+12/45,XY,t(5;17)(p11;q21)	Hagemeijer et al 1981b
50,XY,+Y,t(9;22)(q34;q11),t(6;8)(q34;q11),+8,i(17q),+19,+22q-	Alimena et al 1982b
50,XY,t(9;22)(q34;q11),+8,+17,+21,+del(22)(q11)/51,XY,t(9;22),+8,+8,+17,+21,+del(22)(q11)/52,XY,t(9;22),+8,+8,+8,+17,+21,+del(22)(q11)	Sonta and Sandberg 1978b
50,XY,t(9;22),+8,+15,+17,+del(22q)	Sharp et al 1976
50,XY,t(9;22),+8,+17,+21,+22q-/51,XY,t(9;22),+8,+8,+17,+21,+22q-/54,XY,t(9;22),+8,+8,+8,+17,+21,+22q-,+22q-	Sonta and Sandberg 1977b
50-54,XY,t(9;22),+1,+5,+8,+9,-16,+17,+18,+19,+21,+22q-	Alimena et al 1980
50-55,XY,t(9;22)(q34;q11),+1,+5,+8,+9,-16,+17,+18,+19,+21,+22q-,+mar	Alimena et al 1982b
51,XX,t(9;22)(q34;q11),+6,+8,i(17q),+19,+20,+22q-	Alimena et al 1982b
51,XX,t(9;22),+4,+17,+19,+del(22q),+mar	Hossfeld 1975b
51,XY,t(9;22),+8,+14,+17,+19,+19	Borgström et al 1982
52,XY,t(9;22)(q34;q11),1p+,+8,+9,i(17q),+19,+20,+21,+22q-	Alimena et al 1982b
53,XX,t(9;22),+8,+11,+12,+16,+17,+19,+del(22q)/52,XX,t(9;22),+8,+11,+17,+19,+del(22q)	Gahrton et al 1974b
59,X,-X,t(9;22),+4,+6,+6,t(7;?)(p15;?),+der(7),+8,+8,+12q+,+17,+19,+19,+21,+22,+22q-,+22q-	Miyamoto 1980

Chronic myeloid leukemia, aberrant translocation

??,X?,t(17;22)(q25;q11)	Potter et al 1981
44,Y,-X,t(9;17;22)(q34;p11;q11),t(X;9)(?;p24),i(17q)	Engel et al 1977
45,XY,-22,t(17;17;22)(p1?;q24;q11)	Engel et al 1977
46,X,t(X;9;22)(q27;q34;q12)	Dallapiccola and Alimena 1979
46,X,t(X;9;22)/47-51,X,t(X;9;22),+1,+4,+8,+9,-14,i(17q),-19,+22q-	
46,X?,t(17;22)(p13;q11)	Hagemeijer et al 1980b
46,X?,t(17;22)(p13;q11)	Ishihara 1983
46,X?,t(17;22)(q25;q11)	Hagemeijer et al 1980b
46,X?,t(9;17;22)(q34;p13;q11)	Ishihara and Minamihisamatsu 1981
46,XX,inv(5)(p14q12),t(9;13;17;22)(q31;q11;q12;q12)	Fraisse et al 1980
46,XX,t(12;22)(p13;q11),t(8;17)(q13;q23)	Swolin et al 1983
46,XX,t(17;22)(p11;q11)	Borgström 1981
46,XX,t(17;22)(q21;q11)	Seabright 1983
46,XX,t(17;22)(q25;q11)	Panani et al 1979
46,XX,t(17;22)(q25;q11)/47,XX,t(17;22),+8	Bernstein et al 1980b
46,XX,t(17;22)(q25;q11)	Borgström et al 1982
46,XX,t(17;22)(q25;q11),i(17q+)	

46,XX,t(17;22),inv(5),t(9;13)	Freycon et al 1982
46,XX,t(5;22)(q35;q11)/46,XX,t(5;22),i(17q)/47,XX,t(5;22), +8,i(17q)	Pasquali et al 1979b
46,XX,t(7;9)(q22;q34),t(17;22)(p13;q11)	Mitelman 1983
46,XX,t(7;9)(q22;q34),t(17;22)(p13;q11)	
46,XX,t(9;15;22)(q34;q11;q11)	Hays et al 1981
54-57,XX,t(9;15;22),+5,+7,+8,+10,+11,+12,+17,+22q-	
46,XX,t(9;17;22)(q34;q21;q11)	Sonta and Sandberg 1978b
46,XX,t(9;17;22)(q34;q21;q11)	Rowley et al 1979
46,XX,t(9;17;22)(q34;q21;q11)	Sonta and Sandberg 1977a
46,XY,22q-"c"	Mitelman 1983
46,XY,22q-"c"	
46,XY,22q-"c"	
46,XY,t(13;17;22)(p13;q12;q11)"c"	
46,XY,t(13;17;22)(p13;q12;q11)"c"	
46,XY,t(13;17;22)(p13;q12;q11)"c"	
46,XY,t(13;17;22)(p13;q12;q11)"c"	
46,XY,t(13;17;22)(p13;q12;q11)"c"	
46,XY,t(16;22)(q24;q11),t(9;17)(q34;q21)/46,XY,t(16;22), t(9;17),del(7)(q22)	Mitelman 1983
46,XY,t(16;22)(q24;q11),t(9;17)(q34;q21)	
46,XY,t(17;22)(q25;q11)	Matsunaga et al 1976
46,XY,t(1;7;19;22)(q?;q?;p?;q11)	Borgström et al 1982
46,XY,t(1;7;19;22),i(17q)	
46,XY,t(3;9)(q21;q34),t(17;22)(q21;q11)	Oshimura et al 1982a
46,XY,t(3;9;10;22)(p13;q34;q22;q11),17q+	Seabright 1983
46,XY,t(4;22)(q35;q21),t(3;5)(q27;q22)	Sessarego et al 1981b
46,XY,t(4;22)(q35;q21),t(3;5)(q27;q22),i(17q)/50,XY, t(4;22),t(3;5),i(17q),+8,+15,+21,+22q-	
46,XY,t(7;22)(p22;q12)	Gahrton et al 1979
48,XY,t(7;22)(p22;q12),+8,+8,i(17q)	
46,XY,t(9;17;22)(q34;q21;q11)	Pasquali 1983
46,XY,t(9;20;22)(q34;p11p13;q11),i(17q)	Pasquali et al 1982b
46,XY,t(9;20;22)(q34;p11;q11),i(17q)	Pasquali 1983
46,XY,t(9;20;22)(q34;p11p13;q11),i(17q)	Casalone et al 1983
47,XY,t(10;22)(q26;q11),+8,i(17q)	Prigogina et al 1978

Chronic myeloid leukemia, Ph1 negative

46,XY,i(17q)	Bernstein et al 1980b
46,XY,i(17q)	Bernstein et al 1980b
46,XY,i(17q)	Pasquali et al 1982b
46,XY,t(1;6)(p36;q15),del(3)(q25),del(17)(p11)	Srivastava et al 1981
46,XY,t(3;6)(q?;p?),t(17;21)	Borgström et al 1982
47,XX,+8,17q-	Kohno et al 1979a
47,XX,i(17q),+21	Borgström et al 1982

Eosinophilic leukemia

46,XY,i(17)(q11)	Mitelman et al 1975b
46,XY,i(17q)	Lönnqvist et al 1979
46,XY,t(17;?)(q11;?)	Cabrol 1979

MYELOPROLIFERATIVE DISORDERS

Polycythemia vera

44-45,XY,-10,-13,-16,-17,-20	Wurster-Hill et al 1976
45,X,-Y	Berger and Bernheim 1979
54-57,X,-Y,+6,+8,+11,+17,+20,+21,t(1;22)(p12;q13)	
45,XY,-16,-17,+21/48,XY,+Y,+11,+16,-17,+22	Wurster-Hill et al 1976
46,X?,17q-	Shabtai 1983
46,XY,t(12;17)(q13;p11)/47,X,-Y,+t(Y;1)(q12;q21),+9	Testa et al 1981b

Myelosclerosis / Myelofibrosis

44,XY,-5,-7,-11,-12,t(7;12)(p?;q?),t(17;18)(p11;p11), +mar"s"	Nowell and Finan 1978
44,XY,5p-,-7,-10,11q-,i(17q)/45,XY,-7,18q-/45,XY,2p-,2q-, 5p-,6q-,9q-,10q-,-11,+i(17q)	Whang-Peng et al 1978
46,XY,-7,t(1;7)(p1?;p11),17p+	Geraedts et al 1980b
46,XY,i(17q)	Najfeld 1983
46,XY,t(1;2)(q?;q?),-8,inv(13)(q12q31),i(17q)	Whang-Peng et al 1978
51,XX,+1,del(2)(q33),+6,+9,+11,+17,-19,+20q+	Najfeld et al 1978b

Chronic myeloproliferative disease, NOS

46,XY,t(4;12)(q35;q11),i(17q),del(20)(q11)	Berger et al 1975

Myeloproliferative disorder, special type

45,XX,t(1;?)(p36;?),-5,-7,-12,t(13;?)(q34;?),-17,-22, +4mar/45,XX,t(1;?),-2,-3,-5,-7,-12,-17,-22,+8mar"s"	Rowley et al 1977a
46,XX,-5,del(6)(q13),-7,+8,-17,+2mar"s"	Rowley et al 1977a
47,XX,+8,t(8;17)(q24.2;q22.1)	Hagemeijer et al 1980a

DYSMYELOPOIETIC SYNDROMES

Preleukemia, NOS

44,XY,+Y,-3,-5,+del(6)(q21),-8,t(17;?)(p13;?),+18,-20,+22	Mitelman 1983
44,XY,-5,-6,-7,-8,t(17;?),+2mar"s"	Pedersen-Bjergaard et al 1982
44,XY,t(3;17),-5,-20,+mar"s"	Pedersen-Bjergaard et al 1981
45,XX,-2,-3,5q-,-6,-9,16p-,17p+,+3mar	Swansbury and Lawler 198■
45,XY,-2,-4,-6,-8,-10,-10,+17,+18,+3mar	Panani et al 1980
45,XY,-2,-5,-7,-8,-11,-12,-13,-14,+t(2;5),+t(11;12), +t(16;17),+17,+3mar	Watt et al 1982
45,XY,-5,-17,t(5;17)(p?;p?)	Borgström et al 1982
46,XX,i(17q)	Borgström et al 1982
46,XX,t(5;12)(?q15;q13),-17,+mar"s"	Anderson and Bagby 1982
46,XY,i(17q)	Ruutu et al 1977b
46,XY,i(17q)	Borgström et al 1982
46,XY,t(1;17)(q24;p13),-2,-5,-7,+r,+2mar"s"	Anderson and Bagby 1982
50,XY,-5,-21,+1,+9,+11,17p+,+1-3mar	Borgström et al 1982
53,XX,+1,+2,t(4;19),+6,i(17q),+21,+21,+22	Seabright 1983

Refractory anemia, NOS

Karyotype	Reference
46,XX,i(17q)	Nowell and Finan 1978
47,XY,+13,i(17q)	Prieto et al 1976

Refractory anemia without excess of blasts

Karyotype	Reference
44,XX,5q-,-7,+8,-12,-13,-17,-20,+2mar	Teerenhovi et al 1981

Refractory anemia with excess of blasts

Karyotype	Reference
43,XX,3q-,del(5)(q13q33),7q-,-12,14p+,16q+,-17,-18,18q+	Kerkhofs et al 1982
44,XX,5q-,-13,-17,-20,-22,+2mar	Teerenhovi et al 1981
45,XX,-9,-17,-20,del(4)(q21),del(5)(q12q31), +t(9;17)(p13?;p11)	Streuli et al 1980

Thrombocytopenia

Karyotype	Reference
46,XX,?dup(17)(p11)	Mitelman 1983
46,XX,?dup(17)(p11)	
46,XX,?dup(17)(p11)	

Pancytopenia

Karyotype	Reference
46,XX,-7,+8,-12,-16,-17,t(6;12)(q?;q?),5q-,6q-,+mar	Nowell and Finan 1978

Dysmyelopoietic syndrome, special type

Karyotype	Reference
44,XY,-5,-7,-17,del(3)(q13),+del(3)(p13),del(6)(q13)"s"	Rowley et al 1981a

Hematopoietic proliferation disorder, NOS

Karyotype	Reference
46,X,t(Y;17)(q11;q25)"c"	Mitelman 1983
46,X,t(Y;17)(q11;q25)/45,X,-Y"c"	

SPECIAL LEUKEMIAS

LYMPHOCYTIC LEUKEMIAS

Acute lymphocytic leukemia (ALL)

Karyotype	Reference
28,XX,-1,-2,-3,-4,-5,-7,-8,-9,-11,-12,-13,-14,-15,-16,-17, -19,-20,-22	Kaneko and Sakurai 1980
32,XY,-2,-3,-4,-5,-7,-8,-11,-12,-13,-14,-15,-16,-17,-20	Shabtai and Halbrecht 1981b
44-45,XY,-14,-17,-21	Alimena et al 1977
45,X,-X,dup(1)(q31q41),del(5)(q22),del(17)(p11)	Humbert et al 1978
45,XX,t(14;14)(q34;q11),t(13;15)(q11;p11), t(13;17)(p11;q11),-16,18q+,+3mar	Levitt et al 1978
45,XY,t(9;22),-17,-20,+mar	Rausen et al 1977

46,X,Xp+,del(1)(q21),del(6)(q21),i(17q)	Oshimura et al 1977a
46,XX,t(4;11)(q21;q23)	Arthur et al 1982
46,XX,t(4;11)(q21;q23),del(17)(p11)	
46,XY,+2,t(6;?)(p21;?),+7,-12,-13,14q+,17p+	Oshimura et al 1977a
46,XY,t(12;17),14q+	Minowada et al 1977
46,XY,t(4;11)(q21;q23)	Oshimura et al 1977a
46-49,XY,t(4;11)(q21;q23),+6,+8,+13,+17,t(1;13)(p22;q12), t(7;9)(q11;q34)	
46-47,XX,17p+,+22	Prigogina et al 1979
46-47,XY,del(5)(q15),del(12)(p12),del(17)(p11),+8	Prigogina et al 1979
47,X,-X,del(5)(p23),+t(17;?)(p13;?),+del(18)(p11)	Oshimura et al 1977a
47,XX,+21,i(17q)	Yamada and Furusawa 197(
47,XY,4q-,+17q-	Whang-Peng et al 1976c
49,XX,+14,+17,+21,7q-	Shabtai and Halbrecht 1981b
53,XX,+2,+5,i(7q),+8,+13,+21,+21,+22,+17q+	Oshimura et al 1977a
53,XY,+X,+4,+8,+14,+15,+17,+21	Oshimura et al 1977a
54,XX,+6,+7,+8,+14,+17,+18,+21,+22	Prigogina et al 1979
54,XX,7p+,+10,+10,+14,+14,+18,+18,+21,+21/27,X,-X,-1,-2,-3, -4,-5,-6,-7,7p+,-8,-9,-11,-12,-13,-15,-16,-17,-19,-20, -22	Oshimura et al 1977a
54-55,XX,+5,+12,+14,+16,+17,+18,+21	Prigogina et al 1979
55,XX,+4,+5,+6,+10,+17,+21,+3mar	Humbert et al 1978
57,XX,+3,+4,+8,11p+,+14,+16,+17,+18,+21,+22,+2mar/47,XX,+8	Morse et al 1978

<u>ALL, FAB type L1</u>

26,XX,-1,-2,-3,-4,-5,-6,-7,-8,-9,-10,-11,-12,-13,-14,-15, -16,-17,-19,-20,-22	Hoeltge et al 1982
28,XX,-1,-2,-3,-4,-5,-6,-7,-8,-9,-11,-12,-13,-15,-16,-17, -19,-20,-22/56,XX,+X,+X,+10,+10,+14,+14,+18,+18,+21,+21	Brodeur et al 1981a
34-36,XY,-2,-3,-4,-7,-9,-12,-13,-14,-16,-17,-20,+22	Sandberg et al 1982c
46,XX,t(12;17)(p13;q12)/47,XX,t(12;17),+mar	Kaneko et al 1982e
46,XX,del(9)(p21),t(12;17)(p13;q12)	
47,XY,-16,-17,-18,-20,+5mar	Kaneko et al 1982e
47,XY,-7,-10,-12,+17,del(11)(p11),del(16)(q22), +t(10;?)(q22;?),+t(12;?)(q15;?),+mar	Kaneko et al 1982e
47,XY,del(17)(q11q21),t(1;13)(p34;q14),+mar	Kaneko et al 1981b
56,XY,+X,+4,+6,+10,+15,+17,+18,+21,+21,+mar	Kaneko et al 1982e

<u>ALL, FAB type L2</u>

26,XX,-1,-2,-3,-4,-5,-6,-7,-8,-9,-10,-11,-12,-13,-15,-16, -17,-18,-19,-20,-22	Brodeur et al 1981a
45,X,-X,-1,-14,-17,+21,t(11;12)(p15;p11),+i(17q), +t(1;?)(p34;?)	Kaneko et al 1981b
46,XY,-4,+6,-8,-10,-11,-15,-17,+22,del(6)(q21),+i(17q), +3mar	Kaneko et al 1981b
46,XY,t(16;17)(p13;q23)	Kaneko et al 1982e
47,XY,-11,-17,+3mar	Mitelman 1983
48,XX,t(9;22),+8,+18,/46,XX,t(9;22)	Sandberg et al 1980
48,XX,t(9;22),i(17q),+18,+22q-	
53,XY,t(9;22),+4,+5,+6,+7,+8,+17,+21	Sandberg et al 1980
56,XX,+1,+6,+10,+13,+15,+16,+17,+18,+20,+21	Mitelman 1983
56,XY,t(9;22)(q34;q11),+4,+5,+8,+9,+11,+16,+17,+18,+19,+20	Mitelman 1983

ALL, FAB type L3

Karyotype	Reference
46,XX,-16,-17,-22,+8,11q-,t(8;14),+1-2mar	Borgström et al 1981

Chronic lymphocytic leukemia

Karyotype	Reference
44,XX,-3,14q+,-17	Najfeld et al 1980
46,XX,-3,t(2;3),t(14;18)(q32;q?),2q-,17q+	Finan et al 1978
46,XX,t(6;7)(q?;q?),t(7;13)(q?;q?),t(11;14)(q11;q32),17p+	Schröder et al 1981
46,XY,+12,-17/47,XY,+12,-17,+mar	Robert et al 1982
46,XY,i(17q)	Vahdati et al 1983a
46,XY,i(17q)	Vahdati et al 1983a
46,XY,i(17q)/45,XY,-11,i(17q)	Morita et al 1981
46,XY,t(11;17)(q21;q25)	Fleischman and Prigogina 1977
47,XX,+7/49,XX,+14,+17,+21	Shabtai 1983
47,XY,+2,t(9;13),t(9;19),i(17q)	Finan et al 1978
47,XY,+del(17)(q11)	Shabtai 1983

Prolymphocytic leukemia

Karyotype	Reference
44,XX,-7,-8,t(9;17)(q22;q25),-11	Pittman et al 1982
47,XX,dup(7)(q11q36),+17	Pittman et al 1982
47,XY,16q-,17p-,+mar	Pittman et al 1982

Lymphocytic leukemia, special type

Karyotype	Reference
45,Y,-X,t(Y;14)(q12;q32),-13,-17,del(2)(q33), t(3;13)(q21;q34),del(6)(q16orq22),9p+,9q+,10q+,16q+,18q+	Miyoshi et al 1981a

MONOCLONAL GAMMOPATHIES

Multiple myeloma

Karyotype	Reference
45,XY,-17,17p+	Manolova et al 1979a
46,XX,17p+	Manolova et al 1979a
46,XY,17p+	Manolova et al 1979a
46,XY,17p+	Manolova et al 1979a
47,X,-X,inv(1)(p22q12),t(7;?)(q11;?),+9,t(11;?)(p15;?),-13, -14,t(14;?)(q11;?),+15,17q+	Philip et al 1980

Plasma cell leukemia / Plasmocytoma

Karyotype	Reference
44-45,XY,+10,-17,-21	Wetter et al 1980
45,XY,-5,del(6)(q11),t(6;7)(q?;q?),+9,-12,-13,-16,-17, t(9;22)(q34;q11),+2mar	Karpas et al 1982
78,XY,cx,del(1)(q21),t(1;17)(q12;p13),3q+,3p-,9p+,11p+, 14q+,16p+	Ueshima et al 1983

SOLID TUMORS

UNCLASSIFIED NEOPLASMS

EPITHELIAL NEOPLASMS

Carcinoma, NOS

43,X,-X,t(1;11)(q21;q23),3q+,4q+,5q+,-11,-13,-16,17p+ Atkin and Baker 1982a
44,XX,ins(1;1)(q32;q12q31),i(2q),+3,t(9;11)(q13;p15),-12, -15,17p+,-18 Atkin and Baker 1982a
44-45,XX,t(1;4)(q11;q35),+3,5q-,-14,i(17q),-21 Atkin and Baker 1982a
44-50,XX,+3,dup(1)(q21q32),t(10;11)(q26;q25),t(2;17), t(6;13),t(9;15),del(10)(q24) Riet-Fox et al 1979
45,XY,-4,-7,+8,-9,-12,+13,-17,-18,-19,-20,+21,+3mar Reichmann et al 1981
46,XX,+2,-7,-10,+13,-16,-17,+20,+5q-,15p+,15p+,18p+ Reichmann et al 1981
46,XX,del(2)(p24),+5,+8,t(9;17)(p?;p?),del(16)(q21),-20, -21 Mark 1975b
46,XX,t(3;17)(p24;p13)/46,XX,t(5;17)(q22;p13)/46,XX, t(9;18)(q13;q21)/46,XX,t(12;15)(q13;q22) Stenman et al 1982
46,XY,-4,-5,-11,-13,-17,+20,+21,1p+,9p-,+3mar Reichmann et al 1981
47,XX,-7,+15,+17,+mar Sonta and Sandberg 1978a
48,XX,-7,+15,+17,+mar Sandberg 1977
48,XY,+1,-5,+8,+17 Sonta and Sandberg 1978a
48,XY,+8,-17,-18,+19,+20,t(10;18),t(15;17) Reichmann et al 1981
49,X,-Y,-3,-5,-14,+15,-16,-17,+19,+21,+22,1q+,5q-,6q-,8p-, 9p-,11p-,18p+,20q-,22q-,t(Y;14),+3mar Reichmann et al 1981
49,XX,+8,+17,+20 Sonta and Sandberg 1978a
52,XY,+X,+Y,-2,+7,+8,+9,+10,+13,-17,+i(1q) Reichmann et al 1981
54,XY,+1,+5,+9,+17,+19,+20,+21,+21 Sonta and Sandberg 1978a
55,XY,t(1;17)(p36;q21),+2,+2,+4,+8,+8,-10,+15,+18,+20,+21, +2mar Sonta and Sandberg 1978a
58,XY,del(3)(p14),+4,+7,+8,+10,+11,+12,+14,+15,+17,+18,+19, +21 Kovacs 1978a
63,XX,cx,t(1;4)(q21;q35),7p+,i(17q) Kovacs 1978a
71-75,XX,cx,11p+,17q+,del(1)(p31),del(3)(p13),del(4)(p1?) Riet-Fox et al 1979
73,XY,+1,+7,+11,+13,-16,+17,+18,+19 Sonta and Sandberg 1978a
82,XX,cx,2q+,5q-,9q+,i(17q) Reichmann et al 1981

Carcinoma NOS, metastatic

35,XX,-1,-3,-4,-5,-7,-8,-10,-11,-12,-13,-14,-15,-16,-17, -18,-19,-20,-21,-22,+8mar Pathak et al 1979
46,XX,t(9;22)(q34;q11)"s" Togawa et al 1981
52,XX,+3,+5,+7,+8,+9,+11,12q+,13q-,-14,+16,i(17q)"s"
60-70,XY,i(1)(q11),i(17q),cx Kakati et al 1976b

Small cell carcinoma

40,XY,cx,t(1;16)(q21;q24),t(1;19)(q23;q13),del(3)(p21p24), del(3)(p23q26),del(10)(q22),del(11)(q23), t(5;13)(q13;q32),del(17)(p12),dmin Whang-Peng et al 1982b

Papillary carcinoma

Karyotype	Reference
54,XY,+del(2)(q11),t(2;14)(q11;q32),t(3;11)(p13;p15),+i(5p),+8,+9,+12,+16,+17,+21,+2mar	Pathak et al 1982

Adenocarcinoma

Karyotype	Reference
38,X,-X,cx,-17,-22,del(1)(q25),del(2)(q23),del(6)(q15)	Trent and Salmon 1980
43-46,XY,i(10q),15q+,-17,+19,-21,-22,dmin	Couturier-Turpin et al 1982
44,XX,+5,-9,-13,-13,-15,-17,-18,+22,+2mar	Martin et al 1979
44-50,X,-X,-1,t(1;?)(q21;?),del(2)(q33),4p+,-5,-7,+8,del(10)(q12),t(10;13)(q12;q34),del(17)(q25),-20	Kusyk et al 1982
47,XX,-17,+20,+22	Martin et al 1979
47,XY,-10,-12,+17,+18,+20	Couturier-Turpin et al 1982
48-48,XX,+3,+7,+8,-1,-5,-14,-17,5q+,6q+,del(1)(q42),del(5)(q3?),del(10)(q24),del(14)(q2?)	Riet-Fox et al 1979
68-74,XX,cx,t(13;12),17q+,t(16;16),1q-,i(12p)	Riet-Fox et al 1979
71,XX,cx,1q-,4q+,6q-,9q+,10q-,12q+,17q+,14q+	Wake et al 1981

Adenocarcinoma, metastatic

Karyotype	Reference
49,XX,1q+,+3,4q+,+7,+8,12q+,17q+	Granberg et al 1973
60-65,XX,del(1)(p13),i(1)(q11),i(17q),cx	Kakati et al 1976b
68-72,XY,i(17q),cx	Kakati et al 1976b

Adenomatous polyp

Karyotype	Reference
47,XX,+8/48,XX,+8,+17	Mitelman et al 1974b
50,XX,+1,+5,+17,+19	Sonta and Sandberg 1978a

MESENCHYMAL NEOPLASMS

Liposarcoma

Karyotype	Reference
50,XY,+3,+6,+8,+11,+14,-22,17p+	Sonta et al 1977

Mesothelioma, malignant

Karyotype	Reference
41,XY,del(1)(p13),t(1;3)(p13;q11),t(7;7)(q11;q36),t(9;12)(p24;q13),del(12)(q13),del(13)(q12q14),-14,-15,t(17;?)(p13;?),i(19p),-20,-21,-22	Mark 1978

MELANOMAS

Malignant melanoma

78-82,XY,t(4;9)(q21;q21),del(9)(p13),t(14;14)(p11;q12), del(2)(q24),t(14;?)(q11;?),t(21;?)(q11;?),del(17)(q11) — Kakati et al 1977

Malignant melanoma, metastatic

41-43,XY,del(1)(q21q25),i(1q),t(1;9)(p22;q34), t(11;12)(q11;p11),t(10;12)(q26;q13),t(4;?)(q35;?), del(5)(p13),t(14;?)(p12;?),i(17q),t(18;?)(p11;?), del(6)(q21) — Kakati et al 1977

NEUROGENIC NEOPLASMS

Astrocytoma, grade III, IV

44,XY,-4,-17,-22,1q-,5p-,-11q+,+mar — Yamada et al 1980

Neuroblastoma

46,XX,t(1;?)(p32;?),11p+,17q+,dmin — Gilbert et al 1982
82,XXY,cx,del(1)(p22),del(10)(q22),del(17)(p11) — Brodeur et al 1981b

Retinoblastoma

45,X,-X,+1p+,+i(6p),-16,17q+,-19,22p+ — Kusnetsova et al 1982
46,X,-X,+i(6p),17q+,20q+ — Kusnetsova et al 1982
46,X,-Y,1p+,+1p-,5q+,7p+,10q-,16q-,17q+,21p+ — Kusnetsova et al 1982
46,XX,inv(1),inv(2),inv(3),inv(4),t(6;7),inv(9),t(11;?), 11p-,inv(12),i(17q),-20 — Gardner et al 1982
46,XX,t(1;15)(q11q32;q12),t(1;16)(q21;q11),dup(17)(p11q25) — Gardner et al 1982
46,XY,+4,t(5;13)(q35;q21),t(7;17)(p22;q12),t(1;9)(q22;p24), -10,del(16)(q11) — Gardner et al 1982
47,XX,12p+,12q+,+i(17q),t(4;16)(q35;q24),-16 — Gardner et al 1982
47,XX,i(17q) — Gardner et al 1982
47,XY,del(13)(q14),t(20;?)(q13;?),2q+,i(17q),dmin — Balaban et al 1982
47,XY,dup(1)(q12q44),del(2)(q13),+del(6)(q16), t(14;?)(q32;?),i(17q),+20,-22 — Gardner et al 1982
47,XY,t(X;1)(q28;p11),t(2;?)(q11;?),t(2;15)(p25;q11), t(3;?)(q27;?),4p+,t(5;?)(q35;?),i(17q),+mar — Gardner et al 1982

Meningioma, benign

38,X,-Y,-22,-21,-17,-14,-13,-10,-9,-4,-1p-,+mar — Zankl et al 1975a
41,XY,+del(1)(p?),-5,-10,-17,-20,-22 — Zang and Zankl 1983
43,XY,-1,-8,-17,-19,-22,+2mar/44,XY,-1,-17,-22,+mar — Mark et al 1972a
44,X,-Y,-22/42,X,-Y,-17,-21,-22 — Yamada et al 1980
45,XY,-22/42-45,XY,-8,-15,-17,-20,-22 — Mark et al 1972b
45,XY,-22/44,X,-Y,-22/43,X,-Y,-22,-17 — Yamada et al 1980
46,XX,22q-/43-45,XX,22q-,-15,-17 — Mark et al 1972b

Karyotype	Reference
47,XX,+12,del(22)(q11)/46,XX,-8,+12,-17,del(22)(q11)	Mark et al 1972a
51,XY,+5,6q+,+12,+16,+17,+18,+20,-22/52,XY,+5,+8,+10,+12, +17,+18,+20,-22	Zang and Zankl 1983
52,X,-Y,-22,+20,+19,+17,+15,+11,+9,+7,+5	Zankl et al 1975a

EMBRYONAL AND MISCELLANEOUS NEOPLASMS

Teratoma, malignant

Karyotype	Reference
67,XY,+1,+3,+4,+5,+6,+7,+7,+8,+10,+11,+13,+13,+14,+16,+17, +18,+20,+20,+21,+21	Sonta et al 1977

LYMPHOMAS

1. UNCLASSIFIED AND MISCELLANEOUS LYMPHOMAS

Benign lymphomatous tumor, NOS

Karyotype	Reference
48,XY,t(8;14)(q24;q32),+17,+17q+"s"	Mitelman et al 1979a

Non-Hodgkin's lymphoma, NOS

Karyotype	Reference
46,X,-X,del(1)(q22),t(2;11)(p21;q21),t(9;15)(q11;p12), del(9)(q11),-14,t(14;17)(q23;q23),-16,+mar	Kristoffersson 1983
47-48,XY,t(14;?)(q32;?),16q-,-17,18q-,13q-,+7,+8,8q-	Catovsky et al 1977

Non-Hodgkin's lymphoma, lymphocytic, NOS

Karyotype	Reference
45,XX,i(17q),t(17;21)(q?;q?)	Fleischman and Prigogina 1977
45-46,XY,-1,3q+,del(9)(q22),+14,+17	Reeves 1973
47,XX,14q+,+17p+	Prieto et al 1978a
47-49,XX,+X,-14,i(17q),t(1;17),1p+	Fleischman and Prigogina 1977
48-49,XY,1q-,+3,+del(3)(p?q?),-4,-6,+7,-9,-17,-18,-19, +5mar	Fleischmann et al 1976
51,XX,+5,+9,+11,+del(11)(q?),+del(12)(q?),14q+,i(17q)	Prieto et al 1978a

Non-Hodgkin's lymphoma, histiocytic, NOS

Karyotype	Reference
45,X,-X,del(1)(q22),t(1;17)(q?;q?),del(2)(q33), del(3)(p21p25),+ins(3)(p14p21p25),+5,del(6)(q14),t(1;7)	Mark et al 1978
46,XX,del(1)(p31p35),+3,del(9)(p11),t(10;?)(q26;?), del(11)(q13),-13,t(11;14)(q13;q32),t(17;21), t(9;22)(q11;p13),+r	Mark et al 1977
47-50,XY,t(1;11)(q?;p?),1q-,1p-,3q+,11p+,t(11;17)	Fleischman and Prigogina 1977
50,XY,dup(1)(q12q32),+2,4p+,del(4)(q21),t(8;14)(q24;q32), +17p+,18q+,+20	Brynes et al 1978
89,XX,i(1q),del(3)(p15p23),del(10)(q24),del(12)(p12), i(17q)	Mark 1977

Non-Hodgkin's lymphoma, T-cell, NOS

47,XX,t(2;?),inv(3),4p-,del(5)(q14q32),6p+,6q+,t(7;7),9q+, 10p-,13q-,t(14;14),-15,-15,16q-,17p+,+21,+2mar — Gaeke et al 1981

2. HODGKIN'S LYMPHOMAS

Hodgkin's disease, NOS

47,XY,-2,-5,+16,-17,+3mar — Fleischmann et al 1976

Hodgkin's disease, lymphocytic depletion

45,X,t(X;1)(q28;q11),t(3;5)(q29;q33),del(8)(q22),-9, del(11)(p13),-17 — Hossfeld and Schmidt 1978
48,X,-Y,t(1;?)(q44;?),i(1q),del(1)(q21),del(3)(q21),-10, t(3;11)(q23;q21),+14,t(14;?)(q32;?),+15,-17,+19,+20,-22 — Hossfeld and Schmidt 1978
60,XY,+del(1)(p21),t(1;5)(q44;q33),dup(5)(q12), t(1;5)(p21q23;q13),del(6)(q22),t(9;?)(q12;?), t(3;12)(q11;q24),t(14;?)(p11;?),i(17q) — Hossfeld and Schmidt 1978

Hodgkin's disease, nodular sclerosis

47,XY,t(1;?)(p36;?),+2,t(2;6)(q31;q27),+8,del(12)(p11), t(17;?)(p13;?),-14,-20,+mar — Reeves and Pickup 1980

3A. NON-HODGKIN'S LYMPHOMAS - RAPPAPORT CLASSIFICATION

Lymphocytic, well differentiated, diffuse

48,XY,1q+,3q+,+i(3p),-6,-10,t(14;?)(q12;?),+16,17p-,+18, +mar — Miyamoto et al 1981a

Lymphocytic, poorly differentiated, diffuse

43,XX,-4,-5,-9,i(17q),-18,-21,+2mar — Goh et al 1980
46,XY,t(14;18)(q32;q21),t(10;14)(q24;q32),-1,-17,-22,+3mar — Fukuhara and Rowley 1978
47,XY,-1,+7,-9,+9q+,+9p+,+10,14q+,-17,t(17;18)(p11;q11) — Kristoffersson et al 1981
87,XY,del(1)(q23),del(3)(p14),del(6)(q21),del(7)(p11), del(11)(q21),t(17;?)(p13;?) — Mark et al 1979

Lymphocytic, poorly differentiated, nodular

52,X,-X,+2,-4,+5,+7,+8,-14,-15,-16,+20,+21, +t(1;15)(p11;q11),+t(12;17)(q24;q11),+t(14;?)(q32;?), +mar — Kaneko et al 1982a
92,XY,t(2;9)(q11;p11),t(5;?)(p15;?),t(17;?)(q25;?),17q+, del(1)(p22),del(3)(p21),del(X)(q24),del(11)(q23) — Reeves 1973

Mixed cell, diffuse

Karyotype	Reference
50,XY,+X,-1,+5,+7,+12,-14,-17,del(13)(q12q14), ins(1;1)(q21;q21q32),+t(14;?)(q32;?),+t(1;17)(q21;q25)	Kaneko et al 1982a

Histiocytic, diffuse

Karyotype	Reference
45,X,-Y,t(14;15)(q32;q21orq22),-15,+18,-22,1q+,3p+,5p-,7p-, 9p+,12p+,16q+,17p-,+2mar	Fukuhara and Rowley 1978
45,X,del(X)(q24),del(1)(q21),+del(1)(p11),del(2)(q22), del(3)(q21),del(6)(p11),del(6)(q15),t(7;9)(p22;q13), t(8;10)(p23;q21),del(9)(q13),del(10)(q21), t(2;11)(q22;q23),-13,del(14)(q24),del(15)(q22), del(16)(q22),+17,-19	Mark et al 1979
46,X,-X,-2,-10,+11,-13,-14,-17,del(1)(p22p32), del(3)(p13p21),del(6)(q21),del(9)(p13),+t(2;?)(p23;?), +t(14;?)(q32;?),t(18;?)(q23;?),+t(X;13)(q26;q12),+mar	Kaneko et al 1982a
46,X,-X,-4,-11,-17,+18,-19,del(7)(p15),del(8)(q22),+i(17q), +t(4;19)(p11;q13),t(11;?)(q23;?)	Kaneko et al 1982a
47,X,-X,-7,-17,+19,+i(7q),+i(17q),+mar	Kaneko et al 1982a
49,XY,+del(2)(q31),+3,del(10)(q22),+17,+t(10;17)(q22;p13), -19	Mark et al 1979
50,XX,+3,+3,-4,-6,-9,+12,-17,+18,+i(6p),+3mar	Kaneko et al 1982a
50,XY,t(8;14)(q24;q32),+2,+20,1q+,4p+,+4q+,17p+,18q+	Fukuhara and Rowley 1978
73,XX,cx,del(7)(q32),del(9)(q22),del(10)(p14), t(7;?)(p22;?),t(10;?)(p15;?),t(11;?)(q25;?), t(17;?)(q25;?),t(19;?)(q13;?)	Kaneko et al 1982a
82,XX,del(X)(q13),i(1q),i(1p),del(7)(p11),i(9p), del(14)(q22),t(1;14)(q23;q32),i(15q),i(17q)	Mark et al 1979
89,XY,del(1)(p22),t(5;13)(q15;q34),t(14;20)(q32;q12), t(1;17)(p22;q25),t(1;19)(q21;p13),del(20)(q12)	Mark et al 1979

3B. NON-HODGKIN'S LYMPHOMAS - KIEL CLASSIFICATION

Immunocytoma

Karyotype	Reference
46,XX,1p+,+2,del(17)(p12),17p+	Kristoffersson 1983
46,XX,1p+	
47,XX,1p+,+12,del(17)(p12)	

Centroblastic-centrocytic, diffuse

Karyotype	Reference
48-50,XY,-1,+7,9q+,+9p+,+10,+14q+,-17,-18, +t(17;18)(p11;q11),-19,+20,+22,+mar	Kristoffersson 1983

Centroblastic-centrocytic, follicular

Karyotype	Reference
46-49,XX,1p+,3q-,4q+,+6,+13,-14,14q+,17p+	Kristoffersson 1983
48,XX,+X,t(1;?)(p36;?),del(2)(p21),t(3;?)(q27orq29;?), t(14;?)(q32;?),t(17;18)	Reeves and Pickup 1980

Centroblastic, diffuse

42-45,XY,1p+,-6,-8,14q+,14q-,-16,i(17q),+18,-20,+mar Kristoffersson 1983
46-49,XY,1p+,+2,-6,+del(9)(p21),+12,+13,14q+?,i(17q),+18 Kristoffersson 1983
47,XY,14q+,+22
46-51,XX,+3,del(6)(q21),+del(6)(q21),+7,-8,del(9)(q11),+13, Kristoffersson 1983
-16,-17,+19,-20,-21,+1-5mar

Centroblastic, follicular

47,XY,del(6)(q23),+8,14q+?,i(17q) Kristoffersson 1983

Lymphoblastic, Burkitt's type

45,X?,t(8;14)(q24;q32),t(7;16),t(6;17) Zech et al 1976a
46,X,t(X;1)(p11;q25),del(1)(q25),7p+,17p-,t(2;8)(p11;q24) Slater et al 1982
46,XY,t(8;14)(q24;q32),17p-,17q+ Miyoshi et al 1981c
46-52,X?,t(8;14)(q24;q32),+1,+3,+5,+7,i(17q),+X Zech et al 1976a
47,XX,+12,t(8;14)(q24;q32),del(12)(p11q22),del(17)(q21) Biggar et al 1981

Lymphoblastic, convoluted type

43-46,XY,del(1)(p32),del(2)(p21),-6,t(7;9)(p11;q11),-9,-10, Kristoffersson 1983
t(11;15)(p11;q11),-11,del(12)(q14),-13,-14,i(17q),-17,
+18,del(19)(p11),-22,+2mar
45,XY,t(3;?)(q29;?),t(8;?)(q2?;?),t(12;17)(p13;q11) Kaneko et al 1982d

Immunoblastic

47-48,XX,+del(3)(q11),+17/49,XX,+del(3)(q11),+16,+17 Kristoffersson 1983
45-47,XX,1p+,-14,-15,+16,+17,-19,+mar

3C. NON-HODGKIN'S LYMPHOMAS - LUKES AND COLLINS CLASSIFICATION

Sezary's syndrome

42,XY,-13,-21,-22,1p+,17q+ Nowell et al 1982
44-46,XX,t(2;9),t(2;13),17p+,19p+,-21,-22 Edelson et al 1979
47,XY,-9,-13,-17,-19,+del(9),+t(9;13),+t(9;19),+i(17q) Nowell et al 1982

Follicular center cell, diffuse, large cleaved

45,XY,-6,-9,-17,2q-,3p-,3q-,4q-,5q-,6p+,7p+,7q+,+7q-,9p-, Fukuhara et al 1978b
12q+,13p+,20p-,+mar

Follicular center cell, diffuse, small non-cleaved

44,X,-Y,t(1;14)(p21;q13),5p+,del(6)(q13),del(8)(p21),9p+, Fukuhara et al 1979a
9p-,t(11;21)(q23;q22),13q+,-10,+12,-17p+,der(19),22q+

3D. NON-HODGKIN'S LYMPHOMAS - WORKING FORMULATION

Small lymphocytic

47,XY,+2,2q-,17p-,t(18;22)	Yunis et al 1982
63,XY,cx,t(14;17)	Yunis et al 1982

Follicular, small cleaved cell

44,XY,+der(1),+2,t(2;13),-3,5q-,-12,t(14;18)(q32;q21), i(17q)	Yunis et al 1982
47,XX,t(X;1),ins(1;8),t(2;8),t(14;18)(q32;q21),+17	Yunis et al 1982
47,XY,+X,t(2;3;17),6q-,t(14;18)(q32;q21)	Yunis et al 1982

Follicular, large cell

75,XX,cx,i(6p),t(14;18)(q32;q21),i(17q)	Yunis et al 1982

Diffuse, small cleaved cell

49,XY,dup(1)(q?),8p-,-9,+14,+17,+20,+mar	Yunis et al 1982

Diffuse, large cell

47,XX,+X,6q-,t(14;17)	Yunis et al 1982

Large cell, immunoblastic

45,XY,ins(2;21),t(8;14)(q24;q32),t(9;19),t(17;21),-21	Yunis et al 1982
48,Y,t(X;?),t(1;?),t(3;6),+5,-6,t(8;14)(q24;q32),i(17q), +mar	Yunis et al 1982

Chromosome 18

HEMATOLOGICAL DISORDERS

UNCLASSIFIED LEUKEMIAS

NONLYMPHOCYTIC LEUKEMIAS

Acute nonlymphocytic leukemia (ANLL)

Karyotype	Reference
42,XX,-5,-12,-14,-16,-18,t(12;16)(q11;q22q24)"s"	Oshimura et al 1976b
42,XY,-5,t(12;?)(p11;?),-18,-22,-22	Hustinx and Rutten 1983
43,XX,-1,3p+,3q-,4q+,-5,-5,-7,+9,+9,10p+,+10,+10,10p-,10q-, +11q-,-13,-15,-16,-17,-18,-21"s"	Whang-Peng et al 1979
43,XX,t(1;20),-5,-6,7q-,11q-,-12,17q+,-18,19q+,-21	Mitelman 1983
43-44,XY,-5,-5,-14,-15,-16,-17,-18,-20,+5mar	Mitelman 1983
44-45,XX,-3,-5,-6,-18,-19,t(1;2)(q2?;q11),t(5;?)(q13;?)	Philip et al 1978a
45,XY,-18,-20,t(18;20)(p11;p11)	Philip et al 1978a
46,X?,18q-	Lawler et al 1975
46,XX,del(5)(q12q31),-9,15p+,18p-,-21,+2mar	Kerkhofs et al 1982
46,XY,t(18;?)(q23;?)"s"	Mitelman 1983
46-49,XY,+8,+14,-17,+18,-21"s"	Cavallin-Ståhl et al 197
47,X,Xq-,t(3;5),-5,t(11;?),t(12;17),-18,-21,+6mar	Pedersen-Bjergaard et al 1980
47,XY,+18	Bernard et al 1982a
51,X,-Y,+1,+18,+20,+22,+inv(4),+mar	Zuelzer et al 1976

ANLL, FAB type M1

Karyotype	Reference
41,XY,-1,-2,-4,-5,-12,-15,-17,-17,-18,-22,+5mar	Sessarego et al 1981a
46,XX,-1,inv(1),3q+,5q-,t(13;17),-18,+mar/51,XX,-1,inv(1), 3q+,5q-,t(13;17),-18,+6,+9,+9,+11,+11,+mar	Hagemeijer et al 1981a
46,XX,dup(1)(q21q33),7q+,18p+	Alimena 1983
47,XY,t(9;22)(q34;q11),+18	Sasaki et al 1983

ANLL, FAB type M2

Karyotype	Reference
44,XX,-5,-7,-18,+mar"s"	Rowley et al 1977a
45,,del(X)(q26),-Y,t(8;21)(q22;q22),del(9)(q24), del(18)(p11)	Brodeur et al 1983
45,X,-X,-9,-18,+2mar	Prieto et al 1981
45,XX,5q-,-7,-18,+mar	Teerenhovi et al 1981

46,X,-X,t(8;21)(q22;q22),+18p+	Berger et al 1982b
48-50,XX,+del(6)(q16),+9,-18,+mar"s"	Mitelman 1983
47,XX,del(6)(q16),+mar"s"	
55,XX,+16,+17,+18,+20,+21,+22,+3mar	Prieto et al 1981

ANLL, FAB type M1+M2

37-45,XY,-7,-12,-13,-14,-16,-18,-19,-21,del(4)(q29), del(5)(q24),der(9),der(17),11q+,12p+,t(14;?)	Fitzgerald et al 1983
42-43,XX,cx,t(11;15)(q11;q11),t(9;?)(p?;?),-15,-17,-18,-20, -21	Fitzgerald et al 1983
43,XX,1p+,-4,-5,-6,-7,-13,-17,del(18)(q21),+3mar	Rowley and Potter 1976
43,XX,del(5)(q12q32),-7,-12,-18,t(7;12;21)(q11;q12;q22)	Petit and Van Den Berghe 1979b
44,XX,-3,-4,-5,+9,-17,-18,-21,-22,+5mar	Garson 1980
44,XY,-5,-12,-18,+mar,+dmin	Prigogina et al 1979
44,XY,-5,t(5;12)(q12;q24),i(18)(q11),-19	Mitelman et al 1981
45,X,-Y,+del(5)(q11),del(11)(q21),-14,-16,-18,-20,+2mar,+r	Mitelman et al 1981
45,XX,-5,-11,-13,-18,-21,+4mar,dmin"s"	Sandberg et al 1982a
45,XY,-21,18p+	Yamada and Furusawa 1976
47,XX,+8/51,XX,+X,+8,+18,+20,+21	Testa et al 1979
48,XY,+8,+21/49,XY,+8,+18,+21	Testa et al 1979
45,XY,-18	Testa et al 1979
46-47,XX,del(1)(q32),+8,t(15;17)(q25or26;q22or23),+18	Mitelman et al 1981
46-47,XX,t(1;?)(p36;?),del(18)(q12),del(20)(q11), +1-2mar"s"	Mitelman 1983
46-47,XY,+8,-18,21q+	Mitelman et al 1981
47,X,-Y,t(1;13)(q32p32q11;q11),del(3)(q21), t(8;21)(q22;q22),del(9)(q22),+8,+18	Sakurai et al 1982b
47,XX,del(1)(q?),+18	Benedict et al 1979
47,XX,r(18),+21	Fitzgerald et al 1983

ANLL, FAB type M3

44,XX,-5,-8,-17,-18,+2mar	Gallagher et al 1979
45-51,XY,t(3;12)(p14;q24),t(13;18)(q11;p11),+1,+6,+16,+19, +21,+22	Mitelman et al 1981
46,XY,t(3;18),t(3;12),3p-	Teerenhovi et al 1978

ANLL, FAB type M4

42,XY,-7,i(8)(q11),-16,-17,-18,t(21;21)(q12;q11),+mar	Rowley and Potter 1976
43-46,X,-Y,del(5)(q31),-7,+18,-20	Swolin et al 1981
44,XY,t(3;5)(q29;q13),t(7;12)(q12;q14),t(7;10)(q12;q21),-8, -17,-18,-20	Mitelman et al 1981
44-46,XY,+1,-5,-17,-18,-20,del(6)(q21),-15,t(1;6)(q11;q24), +3mar	Mitelman et al 1981
44-46,XY,+1,t(1;6)(q11;q27),-5,del(6)(q21),-15,-17,-18,-20, +mar	Alimena 1983
45,X,-X,-9,-17,-18,-21,+5mar	Prieto et al 1981
46,XX,t(3;13)(q29;q12),del(5)(q13q31)/46,XX, t(1;18)(q44;q21),t(3;13),del(5)	Alimena 1983
46,XY,2q+,del(18)(p11),del(22)(q11)/47,XY,2q+,+8, del(18)(p11),del(22)(q11)"s"	Alimena et al 1981
45-47,XY,2q+,+8,del(18)(p11),-21,del(22)(q11)"s"	

Karyotype	Reference
46,X,-Y,2q+,+8,del(21)(q21)"s"	
47,XY,t(9;22)(q34;q11),t(16;?)(p13;?),-18,del(20)(p12q12), +22,+mar	Li et al 1981b
53,XX,+15,+18,+21,+21,+mar"c"	Benedict et al 1979

ANLL, FAB type M5

Karyotype	Reference
46,XX,-18,-22,+2r/46,XX,t(2;9)(q23;p22),-3,-22,+2r/49,XX, t(2;9),-22,+4r	Alimena 1983
46,XY,t(6;11)(q26;q22)/52-53,XY,t(6;11)(q26;q22),+der(6), +3,+10,+13,+18,+19,+21,dmin	Hagemeijer et al 1981a
46,XY,t(6;11)(q27;q23)/51-52,XY,t(6;11),+der(6),+3,+19,+19, +21,dmin/53,XY,t(6;11),+der(6),+4,+10,+13,+18,+19,+21	Löwenberg et al 1982
46-51,XX,-3,-18,-22,t(2;9)(q23;p22),+2-4r	Mitelman et al 1981
47,XY,+8,t(1;13)(p31;q11)/48,XY,+8,+18q-	Hagemeijer et al 1981a
51-52,XX,+3,+6,+9,-10,+18,+19,del(1)(p22),del(1)(p22), dup(11)(q11q21),+t(10;?)(p13;?)	Kaneko et al 1982b

ANLL, FAB type M4+M5

Karyotype	Reference
42,XX,-5,-7,-12,-15,-16,-18,+2mar	Li et al 1983
45,XX,-20/45,XX,-18,-20,+mar/46,XX,-15,+mar	Li et al 1983
45,XY,-18,-18,+mar	Li et al 1983
66,XY,+1,+1,+2,+4,+6,+6,+7,+8,+8,+10,+11,+11,+12,+12,+18, +19,+21,+21,+22,+22	Li et al 1983

ANLL, FAB type M6

Karyotype	Reference
43,XX,5q-,-7,-17,-18	Garson 1980
44,X,del(Y)(q?),+Xq-,3p-,3q-,4q-,-6,+7q-,11q+,-15,-18"s"	Bradley et al 1982
44,XY,-3,-18	Yamada and Furusawa 1976
45,XX,-18	Yamada and Furusawa 1976
45,XY,-18	Li et al 1983
45-47,XY,del(6)(q22),-9,+18,+21	Mitelman et al 1981
47,XX,+8/48,XX,+15,+18	Testa et al 1979
46,XX,-8,+18	Li et al 1983
46,XY,t(18;20)(q23;p13)	Yunis et al 1981
72,XY,cx,t(2;15)(q31;q15),del(6)(p22q15),t(8;13)(q23?;q13), t(11;14)(p12;q13),i(16p),t(15;18)(q15;p11)	Douglass and Freeman 1983

Chronic myeloid leukemia, t(9;22)

Karyotype	Reference
26,XX,t(9;22)(q34;q11),-1,-2,-3,-4,-5,-6,-7,-9,-10,-11,-12, -13,-14,-15,-16,-17,-18,-19,-20	Hartley and McBeath 1981
44,XY,t(9;22),-17,-20	Stoll and Oberling 1978
45,XY,t(9;22),-8,-11,-18,+2mar	
47,XY,t(9;22),+7,i(17q)	
44,XY,t(9;22),17p+,-17,-18	Hagemeijer et al 1980b
45,XY,t(9;22)(q34;q11),+8,-18,-19	Alimena et al 1982b
46,X?,t(9;22)/46,X?,t(9;22),18q+	Carbonell et al 1982a
46,XX,t(9;22)(q34;q11)/47,XX,t(9;22),+22q-/49,XX,t(9;22), +8,+17,+22q-/49,XX,t(9;22),+18,+19,+22q-/50,XX,t(9;22), +8,+17,+18,+22q-	Mitelman 1983
46,XX,t(9;22),t(17;18)(p?;q?)	Sadamori et al 1980
46,XY,t(9;22)	Hagemeijer et al 1981b

Karyotype	Reference
46,XY,t(9;22),r(18)	
44,XY,t(9;22),r(18),t(5;17)(q11;p11)	
46,XY,t(9;22)(q34;q11),+t(1;17)(q21;p11)/47,XY,t(9;22), +t(1;17),+18/48,XY,t(9;22),+t(1;17),+8,+19	Mamaeva et al 1983
46,XY,t(9;22)(q34;q11)	Ishihara et al 1982
27,X,-Y,t(9;22),-1,-2,-3,-4,-5,-6,-7,-9,-11,-13,-14,-15, -16,-17,-18,-19,-20,-22	
46,XY,t(9;22),-18,+del(22q)	Hossfeld 1975b
46,XY,t(9;22),i(17q)/47,XY,t(9;22),i(17q),+18	Chrz et al 1983
46,XY,t(9;22),r(18)/45,XY,t(9;22),-5,-17,t(5;17)(q11;p11), r(18)	Hagemeijer et al 1980b
47,XXY,t(9;22)(q34;q11)/53,XXY,+X,t(9;22),+4,+8,+18,+18, +22q-"c"	Mitelman 1983
47,XXY,t(9;22)/53,XXY,+X,t(9;22),+4,+8,+18,+18,+22q-"c"	
47,XXY,t(9;22)"c"	
47,XY,t(9;22),+8,-17,-18,t(17;18)(p11;p11),+del(22q)	Sharp et al 1976
48,XY,t(9;22),+8,+22q-/61,XY,+Y,t(9;22),+4,+5,+6,+8,+8,+8, +8,+15,+17,+18,+19,+20,+22q-,+22q-	Wurster-Hill 1983
50-54,XY,t(9;22),+1,+5,+8,+9,-16,+17,+18,+19,+21,+22q-	Alimena et al 1980
50-55,XY,t(9;22)(q34;q11),+1,+5,+8,+9,-16,+17,+18,+19,+21, +22q-,+mar	Alimena et al 1982b
50-58,XY,+X,t(9;22)(q34;q11),+7,+8,+9,+10,+11,+12, +del(15)(q22),+16,+18,+19,+22q-	Kwan et al 1977
46,XY,t(9;22)	

Chronic myeloid leukemia, aberrant translocation

Karyotype	Reference
45,XY,t(18;22)(q23;q11),+del(1)(q11),6q-,-9,-9	Nowell 1983
46,XX,t(18;22)(p11;q11)	Borgström 1981
46,XY,t(9;11;22)(q34;q13;q11)/52,XY,t(9;11;22),+6,+9q+, +11q-,+18,+19,+del(22q),1q+	Lawler et al 1976

Chronic myeloid leukemia, Ph1 negative

Karyotype	Reference
47,XY,t(11;18)(q23;q12),+del(18)(p11q12)	Bagby Jr et al 1978

MYELOPROLIFERATIVE DISORDERS

Polycythemia vera

Karyotype	Reference
43-45,XX,-9,-18,-22	Wurster-Hill et al 1976
44,X,-X,del(5)(q14q32),-7	Van Den Berghe et al 1979g
44,XX,del(5)(q14q32),-6,-7,-8,t(12;?)(p12;?), t(16;18)(q11;q12)	
47,XY,+9	Van Den Berghe et al 1979g
47,XY,del(1)(q12),t(1;18)(q12;q23),+9,del(20)(q12)/47,XY, +9,del(11)(q21),del(20)(q12)/47,XY,+9,del(20)(q12)	
47,XY,+9/47,XY,del(5)(q14q32),del(8)(q?),+9,del(11)(q21), del(13)(q21)/47,XY,del(5)(q14q32),del(6)(q?),+9, del(11)(q21),del(13)(q21),del(1)(q12),del(20)(q12)	

Myelosclerosis / Myelofibrosis

Karyotype	Reference
44,XY,-5,-7,-11,-12,t(7;12)(p?;q?),t(17;18)(p11;p11), +mar"s"	Nowell and Finan 1978
44,XY,5p-,-7,-10,11q-,i(17q)/45,XY,-7,18q-/45,XY,2p-,2q-, 5p-,6q-,9q-,10q-,-11,+i(17q)	Whang-Peng et al 1978
46,XY,-16,+18	Whang-Peng et al 1978
46,XY,7q-,8q-,10q-,11p-,11q+,12q-,-16,+18	

Chronic myeloproliferative disease, NOS

Karyotype	Reference
47,XX,20q-,18p+	Nowell and Finan 1978

Myeloproliferative disorder, special type

Karyotype	Reference
43,XX,-5,-13,-18/44,XX,-5,t(13;?)(p11;?),-18"s"	Rowley et al 1977a

DYSMYELOPOIETIC SYNDROMES

Preleukemia, NOS

Karyotype	Reference
44,XY,+Y,-3,-5,+del(6)(q21),-8,t(17;?)(p13;?),+18,-20,+22	Mitelman 1983
45,XX,-1,-4,5q-,-7,-12,-18,14q+,+4mar"s"	Pedersen-Bjergaard et al 1981
45,XY,-2,-4,-6,-8,-10,-10,+17,+18,+3mar	Panani et al 1980
46,XX,-7,+8,t(18;22)(p11;q11)"s"	Michalski et al 1982
46,XX,1p+,9q-,18q+,r(20)	Geraedts et al 1980a
46,XY,5q-,-11,+18,18p+,18p+	Swansbury and Lawler 1980
47,XX,+6,t(18;?)(q23;?)	Geraedts et al 1980a

Refractory anemia with excess of blasts

Karyotype	Reference
43,XX,3q-,del(5)(q13q33),7q-,-12,14p+,16q+,-17,-18,18q+	Kerkhofs et al 1982
44-45,XX,del(3)(q?),-7,-16,-18,+1-2mar	Mitelman 1983
45,XY,-5,11q+,-11,-18,+mar	Berger et al 1981a

Dysmyelopoietic syndrome, special type

Karyotype	Reference
44,XY,-18,-21	Yamada and Furusawa 1976

SPECIAL LEUKEMIAS

LYMPHOCYTIC LEUKEMIAS

Acute lymphocytic leukemia (ALL)

Karyotype	Reference
45,XX,t(14;14)(q34;q11),t(13;15)(q11;p11), t(13;17)(p11;q11),-16,18q+,+3mar	Levitt et al 1978
45,XY,-7,-18,+mar	Prigogina et al 1979
46,X?,t(14;18)(q32;q22)	Shabtai 1983
47,X,-X,del(5)(p23),+t(17;?)(p13;?),+del(18)(p11)	Oshimura et al 1977a
54,XX,+6,+7,+8,+14,+17,+18,+21,+22	Prigogina et al 1979
54,XX,7p+,+10,+10,+14,+14,+18,+18,+21,+21/27,X,-X,-1,-2,-3, -4,-5,-6,-7,7p+,-8,-9,-11,-12,-13,-15,-16,-17,-19,-20, -22	Oshimura et al 1977a
54-55,XX,+5,+12,+14,+16,+17,+18,+21	Prigogina et al 1979
56,XX,+3,+6,+12,+13,+14,+15,+18,+21,+r	Prigogina et al 1979
57,XX,+3,+4,+8,11p+,+14,+16,+17,+18,+21,+22,+2mar/47,XX,+8	Morse et al 1978
57,XY,+X,+5,+6,i(7q),+10,+12,+13,+15,+18,+21,+21,+22	Oshimura et al 1977a
61,XX,+X,+1,+2,+3,+4,+5,+6,+8,del(6)(q21),+13,+14,+15,+16, +18,+21,+22	Oshimura et al 1977a

ALL, FAB type L1

Karyotype	Reference
28,XX,-1,-2,-3,-4,-5,-6,-7,-8,-9,-11,-12,-13,-15,-16,-17, -19,-20,-22/56,XX,+X,+X,+10,+10,+14,+14,+18,+18,+21,+21	Brodeur et al 1981a
45,XX,4p-,-18,t(22;?)(q11;?)	Priest et al 1980
47,XY,-16,-17,-18,-20,+5mar	Kaneko et al 1982e
56,XY,+X,+4,+6,+10,+15,+17,+18,+21,+21,+mar	Kaneko et al 1982e
58,XX,+4,+6,+7,+10,+14,+14,+21,+del(18)(q21),+4mar	Kaneko et al 1982e

ALL, FAB type L2

Karyotype	Reference
26,XX,-1,-2,-3,-4,-5,-6,-7,-8,-9,-10,-11,-12,-13,-15,-16, -17,-18,-19,-20,-22	Brodeur et al 1981a
46,XX,-9,-10,t(9;18)(p24;q12),+2mar	Kaneko et al 1981b
46,XX,t(5;8;14)(q11;q24;q32),t(12;22)(p13;q11),-18, +t(18;?)(q23;?)	Kaneko et al 1982e
47,XY,+18	Takeuchi et al 1981
47,XY,+18	Kaneko et al 1982e
46,XY,-2,-5,-6,-7,+8,-9,-10,-11,-12,-16,+18,+t(2;?)(q35;?), +t(11;?)(q21;?),+5mar	
47,XY,-10,-18,+3mar	Takeuchi et al 1981
48,XX,t(9;22),+8,+18,/46,XX,t(9;22)	Sandberg et al 1980
48,XX,t(9;22),i(17q),+18,+22q-	
49,XY,+7,+12,-13,t(6;18)(p25;q21),t(11;14)(q23;q32), +t(9;?)(p24;?),+t(1;13)(q12;p13)	Kaneko et al 1982e
56,XX,+1,+6,+10,+13,+15,+16,+17,+18,+20,+21	Mitelman 1983
56,XY,t(9;22)(q34;q11),+4,+5,+8,+9,+11,+16,+17,+18,+19,+20	Mitelman 1983

ALL, FAB type L3

46,XY,t(8;14)(q24;q32),del(3)(p24),18p+	Berger and Bernheim 1982

Chronic lymphocytic leukemia

41,XX,t(14;14)(q11;q32),t(1;13)(q44;q11),t(15;18)(q11;q23), 10q+,-16,-20	Kaiser McCaw et al 1975
44,XY,-8,del(16)(q21),-18	Pittman et al 1982
46,XX,-3,t(2;3),t(14;18)(q32;q?),2q-,17q+	Finan et al 1978
46,XY,del(6)(p12)/46,XY,t(14;18)(q32;q21)	Robert et al 1982
47,XXY,t(3;8),t(11;14)(q11;q32),+i(18q)"c"	Finan et al 1978
48,XX,+3,+18,t(7;10)(p?;q?)	Hurley et al 1980
49,XY,+12,+18,+19	Finan et al 1978

Prolymphocytic leukemia

44,X,-Y,-11,-12,-13,-13,-14,-15,-15,-18,-18,+8mar	Diamond et al 1980

Lymphocytic leukemia, special type

45,Y,-X,t(Y;14)(q12;q32),-13,-17,del(2)(q33), t(3;13)(q21;q34),del(6)(q16orq22),9p+,9q+,10q+,16q+,18q+	Miyoshi et al 1981a
47,X,-Y,-4,+7,del(1)(q32),i(18q),21q+,i(22q),+2mar	Ueshima et al 1981
47,X,-Y,dup(Y),1q-,4q+,6q-,7q+,+18q+	Miyamoto et al 1982b
48,XX,-6,-11,14q+,+18,+20,+22	Imamura et al 1981

MONOCLONAL GAMMOPATHIES

Multiple myeloma

44,XY,del(3)(q25),t(5;12)(q15;q14),del(7)(q22), del(11)(q24),-14,-18,+mar	Mitelman 1983
46,XY,t(18;?)(q23;?)	Nordenson 1983
52,X,-X,+t(1;?)(q12;?),+3,+7,+8,+9,+11,-13,+18,-21,+2mar	Philip et al 1980
52,XY,-1,+t(1;15)(q12;p1?),+t(1;16)(p36;p13), t(3;5)(q2?;q13),+3,+5,+7,+8,+9,+11,-15,+18,-20	Philip et al 1980

Macroglobulinemia

46,Y,-X,del(1)(q21),del(5)(q23),+del(5)(q23), +t(7;19)(p22;q13),del(8)(p12),+t(1;16)(q12;p13), +t(1;18)(q12;p11),-7,-16,-18,-19,+mar	Ueshima et al 1983

Plasma cell leukemia / Plasmocytoma

46,XY,+t(1;16)(q21;p13),del(6)(q21),-16,-18,+mar	Ueshima et al 1983
47,XY,+t(1;16),del(6)(q21),+t(8;15)(p23;q12),-8,14q+,-15, -16,-18,+3mar	
46,XY,t(8;14)(q24;q32)/47,XY,t(8;14),+7/48,XY,t(8;14),+7, +18	Yamada et al 1983b

SOLID TUMORS

UNCLASSIFIED NEOPLASMS

Malignant neoplasm, NOS

Karyotype	Reference
47,XX,-2,-5,+6,-10,11q+,-13,+18,-21,t(1;?)(q21;?), t(1;13)(q21;p11),t(1;6)(q11;q23)	Kovacs 1978a

EPITHELIAL NEOPLASMS

Carcinoma, NOS

Karyotype	Reference
44,XX,ins(1;1)(q32;q12q31),i(2q),+3,t(9;11)(q13;p15),-12, -15,17p+,-18	Atkin and Baker 1982a
45,XY,-4,-7,+8,-9,-12,+13,-17,-18,-19,-20,+21,+3mar	Reichmann et al 1981
46,XX,+2,-7,-10,+13,-16,-17,+20,+5q-,15p+,15p+,18p+	Reichmann et al 1981
46,XX,t(1;?)(p32;?),-2,3q-,5q-,6q+,-18,-21,+mar	Atkin and Baker 1982a
46,XX,t(3;17)(p24;p13)/46,XX,t(5;17)(q22;p13)/46,XX, t(9;18)(q13;q21)/46,XX,t(12;15)(q13;q22)	Stenman et al 1982
47-48,Y,-X,-5,-8,+13,-14,-16,-18,-22,+7mar	Reichmann et al 1981
48,XY,+8,-17,-18,+19,+20,t(10;18),t(15;17)	Reichmann et al 1981
49,X,-Y,-3,-5,-14,+15,-16,-17,+19,+21,+22,1q+,5q-,6q-,8p-, 9p-,11p-,18p+,20q-,22q-,t(Y;14),+3mar	Reichmann et al 1981
55,XY,t(1;17)(p36;q21),+2,+2,+4,+8,+8,-10,+15,+18,+20,+21, +2mar	Sonta and Sandberg 1978a
58,XY,del(3)(p14),+4,+7,+8,+10,+11,+12,+14,+15,+17,+18,+19, +21	Kovacs 1978a
60-63,XX,cx,t(5;18),t(3;6),t(3;22),del(10)(q24), del(4)(p1?)	Riet-Fox et al 1979
73,XY,+1,+7,+11,+13,-16,+17,+18,+19	Sonta and Sandberg 1978a
80-83,XX,cx,t(2;5;16),t(2;5;18),t(11;13),11q+,t(8;10), del(2)(q24),del(6)(q21),i(14q),dup(9)(q11q12)	Riet-Fox et al 1979

Carcinoma NOS, metastatic

Karyotype	Reference
35,XX,-1,-3,-4,-5,-7,-8,-10,-11,-12,-13,-14,-15,-16,-17, -18,-19,-20,-21,-22,+8mar	Pathak et al 1979
41-42,XX,t(1;18)(p13orp21;q11),3p-,t(3;?)(p25;?), t(8;?)(p2?;?),t(9;?)(q34;?),cx	Bartnitzke and Bullerdiek 1983
45,X?,+16,-18,1p-,3p-,16p-	Wake et al 1981
62-64,XY,del(2)(q34),del(6)(q14),ins(7)(q?),20q+,t(5;18), i(5p)	Ayraud 1975

Large cell carcinoma

Karyotype	Reference
60,XY,del(5)(p11),t(6;10)(q11;p11),t(7;11)(p22;q13), t(14;18)(q11;q11),del(2)(q21),t(6;?)(p11;?)	Pickthall 1976

Small cell carcinoma

Karyotype	Reference
47,XY,+del(1)(p?),+del(3)(q?),+del(3)(p?),-4,+8,-15,+18,-22	Wurster-Hill and Maurer 1978

Adenocarcinoma

Karyotype	Reference
??,XX,cx,del(1)(q42),del(7)(q2?),del(18)(p11),i(8p),i(8q)	Riet-Fox et al 1979
44,XX,+5,-9,-13,-13,-15,-17,-18,+22,+2mar	Martin et al 1979
44,XX,-1,+5,-15,-18,2q+/44,XX,-1,+5,-6,-15,-18,2q+,10p-,+mar	Martin et al 1979
45-47,XX,2q+,-3,-18,+mar	Couturier-Turpin et al 1982
47,XY,-10,-12,+17,+18,+20	Couturier-Turpin et al 1982
50,XY,2q+,+5,+10,+11,-16,-18,+20,+20,+mar	Couturier-Turpin et al 1982
74-75,XY,cx,t(18;19),del(1)(p22),del(6)(q21)	Riet-Fox et al 1979

Adenocarcinoma, metastatic

Karyotype	Reference
40-45,XY,-3,-8,-18,+5,+10,11p+,14q+,del(4)(q32)	Ayraud 1975
56-64,XX,del(X)(q22),del(1)(p12),del(5)(p21),del(7)(q31),t(14;18)(q11;q11),18q+,i(22q)	Berger and Lacour 1974

Adenomatous polyp

Karyotype	Reference
48-52,X,-Y,+7,+8,+13,+14,+16,+18,-20,+12q-,1q+	Reichmann et al 1982b

MESENCHYMAL NEOPLASMS

Leiomyosarcoma

Karyotype	Reference
42,XX,del(1)(p12orp13),del(11)(q13orq14),+7,-9,-13,-14,-15,-18,+19,-22	Mark 1976

Mesothelioma, malignant

Karyotype	Reference
43,XY,t(1;1)(q42;q32p22),t(2;6)(p25;p21),inv(3)(p24q13),t(5;18)(p15;q11),-6,+9,del(13)(q24q14),-14,-18,-22	Mark 1978

MELANOMAS

Malignant melanoma, metastatic

Karyotype	Reference
41,XY,dup(1)(q21q32),t(1;7)(q25;q36),t(7;9)(q11;q11),del(9)(q13),-3,-4,-5,-8,-10,-11,-16,+18,-20,-22	Kakati et al 1977
41-43,XY,del(1)(q21q25),i(1q),t(1;9)(p22;q34),t(11;12)(q11;p11),t(10;12)(q26;q13),t(4;?)(q35;?),del(5)(p13),t(14;?)(p12;?),i(17q),t(18;?)(p11;?),del(6)(q21)	Kakati et al 1977

NEUROGENIC NEOPLASMS

Neuroblastoma

Karyotype	Reference
45,XX,+14,-15,-18,del(1)(p32),dup(1)(p13p31),del(6)(q23), del(11)(q23)	Brodeur et al 1981b

Retinoblastoma

Karyotype	Reference
46,XX,1p+,18p+	Kusnetsova et al 1982
46,XY,dup(1)(q25q44),t(7;10)(q36;q11),-8,-10, t(12;13)(q11;p11),dup(13)(q22q34),+18,+22	Gardner et al 1982
51,XY,del(6)(q21),+8,+11,-12,t(13;18)(p11;q11),14q+,-16, +19,+22,+2mar	Hossfeld 1978

Meningioma, benign

Karyotype	Reference
41,X,-Y,-22,-18,-15,-10,1p-	Zankl et al 1975a
41-42,XY,-1,-5,-8,9q+,-14,-18,-22,+r	Mark 1973b
45,XX,-22/43-44,XX,-8,-18,-22	Mark 1973a
51,XY,+5,6q+,+12,+16,+17,+18,+20,-22/52,XY,+5,+8,+10,+12, +17,+18,+20,-22	Zang and Zankl 1983

EMBRYONAL AND MISCELLANEOUS NEOPLASMS

Teratoma, malignant

Karyotype	Reference
67,XY,+1,+3,+4,+5,+6,+7,+7,+8,+10,+11,+13,+13,+14,+16,+17, +18,+20,+20,+21,+21	Sonta et al 1977

LYMPHOMAS

1. UNCLASSIFIED AND MISCELLANEOUS LYMPHOMAS

Non-Hodgkin's lymphoma, NOS

Karyotype	Reference
47-48,XY,t(14;?)(q32;?),16q-,-17,18q-,13q-,+7,+8,8q-	Catovsky et al 1977
50,XX,+X,t(1;11)(q25;q23),+3,+7,-13,t(14;18)(p?;p?),-15, t(19;?)(q13;?),+3mar	Reeves and Pickup 1980

Non-Hodgkin's lymphoma, lymphocytic, NOS

Karyotype	Reference
47-48,XX,+3,+18	Fleischman and Prigogina 1977
47-50,XX,+3,+18,7q-	Fleischman and Prigogina 1977
48-49,XY,1q-,+3,+del(3)(p?q?),-4,-6,+7,-9,-17,-18,-19, +5mar	Fleischmann et al 1976
50,XY,+3,+3,+12q-,+18,+14q+	Fleischman and Prigogina 1977

Non-Hodgkin's lymphoma, histiocytic, NOS

Karyotype	Reference
45-47,XX,t(10;14)(q21;q32),+8,-18,-21	Mark et al 1977
46,XY,1q+,t(14;?)(q32;?),18p-	Kakati et al 1980
47,XX,t(1;18)(p36;q21),+7,t(8;12)(q11;q15), t(8;12;13)(q11q22;q15;q14),del(13)(q14),t(8;14)(q22;q32)	Mark et al 1978
47,XY,-6,-14,1q-,+1p-,2q+,3q+,9p+,10p-,10p+,12p-,18q+,22p+, t(2;14)(q37?;q13?)	Fukuhara et al 1979b
47,XY,t(11;11)(p23;q13),+18,del(19)	Fleischman and Prigogina 1977
50,XY,dup(1)(q12q32),+2,4p+,del(4)(q21),t(8;14)(q24;q32), +17p+,18q+,+20	Brynes et al 1978

Angioimmunoblastic lymphadenopathy

Karyotype	Reference
48,XY,+3,+18/51,XY,+5,+15,+19,+21,+22	Kaneko et al 1982c

2. HODGKIN'S LYMPHOMAS

Hodgkin's disease, mixed cellularity

Karyotype	Reference
48,XY,+9,+12,t(18;?)(p11;?)	Hossfeld and Schmidt 1978
81,XX,14q+,i(18q),i(18p),del(6)(q21),t(9;15)(q11;p12), del(16)(q13)	Reeves 1973

3A. NON-HODGKIN'S LYMPHOMAS - RAPPAPORT CLASSIFICATION

Lymphocytic, well differentiated, diffuse

Karyotype	Reference
48,XY,1q+,3q+,+i(3p),-6,-10,t(14;?)(q12;?),+16,17p-,+18, +mar	Miyamoto et al 1981a

Lymphocytic, poorly differentiated, diffuse

Karyotype	Reference
43,XX,-4,-5,-9,i(17q),-18,-21,+2mar	Goh et al 1980
45,XY,t(14;18)(q32;q21),+2,-5,-6,-11	Fukuhara and Rowley 1978
46,XY,t(14;18)(q32;q21),t(10;14)(q24;q32),-1,-17,-22,+3mar	Fukuhara and Rowley 1978
47,XY,-1,+7,-9,+9q+,+9p+,+10,14q+,-17,t(17;18)(p11;q11)	Kristoffersson et al 198[illegible]
47,XY,t(14;18)(q32;q21),-9,-10,+12,-16,8p-,13q+,+3mar	Fukuhara and Rowley 1978
48,XX,t(14;18)(q32;q21),1p+,12q+,+mar	Fukuhara and Rowley 1978
48-49,XX,2p-,+7,+i(8q),+12,14q+,15q-,18q-	Kakati et al 1980
50,XX,+3,+4,+10,+18/51,XX,+3,+4,+9,+10,+18	Mark et al 1979

Lymphocytic, poorly differentiated, nodular

Karyotype	Reference
45-47,XX,14q+,del(18)(q21),+del(1)(p34q32)	Reeves 1973
46,XX,del(20)(q11),t(14;18)(q32;q21)	Kaneko et al 1982a
46,XY,12q+/46,XY,t(14;18)(q32;q21)	Kristoffersson et al 198[illegible]
48,XY,t(14;18)(q32;q21),-6,-11,+12,+del(1)(p?),3q+,+2mar	Fukuhara et al 1979a

Mixed cell, diffuse

Karyotype	Reference
53,XY,t(8;14)(q24;q32),t(14;18)(q32;q21),+20,1q-,+1p-,2q-, 4q+,5q-,5p+,6q-,+7q+,10q+,11q+,11q-,13p+,+4mar	Fukuhara and Rowley 1978

Histiocytic, diffuse

Karyotype	Reference
45,X,-Y,t(14;15)(q32;q21orq22),-15,+18,-22,1q+,3p+,5p-,7p-, 9p+,12p+,16q+,17p-,+2mar	Fukuhara and Rowley 1978
45-49,XX,+3,+7p+,-18	Kakati et al 1980
46,X,-X,-11,-18,+t(11;?)(q23;?),t(18;?)(q23;?),+mar	Kaneko et al 1982a
46,X,-X,-2,-10,+11,-13,-14,-17,del(1)(p22p32), del(3)(p13p21),del(6)(q21),del(9)(p13),+t(2;?)(p23;?), +t(14;?)(q32;?),t(18;?)(q23;?),+t(X;13)(q26;q12),+mar	Kaneko et al 1982a
46,X,-X,-4,-11,-17,+18,-19,del(7)(p15),del(8)(q22),+i(17q), +t(4;19)(p11;q13),t(11;?)(q23;?)	Kaneko et al 1982a
47,X,-X,t(4;14)(q?;q32),-18,1q+,2p+,6q-,7q+,9q-,12q+,+3mar	Fukuhara and Rowley 1978
48,XY,-18,+19,+t(18;?)(q23;?),+mar	Kaneko et al 1982a
50,XX,+3,+3,-4,-6,-9,+12,-17,+18,+i(6p),+3mar	Kaneko et al 1982a
50,XX,del(1)(p22),+3,+7,+12,t(1;14)(p22;q32),+18	Mark et al 1979
50,XY,t(8;14)(q24;q32),+2,+20,1q+,4p+,+4q+,17p+,18q+	Fukuhara and Rowley 1978

3B. NON-HODGKIN'S LYMPHOMAS - KIEL CLASSIFICATION

Immunocytoma

Karyotype	Reference
45,X,-Y,t(Y;18)(q11;p11),t(3;3)(q21;q29),5q+	Kristoffersson 1983
49,XX,+3,del(6)(q14),+7,+18/49,XX,+3,del(6)(q14), del(9)(q22),+12,+18	Kristoffersson 1983

Centroblastic-centrocytic, diffuse

Karyotype	Reference
48-50,XY,-1,+7,9q+,+9p+,+10,+14q+,-17,-18, +t(17;18)(p11;q11),-19,+20,+22,+mar	Kristoffersson 1983

Centroblastic-centrocytic, follicular

Karyotype	Reference
45-46,XY,-7,t(14;18)(q32;q21)	Kristoffersson 1983
46,XY,12q+	Kristoffersson 1983
46,XY,t(14;18)(q32;q21)	
46-47,XX,t(8;9)(q13;p24),t(14;18)(q32;q21),+22	Kristoffersson 1983
46-47,XY,del(4)(q28),-10,+12,-18,del(18)(q22)	Kristoffersson 1983
48,XX,+X,t(1;?)(p36;?),del(2)(p21),t(3;?)(q27orq29;?), t(14;?)(q32;?),t(17;18)	Reeves and Pickup 1980
52,XX,+X,+7,+8,del(10)(q24),+12,t(13;?)(q22;?), t(14;?)(q32;?),del(18)(q21),+19,+mar	Reeves and Pickup 1980

Centroblastic, diffuse

42-45,XY,1p+,-6,-8,14q+,14q-,-16,i(17q),+18,-20,+mar Kristoffersson 1983
46-49,XY,1p+,+2,-6,+del(9)(p21),+12,+13,14q+?,i(17q),+18 Kristoffersson 1983
47,XY,14q+,+22

Lymphoblastic, Burkitt's type

??,XY,dup(1)(q12orq21q31),3p-,3q-,+6p-,+8,+14,+15,+18,-21 Douglass et al 1980a
45,X,-Y,t(1;14)(q22;q32),t(8;22)(q24;q11),3q+,6q-,6q+,7q+, 9p-,14q-,18q- Berger et al 1981c
46-47,XY,t(14;?)(q32;?),dup(1)(q21q32),t(2;3)(q13;q29),+5, 6p-,-8,del(9)(q?),+12,+18,+mar Miyoshi et al 1981c

Lymphoblastic, convoluted type

43-46,XY,del(1)(p32),del(2)(p21),-6,t(7;9)(p11;q11),-9,-10, t(11;15)(p11;q11),-11,del(12)(q14),-13,-14,i(17q),-17, +18,del(19)(p11),-22,+2mar Kristoffersson 1983
46,X,-Y,t(1;?)(q21orq23;?),+mar/46,X,-Y,-1,-2,+der(1), +t(2;?)(q33orq35;?),+mar/48,X,-Y,-5,-14,+18,+18,+20, del(9)(p13),t(1;14),+dup(5)(q13q35) Kaneko et al 1982d
47,XY,-7,-9,del(6)(q23?),t(14;18)(p11;q11),+t(9;?)(q34;?), +2mar/47,XY,-7,-9,del(6),del(1)(q32),t(14;18),t(9;?), +t(1;?)(q23orq25;?) Kaneko et al 1982d
47,XY,t(1;2)(p34;p21),+4,t(13;18)(p13;p11)/47,XY,+3 Kaneko et al 1982d

Immunoblastic

78-83,XY,+del(1)(q11),+5,+7,+12,+13q+,+14q+,+14q+,+20,+mar Kristoffersson 1983
80-86,XY,del(1)(q11),+3,+7,+9,+12,+13q+,+14q+,+14q+,+15q+, +18,+20,+mar

3C. NON-HODGKIN'S LYMPHOMAS - LUKES AND COLLINS CLASSIFICATION

Mycosis fungoides

48,XX,del(1)(p22),t(2;8;14)(q37;q24;q24),+5,del(7)(p13), del(9)(q22),t(9;18)(q11;q24),del(10)(p13), del(13)(q12q14),del(5)(q15q22),del(5)(q13),del(5)(q15) Fukuhara et al 1978a
46,XX,t(1;14)(q32;q32)

Follicular center cell, diffuse, large cleaved

47,XY,-6,-14,1q-,+1p-,2q+,3q+,9p+,10p-,10p+,12p-,18q+,19p+, 19q+,+22p+,+mar Fukuhara et al 1978b

3D. NON-HODGKIN'S LYMPHOMAS - WORKING FORMULATION

Small lymphocytic

46,XY,t(10;18)	Yunis et al 1982
47,XY,+2,2q-,17p-,t(18;22)	Yunis et al 1982

Follicular, small cleaved cell

44,XY,+der(1),+2,t(2;13),-3,5q-,-12,t(14;18)(q32;q21), i(17q)	Yunis et al 1982
46,XX,t(2;3),i(6p),t(14;18)(q32;q21)	Yunis et al 1982
46,XY,t(14;18)(q32;q21)	Yunis et al 1982
46,XY,t(14;18)(q32;q21)	Yunis et al 1982
46,XY,t(14;18)(q32;q21)	Yunis et al 1982
46,XY,t(14;18)(q32;q21)	Yunis et al 1982
46,XY,t(1;2),t(4;10),t(14;18)(q32;q21),t(1;22)	Yunis et al 1982
47,XX,t(X;1),ins(1;8),t(2;8),t(14;18)(q32;q21),+17	Yunis et al 1982
47,XY,+7,t(14;18)(q32;q21)	Yunis et al 1982
47,XY,+X,t(2;3;17),6q-,t(14;18)(q32;q21)	Yunis et al 1982
48,X,r(Y),+2,+3,t(14;18)(q32;q21),t(1;22)	Yunis et al 1982
49,XY,+X,+3,ins(11;3),t(14;18)(q32;q21),+18	Yunis et al 1982

Follicular, mixed small cleaved and large cell

50,Y,inv(X),+3,+7,+10,t(12;12),+12,t(14;18)(q32;q21),+18, 18q-,-21	Yunis et al 1982
79,XY,cx,t(1;1),9p-,t(14;18)(q32;q21)	Yunis et al 1982

Follicular, large cell

47,XX,t(2;13),t(14;18)(q32;q21),+21	Yunis et al 1982
75,XX,cx,i(6p),t(14;18)(q32;q21),i(17q)	Yunis et al 1982

Diffuse, large cell

54,XY,dup(1)(q?),inv(5),+3,+3,+9,+13,+18,+18,+20,+mar	Yunis et al 1982

Chromosome 19

HEMATOLOGICAL DISORDERS

UNCLASSIFIED LEUKEMIAS

Acute leukemia, NOS

46,XX,+t(1;19)(q25;q13),der(8),19q+	Prigogina et al 1979
46,XY,t(2;19)(p23;p13)	Prigogina et al 1979
47,XX,+19	Hartley and Sainsbury 1981
47,XX,+19/46,XX,-11,+mar	Hartley and Sainsbury 1981
54,XX,+X,+3,+6,+8,+8,+13,+19,+22	Muir et al 1977

NONLYMPHOCYTIC LEUKEMIAS

Acute nonlymphocytic leukemia (ANLL)

43,XX,2q-,inv(3),-4,del(5)(q?),-7,17p+,-19	Rowley 1976
43,XX,t(1;20),-5,-6,7q-,11q-,-12,17q+,-18,19q+,-21	Mitelman 1983
43-45,XY,-5,5q-,-7,-19"s"	Whang-Peng et al 1979
44-45,XX,-3,-5,-6,-18,-19,t(1;2)(q2?;q11),t(5;?)(q13;?)	Philip et al 1978a
45,XY,-7/50,XY,+8,+15,+19,+20	Berger et al 1981a
46,XX,del(1)(p32),del(5)(q13),+8,t(9;?)(p11;?), t(11;22)(q11;q13),-16,-17,-19,+3mar	Muir et al 1977
46,XY,t(9;22)(q34;q11)	Mitelman 1983
46,XY,t(9;22)(q34;q11)	
46,XY,t(9;22)(q34;q11)	
46,XY,t(9;22)(q34;q11)/47,XY,t(9;22),7p+,+19	
46,XY,t(9;22)	
48,XX,3q+,-5,+9,+16,+19	Nordenson 1983
48,XY,+del(3)(q11),+7,t(6;19)(q15;q13)/50,XY,+del(3)(q11), +7,t(6;19)(q15;q13),+2mar/51,XY,+del(3),+7,+9,t(6;19), +2mar	Mitelman 1983
48-49,XX,-5,del(15)(q?),+19,t(5;?)(q13;?)	Philip et al 1978a
51,XY,+3,+5,-9,+19,+20,+21,t(6;14)(p25;q11),t(9;?)(q22;?)	Philip et al 1978a
52,XX,1p+,+10,+13,+19,+21,+21,+22"c"	Berger et al 1973

ANLL, FAB type M1

Karyotype	Reference
42,XX,-7,-17,-19,-21"s"	Papa et al 1979
45,XX,-10,t(14;19)(p?;p?)	Takeuchi et al 1981
45,XY,t(2;5)(q22orq23;q14q15),6p-,-7,t(6;7)(p21;p22q21), t(7;10)(p21;q21),t(14;19)(q11;q13)"s"	Hagemeijer et al 1981a
46,XX,t(4;19)(p15;p13)	Yunis et al 1981
48,XY,+8,+8,t(11;19)(q23;p12orp13)	Hagemeijer et al 1981a
50,XX,+3,+19,+t(1;6)(q21;p22)	Yunis et al 1981
50,XX,+6,+19,+21"c"	Kaneko et al 1982b

ANLL, FAB type M2

Karyotype	Reference
??,X?,t(8;21),t(Y;19)	Trujillo et al 1979
43,XX,-3,-4,-7,-19,del(5)(q14),17p+	Testa et al 1979
43,XX,-3,-4,-7,-19,del(2)(q31?),del(5)(q14)	
44,X,t(X;8),del(5)(q12q21),t(1;6),t(7;19),-7,-8,t(7;8),-21	Kerkhofs et al 1982
45,X,-X,t(8;21)(q22;q22)/46,X,-X,t(8;21),1p+,7q+,+19/47,X, -X,t(8;21),1p+,+19,+der(21)	Sakurai et al 1982b
46,XX,t(11;19)(q23;p13orq13)	Kaneko et al 1982b
46,XY,+19,-22	Brodeur et al 1983
46,XY,del(1)(p22p32),+19,-21	Hagemeijer et al 1981a
47,XY,del(9)(q22),+19	Kaneko et al 1982b
49,XY,+4,+r(7),+19	Kaneko et al 1982b

ANLL, FAB type M1+M2

Karyotype	Reference
37-45,XY,-7,-12,-13,-14,-16,-18,-19,-21,del(4)(q29), del(5)(q24),der(9),der(17),11q+,12p+,t(14;?)	Fitzgerald et al 1983
44,XY,-5,t(5;12)(q12;q24),i(18)(q11),-19	Mitelman et al 1981
44,XY,-7,-16,-19,-21,-22,+3mar"s"	Sandberg et al 1982a
45,XY,-5,-19,-20,-22,+t(7;12)(q11;q13),+r,+3mar	Fitzgerald et al 1983
45,XY,t(19;22)(p13;q11),t(1;7)(q21;q22),t(7;9),+del(8)(p?), -15	Oshimura and Sandberg 1977
47,XX,+19,t(1;11)	Benedict et al 1979
46,XX,-7,+19	Mitelman et al 1981
46,XX,-7,+19	Alimena et al 1977
47,XX,+8,+10,-19	Mitelman et al 1981
47,XY,+19,t(7;12)(q21;p13),del(4)(q26),del(5)(p12)	Mitelman et al 1981
47,XY,+19	Prigogina et al 1979
48,XX,+t(1;22)(q11;p12),+del(19),-22,+mar	Fitzgerald et al 1983
48,XX,-5,del(5)(q13q31),+8,+t(11;?)(p14;?),-17,-19,+21, +2mar,dmin"s"	Testa et al 1981b
50,XY,+8,+14,+19,+mar	Fitzgerald et al 1983
50-52,XY,+9,+14,+17,+19,+20,+21	Alimena et al 1977
50-52,XY,-5,-7,+9,+14,+17,+19,+20,+21,+mar	Mitelman et al 1981
57,X,-Y,-6,+8,+15,+15,+19,+19,+20,+20,+21,+21,+22,+3mar	Fitzgerald et al 1983

ANLL, FAB type M3

Karyotype	Reference
45,XY,t(15;17),-19	Ferro et al 1982
45-51,XY,t(3;12)(p14;q24),t(13;18)(q11;p11),+1,+6,+16,+19, +21,+22	Mitelman et al 1981
46,XY,t(15;17)(q?;q?),t(3;11)(q?;q?),t(7;19)(q11;p13)	Trujillo and Cork 1981

ANLL, FAB type M4

Karyotype	Reference
46,XY,+19,-22	Bernstein et al 1982b
46,XY,del(19)(p?q?)	Whang-Peng et al 1977
46,XY,t(11;17)(q23;q27),del(11)(q13q23),t(15;19)(q21;p13)	Yunis et al 1981
46,XY,t(11;19)(q23;q13)	Prigogina et al 1979
46,XY,t(5;19),16p+,19p+	Golomb et al 1978c
47,XX,+19	Bernstein et al 1982b
47,XX,+19,20p+/50,XX,+5,+8,+12,+mar	Morse et al 1978
49,XX,del(3)(q25),del(7)(q22),+8,+17,+19	Alimena 1983

ANLL, FAB type M5

Karyotype	Reference
46,XX,ins(10;11)(p11;q23q24)/52,XX,+4,+8,ins(10;11),+12, +16,+19,+20	Kaneko et al 1982b
46,XY,t(6;11)(q26;q22)/52-53,XY,t(6;11)(q26;q22),+der(6), +3,+10,+13,+18,+19,+21,dmin	Hagemeijer et al 1981a
46,XY,t(6;11)(q27;q23)/51-52,XY,t(6;11),+der(6),+3,+19,+19, +21,dmin/53,XY,t(6;11),+der(6),+4,+10,+13,+18,+19,+21	Löwenberg et al 1982
47,XX,+19	Benedict et al 1979
47,XX,+19	Alimena 1983
48,XX,+19,+21	Mitelman 1983
49-50,XY,-21,11q+,19q+,+3-4mar"s"	Prigogina et al 1979
51,XX,+6,+8,+14,+19,+21,t(11;?)(q21orq24;?)	Soukup and Neely 1981
51-52,XX,+3,+6,+9,-10,+18,+19,del(1)(p22),del(1)(p22), dup(11)(q11q21),+t(10;?)(p13;?)	Kaneko et al 1982b
57-58,XX,+1,+3,-4,-5,+6,-8,+11,+12,-17,+19,-20, del(10)(q24),+8-9mar	Testa et al 1979
60-61,XX,Xp+,+1,+2,-4,+6,-7,+19,+20,del(2)(q31), del(10)(q24),+12-13mar	

ANLL, FAB type M5a

Karyotype	Reference
47,XY,+6,t(11;19)(q24;p13)	Berger et al 1982a

ANLL, FAB type M5b

Karyotype	Reference
48,XY,17p+,+19,+mar	Berger et al 1982a

ANLL, FAB type M4+M5

Karyotype	Reference
53,XY,+13,-14,+16,+16,+19,+20,+3mar	Li et al 1983
66,XY,+1,+1,+2,+4,+6,+6,+7,+8,+8,+10,+11,+11,+12,+12,+18, +19,+21,+21,+22,+22	Li et al 1983

ANLL, FAB type M6

Karyotype	Reference
43,XY,-2,4q-,-9,11q-,-15,-19,-20,+2mar/46,XY,-15,-20,+2mar	Li et al 1983
44-46,XY,-5,-17,-19,+mar	Mitelman et al 1981
44-47,XY,+2,-3,-5,-7,+8,-14,-19,+2mar	Mitelman et al 1981
46-47,XX,del(19)(p?q?),11q+,+21"s"	Whang-Peng et al 1977
47,XX,-5,i(17q),-19,+3mar	Najfeld 1976
48,XX,t(1;8)(p11;q11),+8,+19	Morse et al 1979a

Chronic myeloid leukemia, t(9;22)

Karyotype	Reference
26,XX,t(9;22)(q34;q11),-1,-2,-3,-4,-5,-6,-7,-9,-10,-11,-12, -13,-14,-15,-16,-17,-18,-19,-20	Hartley and McBeath 1981
43-46,XX,t(9;22),-5,-7,-10,-17,-19,5p+,14q+,+r(?),+mar	Sadamori et al 1981b
45,X,-Y,t(9;22)	Hartley and McBeath 1981
46,XY,t(9;22)/50,XY,t(9;22),+19,+22q-,+C,+C	
45,XY,t(9;22)(q34;q11),+8,-18,-19	Alimena et al 1982b
45,XY,t(9;22),+del(1)(q12),-3,-7,del(11)(q21),-19,+mar	Lyall and Garson 1978
46,X?,t(9;22),19q+/46,X?,t(9;22),4p+,10q+,19q+	Fleischman et al 1981
46,XX,t(9;22)(q34;q11)/51,XX,t(9;22),+5,+8,+17,+19,+22q-	Misawa 1978
46,XX,t(9;22)(q34;q11)/48,XX,t(9;22),+8,+19	Misawa 1978
46,XX,t(9;22)(q34;q11),5q+,t(8;19),17p+	Alimena et al 1982b
46,XX,t(9;22)(q34;q11)/47,XX,t(9;22),+22q-/49,XX,t(9;22), +8,+17,+22q-/49,XX,t(9;22),+18,+19,+22q-/50,XX,t(9;22), +8,+17,+18,+22q-	Mitelman 1983
46,XX,t(9;22)(q34;q11)/51,XX,t(9;22),+6,+8,+10,+19,+22q-	Mitelman 1983
46,XX,t(9;22),-10,+12,-19,+22q-	Stoll and Oberling 1978
46,XX,t(9;22),19q+	Prigogina et al 1978
46,XX,t(9;22),2p+/48,XX,t(9;22),2p+,+19,+22q-	Sonta and Sandberg 1977b
49,XX,t(9;22),2p+,+19,+22q-,+22q-	
46,XX,t(9;22),2q+,i(17q),19p-	Olah and Rak 1981
46,XX,t(9;22)/47,XX,t(9;22),+22q-/46,XX,t(9;22),-7, +22q-/47,XX,t(9;22),-7,+19,+22q-	Hagemeijer et al 1980b
46,XX,t(9;22)/48,XX,t(9;22),+19,+22q-/50,XX,t(9;22),+8,+12, +19,+22q-/50,XX,t(9;22),+8,+14,+17,+19,+21	Hagemeijer et al 1980b
46,XX,t(9;22)/48,XX,t(9;22),+8,+19	Kohno and Sandberg 1980
46,XY,t(9;22)"s"	Hossfeld 1975b
46,XY,t(9;22)/51,XY,t(9;22),+10,+14,+19,+22q-,+mar"s"	
51,XY,t(9;22),+10,+14,+19,+22q-,+mar"s"	
51,XY,t(9;22),+10,+14,+19,+22q-,+mar"s"	
46,XY,t(9;22)(q34;q11)	Mitelman 1983
46,XY,t(9;22)	
46,XY,t(9;22)	
46,XY,t(9;22)	
46,XY,t(9;22)	
46,XY,t(9;22)/49,XY,t(9;22),+8,+19,+22q-	
46,XY,t(9;22)/47,XY,t(9;22),+22q-/48,XY,t(9;22),+19, +22q-/49,XY,t(9;22),+8,+19,+22q-	
46,XY,t(9;22)(q34;q11)	Mitelman et al 1975a
46,XY,t(9;22)(q34;q11)	
46,XY,t(9;22)(q34;q11)	
46,XY,t(9;22)(q34;q11)	
46,XY,t(9;22)(q34;q11)/49,XY,t(9;22),+8,+19,+del(22)(q11)	

46,XY,t(9;22)(q34;q11)/47,XY,t(9;22),+del(22)(q11)/48,XY,t(9;22),+19,+del(22)(q11)/49,XY,t(9;22),+8,+19,+del(22)(q11)	
46,XY,t(9;22)(q34;q11),-19,3q+,6q-,+17	Greenberg et al 1978
46,XY,t(9;22)(q34;q11)/50,XY,t(9;22),t(1;15)(p36;q22),+8,+10,+19,+22q-	Misawa 1978
46,XY,t(9;22)(q34;q11),+t(1;17)(q21;p11)/47,XY,t(9;22),+t(1;17),+18/48,XY,t(9;22),+t(1;17),+8,+19	Mamaeva et al 1983
46,XY,t(9;22)(q34;q11),i(17q)/47,XY,t(9;22),+8,i(17q)/47,XY,t(9;22),+22q-/48,XY,t(9;22),+8,i(17q),+19	Tomiyasu et al 1982
46,XY,t(9;22)(q34;q11)	Ishihara et al 1982
27,X,-Y,t(9;22),-1,-2,-3,-4,-5,-6,-7,-9,-11,-13,-14,-15,-16,-17,-18,-19,-20,-22	
46,XY,t(9;22),i(17q)/47,XY,t(9;22),+8/48,XY,t(9;22),+8,+19	Hagemeijer et al 1980b
46,XY,t(9;22),i(17q)/47,XY,t(9;22),i(17q),+8/48,XY,t(9;22),i(17q),+8,+8/49,XY,t(9;22),i(17q),+8,+8,+22q-/50,XY,t(9;22),i(17q),+8,+8,+22q-,+19	Hagemeijer et al 1980b
46,XY,t(9;22),t(12;19)(q13;q13)/46,XY,t(9;22),t(1;6)(p36;p21)	Seabright 1983
46,XY,t(9;22)/47,XY,t(9;22),+8	Olinici et al 1978b
46,XY,t(9;22)/47,XY,t(9;22),+8	
46,XY,t(9;22)/48,XY,t(9;22),+19,+22q-	
46,XY,t(9;22)/48,XY,t(9;22),+8,+22q-/49,XY,t(9;22),+8,13q-,+19,+22q-	Hagemeijer et al 1980b
46,XY,t(9;22)/49,XY,t(9;22),+9,+19,+22q-	Gall et al 1976
46,XY,t(9;22)/49,XY,t(9;22),+9,+19,+22q-	
46,XY,t(9;22)/50,XY,t(9;22),+8,+17,+19,+22q-	Hagemeijer et al 1980b
47,X?,t(9;22),+21/48,X?,t(9;22),+21,+21/50-52,X?,t(9;22),+8,+9,+12,+19,+21,+21	Carbonell et al 1982a
47,X?,t(9;22),+22q-/51,X?,t(9;22),+8,+8,+17,+19,+22q-	Fleischman et al 1981
47,X?,t(9;22),+8,i(17q+),19p-	Carbonell et al 1982a
47,X?,t(9;22),r(19),+22q-/48,X?,t(9;22),+17,+22q-/49,X?,t(9;22),+8,+17,+22q-/50,X?,t(9;22),+8,+17,+r(19),+22q-/49,X?,t(9;22),1q+,+8,+17,+22q-	Carbonell et al 1982a
47,XX,t(9;22)(q34;q11),+19	Tomiyasu et al 1982
47,XX,t(9;22),+17/48,XX,t(9;22),+17,+19	Stoll and Oberling 1978
47,XX,t(9;22),i(17q),+8/48,XX,t(9;22),i(17q),+8,+19/49,XX,t(9;22),i(17q),+8,+19,+del(22)(q11)	Prigogina et al 1978
47,XY,t(9;22)(q34;q11),+8/47,XY,t(9;22),+19/47,XY,t(9;22),+21/47,XY,t(9;22),+22q-	Tomiyasu et al 1982
47,XY,t(9;22)(q34;q11),+22q-/47,XY,t(9;22),+19/47,XY,t(9;22),+21	Tomiyasu et al 1982
47,XY,t(9;22),+19	Seabright 1983
47,XY,t(9;22),+8,15q-/48,XY,t(9;22),+8,15q-,+19	Olah and Rak 1981
47,XY,t(9;22),-10,+17,+19,-22,+22q-	Stoll and Oberling 1978
48,X?,t(9;22)(q34;q11),+19,+del(22q)	Lawler et al 1976
48,XX,t(9;22)(q34;q11),t(13;14)(q14;q24orq32),+19,+22q-	Carbone et al 1982a
48,XX,t(9;22)(q34;q11),+8,+19/49,XX,t(9;22),+3,+8,+19/50,XX,t(9;22),+3,+8,+19,+22q-	Mitelman 1983
48,XX,t(9;22)(q34;q11),del(13)(q12q32),+19,+22q-	Tomiyasu et al 1982
48,XX,t(9;22),+19,+21	Hartley and McBeath 1981
48,XX,t(9;22),+8,+19	Wurster-Hill 1983
48,XX,t(9;22),+8,i(17q),+19/46,XX,t(9;22),i(17q)/47,XX,t(9;22),i(17q),+19	Hartley and McBeath 1981
48,XY,t(9;22),+19,+21	Seabright 1983

Karyotype	Reference
48,XY,t(9;22),+8,+22q-/61,XY,+Y,t(9;22),+4,+5,+6,+8,+8,+8, +8,+15,+17,+18,+19,+20,+22q-,+22q-	Wurster-Hill 1983
48,XY,t(9;22),+8,-9,-11,-12,-14,-19,+del(22q),+5mar	Sharp et al 1976
48,XY,t(9;22),+8,i(17q),+19	Lilleyman et al 1978
48-49,XX,t(9;22)(q34;q11),t(2;4),+19,+del(22q)	Sonta and Sandberg 1978b
48-53,X?,t(9;22),+8,+8,+13,+14,+19,+21,+22q-	Carbonell et al 1982a
49,X?,t(9;22),+17,+19,+22q-/50,X?,t(9;22),+8,+17,+19, +22q-/51,X?,t(9;22),+8,+9,+17,+19,+22q-	Carbonell et al 1982a
49,XX,t(9;22),+12,+13,i(17q),+19	Seabright 1983
49,XX,t(9;22),+17,+19,+22q-	Sadamori et al 1980
49,XX,t(9;22),+19,+del(22q),+mar	Hossfeld 1975b
49,XX,t(9;22),+8,+17,+21/49,XX,t(9;22),+8,+17,+17,-19,+21	Olah and Rak 1981
49,XX,t(9;22),+8,+19,+22	Lilleyman et al 1978
49,XX,t(9;22),+8,17q-,+19,+22q-/49,XX,t(9;22),+8,i(17q), +19,+22q-	Olah and Rak 1981
49,XY,t(9;22)(q34;q11),+8,dup(9)(q13q22),+17,+19	Tomiyasu et al 1982
49,XY,t(9;22),+10,-17,+19,+del(22q),+mar	Hossfeld 1975b
49,XY,t(9;22),+17,+19,+22q-/50,XY,t(9;22),+8,+15,+19, +22q-/50,XY,t(9;22),+8,+9,+19,+22q-	Hartley and McBeath 1981
49,XY,t(9;22),+8,+19,+22q-	Seabright 1983
50,XX,t(9;22),+8,+8,+19,+mar	Lilleyman et al 1978
50,XY,+Y,t(9;22)(q34;q11),t(6;8)(q34;q11),+8,i(17q),+19, +22q-	Alimena et al 1982b
50,XY,t(9;22),+C,+19,+21,+22q-	Lessard and Le Prise 1982
50,XY,t(9;22),+Y,+8,+19,+22q-	Sadamori et al 1980
50,XY,t(9;22),+del(22q),+6,+10,+19	Prigogina et al 1978
50-54,XY,t(9;22),+1,+5,+8,+9,-16,+17,+18,+19,+21,+22q-	Alimena et al 1980
50-55,XY,t(9;22)(q34;q11),+1,+5,+8,+9,-16,+17,+18,+19,+21, +22q-,+mar	Alimena et al 1982b
50-58,XY,+X,t(9;22)(q34;q11),+7,+8,+9,+10,+11,+12, +del(15)(q22),+16,+18,+19,+22q-	Kwan et al 1977
46,XY,t(9;22)	
51,XX,t(9;22)(q34;q11),+6,+8,i(17q),+19,+20,+22q-	Alimena et al 1982b
51,XX,t(9;22),+4,+17,+19,+del(22q),+mar	Hossfeld 1975b
51,XX,t(9;22),+8,+9,+14,+19,+22q-	Miyamoto 1980
51,XX,t(9;22),t(8;22)(p12;p11),+del(8)(p12),+19	Lyall and Garson 1978
51,XY,t(9;22)(q34;q11),+6,+8,+19,+21,+22q-	Tomiyasu et al 1982
51,XY,t(9;22),+8,+14,+17,+19,+19	Borgström et al 1982
51-52,XY,t(9;22),+del(22q),+7,+8,+12,+19,+21	Prigogina et al 1978
52,XY,t(9;22)(q34;q11),1p+,+8,+9,i(17q),+19,+20,+21,+22q-	Alimena et al 1982b
52,XY,t(9;22),+8,+14,+15,+19,+21,+22q-	Sadamori et al 1980
52,XY,t(9;22),+8,+9,+12,+19,+21,+22q-	Nordenson 1983
52-54,XX,t(9;22)(q34;q11),+6,+8,+8,+19,+19,+del(22)(q11), +del(22)(q11),+X	Hayata et al 1975b
52-54,XY,t(9;22)(q34;q11),+3,+6,+7,+8,+19,+21,+22q-,+22q-, +1-3mar	Ondreyco et al 1981
53,XX,t(9;22)(q34;q11),+8,+10,+19,+20,+21,+22q-	Greenberg et al 1978
53,XX,t(9;22),+8,+11,+12,+16,+17,+19,+del(22q)/52,XX, t(9;22),+8,+11,+17,+19,+del(22q)	Gahrton et al 1974b
53,XY,+Y,t(9;22)(q34;q11),+3,+6,+8,+9,+10,+12,-15,+22q-/56, XY,+Y,t(9;22),+3,+6,+8,+8,+9,+10,+16,+19,+22q- 56,XY,+Y,t(9;22),+3,+6,+8,+8,+9,+10,+12,+19,+22q-	Stoll and Oberling 1982
53,XY,t(9;22),+6,+9q+,+10,+12,+19,+21,+22q-/54,XY,t(9;22), +6,+8,+9q+,+10,+12,+19,+21,+22q-	Sonta and Sandberg 1977b
53,XY,t(9;22),+8,+9,+19,+22q-,+mar	Lyall and Garson 1978

Karyotype	Reference
53-56,XY,+Y,t(9;22),+del(22q),+6,+10,+19,+8,+11,+21	Prigogina et al 1978
54,XY,t(9;22)(q34;q11),+8,+10,+11,+13,+14,+19,+21,+21	Hayata et al 1975b
55,XX,t(9;22),+3,+8,+12,+14,+15,+19,+20,+22,+22q-	Lilleyman et al 1978
55,XY,t(9;22),+3,+6,+8,+8,+9,+10,+16,+19,+22q-/53,XY, t(9;22),-15,+3,+6,+8,+8,+9,+10,+12,+22q-	Stoll and Oberling 1978
55,XY,t(9;22),+6,+10,+10,+11,+13,+14,+14,+19,+21	Seabright 1983
59,X,-X,t(9;22),+4,+6,+6,t(7;?)(p15;?),+der(7),+8,+8,+12q+, +17,+19,+19,+21,+22,+22q-,+22q-	Miyamoto 1980

Chronic myeloid leukemia, aberrant translocation

Karyotype	Reference
46,X,t(X;9;22)(q27;q34;q12)	Dallapiccola and Alimena 1979
46,X,t(X;9;22)/47-51,X,t(X;9;22),+1,+4,+8,+9,-14,i(17q), -19,+22q-	
46,X?,t(19;22)(p11;q11)	Van Den Berghe 1983
46,X?,t(19;22)(q13;q11)	Lawler et al 1976
46,XX,t(19;22)(q13;q11)	Gahrton et al 1974b
46,XX,t(19;22)(q13;q11)	Petit et al 1978
46,XX,t(2;9;22)(q23;q34;q11)	Pasquali et al 1979b
46,XX,t(2;9;22)/50,XX,t(2;9;22),+6,+10,+19,+der(22)	
46,XX,t(3;22)(p21;q11)/48,XX,t(3;22),+19,+21	Pravtcheva et al 1976
46,XY,t(19;22)(q13;q11),t(3;21)(p11;p11)	Seabright 1983
46,XY,t(1;7;19;22)(q?;q?;p?;q11)	Borgström et al 1982
46,XY,t(1;7;19;22),i(17q)	
46,XY,t(9;10;15;19;22)(q34;q22;q21;q13;q12)	Tomiyasu et al 1982
46,XY,t(9;11;22)(q34;q13;q11)/52,XY,t(9;11;22),+6,+9q+, +11q-,+18,+19,+del(22q),1q+	Lawler et al 1976
47,Y,t(X;22)(p22;q11),+19	Hossfeld and Köhler 1979

Chronic myeloid leukemia, Ph1 negative

Karyotype	Reference
46,XX,t(3;11)(q21;p15)	Mitelman 1983
46,XX,t(3;11)(q21;p15)/49,XX,t(3;11)(q21;p15),+8,+19,+21	
46,XX,t(3;11)/49,XX,t(3;11),+8,+19,+21	
47,XX,+19	Verhest 1983

Eosinophilic leukemia

Karyotype	Reference
49,XY,+10,+15,+19,del(3)(q?)	Huang et al 1979

MYELOPROLIFERATIVE DISORDERS

Polycythemia vera

Karyotype	Reference
45,XX,-19	Kirkland et al 1980
46,XX,t(2;11)(p13;q21)/46,XX,t(1;15)(p1?;q1?), t(2;11)(p13;q21),del(16)(p12)/51,XX,+3,+8,+8,+19,t(1;15), t(2;11),del(16)	Testa et al 1981b

Myelosclerosis / Myelofibrosis

Karyotype	Reference
48,XX,+8,+21/49,XX,+8,+19,+21/50,XX,+8,+19,+19,+21"c"	Ueda et al 1981
51,XX,+1,del(2)(q33),+6,+9,+11,+17,-19,+20q+	Najfeld et al 1978b

Chronic myeloproliferative disease, NOS

Karyotype	Reference
46,XY,-7,+19/47,XY,+19/52,XY,+16,+19,+19,+21,+22,+22	Chrz et al 1983

DYSMYELOPOIETIC SYNDROMES

Preleukemia, NOS

Karyotype	Reference
46,XY,del(7)(p12),t(19;?)(q13;?)"s"	Anderson and Bagby 1982
53,XX,+1,+2,t(4;19),+6,i(17q),+21,+21,+22	Seabright 1983

Chronic myelomonocytic leukemia

Karyotype	Reference
47,XX,+19	Mitelman 1983
47,XX,+19	

Refractory anemia with excess of blasts

Karyotype	Reference
47,XY,+19	Kardon et al 1982a
56-64,XY,+1,-2,+6,+9,+13,+15,+19,+21,10-13mar	Streuli et al 1980

Pancytopenia

Karyotype	Reference
46,XX,-4,-5,-6,-7,-11,-13,-16,+19,+20,t(7;13)(q?;q?), +4mar/43,XX,-2,-3,-5,-7,-16,+19,15q+,+mar	Nowell and Finan 1978

Dysmyelopoietic syndrome, special type

Karyotype	Reference
45,XY,del(2)(q33),-5,del(7)(q22q32),-12,-15, t(12;15)(q11;q26),del(19)(q12),+r"s"	Albain et al 1983

SPECIAL LEUKEMIAS

LYMPHOCYTIC LEUKEMIAS

Acute lymphocytic leukemia (ALL)

Karyotype	Reference
28,XX,-1,-2,-3,-4,-5,-7,-8,-9,-11,-12,-13,-14,-15,-16,-17, -19,-20,-22	Kaneko and Sakurai 1980
45,XX,-5,19q+	Prigogina et al 1979
45,XX,-8,t(13;?)(q?;?),19q+	Prigogina et al 1979
46,X?,19p+	Lawler et al 1975
46,XX,1q-,7q-,t(1;19)	Goldstone et al 1979
55,XX,+1q-,+7q-,+t(1;19),+19,+22,+4mar	
46,XX,del(11)(q13q23),ins(19;11)(p13;q13q23)	Abe et al 1983
46,XX,t(4;11)(q21;q23)/49-50,XX,t(4;11),+8,+13,+19,+4q-	Prigogina et al 1979
46,XY,3q-,6q-,19q+	Dewald et al 1978
47,XY,t(7;12)(q22;p13),+19	Morse et al 1979b
48,XX,+19,+mar	Mitelman 1983
48,XX,+5,-19,+21,+mar	Seabright 1983
48,XY,t(1;22)(q21;p11),+5,+8,19p+	Humbert et al 1978
52-56,XY,+5,+8,+11,+19,+20,+21,del(6)(q15),+mar	Prigogina et al 1979

Karyotype	Reference
54,XX,7p+,+10,+10,+14,+14,+18,+18,+21,+21/27,X,-X,-1,-2,-3,-4,-5,-6,-7,7p+,-8,-9,-11,-12,-13,-15,-16,-17,-19,-20,-22	Oshimura et al 1977a
55,XX,+2,+4,+9,+14,+15,+19,+21,+21	Oshimura et al 1977a

ALL, FAB type L1

Karyotype	Reference
26,XX,-1,-2,-3,-4,-5,-6,-7,-8,-9,-10,-11,-12,-13,-14,-15,-16,-17,-19,-20,-22	Hoeltge et al 1982
28,XX,-1,-2,-3,-4,-5,-6,-7,-8,-9,-11,-12,-13,-15,-16,-17,-19,-20,-22/56,XX,+X,+X,+10,+10,+14,+14,+18,+18,+21,+21	Brodeur et al 1981a
46,XX,t(1;19)(q21;q13),del(3)(q23),del(11)(q21)	Kaneko et al 1982e
46,XY,t(7;19)(q11;q13)	Green et al 1982
47,XY,-19,+2mar	Kaneko et al 1982e
48,XX,+13,+19,/47,XX,+19	Kaneko et al 1982e
53,XX,+1,+5,+14,+15,+19,+20,+22	Kaneko et al 1981b

ALL, FAB type L2

Karyotype	Reference
26,XX,-1,-2,-3,-4,-5,-6,-7,-8,-9,-10,-11,-12,-13,-15,-16,-17,-18,-19,-20,-22	Brodeur et al 1981a
56,XY,t(9;22)(q34;q11),+4,+5,+8,+9,+11,+16,+17,+18,+19,+20	Mitelman 1983

Chronic lymphocytic leukemia

Karyotype	Reference
45-46,XY,-8,9q+,-10,13p+,-14,-15,-19,-20,+mar	Vahdati et al 1983a
46,XY,t(8;14)(q24;q32),19p-	Fleischman and Prigogina 1977
47,XY,+2,t(9;13),t(9;19),i(17q)	Finan et al 1978
48,XY,+12,+19/48,XY,t(1;11)(q31;q22),+12,+19	Vahdati et al 1983a
49,XY,+12,+18,+19	Finan et al 1978

MONOCLONAL GAMMOPATHIES

Multiple myeloma

Karyotype	Reference
45,X,del(X)(q25),t(3;8;22)(q13;q34;q11),-5,-7,-10,-11,del(13)(q13?),+16,19q+,+3mar	Van Den Berghe et al 1979f
46,XY,+t(1;?)(p1?;?),+t(1;12)(q21;q24),del(6)(p11),-8,+9,-12,-13,-16,-19,22q+,+mar	Philip et al 1980
47,X,-Y,+t(1;15)(q12;p11),+3,+11,-15,+19,-22/47,X,-Y,+3,+11,+19,-22	Philip et al 1980
55,XY,+1,+5,+9,+11,-12,+13,+15,del(1)(q32),del(6)(q25),t(9;19)(q13;q13),14q+,16p+,+4mar	Liang et al 1979
45,XY,-2,-2,-3,+5,-8,-10,+14,-15,-16,del(6)(q25),+4mar	
55,XY,+3,+5,+7,+9,+11,+14,+15,+21,+21/53,XY,+3,+5,-8,+9,+11,+14,+15,-19,+21,14q+,+3mar	Liang et al 1979

Macroglobulinemia

Karyotype	Reference
46,Y,-X,del(1)(q21),del(5)(q23),+del(5)(q23),+t(7;19)(p22;q13),del(8)(p12),+t(1;16)(q12;p13),+t(1;18)(q12;p11),-7,-16,-18,-19,+mar	Ueshima et al 1983

SOLID TUMORS

UNCLASSIFIED NEOPLASMS

EPITHELIAL NEOPLASMS

Carcinoma, NOS

Karyotype	Reference
44,XY,+5,-8,-11,-13,-13,-14,-14,-15,+19,-22,-22,t(1;11), 5q-,9q+,+5mar	Reichmann et al 1981
45,XY,-4,-7,+8,-9,-12,+13,-17,-18,-19,-20,+21,+3mar	Reichmann et al 1981
48,XY,+8,-17,-18,+19,+20,t(10;18),t(15;17)	Reichmann et al 1981
49,X,-Y,-3,-5,-14,+15,-16,-17,+19,+21,+22,1q+,5q-,6q-,8p-, 9p-,11p-,18p+,20q-,22q-,t(Y;14),+3mar	Reichmann et al 1981
49,XX,+7,+8,+19	Reichmann et al 1981
54,XY,+1,+5,+9,+17,+19,+20,+21,+21	Sonta and Sandberg 1978a
58,XX,+X,+1,+2,+3,5p-,+6,+7,+9,11q+,-13,-14,-15,+16,+19, +20,+6mar	Kovacs 1978a
58,XY,del(3)(p14),+4,+7,+8,+10,+11,+12,+14,+15,+17,+18,+19, +21	Kovacs 1978a
73,XY,+1,+7,+11,+13,-16,+17,+18,+19	Sonta and Sandberg 1978a

Carcinoma NOS, metastatic

Karyotype	Reference
35,XX,-1,-3,-4,-5,-7,-8,-10,-11,-12,-13,-14,-15,-16,-17, -18,-19,-20,-21,-22,+8mar	Pathak et al 1979
63,XY,t(19;?),cx	Bartnitzke and Bullerdiek 1983

Small cell carcinoma

Karyotype	Reference
40,XY,cx,t(1;16)(q21;q24),t(1;19)(q23;q13),del(3)(p21p24), del(3)(p23q26),del(10)(q22),del(11)(q23), t(5;13)(q13;q32),del(17)(p12),dmin	Whang-Peng et al 1982b
44,XX,cx,del(3)(p14p23),t(3;19)(p13;p11)	Whang-Peng et al 1982b
67,XY,cx,inv(1)(q32p36),dup(1)(q32q44),del(3)(p14q13), del(3)(p14p23),t(7;13)(p11;q11),del(9)(q11),del(11)(p11), t(12;19)(q24;p13)	Whang-Peng et al 1982b

Adenocarcinoma

Karyotype	Reference
43-46,XY,i(10q),15q+,-17,+19,-21,-22,dmin	Couturier-Turpin et al 1982
74-75,XY,cx,t(18;19),del(1)(p22),del(6)(q21)	Riet-Fox et al 1979

Adenomatous polyp

Karyotype	Reference
50,XX,+1,+5,+17,+19	Sonta and Sandberg 1978a

MESENCHYMAL NEOPLASMS

Leiomyosarcoma

Karyotype	Reference
42,XX,del(1)(p12orp13),del(11)(q13orq14),+7,-9,-13,-14,-15, -18,+19,-22	Mark 1976

Mesothelioma, malignant

Karyotype	Reference
41,XY,del(1)(p13),t(1;3)(p13;q11),t(7;7)(q11;q36), t(9;12)(p24;q13),del(12)(q13),del(13)(q12q14),-14,-15, t(17;?)(p13;?),i(19p),-20,-21,-22	Mark 1978

MELANOMAS

Malignant melanoma

Karyotype	Reference
44,X?,+2,+9,-6,-8,-15,-21,9p-,19q+	Wake et al 1981

NEUROGENIC NEOPLASMS

Neuroblastoma

Karyotype	Reference
46,XY,t(1;?)(p32;?),6p+,hsr(6q),7q-,hsr(7p),19q+	Gilbert et al 1982
46,XY,t(1;?)(p?;?),19q+,20q+,-22	Gilbert et al 1982
47,XY,+1,11q-,16q+,19q+	Gilbert et al 1982
47,XY,1p+,4p+,6q+,10q+,12q+,+19	Brodeur et al 1977

Retinoblastoma

Karyotype	Reference
45,X,-X,+1p+,+i(6p),-16,17q+,-19,22p+	Kusnetsova et al 1982
46,XX,+i(6p),12q+,19p+,-22	Kusnetsova et al 1982
46,XY,+del(1)(p13),del(16)(p11),-19	Gardner et al 1982
51,XY,del(6)(q21),+8,+11,-12,t(13;18)(p11;q11),14q+,-16, +19,+22,+2mar	Hossfeld 1978

Meningioma, benign

Karyotype	Reference
43,XX,-1,-6,-7,-11,-22,+2mar/40-45,XX,-1,-6,-7,-9,-11,-19, -22,+2mar	Mark et al 1972b
43,XY,-1,-8,-17,-19,-22,+2mar/44,XY,-1,-17,-22,+mar	Mark et al 1972a
44,XX,-19,-22	Zankl et al 1979
44,XX,-8,-22/40-44,XX,-8,-9,-19,-20,-21,-22	Mark et al 1972a
45,XY,-22/42-44,XY,-8,-16,-19,-22	Mark 1973a
52,X,-Y,-22,+20,+19,+17,+15,+11,+9,+7,+5	Zankl et al 1975a

EMBRYONAL AND MISCELLANEOUS NEOPLASMS

Seminoma

53,XY,+3,+4,+6,+7,+11,+13,+19 — Sonta et al 1977

LYMPHOMAS

1. UNCLASSIFIED AND MISCELLANEOUS LYMPHOMAS

Malignant lymphoma, NOS

46,XX,del(19)(p?q?) — Whang-Peng et al 1977

Non-Hodgkin's lymphoma, NOS

50,XX,+X,t(1;11)(q25;q23),+3,+7,-13,t(14;18)(p?;p?),-15, t(19;?)(q13;?),+3mar — Reeves and Pickup 1980

Non-Hodgkin's lymphoma, lymphocytic, NOS

45,X,-Y,t(11;14)(q13;q32),t(11;19)(q13;q13) — Fleischman and Prigogina 1977

48-49,XY,1q-,+3,+del(3)(p?q?),-4,-6,+7,-9,-17,-18,-19, +5mar — Fleischmann et al 1976

Non-Hodgkin's lymphoma, histiocytic, NOS

47,XY,t(11;11)(p23;q13),+18,del(19) — Fleischman and Prigogina 1977

48,XY,t(3;8)(q22;q11),t(5;10)(q33;q22),del(7)(q21), t(4;9)(q11;p24),t(10;13)(q22;q21),r(12),del(19)(p13), del(22)(q11),+7 — Mark et al 1978

Angioimmunoblastic lymphadenopathy

48,XY,+3,+18/51,XY,+5,+15,+19,+21,+22 — Kaneko et al 1982c

56,XY,+5,+5,+6,+10,+15,+15,+19,+20,+21,+22 — Kaneko et al 1982c

2. HODGKIN'S LYMPHOMAS

Hodgkin's disease, lymphocytic depletion

48,X,-Y,t(1;?)(q44;?),i(1q),del(1)(q21),del(3)(q21),-10, t(3;11)(q23;q21),+14,t(14;?)(q32;?),+15,-17,+19,+20,-22 — Hossfeld and Schmidt 1978

3A. NON-HODGKIN'S LYMPHOMAS - RAPPAPORT CLASSIFICATION

Histiocytic, diffuse

45,X,del(X)(q24),del(1)(q21),+del(1)(p11),del(2)(q22), del(3)(q21),del(6)(p11),del(6)(q15),t(7;9)(p22;q13), t(8;10)(p23;q21),del(9)(q13),del(10)(q21), t(2;11)(q22;q23),-13,del(14)(q24),del(15)(q22), del(16)(q22),+17,-19 — Mark et al 1979

46,X,-X,-4,-11,-17,+18,-19,del(7)(p15),del(8)(q22),+i(17q), +t(4;19)(p11;q13),t(11;?)(q23;?) — Kaneko et al 1982a

47,X,-X,-7,-17,+19,+i(7q),+i(17q),+mar — Kaneko et al 1982a

47,XX,-1,-2,-6,+8,-14,+16,-19,del(11)(q23), t(8;14)(q22;q32),+t(2;6)(q21;p23),del(6)(q21), +t(1;14)(p11;q11),+t(19;?)(q13;?),+t(14;?)(q11;?) — Kaneko et al 1982a

47,XY,-10,-13,+15,-16,+19,+20,del(6)(q21),del(8)(q22), t(2;14)(q21;q32),t(4;4)(q31;q35),+t(10;?)(p11;?) — Kaneko et al 1982a

48,XY,-18,+19,+t(18;?)(q23;?),+mar — Kaneko et al 1982a

48,XY,-6,-8,+21,i(6p),+t(8;?)(q11;?),+t(19;?)(q13;?) — Kaneko et al 1982a

49,XY,+del(2)(q31),+3,del(10)(q22),+17,+t(10;17)(q22;p13), -19 — Mark et al 1979

73,XX,cx,del(7)(q32),del(9)(q22),del(10)(p14), t(7;?)(p22;?),t(10;?)(p15;?),t(11;?)(q25;?), t(17;?)(q25;?),t(19;?)(q13;?) — Kaneko et al 1982a

89,XY,del(1)(p22),t(5;13)(q15;q34),t(14;20)(q32;q12), t(1;17)(p22;q25),t(1;19)(q21;p13),del(20)(q12) — Mark et al 1979

3B. NON-HODGKIN'S LYMPHOMAS - KIEL CLASSIFICATION

Centroblastic-centrocytic, diffuse

48-50,XY,-1,+7,9q+,+9p+,+10,+14q+,-17,-18, +t(17;18)(p11;q11),-19,+20,+22,+mar — Kristoffersson 1983

Centroblastic-centrocytic, follicular

47,XY,t(1;14)(q21;q32),+12,-16,+19,t(14;?)(q32;?) — Reeves and Pickup 1980

52,XX,+X,+7,+8,del(10)(q24),+12,t(13;?)(q22;?), t(14;?)(q32;?),del(18)(q21),+19,+mar — Reeves and Pickup 1980

Centroblastic, diffuse

46-51,XX,+3,del(6)(q21),+del(6)(q21),+7,-8,del(9)(q11),+13, -16,-17,+19,-20,-21,+1-5mar — Kristoffersson 1983

Lymphoblastic, convoluted type

43-46,XY,del(1)(p32),del(2)(p21),-6,t(7;9)(p11;q11),-9,-10, t(11;15)(p11;q11),-11,del(12)(q14),-13,-14,i(17q),-17, +18,del(19)(p11),-22,+2mar — Kristoffersson 1983

Immunoblastic

47-48,XX,+del(3)(q11),+17/49,XX,+del(3)(q11),+16,+17	Kristoffersson 1983
45-47,XX,1p+,-14,-15,+16,+17,-19,+mar	
47-48,XY,+4,del(4)(q21),+5,t(15;?)(q25;?),+19,-21	Reeves and Stathopoulos 1976
49,XX,+5,+19,+22	Kaneko et al 1982a

3C. NON-HODGKIN'S LYMPHOMAS - LUKES AND COLLINS CLASSIFICATION

Sezary's syndrome

44-46,XX,t(2;9),t(2;13),17p+,19p+,-21,-22	Edelson et al 1979
47,XY,-9,-13,-17,-19,+del(9),+t(9;13),+t(9;19),+i(17q)	Nowell et al 1982

Follicular center cell, diffuse, large cleaved

47,XY,-6,-14,1q-,+1p-,2q+,3q+,9p+,10p-,10p+,12p-,18q+,19p+, 19q+,+22p+,+mar	Fukuhara et al 1978b

Follicular center cell, diffuse, small non-cleaved

44,X,-Y,t(1;14)(p21;q13),5p+,del(6)(q13),del(8)(p21),9p+, 9p-,t(11;21)(q23;q22),13q+,-10,+12,-17p+,der(19),22q+	Fukuhara et al 1979a

3D. NON-HODGKIN'S LYMPHOMAS - WORKING FORMULATION

Follicular, large cell

50,XY,1q-,t(2;3),+3,+10,t(1;12),13q-,+15,-19,t(3;19),+21	Yunis et al 1982

Large cell, immunoblastic

45,XY,ins(2;21),t(8;14)(q24;q32),t(9;19),t(17;21),-21	Yunis et al 1982

Chromosome 20

HEMATOLOGICAL DISORDERS

UNCLASSIFIED LEUKEMIAS

NONLYMPHOCYTIC LEUKEMIAS

Acute nonlymphocytic leukemia (ANLL)

Karyotype	Reference
43,XX,t(1;20),-5,-6,7q-,11q-,-12,17q+,-18,19q+,-21	Mitelman 1983
43-44,XY,-5,-5,-14,-15,-16,-17,-18,-20,+5mar	Mitelman 1983
44,X,-Y,-3,+4,-12,+15,-20	Bernard et al 1982a
44,XX,t(8;11)(p23;p13 or p15),-20/46,XX,del(7)(q11), t(1;3)(q2?;q2?),-20,-21,-22,+3mar	Philip et al 1978a
44,XY,-7,-17,-20,-22,+2mar"s"	Pedersen-Bjergaard et al 1982
44,XY,del(5)(q21),del(7)(q11),del(11)(q11), t(7;11;12)(q11;q11;p13),del(13)(q11),-20,-21/45,XY, t(3;12)(q11;p13),del(5)(q21),del(7)(q11),del(13)(q11), -20	Mitelman 1983
45,X,-X,t(8;16)(q22;q24),+20,-21	Trent et al 1983a
45,XY,-18,-20,t(18;20)(p11;p11)	Philip et al 1978a
45,XY,-5,t(20;?)(q13;?)	Hustinx and Rutten 1983
45,XY,-7/50,XY,+8,+15,+19,+20	Berger et al 1981a
45,XY,5q+,-20	Bernard et al 1982a
46,XX,20p-	Bernard et al 1982a
46,XX,del(20)(q11)"s"	Weinfeld et al 1977a
46,XY,20p-	Bernard et al 1982a
46,XY,del(20)(q11),del(7)(q22)	Philip et al 1978a
47,XX,+t(1;16),-16,-16,+13,+20	Prieto et al 1978b
47,XY,t(7;20)(p13;p12),+8"c"	Riccardi et al 1978
51,X,-Y,+1,+18,+20,+22,+inv(4),+mar	Zuelzer et al 1976
51,XY,+3,+5,-9,+19,+20,+21,t(6;14)(p25;q11),t(9;?)(q22;?)	Philip et al 1978a
57,XY,+6,+11,+17,+20,+21"s"	Berger et al 1981a

ANLL, FAB type M1

39,XY,-5,-7,-12,-16,-17,-20,-21	Berger et al 1981a
44,XX,-5,-7,-20,-22,+2mar/44,XX,-3,-11,-11p+,-22	Prieto et al 1981
46,XY,t(9;22)(q34;q11)	Sasaki et al 1983
46,XY,t(9;20;22)(q34;q12;q11)/46,XY,t(9;22)(q34;q11),-6,-7,-7,-8,10q+,+t(7;8),i(8q)	
47,X,-X,+i(Xp)/48,X,-X,+i(Xp),+8/48,X,-X,+i(Xp),+20	Hagemeijer et al 1981a
47-48,XY,+8,t(9;22)(q34;q11),-20,+t(1;20)(q21;q13),t(10;21)(p11;q22),+22q-	Sasaki et al 1983

ANLL, FAB type M2

43,XX,+2,3q+,-5,-20,-22	Berger et al 1981a
43,XY,-3,-4,-7,-7,-8,-8,-17,-20,+5mar	Berger et al 1981a
43,XY,-5,-7,-12,-16,4p+,17p+,20q-,+mar"s"	Rowley et al 1981a
44,XY,-7,17p+,-20,21q+"s"	Pedersen-Bjergaard et al 1981
45-47,XX,-9,-17,-20,del(4)(q21),del(5)(q12q31),t(9;17)(p13;p11),+mar	Testa et al 1979
46,XX,-4,-6,-7,-20,+mar	Golomb et al 1978c
46,XX,del(20)(q11)/48,XX,+9,del(20)(q11),+21	Golomb et al 1978c
55,XX,+16,+17,+18,+20,+21,+22,+3mar	Prieto et al 1981

ANLL, FAB type M1+M2

42-43,XX,cx,t(11;15)(q11;q11),t(9;?)(p?;?),-15,-17,-18,-20,-21	Fitzgerald et al 1983
44,XX,-15,-17,dmin/43,XX,-15,-17,-20,dmin"s"	Sandberg et al 1982a
44,XX,del(3)(p21),del(5)(q12),-7,-20,i(22)(q11)	Mitelman et al 1981
45,X,-Y,+del(5)(q11),del(11)(q21),-14,-16,-18,-20,+2mar,+r	Mitelman et al 1981
45,XX,-20	Li et al 1983
45,XY,-20/47,XY,+21	Li et al 1983
45,XY,-5,-19,-20,-22,+t(7;12)(q11;q13),+r,+3mar	Fitzgerald et al 1983
45-46,XX,+14,-7,-22,+16,-20	Mitelman et al 1981
47,XX,+8/51,XX,+X,+8,+18,+20,+21	Testa et al 1979
46,XX,+14,-22/46,XX,-7,+14,+16,-20,-22	Alimena et al 1977
46,XX,t(8;9)(q23or24;q31),-17,+20	Mitelman et al 1981
46,XY,del(20)(q12)	Fitzgerald et al 1983
46,XY,del(7)(q22),del(7)(p?),del(20)(q12)"s"	Whang-Peng et al 1977
46-47,XX,t(1;?)(p36;?),del(18)(q12),del(20)(q11),+1-2mar"s"	Mitelman 1983
50-52,XY,+9,+14,+17,+19,+20,+21	Alimena et al 1977
50-52,XY,-5,-7,+9,+14,+17,+19,+20,+21,+mar	Mitelman et al 1981
51,XY,+1,+4,+5,+8,+20	Nordenson 1983
57,X,-Y,-6,+8,+15,+15,+19,+19,+20,+20,+21,+21,+22,+3mar	Fitzgerald et al 1983

ANLL, FAB type M3

47,XY,del(12)(p11),del(20)(q11),+9	Mitelman et al 1981

ANLL, FAB type M4

Karyotype	Reference
42,XY,-5,-9,-10,-15,-16,-17,-20,-21,-22,+5mar	Testa et al 1979
43,XY,+1,-5,+6,-9,-10,-15,-16,-17,-20,-21,-22,+6mar	
43-46,X,-Y,del(5)(q31),-7,+18,-20	Swolin et al 1981
44,XY,-7,+20,-21,-22"s"	Berger et al 1981a
44,XY,t(3;5)(q29;q13),t(7;12)(q12;q14),t(7;10)(q12;q21),-8, -17,-18,-20	Mitelman et al 1981
44-46,XY,+1,-5,-17,-18,-20,del(6)(q21),-15,t(1;6)(q11;q24), +3mar	Mitelman et al 1981
44-46,XY,+1,t(1;6)(q11;q27),-5,del(6)(q21),-15,-17,-18,-20, +mar	Alimena 1983
46,XX,4q+,-20,+mar	Takeuchi et al 1981
46,XY,del(20)(q11)"c"	Golomb et al 1978c
46,XY,del(20)(q21)	Alimena 1983
47,XX,+19,20p+/50,XX,+5,+8,+12,+mar	Morse et al 1978
47,XY,t(9;22)(q34;q11),t(16;?)(p13;?),-18,del(20)(p12q12), +22,+mar	Li et al 1981b

ANLL, FAB type M5

Karyotype	Reference
46,XX,ins(10;11)(p11;q23q24)/52,XX,+4,+8,ins(10;11),+12, +16,+19,+20	Kaneko et al 1982b
46,Y,-X,-1,-7,-8,-14,-15,-17,-20,inv(6)(p23q13), del(11)(q23),+t(X;?)(p11;?),+t(1;?)(p24;?), +t(12;?)(q22;?),+t(17;?)(p11;?),+4mar	Kaneko et al 1982b
46-47,XX,-7,+8,+20	Mitelman et al 1981
47,XX,+8/48,XX,+8,+17/48,XX,+8,+20	Garson 1980
57-58,XX,+1,+3,-4,-5,+6,-8,+11,+12,-17,+19,-20, del(10)(q24),+8-9mar	Testa et al 1979
60-61,XX,Xp+,+1,+2,-4,+6,-7,+19,+20,del(2)(q31), del(10)(q24),+12-13mar	

ANLL, FAB type M4+M5

Karyotype	Reference
45,XX,-20/45,XX,-18,-20,+mar/46,XX,-15,+mar	Li et al 1983
53,XY,+13,-14,+16,+16,+19,+20,+3mar	Li et al 1983

ANLL, FAB type M6

Karyotype	Reference
40-45,XY,-4,-6,-7,-10,-11,-20,+1-3mar	Mitelman et al 1981
43,X,-X,del(5)(q14q34),-7,11q-,-12,12p+,t(12;17),t(20;21)	Kerkhofs et al 1982
43,XY,-2,4q-,-9,11q-,-15,-19,-20,+2mar/46,XY,-15,-20,+2mar	Li et al 1983
44,XY,-3,5q-,-7,-D,+20	Li et al 1983
44,XY,3p-,t(2;6),t(5;6),t(17;20)"s"	Sandberg et al 1982a
46,XY,t(18;20)(q23;p13)	Yunis et al 1981

Chronic myeloid leukemia, t(9;22)

Karyotype	Reference
26,XX,t(9;22)(q34;q11),-1,-2,-3,-4,-5,-6,-7,-9,-10,-11,-12, -13,-14,-15,-16,-17,-18,-19,-20	Hartley and McBeath 1981
44,XY,t(9;22),-17,-20	Stoll and Oberling 1978
45,XY,t(9;22),-8,-11,-18,+2mar	
47,XY,t(9;22),+7,i(17q)	

45,X?,t(9;22),-13/45,X?,t(9;22),-13,-20,+mar   Carbonell et al 1982a
45,XY,t(9;22),-20   Stoll and Oberling 1978
45,XY,t(9;22),-7,-8,+17
45,XY,t(9;22),-20,-22,+22q-/46,XY,t(9;22),-20,-22,+8,+22q-   Stoll and Oberling 1978
44,XY,t(9;22),-17,-22,-22,+22q-/46,XY,t(9;22),-20,+22q-/47,
XY,t(9;22),+11,-22,+22q-
45,XY,t(9;22),-9,t(20;21)(p13;q22)   Miyamoto 1980
45,XY,t(9;22),t(17;20)(q12;q11),del(11)(p13)   Fleischman et al 1981
46,X,del(X)(q?),t(9;22),5q-,inv(11),16q+,20q+   Hagemeijer et al 1980b
46,X?,t(9;22),20q+   Carbonell et al 1982a
46,X?,t(9;22),t(3;20)   Carbonell et al 1982a
46,XX,t(9;22)(q34;q11),t(1;20)(q21;q13)   Norman and Boucher 1978
46,XX,t(9;22)(q34;q11)   Mitelman 1983
46,XX,t(9;22)(q34;q11),t(20;?)(p13;?)
46,XX,t(9;22)(q34;q11),t(20;?)(p13;?)
46,XX,t(9;22)(q34;q11)/46,XX,t(9;22),del(6)(q16)/46,XX,   Mitelman 1983
t(9;22),del(20)(q11)/45,XX,t(9;22),del(20)(q11),
t(1;2)(p31;q37)
46,XX,t(9;22)/46,XX,t(9;22),del(6)(q16)/46,XX,t(9;22),
del(20)(q11)
46,XX,t(9;22),-20,+mar   Hossfeld 1975b
46,XX,t(9;22),-7,-15,+20,+i(17q)   Stoll and Oberling 1978
46,XX,t(9;22),i(17q)/47,XX,t(9;22),i(17q),+20/46,XX,   Olah and Rak 1981
t(9;22),-1,+20
46,XX,t(9;22),t(8;13)(p12;q34),+der(8),-20   Lyall and Garson 1978
46,XY,t(9;22)(q34;q11)/47,XY,t(9;22),+22q-/47,XY,t(9;22),   Misawa 1978
-20,t(11;?),+22q-/48,XY,t(9;22),+8,-20,+t(11;?),+22q-
46,XY,t(9;22)(q34;q11)   Ishihara et al 1982
27,X,-Y,t(9;22),-1,-2,-3,-4,-5,-6,-7,-9,-11,-13,-14,-15,
-16,-17,-18,-19,-20,-22
46,XY,t(9;22),i(17q),20q-   Prigogina et al 1978
47,XY,t(9;22)(q34;q11),-7,13q+,20p+,+21,+mar   Alimena et al 1982b
48,XY,t(9;22),+8,+22q-/61,XY,+Y,t(9;22),+4,+5,+6,+8,+8,+8,   Wurster-Hill 1983
+8,+15,+17,+18,+19,+20,+22q-,+22q-
48,XY,t(9;22),-22,+17,+10,+22q-   Stoll and Oberling 1978
48,XY,t(9;22),-22,+17,+20,+22q-
51,XX,t(9;22)(q34;q11),+6,+8,i(17q),+19,+20,+22q-   Alimena et al 1982b
52,XY,t(9;22)(q34;q11),1p+,+8,+9,i(17q),+19,+20,+21,+22q-   Alimena et al 1982b
53,XX,t(9;22)(q34;q11),+8,+10,+19,+20,+21,+22q-   Greenberg et al 1978
55,XX,t(9;22),+3,+8,+12,+14,+15,+19,+20,+22,+22q-   Lilleyman et al 1978

Chronic myeloid leukemia, aberrant translocation

??,X?,t(1;4;20;22)   Geraedts et al 1977
46,XX,t(9;20;22)(q31orq33;q11;q11)   Oshimura et al 1982a
46,XY,t(20;22)(p12;q11)   Lai et al 1982
46,XY,t(9;20;22)(q34;p11p13;q11),i(17q)   Pasquali et al 1982b
46,XY,t(9;20;22)(q34;p11;q11),i(17q)   Pasquali 1983
46,XY,t(9;20;22)(q34;p11p13;q11),i(17q)   Casalone et al 1983

Chronic myeloid leukemia, Ph1 negative

46,XX,del(20)(q11)	Hustinx and Rutten 1983
46,XX,del(20)(q11)	Testa et al 1978a
47,XY,20q-,+21/46,XY,20q-	Hossfeld 1975b

Megakaryocytic leukemia

46,XY,del(20)(q11)	Hustinx and Rutten 1983

MYELOPROLIFERATIVE DISORDERS

Polycythemia vera

??,X?,4q+,+13,15q-,20q-	Vykoupil et al 1980
44-45,XY,-10,-13,-16,-17,-20	Wurster-Hill et al 1976
45,X,-Y	Berger and Bernheim 1979
54-57,X,-Y,+6,+8,+11,+17,+20,+21,t(1;22)(p12;q13)	
45,X,-Y,20q-	Zech et al 1976b
46,X?,20q-	Lessard and Le Prise 1983
46,X?,del(20)(q11)	Reeves et al 1972
46,X?,del(20)(q11)	Reeves et al 1972
46,XX,del(12)(p11)/46,XX,del(12)(p11),del(20)(q11)	Westin 1976
46,XX,20q-	Van Der Weyden et al 1978
46,XX,20q-	Kirkland et al 1980
46,XX,20q-,t(14;20)	Wurster-Hill et al 1976
46,XX,del(20)(q11)	Testa et al 1981b
46,XY,20q-	Van Der Weyden et al 1978
46,XY,20q-	Kirkland et al 1980
46,XY,20q-	Kirkland et al 1980
46,XY,20q-	Kirkland et al 1980
46,XY,20q-	Knuutila et al 1981
46,XY,20q-	Van Der Weyden et al 1978
46,XY,del(20)(q11)	Shiraishi et al 1975
46,XY,del(20)(q11)	Testa et al 1981b
46,XY,del(20)(q11),t(5;20)(q35;q11)	Zech et al 1976b
46,XY,del(5)(q15),del(20)(q11)/46,XY,del(11)(q22)/46,XY, del(13)(q14q32)	Testa et al 1981b
46,XY,t(11;20)(p15;q11)	Berger 1975
46-47,X?,del(20)(q11)	Reeves et al 1972
46-47,X?,del(20)(q11)	Reeves et al 1972
47,XX,+20	Shabtai et al 1978
47,XX,+20	Shabtai et al 1978
47,XX,+9/49,XX,+9,+20,+20,t(4;5),-7,-11,-12,+3mar	Zech et al 1976b
47,XX,t(1;9)(p2?;p2?),t(1;9),del(7)(q2?),del(20)(q11)"s"	Nowell and Finan 1978
47,XY,+9,del(20)(q11)	Mitelman 1983
47,XY,+9	Van Den Berghe et al 1979g
47,XY,del(1)(q12),t(1;18)(q12;q23),+9,del(20)(q12)/47,XY, +9,del(11)(q21),del(20)(q12)/47,XY,+9,del(20)(q12)	
47,XY,+9/47,XY,del(5)(q14q32),del(8)(q?),+9,del(11)(q21), del(13)(q21)/47,XY,del(5)(q14q32),del(6)(q?),+9, del(11)(q21),del(13)(q21),del(1)(q12),del(20)(q12)	

47,XY,+del(1)(p21),del(20)(q11)/47,XY,+9,del(20)(q11)	Westin et al 1976
47,XY,+del(1)(p21),del(5)(q31),del(20)(q11)/47,XY,+9, del(20)(q11)	Swolin et al 1981
47,XY,+del(1)(p?),-20,+mar	Nowell and Finan 1978

Myelosclerosis / Myelofibrosis

46,XX,del(11)(q14),del(20)(q11)	Mitelman 1983
46,XX,del(11)(q14),del(20)(q11)	
46,XX,del(20)(q11)	Findley et al 1979
46,XY,9q+,13q-,20q-/47,XY,-7,+2mar	Chrz et al 1983
46,XY,del(20)(q11)	Fleischman et al 1979
47,XXY,del(20)(q11)"c"	Testa et al 1978a
51,XX,+1,del(2)(q33),+6,+9,+11,+17,-19,+20q+	Najfeld et al 1978b

Chronic myeloproliferative disease, NOS

46,XY,20q-	Lessard and Le Prise 1983
46,XY,del(20)(q11)"s"	Mitelman 1983
46,XY,t(4;12)(q35;q11),i(17q),del(20)(q11)	Berger et al 1975
47,XX,20q-,18p+	Nowell and Finan 1978

Myeloproliferative disorder, special type

46,XX,del(20)(q11)"s"	Testa et al 1978a

DYSMYELOPOIETIC SYNDROMES

Preleukemia, NOS

44,XY,+Y,-3,-5,+del(6)(q21),-8,t(17;?)(p13;?),+18,-20,+22	Mitelman 1983
44,XY,t(3;17),-5,-20,+mar"s"	Pedersen-Bjergaard et al 1981
45,XY,-7,del(20)(q11)	Ruutu et al 1977b
45,XY,t(20;21)(p13;q11),-21"s"	Anderson and Bagby 1982
46,XX,1p+,9q-,18q+,r(20)	Geraedts et al 1980a
46,XX,t(1;16)(q11;p11),t(1;20)(p?;p?)	Anderson and Bagby 1982
46,XY,del(12)(p11)/46,XY,del(12)(p11),del(20)(q11)	Mitelman 1983
46,XY,del(12)(p11)/46,XY,del(12)(p11),del(20)(q11)	
46,XY,del(12)(p11),del(20)(q11)	
46,XY,del(20)(q11)	Seabright 1983
46,XY,del(20)(q11)	Seabright 1983

Chronic monocytic leukemia

47,XY,+20	Nordenson 1983

Sideroblastic anemia, NOS

46,XY,del(20)(q11)	Cohen et al 1974

Refractory anemia without excess of blasts

Karyotype	Reference
44,XX,5q-,-7,+8,-12,-13,-17,-20,+2mar	Teerenhovi et al 1981

Refractory anemia with excess of blasts

Karyotype	Reference
44,XX,5q-,-13,-17,-20,-22,+2mar	Teerenhovi et al 1981
44-47,XY,-3,-5,-7,-12,-16,-20,del(1)(q32),3p+,del(11)(q22), 3-6mar,+r	Streuli et al 1980
45,XX,-9,-17,-20,del(4)(q21),del(5)(q12q31), +t(9;17)(p13?;p11)	Streuli et al 1980
46,XY,del(5)(q12q31),-7,20q+,+22	Kerkhofs et al 1982

Thrombocytopenia

Karyotype	Reference
46,XX,20q-	Tricot et al 1982

Pancytopenia

Karyotype	Reference
46,XX,-4,-5,-6,-7,-11,-13,-16,+19,+20,t(7;13)(q?;q?), +4mar/43,XX,-2,-3,-5,-7,-16,+19,15q+,+mar	Nowell and Finan 1978

SPECIAL LEUKEMIAS

LYMPHOCYTIC LEUKEMIAS

Acute lymphocytic leukemia (ALL)

Karyotype	Reference
28,XX,-1,-2,-3,-4,-5,-7,-8,-9,-11,-12,-13,-14,-15,-16,-17, -19,-20,-22	Kaneko and Sakurai 1980
32,XY,-2,-3,-4,-5,-7,-8,-11,-12,-13,-14,-15,-16,-17,-20	Shabtai and Halbrecht 1981b
45,XY,-20	Prigogina et al 1979
45,XY,t(9;22),-17,-20,+mar	Rausen et al 1977
46,X,-Y,+1,-7,+20,t(1;22)(q24;q12)	Vercherat et al 1980
46,XX,2q+,4q-,12p+,13q+,14q+,-20,+21	Whang-Peng et al 1976c
48,XY,t(4;12)(q?;q?),t(6;15)(q?;q?),t(9;22)(q34;q11),11q-, +20,+21	Nordenson 1983
52-56,XY,+5,+8,+11,+19,+20,+21,del(6)(q15),+mar	Prigogina et al 1979
53-55,XY,+4,+5,+6,+20,+22	Lessard and Le Prise 1983
54,XX,7p+,+10,+10,+14,+14,+18,+18,+21,+21/27,X,-X,-1,-2,-3, -4,-5,-6,-7,7p+,-8,-9,-11,-12,-13,-15,-16,-17,-19,-20, -22	Oshimura et al 1977a

ALL, FAB type L1

Karyotype	Reference
26,XX,-1,-2,-3,-4,-5,-6,-7,-8,-9,-10,-11,-12,-13,-14,-15, -16,-17,-19,-20,-22	Hoeltge et al 1982
28,XX,-1,-2,-3,-4,-5,-6,-7,-8,-9,-11,-12,-13,-15,-16,-17, -19,-20,-22/56,XX,+X,+X,+10,+10,+14,+14,+18,+18,+21,+21	Brodeur et al 1981a
34-36,XY,-2,-3,-4,-7,-9,-12,-13,-14,-16,-17,-20,+22	Sandberg et al 1982c
46,XY,del(9)(p22),del(20)(p12)	Kaneko et al 1982e

47,XY,-16,-17,-18,-20,+5mar	Kaneko et al 1982e
52-56,XY,t(9;22),+4,+6,+9,+10,+14,+15,+20,+21,+22q-	Sandberg et al 1980
53,XX,+1,+5,+14,+15,+19,+20,+22	Kaneko et al 1981b

ALL, FAB type L2

26,XX,-1,-2,-3,-4,-5,-6,-7,-8,-9,-10,-11,-12,-13,-15,-16, -17,-18,-19,-20,-22	Brodeur et al 1981a
46,XX,t(9;22),-20,+mar	Borgström et al 1981
56,XX,+1,+6,+10,+13,+15,+16,+17,+18,+20,+21	Mitelman 1983
56,XY,t(9;22)(q34;q11),+4,+5,+8,+9,+11,+16,+17,+18,+19,+20	Mitelman 1983

Chronic lymphocytic leukemia

41,XX,t(14;14)(q11;q32),t(1;13)(q44;q11),t(15;18)(q11;q23), 10q+,-16,-20	Kaiser McCaw et al 1975
44,XX,6q-,i(8q),12p-,-14,t(14;14)(q31;q12),-20,20p+	Sparkes et al 1980
45-46,XY,-8,9q+,-10,13p+,-14,-15,-19,-20,+mar	Vahdati et al 1983a
46,XY,+12,-20/45,XY,-20	Robert et al 1982
46,XY,t(6;20)(q?;q?)	Schröder et al 1981
47,XY,+3,+12,-20,/47,XY,+9,+12,-20/45,XY,-20/46,XY,+12,-20	Morita et al 1981

Lymphocytic leukemia, special type

48,XX,-6,-11,14q+,+18,+20,+22	Imamura et al 1981

MONOCLONAL GAMMOPATHIES

Multiple myeloma

45,XY,del(1),+del(1),-10,-11,-11,+t(11;?),-13,-14,-20, +3mar	Philip et al 1980
52,XY,-1,+t(1;15)(q12;p1?),+t(1;16)(p36;p13), t(3;5)(q2?;q13),+3,+5,+7,+8,+9,+11,-15,+18,-20	Philip et al 1980
53,XY,+3,+5,-7,-8,+9,+11,-14,+20,del(1)(p34),del(6)(q25), +5mar	Liang et al 1979

SOLID TUMORS

UNCLASSIFIED NEOPLASMS

EPITHELIAL NEOPLASMS

Carcinoma, NOS

Karyotype	Reference
45,XY,-4,-7,+8,-9,-12,+13,-17,-18,-19,-20,+21,+3mar	Reichmann et al 1981
46,XX,+2,-7,-10,+13,-16,-17,+20,+5q-,15p+,15p+,18p+	Reichmann et al 1981
46,XX,del(2)(p24),+5,+8,t(9;17)(p?;p?),del(16)(q21),-20,-21	Mark 1975b
46,XY,-4,-5,-11,-13,-17,+20,+21,1p+,9p-,+3mar	Reichmann et al 1981
48,XX,+3,t(1;20)(p13;p13),+i(1q)	Kovacs 1978a
48,XY,+8,-17,-18,+19,+20,t(10;18),t(15;17)	Reichmann et al 1981
49,X,-Y,-3,-5,-14,+15,-16,-17,+19,+21,+22,1q+,5q-,6q-,8p-,9p-,11p-,18p+,20q-,22q-,t(Y;14),+3mar	Reichmann et al 1981
49,XX,+8,+17,+20	Sonta and Sandberg 1978a
54,XY,+1,+5,+9,+17,+19,+20,+21,+21	Sonta and Sandberg 1978a
55,XY,t(1;17)(p36;q21),+2,+2,+4,+8,+8,-10,+15,+18,+20,+21,+2mar	Sonta and Sandberg 1978a
58,XX,+X,+1,+2,+3,5p-,+6,+7,+9,11q+,-13,-14,-15,+16,+19,+20,+6mar	Kovacs 1978a
63,XX,-1,+3,+5,+6,+7,+8,+10,+11,+12,-13,-15,+20,+21,+22,+9mar	Sandberg 1977

Carcinoma NOS, metastatic

Karyotype	Reference
35,XX,-1,-3,-4,-5,-7,-8,-10,-11,-12,-13,-14,-15,-16,-17,-18,-19,-20,-21,-22,+8mar	Pathak et al 1979
62-64,XY,del(2)(q34),del(6)(q14),ins(7)(q?),20q+,t(5;18),i(5p)	Ayraud 1975

Small cell carcinoma

Karyotype	Reference
80,XX,cx,del(1)(q32),del(3)(p14p21),t(3;4)(p23q11;q11),t(5;20)(q11;q13),dmin	Whang-Peng et al 1982b

Papillary carcinoma

Karyotype	Reference
42-54,X,-X,-14,-16,-16,+1,+3,+12,+20,1p-,6q-	Wake et al 1981
50,X,-X,+1,+3,+12,+20,-14,-16,1p-,6q-	Wake et al 1980

Adenoma

Karyotype	Reference
46,XX,t(1;3)(p21;p21),t(1;13)(p36;q14),t(3;8)(p21;q12),t(1;5;20)(p21;p14q12;p13),t(5;8)(p14;q12),t(X;9)(q27;q12)	Mark et al 1982a

Adenocarcinoma

Karyotype	Reference
38,X,-X,-8,-14,-20,+3-5mar	Trent and Salmon 1981
44-50,X,-X,-1,t(1;?)(q21;?),del(2)(q33),4p+,-5,-7,+8, del(10)(q12),t(10;13)(q12;q34),del(17)(q25),-20	Kusyk et al 1982
47,XX,-17,+20,+22	Martin et al 1979
47,XY,-10,-12,+17,+18,+20	Couturier-Turpin et al 1982
50,XY,2q+,+5,+10,+11,-16,-18,+20,+20,+mar	Couturier-Turpin et al 1982

Adenomatous polyp

Karyotype	Reference
48-52,X,-Y,+7,+8,+13,+14,+16,+18,-20,+12q-,1q+	Reichmann et al 1982b

MESENCHYMAL NEOPLASMS

Mesothelioma, malignant

Karyotype	Reference
40-46,XY,+16,+20,-22,t(14;15),del(2)(p?),i(5p)	Ayraud 1975
41,XY,del(1)(p13),t(1;3)(p13;q11),t(7;7)(q11;q36), t(9;12)(p24;q13),del(12)(q13),del(13)(q12q14),-14,-15, t(17;?)(p13;?),i(19p),-20,-21,-22	Mark 1978

MELANOMAS

Malignant melanoma

Karyotype	Reference
46,XY,5q+,+7,-8,-9,-10,-13,-20,+21,+3mar	McCulloch et al 1976

Malignant melanoma, metastatic

Karyotype	Reference
41,XY,dup(1)(q21q32),t(1;7)(q25;q36),t(7;9)(q11;q11), del(9)(q13),-3,-4,-5,-8,-10,-11,-16,+18,-20,-22	Kakati et al 1977

NEUROGENIC NEOPLASMS

Neuroblastoma

Karyotype	Reference
46,XY,t(1;?)(p?;?),19q+,20q+,-22	Gilbert et al 1982
53,XX,+1,+4,+15,+20,dup(1)(p13p32),t(3;15)(q13;q26), del(4)(q31),t(7;11)(q36;q13)	Brodeur et al 1981b

Retinoblastoma

Karyotype	Reference
46,X,-X,+i(6p),17q+,20q+	Kusnetsova et al 1982
46,XX,inv(1),inv(2),inv(3),inv(4),t(6;7),inv(9),t(11;?), 11p-,inv(12),i(17q),-20	Gardner et al 1982
47,XY,del(13)(q14),t(20;?)(q13;?),2q+,i(17q),dmin	Balaban et al 1982
47,XY,dup(1)(q12q44),del(2)(q13),+del(6)(q16), t(14;?)(q32;?),i(17q),+20,-22	Gardner et al 1982

Meningioma, benign

41,XY,+del(1)(p?),-5,-10,-17,-20,-22	Zang and Zankl 1983
44,XX,-8,-22/40-44,XX,-8,-9,-19,-20,-21,-22	Mark et al 1972a
45,XX,-22,del(20)(q11)/44,XX,-22,-20	Zankl et al 1975a
45,XY,-22/42-45,XY,-8,-15,-17,-20,-22	Mark et al 1972b
50,XX,+9,+10,+12,+15,+20,-22	Mark 1973c
51,XY,+5,6q+,+12,+16,+17,+18,+20,-22/52,XY,+5,+8,+10,+12, +17,+18,+20,-22	Zang and Zankl 1983
52,X,-Y,-22,+20,+19,+17,+15,+11,+9,+7,+5	Zankl et al 1975a

EMBRYONAL AND MISCELLANEOUS NEOPLASMS

Teratoma, malignant

67,XY,+1,+3,+4,+5,+6,+7,+7,+8,+10,+11,+13,+13,+14,+16,+17, +18,+20,+20,+21,+21	Sonta et al 1977

LYMPHOMAS

1. UNCLASSIFIED AND MISCELLANEOUS LYMPHOMAS

Non-Hodgkin's lymphoma, NOS

46-48,XX,del(1)(q32),+1,+3,r(7),14q+,-20,+mar	Kristoffersson 1983
52,XY,+2,+3,t(5;9)(q22;q32),+6,+12,+14,+20	Reeves and Pickup 1980

Non-Hodgkin's lymphoma, histiocytic, NOS

50,XY,dup(1)(q12q32),+2,4p+,del(4)(q21),t(8;14)(q24;q32), +17p+,18q+,+20	Brynes et al 1978
50,XY,t(2;?)(p25;?),del(2)(p13),+i(3p),del(6)(p21),+7,+8, +9,+9,-11,-13,+14,-15,+20,+21,-22	Mark et al 1978
44,XY,del(2)(p13),t(6;?)(p21;?),del(11)(q13), t(11;14)(q13;q31),i(15q),-20,-22	

Angioimmunoblastic lymphadenopathy

47,XY,+3/47,XY,+3,-9,+21/48,XY,+3,+8,+9,-20	Hossfeld et al 1976
56,XY,+5,+5,+6,+10,+15,+15,+19,+20,+21,+22	Kaneko et al 1982c

2. HODGKIN'S LYMPHOMAS

Hodgkin's disease, lymphocytic depletion

48,X,-Y,t(1;?)(q44;?),i(1q),del(1)(q21),del(3)(q21),-10, t(3;11)(q23;q21),+14,t(14;?)(q32;?),+15,-17,+19,+20,-22	Hossfeld and Schmidt 197[illegible]

Hodgkin's disease, nodular sclerosis

47,XY,t(1;?)(p36;?),+2,t(2;6)(q31;q27),+8,del(12)(p11), t(17;?)(p13;?),-14,-20,+mar	Reeves and Pickup 1980

3A. NON-HODGKIN'S LYMPHOMAS - RAPPAPORT CLASSIFICATION

Lymphocytic, poorly differentiated, NOS

46,XY,del(6)(q21),-8,-9,t(6;14)(q21;q32),+16,+20	Mark et al 1979

Lymphocytic, poorly differentiated, nodular

46,XX,del(20)(q11),t(14;18)(q32;q21)	Kaneko et al 1982a
52,X,-X,+2,-4,+5,+7,+8,-14,-15,-16,+20,+21, +t(1;15)(p11;q11),+t(12;17)(q24;q11),+t(14;?)(q32;?), +mar	Kaneko et al 1982a

Mixed cell, diffuse

53,XY,t(8;14)(q24;q32),t(14;18)(q32;q21),+20,1q-,+1p-,2q-, 4q+,5q-,5p+,6q-,+7q+,10q+,11q+,11q-,13p+,+4mar	Fukuhara and Rowley 1978

Histiocytic, diffuse

47,XY,-10,-13,+15,-16,+19,+20,del(6)(q21),del(8)(q22), t(2;14)(q21;q32),t(4;4)(q31;q35),+t(10;?)(p11;?)	Kaneko et al 1982a
50,XY,del(3)(p24),+7,+12,+14,+20	Kristoffersson et al 1981
50,XY,t(8;14)(q24;q32),+2,+20,1q+,4p+,+4q+,17p+,18q+	Fukuhara and Rowley 1978
89,XY,del(1)(p22),t(5;13)(q15;q34),t(14;20)(q32;q12), t(1;17)(p22;q25),t(1;19)(q21;p13),del(20)(q12)	Mark et al 1979

3B. NON-HODGKIN'S LYMPHOMAS - KIEL CLASSIFICATION

Immunocytoma

45-46,XY,-8	Kristoffersson 1983
47,XY,15p+,+20/46,XY,del(5)(q23),del(6)(p22),+del(7)(p11), -21	

Centroblastic-centrocytic, diffuse

48-50,XY,-1,+7,9q+,+9p+,+10,+14q+,-17,-18, +t(17;18)(p11;q11),-19,+20,+22,+mar	Kristoffersson 1983
49,XY,+1,+20,+r	Kristoffersson 1983

Centroblastic, diffuse

42-45,XY,1p+,-6,-8,14q+,14q-,-16,i(17q),+18,-20,+mar	Kristoffersson 1983
46-51,XX,+3,del(6)(q21),+del(6)(q21),+7,-8,del(9)(q11),+13, -16,-17,+19,-20,-21,+1-5mar	Kristoffersson 1983

Lymphoblastic, Burkitt's type

48-53,X?,t(8;14)(q24;q32),+X,+2,+3,+9,+10,+11,+12,+20 — Zech et al 1976a

Lymphoblastic, convoluted type

46,X,-Y,t(1;?)(q21orq23;?),+mar/46,X,-Y,-1,-2,+der(1), +t(2;?)(q33orq35;?),+mar/48,X,-Y,-5,-14,+18,+18,+20, del(9)(p13),t(1;14),+dup(5)(q13q35) — Kaneko et al 1982d

Immunoblastic

48-49,XY,+3,+7,+11,t(12;?)(p13;?),t(13;?)(p13;?),-16,-20, +21,+22 — Reeves and Pickup 1980
48-49,XY,del(3)(p23),+7,+12,+14,+20 — Kristoffersson 1983
51,XY,t(1;6)(p22;q13orq15),t(2;13)(q21;q14), t(8;16)(q24;p11),+11,+12,del(15)(q22),+16,+20,+mar — Reeves and Pickup 1980
78-83,XY,+del(1)(q11),+5,+7,+12,+13q+,+14q+,+14q+,+20,+mar — Kristoffersson 1983
80-86,XY,del(1)(q11),+3,+7,+9,+12,+13q+,+14q+,+14q+,+15q+, +18,+20,+mar

3C. NON-HODGKIN'S LYMPHOMAS - LUKES AND COLLINS CLASSIFICATION

Follicular center cell, diffuse, large cleaved

45,XY,-6,-9,-17,2q-,3p-,3q-,4q-,5q-,6p+,7p+,7q+,+7q-,9p-, 12q+,13p+,20p-,+mar — Fukuhara et al 1978b

Immunoblastic type (B-cell)

48,XY,t(4;12)(q?;q?),t(6;15)(q?;q?),t(9;22)(q34;q11),11q-, +20,+21 — Nordenson et al 1983

3D. NON-HODGKIN'S LYMPHOMAS - WORKING FORMULATION

Diffuse, small cleaved cell

49,XY,dup(1)(q?),8p-,-9,+14,+17,+20,+mar — Yunis et al 1982

Diffuse, large cell

54,XY,dup(1)(q?),inv(5),+3,+3,+9,+13,+18,+18,+20,+mar — Yunis et al 1982

Chromosome 21

HEMATOLOGICAL DISORDERS

UNCLASSIFIED LEUKEMIAS

Acute leukemia, NOS

48,XX,+8,+21"c" — Debiec-Rychter et al 1982

Leukemoid reaction

46,XY,-21,+t(21;21)(p11;q11)"c" — Heaton et al 1981
49,XY,+21,+21,+21 — Van Den Berghe et al 1983

NONLYMPHOCYTIC LEUKEMIAS

Acute nonlymphocytic leukemia (ANLL)

Karyotype	Reference
39,X,-X,-3,-5,-13,-14,-15,-16,-17,-21,-22,del(7)(p11), t(12;?)(p13;?),t(16;?)(p13;?),+3mar	Mitelman 1983
39,X,-X,-3,-5,-13,-14,-15,-16,-17,-21,-22,del(7)(p11), t(12;?)(p13;?),t(16;?)(p13;?),+3mar	
43,XX,-1,3p+,3q-,4q+,-5,-5,-7,+9,+9,10p+,+10,+10,10p-,10q-, +11q-,-13,-15,-16,-17,-18,-21"s"	Whang-Peng et al 1979
43,XX,-4,-5,-21,t(8;21)(q22;q22)	Oshimura et al 1976b
43,XX,t(1;20),-5,-6,7q-,11q-,-12,17q+,-18,19q+,-21	Mitelman 1983
44,X,-X,-21	Mitelman 1983
44,XX,t(8;11)(p23;p13 or p15),-20/46,XX,del(7)(q11), t(1;3)(q2?;q2?),-20,-21,-22,+3mar	Philip et al 1978a
44,XY,del(5)(q21),del(7)(q11),del(11)(q11), t(7;11;12)(q11;q11;p13),del(13)(q11),-20,-21/45,XY, t(3;12)(q11;p13),del(5)(q21),del(7)(q11),del(13)(q11), -20	Mitelman 1983
45,X,-X,t(8;16)(q22;q24),+20,-21	Trent et al 1983a
45,X,-X,t(8;21)(q22;q22)"s"	Elfenbein et al 1978
45,X,-X,t(8;21)(q22;q22)	Sasaki et al 1976
45,X,-Y,t(8;21)(q22;q22)	Oshimura et al 1976b
45,X,-Y,t(8;21)(q22;q22)	Philip et al 1978a
45,X,-Y,t(8;21)(q22;q22),t(9;22)(q34;q11)	Francesconi and Pasquali 1978a
45,X,-Y,t(8;21)(q22;q22)	Sasaki et al 1976

45,X,-Y,t(8;21)(q22;q22)	Sasaki et al 1976
45,X,-Y,t(8;21)(q22;q22)	Seabright 1983
45,XX,-21	Oshimura et al 1976b
45,XX,-21	Oshimura et al 1976b
46,X?,t(8;21)(q22;q22)	Hartley 1983
46,XX,5q-/47,XX,5q-,+21	Rowley 1976
46,XX,del(5)(q12q31),-9,15p+,18p-,-21,+2mar	Kerkhofs et al 1982
46,XX,del(5),del(6),del(7),21q+"s"	Geraedts et al 1980a
46,XX,t(4;8),dic(21)	Seabright 1983
46,XX,t(8;21)(q22;q22)	Oshimura et al 1976b
46,XX,t(8;21)(q22;q22)	Philip et al 1978a
46,XX,t(8;21)(q22;q22)	Philip et al 1978a
46,XX,t(8;21)(q22;q22),del(9)(q?)	Rozynkowa et al 1977
46,XX,t(8;21)(q22;q22)	Sasaki et al 1976
46,XX,t(8;21)(q22;q22)	Seabright 1983
46,XY,+del(7)(p11),-8,del(12)(p11),t(21;?)(q22;?)/50,XY,+Y, +del(7)(p11),del(12)(p11),+17,t(17;?)(q25;?), t(21;?)(q22;?),+22	Mitelman 1983
46,XY,+del(7)(p11),-8,del(12)(p11),t(21;?)(q22;?)/47,XY,+9	
46,XY,-7,+21/47,XY,+21	Philip et al 1978a
46,XY,del(2)(q31),-7,del(9)(p13),t(2;21)(q21;q22), t(2;?)(p21;?)/46,XY,t(1;?)(p22;?),t(17;?)(q25;?)	Philip et al 1978a
46,XY,t(21;?)(q22;?)"s"	Cavallin-Ståhl et al 1977
46,XY,t(8;21)	Bernard et al 1982a
46,XY,t(8;21)	Knuutila et al 1981
46,XY,t(8;21)	Carbonell 1983
46,XY,t(8;21)(q22;q22)	Sasaki et al 1976
46,XY,t(8;21)(q22;q22)	Seabright 1983
46,XY,t(8;21)(q22;q22)	Watt 1983
46,XY,t(8;21)	Bernard et al 1982a
46,XY,t(8;21)	Bernard et al 1982a
46,XY,t(9;22)(q34;q11),21q+/46,X,-Y,t(9;22),+8	Abe and Sandberg 1979
46-49,XY,+8,+14,-17,+18,-21"s"	Cavallin-Ståhl et al 1977
47,X,Xq-,t(3;5),-5,t(11;?),t(12;17),-18,-21,+6mar	Pedersen-Bjergaard et al 1980
47,XX,+21	Bernard et al 1982a
47,XX,+21/48,XX,+8,+21	Sikand et al 1980
47,XX,del(5)(q?),+21	Van Den Berghe et al 1976
47,XY,+21	Zuelzer et al 1976
47,XY,+21"c"	Mitelman 1983
47,XY,+21/47,XY,+21,r(7)"c"	
47,XY,+21,r(7)"c"	
47,XY,+21,r(7)"c"	
47,XY,+21	Mitelman et al 1976b
47,XY,+?21	Mitelman 1983
48,XX,+8,+21"c"	Mitelman 1983
48,XX,+8,+21"c"	
48,XX,t(9;22)(q34;q11),+del(22)(q11),+21	Abe and Sandberg 1979
48,XY,+15,+21	Bernard et al 1982a
48,XY,+9,+21	Mitelman 1983
51,XY,+3,+5,-9,+19,+20,+21,t(6;14)(p25;q11),t(9;?)(q22;?)	Philip et al 1978a
51,XY,+8,+9,+13,+14,+21	Oshimura et al 1976b
52,XX,1p+,+10,+13,+19,+21,+21,+22"c"	Berger et al 1973

56,XY,+Y,+6,+14,+21,+t(5;6)(q?;p?),+del(1)(p22q25), +del(2)(q12),+del(3)(p12),+del(4)(q13),+del(5)(p14q14), +mar	Zuelzer et al 1976
57,XY,+6,+11,+17,+20,+21"s"	Berger et al 1981a

ANLL, FAB type M1

39,XY,-5,-7,-12,-16,-17,-20,-21	Berger et al 1981a
42,XX,-7,-17,-19,-21"s"	Papa et al 1979
45,X,-X,-5,-7,-11,-17,+21,+3mar"s"	Zaccaria et al 1983b
47,XX,+21"s"	Mitelman 1983
45,X,-X"s"	
46,XY,-2,-5,-7,-13,+16,-21,-21,+5mar	Sessarego et al 1981a
46,XY,t(8;21)(q22;q22)	Brodeur et al 1983
47,XX,+21	Prieto et al 1981
47,XY,+21	Takeuchi et al 1981
47,XY,+21"s"	Berger et al 1981a
47,XY,+8,-21,+dic(21;21)(p13;p11)"c"	Kaneko et al 1982b
47-48,XY,+8,t(9;22)(q34;q11),-20,+t(1;20)(q21;q13), t(10;21)(p11;q22),+22q-	Sasaki et al 1983
48,XX,+8,+21	Prieto et al 1981
48,XX,+8,+21"c"	Kaneko et al 1982b
50,XX,+6,+19,+21"c"	Kaneko et al 1982b
58,XY,+1,+4,+5,+6,+8,+11,+13,+13,+14,+14,+15,+21,+21, t(9;22)(q34;q11)	Alimena 1983

ANLL, FAB type M2

??,X?,t(8;21)	Trujillo et al 1979
??,X?,t(8;21)	Trujillo et al 1979
??,X?,t(8;21)	Trujillo et al 1979
??,X?,t(8;21)	Trujillo et al 1979
??,X?,t(8;21)	Trujillo et al 1979
??,X?,t(8;21)	Trujillo et al 1979
??,X?,t(8;21)	Trujillo et al 1979
??,X?,t(8;21)	Trujillo et al 1979
??,X?,t(8;21)	Trujillo et al 1979
??,X?,t(8;21)	Trujillo et al 1979
??,X?,t(8;21)	Trujillo et al 1979
??,X?,t(8;21),t(Y;19)	Trujillo et al 1979
42,XY,5q-,7q-,-8,-12,-17,-21,-22,+i(17q),+2mar	Berger et al 1981a
44,X,t(X;8),del(5)(q12q21),t(1;6),t(7;19),-7,-8,t(7;8),-21	Kerkhofs et al 1982
44,XY,-7,17p+,-20,21q+"s"	Pedersen-Bjergaard et al 1981
45,,del(X)(q26),-Y,t(8;21)(q22;q22),del(9)(q24), del(18)(p11)	Brodeur et al 1983
45,X,-X,t(8;21)	Tricot et al 1981
45,X,-X,t(8;21)	Takeuchi et al 1981
45,X,-X,t(8;21)	Takeuchi et al 1981
45,X,-X,t(8;21)(q21;q21)	Hagemeijer et al 1981a
45,X,-X,t(8;21)(q21;q21)	Hagemeijer et al 1981a
45,X,-X,t(8;21)(q21;q21),1p+,17p-	
45,X,-X,t(8;21)(q22;q22)	Levan and Mitelman 1979
45,X,-X,t(8;21)(q22;q22)/46,X,-X,t(8;21),1p+,7q+,+19/47,X, -X,t(8;21),1p+,+19,+der(21)	Sakurai et al 1982b

45,X,-X,t(8;21)(q22;q22)	Oshimura et al 1982b
45,X,-X,t(8;21)(q22;q22)/46,XX,t(8;21)	Berger et al 1982b
45,X,-X,t(8;21)(q22;q22)/46,XX,t(8;21)	Sakurai et al 1982b
45,X,-X,t(8;21)(q22;q22)/46,XX,t(8;21)	Alimena 1983
45,X,-X,t(8;21)(q22;q22),del(9)(q31)	Brodeur et al 1983
45,X,-Y,t(8;21)	Tricot et al 1981
45,X,-Y,t(8;21)	Tricot et al 1981
45,X,-Y,t(8;21)	Tricot et al 1981
45,X,-Y,t(8;21)	Takeuchi et al 1981
45,X,-Y,t(8;21)	Hagemeijer et al 1981a
45,X,-Y,t(8;21)(q21;q22)	Prieto et al 1981
45,X,-Y,t(8;21)(q21;q21)	Hagemeijer et al 1981a
45,X,-Y,t(8;21)(q22;q22)	Kamada et al 1976
45,X,-Y,t(8;21)(q22;q22)	Bernstein et al 1982b
45,X,-Y,t(8;21)(q22;q22),del(9)(q13q22)	Hossfeld et al 1980
45,X,-Y,t(8;21)(q22;q22)/46,XY,t(8;21)	Oshimura et al 1982b
45,X,-Y,t(8;21)(q22;q22)	Oshimura et al 1982b
45,X,-Y,t(8;21)(q22;q22)	Berger et al 1982b
45,X,-Y,t(8;21)(q22;q22)/46,XY,t(8;21)	Berger et al 1982b
45,X,-Y,t(8;21)(q22;q22),t(1;5)(q43;q11)	Berger et al 1982b
45,X,-Y,t(8;21)(q22;q22)	Berger et al 1982b
45,X,-Y,t(8;21)(q22;q22),+1,-B/46,X,-Y,t(8;21),+8	Berger et al 1982b
45,X,-Y,t(8;21)(q22;q22)	Kamada et al 1981
45,X,-Y,t(8;21)(q22;q22)/46,X,-Y,t(8;21)(q22;q22), +t(11;12)(p11;q13)	Sakurai et al 1982b
45,X,-Y,t(8;21)(q22;q22)"s"	Mitelman 1983
45,X,-Y,t(8;21)(q22;q22)/46,X,-Y,t(8;21)(q22;q22),+der(21)	Mitelman 1983
45,X,-Y,t(8;21)	
45,X,-Y,t(8;21)	Golomb et al 1978c
45,XX,-17,-21,+mar	Oshimura et al 1982b
45,XY,-21/46,XY,+3,-21	Prieto et al 1981
46,X,-X,t(8;21)(q22;q22),+18p+	Berger et al 1982b
46,X,-Y,t(8;21)(q22;q22)	Kamada et al 1976
46,X,-Y,t(8;21)(q22;q22)	Bernstein et al 1982b
46,XX,4q+,del(7)(q31),t(8;21)(q21;q21),del(9)(q12q21)	Hagemeijer et al 1981a
46,XX,del(20)(q11)/48,XX,+9,del(20)(q11),+21	Golomb et al 1978c
46,XX,t((8;21)(q22;q22)	Berger et al 1982b
46,XX,t(14;21)(q24;q21)	Alimena 1983
46,XX,t(1;8;21)	Tricot et al 1981
46,XX,t(2;8;21)(p11q13;q22;q22)	Pasquali and Casalone 1981
46,XX,t(8;21)	Tricot et al 1981
46,XX,t(8;21)	Tricot et al 1981
46,XX,t(8;21)	Tricot et al 1981
46,XX,t(8;21)	Takeuchi et al 1981
46,XX,t(8;21)	Takeuchi et al 1981
46,XX,t(8;21)	Takeuchi et al 1981
46,XX,t(8;21)(q21;q21),del(9)(q11q13)	Hagemeijer et al 1981a
46,XX,t(8;21)(q22;q22),del(5)(q11?;q13?)	Bernstein et al 1982b
46,XX,t(8;21)(q22;q22),del(9)(q13q22)	Hossfeld et al 1980
46,XX,t(8;21)(q22;q22)	Oshimura et al 1982b
46,XX,t(8;21)(q22;q22)	Oshimura et al 1982b
46,XX,t(8;21)(q22;q22)	Berger et al 1982b
46,XX,t(8;21)(q22;q22)	Kamada et al 1981
46,XX,t(8;21)(q22;q22)	Kamada et al 1981

46,XX,t(8;21)(q22;q22)	Mitelman 1983
46,XX,t(8;21)(q22;q22)/45,X,-X,t(8;21)	Kaneko et al 1982b
46,XX,t(8;21)(q22;q22)	Brodeur et al 1983
46,XX,t(8;21)(q22;q22)	Brodeur et al 1983
46,XX,t(8;21),t(14;21)"c"	Tricot et al 1981
46,XY,21q-	Tadano et al 1980
46,XY,del(1)(p22p32),+19,-21	Hagemeijer et al 1981a
46,XY,t(8;21)	Tricot et al 1981
46,XY,t(8;21)	Hagemeijer et al 1981a
46,XY,t(8;21)(q22;q22),del(9)(q21)/45,X,-Y,t(8;21), del(9)/45,XY,t(8;21),-9	Bernstein et al 1982b
46,XY,t(8;21)(q22;q22)/45,X,-Y,t(8;21)/48,XY,t(8;21), +der(21),t(10;15)(p15;q22),+r	Sakurai et al 1982b
46,XY,t(8;21)(q22;q22)	Sakurai et al 1982b
46,XY,t(8;21)(q22;q22)	Oshimura et al 1982b
46,XY,t(8;21)(q22;q22)/46,XY,t(8;21),9q-/47,XY,t(8;21),+9	Berger et al 1982b
46,XY,t(8;21)(q22;q22)	Kamada et al 1981
46,XY,t(8;21)(q22;q22)	Alimena 1983
46,XY,t(8;21)(q22;q22)	Brodeur et al 1983
47,X,-X,+8,t(8;21)(q22;q22),+14	Brodeur et al 1983
47,X?,t(8;21),+21	Trujillo et al 1979
47,X?,t(8;21),+8	Trujillo et al 1979
47,X?,t(8;21),+8	Trujillo et al 1979
47,XX,+13/47,XX,+21/48,XX,+13,+21	Prieto et al 1981
47,XX,+21	Prieto et al 1981
47,XX,+21	Mitelman 1983
47,XX,1p+,der(5),-6,12p+,-13,-14,-16,-17,-17,-21,+22,+7mar, dmin	Cooperman and Klinger 1981
47,XX,t(8;11;21),+8	Golomb et al 1978c
47,XX,t(8;21)(q22;q22),+8	Sakurai et al 1982b
47,XY,+21	Takeuchi et al 1981
47,XY,+21	Alimena 1983
47,XY,del(5)(q15),+21	Swolin et al 1981
48,XX,-5,t(8;?)(p23;?),del(11)(q23),t(15;?)(p11;?),-21, +4mar"s"	Rowley et al 1977a
53,XX,+1,-4,-5,+6,-7,+8,+11,+12,+21,+21,+3mar"s"	Zaccaria et al 1983b
55,XX,+16,+17,+18,+20,+21,+22,+3mar	Prieto et al 1981
90,XX,-Y,-Y,t(8;21)(q22;q22),t(8;21)(q22;q22)	Testa et al 1983

ANLL, FAB type M1+M2

45,XY,-21,t(4;9)(p16;q22)	Yamada and Furusawa 1976
37-45,XY,-7,-12,-13,-14,-16,-18,-19,-21,del(4)(q29), del(5)(q24),der(9),der(17),11q+,12p+,t(14;?)	Fitzgerald et al 1983
42,X,-Y,4q+,-8,-16,17p+,-21,-22,+Yp+"s"	Sandberg et al 1982a
42-43,XX,cx,t(11;15)(q11;q11),t(9;?)(p?;?),-15,-17,-18,-20, -21	Fitzgerald et al 1983
43,XX,del(5)(q12q32),-7,-12,-18,t(7;12;21)(q11;q12;q22)	Petit and Van Den Berghe 1979b
43,XY,-3,-5,-21	Rowley and Potter 1976
44,XX,-3,-4,-5,+9,-17,-18,-21,-22,+5mar	Garson 1980
44,XY,-7,-16,-19,-21,-22,+3mar"s"	Sandberg et al 1982a
45,X,-X,t(8;21)(q21;q22)	Mitelman et al 1981
45,X,-X,t(8;21)(q22;q22)	Mitelman et al 1981
45,X,-X,t(8;21)(q22;q22)	Rowley and Potter 1976

45,X,-X,t(8;21)(q22;q22)	Rowley and Potter 1976
45,X,-X,t(8;21)(q22;q22)	Prigogina et al 1979
45,X,-Y,t(8;21)	Li et al 1983
45,X,-Y,t(8;21)(q21;q22)	Mitelman et al 1981
45,X,-Y,t(8;21)(q22;q22)	Yamada and Furusawa 1976
45,X,-Y,t(8;21)(q22;q22)	Prigogina et al 1979
45,X,-Y,t(8;21)(q22;q22)	Prigogina et al 1979
45,X,-Y,t(8;21)(q22;q22)	Prigogina et al 1979
45,X,-Y,t(8;21)(q22;q22)	Prigogina et al 1979
45,X,-Y,t(8;21)(q22;q22),del(9)(q13)	Shiraishi et al 1982
45,XX,-21/47,XX,+8	Li et al 1983
45,XX,-5,-11,-13,-18,-21,+4mar,dmin"s"	Sandberg et al 1982a
45,XY,+t(1;10)(q23;q26),t(8;21)(q22;q22),-10/45,X,-Y, +t(1;10),t(8;21),-10	Fitzgerald et al 1983
45,XY,-20/47,XY,+21	Li et al 1983
45,XY,-21,18p+	Yamada and Furusawa 1976
45,XY,-7,t(11;21)(q25;q11)	Fitzgerald et al 1983
45,XY,-8,-10,+21/47,XY,+21	Alimena et al 1977
45-46,XX,t(5;11),4q+,-21	Alimena et al 1977
46,X,del(X)(q21),t(5;21)(q13;q22)	Mitelman et al 1981
46,X,i(Xq),t(8;21)(q22;q22)	Shiraishi et al 1982
46,X,r(X),t(8;21)	Garson 1980
47,XX,+8/51,XX,+X,+8,+18,+20,+21	Testa et al 1979
46,XX,t(8;21)(q22;q22)	Rowley and Potter 1976
46,XX,7q+,t(8;21)	Li et al 1983
46,XX,t(8;21)	Garson 1980
46,XX,t(8;21)	Li et al 1983
46,XX,t(8;21)(q22;q22)	Mitelman et al 1981
46,XX,t(8;21)(q22;q22)	Fitzgerald et al 1983
48,XY,+8,+21/49,XY,+8,+18,+21	Testa et al 1979
46,XY,21q+	Nordenson 1983
46,XY,t(5;8;21)(q31;q22;q22)/47,XY,t(5;8;21),+mar	Fitzgerald et al 1983
46,XY,t(8;21)	Garson 1980
46,XY,t(8;21)(q22;q22)	Prigogina et al 1979
46-47,XY,+8,-18,21q+	Mitelman et al 1981
46-50,XY,+Y,+del(7)(p11),-8,+9,del(12)(p11),+17,21q+,+22	Mitelman et al 1981
47,X,-X,+del(1)(p11),del(5)(q13),t(5;12)(q21;q24), t(11;12)(q25;q13),t(13;14)(q11;p11),+16,+21	Mitelman et al 1981
47,X,-Y,t(1;13)(q32p32q11;q11),del(3)(q21), t(8;21)(q22;q22),del(9)(q22),+8,+18	Sakurai et al 1982b
47,XX,+21	Mitelman et al 1981
47,XX,r(18),+21	Fitzgerald et al 1983
47,XY,+21	Fitzgerald et al 1983
48,XX,-5,del(5)(q13q31),+8,+t(11;?)(p14;?),-17,-19,+21, +2mar,dmin"s"	Testa et al 1981b
48,XX,t(1;3)(q44;p14),del(3)(q21p14),-5,+10,+11,+21	Mitelman et al 1981
48,XY,+11,+21	Fitzgerald et al 1983
49,XX,+6,t(9;22)(q34;q11),+17,+21	Nordenson 1983
50-52,XY,+9,+14,+17,+19,+20,+21	Alimena et al 1977
50-52,XY,-5,-7,+9,+14,+17,+19,+20,+21,+mar	Mitelman et al 1981
57,X,-Y,-6,+8,+15,+15,+19,+19,+20,+20,+21,+21,+22,+3mar	Fitzgerald et al 1983
57,XY,+6,+11,+13,+15,+21,+21,+21,+21,+21,+21,+22"c"	Shiraishi et al 1982

ANLL, FAB type M3

Karyotype	Reference
45,XX,-4,-5,-14,t(21;?)(q22;?),+2mar"s"	Rowley et al 1977a
45-51,XY,t(3;12)(p14;q24),t(13;18)(q11;p11),+1,+6,+16,+19, +21,+22	Mitelman et al 1981
46,XY,t(14;21)"c"	Alimena 1983
46,XY,t(15;17),t(14;21)"c"	
46,XY,t(15;17)(q22;q12)/46,XY,t(15;17),t(21;?)(p11;?)/46, XY,del(4)(q22),t(15;17)	Kaneko et al 1982b
47,XY,t(15;17)(q24;q21),+21	Scheres et al 1978

ANLL, FAB type M4

Karyotype	Reference
42,XY,-5,-9,-10,-15,-16,-17,-20,-21,-22,+5mar	Testa et al 1979
43,XY,+1,-5,+6,-9,-10,-15,-16,-17,-20,-21,-22,+6mar	
42,XY,-7,i(8)(q11),-16,-17,-18,t(21;21)(q12;q11),+mar	Rowley and Potter 1976
43,XY,-2,-3,-6,-7,-12,-15,-21,-22,del(2)(p11),del(11)(p14), t(14;?)(q32;?),t(17;?)(p13;?),del(22)(q11),+5mar"s"	Rowley et al 1981a
44,XX,-16,-21	Garson 1980
44,XY,-7,+20,-21,-22"s"	Berger et al 1981a
45,X,-X,-9,-17,-18,-21,+5mar	Prieto et al 1981
45,X,-X,t(8;17;21)(q22;q23;q22)	Testa et al 1979
45,X,-X,t(8;17;21)(q22;q23;q22)/45,X,-X, t(1;11)(p36q32;q13),t(8;17;21)	
45,X,-Y,del(7)(q32),t(8;21)(q22;q22)	Hustinx et al 1980
45,X,-Y,r(21)"c"	Testa et al 1979
45,X,-Y,r(21)/46,X,-Y,r(21),+8"c"	
45,XY,7p-,21	Garson 1980
46,XX,-1,-3,-5,+10,+11,-13,-14,+21,+del(3?)(p?q?), ?t(1;3;5)(p?;p?q?;q?)	Mitelman 1983
46,XX,-12,+21"s"	Berger et al 1981a
46,XX,del(7)(q22)/45,XX,del(7)(q22),-21	Hagemeijer et al 1981a
46,XY,-17,+21/47,XY,+21"s"	Papa et al 1979
46,XY,2q+,del(18)(p11),del(22)(q11)/47,XY,2q+,+8, del(18)(p11),del(22)(q11)"s"	Alimena et al 1981
45-47,XY,2q+,+8,del(18)(p11),-21,del(22)(q11)"s"	
46,X,-Y,2q+,+8,del(21)(q21)"s"	
46,XY,7q-,-21,+i(?21q)"s"	Berger et al 1981a
47,XX,+21	Mitelman 1983
47,XX,-7,+21,+22/45,XX,-7	Prieto et al 1981
47,XY,+21	Mitelman 1983
47,XY,+21	Mitelman et al 1981
47-48,XY,+8,+10,+21,+22	Mitelman et al 1981
48,XY,+9,+21"s"	Berger et al 1981a
49,XX,+8,+13,+21	Prigogina et al 1979
53,XX,+15,+18,+21,+21,+mar"c"	Benedict et al 1979

ANLL, FAB type M5

Karyotype	Reference
44,X,-Y,del(2)(p11),-15,-21,+mar	Weber et al 1979
46,XY,t(17;21)(q11;q11orq21),+i(8q),del(5)(q11q13),9p+, 11q+	Fitzgerald et al 1983
46,XY,t(6;11)(q26;q22)/52-53,XY,t(6;11)(q26;q22),+der(6), +3,+10,+13,+18,+19,+21,dmin	Hagemeijer et al 1981a

46,XY,t(6;11)(q27;q23)/51-52,XY,t(6;11),+der(6),+3,+19,+19, +21,dmin/53,XY,t(6;11),+der(6),+4,+10,+13,+18,+19,+21	Löwenberg et al 1982
47,XY,+21/48,XY,+11,+21	Brodeur et al 1983
48,XX,+19,+21	Mitelman 1983
49-50,XY,-21,11q+,19q+,+3-4mar"s"	Prigogina et al 1979
51,XX,+6,+8,+14,+19,+21,t(11;?)(q21orq24;?)	Soukup and Neely 1981

ANLL, FAB type M5a

47,XY,+21	Berger et al 1982a

ANLL, FAB type M4+M5

45,X,-X,r(21)	Olinici et al 1978a
45,XY,-21	Li et al 1983
46,XY,-17,+21/47,XY,+21	Alimena et al 1977
47,XY,+8/48,XY,+8,+10/48,XY,+21,+22	Alimena et al 1977
66,XY,+1,+1,+2,+4,+6,+6,+7,+8,+8,+10,+11,+11,+12,+12,+18, +19,+21,+21,+22,+22	Li et al 1983

ANLL, FAB type M6

43,X,-X,del(5)(q14q34),-7,11q-,-12,12p+,t(12;17),t(20;21)	Kerkhofs et al 1982
43-45,XX,der(1),inv(2)(p25q14),del(3)(q25),-4,-5,-6,6q-,-7, 7p+,11p+,-12,-17,-21,+r,+1-4mar"s"	Smadja et al 1983b
44,XX,del(5)(q13q31orq33),del(11)(q21),-17,-21"c""s"	Papa et al 1979
45,X,-X,-5,-21,22q-,+3mar"s"	Sandberg et al 1982a
45,X,-Y/44,X,-Y,-21"s"	Sandberg et al 1982a
45,XY,del(5)(q13q31),r(8),-17,-21,+mar	Kerkhofs et al 1982
45-47,XY,del(6)(q22),-9,+18,+21	Mitelman et al 1981
45-49,XY,-4,-7,-8,-10,+14,+16,-21,-22,+2mar	Mitelman et al 1981
46,XX,-7,-16,-21,+t(2;7)(p11;q11),del(2)(q33),del(11)(q22), +11p+,14p+,+t(16;21;21;21)(p13;q22;q11;q22)"s"	Rowley et al 1981a
46-47,XX,del(19)(p?q?),11q+,+21"s"	Whang-Peng et al 1977
47,XY,+21	Prieto et al 1981
47,XY,+21	Li et al 1983
48,XX,+t(1;12)(p22;p13),+21,del(22)(q11)	Brodeur et al 1983
48,XX,del(6)(q21orq22q24orq25),+21,+21"c"	Hagemeijer et al 1981a
49,XX,+6,+6,+21,der(13)(q12q21)	Bernstein et al 1982b
49,XY,+8,+12,+17,t(17;21)(p11;q22)"s"	Yamada and Furusawa 1976
51,XY,+1,+2,+6,-7,+8,-14,+15,+21,+mar"s"	Rowley et al 1981a

Chronic myeloid leukemia, t(9;22)

45,X,-Y,t(9;22)	Sharp et al 1976
45-46,X,-Y,t(9;22),21q+,+mar	
45,XX,t(9;22),-21	Stoll and Oberling 1978
45,XY,t(9;22),-9,t(20;21)(p13;q22)	Miyamoto 1980
46,X,-Y,t(9;22),21q+,+mar	Lilleyman et al 1978
46,X?,t(9;22),21q+	Carbonell et al 1982a
46,X?,t(9;22)/47,X?,t(9;22),+21	Carbonell et al 1982a
46,XX,t(9;22),-21,+mar	Hossfeld 1975b
46,XX,t(9;22),t(12;21)(q13;q22)	Fleischman et al 1981
46,XX,t(9;22),t(5;21)(q?;q?)	Sadamori et al 1980

46,XX,t(9;22)/48,XX,t(9;22),+19,+22q-/50,XX,t(9;22),+8,+12, +19,+22q-/50,XX,t(9;22),+8,+14,+17,+19,+21	Hagemeijer et al 1980b
46,XY,t(9;22)(q34;q11)/47,XY,t(9;22),i(17q),+del(22q)/49, XY,t(9;22),+8,+21,i(17q),+del(22q)	Sonta and Sandberg 1978b
46,XY,t(9;22)(q34;q11),21q+/50,XY,t(9;22),+7,+8,+12,+15, 21q+	Tomiyasu et al 1982
46,XY,t(9;22),-8,12q+,i(17q),+21/46,XY,t(9;22),-8,-8,12q+, +i(17q),+21	Olah and Rak 1981
47,X?,t(9;22),+21	Carbonell et al 1982a
47,X?,t(9;22),+21/48,X?,t(9;22),+21,+21/50-52,X?,t(9;22), +8,+9,+12,+19,+21,+21	Carbonell et al 1982a
47,X?,t(9;22),+22q-/48,X?,t(9;22),+8,21q+,+22q-/49,X?, t(9;22),+8,t(14;?),21q+,+22q-	Carbonell et al 1982a
47,XX,t(9;22),+21	Vilpo et al 1979
47,XX,t(9;22),+mar	
47,XY,t(9;22)(q34;q11),-7,13q+,20p+,+21,+mar	Alimena et al 1982b
47,XY,t(9;22)(q34;q11),+8/47,XY,t(9;22),+19/47,XY,t(9;22), +21/47,XY,t(9;22),+22q-	Tomiyasu et al 1982
47,XY,t(9;22)(q34;q11),+22q-/47,XY,t(9;22),+19/47,XY, t(9;22),+21	Tomiyasu et al 1982
47,XY,t(9;22),+17/47,XY,t(9;22),-22,+5,+22q-/49,XY,t(9;22), +5,+21,+22q-	Stoll and Oberling 1978
47,XY,t(9;22),i(17q)/49,XY,t(9;22),i(1q),+8,+21,+22q-	Sonta and Sandberg 1977b
48,XX,t(9;22),+19,+21	Hartley and McBeath 1981
48,XY,+Y,t(9;22)(q34;q11),+21	Alimena et al 1979
48,XY,t(9;22),+12,+21	Lilleyman et al 1978
48,XY,t(9;22),+19,+21	Seabright 1983
48,XY,t(9;22),+22q-,+22q-/47,XY,t(9;22),+17/46,XY,t(9;22), -4,-7,+6,+17	Stoll and Oberling 1978
45,XY,t(9;22),-21/46,XY,t(9;22),-10,+17	
48,XY,t(9;22),t(3;4)(p?;p?),+21,+22q-	Sadamori et al 1980
48-53,X?,t(9;22),+8,+8,+13,+14,+19,+21,+22q-	Carbonell et al 1982a
49,XX,t(9;22),+10,+21,del(22q)	Castleman et al 1973
49,XX,t(9;22),+8,+17,+21/49,XX,t(9;22),+8,+17,+17,-19,+21	Olah and Rak 1981
49,XY,t(9;22)(q34;q11),+4,+21,+22q-	Alimena et al 1982b
50,XY,t(9;22)(q34;q11),+8,+17,+21,+del(22)(q11)/51,XY, t(9;22),+8,+8,+17,+21,+del(22)(q11)/52,XY,t(9;22),+8,+8, +8,+17,+21,+del(22)(q11)	Sonta and Sandberg 1978b
50,XY,t(9;22),+8,+17,+21,+22q-/51,XY,t(9;22),+8,+8,+17,+21, +22q-/54,XY,t(9;22),+8,+8,+8,+17,+21,+22q-,+22q-	Sonta and Sandberg 1977b
50,XY,t(9;22),+C,+19,+21,+22q-	Lessard and Le Prise 1982
50-54,XY,t(9;22),+1,+5,+8,+9,-16,+17,+18,+19,+21,+22q-	Alimena et al 1980
50-55,XY,t(9;22)(q34;q11),+1,+5,+8,+9,-16,+17,+18,+19,+21, +22q-,+mar	Alimena et al 1982b
51,XY,t(9;22)(q34;q11),+6,+8,+19,+21,+22q-	Tomiyasu et al 1982
51-52,XY,t(9;22),+del(22q),+7,+8,+12,+19,+21	Prigogina et al 1978
52,XY,t(9;22)(q34;q11),1p+,+8,+9,i(17q),+19,+20,+21,+22q-	Alimena et al 1982b
52,XY,t(9;22),+1,+8,+10,+11,+21,+22q-	Sadamori et al 1980
52,XY,t(9;22),+8,+14,+15,+19,+21,+22q-	Sadamori et al 1980
52,XY,t(9;22),+8,+9,+12,+19,+21,+22q-	Nordenson 1983
52-54,XY,t(9;22)(q34;q11),+3,+6,+7,+8,+19,+21,+22q-,+22q-, +1-3mar	Ondreyco et al 1981
53,XX,t(9;22)(q34;q11),+8,+10,+19,+20,+21,+22q-	Greenberg et al 1978
53,XY,t(9;22),+6,+9q+,+10,+12,+19,+21,+22q-/54,XY,t(9;22), +6,+8,+9q+,+10,+12,+19,+21,+22q-	Sonta and Sandberg 1977b

53-56,XY,+Y,t(9;22),+del(22q),+6,+10,+19,+8,+11,+21	Prigogina et al 1978
54,XY,t(9;22)(q34;q11),+8,+10,+11,+13,+14,+19,+21,+21	Hayata et al 1975b
55,XY,t(9;22),+6,+10,+10,+11,+13,+14,+14,+19,+21	Seabright 1983
59,X,-X,t(9;22),+4,+6,+6,t(7;?)(p15;?),+der(7),+8,+8,+12q+, +17,+19,+19,+21,+22,+22q-,+22q-	Miyamoto 1980

<u>Chronic myeloid leukemia, aberrant translocation</u>

46,X?,t(21;22)(q22;q11)	Bottura and Couthinho 1974
46,XX,t(21;22)(p1?;q11)	Seabright 1983
46,XX,t(3;22)(p21;q11)/48,XX,t(3;22),+19,+21	Pravtcheva et al 1976
46,XY,t(19;22)(q13;q11),t(3;21)(p11;p11)	Seabright 1983
46,XY,t(21;22;22)(p13;p13;q11)	Ishihara et al 1974
46,XY,t(4;22)(q35;q21),t(3;5)(q27;q22)	Sessarego et al 1981b
46,XY,t(4;22)(q35;q21),t(3;5)(q27;q22),i(17q)/50,XY, t(4;22),t(3;5),i(17q),+8,+15,+21,+22q-	
50,XY,del(22)(q11),+8,+11,+21,+21/51,XY,del(22)(q11),+8, +11,+13,+21,+21	Sonta and Sandberg 1978b
50,XY,del(22)(q11),+8,+11,+21,+21,/51,XY,del(22)(q11),+8, +11,+13,+21,+21	

<u>Chronic myeloid leukemia, Ph1 negative</u>

46,X,del(Xq),21q+	Hossfeld 1975b
46,XX,t(3;11)(q21;p15)	Mitelman 1983
46,XX,t(3;11)(q21;p15)/49,XX,t(3;11)(q21;p15),+8,+19,+21	
46,XX,t(3;11)/49,XX,t(3;11),+8,+19,+21	
46,XY,t(3;6)(q?;p?),t(17;21)	Borgström et al 1982
47,XX,i(17q),+21	Borgström et al 1982
47,XY,20q-,+21/46,XY,20q-	Hossfeld 1975b

<u>Eosinophilic leukemia</u>

45,X,-X,t(8;21)(q22;q22)	Kaneko et al 1983

<u>Megakaryocytic leukemia</u>

46,XY,+C,-D,-E,r(21),+r"c"	Pui et al 1982

MYELOPROLIFERATIVE DISORDERS

<u>Polycythemia vera</u>

??,X?,+15,+21	Vykoupil et al 1980
45,X,-Y	Berger and Bernheim 1979
54-57,X,-Y,+6,+8,+11,+17,+20,+21,t(1;22)(p12;q13)	
45,XY,-16,-17,+21/48,XY,+Y,+11,+16,-17,+22	Wurster-Hill et al 1976

Myelosclerosis / Myelofibrosis

Karyotype	Reference
47,XX,+21	Whang-Peng et al 1978
47,XY,+21	Whang-Peng et al 1978
47,XY,-7,+21,+t(1;7)(p1?;p11)"s"	Geraedts et al 1980b
48,XX,+8,+21/49,XX,+8,+19,+21/50,XX,+8,+19,+19,+21"c"	Ueda et al 1981
50,XY,+del(1)(p?),+8,+9,+21	Nowell and Finan 1978

Idiopathic thrombocythemia

Karyotype	Reference
46,XX,t(11;21)(q25;q21)	Zaccaria et al 1980
46,XX,t(11;21)(q25;q11)	Petit and Van Den Berghe 1979a

Chronic myeloproliferative disease, NOS

Karyotype	Reference
46,XX,del(21)(q21)	Mitelman 1983
46,XY,-7,+19/47,XY,+19/52,XY,+16,+19,+19,+21,+22,+22	Chrz et al 1983
47,XX,+21	Tabachnik et al 1979

Myeloproliferative disorder, special type

Karyotype	Reference
48,XY,+8,+21	Whaun et al 1981

DYSMYELOPOIETIC SYNDROMES

Preleukemia, NOS

Karyotype	Reference
45,XY,t(20;21)(p13;q11),-21"s"	Anderson and Bagby 1982
46,XX,-21,+mar"s"	Berger et al 1980a
50,XY,-5,-21,+1,+9,+11,17p+,+1-3mar	Borgström et al 1982
52,XY,+1,+3,+21,+3mar	Panani et al 1980
53,XX,+1,+2,t(4;19),+6,i(17q),+21,+21,+22	Seabright 1983

Chronic myelomonocytic leukemia

Karyotype	Reference
47,XX,+8/48,XX,+8,+21	Kardon et al 1982a

Refractory anemia, NOS

Karyotype	Reference
45,XX,-21	Lessard and Le Prise 1983
46,XX,del(5)(q14q32),t(11;21)(q25;q21)	Van Den Berghe et al 1979e

Sideroblastic anemia, NOS

Karyotype	Reference
44,X,-Y,-21/45,XY,-21/46,XY,1q+	Yamada and Furusawa 1976

Refractory anemia with excess of blasts

46,XY,del(5)(q12)/47,XY,del(5)(q21),+21/49-50,XY, del(5)(q12),+9,+21,+2mar	Kardon et al 1982a
56-64,XY,+1,-2,+6,+9,+13,+15,+19,+21,10-13mar	Streuli et al 1980

Thrombocytopenia

46,XY,t(11;21)(q22;q21)	Tricot and Van Den Berghe 1982
46,XY,t(11;21)(q25;q?)	Tricot et al 1982

Pancytopenia

47,XX,+del(11)(q?),21q-	Nowell and Finan 1978

Dysmyelopoietic syndrome, special type

44,XY,-18,-21	Yamada and Furusawa 1976

SPECIAL LEUKEMIAS

Hairy cell leukemia

44,X,del(X)(q22),del(6)(q23),-7,-8,-10,del(11)(p15q21),-21, 14q+,+r,+mar	Sadamori and Sandberg 1983a

LYMPHOCYTIC LEUKEMIAS

Acute lymphocytic leukemia (ALL)

44,XY,-14,-21	Alimena et al 1977
44-45,XY,-14,-17,-21	Alimena et al 1977
45,XX,-21/52,XX,cx	Yamada and Furusawa 1976
46,XX,21q-	Tosato et al 1978
46,XX,21q-	Whang-Peng et al 1976c
46,XX,2q+,4q-,12p+,13q+,14q+,-20,+21	Whang-Peng et al 1976c
46,XY,13q+,21q-	Whang-Peng et al 1976c
46,XY,t(8;9)(q22;p24)/45,XY,t(8;9),-10,+12,-21	Morse et al 1982a
47,XX,+21"c"	Mitelman 1983
47,XX,+21,i(17q)	Yamada and Furusawa 1976
47,XX,+21	Alimena et al 1977
47,XY,+21	Prigogina et al 1979
48,XX,+5,-19,+21,+mar	Seabright 1983
48,XY,t(4;12)(q?;q?),t(6;15)(q?;q?),t(9;22)(q34;q11),11q-, +20,+21	Nordenson 1983
48-53,XX,t(4;11)(q21;q23),+6,+7,+9,+13,+21	Prigogina et al 1979
49,XX,+10,-21,+3r"c"	Stern et al 1979b
49,XX,+14,+17,+21,7q-	Shabtai and Halbrecht 1981b
52-56,XY,+5,+8,+11,+19,+20,+21,del(6)(q15),+mar	Prigogina et al 1979
53,XX,+2,+5,i(7q),+8,+13,+21,+21,+22,+17q+	Oshimura et al 1977a

Karyotype	Reference
53,XY,+X,+4,+8,+14,+15,+17,+21	Oshimura et al 1977a
54,XX,+6,+7,+8,+14,+17,+18,+21,+22	Prigogina et al 1979
54,XX,7p+,+10,+10,+14,+14,+18,+18,+21,+21/27,X,-X,-1,-2,-3,-4,-5,-6,-7,7p+,-8,-9,-11,-12,-13,-15,-16,-17,-19,-20,-22	Oshimura et al 1977a
54-55,XX,+5,+12,+14,+16,+17,+18,+21	Prigogina et al 1979
54-55,XY,+5,+8,+9,+14,+21,+mar	Prigogina et al 1979
54-57,XY,+3,+5,+6,+7,+12,+13,+14,+21,+2mar	Prigogina et al 1979
55,XX,+2,+4,+9,+14,+15,+19,+21,+21	Oshimura et al 1977a
55,XX,+4,+5,+6,+10,+17,+21,+3mar	Humbert et al 1978
55,XY,+X,+3,+6,+10,+13,+14,+16,+21,+21	Oshimura et al 1977a
56,XX,+3,+6,+12,+13,+14,+15,+18,+21,+r	Prigogina et al 1979
57,XX,+3,+4,+8,11p+,+14,+16,+17,+18,+21,+22,+2mar/47,XX,+8	Morse et al 1978
57,XY,+X,+5,+6,i(7q),+10,+12,+13,+15,+18,+21,+21,+22	Oshimura et al 1977a
61,XX,+X,+1,+2,+3,+4,+5,+6,+8,del(6)(q21),+13,+14,+15,+16,+18,+21,+22	Oshimura et al 1977a

ALL, FAB type L1

Karyotype	Reference
28,XX,-1,-2,-3,-4,-5,-6,-7,-8,-9,-11,-12,-13,-15,-16,-17,-19,-20,-22/56,XX,+X,+X,+10,+10,+14,+14,+18,+18,+21,+21	Brodeur et al 1981a
46,XY,del(2)(q32),-11,-12,+2mar/46,X,-Y,del(2),-4,-5,-11,-12,-21,+6	Kaneko et al 1981b
46,XY,t(21;22)(q22;q11)	Kaneko et al 1982e
52-56,XY,t(9;22),+4,+6,+9,+10,+14,+15,+20,+21,+22q-	Sandberg et al 1980
56,XY,+X,+4,+6,+10,+15,+17,+18,+21,+21,+mar	Kaneko et al 1982e
58,XX,+4,+6,+7,+10,+14,+14,+21,+del(18)(q21),+4mar	Kaneko et al 1982e

ALL, FAB type L2

Karyotype	Reference
45,X,-X,-1,-14,-17,+21,t(11;12)(p15;p11),+i(17q),+t(1;?)(p34;?)	Kaneko et al 1981b
48,XY,+8,+21	Prieto et al 1981
53,XY,t(9;22),+4,+5,+6,+7,+8,+17,+21	Sandberg et al 1980
56,XX,+1,+6,+10,+13,+15,+16,+17,+18,+20,+21	Mitelman 1983

ALL, FAB type L3

Karyotype	Reference
46,XX,t(2;8;14)(p11orp12;q24;q32),t(21;21)(p11;q11),+t(21;21)	Ekblom et al 1982
47,XX,t(8;14)(q22;q32),del(21)(q21),+mar	Alimena et al 1981
47,XX,t(8;14)(q24;q32),+mar/47,XX,t(8;14),del(21)(q21),+mar	Rossi et al 1982

Chronic lymphocytic leukemia

Karyotype	Reference
46,XX,t(6;21)(p21;q22)/46,XX,del(6)(q15)/46,XX,del(6)(q21)	Robert et al 1982
46,XY,del(11)(q22),+12,-13,-21,-22,+3mar/45,XY,del(11),+12,-13,-15,-21,-22,+2mar/46,XY,del(11),+12,+12,-13,-21,-22,+2mar/46,XY,t(1;8)(p22;q24)	Robert et al 1982
47,XX,+7/49,XX,+14,+17,+21	Shabtai 1983

Lymphocytic leukemia, special type

Karyotype	Reference
46,XY,-10,-13,5p+,+t(10;13)(p13;q11q32),21p+	Ueshima et al 1981
47,X,-Y,-4,+7,del(1)(q32),i(18q),21q+,i(22q),+2mar	Ueshima et al 1981
49,XX,+7,+8,+12,8q+,8p+,del(10)(q23),21q+/49,XX,+8,+12, +del(7)(p13),8q+,8p+,10q-,21q+	Ueshima et al 1981

MONOCLONAL GAMMOPATHIES

Multiple myeloma

Karyotype	Reference
47,XY,dup(1)(q11q32),t(1;10)(p22;p15),inv(6)(p21q13), t(11;14)(q23;q32),21p+,+mar	Liang et al 1979
48,X,-X,+1,+7,-8,+9,+21,del(1)(p36),4p+,del(5)(q22),9p+, 14q+	Liang et al 1979
52,X,-X,+t(1;?)(q12;?),+3,+7,+8,+9,+11,-13,+18,-21,+2mar	Philip et al 1980
55,XY,+3,+5,+7,+9,+11,+14,+15,+21,+21/53,XY,+3,+5,-8,+9, +11,+14,+15,-19,+21,14q+,+3mar	Liang et al 1979

Plasma cell leukemia / Plasmocytoma

Karyotype	Reference
44,X,-Y,del(1)(p11),+del(1)(q11),t(11;14)(q11;q32),14q+, -21,-22	Ueshima et al 1983
44-45,XY,+10,-17,-21	Wetter et al 1980

SOLID TUMORS

UNCLASSIFIED NEOPLASMS

Malignant neoplasm, NOS

47,XX,-2,-5,+6,-10,11q+,-13,+18,-21,t(1;?)(q21;?), t(1;13)(q21;p11),t(1;6)(q11;q23) — Kovacs 1978a

EPITHELIAL NEOPLASMS

Carcinoma, NOS

44-45,XX,t(1;4)(q11;q35),+3,5q-,-14,i(17q),-21 — Atkin and Baker 1982a

45,XY,-1,-6,-14,+21,+2mar/45,XY,t(1;14)(q11?;q24?),-6,-16, +21 — Sonta and Sandberg 1978a

45,XY,-4,-7,+8,-9,-12,+13,-17,-18,-19,-20,+21,+3mar — Reichmann et al 1981

46,XX,+8,+12,+21,-2,-15,-16 — Wake et al 1981

46,XX,del(2)(p24),+5,+8,t(9;17)(p?;p?),del(16)(q21),-20, -21 — Mark 1975b

46,XX,t(1;?)(p32;?),-2,3q-,5q-,6q+,-18,-21,+mar — Atkin and Baker 1982a

46,XY,-4,-5,-11,-13,-17,+20,+21,1p+,9p-,+3mar — Reichmann et al 1981

47,XX,+1,+3,t(11;14)(q?;q?),-12,-21,+r — Atkin and Baker 1982a

48,XX,+del(1)(p32),-10,-21,+mar — Atkin and Baker 1982a

49,X,-Y,-3,-5,-14,+15,-16,-17,+19,+21,+22,1q+,5q-,6q-,8p-, 9p-,11p-,18p+,20q-,22q-,t(Y;14),+3mar — Reichmann et al 1981

49,XY,+3,+8,+21 — Sonta and Sandberg 1978a

50,XX,+X,+8,+12,+21 — Sonta et al 1977

50,XX,inv(3)(p13q25),+8,+14,+21/63,XX,inv(3),del(11)(p13), cx — Sonta and Sandberg 1978a

54,XX,+1,+2,+8,+11,+13,+14,+21,+21 — Sonta et al 1977

54,XY,+1,+5,+9,+17,+19,+20,+21,+21 — Sonta and Sandberg 1978a

55,XY,t(1;17)(p36;q21),+2,+2,+4,+8,+8,-10,+15,+18,+20,+21, +2mar — Sonta and Sandberg 1978a

58,XY,del(3)(p14),+4,+7,+8,+10,+11,+12,+14,+15,+17,+18,+19, +21 — Kovacs 1978a

63,XX,-1,+3,+5,+6,+7,+8,+10,+11,+12,-13,-15,+20,+21,+22, +9mar — Sandberg 1977

79,XX,cx,i(1q),t(1;3)(p13;q29),t(1;14)(p13;p13), t(12;?)(q24;?),t(15;21)(p11;q11),i(5p),r — Kovacs 1981

Carcinoma NOS, metastatic

35,XX,-1,-3,-4,-5,-7,-8,-10,-11,-12,-13,-14,-15,-16,-17, -18,-19,-20,-21,-22,+8mar — Pathak et al 1979

47,XX,+4,+16,-21 — Musilova et al 1981

60,XX,t(1;?),t(7;?)(q3?;?),i(10q),der(11),t(14;21),cx — Bartnitzke and Bullerdiek 1983

65,XX,i(1q),t(1;9),der(5),t(6;10)(p?;q?),t(14;?),der(21), cx — Bartnitzke and Bullerdiek 1983

85-90,XY,i(9q),del(1)(q2?),t(7;?)(q35;?),t(3;?)(p11;?), t(21;?)(q11;?) — Kakati et al 1975

Large cell carcinoma

55-70,XY,t(4;5;14),del(1)(q11q21),t(7;11)(p22;q13), t(7;21)(q11;q22),t(3;8)(p?;q?),t(6;11)(p11;p11),i(16q), t(3;?)(q11;?),t(21;?)(q11;?),t(13;?) — Kakati et al 1975

Papillary carcinoma

54,XY,+del(2)(q11),t(2;14)(q11;q32),t(3;11)(p13;p15), +i(5p),+8,+9,+12,+16,+17,+21,+2mar — Pathak et al 1982

Adenoma

46,XY,t(6;21)(q13;q22)/46,XY,t(1;14)(q42;p12)/45,X,-Y — Mark et al 1981b
45,X,-Y
46,Y,t(X;15)(q24;q15)/45,X,-Y

Adenocarcinoma

43-46,XY,i(10q),15q+,-17,+19,-21,-22,dmin — Couturier-Turpin et al 1982

54,XX,+5,+6,+7,+9,+11,+21,+21,+del(1)(q32) — Sonta et al 1977

MESENCHYMAL NEOPLASMS

Mesothelioma, malignant

41,XY,del(1)(p13),t(1;3)(p13;q11),t(7;7)(q11;q36), t(9;12)(p24;q13),del(12)(q13),del(13)(q12q14),-14,-15, t(17;?)(p13;?),i(19p),-20,-21,-22 — Mark 1978

MELANOMAS

Malignant melanoma

44,X?,+2,+9,-6,-8,-15,-21,9p-,19q+ — Wake et al 1981
46,XY,5q+,+7,-8,-9,-10,-13,-20,+21,+3mar — McCulloch et al 1976
78-82,XY,t(4;9)(q21;q21),del(9)(p13),t(14;14)(p11;q12), del(2)(q24),t(14;?)(q11;?),t(21;?)(q11;?),del(17)(q11) — Kakati et al 1977

Malignant melanoma, metastatic

24,X,cx,t(1;15),t(1;14)(q12;p11),del(7)(p13), t(6;7)(p11;q11),t(13;21)(p11;q11),+r — Atkin and Baker 1981

NEUROGENIC NEOPLASMS

Oligodendroglioma

45,XY,t(21;22)(p11;q11),7q-,9q-,14q+ Yamada et al 1980

Neuroblastoma

45,XY,-21,t(1;14)(p32;q31),t(8;?)(q24;?),dup(14)(q11q24) Brodeur et al 1981b

Retinoblastoma

46,X,-Y,1p+,+1p-,5q+,7p+,10q-,16q-,17q+,21p+ Kusnetsova et al 1982

Meningioma, benign

38,X,-Y,-22,-21,-17,-14,-13,-10,-9,-4,-1p-,+mar Zankl et al 1975a
44,X,-Y,-22/42,X,-Y,-17,-21,-22 Yamada et al 1980
44,XX,-8,-22/40-44,XX,-8,-9,-19,-20,-21,-22 Mark et al 1972a
45,XX,-22/40-44,X,-X,-8,-9,-15,-21,-22 Mark et al 1972b
45,XX,-22/44-45,XX,-8,-12,-21,-22 Mark et al 1972b

EMBRYONAL AND MISCELLANEOUS NEOPLASMS

Embryonal carcinoma, NOS

50,XXY,+7,+21,+mar"c" Mann et al 1983

Teratoma, benign

47,XX,+21"c" Linder et al 1975b

Teratoma, malignant

67,XY,+1,+3,+4,+5,+6,+7,+7,+8,+10,+11,+13,+13,+14,+16,+17,
+18,+20,+20,+21,+21 Sonta et al 1977

LYMPHOMAS

1. UNCLASSIFIED AND MISCELLANEOUS LYMPHOMAS

Malignant lymphoma, NOS

47,XX,t(1;14)(q23;q32),+3,t(6;21)(p25;q11),+8,
t(7;11)(q11;q25),-21 Yamada et al 1980
48,XY,+7,+21"c" Kristoffersson 1983

Non-Hodgkin's lymphoma, lymphocytic, NOS

45,XX,i(17q),t(17;21)(q?;q?) Fleischman and Prigogina 1977

47,XY,+21,14q+"c" Oshimura et al 1981b

Non-Hodgkin's lymphoma, histiocytic, NOS

45-47,XX,t(10;14)(q21;q32),+8,-18,-21 Mark et al 1977

46,XX,del(1)(p31p35),+3,del(9)(p11),t(10;?)(q26;?), del(11)(q13),-13,t(11;14)(q13;q32),t(17;21), t(9;22)(q11;p13),+r Mark et al 1977

50,XY,t(2;?)(p25;?),del(2)(p13),+i(3p),del(6)(p21),+7,+8, +9,+9,-11,-13,+14,-15,+20,+21,-22 Mark et al 1978

44,XY,del(2)(p13),t(6;?)(p21;?),del(11)(q13), t(11;14)(q13;q31),i(15q),-20,-22

Non-Hodgkin's lymphoma, T-cell, NOS

47,XX,t(2;?),inv(3),4p-,del(5)(q14q32),6p+,6q+,t(7;7),9q+, 10p-,13q-,t(14;14),-15,-15,16q-,17p+,+21,+2mar Gaeke et al 1981

Malignant histiocytosis

50,XY,+5,+12,+13,+21 Sonta et al 1977

Angioimmunoblastic lymphadenopathy

47,XY,+3/47,XY,+3,-9,+21/48,XY,+3,+8,+9,-20 Hossfeld et al 1976

48,XY,+3,+18/51,XY,+5,+15,+19,+21,+22 Kaneko et al 1982c

56,XY,+5,+5,+6,+10,+15,+15,+19,+20,+21,+22 Kaneko et al 1982c

2. HODGKIN'S LYMPHOMAS

3A. NON-HODGKIN'S LYMPHOMAS - RAPPAPORT CLASSIFICATION

Lymphocytic, poorly differentiated, diffuse

43,XX,-4,-5,-9,i(17q),-18,-21,+2mar Goh et al 1980

48,XY,t(14;?)(q32;?),21q+,+mar Fukuhara et al 1979a

Lymphocytic, poorly differentiated, nodular

52,X,-X,+2,-4,+5,+7,+8,-14,-15,-16,+20,+21, +t(1;15)(p11;q11),+t(12;17)(q24;q11),+t(14;?)(q32;?), +mar Kaneko et al 1982a

Mixed cell, nodular

50,XX,t(3;6)(q29;p21),del(9)(p13q22),+11,+12,+21 — Mark et al 1976

Histiocytic, diffuse

47,XY,-11,-22,del(1)(q42),del(2)(q33),del(5)(q22q31), del(7)(q32),del(9)(p22),t(21;22)(q22;q11), +t(11;?)(q23;?),+t(21;22)(q22;q11),+mar — Kaneko et al 1982a
48,XY,-6,-8,+21,i(6p),+t(8;?)(q11;?),+t(19;?)(q13;?) — Kaneko et al 1982a
51,XY,+X,-2,+5,+11,-12,+21,del(6)(q21),+t(2;?)(q33orq35;?), +t(12;?)(p11;?),+mar — Kaneko et al 1982a

3B. NON-HODGKIN'S LYMPHOMAS - KIEL CLASSIFICATION

Immunocytoma

47,XX,+3,+del(3)(q11),del(10)(q11),-21 — Kristoffersson 1983
45-46,XY,-8 — Kristoffersson 1983
47,XY,15p+,+20/46,XY,del(5)(q23),del(6)(p22),+del(7)(p11), -21

Centroblastic-centrocytic, diffuse

48-51,XY,del(1)(p33q23q25),-2,+del(3)(p22),+4,4p+,+5,+6p+, +7,+8,+9,+21,+22 — Kristoffersson 1983
45-51,XY,del(1)(p33q23q25),del(2)(q14),+del(3)(p22),+4,+5, +7,+8,-12

Centroblastic, diffuse

46-51,XX,+3,del(6)(q21),+del(6)(q21),+7,-8,del(9)(q11),+13, -16,-17,+19,-20,-21,+1-5mar — Kristoffersson 1983

Lymphoblastic, Burkitt's type

??,XY,dup(1)(q12orq21q31),3p-,3q-,+6p-,+8,+14,+15,+18,-21 — Douglass et al 1980a
??,XY,t(8;14),+15,+21 — Douglass et al 1980a
47,XX,t(8;22)(q24;q11),+8,t(11;21)(q14;q22) — Berger and Bernheim 1982
47,XY,+11,-14,+21,inv(1)(p11p31),11q+,21p+ — Biggar et al 1981

Immunoblastic

45-46,XX,+3,del(6)(q21),+4,+16,+21,-C,-C — Kristoffersson 1983
47-48,XY,+4,del(4)(q21),+5,t(15;?)(q25;?),+19,-21 — Reeves and Stathopoulos 1976
48-49,XY,+3,+7,+11,t(12;?)(p13;?),t(13;?)(p13;?),-16,-20, +21,+22 — Reeves and Pickup 1980
51,XY,+X,t(3;3)(q11;q27),t(8;?)(q2?;?),+7,+12,+21,+21 — Reeves and Pickup 1980

3C. NON-HODGKIN'S LYMPHOMAS - LUKES AND COLLINS CLASSIFICATION

Sezary's syndrome

Karyotype	Reference
42,XY,-13,-21,-22,1p+,17q+	Nowell et al 1982
44-46,XX,t(2;9),t(2;13),17p+,19p+,-21,-22	Edelson et al 1979
50,XX,1q+,2p-,+3q-,+6q-,+9,-15,+21,+mar	Nowell et al 1982

Follicular center cell, diffuse, small non-cleaved

Karyotype	Reference
44,X,-Y,t(1;14)(p21;q13),5p+,del(6)(q13),del(8)(p21),9p+, 9p-,t(11;21)(q23;q22),13q+,-10,+12,-17p+,der(19),22q+	Fukuhara et al 1979a

Immunoblastic type (B-cell)

Karyotype	Reference
48,XY,t(4;12)(q?;q?),t(6;15)(q?;q?),t(9;22)(q34;q11),11q-, +20,+21	Nordenson et al 1983

3D. NON-HODGKIN'S LYMPHOMAS - WORKING FORMULATION

Follicular, mixed small cleaved and large cell

Karyotype	Reference
50,Y,inv(X),+3,+7,+10,t(12;12),+12,t(14;18)(q32;q21),+18, 18q-,-21	Yunis et al 1982

Follicular, large cell

Karyotype	Reference
47,XX,t(2;13),t(14;18)(q32;q21),+21	Yunis et al 1982
50,XY,1q-,t(2;3),+3,+10,t(1;12),13q-,+15,-19,t(3;19),+21	Yunis et al 1982

Large cell, immunoblastic

Karyotype	Reference
45,XY,ins(2;21),t(8;14)(q24;q32),t(9;19),t(17;21),-21	Yunis et al 1982

Chromosome 22

HEMATOLOGICAL DISORDERS

UNCLASSIFIED LEUKEMIAS

Acute leukemia, NOS

Karyotype	Reference
46,XY,+del(1)(q32),-9,del(9)(q13),del(22)(q11)	Prigogina et al 1979
54,XX,+X,+3,+6,+8,+8,+13,+19,+22	Muir et al 1977

NONLYMPHOCYTIC LEUKEMIAS

Acute nonlymphocytic leukemia (ANLL)

Karyotype	Reference
??,X?,22q-	Bloomfield et al 1977
39,X,-X,-3,-5,-13,-14,-15,-16,-17,-21,-22,del(7)(p11), t(12;?)(p13;?),t(16;?)(p13;?),+3mar	Mitelman 1983
39,X,-X,-3,-5,-13,-14,-15,-16,-17,-21,-22,del(7)(p11), t(12;?)(p13;?),t(16;?)(p13;?),+3mar	
42,XY,-5,t(12;?)(p11;?),-18,-22,-22	Hustinx and Rutten 1983
43,X,-Y,t(1;8)(p22;q24),-1,-11,-11,-13,-16,-17,-22, t(1;22)(q11;p11),t(11;13)(q23q25;q12q14),t(17;?)(p11;?), +mar	Oshimura et al 1976b
44,XX,t(8;11)(p23;p13 or p15),-20/46,XX,del(7)(q11), t(1;3)(q2?;q2?),-20,-21,-22,+3mar	Philip et al 1978a
44,XY,-7,-17,-20,-22,+2mar"s"	Pedersen-Bjergaard et al 1982
44,XY,5q-,-7,12p-,-22"s"	Whang-Peng et al 1979
45,X,-Y,t(8;21)(q22;q22),t(9;22)(q34;q11)	Francesconi and Pasquali 1978a
45-47,XX,-2,-7,7q-,-8,+15,+22,+2mar"s"	Whang-Peng et al 1979
46,X?,t(9;22)(q34;q12)	Rozynkowa et al 1977
46,XX,12q+,-22,+mar	Chrz et al 1983
46,XX,del(1)(p32),del(5)(q13),+8,t(9;?)(p11;?), t(11;22)(q11;q13),-16,-17,-19,+3mar	Muir et al 1977
46,XX,t(1;22)(p36;q11),del(5)(q14q34),del(7)(q22q36),r(22)	Yunis 1982
46,XX,t(9;22)(q34;q11)	Abe and Sandberg 1979
46,XY,+del(7)(p11),-8,del(12)(p11),t(21;?)(q22;?)/50,XY,+Y, +del(7)(p11),del(12)(p11),+17,t(17;?)(q25;?), t(21;?)(q22;?),+22	Mitelman 1983
46,XY,+del(7)(p11),-8,del(12)(p11),t(21;?)(q22;?)/47,XY,+9	

Karyotype	Reference
46,XY,t(11;22)(q13;p11)"s"	Oshimura et al 1976b
46,XY,t(9;22)	Lessard and Le Prise 198
46,XY,t(9;22)	Lessard and Le Prise 198
46,XY,t(9;22)	Carbonell 1983
46,XY,t(9;22)(q34;q11),21q+/46,X,-Y,t(9;22),+8	Abe and Sandberg 1979
46,XY,t(9;22)(q34;q11)	Abe and Sandberg 1979
46,XY,t(9;22)(q34;q11)	Mitelman 1983
46,XY,t(9;22)(q34;q11)	
46,XY,t(9;22)(q34;q11)	
46,XY,t(9;22)(q34;q11)/47,XY,t(9;22),7p+,+19	
46,XY,t(9;22)	
47,XX,+22	Bernard et al 1982a
47,XX,t(3;5)(q21;q31),t(1;13)(p36;q14),+8,+22	Oshimura et al 1976b
47,XX,t(9;22)(q34;q11),+22q-/48,XX,t(9;22),+8,+22q-	Mitelman 1983
49,XX,t(9;22)(q34;q11),+8,+13,+22q-	
49,XX,t(9;22)(q34;q11),+8,+22q-,+mar	
50,XX,t(9;22)(q34;q11),dup(1)(q25q44),+del(3)(p11),+8, +22q-,+mar	
47,XY,+22	Oshimura et al 1976b
48,XX,+8,t(9;22),+10	Chrz et al 1983
48,XX,t(9;22)(q34;q11),+del(22)(q11),+21	Abe and Sandberg 1979
50,XX,t(9;22)(q34;q11),+6,+8,+10,+del(9)(p13), +del(22)(q11)	Abe and Sandberg 1979
51,X,-Y,+1,+18,+20,+22,+inv(4),+mar	Zuelzer et al 1976
51,XX,del(5)(q?),+7,+8,+13,+17,+22	Van Den Berghe et al 197
52,XX,1p+,+10,+13,+19,+21,+21,+22"c"	Berger et al 1973

ANLL, FAB type M1

Karyotype	Reference
41,XY,-1,-2,-4,-5,-12,-15,-17,-17,-18,-22,+5mar	Sessarego et al 1981a
44,XX,-5,-7,-20,-22,+2mar/44,XX,-3,-11,-11p+,-22	Prieto et al 1981
45,XX,-7,-9,+t(7;9)(q11;p11),22q-/46,XX,t(9;22)(q34;q11)	Sasaki et al 1983
45,XX,t(9;22)(q34;q11),-7	Sasaki et al 1983
45,XY,-7,t(9;22)(q34;q11)	Sasaki et al 1983
46,XX,t(9;22)/45,XX,-7,t(9;22)	Hagemeijer et al 1981a
46,XY,t(12;22)(p12;q11)	Hagemeijer et al 1981a
46,XY,t(9;22)(q34;q11)	Sasaki et al 1983
46,XY,t(9;20;22)(q34;q12;q11)/46,XY,t(9;22)(q34;q11),-6,-7, -7,-8,10q+,+t(7;8),i(8q)	
47,XY,t(9;22)(q34;q11),+18	Sasaki et al 1983
47-48,XY,+8,t(9;22)(q34;q11),-20,+t(1;20)(q21;q13), t(10;21)(p11;q22),+22q-	Sasaki et al 1983
49,XX,+X,-1,+der(1)(q21q31),+del(1)(p22),+t(1;5)(p11;q35), -5,t(9;22)(q34;q11),+22q-	Sasaki et al 1983
58,XY,+1,+4,+5,+6,+8,+11,+13,+13,+14,+14,+15,+21,+21, t(9;22)(q34;q11)	Alimena 1983

ANLL, FAB type M2

Karyotype	Reference
42,XY,5q-,7q-,-8,-12,-17,-21,-22,+i(17q),+2mar	Berger et al 1981a
43,XX,+2,3q+,-5,-20,-22	Berger et al 1981a
45,XY,-7,t(9;22)(q34;q11)	Sasaki et al 1983
46,XX,t(9;22)(q34;q11),i(17q)	Sasaki et al 1983
46,XY,+19,-22	Brodeur et al 1983
46,XY,del(5)(q13?q22?),t(9;22)(q34;q11)	Sasaki et al 1983

46,XY,t(9;22)(q34;q11)	Bernstein et al 1982b
47,XX,+8,t(12;22)(p13;q12)	Mitelman 1983
47,XX,+8,t(12;22)(p13;q12)	
47,XX,+8,t(12;22)(p13;q12)	
47,XX,1p+,der(5),-6,12p+,-13,-14,-16,-17,-17,-21,+22,+7mar, dmin	Cooperman and Klinger 1981
55,XX,+16,+17,+18,+20,+21,+22,+3mar	Prieto et al 1981

ANLL, FAB type M1+M2

42,X,-Y,4q+,-8,-16,17p+,-21,-22,+Yp+"s"	Sandberg et al 1982a
44,XX,-3,-4,-5,+9,-17,-18,-21,-22,+5mar	Garson 1980
44,XX,del(3)(p21),del(5)(q12),-7,-20,i(22)(q11)	Mitelman et al 1981
44,XY,-7,-16,-19,-21,-22,+3mar"s"	Sandberg et al 1982a
44-45,XX,-7,-22	Mitelman et al 1981
45,XY,-5,-19,-20,-22,+t(7;12)(q11;q13),+r,+3mar	Fitzgerald et al 1983
45,XY,t(19;22)(p13;q11),t(1;7)(q21;q22),t(7;9),+del(8)(p?), -15	Oshimura and Sandberg 1977
45-46,XX,+14,-7,-22,+16,-20	Mitelman et al 1981
45-46,XY,t(2;6)(q37;q23),-7,del(11)(q21),del(22)(q11)	Mitelman et al 1981
46,XX,+14,-22/46,XX,-7,+14,+16,-20,-22	Alimena et al 1977
46,XX,t(9;22)	Garson 1980
46,XX,t(9;22)	Garson 1980
46,XX,t(9;22)(q34;q11)	Sasaki et al 1975
46,XX,t(9;22)(q34;q11)	Sasaki et al 1975
46,XY,t(9;22)(q34;q11)	Mitelman et al 1981
46,XY,t(9;22)(q34;q11),t(2;15)(q37;q12)	Mitelman et al 1981
46,XY,t(9;22),del(10)(q22)	Wayne et al 1979
47,XY,t(9;22),+8,del(10)(q22)/48-50,XY,t(9;22),+8,+9, del(10)(q22),+10,+22q-	
46,XY,t(9;22),t(7;10)(q34;q22)	Wayne et al 1979
46,XY,t(9;22)/47,XY,t(9;22),+17/47,XY,t(9;22),+22q-	Nagao et al 1977
46-50,XY,+Y,+del(7)(p11),-8,+9,del(12)(p11),+17,21q+,+22	Mitelman et al 1981
47-49,XY,-5,del(5)(q15),-7,i(8q),dic(9),-13,-16,-22, +6-9mar"s"	Testa et al 1981b
48,XX,+t(1;22)(q11;p12),+del(19),-22,+mar	Fitzgerald et al 1983
49,XX,+6,t(9;22)(q34;q11),+17,+21	Nordenson 1983
57,X,-Y,-6,+8,+15,+15,+19,+19,+20,+20,+21,+21,+22,+3mar	Fitzgerald et al 1983
57,XY,+6,+11,+13,+15,+21,+21,+21,+21,+21,+21,+22"c"	Shiraishi et al 1982

ANLL, FAB type M3

45-51,XY,t(3;12)(p14;q24),t(13;18)(q11;p11),+1,+6,+16,+19, +21,+22	Mitelman et al 1981

ANLL, FAB type M4

42,XY,-4,-15,-16,-22	Prieto et al 1981
42,XY,-5,-9,-10,-15,-16,-17,-20,-21,-22,+5mar	Testa et al 1979
43,XY,+1,-5,+6,-9,-10,-15,-16,-17,-20,-21,-22,+6mar	
43,XY,-2,-3,-6,-7,-12,-15,-21,-22,del(2)(p11),del(11)(p14), t(14;?)(q32;?),t(17;?)(p13;?),del(22)(q11),+5mar"s"	Rowley et al 1981a
44,XY,-7,+20,-21,-22"s"	Berger et al 1981a
46,XY,+19,-22	Bernstein et al 1982b

Karyotype	Reference
46,XY,2q+,del(18)(p11),del(22)(q11)/47,XY,2q+,+8, del(18)(p11),del(22)(q11)"s"	Alimena et al 1981
45-47,XY,2q+,+8,del(18)(p11),-21,del(22)(q11)"s"	
46,X,-Y,2q+,+8,del(21)(q21)"s"	
46,XY,del(22)(q?)	Padre-Mendoza et al 1978
46,XY,t(3;22)(q21;q13)	Alimena 1983
46,XY,t(9;22)(q34;q11)/48,XY,t(9;22),+8,+22q-	Bernstein et al 1982b
47,XX,+22	Mitelman et al 1981
47,XX,-7,+21,+22/45,XX,-7	Prieto et al 1981
47,XY,+22	Shiloh et al 1979
47,XY,+22	Brodeur et al 1983
47,XY,+8,t(9;22)	Prieto et al 1981
47,XY,del(7)(q32),+22	Prigogina et al 1979
47,XY,t(9;22)(q34;q11),t(16;?)(p13;?),-18,del(20)(p12q12), +22,+mar	Li et al 1981b
47-48,XY,+8,+10,+21,+22	Mitelman et al 1981

ANLL, FAB type M5

Karyotype	Reference
46,XX,-18,-22,+2r/46,XX,t(2;9)(q23;p22),-3,-22,+2r/49,XX, t(2;9),-22,+4r	Alimena 1983
46,XY,t(9;22)(q34;q11)/47,XY,+8,t(9;22)	Hagemeijer et al 1981a
46-51,XX,-3,-18,-22,t(2;9)(q23;p22),+2-4r	Mitelman et al 1981
47,XY,+8/48,XY,+8,+17/48,XY,+8,+22	Garson 1980

ANLL, FAB type M4+M5

Karyotype	Reference
45,XY,+9,-15,-22	Li et al 1983
46,XY,t(9;22)	Li et al 1983
47,XY,+22	Zuelzer et al 1976
47,XY,+8/48,XY,+8,+10/48,XY,+21,+22	Alimena et al 1977
48,XY,t(1;6)(q21;q27),+del(8)(q22),+22	Shiraishi et al 1982
66,XY,+1,+1,+2,+4,+6,+6,+7,+8,+8,+10,+11,+11,+12,+12,+18, +19,+21,+21,+22,+22	Li et al 1983

ANLL, FAB type M6

Karyotype	Reference
44,XY,t(3;5;11),17p+,-22	Garson 1980
44-45,XY,-2,-4,-7,-10,-13,-16,-17,-22,+3mar	Mitelman et al 1981
45,X,-X,-5,-21,22q-,+3mar"s"	Sandberg et al 1982a
45,XX,t(4;22)(p16;q11),-7/46,XX,t(4;22)(p16;q11)	Oshimura and Sandberg 1977
45-49,XY,-4,-7,-8,-10,+14,+16,-21,-22,+2mar	Mitelman et al 1981
48,XX,+t(1;12)(p22;p13),+21,del(22)(q11)	Brodeur et al 1983

Chronic myeloid leukemia, t(9;22)

Karyotype	Reference
??,X?,t(9;22),1p+,+22q-,cx	Carbonell et al 1982a
44,XX,t(9;22),-16,-17/46,XX,t(9;22),-22,+22q-	Stoll and Oberling 1978
44,XX,t(9;22),-16,-17/46,XX,t(9;22),-22,+22q-	
44,XX,t(9;22),-16,-17/46,XX,t(9;22),-22,+22q-	
45,X,-Y,t(9;22)	Hartley and McBeath 198[illegible]
46,XY,t(9;22)/50,XY,t(9;22),+19,+22q-,+C,+C	
45,XY,t(9;22),-17	Stoll and Oberling 1978
47,XY,t(9;22),+8,-17,+22q-	

45,XY,t(9;22),-20,-22,+22q-/46,XY,t(9;22),-20,-22,+8,+22q-	Stoll and Oberling 1978
44,XY,t(9;22),-17,-22,-22,+22q-/46,XY,t(9;22),-20,+22q-/47, XY,t(9;22),+11,-22,+22q-	
46,X,-Y,t(9;22)(q34;q12),+13/47-48,X,-Y,t(9;22),+8,+13, +22q-/47,XY,t(9;22),+22q-	Izakovic et al 1982
46,X,-Y,t(9;22),+22q-/47,X,-Y,t(9;22),+8,i(17q),+22q-	Bernstein et al 1980b
46,X,i(Xq),t(9;22)/47,X,i(Xq),t(9;22),+22q-	Lyall and Garson 1978
46,X?,t(9;22)(q34;q11),i(17q)	Lawler et al 1976
46,X?,t(9;22),-16,i(17q),+del(22)(q11)	
46,X?,t(9;22)(q34;q11)/47,X?,t(9;22),+del(22q)/48,X?, t(9;22),+C,+del(22q)	Lawler et al 1976
46,XX,t(9;22)(q34;q11)	Whang-Peng et al 1973
47,XX,t(9;22)(q34;q11),+del(22)(q11)/46,XX, t(9;22)(q34;q11),dic(22)(q11)	
46,XX,t(9;22)(q34;q11)/47,XX,t(9;22)(q34;q11), +del(22)(q11)	Hayata et al 1975b
46,XX,t(9;22)(q34;q11)	Mitelman et al 1974a
46,XX,t(9;22)(q34;q11)	
46,XX,t(9;22)(q34;q11)	
46,XX,t(9;22)(q34;q11)	
46,XX,t(9;22)(q34;q11)	
46,XX,t(9;22)(q34;q11)/48,XX,t(9;22)(q34;q11),+17, del(22)(q11)	
46,XX,t(9;22)(q34;q11)/48,XX,t(9;22)(q34;q11),+17, del(22)(q11)	
46,XX,t(9;22)(q34;q11)/48,XX,t(9;22)(q34;q11),+17, +del(22)(q11)	
46,XX,t(9;22)(q34;q11)/47,XX,t(9;22),+del(22q)	Sonta and Sandberg 1978b
46,XX,t(9;22)(q34;q11)/47,XX,t(9;22),+22q-	Shiloh et al 1979
46,XX,t(9;22)(q34;q11)/51,XX,t(9;22),+5,+8,+17,+19,+22q-	Misawa 1978
46,XX,t(9;22)(q34;q11)/47,XX,t(9;22),+22q-	Misawa 1978
46,XX,t(9;22)(q34;q11)	Rajasekariah et al 1982
50,XX,t(9;22)(q34;q11),t(9;22)(q34;q11),+8,+11,+13, t(16;17)(p11;p12),+22	
46,XX,t(9;22)(q34;q11),+8,+22q-	Alimena et al 1982b
46,XX,t(9;22)(q34;q11)/47,XX,t(9;22),+22q-/49,XX,t(9;22), +8,+17,+22q-/49,XX,t(9;22),+18,+19,+22q-/50,XX,t(9;22), +8,+17,+18,+22q-	Mitelman 1983
46,XX,t(9;22)(q34;q11)/51,XX,t(9;22),+6,+8,+10,+19,+22q-	Mitelman 1983
46,XX,t(9;22)(q34;q11),i(17q)/47,XX,t(9;22),+8,i(17q), +22q-	Tomiyasu et al 1982
46,XX,t(9;22)(q34;q11)	Mitelman 1983
46,XX,t(9;22)	
46,XX,t(9;22)	
46,XX,t(9;22)	
46,XX,t(9;22)	
46,XX,t(9;22)	
46,XX,t(9;22)	
46,XX,t(9;22)/48,XX,t(9;22),+17,+22q-	
46,XX,t(9;22)/48,XX,t(9;22),+17,+22q-	
46,XX,t(9;22)/48,XX,t(9;22),+17,+22q-	
46,XX,t(9;22),-10,+12,-19,+22q-	Stoll and Oberling 1978
46,XX,t(9;22),2p+/48,XX,t(9;22),2p+,+19,+22q-	Sonta and Sandberg 1977b
49,XX,t(9;22),2p+,+19,+22q-,+22q-	
46,XX,t(9;22),dic(22q-),-17,+mar/46,XX,t(9;22)	Hossfeld 1975b

47,XX,t(9;22),dic(22q-),-17,+2mar	
47,XX,t(9;22),dic(22q-),-17,+2mar	
46,XX,t(9;22),t(X;17)(q?;q?),-17,+22q-	Borgström et al 1982
46,XX,t(9;22)/46,XX,t(9;22),+i(22q-)	Hagemeijer et al 1980b
46,XX,t(9;22)/47,XX,t(9;22),+11/48,XX,t(9;22),+11, +del(22)(q?)	Philip 1975a
46,XX,t(9;22)/47,XX,t(9;22),+22q-/46,XX,t(9;22),-7, +22q-/47,XX,t(9;22),-7,+19,+22q-	Hagemeijer et al 1980b
46,XX,t(9;22)/47,XX,t(9;22),+22q-/48,XX,t(9;22),+8,+22q-	Bernstein et al 1980b
46,XX,t(9;22)/47,XX,t(9;22),+22q-	Kohno and Sandberg 1980
46,XX,t(9;22)/48,XX,t(9;22),+19,+22q-/50,XX,t(9;22),+8,+12, +19,+22q-/50,XX,t(9;22),+8,+14,+17,+19,+21	Hagemeijer et al 1980b
46,XY,t(9;22)"s"	Hossfeld 1975b
46,XY,t(9;22)/51,XY,t(9;22),+10,+14,+19,+22q-,+mar"s"	
51,XY,t(9;22),+10,+14,+19,+22q-,+mar"s"	
51,XY,t(9;22),+10,+14,+19,+22q-,+mar"s"	
46,XY,t(9;22)	Mitelman 1983
50,XY,t(9;22)(q34;q11),del(3)(q11),+4,+5,+12,+22q-	
46,XY,t(9;22)	Mitelman 1983
47,XY,t(9;22)(q34;q11),+22q-	
46,XY,t(9;22)(q34;q11)	Mitelman 1983
46,XY,t(9;22)	
46,XY,t(9;22)	
46,XY,t(9;22)	
46,XY,t(9;22)	
46,XY,t(9;22)/49,XY,t(9;22),+8,+19,+22q-	
46,XY,t(9;22)/47,XY,t(9;22),+22q-/48,XY,t(9;22),+19, +22q-/49,XY,t(9;22),+8,+19,+22q-	
46,XY,t(9;22)(q34;q11)	Whang-Peng et al 1973
46,XY,t(9;22)(q34;q11),dic(22)(q11)/46,XY,t(9;22)(q34;q11)	
46,XY,t(9;22)(q34;q11)	Whang-Peng et al 1973
46,XY,t(9;22)(q34;q11)/47,XY,t(9;22)(q34;q11), +del(22)(q11)/46,XY,t(9;22)(q34;q11),dic(22)(q11)	
46,XY,t(9;22)(q34;q11)	Whang-Peng et al 1973
46,XY,t(9;22)(q34;q11)/47,XY,t(9;22)(q34;q11), +del(22)(q11)/46,XY,t(9;22)(q34;q11),dic(22)(q11)	
46,XY,t(9;22)(q34;q11)	Whang-Peng et al 1973
46,XY,t(9;22)(q34;q11)/47,XY,t(9;22)(q34;q11), +del(22)(q11)/46,XY,t(9;22)(q34;q11),dic(22)(q11)	
46,XY,t(9;22)(q34;q12)	Whang-Peng et al 1976b
45,XY,t(9;22)(q34;q12),t(7;22)(p12;p13)	
46,XY,t(9;22)(q34;q12),del(8)(q23)	
46,XY,t(9;22)(q34;q11)	Mitelman et al 1975a
46,XY,t(9;22)(q34;q11)	
46,XY,t(9;22)(q34;q11)	
46,XY,t(9;22)(q34;q11)	
46,XY,t(9;22)(q34;q11)/49,XY,t(9;22),+8,+19,+del(22)(q11)	
46,XY,t(9;22)(q34;q11)/47,XY,t(9;22),+del(22)(q11)/48,XY, t(9;22),+19,+del(22)(q11)/49,XY,t(9;22),+8,+19, +del(22)(q11)	
46,XY,t(9;22)(q34;q11)/47,XY,t(9;22),i(17q),+del(22q)/49, XY,t(9;22),+8,+21,i(17q),+del(22q)	Sonta and Sandberg 1978b
46,XY,t(9;22)(q34;q11)/47,XY,t(9;22),+del(22q)	Sonta and Sandberg 1978b
46,XY,t(9;22)(q34;q11),11p-/47-48,XY,t(9;22),+del(22)(q11), 11p-,+8/46,XY,t(9;22),11p-,dic(22q-)	Prigogina et al 1978

46,XY,t(9;22),dic(22q-)	
46,XY,t(9;22),dic(22q-)	
46,XY,t(9;22)(q34;q11)	Daniel et al 1978
24,XY,t(9;22)(q34;q11)/46,XY,t(9;22)/47,XY,t(9;22),+22q-	
46,XY,t(9;22)(q34;q11)/47,XY,t(9;22),+22q-/47,XY,t(9;22),-20,t(11;?),+22q-/48,XY,t(9;22),+8,-20,+t(11;?),+22q-	Misawa 1978
46,XY,t(9;22)(q34;q11),i(17q)/47,XY,t(9;22),+8,i(17q)/47,XY,t(9;22),+22q-/48,XY,t(9;22),+8,i(17q),+19	Tomiyasu et al 1982
46,XY,t(9;22)(q34;q11),t(7;11)(p14;p15)/47,XY,t(9;22),t(7;11),+22q-	Tomiyasu et al 1982
46,XY,t(9;22)(q34;q11)	Ishihara et al 1982
27,X,-Y,t(9;22),-1,-2,-3,-4,-5,-6,-7,-9,-11,-13,-14,-15,-16,-17,-18,-19,-20,-22	
46,XY,t(9;22),-18,+del(22q)	Hossfeld 1975b
46,XY,t(9;22),-22,+22q-	Kamada et al 1981
46,XY,t(9;22),-5,-17,t(5;17),+22q-	Borgström et al 1982
46,XY,t(9;22),i(17q)/47,XY,t(9;22),i(17q),+8/48,XY,t(9;22),i(17q),+8,+8/49,XY,t(9;22),i(17q),+8,+8,+22q-/50,XY,t(9;22),i(17q),+8,+8,+22q-,+19	Hagemeijer et al 1980b
46,XY,t(9;22),i(17q)/47,XY,t(9;22),+8,i(17q)/48,XY,t(9;22),+8,+22q-	Bernstein et al 1980b
46,XY,t(9;22),i(17q)/47,XY,t(9;22),+22q-/48,XY,t(9;22),+8,i(17q),+22q-	Miyamoto 1980
46,XY,t(9;22)/46,XY,t(9;22),i(9q),-22,+mar	Prigogina et al 1978
46,XY,t(9;22)/47,XY,t(9;22),+22q-	Hagemeijer et al 1980b
46,XY,t(9;22)/47,XY,t(9;22),+dic(22q-)	
46,XY,t(9;22)/47,XY,t(9;22),+8	Olinici et al 1978b
46,XY,t(9;22)/47,XY,t(9;22),+8	
46,XY,t(9;22)/48,XY,t(9;22),+19,+22q-	
46,XY,t(9;22)/47,XY,t(9;22),+22q-/48,XY,t(9;22),+8,+22q-/48,XY,t(9;22),+22q-,+22q-	Sonta and Sandberg 1977b
46,XY,t(9;22)/47,XY,t(9;22),+22q-/48,XY,t(9;22),+8,+22q-	Bernstein et al 1980b
46,XY,t(9;22)/47,XY,t(9;22),+22q-	Kohno and Sandberg 1980
46,XY,t(9;22)/47,XY,t(9;22),+22q-	Kohno and Sandberg 1980
46,XY,t(9;22)/47,XY,t(9;22),+8/47,XY,t(9;22),+22q-	Kohno and Sandberg 1980
46,XY,t(9;22)/48,XY,t(9;22),+8,+22q-/49,XY,t(9;22),+8,13q-,+19,+22q-	Hagemeijer et al 1980b
46,XY,t(9;22)/48,XY,t(9;22),+8,+22q-	Bernstein et al 1980b
46,XY,t(9;22)/49,XY,t(9;22),+9,+19,+22q-	Gall et al 1976
46,XY,t(9;22)/49,XY,t(9;22),+9,+19,+22q-	
46,XY,t(9;22)/50,XY,t(9;22),+8,+17,+19,+22q-	Hagemeijer et al 1980b
46-49,XX,t(9;22),+del(22q),+5,+6,+9	Prigogina et al 1978
47,X,-Y,t(9;22)(q34;q11),+8,i(17q),+22q-	Pasquali et al 1982b
47,X?,t(9;22),+22q-	Fleischman et al 1981
47,X?,t(9;22),+22q-	Carbonell et al 1982a
47,X?,t(9;22),+22q-/51,X?,t(9;22),+8,+8,+17,+19,+22q-	Fleischman et al 1981
47,X?,t(9;22),+22q-/48,X?,t(9;22),+8,21q+,+22q-/49,X?,t(9;22),+8,t(14;?),21q+,+22q-	Carbonell et al 1982a
47,X?,t(9;22),+22q-/49,X?,t(9;22),+8,+9,+22q-	Carbonell et al 1982a
47,X?,t(9;22),+8,i(17q)/48,X?,t(9;22),+8,i(17q),+22q-	Carbonell et al 1982a
47,X?,t(9;22),r(19),+22q-/48,X?,t(9;22),+17,+22q-/49,X?,t(9;22),+8,+17,+22q-/50,X?,t(9;22),+8,+17,+r(19),+22q-/49,X?,t(9;22),1q+,+8,+17,+22q-	Carbonell et al 1982a

47,XX,t(9;22)(q34;q12)/62,XX,cx,t(9;22)(q34;q12), del(1)(q25),del(1)(q11),del(1)(q21),del(1)(q11q25), del(2)(p12),del(3)(q21),del(6)(q23),i(13q),+22q-,+22q-, +22q-,+22q-	Hjorth et al 1980
47,XX,t(9;22)(q34;q11),+22q-	Alimena et al 1982b
47,XX,t(9;22)(q34;q11),t(7;11)(q21;p15),+22q-	Mitelman 1983
47,XX,t(9;22)(q34;q11),t(1;1)(q25;p36),+22	Mamaeva et al 1983
47,XX,t(9;22)(q34;q11),+8,i(17q)/48,XX,t(9;22),+8,i(17q), +22q-	Tomiyasu et al 1982
47,XX,t(9;22)(q34;q11),+t(1;17)(q21;p11),-17,+22q-	Tomiyasu et al 1982
47,XX,t(9;22),+22q-	Lilleyman et al 1978
47,XX,t(9;22),+22q-	Lilleyman et al 1978
47,XX,t(9;22),+22q-	Stoll and Oberling 1978
47,XX,t(9;22),+22q-	Stoll and Oberling 1978
47,XX,t(9;22),+22q-	Seabright 1983
47,XX,t(9;22),+8,i(17q)	Stoll and Oberling 1978
47,XX,t(9;22),+8,i(17q)/48,XX,t(9;22),+i(17q),+22q-	
47,XX,t(9;22),+del(22q)	Prigogina et al 1978
47,XX,t(9;22),i(17q),+8/48,XX,t(9;22),i(17q),+8,+19/49,XX, t(9;22),i(17q),+8,+19,+del(22)(q11)	Prigogina et al 1978
47,XX,t(9;22),i(17q),+22q-	Stoll and Oberling 1978
47,XX,t(9;22),i(17q),+22q-	Stoll and Oberling 1978
47,XX,t(9;22),i(17q),+22q-	Seabright 1983
47,XXY,t(9;22)(q34;q11)/53,XXY,+X,t(9;22),+4,+8,+18,+18, +22q-"c"	Mitelman 1983
47,XXY,t(9;22)/53,XXY,+X,t(9;22),+4,+8,+18,+18,+22q-"c"	
47,XXY,t(9;22)"c"	
47,XY,t(9;22)(q34;q11),+del(22q)	Sonta and Sandberg 1978b
47,XY,t(9;22)(q34;q11),+del(22q)/46,XY,t(9;22)/46,XY, t(9;22),dic(22)(q11)	Lawler et al 1976
47,XY,t(9;22)(q34;q11),-17,+t(1;7)(p11;q11),+22	Pasquali et al 1979a
47,XY,t(9;22)(q34;q11),+22q-	Bernstein et al 1980b
47,XY,t(9;22)(q34;q11),+del(22)(q11)	McGlave et al 1982
47,XY,t(9;22)(q34;q11),+22q-	Alimena et al 1979
47,XY,t(9;22)(q34;q11),+22q-	Alimena et al 1982b
47,XY,t(9;22)(q34;q11),del(1)(q21),dup(1)(q?),+8,+9,-14, -16,-21,+22q-,+mar	Alimena et al 1982b
47,XY,t(9;22)(q34;q11),t(3;5),-8,-10,+17,+22q-,+mar	Alimena et al 1982b
47,XY,t(9;22)(q34;q11),+8/47,XY,t(9;22),+19/47,XY,t(9;22), +21/47,XY,t(9;22),+22q-	Tomiyasu et al 1982
47,XY,t(9;22)(q34;q11),+22q-/47,XY,t(9;22),+19/47,XY, t(9;22),+21	Tomiyasu et al 1982
47,XY,t(9;22),+15/48,XY,t(9;22),+17,+22q-	Stoll and Oberling 1978
47,XY,t(9;22),+17/47,XY,t(9;22),-22,+5,+22q-/49,XY,t(9;22), +5,+21,+22q-	Stoll and Oberling 1978
47,XY,t(9;22),+22q-	Lilleyman et al 1978
47,XY,t(9;22),+22q-	Lilleyman et al 1978
47,XY,t(9;22),+22q-	Lyall and Garson 1978
47,XY,t(9;22),+22q-	Bernstein et al 1980b
47,XY,t(9;22),+22q-	Sadamori et al 1980
47,XY,t(9;22),+22q-	Kohno and Sandberg 1980
47,XY,t(9;22),+22q-	Carbonell et al 1981
47,XY,t(9;22),+22q-	Seabright 1983
47,XY,t(9;22),+22q-	Seabright 1983
47,XY,t(9;22),+22q-,+8,t(12;13)(q?;q?),-13,i(17q)	Rozynkowa et al 1977

47,XY,t(9;22),+22q-/46,XY,t(9;22),i(17q)	Stoll and Oberling 1978
47,XY,t(9;22),+22q-/47,XY,t(9;22),+17	Stoll and Oberling 1978
47,XY,t(9;22),+22q-/48,XY,t(9;22),+22q-,+22q-	Stoll and Oberling 1978
47,XY,t(9;22),+22q-/56,XY,t(9;22),+1,+2,+5,+6,+7,+8,+10, +12,+17,+22q-	Stoll and Oberling 1978
47,XY,t(9;22),+22q-/48,XY,t(9;22),+8,+22q-/48,XY,t(9;22), +17,+22	Stoll and Oberling 1978
47,XY,t(9;22),+22q-/46,XY,t(9;22),-9,+22q-/46,XY,t(9;22), t(3;5)(p?;q?),t(6;13)(p?;q?)	Kohno and Sandberg 1980
47,XY,t(9;22),+8,-17,-18,t(17;18)(p11;p11),+del(22q)	Sharp et al 1976
47,XY,t(9;22),+8,i(17q)/48,XY,t(9;22),+8,i(17q),+22q-	Lyall and Garson 1978
47,XY,t(9;22),+8/48,XY,t(9;22),+8,+22q-	Stoll and Oberling 1978
47,XY,t(9;22),+8/48,XY,t(9;22),+8,+22q-	Olah and Rak 1981
47,XY,t(9;22),+8/50,XY,t(9;22),+8,+13,+i(17q),+22q-	Miyamoto 1980
47,XY,t(9;22),+del(22)(q?)	Raposa et al 1974
47,XY,t(9;22),-10,+17,+19,-22,+22q-	Stoll and Oberling 1978
47,XY,t(9;22),4p+,+22q-	Sadamori et al 1980
47,XY,t(9;22),i(17q)/49,XY,t(9;22),i(1q),+8,+21,+22q-	Sonta and Sandberg 1977b
47,XY,t(9;22),inv(1)(p13q44),+22q-	Seabright 1983
47-48,XY,t(9;22),+8,+22q-	Williams and Weiss 1982
48,X?,t(9;22)(q34;q11),+8,+del(22q)	Lawler et al 1976
48,X?,t(9;22)(q34;q11),+19,+del(22q)	Lawler et al 1976
48,XX,t(9;22)(q34;q11),t(13;14)(q14;q24orq32),+19,+22q-	Carbone et al 1982a
48,XX,t(9;22)(q34;q11),+8,+19/49,XX,t(9;22),+3,+8,+19/50, XX,t(9;22),+3,+8,+19,+22q-	Mitelman 1983
48,XX,t(9;22)(q34;q11),+8,+22q-	Tomiyasu et al 1982
48,XX,t(9;22)(q34;q11),del(13)(q12q32),+19,+22q-	Tomiyasu et al 1982
48,XX,t(9;22),+17,+22q-	Stoll and Oberling 1978
48,XY,t(9;22)(q34;q11),+8,+del(22)(q11)	Ishihara et al 1974
48,XY,t(9;22)(q34;q11),+del(22)(q11),+mar	Rowley 1973a
48,XY,t(9;22)(q34;q11),+8,+22q-	Misawa 1978
48,XY,t(9;22)(q34;q11),+8,+22q-	Alimena et al 1982b
48,XY,t(9;22)(q34;q11),+8,i(17q),+22q-	Tomiyasu et al 1982
48,XY,t(9;22)(q34;q11),+8,+22q-	Tomiyasu et al 1982
48,XY,t(9;22)(q34;q11),+8,+22q-	Tomiyasu et al 1982
48,XY,t(9;22)(q34;q11),+7,+22q-/47,XY,t(9;22),+22q-	Tomiyasu et al 1982
48,XY,t(9;22),+17,+22q-	Stoll and Oberling 1978
48,XY,t(9;22),+17,+22q-	Stoll and Oberling 1978
48,XY,t(9;22),+22q-,+22q-/47,XY,t(9;22),+17/46,XY,t(9;22), -4,-7,+6,+17	Stoll and Oberling 1978
45,XY,t(9;22),-21/46,XY,t(9;22),-10,+17	
48,XY,t(9;22),+8,+13,-15,t(9;15),+i(22q)	Stoll and Oberling 1978
50,XY,t(9;22),+5,+8,+13,-15,t(9;15),+i(22q),+22q-	
48,XY,t(9;22),+8,+22q-	Lilleyman et al 1978
48,XY,t(9;22),+8,+22q-	Miyamoto 1980
48,XY,t(9;22),+8,+22q-/61,XY,+Y,t(9;22),+4,+5,+6,+8,+8,+8, +8,+15,+17,+18,+19,+20,+22q-,+22q-	Wurster-Hill 1983
48,XY,t(9;22),+8,-9,-11,-12,-14,-19,+del(22q),+5mar	Sharp et al 1976
48,XY,t(9;22),+8,i(17q),+22q-	Stoll and Oberling 1978
48,XY,t(9;22),+8,i(17q),+22	Stoll and Oberling 1978
48,XY,t(9;22),-22,+17,+10,+22q-	Stoll and Oberling 1978
48,XY,t(9;22),-22,+17,+20,+22q-	
48,XY,t(9;22),t(3;4)(p?;p?),+21,+22q-	Sadamori et al 1980
48-49,XX,t(9;22)(q34;q11),t(2;4),+19,+del(22q)	Sonta and Sandberg 1978b
48-53,X?,t(9;22),+8,+8,+13,+14,+19,+21,+22q-	Carbonell et al 1982a

49,X?,t(9;22),+17,+19,+22q-/50,X?,t(9;22),+8,+17,+19,+22q-/51,X?,t(9;22),+8,+9,+17,+19,+22q- Carbonell et al 1982a
49,XX,t(9;22)(q34;q11),8q-,+10,+14,+22q- Alimena et al 1982b
49,XX,t(9;22),+10,+21,del(22q) Castleman et al 1973
49,XX,t(9;22),+17,+19,+22q- Sadamori et al 1980
49,XX,t(9;22),+19,+del(22q),+mar Hossfeld 1975b
49,XX,t(9;22),+3,+8,+22q- Seabright 1983
49,XX,t(9;22),+8,+19,+22 Lilleyman et al 1978
49,XX,t(9;22),+8,17q-,+19,+22q-/49,XX,t(9;22),+8,i(17q),+19,+22q- Olah and Rak 1981
49,XY,t(9;22)(q34;q11),+6,+8,t(22;?)(p11;?),+der(22) Oshimura et al 1982a
49,XY,t(9;22)(q34;q11),+4,+21,+22q- Alimena et al 1982b
49,XY,t(9;22),+10,-17,+19,+del(22q),+mar Hossfeld 1975b
49,XY,t(9;22),+13,+22q-,+22q-/52,XY,t(9;22),+13,+13,+17,+17,+22q-,+22q- Stoll and Oberling 1978
49,XY,t(9;22),+17,+19,+22q-/50,XY,t(9;22),+8,+15,+19,+22q-/50,XY,t(9;22),+8,+9,+19,+22q- Hartley and McBeath 1981
49,XY,t(9;22),+22q-,+8,+10 Rozynkowa et al 1977
49,XY,t(9;22),+6,+8,1p+,14q+,+22q- Sadamori et al 1980
49,XY,t(9;22),+8,+19,+22q- Seabright 1983
49,XY,t(9;22),+8,+i(17q),+22q-/50,XY,t(9;22),+8,+10,+i(17q),+22q- Olah and Rak 1981
49,XY,t(9;22),+9,+14,+22q- Lyall and Garson 1978
50,XX,t(9;22),t(2;12)(p13;q24),+8,+8,+12q+,+12q+,+del(22q) Sharp et al 1976
50,XY,+Y,t(9;22)(q34;q11),t(6;8)(q34;q11),+8,i(17q),+19,+22q- Alimena et al 1982b
50,XY,t(9;22)(q34;q11),+8,+C,+del(22)(q11)+i(17)(q11) Rowley 1973a
50,XY,t(9;22)(q34;q11),+8,+17,+21,+del(22)(q11)/51,XY,t(9;22),+8,+8,+17,+21,+del(22)(q11)/52,XY,t(9;22),+8,+8,+8,+17,+21,+del(22)(q11) Sonta and Sandberg 1978b
50,XY,t(9;22),+8,+15,+17,+del(22q) Sharp et al 1976
50,XY,t(9;22),+8,+17,+21,+22q-/51,XY,t(9;22),+8,+8,+17,+21,+22q-/54,XY,t(9;22),+8,+8,+8,+17,+21,+22q-,+22q- Sonta and Sandberg 1977b
50,XY,t(9;22),+C,+19,+21,+22q- Lessard and Le Prise 198
50,XY,t(9;22),+Y,+8,+19,+22q- Sadamori et al 1980
50,XY,t(9;22),+del(22q),+6,+10,+19 Prigogina et al 1978
50-54,XY,t(9;22),+1,+5,+8,+9,-16,+17,+18,+19,+21,+22q- Alimena et al 1980
50-55,XY,t(9;22)(q34;q11),+1,+5,+8,+9,-16,+17,+18,+19,+21,+22q-,+mar Alimena et al 1982b
51,XX,t(9;22)(q34;q11),+6,+8,i(17q),+19,+20,+22q- Alimena et al 1982b
51,XX,t(9;22),+4,+17,+19,+del(22q),+mar Hossfeld 1975b
51,XX,t(9;22),+8,+9,+14,+19,+22q- Miyamoto 1980
51,XX,t(9;22),t(8;22)(p12;p11),+del(8)(p12),+19 Lyall and Garson 1978
51,XY,t(9;22)(q34;q11),+6,+8,+19,+21,+22q- Tomiyasu et al 1982
51-52,XY,t(9;22),+del(22q),+7,+8,+12,+19,+21 Prigogina et al 1978
52,XX,+X,+X,+8,+12,+21,t(9;22)(q34;q11),+del(22q) Sonta and Sandberg 1978b
52,XY,t(9;22)(q34;q11),1p+,+8,+9,i(17q),+19,+20,+21,+22q- Alimena et al 1982b
52,XY,t(9;22),+1,+8,+10,+11,+21,+22q- Sadamori et al 1980
52,XY,t(9;22),+8,+14,+15,+19,+21,+22q- Sadamori et al 1980
52,XY,t(9;22),+8,+9,+12,+19,+21,+22q- Nordenson 1983
52-54,XX,t(9;22)(q34;q11),+6,+8,+8,+19,+19,+del(22)(q11),+del(22)(q11),+X Hayata et al 1975b
52-54,XY,t(9;22)(q34;q11),+3,+6,+7,+8,+19,+21,+22q-,+22q-,+1-3mar Ondreyco et al 1981
53,XX,t(9;22)(q34;q11),+8,+10,+19,+20,+21,+22q- Greenberg et al 1978

53,XX,t(9;22),+8,+11,+12,+16,+17,+19,+del(22q)/52,XX, t(9;22),+8,+11,+17,+19,+del(22q)	Gahrton et al 1974b
53,XY,+Y,t(9;22)(q34;q11),+3,+6,+8,+9,+10,+12,-15,+22q-/56, XY,+Y,t(9;22),+3,+6,+8,+8,+9,+10,+16,+19,+22q-	Stoll and Oberling 1982
56,XY,+Y,t(9;22),+3,+6,+8,+8,+9,+10,+12,+19,+22q-	
53,XY,t(9;22),+6,+9q+,+10,+12,+19,+21,+22q-/54,XY,t(9;22), +6,+8,+9q+,+10,+12,+19,+21,+22q-	Sonta and Sandberg 1977b
53,XY,t(9;22),+8,+9,+19,+22q-,+mar	Lyall and Garson 1978
53-56,XY,+Y,t(9;22),+del(22q),+6,+10,+19,+8,+11,+21	Prigogina et al 1978
55,XX,t(9;22),+3,+8,+12,+14,+15,+19,+20,+22,+22q-	Lilleyman et al 1978
55,XY,t(9;22),+3,+6,+8,+8,+9,+10,+16,+19,+22q-/53,XY, t(9;22),-15,+3,+6,+8,+8,+9,+10,+12,+22q-	Stoll and Oberling 1978
59,X,-X,t(9;22),+4,+6,+6,t(7;?)(p15;?),+der(7),+8,+8,+12q+, +17,+19,+19,+21,+22,+22q-,+22q-	Miyamoto 1980

Chronic myeloid leukemia, aberrant translocation

??,X?,del(22)(q11)	Potter et al 1981
??,X?,del(22)(q11)	Potter et al 1981
??,X?,t(11;22)(p15;q11)	Potter et al 1981
??,X?,t(12;22)	Geraedts et al 1977
??,X?,t(12;22)(p13;q11)	Potter et al 1981
??,X?,t(12;22)(q24;q11)	Potter et al 1981
??,X?,t(15;22)(p11;q11)	Potter et al 1981
??,X?,t(17;22)(q25;q11)	Potter et al 1981
??,X?,t(1;4;20;22)	Geraedts et al 1977
??,X?,t(2;9;22)(q14;q34;q11)	Smadja et al 1980
??,X?,t(3;22)(p21;q11)	Potter et al 1981
??,X?,t(3;4;9;11;22)	Potter et al 1981
??,X?,t(3;9;22)(p21;q34;q11)	Potter et al 1981
??,X?,t(3;9;22)(q11orq12;q34;q11)	Tanzer et al 1980a
??,X?,t(5;9;22)(q13;q34;q11)	Potter et al 1981
??,X?,t(6;22)(p25;q11)	Potter et al 1981
??,X?,t(7;22)(p22;q11)	Potter et al 1981
??,X?,t(9;10;22)	Geraedts et al 1977
??,X?,t(9;13;22)(q34;q22;q11)	Potter et al 1981
??,X?,t(9;15;22)(q34;q15;q11)	Potter et al 1981
??,X?,t(9;15;22)(q34;q15;q11)	Potter et al 1981
44,Y,-X,t(9;17;22)(q34;p11;q11),t(X;9)(?;p24),i(17q)	Engel et al 1977
45,XY,-22,t(17;17;22)(p1?;q24;q11)	Engel et al 1977
45,XY,t(18;22)(q23;q11),+del(1)(q11),6q-,-9,-9	Nowell 1983
46,X,t(X;9;22)(q27;q34;q12)	Dallapiccola and Alimena 1979
46,X,t(X;9;22)/47-51,X,t(X;9;22),+1,+4,+8,+9,-14,i(17q), -19,+22q-	
46,X?,t(10;22)(q26;q11)	Hossfeld 1983
46,X?,t(11;22)	Stoll 1983
46,X?,t(13;22)(q34;q11),t(8;9)(q22;q34)	Hossfeld 1983
46,X?,t(14;22)	Rowley 1983
46,X?,t(14;22)(q32;q11)	Nowell 1983
46,X?,t(14;22)(q32;q11)	Ishihara 1983
46,X?,t(17;22)(p13;q11)	Hagemeijer et al 1980b
46,X?,t(17;22)(p13;q11)	Ishihara 1983
46,X?,t(17;22)(q25;q11)	Hagemeijer et al 1980b
46,X?,t(19;22)(p11;q11)	Van Den Berghe 1983

46,X?,t(19;22)(q13;q11)	Lawler et al 1976
46,X?,t(1;9;22)(p31;q34;q11)	Ishihara 1983
46,X?,t(1;9;22)(p32;q34;q11)	Hagemeijer et al 1980b
46,X?,t(21;22)(q22;q11)	Bottura and Couthinho 1974
46,X?,t(2;22)(q37;q11)	Van Den Berghe 1983
46,X?,t(3;9;22)	Nowell 1983
46,X?,t(3;9;22)(p21;q34;q12)	Rozynkowa 1983
46,X?,t(3;9;22)(p21;q34;q12)	Rozynkowa 1983
46,X?,t(3;9;22)(q22;q34;q11)	Van Den Berghe 1983
46,X?,t(4;22)(p16;q11)	Van Den Berghe 1983
46,X?,t(5;9;22)(q35;q11;q11)	Ishihara 1983
46,X?,t(7;9;11;22)(p11;q34;q22;q11)	Ishihara 1983
46,X?,t(7;9;22)(q22;q34;q11)	Nowell 1983
46,X?,t(9;11;22)(q34;q13;q11)	Carbonell et al 1982a
46,X?,t(9;11;22)(q34;q13;q11)	Ishihara 1983
46,X?,t(9;14;22)(q34;q2?;q11)	Nowell 1983
46,X?,t(9;17;22)(q34;p13;q11)	Ishihara and Minamihisamatsu 1981
46,X?,t(9;9;22)	Ishihara 1983
46,XX,del(22)(q11)	Mitelman 1974
46,XX,inv(5)(p14q12),t(9;13;17;22)(q31;q11;q12;q12)	Fraisse et al 1980
46,XX,t(10;14;22)(q26;q24;q11)	Shabtai et al 1980
46,XX,t(11;14;22)(q13;q32;q11)	Kolitz et al 1981
46,XX,t(12;22)(p13;q13)	Verma and Dosik 1979
46,XX,t(12;22)(p13;q11)	Van Der Blij-Philipsen et al 1977
46,XX,t(12;22)(p13;q11),t(8;17)(q13;q23)	Swolin et al 1983
46,XX,t(12;22)(q12;q11)	Marinello et al 1981
46,XX,t(12;22)(q34;q11),t(9;11)(q34;q11)	Seabright 1983
46,XX,t(14;22)(p13;q11)	Borgström 1981
46,XX,t(15;22)(p13;q11)	Jotterand-Bellomo 1978
46,XX,t(15;22)(q26;q11)	Hossfeld and Köhler 1979
46,XX,t(16;22)(p13;q11)	Lyall et al 1978
46,XX,t(17;22)(p11;q11)	Borgström 1981
46,XX,t(17;22)(q21;q11)	Seabright 1983
46,XX,t(17;22)(q25;q11)	Panani et al 1979
46,XX,t(17;22)(q25;q11)/47,XX,t(17;22),+8	Bernstein et al 1980b
46,XX,t(17;22)(q25;q11)	Borgström et al 1982
46,XX,t(17;22)(q25;q11),i(17q+)	
46,XX,t(17;22),inv(5),t(9;13)	Freycon et al 1982
46,XX,t(18;22)(p11;q11)	Borgström 1981
46,XX,t(19;22)(q13;q11)	Gahrton et al 1974b
46,XX,t(19;22)(q13;q11)	Petit et al 1978
46,XX,t(1;11;22)(q24;p11;q11)	Najfeld et al 1983
46,XX,t(1;9;22)(q22;q34;q11q13)	Fleischman et al 1981
46,XX,t(1;9;22)(q32;q34;q11)	Borgström 1981
46,XX,t(1;9;22)(q32;q34;q11)	Pasquali 1983
46,XX,t(21;22)(p1?;q11)	Seabright 1983
46,XX,t(22;22)(q13;q12)	Foerster et al 1974
46,XX,t(2;9;22)(p11;q34;q11)	Rochon and Vaillancourt 1980
46,XX,t(2;9;22)(p21;q34;q11)	Pasquali et al 1979b
46,XX,t(2;9;22)(p24q24;q34;q11)	Pasquali 1983
46,XX,t(2;9;22)(q11;q34;q11),t(2;7)(q11;q11)	Fleischman et al 1981

46,XX,t(2;9;22)(q11;q34;q11)	Smadja et al 1983a
46 XX,t(2;9;22)(q14q37;q34;q11)	Van Den Akker et al 1980
46,XX,t(2;9;22)(q23;q34;q11)	Pasquali et al 1979b
46,XX,t(2;9;22)/50,XX,t(2;9;22),+6,+10,+19,+der(22)	
46,XX,t(3;22)(p21;q11)/48,XX,t(3;22),+19,+21	Pravtcheva et al 1976
46,XX,t(3;4;9;22)(q21;q31;q34;q11)	Chessells et al 1979
46,XX,t(3;9;22)(p21;q34;q11)	Nowell et al 1975
46,XX,t(3;9;22)(p21;q34;q11)	Anderson et al 1978
46,XX,t(3;9;22)(p21;q34;q12)	Rozynkowa et al 1977
46,XX,t(3;9;22)(p21;q34;q11)	Mohandas et al 1980
46,XX,t(3;9;22)(p21;q34;q11)	Seabright 1983
46,XX,t(3;9;22)(q21;q34;q11)	Pasquali 1983
46,XX,t(3;9;22)(q23;q34;q11)	Borgström 1981
46,XX,t(4;22)(p?;q11)	Sekhon et al 1978
46,XX,t(4;9;22)(p14;q34;q11)	Tomiyasu et al 1982
46,XX,t(4;9;22)(q23;q34;q11)	Ciric and Rolovic 1980
46,XX,t(4;9;22)(q31;q34;q11)	Sudries et al 1980
46,XX,t(4;9;22)(q31;q34;q11)	Kessous et al 1980
46,XX,t(4;9;22)(q32;q34;q12)	Fraisse et al 1980
46,XX,t(5;22)(q35;q11)/46,XX,t(5;22),i(17q)/47,XX,t(5;22), +8,i(17q)	Pasquali et al 1979b
46,XX,t(5;9;22)(q13;q34;q11)	Alimena 1983
46,XX,t(6;22)(p25;q12)	Mammon et al 1976
46,XX,t(6;22)(q26;q11),t(1;9)(q21;q24)	Berger et al 1976
46,XX,t(6;9;11;22)(p21;q34;q13;q11)	Carbonell et al 1980
46,XX,t(6;9;22)(p21;q34;q11)	Pasquali 1983
46,XX,t(6;9;22)(p21;q34;q11)	Pasquali 1983
46,XX,t(7;9)(q22;q34),t(17;22)(p13;q11)	Mitelman 1983
46,XX,t(7;9)(q22;q34),t(17;22)(p13;q11)	
46,XX,t(7;9;22)(p14;q34;q12)	Fraisse et al 1980
46,XX,t(7;9;22)(q11orq21;q34;q11)	Martin et al 1980
46,XX,t(7;9;22)(q22;q32q34;q11)	Pasquali et al 1979b
46,XX,t(9;11;22)(q34;q13;q11)	Sessarego et al 1982
46,XX,t(9;14;22)(q34;q24;q11)	Potter et al 1975
46,XX,t(9;14;22)(q34;q12orq13;q11)	Tanzer et al 1980a
46,XX,t(9;14;22)(q34;q11;q11)	Borgström 1981
46,XX,t(9;14;22)(q34;q13;q11)	Seabright 1983
46,XX,t(9;14;22)(q34;q21;q11)	Pasquali 1983
46,XX,t(9;15;22)(q34;q11;q11)	Hays et al 1981
54-57,XX,t(9;15;22),+5,+7,+8,+10,+11,+12,+17,+22q-	
46,XX,t(9;17;22)(q34;q21;q11)	Sonta and Sandberg 1978b
46,XX,t(9;17;22)(q34;q21;q11)	Rowley et al 1979
46,XX,t(9;17;22)(q34;q21;q11)	Sonta and Sandberg 1977a
46,XX,t(9;20;22)(q31orq33;q11;q11)	Oshimura et al 1982a
46,XX,t(9;22)(q22q34;q11)	Lessard et al 1981
46,XY,22q-"c"	Mitelman 1983
46,XY,22q-"c"	
46,XY,22q-"c"	
46,XY,t(13;17;22)(p13;q12;q11)"c"	
46,XY,t(13;17;22)(p13;q12;q11)"c"	
46,XY,t(13;17;22)(p13;q12;q11)"c"	
46,XY,t(13;17;22)(p13;q12;q11)"c"	
46,XY,t(13;17;22)(p13;q12;q11)"c"	
46,XY,t(10;22)(p15?;q11)/45,X,-Y,t(10;22)	Testa et al 1982
46,XY,t(10;22)(q26;q11)	Misawa 1978

46,XY,t(10;22)(q26;q11)	Fleischman et al 1981
46,XY,t(11;22)(p15;q11)	Muldal et al 1975
46,XY,t(11;22)(p15;q11)	Seabright 1983
46,XY,t(11;22)(q23;q11)	Seabright 1983
46,XY,t(12;22)(p13;q11)	Engel et al 1977
46,XY,t(13;22)(p13;q11)	Hayata et al 1975b
46,XY,t(13;22)(q34;q11)	Seabright 1983
46,XY,t(14;22)(q32;q11)	Seabright 1983
46,XY,t(16;22)(p13;q11)	Engel et al 1977
46,XY,t(16;22)(q24;q11)	Panani et al 1979
46,XY,t(16;22)(q24;q11),t(9;17)(q34;q21)/46,XY,t(16;22), t(9;17),del(7)(q22)	Mitelman 1983
46,XY,t(16;22)(q24;q11),t(9;17)(q34;q21)	
46,XY,t(17;22)(q25;q11)	Matsunaga et al 1976
46,XY,t(19;22)(q13;q11),t(3;21)(p11;p11)	Seabright 1983
46,XY,t(1;7;19;22)(q?;q?;p?;q11)	Borgström et al 1982
46,XY,t(1;7;19;22),i(17q)	
46,XY,t(1;9;22)(p11;q34;q11)	Borgström 1981
46,XY,t(1;9;22)(q21;q34;q11)/46,XY,t(9;22)(q34;q11)	Berger et al 1981d
46,XY,t(1;9;22)(q21;q34;q11)	Lessard and Le Prise 1982
46,XY,t(1;9;22)(q23;q34;q12)	Berger and Bernheim 1978
46,XY,t(1;9;22)(q23;q22;q12)	Verma and Dosik 1977
46,XY,t(1;9;22)(q32;q34;q11)	Fleischman et al 1981
46,XY,t(20;22)(p12;q11)	Lai et al 1982
46,XY,t(21;22;22)(p13;p13;q11)	Ishihara et al 1974
46,XY,t(22;?)(q11;?)	Mitelman 1983
46,XY,t(2;22)(q37;q11)	Hayata et al 1975b
46,XY,t(2;22;22)(q11;q11;q13)	Fraisse et al 1980
46,XY,t(2;9;22)(q11;q34;q11)	Alimena et al 1982b
46,XY,t(2;9;22)(q24orq31;q34;q11)	Tanzer et al 1977
46,XY,t(2;9;22)(q32;q34;q11)	Nowell 1983
46,XY,t(2;9;22)(q37;q22q34;q11)	Testa et al 1982
46,XY,t(3;9)(q21;q34),t(17;22)(q21;q11)	Oshimura et al 1982a
46,XY,t(3;9;10;22)(p13;q34;q22;q11),17q+	Seabright 1983
46,XY,t(3;9;22)(p21;q34;q11)	Lessard and Le Prise 1982
46,XY,t(3;9;22)(q21;q34;q11)	Pasquali et al 1979b
46,XY,t(3;9;22)(q21;q34;q11)	Pasquali 1983
46,XY,t(3;9;22)(q23;q34;q11)	Seabright 1983
46,XY,t(4;22)(q35;q21),t(3;5)(q27;q22)	Sessarego et al 1981b
46,XY,t(4;22)(q35;q21),t(3;5)(q27;q22),i(17q)/50,XY, t(4;22),t(3;5),i(17q),+8,+15,+21,+22q-	
46,XY,t(4;9;22)(q21;q34;q11)	Pasquali et al 1979b
46,XY,t(4;9;22)(q23;q34;q11)	Rowley et al 1976
46,XY,t(4;9;22)(q31;q34;q11)	Seabright 1983
46,XY,t(5;13)(q33;q14),del(22)(q11)	Rolovic et al 1983
46,XY,t(5;9;22)(q13;q34;q11)	Oshimura et al 1982a
46,XY,t(6;9;22)(p21;q34;q11)	Potter et al 1975
46,XY,t(6;9;22)(p21;q34;q11)	Potter et al 1975
46,XY,t(6;9;22)(q21;q34;q11)	Chessells et al 1979
46,XY,t(6;9;22)(q24;q34;q11)	Sudries et al 1980
46,XY,t(7;22)(p22;q12)	Gahrton et al 1979
48,XY,t(7;22)(p22;q12),+8,+8,i(17q)	
46,XY,t(7;22)(p22;q11)	Adler et al 1978
46,XY,t(7;9;22)(q22;q34;q11)	Seabright 1983
46,XY,t(7;9;22)(q35;q31q34;q11)	Lessard and Le Prise 1982

46,XY,t(8;9;22)(q22;q34;q11)/45,X,-Y, t(8;9;22)(q22;q34;q11)	Lawler et al 1976
46,XY,t(9;10;15;19;22)(q34;q22;q21;q13;q12)	Tomiyasu et al 1982
46,XY,t(9;10;22)(q34;q11;q11)	Hayata et al 1975b
46,XY,t(9;11;22)(q34;q13;q11)/52,XY,t(9;11;22),+6,+9q+, +11q-,+18,+19,+del(22q),1q+	Lawler et al 1976
46,XY,t(9;11;22)(q34;q13;q11)	Fleischman et al 1981
46,XY,t(9;12;22)(q34;q15;q11)	Borgström 1981
46,XY,t(9;13;22)(q13-21;q13;q11)	Hayata et al 1975b
46,XY,t(9;13;22)(q34;q14;q11)	Seabright 1983
46,XY,t(9;14;22)(q34;q24;q11)	Borgström 1981
46,XY,t(9;14;22)(q34;q21;q11)	Pasquali 1983
46,XY,t(9;14;22)(q34;q24;q11)	Tomiyasu et al 1982
46,XY,t(9;15;22)(q34;q15;q11)	Seabright 1983
46,XY,t(9;17;22)(q34;q21;q11)	Pasquali 1983
46,XY,t(9;20;22)(q34;p11p13;q11),i(17q)	Pasquali et al 1982b
46,XY,t(9;20;22)(q34;p11;q11),i(17q)	Pasquali 1983
46,XY,t(9;20;22)(q34;p11p13;q11),i(17q)	Casalone et al 1983
46,XY,t(9;22;?)(q34;q11q13;?)	Fleischman et al 1981
46,XY,t(9;9;11;22)(q34;p24;p13;q11)	Gahrton et al 1977
46,XY,t(9;9;22)(q34;q22;q11)	Hayata et al 1975b
47,XY,+Y,t(9;13;15;22)(q34;q12;q22;q11)	Potter et al 1975
47,XY,inv(9)(p11q13),t(5;9;22)(q13;q34;q11), +del(22)(q11)"c"	Nowell et al 1975
47,XY,t(10;22)(q26;q11),+8,i(17q)	Prigogina et al 1978
47,XY,t(9;9;22)(p24;q34;q11),+8	Pasquali 1983
47,Y,t(X;22)(p22;q11),+19	Hossfeld and Köhler 1979
50,XY,del(22)(q11),+8,+11,+21,+21/51,XY,del(22)(q11),+8, +11,+13,+21,+21	Sonta and Sandberg 1978b
50,XY,del(22)(q11),+8,+11,+21,+21,/51,XY,del(22)(q11),+8, +11,+13,+21,+21	

Chronic myeloid leukemia, Ph1 negative

48,XY,+9,+22	Seabright 1983

Eosinophilic leukemia

48,XY,t(9;22),r(12),+22,+22q-	Hartley 1983

MYELOPROLIFERATIVE DISORDERS

Polycythemia vera

43-45,XX,-9,-18,-22	Wurster-Hill et al 1976
??,X?,+22	Vykoupil et al 1980
??,X?,+6,+11,+14,+22	Vykoupil et al 1980
??,X?,+8,+11,+22	Vykoupil et al 1980
??,X?,t(9;22),+8,+10,+12,+16	Vykoupil et al 1980
45,X,-Y	Berger and Bernheim 1979
54-57,X,-Y,+6,+8,+11,+17,+20,+21,t(1;22)(p12;q13)	
45,XY,-16,-17,+21/48,XY,+Y,+11,+16,-17,+22	Wurster-Hill et al 1976
46,XX,22q-	Shabtai et al 1978
46,XY,t(9;22)(q34;q11)	Nicoara et al 1977

47,X?,+9,22p+	Shabtai 1983
48,XX,t(9;22),+5,+8	Armenta et al 1976

Myelosclerosis / Myelofibrosis

47,XX,del(22)(q?),t(1;2),t(6;7),+r,+mar	Page et al 1979

Idiopathic thrombocythemia

46,X,t(X;9;22)(q11;q34;q11)	Fitzgerald et al 1981
46,XX,t(9;22)(q34;q12)	Rajendra et al 1981

Chronic myeloproliferative disease, NOS

46,XX,t(1;22)(q?;q13)	Lessard and Le Prise 1983
46,XY,-7,+19/47,XY,+19/52,XY,+16,+19,+19,+21,+22,+22	Chrz et al 1983

Myeloproliferative disorder, special type

45,XX,t(1;?)(p36;?),-5,-7,-12,t(13;?)(q34;?),-17,-22, +4mar/45,XX,t(1;?),-2,-3,-5,-7,-12,-17,-22,+8mar"s"	Rowley et al 1977a

DYSMYELOPOIETIC SYNDROMES

Preleukemia, NOS

43,X,-Y,-3,5q-,-7,-10,-16,-22,+3mar"s"	Pedersen-Bjergaard et al 1981
44,XY,+Y,-3,-5,+del(6)(q21),-8,t(17;?)(p13;?),+18,-20,+22	Mitelman 1983
46,XX,-7,+8,t(18;22)(p11;q11)"s"	Michalski et al 1982
46,XX,5q-,22q-	Teerenhovi et al 1981
46,XX,del(5)(q13orq14),del(22)(q11)	Ruutu et al 1977b
46,XX,t(9;22)(q34;q11)	Roth et al 1980
53,XX,+1,+2,t(4;19),+6,i(17q),+21,+21,+22	Seabright 1983

Acquired idiopathic sideroblastic anemia

46,XX,del(5)(q12q23)/45,XX,3p-,4q+,6q-,-22"s"	Kerkhofs et al 1982

Refractory anemia with excess of blasts

44,XX,5q-,-13,-17,-20,-22,+2mar	Teerenhovi et al 1981
46,XX,del(5)(q15)/49-51,XX,+1,del(5)(q15),+11,-14,+22,+mar	Swolin et al 1981
46,XY,del(5)(q12q31),-7,20q+,+22	Kerkhofs et al 1982

Erythrocytopenia

46,XX,t(9;22)(q34;q11)	Shiraishi et al 1980

SPECIAL LEUKEMIAS

LYMPHOCYTIC LEUKEMIAS

Acute lymphocytic leukemia (ALL)

??,X?,t(11;22)(p15;q12)	Pittman et al 1979
??,X?,t(1;22)(p36;q11)	Pittman et al 1979
??,X?,t(22;?)(q11;?)	Pittman et al 1979
??,X?,t(9;22)(q34;q11)	Pittman et al 1979
??,X?,t(9;22)(q34;q11)	Pittman et al 1979
??,X?,t(9;22)(q34;q11)	Pittman et al 1979
43,X,-X,-7,-15,t(9;22)	Gibbs et al 1977
43,X,-X,-7,-8,t(9;22)(q34;q11)	Oshimura and Sandberg 1977
45,XY,t(9;22),-17,-20,+mar	Rausen et al 1977
46,X,-Y,+1,-7,+20,t(1;22)(q24;q12)	Vercherat et al 1980
46,X?,22q-	Bloomfield et al 1977
46,X?,22q-	Bloomfield et al 1977
46,X?,t(9;22)	Bloomfield et al 1977
46,X?,t(9;22)	Bloomfield et al 1977
46,X?,t(9;22)(q34;q11)	Philip et al 1976
46,XX,1q-,7q-,t(1;19)	Goldstone et al 1979
55,XX,+1q-,+7q-,+t(1;19),+19,+22,+4mar	
46,XX,t(9;22)	Chessells et al 1979
46,XX,t(9;22)	Forman et al 1977
46,XX,t(9;22)	Olah et al 1981a
46,XX,t(9;22)	Olah et al 1981a
46,XX,t(9;22)	Lessard and Le Prise 1982
46,XX,t(9;22)(q34;q11),t(2;9)(q21;p13),del(9)(p13)	Morse et al 1982a
46,XX,t(9;22)(q34;q11)/46,XX,t(9;22),del(7)(q11)/45,XX,t(9;22),-7	Shabtai and Halbrecht 1981b
46,XY,?22q-	Mitelman 1983
46,XY,del(3)(p?),t(9;22)(q34;q11)	Mitelman 1983
46,XY,del(3)(p?),t(9;22)(q34;q11)	
46,XY,t(14;22)(q32;q11)/46,XY,t(14;22),+14,-15	Ayraud et al 1975
46,XY,t(8;22)(q24;q12),t(3;11)(p12p21;q23),inv(2)(p11q13),del(2)(p21)/46,XY,t(8;22)(q24;q12),dup(1)(q23q44)"s"	Fonatsch et al 1982
46,XY,t(9;22)	Gibbs et al 1977
46,XY,t(9;22)	Chessells et al 1979
46,XY,t(9;22)	Chessells et al 1979
46,XY,t(9;22)	Chessells et al 1979
46,XY,t(9;22)(q34;q11)	Secker Walker and Hardy 1976
46,XY,t(9;22)(q34;q11)	
46,XY,t(9;22)(q34;q11)	Cheson et al 1980
46,XY,t(9;22)(q34;q11)	Nordenson 1983
46,XY,t(9;22)(q34;q11)	Mitelman 1983
46,XY,t(9;22)	Schmidt et al 1975
45,X,-Y,t(9;22)	
46,XY,t(9;22)	
46-47,XX,17p+,+22	Prigogina et al 1979
48,XX,+13,+22	Nordenson 1983

48,XY,t(1;22)(q21;p11),+5,+8,19p+	Humbert et al 1978
48,XY,t(4;12)(q?;q?),t(6;15)(q?;q?),t(9;22)(q34;q11),11q-, +20,+21	Nordenson 1983
53,XX,+2,+5,i(7q),+8,+13,+21,+21,+22,+17q+	Oshimura et al 1977a
53-55,XY,+4,+5,+6,+20,+22	Lessard and Le Prise 1983
54,XX,+6,+7,+8,+14,+17,+18,+21,+22	Prigogina et al 1979
54,XX,7p+,+10,+10,+14,+14,+18,+18,+21,+21/27,X,-X,-1,-2,-3, -4,-5,-6,-7,7p+,-8,-9,-11,-12,-13,-15,-16,-17,-19,-20, -22	Oshimura et al 1977a
57,XX,+3,+4,+8,11p+,+14,+16,+17,+18,+21,+22,+2mar/47,XX,+8	Morse et al 1978
57,XY,+X,+5,+6,i(7q),+10,+12,+13,+15,+18,+21,+21,+22	Oshimura et al 1977a
61,XX,+X,+1,+2,+3,+4,+5,+6,+8,del(6)(q21),+13,+14,+15,+16, +18,+21,+22	Oshimura et al 1977a

ALL, FAB type L1

26,XX,-1,-2,-3,-4,-5,-6,-7,-8,-9,-10,-11,-12,-13,-14,-15, -16,-17,-19,-20,-22	Hoeltge et al 1982
28,XX,-1,-2,-3,-4,-5,-6,-7,-8,-9,-11,-12,-13,-15,-16,-17, -19,-20,-22/56,XX,+X,+X,+10,+10,+14,+14,+18,+18,+21,+21	Brodeur et al 1981a
34-36,XY,-2,-3,-4,-7,-9,-12,-13,-14,-16,-17,-20,+22	Sandberg et al 1982c
45,XX,4p-,-18,t(22;?)(q11;?)	Priest et al 1980
46,XX,t(9;22)	Sandberg et al 1980
46,XY,t(21;22)(q22;q11)	Kaneko et al 1982e
46,XY,t(9;22)	Priest et al 1980
46,XY,t(9;22)	Priest et al 1980
46,XY,t(9;22)	Sandberg et al 1980
47,XY,t(22;?),-1,-1,-2,-3,-9,+6mar	Sandberg et al 1980
52-56,XY,t(9;22),+4,+6,+9,+10,+14,+15,+20,+21,+22q-	Sandberg et al 1980
53,XX,+1,+5,+14,+15,+19,+20,+22	Kaneko et al 1981b

ALL, FAB type L2

26,XX,-1,-2,-3,-4,-5,-6,-7,-8,-9,-10,-11,-12,-13,-15,-16, -17,-18,-19,-20,-22	Brodeur et al 1981a
34-37,XX,cx,t(9;22)	Priest et al 1980
44,XX,-8,t(13;14),t(9;22)"c"	Takeuchi et al 1981
46,XX,t(5;8;14)(q11;q24;q32),t(12;22)(p13;q11),-18, +t(18;?)(q23;?)	Kaneko et al 1982e
46,XX,t(9;22)	Borgström et al 1981
46,XX,t(9;22)(q34;q11)	Mitelman 1983
46,XX,t(9;22),-20,+mar	Borgström et al 1981
46,XY,-4,+6,-8,-10,-11,-15,-17,+22,del(6)(q21),+i(17q), +3mar	Kaneko et al 1981b
46,XY,t(9;22)	Sandberg et al 1980
46,XY,t(9;22)	Sandberg et al 1980
46,XY,t(9;22)(q34;q11)	Roozendaal et al 1981
46,XY,t(9;22)(q34;q11)	
47,XY,-11,+16,+22	Nordenson 1983
47,XY,t(9;22)(q34;q11),+del(7)(q11)/48,XY,t(9;22),+7, +del(7)	Mitelman 1983
47,XY,t(9;22)(q34;q11),+del(7)(q11)	
48,XX,t(9;22),+8,+18,/46,XX,t(9;22)	Sandberg et al 1980
48,XX,t(9;22),i(17q),+18,+22q-	
52,XX,t(9;22),+22q-,+1-5mar	Borgström et al 1981

53,XY,t(9;22),+4,+5,+6,+7,+8,+17,+21	Sandberg et al 1980
56,XY,t(9;22)(q34;q11),+4,+5,+8,+9,+11,+16,+17,+18,+19,+20	Mitelman 1983

ALL, FAB type L3

46,XX,-16,-17,-22,+8,11q-,t(8;14),+1-2mar	Borgström et al 1981
46,XY,t(8;14)(q24;q32),+8,-22	Berger and Bernheim 1982
46,XY,t(8;22)(q24;q12)/46,XY,1q+,+del(1)(p22),-5, t(8;22)(q24;q12)/47,XY,+del(1)(p22),t(8;22)(q24;q12)	Abe et al 1982a
46,XY,t(8;22)(q24;q11)	Berger and Bernheim 1982
47,XY,ins(1),t(3;22)(q?;q?),+7	Penchansky et al 1981

Chronic lymphocytic leukemia

45,XX,-8,-14,-15,-22,+2,+9,+1p+,+mar	Finan et al 1978
46,XX,t(14;22)(q24q32;q11)	Nowell et al 1981
46,XY,9p+/43,X,-Y,-14,-22	Nordenson 1983
46,XY,del(11)(q22),+12,-13,-21,-22,+3mar/45,XY,del(11),+12, -13,-15,-21,-22,+2mar/46,XY,del(11),+12,+12,-13,-21,-22, +2mar/46,XY,t(1;8)(p22;q24)	Robert et al 1982
47,XY,+12,del(22)(q13)	Morita et al 1981
50,XX,+5,+5,t(13;?)(q34;?),+22	Nordenson 1983

Lymphocytic leukemia, special type

47,X,-Y,-4,+7,del(1)(q32),i(18q),21q+,i(22q),+2mar	Ueshima et al 1981
48,XX,-6,-11,14q+,+18,+20,+22	Imamura et al 1981

MONOCLONAL GAMMOPATHIES

Multiple myeloma

45,X,del(X)(q25),t(3;8;22)(q13;q34;q11),-5,-7,-10,-11, del(13)(q13?),+16,19q+,+3mar	Van Den Berghe et al 1979f
46,XX,t(22;?)(q11;?)	Van Den Berghe et al 1979f
46,XX,t(9;22)(q34;q11)	Van Den Berghe et al 1979f
46,XY,+t(1;?)(p1?;?),+t(1;12)(q21;q24),del(6)(p11),-8,+9, -12,-13,-16,-19,22q+,+mar	Philip et al 1980
46,XY,t(22;?)(q11;?)	Van Den Berghe et al 1979f
47,X,-Y,+t(1;15)(q12;p11),+3,+11,-15,+19,-22/47,X,-Y,+3, +11,+19,-22	Philip et al 1980
47-48,XY,+8,+22,+mar	Shiloh et al 1979

Plasma cell leukemia / Plasmocytoma

43,X,-X,-1,+t(1;8)(q11;q11),+t(1;9)(q21;p24), +t(1;10)(p31;q26),+t(1;?)(q12;?),-8,-9,-10,-13,-22	Ueshima et al 1983
44,X,-Y,del(1)(p11),+del(1)(q11),t(11;14)(q11;q32),14q+, -21,-22	Ueshima et al 1983
45,XY,-5,del(6)(q11),t(6;7)(q?;q?),+9,-12,-13,-16,-17, t(9;22)(q34;q11),+2mar	Karpas et al 1982

SOLID TUMORS

UNCLASSIFIED NEOPLASMS

EPITHELIAL NEOPLASMS

Carcinoma, NOS

44,XY,+5,-8,-11,-13,-13,-14,-14,-15,+19,-22,-22,t(1;11), 5q-,9q+,+5mar	Reichmann et al 1981
47-48,Y,-X,-5,-8,+13,-14,-16,-18,-22,+7mar	Reichmann et al 1981
49,X,-Y,-3,-5,-14,+15,-16,-17,+19,+21,+22,1q+,5q-,6q-,8p-, 9p-,11p-,18p+,20q-,22q-,t(Y;14),+3mar	Reichmann et al 1981
60-63,XX,cx,t(5;18),t(3;6),t(3;22),del(10)(q24), del(4)(p1?)	Riet-Fox et al 1979
63,XX,-1,+3,+5,+6,+7,+8,+10,+11,+12,-13,-15,+20,+21,+22, +9mar	Sandberg 1977

Carcinoma NOS, metastatic

35,XX,-1,-3,-4,-5,-7,-8,-10,-11,-12,-13,-14,-15,-16,-17, -18,-19,-20,-21,-22,+8mar	Pathak et al 1979
45-50,XX,-2,+7,-22,t(1;12),t(13;14),inv(7)(q?)	Ayraud 1975
46,XX,t(9;22)(q34;q11)"s"	Togawa et al 1981
52,XX,+3,+5,+7,+8,+9,+11,12q+,13q-,-14,+16,i(17q)"s"	
46,XY,+2,-6,-9,del(22)(q11),+mar	Hansson and Korsgaard 1974

Small cell carcinoma

47,XY,+del(1)(p?),+del(3)(q?),+del(3)(p?),-4,+8,-15,+18, -22	Wurster-Hill and Maurer 1978
68,XY,cx,del(3)(p14p23),del(22)(q11)	Whang-Peng et al 1982b

Adenocarcinoma

38,X,-X,-2,-3,-10,-14,-22,del(1)(q25),del(6)(q15), del(14)(p11),dmin	Trent and Salmon 1981
38,X,-X,cx,-17,-22,del(1)(q25),del(2)(q23),del(6)(q15)	Trent and Salmon 1980
40-42,X,-X,t(1;?)(p13;?),del(1)(p12),2q+,del(3)(q21), del(6)(q13),del(7)(q11),t(11;11)(p15;q13), t(12;12)(q24;?),13p+,15p+,i(22q)	Kusyk et al 1981
43-46,XY,i(10q),15q+,-17,+19,-21,-22,dmin	Couturier-Turpin et al 1982
44,XX,+5,-9,-13,-13,-15,-17,-18,+22,+2mar	Martin et al 1979
47,XX,-17,+20,+22	Martin et al 1979

Adenocarcinoma, metastatic

43-46,XX,t(1;16),t(1;22),+r	Bertrand et al 1979
56-64,XX,del(X)(q22),del(1)(p12),del(5)(p21),del(7)(q31), t(14;18)(q11;q11),18q+,i(22q)	Berger and Lacour 1974

MESENCHYMAL NEOPLASMS

Liposarcoma

50,XY,+3,+6,+8,+11,+14,-22,17p+ Sonta et al 1977

Leiomyosarcoma

42,XX,del(1)(p12orp13),del(11)(q13orq14),+7,-9,-13,-14,-15, -18,+19,-22 Mark 1976

Mesothelioma, malignant

40-46,XY,+16,+20,-22,t(14;15),del(2)(p?),i(5p) Ayraud 1975
41,XY,del(1)(p13),t(1;3)(p13;q11),t(7;7)(q11;q36), t(9;12)(p24;q13),del(12)(q13),del(13)(q12q14),-14,-15, t(17;?)(p13;?),i(19p),-20,-21,-22 Mark 1978
43,XY,t(1;1)(q42;q32p22),t(2;6)(p25;p21),inv(3)(p24q13), t(5;18)(p15;q11),-6,+9,del(13)(q24q14),-14,-18,-22 Mark 1978

MELANOMAS

Malignant melanoma, metastatic

41,XY,dup(1)(q21q32),t(1;7)(q25;q36),t(7;9)(q11;q11), del(9)(q13),-3,-4,-5,-8,-10,-11,-16,+18,-20,-22 Kakati et al 1977

NEUROGENIC NEOPLASMS

Astrocytoma, grade III, IV

44,XY,-4,-17,-22,1q-,5p-,-11q+,+mar Yamada et al 1980

Oligodendroglioma

45,XY,t(21;22)(p11;q11),7q-,9q-,14q+ Yamada et al 1980

Neuroblastoma

46,XX,del(1)(p31),22p+ Brodeur et al 1977
46,XX,del(22)(q13) Douglass et al 1980b
46,XY,t(1;?)(p?;?),19q+,20q+,-22 Gilbert et al 1982
47,XX,+7,9q+,22q+ Brodeur et al 1977

Retinoblastoma

45,X,-X,+1p+,+i(6p),-16,17q+,-19,22p+ Kusnetsova et al 1982
46,XX,+i(6p),12q+,19p+,-22 Kusnetsova et al 1982
46,XY,dup(1)(q25q44),t(7;10)(q36;q11),-8,-10, t(12;13)(q11;p11),dup(13)(q22q34),+18,+22 Gardner et al 1982
47,XX,+t(1;22)(q?;q?),+i(6p),-22 Kusnetsova et al 1982

47,XY,dup(1)(q12q44),del(2)(q13),+del(6)(q16), t(14;?)(q32;?),i(17q),+20,-22	Gardner et al 1982
51,XY,del(6)(q21),+8,+11,-12,t(13;18)(p11;q11),14q+,-16, +19,+22,+2mar	Hossfeld 1978

Meningioma, benign

38,X,-Y,-22,-21,-17,-14,-13,-10,-9,-4,-1p-,+mar	Zankl et al 1975a
41,X,-Y,-22,-18,-15,-10,1p-	Zankl et al 1975a
41,XY,+del(1)(p?),-5,-10,-17,-20,-22	Zang and Zankl 1983
41-42,XY,-1,-5,-8,9q+,-14,-18,-22,+r	Mark 1973b
43,X,-Y,-22,-15	Zankl et al 1975a
43,XX,-1,-6,-7,-11,-22,+2mar/40-45,XX,-1,-6,-7,-9,-11,-19, -22,+2mar	Mark et al 1972b
43,XY,-1,-8,-17,-19,-22,+2mar/44,XY,-1,-17,-22,+mar	Mark et al 1972a
43,XY,-22	Zankl and Zang 1972
43-45,XX,-8,-16,-22	Mark et al 1972a
44,X,-Y,-22/42,X,-Y,-17,-21,-22	Yamada et al 1980
44,XX,-19,-22	Zankl et al 1979
44,XX,-6,-22/45,XX,-22	Zang and Zankl 1983
44,XX,-8,-22/40-44,XX,-8,-9,-19,-20,-21,-22	Mark et al 1972a
44,XY,-22,-15	Zankl et al 1975a
44,XY,-9,-22,+del(22q)	Mark 1973a
44,Y,-X,-22,-14,+mar/43,Y,-X,-22,-14,-15,-11,-10,+3mar	Zankl et al 1975a
44-45,XX,-8,-22,+del(22q)	Mark 1973a
45,X,-Y/44,X,-Y,-22	Yamada et al 1980
45,X,-Y/44,X,-Y,-22	Yamada et al 1980
45,XX,-15,-22,1p-,2p-,4p-,+mar	Zankl et al 1979
45,XX,-22	Mark et al 1972b
45,XX,-22	Zankl et al 1975a
45,XX,-22	Zankl et al 1975a
45,XX,-22	Zankl and Zang 1972
45,XX,-22	Weiss et al 1975
45,XX,-22	Zankl et al 1979
45,XX,-22	Zankl et al 1979
45,XX,-22	Zang and Zankl 1983
45,XX,-22	Zang and Zankl 1983
45,XX,-22,+13,-14/44,XX,-22,+13,-14,-1	Mark 1973a
45,XX,-22,del(20)(q11)/44,XX,-22,-20	Zankl et al 1975a
45,XX,-22/40-44,X,-X,-8,-9,-15,-21,-22	Mark et al 1972b
45,XX,-22/43-44,XX,-8,-16	Mark 1973a
45,XX,-22/43-44,XX,-8,-18,-22	Mark 1973a
45,XX,-22/43-44,XX,-2,-16,-22	Mark 1973a
45,XX,-22/44-45,XX,-8,-12,-21,-22	Mark et al 1972b
45,XX,-22/45,XX,3p-,-22	Zang and Zankl 1983
45,XX,-22/45-46,XX,-22,-9,+6,+14,+del(22q)	Mark 1973a
45,XX,-22	Mark et al 1972b
45,XX,-22	Zankl et al 1982
45,XX,-22	Zankl et al 1982
45,XX,-22	Yamada et al 1980
45,XX,-22	Yamada et al 1980
45,XX,-22/46,XX,-22,+mar	Zang and Zankl 1983
45,XX,-22/46,XX,22q-	Zankl et al 1975a
45,XX,t(6;15)(p12;q26),-22	Zang and Zankl 1983
45,XY,-22	Zankl et al 1975a

45,XY,-22/42-44,XY,-8,-16,-19,-22	Mark 1973a
45,XY,-22/42-45,XY,-8,-15,-17,-20,-22	Mark et al 1972b
45,XY,-22/44,X,-Y,-22/43,X,-Y,-22,-17	Yamada et al 1980
45,XY,-22/44,XY,-12,-22	Mark 1973a
45,XY,-22/44,XY,-4,-22	Mark 1973a
45,XY,-22	Zankl et al 1979
45,XY,-22/46,XY,22q-	Zang and Zankl 1983
46,XX,-22,+mar/44-45,XX,-8,-9,-22,+mar	Mark et al 1972b
46,XX,22q-/43-45,XX,22q-,-15,-17	Mark et al 1972b
47,XX,+12,del(22)(q11)/46,XX,-8,+12,-17,del(22)(q11)	Mark et al 1972a
47,XX,-22,+7,+9	Zankl et al 1975a
50,XX,+9,+10,+12,+15,+20,-22	Mark 1973c
51,XY,+5,6q+,+12,+16,+17,+18,+20,-22/52,XY,+5,+8,+10,+12, +17,+18,+20,-22	Zang and Zankl 1983
52,X,-Y,-22,+20,+19,+17,+15,+11,+9,+7,+5	Zankl et al 1975a
64,XX,-22,cx	Mark 1973c
67,XX,-22,cx	Mark 1973c

EMBRYONAL AND MISCELLANEOUS NEOPLASMS

LYMPHOMAS

1. UNCLASSIFIED AND MISCELLANEOUS LYMPHOMAS

Non-Hodgkin's lymphoma, lymphocytic, NOS

46,XX,+3,t(13;22)(p11;p11)	Fleischman and Prigogina 1977
47,XY,+7,-9,-11,1p+,6q-,10p-,14q-, t(14;?;14)(q24orq32;?;q13),15q-,22q-	Fukuhara et al 1979b

Non-Hodgkin's lymphoma, histiocytic, NOS

46,XX,del(1)(p31p35),+3,del(9)(p11),t(10;?)(q26;?), del(11)(q13),-13,t(11;14)(q13;q32),t(17;21), t(9;22)(q11;p13),+r	Mark et al 1977
47,XY,-6,-14,1q-,+1p-,2q+,3q+,9p+,10p-,10p+,12p-,18q+,22p+, t(2;14)(q37?;q13?)	Fukuhara et al 1979b
48,XY,t(3;8)(q22;q11),t(5;10)(q33;q22),del(7)(q21), t(4;9)(q11;p24),t(10;13)(q22;q21),r(12),del(19)(p13), del(22)(q11),+7	Mark et al 1978
50,XY,t(2;?)(p25;?),del(2)(p13),+i(3p),del(6)(p21),+7,+8, +9,+9,-11,-13,+14,-15,+20,+21,-22	Mark et al 1978
44,XY,del(2)(p13),t(6;?)(p21;?),del(11)(q13), t(11;14)(q13;q31),i(15q),-20,-22	

Angioimmunoblastic lymphadenopathy

48,XY,+3,+18/51,XY,+5,+15,+19,+21,+22 Kaneko et al 1982c
56,XY,+5,+5,+6,+10,+15,+15,+19,+20,+21,+22 Kaneko et al 1982c

Lennert's lymphoma

45,XX,-22 Kakati et al 1980

2. HODGKIN'S LYMPHOMAS

Hodgkin's disease, lymphocytic depletion

48,X,-Y,t(1;?)(q44;?),i(1q),del(1)(q21),del(3)(q21),-10, t(3;11)(q23;q21),+14,t(14;?)(q32;?),+15,-17,+19,+20,-22 Hossfeld and Schmidt 1978

3A. NON-HODGKIN'S LYMPHOMAS - RAPPAPORT CLASSIFICATION

Lymphocytic, poorly differentiated, diffuse

46,XY,t(14;18)(q32;q21),t(10;14)(q24;q32),-1,-17,-22,+3mar Fukuhara and Rowley 1978
47,XY,t(14;14)(q32;q?),+7,-9,-11,1p+,6q-,10p-,15q-,22q-, +2mar Fukuhara and Rowley 1978
88,XX,del(1)(p22),del(3)(q21),t(3;9)(q?;q?),del(6)(q21), t(22;?)(q13;?) Mark et al 1979

Histiocytic, diffuse

45,X,-Y,t(14;15)(q32;q21orq22),-15,+18,-22,1q+,3p+,5p-,7p-, 9p+,12p+,16q+,17p-,+2mar Fukuhara and Rowley 1978
47,XY,-11,-22,del(1)(q42),del(2)(q33),del(5)(q22q31), del(7)(q32),del(9)(p22),t(21;22)(q22;q11), +t(11;?)(q23;?),+t(21;22)(q22;q11),+mar Kaneko et al 1982a

3B. NON-HODGKIN'S LYMPHOMAS - KIEL CLASSIFICATION

Centroblastic-centrocytic, diffuse

48-50,XY,-1,+7,9q+,+9p+,+10,+14q+,-17,-18, +t(17;18)(p11;q11),-19,+20,+22,+mar Kristoffersson 1983
48-51,XY,del(1)(p33q23q25),-2,+del(3)(p22),+4,4p+,+5,+6p+, +7,+8,+9,+21,+22 Kristoffersson 1983
45-51,XY,del(1)(p33q23q25),del(2)(q14),+del(3)(p22),+4,+5, +7,+8,-12

Centroblastic-centrocytic, follicular

46-47,XX,t(8;9)(q13;p24),t(14;18)(q32;q21),+22	Kristoffersson 1983

Centroblastic, diffuse

46-49,XY,1p+,+2,-6,+del(9)(p21),+12,+13,14q+?,i(17q),+18	Kristoffersson 1983
47,XY,14q+,+22	

Lymphoblastic, Burkitt's type

43,XY,+1,+2,+3,-7,-8,-9,-12,-16,-22,del(1)(q25), del(2)(p11),del(10)(q22),14q+	Biggar et al 1981
45,X,-Y,t(1;14)(q22;q32),t(8;22)(q24;q11),3q+,6q-,6q+,7q+, 9p-,14q-,18q-	Berger et al 1981c
45-47,XY,t(14;?)(q32;?),del(3)(p?),-9,-22	Philip et al 1977b
46,XX,t(8;22)(q24;q13)	Miyoshi et al 1981c
46,XY,t(8;22)(q23;q12)/46,XY,t(8;22)(q23;q12), t(1;6)(q21;q27)	Berger et al 1981c
46,XY,t(8;22)(q23;q12)/46,XY,t(8;22)(q23;q12), +t(1;6)(q23;q26)	Berger et al 1981c
46,XY,t(8;22)(q23;q11)/46,XY,t(8;22)(q23;q11), t(1;4)(p32;q32)	Berger et al 1981c
47,XX,t(8;22)(q24;q11),+8,t(11;21)(q14;q22)	Berger and Bernheim 1982
47,XY,t(8;22)(q24;q11),+22/47,XY,t(8;22)(q24;q11),+22, dup(1)(q23q24)	Berger and Bernheim 1982

Lymphoblastic, convoluted type

43-46,XY,del(1)(p32),del(2)(p21),-6,t(7;9)(p11;q11),-9,-10, t(11;15)(p11;q11),-11,del(12)(q14),-13,-14,i(17q),-17, +18,del(19)(p11),-22,+2mar	Kristoffersson 1983

Immunoblastic

48-49,XY,+3,+7,+11,t(12;?)(p13;?),t(13;?)(p13;?),-16,-20, +21,+22	Reeves and Pickup 1980
49,XX,+5,+19,+22	Kaneko et al 1982a

3C. NON-HODGKIN'S LYMPHOMAS - LUKES AND COLLINS CLASSIFICATION

Sezary's syndrome

42,XY,-13,-21,-22,1p+,17q+	Nowell et al 1982
44-46,XX,t(2;9),t(2;13),17p+,19p+,-21,-22	Edelson et al 1979

Follicular center cell, diffuse, large cleaved

47,XY,-6,-14,1q-,+1p-,2q+,3q+,9p+,10p-,10p+,12p-,18q+,19p+, 19q+,+22p+,+mar	Fukuhara et al 1978b

Follicular center cell, diffuse, small non-cleaved

Karyotype	Reference
44,X,-Y,t(1;14)(p21;q13),5p+,del(6)(q13),del(8)(p21),9p+, 9p-,t(11;21)(q23;q22),13q+,-10,+12,-17p+,der(19),22q+	Fukuhara et al 1979a

Immunoblastic type (B-cell)

Karyotype	Reference
48,XY,t(4;12)(q?;q?),t(6;15)(q?;q?),t(9;22)(q34;q11),11q-, +20,+21	Nordenson et al 1983
49,XX,+i(1q),+5,del(8)(q24),t(3;22)(q29;q11),+13	Berger et al 1983a

3D. NON-HODGKIN'S LYMPHOMAS - WORKING FORMULATION

Small lymphocytic

Karyotype	Reference
47,XY,+2,2q-,17p-,t(18;22)	Yunis et al 1982

Follicular, small cleaved cell

Karyotype	Reference
46,XY,t(1;2),t(4;10),t(14;18)(q32;q21),t(1;22)	Yunis et al 1982
48,X,r(Y),+2,+3,t(14;18)(q32;q21),t(1;22)	Yunis et al 1982

Small non-cleaved cell

Karyotype	Reference
46,X,-X,t(8;14)(q24;q32),dup(13),+r/45,X,-X,t(8;14),-12, t(1;15),der(22),+r	Yunis et al 1982

Chromosome X

HEMATOLOGICAL DISORDERS

UNCLASSIFIED LEUKEMIAS

Acute leukemia, NOS

Karyotype	Reference
45,X,-X,+t(X;1)(q13;p12),-7,+8,-9	Mamaeva et al 1983
54,XX,+X,+3,+6,+8,+8,+13,+19,+22	Muir et al 1977

NONLYMPHOCYTIC LEUKEMIAS

Acute nonlymphocytic leukemia (ANLL)

Karyotype	Reference
39,X,-X,-3,-5,-13,-14,-15,-16,-17,-21,-22,del(7)(p11), t(12;?)(p13;?),t(16;?)(p13;?),+3mar	Mitelman 1983
39,X,-X,-3,-5,-13,-14,-15,-16,-17,-21,-22,del(7)(p11), t(12;?)(p13;?),t(16;?)(p13;?),+3mar	
44,X,-X,-21	Mitelman 1983
45,X,-X,-7,t(X;X)(q13;q13)	Philip et al 1978a
45,X,-X,7q-,-8,+i(17q)	Nordenson 1983
45,X,-X,t(8;16)(q22;q24),+20,-21	Trent et al 1983a
45,X,-X,t(8;21)(q22;q22)"s"	Elfenbein et al 1978
45,X,-X,t(8;21)(q22;q22)	Sasaki et al 1976
46,X,-X,+8	Mitelman 1983
46,Y,t(X;11)(q22.3;q23.3),ins(X;11)(q22.3;p13p15), dup(1)(q11q21),t(4;6)(p16;p21),t(7;12)(p13;q15)	Nacheva et al 1982
48,XX,+8,-9,+t(X;9)(q13;q32)	Philip et al 1978a
49,X,t(X;10)(p11;p11),+del(1)(p22),+del(1)(p22), del(5)(q23),+7/49,X,t(X;10)(p11;p11),+del(1)(p22),+7,+8	Mitelman 1983

ANLL, FAB type M1

Karyotype	Reference
45,X,-X,-5,-7,-11,-17,+21,+3mar"s"	Zaccaria et al 1983b
47,XX,+21"s"	Mitelman 1983
45,X,-X"s"	
47,X,-X,+i(Xp)/48,X,-X,+i(Xp),+8/48,X,-X,+i(Xp),+20	Hagemeijer et al 1981a
49,XX,+X,-1,+der(1)(q21q31),+del(1)(p22),+t(1;5)(p11;q35), -5,t(9;22)(q34;q11),+22q-	Sasaki et al 1983

ANLL, FAB type M2

Karyotype	Reference
44,X,t(X;8),del(5)(q12q21),t(1;6),t(7;19),-7,-8,t(7;8),-21	Kerkhofs et al 1982
45,,del(X)(q26),-Y,t(8;21)(q22;q22),del(9)(q24), del(18)(p11)	Brodeur et al 1983
45,X,-X,-9,-18,+2mar	Prieto et al 1981
45,X,-X,t(8;21)	Tricot et al 1981
45,X,-X,t(8;21)	Takeuchi et al 1981
45,X,-X,t(8;21)	Takeuchi et al 1981
45,X,-X,t(8;21)(q21;q21)	Hagemeijer et al 1981a
45,X,-X,t(8;21)(q21;q21)	Hagemeijer et al 1981a
45,X,-X,t(8;21)(q21;q21),1p+,17p-	
45,X,-X,t(8;21)(q22;q22)	Levan and Mitelman 1979
45,X,-X,t(8;21)(q22;q22)/46,X,-X,t(8;21),1p+,7q+,+19/47,X, -X,t(8;21),1p+,+19,+der(21)	Sakurai et al 1982b
45,X,-X,t(8;21)(q22;q22)	Oshimura et al 1982b
45,X,-X,t(8;21)(q22;q22)/46,XX,t(8;21)	Berger et al 1982b
45,X,-X,t(8;21)(q22;q22)/46,XX,t(8;21)	Sakurai et al 1982b
45,X,-X,t(8;21)(q22;q22)/46,XX,t(8;21)	Alimena 1983
45,X,-X,t(8;21)(q22;q22),del(9)(q31)	Brodeur et al 1983
45,X,-X	Alimena 1983
45,XX,-7,t(3;9)(q29;p21)/44,X,-X,-7,t(3;9)"s"	Rowley et al 1981a
46,X,-X,t(8;21)(q22;q22),+18p+	Berger et al 1982b
46,XX,t(8;21)(q22;q22)/45,X,-X,t(8;21)	Kaneko et al 1982b
47,X,-X,+2mar	Golomb et al 1978c
47,X,-X,+8,t(8;21)(q22;q22),+14	Brodeur et al 1983
47,XX,+8,/48,XX,+8,+8/46,X,-X,+8"s"	Hagemeijer et al 1981a

ANLL, FAB type M1+M2

Karyotype	Reference
44,X,-X,-5,t(3;7)(q13;p12)/44,XX,-7,-8/44,XX,-5,-8"s"	Sandberg et al 1982a
45,X,-X,t(8;21)(q21;q22)	Mitelman et al 1981
45,X,-X,t(8;21)(q22;q22)	Mitelman et al 1981
45,X,-X,t(8;21)(q22;q22)	Rowley and Potter 1976
45,X,-X,t(8;21)(q22;q22)	Rowley and Potter 1976
45,X,-X,t(8;21)(q22;q22)	Prigogina et al 1979
46,X,del(X)(q21),t(5;21)(q13;q22)	Mitelman et al 1981
46,X,r(X),t(8;21)	Garson 1980
47,XX,+8/51,XX,+X,+8,+18,+20,+21	Testa et al 1979
46,XY,t(X;12)(p?;q?)	Garson 1980
47,X,-X,+2mar"s"	Testa et al 1981b
47,X,-X,+del(1)(p11),del(5)(q13),t(5;12)(q21;q24), t(11;12)(q25;q13),t(13;14)(q11;p11),+16,+21	Mitelman et al 1981

ANLL, FAB type M3

Karyotype	Reference
46,XX,t(15;17)(q22;q21)/46,XX,t(15;17),t(1;4)(p35;q21), del(X)(q25)	Berger et al 1981e
46,XY,t(15;17)(q22;q21),t(X;9)(q28;q22)	Berger et al 1981e

ANLL, FAB type M4

Karyotype	Reference
45,X,-X	Chrz et al 1983
45,X,-X,-9,-17,-18,-21,+5mar	Prieto et al 1981
45,X,-X,t(8;17;21)(q22;q23;q22)	Testa et al 1979
45,X,-X,t(8;17;21)(q22;q23;q22)/45,X,-X,t(1;11)(p36q32;q13),t(8;17;21)	
46,X,del(X)(p21)	Alimena 1983
46,XX,del(X)(q11),del(8)(q21)	Alimena 1983
46,XX,t(X;11)(q25;q23)	Prigogina et al 1979

ANLL, FAB type M5

Karyotype	Reference
46,Y,-X,-1,-7,-8,-14,-15,-17,-20,inv(6)(p23q13),del(11)(q23),+t(X;?)(p11;?),+t(1;?)(p24;?),+t(12;?)(q22;?),+t(17;?)(p11;?),+4mar	Kaneko et al 1982b

ANLL, FAB type M5a

Karyotype	Reference
44,Y,-X,-5,del(11)(q22),17p+,+r	Berger et al 1982a

ANLL, FAB type M4+M5

Karyotype	Reference
45,X,-X,r(21)	Olinici et al 1978a

ANLL, FAB type M6

Karyotype	Reference
43,X,-X,del(5)(q14q34),-7,11q-,-12,12p+,t(12;17),t(20;21)	Kerkhofs et al 1982
44,X,del(Y)(q?),+Xq-,3p-,3q-,4q-,-6,+7q-,11q+,-15,-18"s"	Bradley et al 1982
45,X,-X,-5,-21,22q-,+3mar"s"	Sandberg et al 1982a

Chronic myeloid leukemia, t(9;22)

Karyotype	Reference
45,X,-X,t(9;22)/46,XX,t(9;22)	Prigogina et al 1978
46,XX,t(9;22)(q34;q11)/48,XX,+X,t(9;22),+8	Sonta and Sandberg 1978b
46,XX,t(9;22)(q34;q11)/46,X,-X,t(9;22),4q+,+mar	McGlave et al 1982
46,XX,t(9;22),t(X;17)(q?;q?),-17,+22q-	Borgström et al 1982
46,XY,t(9;22),t(X;6)(q28?;q22?),t(5;17)(q13?;q25)	Bernstein et al 1980b
46,XY,t(9;22),t(X;9)(p22;q22),t(2;3)(p22;q27)	Fleischman et al 1981
47,XXY,t(9;22)(q34;q11)/53,XXY,+X,t(9;22),+4,+8,+18,+18,+22q-"c"	Mitelman 1983
47,XXY,t(9;22)/53,XXY,+X,t(9;22),+4,+8,+18,+18,+22q-"c"	
47,XXY,t(9;22)"c"	
52-54,XX,t(9;22)(q34;q11),+6,+8,+8,+19,+19,+del(22)(q11),+del(22)(q11),+X	Hayata et al 1975b

Chronic myeloid leukemia, aberrant translocation

Karyotype	Reference
44,Y,-X,t(9;17;22)(q34;p11;q11),t(X;9)(?;p24),i(17q)	Engel et al 1977
46,X,t(X;9;22)(q27;q34;q12)	Dallapiccola and Alimena 1979
46,X,t(X;9;22)/47-51,X,t(X;9;22),+1,+4,+8,+9,-14,i(17q),-19,+22q-	
47,Y,t(X;22)(p22;q11),+19	Hossfeld and Köhler 1979

Eosinophilic leukemia

45,X,-X,t(8;21)(q22;q22)	Kaneko et al 1983

MYELOPROLIFERATIVE DISORDERS

Polycythemia vera

44,X,-X,del(5)(q14q32),-7	Van Den Berghe et al 1979g
44,XX,del(5)(q14q32),-6,-7,-8,t(12;?)(p12;?), t(16;18)(q11;q12)	

Myelosclerosis / Myelofibrosis

46,X,del(X)(p?q?),del(5)(q?),13q+,15q-"s"	Whang-Peng et al 1978

Idiopathic thrombocythemia

46,X,t(X;9;22)(q11;q34;q11)	Fitzgerald et al 1981

DYSMYELOPOIETIC SYNDROMES

Preleukemia, NOS

43,X,-X,-3,-13,4q-,8q-,12q+	Geraedts et al 1980a
45,XX,-7/45,X,-X	Panani et al 1980

Acquired idiopathic sideroblastic anemia

46,X,t(X;11)(q13;p15)	Dewald et al 1982

Refractory anemia without excess of blasts

46,XX,5q-/47,XX,+8/47,XX,+14/46,X,-X,+8	Teerenhovi et al 1981

Refractory anemia with excess of blasts

49,XX,+X,del(1)(p21?),+9,+mar	Mitelman 1983

SPECIAL LEUKEMIAS

Hairy cell leukemia

44,X,del(X)(q22),del(6)(q23),-7,-8,-10,del(11)(p15q21),-21, 14q+,+r,+mar	Sadamori and Sandberg 1983a

LYMPHOCYTIC LEUKEMIAS

Acute lymphocytic leukemia (ALL)

Karyotype	Reference
43,X,-X,-7,-15,t(9;22)	Gibbs et al 1977
43,X,-X,-7,-8,t(9;22)(q34;q11)	Oshimura and Sandberg 1977
45,X,-X	Yamada and Furusawa 1976
45,X,-X,dup(1)(q31q41),del(5)(q22),del(17)(p11)	Humbert et al 1978
47,X,-X,del(5)(p23),+t(17;?)(p13;?),+del(18)(p11)	Oshimura et al 1977a
47,XX,+mar/50,X,-X,-11,+6mar	Morse et al 1978
53,XY,+X,+4,+8,+14,+15,+17,+21	Oshimura et al 1977a
54,XX,7p+,+10,+10,+14,+14,+18,+18,+21,+21/27,X,-X,-1,-2,-3,-4,-5,-6,-7,7p+,-8,-9,-11,-12,-13,-15,-16,-17,-19,-20,-22	Oshimura et al 1977a
55,XY,+X,+3,+6,+10,+13,+14,+16,+21,+21	Oshimura et al 1977a
57,XY,+X,+5,+6,i(7q),+10,+12,+13,+15,+18,+21,+21,+22	Oshimura et al 1977a
61,XX,+X,+1,+2,+3,+4,+5,+6,+8,del(6)(q21),+13,+14,+15,+16,+18,+21,+22	Oshimura et al 1977a

ALL, FAB type L1

Karyotype	Reference
28,XX,-1,-2,-3,-4,-5,-6,-7,-8,-9,-11,-12,-13,-15,-16,-17,-19,-20,-22/56,XX,+X,+X,+10,+10,+14,+14,+18,+18,+21,+21	Brodeur et al 1981a
56,XY,+X,+4,+6,+10,+15,+17,+18,+21,+21,+mar	Kaneko et al 1982e

ALL, FAB type L2

Karyotype	Reference
45,X,-X,-1,-14,-17,+21,t(11;12)(p15;p11),+i(17q),+t(1;?)(p34;?)	Kaneko et al 1981b
46,X,-X,-15,del(11)(q21),del(12)(p11),+2mar	Kaneko et al 1981b
47,XX,+X,t(4;11)(q21;q23)	Weh and Hossfeld 1982
49,XX,+X,+X,+mar	Borgström et al 1981

Lymphocytic leukemia, special type

Karyotype	Reference
45,X,-X,-4,+t(4;7)(p16;q11),15q+	Ueshima et al 1981
45,Y,-X,t(Y;14)(q12;q32),-13,-17,del(2)(q33),t(3;13)(q21;q34),del(6)(q16orq22),9p+,9q+,10q+,16q+,18q+	Miyoshi et al 1981a

MONOCLONAL GAMMOPATHIES

Multiple myeloma

Karyotype	Reference
45,X,-X	Nordenson 1983
45,X,del(X)(q25),t(3;8;22)(q13;q34;q11),-5,-7,-10,-11,del(13)(q13?),+16,19q+,+3mar	Van Den Berghe et al 1979f
47,X,-X,inv(1)(p22q12),t(7;?)(q11;?),+9,t(11;?)(p15;?),-13,-14,t(14;?)(q11;?),+15,17q+	Philip et al 1980
48,X,-X,+1,+7,-8,+9,+21,del(1)(p36),4p+,del(5)(q22),9p+,14q+	Liang et al 1979
52,X,-X,+t(1;?)(q12;?),+3,+7,+8,+9,+11,-13,+18,-21,+2mar	Philip et al 1980

Macroglobulinemia

46,Y,-X,del(1)(q21),del(5)(q23),+del(5)(q23), Ueshima et al 1983
+t(7;19)(p22;q13),del(8)(p12),+t(1;16)(q12;p13),
+t(1;18)(q12;p11),-7,-16,-18,-19,+mar

Plasma cell leukemia / Plasmocytoma

43,X,-X,-1,+t(1;8)(q11;q11),+t(1;9)(q21;p24), Ueshima et al 1983
+t(1;10)(p31;q26),+t(1;?)(q12;?),-8,-9,-10,-13,-22

SOLID TUMORS

UNCLASSIFIED NEOPLASMS

EPITHELIAL NEOPLASMS

Carcinoma, NOS

43,X,-X,t(1;11)(q21;q23),3q+,4q+,5q+,-11,-13,-16,17p+	Atkin and Baker 1982a
46,X,-X,+mar/61,XX,cx,der(2)(q23)	Vang Nielsen et al 1982
47-48,Y,-X,-5,-8,+13,-14,-16,-18,-22,+7mar	Reichmann et al 1981
50,XX,+X,+8,+12,+21	Sonta et al 1977
50,XX,+X,+8,+8,+13	Sonta and Sandberg 1978a
50,XY,+X,+7,+8,+9	Reichmann et al 1981
52,XY,+X,+Y,-2,+7,+8,+9,+10,+13,-17,+i(1q)	Reichmann et al 1981
58,XX,+X,+1,+2,+3,5p-,+6,+7,+9,11q+,-13,-14,-15,+16,+19, +20,+6mar	Kovacs 1978a

Carcinoma NOS, metastatic

??,X,-X,t(1;?),hsr(1),6q-,t(9;?),dic(13),dmin,cx	Bartnitzke and Bullerdiek 1983

Small cell carcinoma

44,XY,cx,del(1)(q31),t(2;3)(p?;q?),del(3)(p14p23), t(9;13)(p11;q11),del(11)(p15),12q+,del(X)(q22)	Whang-Peng et al 1982b
69,XY,cx,del(3)(p14q24),t(11;14)(p11;p11), t(11;13)(p11;p11),del(X)(q23)	Whang-Peng et al 1982b

Papillary carcinoma

42-54,X,-X,-14,-16,-16,+1,+3,+12,+20,1p-,6q-	Wake et al 1981
50,X,-X,+1,+3,+12,+20,-14,-16,1p-,6q-	Wake et al 1980

Adenoma

46,X,t(X;4)(q25;p16),del(2)(p13p21),del(5)(q15q31),-7, t(9;12)(p13;q13),del(11)(q11q14), t(7;9;12)(p11p22q11;p13p24;q13)/45,X,-X	Mark et al 1980
46,XX,t(1;3)(p21;p21),t(1;13)(p36;q14),t(3;8)(p21;q12), t(1;5;20)(p21;p14q12;p13),t(5;8)(p14;q12), t(X;9)(q27;q12)	Mark et al 1982a
46,XY,t(6;21)(q13;q22)/46,XY,t(1;14)(q42;p12)/45,X,-Y	Mark et al 1981b
45,X,-Y	
46,Y,t(X;15)(q24;q15)/45,X,-Y	
46,Y,-X,ins(X;3)(q22;q25q27),del(3)(q25),t(8;14)(q12;q22)	Mark et al 1981a

Adenocarcinoma

37,X,-X,-3,-8,-15,del(2)(p23),del(6)(q15),t(11;?)(p12;?) Trent and Salmon 1981
38,X,-X,-2,-3,-10,-14,-22,del(1)(q25),del(6)(q15), del(14)(p11),dmin Trent and Salmon 1981
38,X,-X,-8,-14,-20,+3-5mar Trent and Salmon 1981
38,X,-X,cx,-17,-22,del(1)(q25),del(2)(q23),del(6)(q15) Trent and Salmon 1980
40-42,X,-X,t(1;?)(p13;?),del(1)(p12),2q+,del(3)(q21), del(6)(q13),del(7)(q11),t(11;11)(p15;q13), t(12;12)(q24;?),13p+,15p+,i(22q) Kusyk et al 1981
44-50,X,-X,-1,t(1;?)(q21;?),del(2)(q33),4p+,-5,-7,+8, del(10)(q12),t(10;13)(q12;q34),del(17)(q25),-20 Kusyk et al 1982

Adenocarcinoma, metastatic

??,X,del(X)(q21),del(1)(q12),i(7q),del(9)(q31), del(16)(q23) Ayraud et al 1977
??,X,del(X)(q21),t(13;13)(q?;q?),del(16)(q21) Ayraud et al 1977
??,X,del(X)(q21),t(1;16)(q?;p?),del(12)(p12),del(16)(q21) Ayraud et al 1977
56-64,XX,del(X)(q22),del(1)(p12),del(5)(p21),del(7)(q31), t(14;18)(q11;q11),18q+,i(22q) Berger and Lacour 1974

MESENCHYMAL NEOPLASMS

MELANOMAS

NEUROGENIC NEOPLASMS

Retinoblastoma

45,X,-X,+1p+,+i(6p),-16,17q+,-19,22p+ Kusnetsova et al 1982
46,X,-X,+i(6p),17q+,20q+ Kusnetsova et al 1982
46,X,t(X;1)(q28;q11) Gardner et al 1982
47,XY,t(X;1)(q28;p11),t(2;?)(q11;?),t(2;15)(p25;q11), t(3;?)(q27;?),4p+,t(5;?)(q35;?),i(17q),+mar Gardner et al 1982

Meningioma, benign

44,Y,-X,-22,-14,+mar/43,Y,-X,-22,-14,-15,-11,-10,+3mar Zankl et al 1975a
45,XX,-22/40-44,X,-X,-8,-9,-15,-21,-22 Mark et al 1972b

EMBRYONAL AND MISCELLANEOUS NEOPLASMS

Blastoma, NOS

Karyotype	Reference
40-47,X,-X,cx,7q+,+del(10)(p13),+del(12)(p11), del(11)(p11p14)	Slater and Kraker 1982

LYMPHOMAS

1. UNCLASSIFIED AND MISCELLANEOUS LYMPHOMAS

Non-Hodgkin's lymphoma, NOS

Karyotype	Reference
46,X,-X,del(1)(q22),t(2;11)(p21;q21),t(9;15)(q11;p12), del(9)(q11),-14,t(14;17)(q23;q23),-16,+mar	Kristoffersson 1983
48-54,XX,+X,3p+,-4,-6,-8,-9,-10,+12,-15,+7-10mar	Pearson et al 1982
50,XX,+X,t(1;11)(q25;q23),+3,+7,-13,t(14;18)(p?;p?),-15, t(19;?)(q13;?),+3mar	Reeves and Pickup 1980

Non-Hodgkin's lymphoma, lymphocytic, NOS

Karyotype	Reference
47-49,XX,+X,-14,i(17q),t(1;17),1p+	Fleischman and Prigogina 1977

Non-Hodgkin's lymphoma, histiocytic, NOS

Karyotype	Reference
45,X,-X,del(1)(q22),t(1;17)(q?;q?),del(2)(q33), del(3)(p21p25),+ins(3)(p14p21p25),+5,del(6)(q14),t(1;7)	Mark et al 1978
47-48,XY,+X,-13,-15,t(6;11)(p?;q?),t(13;14;?)	Fleischman and Prigogina 1977

2. HODGKIN'S LYMPHOMAS

Hodgkin's disease, mixed cellularity

Karyotype	Reference
47,XX,+X,t(6;14)(q15;q32)	Reeves and Pickup 1980

Hodgkin's disease, lymphocytic depletion

Karyotype	Reference
45,X,t(X;1)(q28;q11),t(3;5)(q29;q33),del(8)(q22),-9, del(11)(p13),-17	Hossfeld and Schmidt 1978

3A. NON-HODGKIN'S LYMPHOMAS - RAPPAPORT CLASSIFICATION

Lymphocytic, poorly differentiated, nodular

Karyotype	Reference
52,X,-X,+2,-4,+5,+7,+8,-14,-15,-16,+20,+21, +t(1;15)(p11;q11),+t(12;17)(q24;q11),+t(14;?)(q32;?), +mar	Kaneko et al 1982a
92,XY,t(2;9)(q11;p11),t(5;?)(p15;?),t(17;?)(q25;?),17q+, del(1)(p22),del(3)(p21),del(X)(q24),del(11)(q23)	Reeves 1973

Mixed cell, diffuse

Karyotype	Reference
50,XY,+11,+12,-15,+del(3)(q11),+del(3)(q11), +t(X;1)(p22;q11)	Kaneko et al 1982a
50,XY,+X,-1,+5,+7,+12,-14,-17,del(13)(q12q14), ins(1;1)(q21;q21q32),+t(14;?)(q32;?),+t(1;17)(q21;q25)	Kaneko et al 1982a

Histiocytic, diffuse

Karyotype	Reference
45,X,del(X)(q24),del(1)(q21),+del(1)(p11),del(2)(q22), del(3)(q21),del(6)(p11),del(6)(q15),t(7;9)(p22;q13), t(8;10)(p23;q21),del(9)(q13),del(10)(q21), t(2;11)(q22;q23),-13,del(14)(q24),del(15)(q22), del(16)(q22),+17,-19	Mark et al 1979
46,X,-X,-11,-18,+t(11;?)(q23;?),t(18;?)(q23;?),+mar	Kaneko et al 1982a
46,X,-X,-2,-10,+11,-13,-14,-17,del(1)(p22p32), del(3)(p13p21),del(6)(q21),del(9)(p13),+t(2;?)(p23;?), +t(14;?)(q32;?),t(18;?)(q23;?),+t(X;13)(q26;q12),+mar	Kaneko et al 1982a
46,X,-X,-4,-11,-17,+18,-19,del(7)(p15),del(8)(q22),+i(17q), +t(4;19)(p11;q13),t(11;?)(q23;?)	Kaneko et al 1982a
47,X,+X,-Y,+12,-14,del(11)(q21q23),del(16)(q22), t(14;?)(q32;?)	Kaneko et al 1982a
47,X,-X,-7,-17,+19,+i(7q),+i(17q),+mar	Kaneko et al 1982a
47,X,-X,t(4;14)(q?;q32),-18,1q+,2p+,6q-,7q+,9q-,12q+,+3mar	Fukuhara and Rowley 1978
51,XY,+X,-2,+5,+11,-12,+21,del(6)(q21),+t(2;?)(q33orq35;?), +t(12;?)(p11;?),+mar	Kaneko et al 1982a
82,XX,del(X)(q13),i(1q),i(1p),del(7)(p11),i(9p), del(14)(q22),t(1;14)(q23;q32),i(15q),i(17q)	Mark et al 1979

3B. NON-HODGKIN'S LYMPHOMAS - KIEL CLASSIFICATION

Centroblastic-centrocytic, follicular

Karyotype	Reference
48,XX,+X,t(1;?)(p36;?),del(2)(p21),t(3;?)(q27orq29;?), t(14;?)(q32;?),t(17;18)	Reeves and Pickup 1980
52,XX,+X,+7,+8,del(10)(q24),+12,t(13;?)(q22;?), t(14;?)(q32;?),del(18)(q21),+19,+mar	Reeves and Pickup 1980

Lymphoblastic, Burkitt's type

45,X,-X,t(2;8)(p11;q24)	Abe et al 1982b
46,X,t(X;1)(p11;q25),del(1)(q25),7p+,17p-,t(2;8)(p11;q24)	Slater et al 1982
46-52,X?,t(8;14)(q24;q32),+1,+3,+5,+7,i(17q),+X	Zech et al 1976a
47-49,XY,t(8;14)(q23;q32),+7,+t(X;1)(p21;q21)	Kakati et al 1979
48-53,X?,t(8;14)(q24;q32),+X,+2,+3,+9,+10,+11,+12,+20	Zech et al 1976a

Immunoblastic

51,XY,+X,t(3;3)(q11;q27),t(8;?)(q2?;?),+7,+12,+21,+21	Reeves and Pickup 1980

3C. NON-HODGKIN'S LYMPHOMAS - LUKES AND COLLINS CLASSIFICATION

Mycosis fungoides

46,XX,dmin/47,XX,+X	Vloten et al 1979
47,XY,+Xp-,t(4;14),+14	Edelson et al 1979

Follicular center cell, follicular, small cleaved

45,X,-X,t(14;?)(q?;?)	Fukuhara et al 1979a

3D. NON-HODGKIN'S LYMPHOMAS - WORKING FORMULATION

Follicular, small cleaved cell

47,XX,t(X;1),ins(1;8),t(2;8),t(14;18)(q32;q21),+17	Yunis et al 1982
47,XY,+X,t(2;3;17),6q-,t(14;18)(q32;q21)	Yunis et al 1982
48,X,-X,inv(1),1q-,+3,4q-,ins(5),+7,+mar	Yunis et al 1982
49,XY,+X,+3,ins(11;3),t(14;18)(q32;q21),+18	Yunis et al 1982

Follicular, mixed small cleaved and large cell

50,Y,inv(X),+3,+7,+10,t(12;12),+12,t(14;18)(q32;q21),+18, 18q-,-21	Yunis et al 1982

Diffuse, large cell

47,XX,+X,6q-,t(14;17)	Yunis et al 1982

Large cell, immunoblastic

48,Y,t(X;?),t(1;?),t(3;6),+5,-6,t(8;14)(q24;q32),i(17q), +mar	Yunis et al 1982

Small non-cleaved cell

46,X,-X,t(8;14)(q24;q32),dup(13),+r/45,X,-X,t(8;14),-12, t(1;15),der(22),+r	Yunis et al 1982

Chromosome Y

HEMATOLOGICAL DISORDERS

UNCLASSIFIED LEUKEMIAS

Acute leukemia, NOS

Karyotype	Reference
45,X,-Y/46,X,-Y,+12	Najfeld et al 1981

NONLYMPHOCYTIC LEUKEMIAS

Acute nonlymphocytic leukemia (ANLL)

Karyotype	Reference
43,X,-Y,t(1;8)(p22;q24),-1,-11,-11,-13,-16,-17,-22, t(1;22)(q11;p11),t(11;13)(q23q25;q12q14),t(17;?)(p11;?), +mar	Oshimura et al 1976b
44,X,-Y,-3,+4,-12,+15,-20	Bernard et al 1982a
45,X,-Y	Muir et al 1977
45,X,-Y,t(6;11)(p2?;q13),t(8;11)(q22;q13),del(8)(q22)	Philip et al 1978a
45,X,-Y,t(8;21)(q22;q22)	Oshimura et al 1976b
45,X,-Y,t(8;21)(q22;q22)	Philip et al 1978a
45,X,-Y,t(8;21)(q22;q22),t(9;22)(q34;q11)	Francesconi and Pasquali 1978a
45,X,-Y,t(8;21)(q22;q22)	Sasaki et al 1976
45,X,-Y,t(8;21)(q22;q22)	Sasaki et al 1976
45,X,-Y,t(8;21)(q22;q22)	Seabright 1983
45,X,-Y,t(Y;9)(p11;p11)	Sasaki et al 1976
45,X,-Y	Philip et al 1978a
45,X,-Y	Mitelman 1983
46,XY,+del(7)(p11),-8,del(12)(p11),t(21;?)(q22;?)/50,XY,+Y, +del(7)(p11),del(12)(p11),+17,t(17;?)(q25;?), t(21;?)(q22;?),+22	Mitelman 1983
46,XY,+del(7)(p11),-8,del(12)(p11),t(21;?)(q22;?)/47,XY,+9	
46,XY,t(9;22)(q34;q11),21q+/46,X,-Y,t(9;22),+8	Abe and Sandberg 1979
51,X,-Y,+1,+18,+20,+22,+inv(4),+mar	Zuelzer et al 1976
56,XY,+Y,+6,+14,+21,+t(5;6)(q?;p?),+del(1)(p22q25), +del(2)(q12),+del(3)(p12),+del(4)(q13),+del(5)(p14q14), +mar	Zuelzer et al 1976

ANLL, FAB type M1

Karyotype	Reference
45,X,-Y	Mitelman 1983

ANLL, FAB type M2

Karyotype	Reference
??,X?,t(8;21),t(Y;19)	Trujillo et al 1979
45,,del(X)(q26),-Y,t(8;21)(q22;q22),del(9)(q24), del(18)(p11)	Brodeur et al 1983
45,X,-Y	Yunis et al 1981
45,X,-Y	Oshimura et al 1982b
45,X,-Y,del(8)(q22)/45,X,-Y,del(8)(q22),del(11)(q21)/46,XY, del(8)(q22)	Sakurai et al 1982b
45,X,-Y,t(8;21)	Tricot et al 1981
45,X,-Y,t(8;21)	Tricot et al 1981
45,X,-Y,t(8;21)	Tricot et al 1981
45,X,-Y,t(8;21)	Takeuchi et al 1981
45,X,-Y,t(8;21)	Hagemeijer et al 1981a
45,X,-Y,t(8;21)(q21;q22)	Prieto et al 1981
45,X,-Y,t(8;21)(q21;q21)	Hagemeijer et al 1981a
45,X,-Y,t(8;21)(q22;q22)	Kamada et al 1976
45,X,-Y,t(8;21)(q22;q22)	Bernstein et al 1982b
45,X,-Y,t(8;21)(q22;q22),del(9)(q13q22)	Hossfeld et al 1980
45,X,-Y,t(8;21)(q22;q22)/46,XY,t(8;21)	Oshimura et al 1982b
45,X,-Y,t(8;21)(q22;q22)	Oshimura et al 1982b
45,X,-Y,t(8;21)(q22;q22)	Berger et al 1982b
45,X,-Y,t(8;21)(q22;q22)/46,XY,t(8;21)	Berger et al 1982b
45,X,-Y,t(8;21)(q22;q22),t(1;5)(q43;q11)	Berger et al 1982b
45,X,-Y,t(8;21)(q22;q22)	Berger et al 1982b
45,X,-Y,t(8;21)(q22;q22),+1,-8/46,X,-Y,t(8;21),+8	Berger et al 1982b
45,X,-Y,t(8;21)(q22;q22)	Kamada et al 1981
45,X,-Y,t(8;21)(q22;q22)/46,X,-Y,t(8;21)(q22;q22), +t(11;12)(p11;q13)	Sakurai et al 1982b
45,X,-Y,t(8;21)(q22;q22)"s"	Mitelman 1983
45,X,-Y,t(8;21)(q22;q22)/46,X,-Y,t(8;21)(q22;q22),+der(21)	Mitelman 1983
45,X,-Y,t(8;21)	
45,X,-Y,t(8;21)	Golomb et al 1978c
46,X,-Y,t(8;21)(q22;q22)	Kamada et al 1976
46,X,-Y,t(8;21)(q22;q22)	Bernstein et al 1982b
46,XY,t(8;21)(q22;q22),del(9)(q21)/45,X,-Y,t(8;21), del(9)/45,XY,t(8;21),-9	Bernstein et al 1982b
46,XY,t(8;21)(q22;q22)/45,X,-Y,t(8;21)/48,XY,t(8;21), +der(21),t(10;15)(p15;q22),+r	Sakurai et al 1982b
90,XX,-Y,-Y,t(8;21)(q22;q22),t(8;21)(q22;q22)	Testa et al 1983

ANLL, FAB type M1+M2

Karyotype	Reference
42,X,-Y,4q+,-8,-16,17p+,-21,-22,+Yp+"s"	Sandberg et al 1982a
44,X,-Y,-3,6q-,+mar	Shiraishi et al 1982
45,X,-Y	Prigogina et al 1979
45,X,-Y	Abe et al 1980
45,X,-Y	Abe et al 1980
45,X,-Y,+del(5)(q11),del(11)(q21),-14,-16,-18,-20,+2mar,+r	Mitelman et al 1981
45,X,-Y,del(5)(q22)/45,X,-Y,5q+	Shiraishi et al 1982

Karyotype	Reference
45,X,-Y,t(8;21)	Li et al 1983
45,X,-Y,t(8;21)(q21;q22)	Mitelman et al 1981
45,X,-Y,t(8;21)(q22;q22)	Yamada and Furusawa 1976
45,X,-Y,t(8;21)(q22;q22)	Prigogina et al 1979
45,X,-Y,t(8;21)(q22;q22)	Prigogina et al 1979
45,X,-Y,t(8;21)(q22;q22)	Prigogina et al 1979
45,X,-Y,t(8;21)(q22;q22)	Prigogina et al 1979
45,X,-Y,t(8;21)(q22;q22),del(9)(q13)	Shiraishi et al 1982
45,X,-Y/46,X,-Y,+8	Alimena et al 1977
45,X,-Y	Abe et al 1980
45,XY,+t(1;10)(q23;q26),t(8;21)(q22;q22),-10/45,X,-Y, +t(1;10),t(8;21),-10	Fitzgerald et al 1983
45-46,X,-Y,+8	Mitelman et al 1981
46-50,XY,+Y,+del(7)(p11),-8,+9,del(12)(p11),+17,21q+,+22	Mitelman et al 1981
47,X,-Y,+mar	Benedict et al 1979
47,X,-Y,t(1;13)(q32p32q11;q11),del(3)(q21), t(8;21)(q22;q22),del(9)(q22),+8,+18	Sakurai et al 1982b
48,X,-Y,1p-,+3p-,+15,+mar"s"	Sandberg et al 1982a
57,X,-Y,-6,+8,+15,+15,+19,+19,+20,+20,+21,+21,+22,+3mar	Fitzgerald et al 1983

ANLL, FAB type M3

Karyotype	Reference
45,X,-Y,t(15;17)(q22;q12)	Kaneko et al 1982b
45,X,-Y/46,XY,del(17)(q22)	Van Den Berghe et al 1979c
46,XY,del(12)(q11)	
46,X,inv(Y)	Hurd et al 1982

ANLL, FAB type M4

Karyotype	Reference
43-46,X,-Y,del(5)(q31),-7,+18,-20	Swolin et al 1981
44,XY,del(5)(q15q23),del(7)(q22),-9,-12,-13,del(15)(q22), +mar/46,X,-Y,del(5),del(7),+8,-12,-13,del(15),+mar,dmin	Alimena 1983
45,X,-Y	Rowley and Potter 1976
45,X,-Y	Rowley and Potter 1976
45,X,-Y	Berger and Bernheim 1979
45,X,-Y,del(16)(q21)	Hagemeijer et al 1981a
45,X,-Y,del(7)(q32),t(8;21)(q22;q22)	Hustinx et al 1980
45,X,-Y,r(21)"c"	Testa et al 1979
45,X,-Y,r(21)/46,X,-Y,r(21),+8"c"	
45,X,-Y/46,X,-Y,+7	Chrz et al 1983
45,X,-Y	Mitelman et al 1981
46,XY,2q+,del(18)(p11),del(22)(q11)/47,XY,2q+,+8, del(18)(p11),del(22)(q11)"s"	Alimena et al 1981
45-47,XY,2q+,+8,del(18)(p11),-21,del(22)(q11)"s"	
46,X,-Y,2q+,+8,del(21)(q21)"s"	

ANLL, FAB type M5

Karyotype	Reference
44,X,-Y,del(2)(p11),-15,-21,+mar	Weber et al 1979
47,XY,+Y	Benedict et al 1979

ANLL, FAB type M6

Karyotype	Reference
43,X,-Y,-4,-13,-14,-15,+2mar/46,X,-Y,-3,+9,+mar	Li et al 1983
44,X,del(Y)(q?),+Xq-,3p-,3q-,4q-,-6,+7q-,11q+,-15,-18"s"	Bradley et al 1982
45,X,-Y/44,X,-Y,-21"s"	Sandberg et al 1982a
47,XY,+Y,3p-,5q-,10q-	Garson 1980
51,XY,+Y,+8,+15,+2mar	Hagemeijer et al 1981a

Chronic myeloid leukemia, t(9;22)

Karyotype	Reference
45,X,-Y,t(9;22)	Hartley and McBeath 1981
46,XY,t(9;22)/50,XY,t(9;22),+19,+22q-,+C,+C	
45,X,-Y,t(9;22)/45,X,-Y,t(9;22),t(1;11)(p?;p?)"s"	Hagemeijer et al 1980b
45,X,-Y,t(9;22)/47,XY,t(9;22),+8	Seabright 1983
45,X,-Y,t(9;22)/49,X,-Y,t(9;22),+t(9;22),+8,+12	Miyamoto 1980
45,X,-Y/45,X,-Y,t(9;22)(q34;q11)	Lawler et al 1976
45,XY,t(9;22),-8	Stoll and Oberling 1978
45,X,-Y,t(9;22),i(17q)/45,XY,t(9;22),-8	
46,X,-Y,t(9;22)(q34;q12),+13/47-48,X,-Y,t(9;22),+8,+13, +22q-/47,XY,t(9;22),+22q-	Izakovic et al 1982
46,X,-Y,t(9;22),+22q-/47,X,-Y,t(9;22),+8,i(17q),+22q-	Bernstein et al 1980b
46,XY,t(9;22)(q34;q11)	Berger and Bernheim 1979
46,XY,t(9;22)(q34;q11)/45,X,-Y,t(9;22)	
46,XY,t(9;22)(q34;q11)/45,X,-Y,t(9;22)(q34;q11)	Mitelman 1983
48,XY,t(9;22),+8,+22q-/61,XY,+Y,t(9;22),+4,+5,+6,+8,+8,+8, +8,+15,+17,+18,+19,+20,+22q-,+22q-	Wurster-Hill 1983
50,XY,t(9;22),+Y,+8,+19,+22q-	Sadamori et al 1980
53,XY,+Y,t(9;22)(q34;q11),+3,+6,+8,+9,+10,+12,-15,+22q-/56, XY,+Y,t(9;22),+3,+6,+8,+8,+9,+10,+16,+19,+22q-	Stoll and Oberling 1982
56,XY,+Y,t(9;22),+3,+6,+8,+8,+9,+10,+12,+19,+22q-	

Chronic myeloid leukemia, aberrant translocation

Karyotype	Reference
46,XY,t(10;22)(p15?;q11)/45,X,-Y,t(10;22)	Testa et al 1982
46,XY,t(8;9;22)(q22;q34;q11)/45,X,-Y, t(8;9;22)(q22;q34;q11)	Lawler et al 1976
47,XY,+Y,t(9;13;15;22)(q34;q12;q22;q11)	Potter et al 1975

Chronic myeloid leukemia, Ph1 negative

Karyotype	Reference
45,X,-Y	Engel et al 1977
45,X,-Y	Whittaker et al 1975
45,X,-Y	Kohno et al 1979a
45,X,-Y	Hossfeld 1975b
45,X,-Y	Sonta and Sandberg 1978b
46,X,t(Y;3;9)(q12;q25;q34)	Verhest et al 1980

Eosinophilic leukemia

Karyotype	Reference
45,X,-Y,del(15)(q22)"s"	Goffman et al 1983

MYELOPROLIFERATIVE DISORDERS

Polycythemia vera

Karyotype	Reference
45,X,-Y	Berger and Bernheim 1979
54-57,X,-Y,+6,+8,+11,+17,+20,+21,t(1;22)(p12;q13)	
45,X,-Y	Testa et al 1981b
45,X,-Y	Testa et al 1981b
45,X,-Y,20q-	Zech et al 1976b
45,X,-Y	Shiraishi et al 1975
45,XY,-16,-17,+21/48,XY,+Y,+11,+16,-17,+22	Wurster-Hill et al 1976
46,XY,t(12;17)(q13;p11)/47,X,-Y,+t(Y;1)(q12;q21),+9	Testa et al 1981b
47,XY,+Y	Shabtai 1983

Myelosclerosis / Myelofibrosis

Karyotype	Reference
45,X,-Y	Carbonell et al 1983

Chronic myeloproliferative disease, NOS

Karyotype	Reference
45,X,-Y	Dileo et al 1977
45,X,-Y	Dileo et al 1977
45,X,-Y	Carbonell et al 1983
45,X,-Y	Carbonell et al 1983
45,X,-Y	Mitelman 1983
46-49,X,-Y,+1-5mar	Carbonell et al 1983

Myeloproliferative disorder, special type

Karyotype	Reference
45,X,-Y	Mitelman 1983

DYSMYELOPOIETIC SYNDROMES

Preleukemia, NOS

Karyotype	Reference
43,X,-Y,-3,5q-,-7,-10,-16,-22,+3mar"s"	Pedersen-Bjergaard et al 1981
44,XY,+Y,-3,-5,+del(6)(q21),-8,t(17;?)(p13;?),+18,-20,+22	Mitelman 1983
45,X,-Y	Geraedts et al 1980a
45,X,-Y	Geraedts et al 1980a
45,X,-Y	Geraedts et al 1980a
45,X,-Y	Geraedts et al 1980a
45,X,-Y	Mitelman 1983
46,X,-Y,+8	Geraedts et al 1980a
46,X,-Y,+t(1;?)(p?;?)	Geraedts et al 1980a

Chronic myelomonocytic leukemia

45,X,-Y	Streuli et al 1980
45,XY,-7,dmin/45,XY,-7	
46,X,-Y	Geary et al 1975

Refractory anemia, NOS

47,X,del(Y)(q11),+1	Warburton and Bluming 1973

Sideroblastic anemia, NOS

44,X,-Y,-21/45,XY,-21/46,XY,1q+	Yamada and Furusawa 1976
45,X,-Y	Nordenson 1983

Refractory anemia without excess of blasts

47,XY,+8/46,X,-Y,+8	Mitelman 1983

Pancytopenia

45,X,-Y	Seabright 1983
45,X,-Y	Mitelman 1983

Paroxysmal nocturnal hemoglobinuria

45,X,-Y	Whang-Peng et al 1976a

Hematopoietic proliferation disorder, NOS

46,X,t(Y;17)(q11;q25)"c"	Mitelman 1983
46,X,t(Y;17)(q11;q25)/45,X,-Y"c"	

SPECIAL LEUKEMIAS

Hairy cell leukemia

46,X,-Y,+12	Golomb et al 1978b
46,X,-Y,t(12;?)(q14;?)	Golomb et al 1978b

LYMPHOCYTIC LEUKEMIAS

Acute lymphocytic leukemia (ALL)

45,X,-Y	Berger and Bernheim 1979
45,X,-Y	Humbert et al 1978
45,X,-Y/90,XX,-Y,-Y"s"	Okada et al 1982
46,X,-Y,+1,-7,+20,t(1;22)(q24;q12)	Vercherat et al 1980
46,XY,t(9;22)	Schmidt et al 1975
45,X,-Y,t(9;22)	
46,XY,t(9;22)	

47,X,del(Y)(q11),del(3)(q11),t(Y;12)/50,X,del(Y)(q11), del(3)(q11),del(1)(p31),t(Y;12),+4mar	Zuelzer et al 1976

ALL, FAB type L1

46,XY,del(2)(q32),-11,-12,+2mar/46,X,-Y,del(2),-4,-5,-11, -12,-21,+6	Kaneko et al 1981b

ALL, FAB type L3

??,XY,t(8;14)(q?;q?),t(Y;1)	Prieto et al 1981

Chronic lymphocytic leukemia

44-48,XY,t(Y;9)(q12;q13),del(1)(p22p32),t(4;6)(p16;q15), del(6)(q15),t(12;14)(q15;q32),+mar	Robert et al 1982
46,XY,9p+/43,X,-Y,-14,-22	Nordenson 1983

Prolymphocytic leukemia

44,X,-Y,-11,-12,-13,-13,-14,-15,-15,-18,-18,+8mar	Diamond et al 1980
46,X,-Y,+mar	Diamond et al 1980

Lymphocytic leukemia, special type

45,X,-Y,6q-/46,XY,6q-	Miyamoto et al 1982b
45,Y,-X,t(Y;14)(q12;q32),-13,-17,del(2)(q33), t(3;13)(q21;q34),del(6)(q16orq22),9p+,9q+,10q+,16q+,18q+	Miyoshi et al 1981a
46,X,-Y,-1,1p+,del(9)(q32),12p+,14q+,+2mar	Ueshima et al 1981
47,X,-Y,-4,+7,del(1)(q32),i(18q),21q+,i(22q),+2mar	Ueshima et al 1981
47,X,-Y,dup(Y),1q-,4q+,6q-,7q+,+18q+	Miyamoto et al 1982b

MONOCLONAL GAMMOPATHIES

Multiple myeloma

45,X,-Y	Knuutila et al 1981
45,X,-Y,-1,+t(1;3)(q3?;q2?),-3,+11,14q+/45,X,-Y,-1,-1, +t(1;1)(q2?;q3?),14q+	Philip et al 1980
47,X,-Y,+t(1;15)(q12;p11),+3,+11,-15,+19,-22/47,X,-Y,+3, +11,+19,-22	Philip et al 1980

Macroglobulinemia

45,X,-Y	Wurster-Hill 1983

Plasma cell leukemia / Plasmocytoma

44,X,-Y,+t(1;9)(q12q44;q34),t(6;8)(q13;p11),-9,-13	Ueshima et al 1983
44,X,-Y,del(1)(p11),+del(1)(q11),t(11;14)(q11;q32),14q+, -21,-22	Ueshima et al 1983

SOLID TUMORS

UNCLASSIFIED NEOPLASMS

EPITHELIAL NEOPLASMS

Carcinoma, NOS

Karyotype	Reference
49,X,-Y,-3,-5,-14,+15,-16,-17,+19,+21,+22,1q+,5q-,6q-,8p-, 9p-,11p-,18p+,20q-,22q-,t(Y;14),+3mar	Reichmann et al 1981
52,XY,+X,+Y,-2,+7,+8,+9,+10,+13,-17,+i(1q)	Reichmann et al 1981
58-61,XY,cx,t(Y;15)	Reichmann et al 1981
64-68,XX,cx,t(Y;13),i(9p),i(9q),dmin	Reichmann et al 1981
66,XX,t(Y;13),1q+,i(4q),i(9p),i(9q),cx	Martin et al 1979

Adenoma

Karyotype	Reference
46,XY,t(6;21)(q13;q22)/46,XY,t(1;14)(q42;p12)/45,X,-Y 45,X,-Y 46,Y,t(X;15)(q24;q15)/45,X,-Y	Mark et al 1981b

Adenomatous polyp

Karyotype	Reference
48-52,X,-Y,+7,+8,+13,+14,+16,+18,-20,+12q-,1q+	Reichmann et al 1982b

MESENCHYMAL NEOPLASMS

MELANOMAS

NEUROGENIC NEOPLASMS

Neuroblastoma

Karyotype	Reference
47,X,-Y,del(1)(p?),+6,16q+	Brodeur et al 1977

Retinoblastoma

Karyotype	Reference
46,X,-Y,1p+,+1p-,5q+,7p+,10q-,16q-,17q+,21p+	Kusnetsova et al 1982

Meningioma, benign

Karyotype	Reference
38,X,-Y,-22,-21,-17,-14,-13,-10,-9,-4,-1p-,+mar	Zankl et al 1975a
41,X,-Y,-22,-18,-15,-10,1p-	Zankl et al 1975a
43,X,-Y,-22,-15	Zankl et al 1975a
44,X,-Y,-22/42,X,-Y,-17,-21,-22	Yamada et al 1980
45,X,-Y/44,X,-Y,-22	Yamada et al 1980
45,X,-Y/44,X,-Y,-22	Yamada et al 1980
45,XY,-22/44,X,-Y,-22/43,X,-Y,-22,-17	Yamada et al 1980
52,X,-Y,-22,+20,+19,+17,+15,+11,+9,+7,+5	Zankl et al 1975a

EMBRYONAL AND MISCELLANEOUS NEOPLASMS

Teratoma, malignant

48,X,-Y,+1,del(5)(q12q32),+8,+10 Tricot et al 1983

LYMPHOMAS

1. UNCLASSIFIED AND MISCELLANEOUS LYMPHOMAS

Non-Hodgkin's lymphoma, lymphocytic, NOS

45,X,-Y,t(11;14)(q13;q32),t(11;19)(q13;q13) Fleischman and Prigogina 1977

Non-Hodgkin's lymphoma, histiocytic, NOS

45,X,del(Y)(q12),t(4;10)(q12;q24),t(6;7)(q21;q32), del(8)(p21) Mark et al 1978

Malignant histiocytosis

45,X,-Y Mitelman 1983

2. HODGKIN'S LYMPHOMAS

Hodgkin's disease, lymphocytic depletion

48,X,-Y,t(1;?)(q44;?),i(1q),del(1)(q21),del(3)(q21),-10, t(3;11)(q23;q21),+14,t(14;?)(q32;?),+15,-17,+19,+20,-22 Hossfeld and Schmidt 1978

3A. NON-HODGKIN'S LYMPHOMAS - RAPPAPORT CLASSIFICATION

Histiocytic, diffuse

45,X,-Y,t(14;15)(q32;q21orq22),-15,+18,-22,1q+,3p+,5p-,7p-, 9p+,12p+,16q+,17p-,+2mar Fukuhara and Rowley 1978

47,X,+X,-Y,+12,-14,del(11)(q21q23),del(16)(q22), t(14;?)(q32;?) Kaneko et al 1982a

3B. NON-HODGKIN'S LYMPHOMAS - KIEL CLASSIFICATION

Lymphocytic, T-zone

44-46,X,+t(Y;14)(q24;q12),-3,del(3)(q21),+9,14q+,+mar — Gödde-Salz et al 1981a
46,XY,+t(Y;14)(q12;q24),3q-,-5,-13,+mar — Gödde-Salz et al 1981b

Immunocytoma

45,X,-Y,t(Y;18)(q11;p11),t(3;3)(q21;q29),5q+ — Kristoffersson 1983

Lymphoblastic, Burkitt's type

45,X,-Y,t(1;14)(q22;q32),t(8;22)(q24;q11),3q+,6q-,6q+,7q+, 9p-,14q-,18q- — Berger et al 1981c

Lymphoblastic, convoluted type

46,X,-Y,t(1;?)(q21orq23;?),+mar/46,X,-Y,-1,-2,+der(1), +t(2;?)(q33orq35;?),+mar/48,X,-Y,-5,-14,+18,+18,+20, del(9)(p13),t(1;14),+dup(5)(q13q35) — Kaneko et al 1982d

3C. NON-HODGKIN'S LYMPHOMAS - LUKES AND COLLINS CLASSIFICATION

Follicular center cell, diffuse, large cleaved

47,XY,+Y,t(2;3)(q11;q29) — Fukuhara et al 1978b

Follicular center cell, diffuse, small non-cleaved

44,X,-Y,t(1;14)(p21;q13),5p+,del(6)(q13),del(8)(p21),9p+, 9p-,t(11;21)(q23;q22),13q+,-10,+12,-17p+,der(19),22q+ — Fukuhara et al 1979a

3D. NON-HODGKIN'S LYMPHOMAS - WORKING FORMULATION

Follicular, small cleaved cell

48,X,r(Y),+2,+3,t(14;18)(q32;q21),t(1;22) — Yunis et al 1982

References

Abe, S. and Sandberg, A.A. (1979): Chromosomes and causation of human cancer and leukemia. XXXII. Unusual features of Ph1-positive acute myeloblastic leukemia (AML), including a review of the literature. Cancer 43:2352-2364

Abe, S., Kohno, S., Kubonishi, I., Minowada, J. and Sandberg, A.A. (1979): Chromosomes and causation of human cancer and leukemia. XXXIII. 5q- in a case of acute lymphoblastic leukemia (ALL). Am. J. Hematol. 6:259-266

Abe, S., Golomb, H.M., Rowley, J.D., Mitelman, F. and Sandberg, A.A. (1980): Chromosomes and causation of human cancer and leukemia. XXXV. The missing Y in acute non-lymphocytic leukemia (ANLL). Cancer 45:84-90

Abe, R., Tebbi, C.K., Yasuda, H. and Sandberg, A.A. (1982a): North American Burkitt-type ALL with a variant translocation t(8;22). Cancer Genet. Cytogenet. 7:185-195

Abe, R., Hayashi, Y., Sampi, K. and Sakurai, M. (1982b): Burkitt's lymphoma with 2/8 translocation: a case report with special reference to the clinical features. Cancer Genet. Cytogenet. 6:135-142

Abe, R., Ryan, D., Cecalupo, A., Cohen, H. and Sandberg, A.A. (1983): Cytogenetic findings in congenital leukemia: Case report and review of the literature. Cancer Genet. Cytogenet. 9:139-144

Adler, K.R., Lempert, N. and Scharfman, W.B. (1978): Chronic granulocytic leukemia following successful renal transplantation. Cancer 41:2206-2208

Albain, K.S., Le Beau, M.M., Vardiman, J.W., Golomb, H.M. and Rowley, J.D. (1983): Development of dysmyelopoietic syndrome in a hairy cell leukemia patient treated with chlorambucil: Cytogenetic and morphologic evaluation. Cancer Genet. Cytogenet. 8:107-115

Alimena, G. (1983): Personal communication.

Alimena, G., Annino, L., Balestrazzi, P., Montuoro, A. and Dallapiccola, B. (1977): Cytogenetic studies in acute leukaemias. Prognostic implications of chromosome imbalances. Acta Haematol. 58:234-239

Alimena, G., Brandt, L., Dallapiccola, B., Mitelman, F. and Nilsson, P.G. (1979): Secondary chromosome changes in chronic myeloid leukemia: Relation to treatment. Cancer Genet. Cytogenet. 1:79-85

Alimena, G., Dallapiccola, B., Mitelman, F. and Montuoro, A. (1980): Aberrations of chromosome No. 1 in blastic phase of chronic myeloid leukemia. Hereditas 92:59-63

Alimena, G., Dallapiccola, B., De Cuia, M.R., Mandelli, F. and Mitelman, F. (1981): Acute lymphocytic and myelomonocytic leukemia associated with low platelet counts and a 21q- marker chromosome. Hum. Genet. 57:329-331

Alimena, G., Gastaldi, R., de Cuia, M.R., Gallo, E. and Dallapiccola, B. (1982a): 5q-chromosome in a case of acute non-lymphocytic leukemia (ANLL): the marker of a polyphasic disease?. Haematologica 67:261-266

Alimena, G., Dallapiccola, B., Gastaldi, R., Mandelli, F., Brandt, L., Mitelman, F. and Nilsson, P.G. (1982b): Chromosomal, morphological and clinical correlations in blastic crisis of chronic myeloid leukaemia. A study of 69 cases. Scand. J. Haematol. 28:103-117

Altman, A.J., Palmer, C.G. and Baehner, R.L. (1974): Juvenile "chronic granulocytic" leukemia: A panmyelopathy with prominent monocytic involvement and circulating monocyte colony-forming cells. Blood 43:341-350

Anderson, R.L. and Bagby, G.C. (1982): The prognostic value of chromosome studies in patients with the preleukemic syndrome (hemopoietic dysplasia). Leukemia Res. 6:175-181

Anderson, C., Mohandas, T. and Okun, D. (1978): Chronic myelogenous leukemia with a complex translocation. Am. J. Genet. 30:73A

Andrieu, J.M., Casassus, P., Degos, L., Preud'homme, J.L., Berger, R. and Flandrin, G. (1980): Burkitt's lymphoma occurring 6 years after Hodgkin's disease. Acta Haematol. 63:330-332

Anglani, F., Artifoni, L., Zanesco, L., Baccichetti, C. and Tenconi, R. (1981): Trisomy 1q and hematologic disorders. Cancer Genet. Cytogenet. 4:275-279

Armenta, D., Cadotte, M., Beaulieu, R., Neemeh, J., Long, L., Pretty, H. and Gosselin, G. (1976): Évidence cytogénétique de l'origine splénique de la leucémie myéloide chronique. Union Méd. Canada 105:922-927

Arthur, D.C., Bloomfield, C.D., Lindquist, L.L. and Nesbit, M.E. (1982): Translocation 4;11 in acute lymphoblastic leukemia: clinical characteristics and prognostic significance. Blood 59:96-99

Atkin, N.B. and Baker, M.C. (1977a): Chromosome 1 in cervical carcinoma. Lancet (2):984

Atkin, N.B. and Baker, M.C. (1977b): Abnormal chromosomes and number 1 hetero-chromatin variants revealed in C-banded preparations from 13 bladder carcinomas. Cytobios 18:101-109

Atkin, N.B. and Baker, M.C. (1978): Duplication of the long arm of chromosome 1 in a malignant vaginal tumour. Brit. J. Cancer 38:468-471

Atkin, N.B. and Baker, M.C. (1979): Chromosome 1 in 26 carcinomas of the cervix uteri. Cancer 44:604-613

Atkin, N.B. and Baker, M.C. (1980): Cytogenetic observations on a carcinoma of the cervix uteri with double minute chromatin bodies. Europ. J. Cancer 16:793-797

Atkin, N.B. and Baker, M.C. (1981): A metastatic malignant melanoma with 24 chromosomes. Hum. Genet. 58:217-219

Atkin, N.B. and Baker, M.C. (1982a): Nonrandom chromosome changes in carcinoma of the cervix uteri. I. Nine near-diploid tumors. Cancer Genet. Cytogenet. 7:209-222

Atkin, N.B. and Baker, M.C. (1982b): Specific chromosome change, i(12p), in testicular tumours?. Lancet (2):1349

Atkin, N.B. and Pickthall, V.J. (1977): Chromosomes 1 in 14 ovarian cancers. Heterochromatin variants and structural changes. Hum. Genet. 38:25-33

Atkin, N.B., Amin, S. and Brito-Babapulle, V. (1981): Three or four copies of a dicentric 17q isochromosome in an acute myeloproliferative disorder. Cancer Genet. Cytogenet. 3:75-80

Auerbach, A.D., Weiner, M.A., Warburton, D., Yeboa, K., Lu, L. and Broxmeyer, H.E. (1982): Acute myeloid leukemia as the first hematologic manifestation of Fanconi anemia. Am. J. Hematol. 12:289-300

Autio, K., Turunen, O., Penttilä, O., Erämaa, E., Chapelle, A. de la and Schröder, J. (1979): Trisomy 12 in hematologic disorders. Blood 56:741

Ayraud, N. (1975): Identification par denaturation thermique menagee des anomalies chromosomiques observees dans six tumeurs metastatiques humaines. Biomedicine 23:423-430

Ayraud, N., Dujardin, P. and Audoly, P. (1975): Leucemie aigue lymphoblastique avec chromosome Philadelphie. Role probable d'une translocation 14-22. Nouv. Presse Med. 4:3013

Ayraud, N., Lambert, J.-C., Hufferman-Tribollet, K. and Basteris, B. (1977): Etude de cytogenetique comparative de sept carcinomes d'origin mammaire. Ann. Genet. 20:171-177

Bagby Jr, G.C., Kaiser-McCaw, B., Hecht, F., Koler, R.D. and Linman, J.W. (1978): Clonal evolution in atypical chronic granulocytic leukemia: A non-Philadelphia translocation. Blood 51:997-1004

Balaban, G., Gilbert, F., Nichols, W., Meadows, A.T. and Shields, J. (1982): Abnormalities of chromosome #13 in retinoblastomas from individuals with normal constitutional karyotypes. Cancer Genet. Cytogenet. 6:213-221

Balaban-Malenbaum, G., Gilbert, F., Nichols, W.W., Hill, R., Shields, J. and Meadows, A.T. (1981): A deleted chromosome No. 13 in human retinoblastoma cells: relevance to tumorigenesis. Cancer Genet. Cytogenet. 3:243-250

Bartnitzke, S. and Bullerdiek, J. (1983): Personal communication.

Bartoli, E., Massarelli, G., Soggia, G., Tanda, F. and Vianello, M.G. (1979): Acute agnogenic myeloid metaplasia with chromosomal abnormalities. Acta Haematol. 62:206-213

Becher, R., Gibas, Z. and Sandberg, A.A. (1983): Chromosome 6 in malignant melanoma. Cancer Genet. Cytogenet. 9:173-175

Beltran, G. and Varela, M. (1980): Clonal changes in chronic granulocytic leukemia in blastic transformation and during remission. Cancer 46:1590-1593

Benedict, W.F., Lange, M., Greene, J., Derencsenyi, A. and Alfi, O.S. (1979): Correlation between prognosis and bone marrow chromosomal patterns in children with acute nonlymphocytic leukemia: Similarities and differences compared to adults. Blood 54:818-823

Benitez, I. and Outeirino, I. (1979): Estudio citogenético en medula osea de 21 casos de leucemia mieloide cronica y dos casos de leucemia aguda con Ph1 positivo. Sangre 24:1048-1056

Bennett, J.M., Catovsky, D., Daniel, M.-T., Flandrin, G., Galton, D.A.G., Gralnick, H.R. and Sultan, C. (1976): Proposals for the classification of the acute leukaemias. Brit. J. Haematol. 33:451-458

Berger, R. (1973): Chromosome Philadelphie. Nouv. Presse Med. 2:3121

Berger, R. (1975): Translocation t(11;20) et polyglobulie primitive. Nouv. Presse Med. 4:1972

Berger, R. and Bernheim, A. (1978): Non-randomness in complex translocations of chronic myeloid leukaemia. Scand. J. Haematol. 21:418-420

Berger, R. and Bernheim, A. (1979): Y chromosome loss in leukemias. Cancer Genet. Cytogenet. 1:1-8

Berger, R. and Bernheim, A. (1982): Cytogenetic studies on Burkitt's lymphoma-leukemia. Cancer Genet. Cytogenet. 7:231-244

Berger, R. and Lacour, J. (1974): Anomalies chromosomiques dans un cancer de l'ovaire. Pathol. Biol. 22:603-606

Berger, R., Weisgerber, C. and Bernard, J. (1973): Evolution clonale au cours d'une leucemie aigue chez un enfant mongolien. Nouv. Rev. Fr. Hematol. 13:229-236

Berger, R., Briéere, J. and Clauvel, J.P. (1975): Homozygotie et syndrome myeloproliferatif. Nouv. Rev. Fr. Hematol. 15:667-676

Berger, R., Gyger, M. and Bussel, A. (1976): Anomalie chromosomique nouvelle dans une leucemie myeloide chronique. Nouv. Rev. Fr. Hematol. 16:309-320

Berger, R., Bussel, A. and Schenmetzler, C. (1977): Somatic segregation and Fanconi anemia. Clin. Genet. 11:409-415

Berger, R., Bernheim, A., Daniel, M-T., Valiensi, F. and Flandrin, G. (1979a): Une nouvelle variété de leucémie aigué non promyélocytaire avec translocation t(15;17). C.R. Acad. Sc. Paris 288:177-179

Berger, R., Bernheim, A., Flandrin, G., Daniel, M.T., Schaison, G., Brouet, J.-C. and Bernard, J. (1979b): Translocation t(8;14) dans la leucemie lymphoblastique de type Burkitt. Nouv. Presse Med. 8:181-183

Berger, R., Bernheim, A., Brouet, J.C., Daniel, M.T. and Flandrin, G. (1979c): t(8;14) translocation in a Burkitt's type of lymphopblastic leukaemia (L3). Brit. J. Haematol. 43:87-90

Berger, R., Bernheim, A., Daniel, M.T., Valensi, F., Flandrin, G. and Bernard, J. (1979d): Translocation t(15;17), leucemie aigue promyelocytaire et non promyelocytaire. Nouv. Rev. Fr. Hematol. 21:117-131

Berger, R., Bernheim, A., Weh, H.-J., Flandrin, G., Daniel, M.T., Brouet, J.-C. and Colbert, N. (1979e): A new translocation in Burkitt's tumor cells. Hum. Genet. 53:111-112

Berger, R., Bernheim, A., Coniat, M. le, Vecchione, D. and Schaison, G. (1980a): Chromosomal studies of leukemic and preleukemic Fanconi's anemia patients. Hum. Genet. 56:59-62

Berger, R., Bernheim, A., Weh, H-J., Daniel, M-T. and Flandrin, G. (1980b): Cytogenetic studies on acute monocytic leukemia. Leukemia Res. 4:119-127

Berger, R., Bernheim, A., Daniel, M.T., Valensi, F. and Flandrin, G. (1981a): Leucémies "induites". Aspects cytogénétique et cytologique. Comparaison avec des leucémies primitives. Nouv. Rev. Fr. Hematol. 23:275-284

Berger, R., Bernheim, A., Daniel, M.T., Valensi, F. and Flandrin, G. (1981b): Karyotypes and cell phenotypes in acute leukemia following other diseases. Blood cells 7:293-299

Berger, R., Bernheim, A., Bertrand, S., Fraisse, J., Frocrain, C., Tanzer, J. and Lenoir, G. (1981c): Variant chromosomal t(8;22) translocation in four French cases with Burkitt lymphoma-leukemia. Nouv. Rev. Fr. Hematol. 23:39-41

Berger, R., Bernheim, A. and Bussel, A. (1981d): Chromosome Philadelphie par t(9;22) et t(1;9;22) en mosaique. Nouv. Rev. Fr. Hematol. 23:209-212

Berger, R., Bernheim, A., Daniel, M-T., Valensi, F. and Flandrin, G. (1981e): t(15;17) translocation in acute promyelocytic leukaemia (M3) and cytological "M3-variant". Nouv. Rev. Fr. Hematol. 23:27-38

Berger, R., Bernheim, A. and Schaison, G. (1981f): Discrepancy between G and R bands. Hum. Genet. 59:84-86

Berger, R., Bernheim, A., Sigaux, F., Daniel, M-T., Valensi, F. and Flandrin, G. (1982a): Acute monocytic leukemia chromosome studies. Leukemia Res. 6:17-26

Berger, R., Bernheim, A., Daniel, M-T., Valensi, F., Sigaux, F. and Flandrin, G. (1982b): Cytologic characterization and significance of normal karyotypes in t(8;21) acute myeloblastic leukemia. Blood 59:171-178

Berger, R., Bernheim, A., Valensi, F. and Flandrin, G. (1983a): 22q- and 8q- in a non-Burkitt lymphoma. Cancer Genet. Cytogenet. 8:91-92

Berger, R., Bernheim, A., Daniel, M-T. and Flandrin, G. (1983b): t(15;17) in a promyelocytic form of chronic myeloid leukemia blastic crisis. Cancer Genet. Cytogenet. 8:149-152

Bernard, P., Reiffers, J., Lacombe, F., Dachary, D., David, B., Boisseau, M.R. and Broustet, A. (1982a): Prognostic value of age and bone marrow karyotype in 78 adults with acute myelogenous leukemia. Cancer Genet. Cytogenet. 7:153-163

Bernard, P., Fialon, P., Longy, M., Boisseau, M.R. and Broustet, A. (1982b): Acute promyelocytic leukemia with (15;17) translocation and chromosome No. 11 deletion (q23). Leukemia Res. 6:869-871

Bernstein, R., Mendelow, B., Pinto, M.R., Morcom, G. and Bezwoda, W. (1980a): Complex translocations involving chromosomes 15 and 17 in acute promyelocytic leukaemia. Brit. J. Haematol. 46:311-314

Bernstein, R., Morcom, G., Pinto, M.R., Mendelow, B., Dukes, I., Penfold, G. and Bezwoda, W. (1980b): Cytogenetic findings in chronic myeloid leukemia (CML); evaluation of karyotype, blast morphology, and survival in the acute phase. Cancer Genet. Cytogenet. 2:23-37

Bernstein, R., Pinto, M. and Jenkins, T. (1981): Ataxia telangiectasia with evolution of monosomy 14 and emergence of Hodgkin's disease. Cancer Genet. Cytogenet. 4:31-37

Bernstein, R., Pinto, M.R., Behr, A. and Mendelow, B. (1982a): Chromosome 3 abnormalities in acute nonlymphocytic leukemia (ANLL) with abnormal thrombopoiesis: report of three patients with a "new" inversion anomaly and a further case of homologous translocation. Blood 60:613-617

Bernstein, R., Pinto, M.R., Morcom, G., Macdougall, L.G., Bezwoda, W., Dukes, I., Penfold, G. and Mendelow, B. (1982b): Karyotype analysis in acute nonlymphocytic leukemia (ANLL): comparison with ethnic group, age, morphology, and survival. Cancer Genet. Cytogenet. 6:187-199

Bertrand, S., Branger, M.R. and Cheix, F. (1979): Use of G banding technique in the cytogenetic study of metastatic breast cancer effusion. A case report. Europ. J. Cancer 15:737-743

Bertrand, S., Berger, R., Philip, T., Bernheim, A., Bryon, P-A., Bertoglio, J., Doré, J-F., Brunat-Mentigny, M. and Lenoir, G.M. (1981): Variant translocation in a non endemic case of Burkitt's lymphoma: t(8;22) in an Epstein-Barr virus negative tumour and in a derived cell line. Europ. J. Cancer 17:577-584

Biggar, R.J., Lee, E.C., Nkrumah, F.K. and Whang-Peng, J. (1981): Direct cytogenetic studies by needle aspiration of Burkitt's lymphoma in Ghana, West Africa. J. Natl. Cancer Inst. 67:769-776

Bitran, J., Golomb, H.M. and Rowley, J.D. (1977): Idiopathic acquired refractory sideroblastic anemia: Banded chromosome analysis in six patients. Acta Haematol. 57:15-23

Bloomfield, C.D., Peterson, L.C., Yunis, J.J. and Brunning, R.D. (1977): The Philadelphia chromosome (Ph1) in adults presenting with acute leukaemia: a comparison of Ph1+ and Ph1- patients. Brit. J. Haematol. 36:347-358

Bloomfield, C.D., Lindquist, L.L., Arthur, D., McKenna, R.W., LeBien, T.W., Peterson, B.A. and Nesbit, M.E. (1981): Chromosomal abnormalities in acute lymphoblastic leukemia. Cancer Res. 41:4838-4843

Boetius, G., Hustinx, T.W.J., Smits, A.P.T., Scheres, J.M.J.C., Rutten, F.J. and Haanen, C. (1977): Monosomy 7 in two patients with a myeloproliferative disorder. Brit. J. Haematol. 37:101-109

Borgström, G.H. (1981): New types of unusual and complex Philadelphia chromosome (Ph1) translocations in chronic myeloid leukemia. Cancer Genet. Cytogenet. 3:19-31

Borgström, G.H., Teerenhovi, L., Vuopio, P., Chapelle, A. de la , Van Den Berghe, H., Brandt, L., Golomb, H.M., Louwagie, A., Mitelman, F., Rowley, J.D. and Sandberg, A.A. (1980): Clinical implications of monosomy 7 in acute nonlymphocytic leukemia. Cancer Genet. Cytogenet. 2:115-126

Borgström, G.H., Teerenhovi, L., Vuopio, P., Andersson, L.C., Knuutila, S., Elonen, E. and Chapelle, A. de la (1981): Chromosome studies in acute lymphoblastic leukaemia (ALL). Scand. J. Haematol. 26:241-251

Borgström, G.H., Vuopio, P. and Chapelle, A. de la (1982): Abnormalities of chromosome No. 17 in myeloproliferative disorders. Cancer Genet. Cytogenet. 5:123-135

Bornkamm, G.W., Kaduk, B., Kachel, G., Schneider, U., Fresen, K.O., Schwanitz, G. and Hermanek, P. (1980): Epstein-Barr virus-positive Burkitt's lymphoma in a German woman during pregnancy. Blut 40:167-177

Bottura, C. and Couthinho, V. (1974): G/G translocation and chronic myelocytic leukaemia. Blut 29:216-218

Bradley, E.C., Schechter, G.P., Matthew, M.J., Whang-Peng, J., Cohen, M.H., Bunn, P.A., Ihde, D.C. and Minna, J.D. (1982): Erythroleukemia and other hematologic complications of intensive therapy in long-term survivors of small cell lung cancer. Cancer 49:221-223

Brandt, L., Levan, G., Mitelman, F., Olsson, I. and Sjögren, U. (1974a): Trisomy G-21 in adult myelomonocytic leukaemia. An abnormality common to granulocytic and monocytic cells. Scand. J. Haematol. 12:117-122

Brandt, L., Levan, G., Mitelman, F. and Sjögren, U. (1974b): Defective differentiation of megakaryocytes in acute myeloid leukemia. Acta Med. Scand. 196:227-230

Brandt, L., Forssman, O., Mitelman, F., Odeberg, H., Olofsson, T., Olsson, I. and Svensson, B. (1975a): Cell production and cell function in human cyclic neutropenia. Scand. J. Haematol. 15:228-240

Brandt, L., Mitelman, F. and Panani, A. (1975b): Cytogenetic differences between bone marrow and spleen in a case of agnogenic myeloid metaplasia developing blast crisis. Scand. J. Haematol. 15:187-191

Brandt, L., Mitelman, F. and Sjögren, U. (1975c): Megaloblastic changes and chromosome abnormalities of erythropoietic cells in acute myeloid leukaemia. Acta Haematol. 54:280-283

Brandt, L., Mitelman, F., Panani, A. and Lenner, H.C. (1976): Extremely long duration of chronic myeloid leukaemia with Ph1 negative and Ph1 positive bone marrow cells. Scand. J. Haematol. 16:321-325

Brandt, L., Mitelman, F., Beckman, G., Laurell, H. and Nordenson, I. (1977): Different composition of the eosinophilic bone marrow pool in reactive eosinophilia and eosinophilic leukaemia. Acta Med. Scand. 201:177-180

Brodeur, G.M., Sekhon, G.S. and Goldstein, M.N. (1977): Chromosomal aberrations in human neuroblastomas. Cancer 40:2256-2263

Brodeur, G.M., Dow, L.W. and Williams, D.L. (1979): Cytogenetic features of juvenile chronic myelogenous leukemia. Blood 53:812-819

Brodeur, G.M., Williams, D.L., Look, A.T., Bowman, W.P. and Kalwinsky, D.K. (1981a): Near-haploid acute lymphoblastic leukemia: a unique subgroup with a poor prognosis?. Blood 58:14-19

Brodeur, G.M., Green, A.A., Hayes, F.A., Williams, K.J., Williams, D.L. and Tsiatis, A.A. (1981b): Cytogenetic features of human neuroblastomas and cell lines. Cancer Res. 41:4678-4686

Brodeur, G.M., Williams, D.L., Kalwinsky, D.K., Williams, K.J. and Dahl, G.V. (1983): Cytogenetic features of acute nonlymphoblastic leukemia in 73 children and adolescents. Cancer Genet. Cytogenet. 8:93-105

Brynes, R.K., Golomb, H.M., Desser, R.K., Recant, W., Reese, C. and Rowley, J. (1976): Acute monocytic leukemia. Cytologic, histologic, cytochemical, ultrastructural, and cytogenetic observations. Am. J. Clin. Pathol. 65:471-482

Brynes, R.K., Golomb, H.M., Gelder, F., Desser, R.K. and Rowley, J. (1978): The leukemic phase of histiocytic lymphoma. Histologic, cytologic, cytochemical, ultrastructural, immunologic and cytogenetic observations in a case. Am. J. Clin. Pathol. 69:550-558

Cabrol, C. (1979): Chromosomal anomaly in eosinophilic leukemia. New Engl. J. Med. 301:439

Cabrol, C. and Abele, R. (1978): Chromosome 5q- dans les cellules medullaires d'un sujet atteint d'anemie ayant evolue en leucemie aigue indifferenciee. J. Genet. Hum. 26:195-202

Carbone, P., Granata, G., Margiotta, G., Barbata, G. and Majolino, I. (1982a): Ph1 duplication, t(13q-;14q+) and trisomy 19 in a case with chronic myeloid leukemia in lymphoid blast crisis at presentation. Haematologica 67:595-604

Carbone, P., Barbata, G. and Caronia, F. (1982b): Marker chromosome 14q+ in a case of Ph1-positive chronic myeloid leukemia in lymphoid blastic crisis at presentation. Acta Haematol. 67:230-231

Carbonell, F. (1983): Personal communication.

Carbonell, F., Kratt, E. and Neuhaus, K. (1980): Complex translocations between chromosomes #6, #9, #22, and #11 in a patient with chronic myelocytic leukemia: 46,XX,t(6;9;22;11)(p21;q34;q11;q13). Cancer Genet. Cytogenet. 2:139-143

Carbonell, F., Grilli, G. and Fliedner, T.M. (1981): Cytogenetic evidence for a clonal selection of leukemic cells in culture. Leukemia Res. 5:395-398

Carbonell, F., Benitez, J., Prieto, F., Badia, L. and Sanchez-Fayos, J. (1982a): Chromosome banding patterns in patients with chronic myelocytic leukemia. Cancer Genet. Cytogenet. 7:287-297

Carbonell, F., Hoelzer, D., Thiel, E. and Bartl, R. (1982b): Ph1-positive CML associated with megakaryocytic hyperplasia and thrombocythemia and an abnormality of chromosome No. 3. Cancer Genet. Cytogenet. 6:153-161

Carbonell, F., Ganser, A. and Heimpel, H. (1983): Cytogenetic studies in chronic myeloproliferative disorders. Acta Haematol. 69:145-151

Carrano, A.V., Mayall, B.H., Testa, J.R., Ashworth, L.K. and Rowley, J.D. (1979): Chromosomal DNA cytophotometry in 20q- nonspecific myeloid disorders. Cancer Res. 39:2984-2987

Casalone, R., Francesconi, D., Pasquali, F., Comotti, B. and Vaccari, F. (1981): Isochromosome (17q) in Philadelphia chromosome (Ph1)-negative juvenile chronic myelocytic leukemia. Cancer Genet. Cytogenet. 3:145-148

Casalone, R., Bernasconi, P. and Pasquali, F. (1983): Involvement of chromosome No. 20 in a complex Ph1 translocation. Cancer Genet. Cytogenet. 8:181-182

Castleman, B., Scully, R.E. and McNeely, B.U. (1973): Case records of the Massachusetts General Hospital (Case 18-1973). New Engl. J. Med. 288:957-963

Catovsky, D., Pittman, S., Lewis, D. and Pearse, E. (1977): Marker chromosome 14q+ in follicular lymphoma in transformation. Lancet (2):934

Cavallin-Ståhl, E., Landberg, T., Ottow, Z. and Mitelman, F. (1977): Hodgkin's disease and acute leukaemia. Scand. J. Haematol. 19:273-280

Chaganti, R.S.K., Miller, D.R., Meyers, P.A. and German, J. (1979): Cytogenetic evidence of the intrauterine origin of acute leukemia in monozygotic twins. New Engl. J. Med. 300:1032-1034

Chaganti, R.S.K., Bailey, R.B., Jhanwar, S.C., Arlin, Z.A. and Clarkson, B.D. (1982a): Chronic myelogenous leukemia in the monosomic cell line of a fertile Turner syndrome mosaic (45,X/46,XX). Cancer Genet. Cytogenet. 5:215-221

Chaganti, R.S.K., Jhanwar, S.C., Arlin, Z.A. and Clarkson, B.D. (1982b): Chronic myelogenous leukemia in an XYY male. Cancer Genet. Cytogenet. 5:223-226

Chapelle, A. de la, Schröder, J. and Vuopio, P. (1972): 8-trisomy in the bone marrow. Report of two cases. Clin. Genet. 3:470-476

Chapelle, A. de la, Vuopio, P., Sanger, R. and Teesdale, P. (1975): Monosomy-7 and the Colton blood-groups. Lancet (2):817

Chapelle, A. de la, Vuopio, P. and Icen, A. (1976): Trisomy 8 in the bone marrow associated with high red cell glutathione reductase activity. Blood 47:815-826

Chapelle, A. de la, Knuutila, S., Elonen, E. and Vuopio, P. (1981): Chromosomal abnormalities in acute promyelocytic leukaemia. Scand. J. Haematol. 26:57-60

Chen, T.R. and Shaw, M.W. (1973): Stable chromosome changes in a human malignant melanoma. Cancer Res. 33:2042-2047

Cheson, B.D., Vananetti, S.M., Buskjaer, L. and Fineman, R.M. (1980): Philadelphia chromosome (Ph1)-associated leukemias: a family study. Cancer Genet. Cytogenet. 1:321-328

Chessells, J.M., Janossy, G., Lawler, S.D. and Secker-Walker, L.M. (1979): The Ph1 chromosome in childhood leukaemia. Brit. J. Haematol. 41:25-41

Chilcote, R.R., Pergament, E., Kretschmer, R. and Mikuta, J.C. (1982): The hypereosinophilic syndrome and lymphoblastic leukemia with extra C-group chromosome and q14+ marker. J. Pediatr. 101:57-60

Chrz, R., Michalova, J. and Musilova, J. (1983): Personal communication.

Cimino, M.C., Roth, D.G., Golomb, H.M. and Rowley, J.D. (1978): A chromosome marker for B-cell cancers. New Engl. J. Med. 298:1422

Cimino, M.C., Rowley, J.D., Kinnealey, A., Variakojis, D. and Golomb, H.M. (1979): Banding studies of chromosomal abnormalities in patients with acute lymphocytic leukemia. Cancer Res. 39:227-238

Ciric, M. and Rolovic, Z. (1980): Another case of t(4;9;22) in chronic myeloid leukemia. Blood 55:871

Clare, N., Elson, D. and Manhoff, L. (1982): Cytogenetic studies of peripheral myeloblasts and bone marrow fibroblasts in acute myelofibrosis. Am. J. Clin. Pathol. 77:762-766

Cohen, M.M., Ariel, I. and Dagan, J. (1974): Chromosome deletion (46,XY,20q-) in sideroblastic anemia. Israel J. Med. Sci. 10:1393-1396

Cohen, A.M., Shabtai, F., Lewinski, U., Klein, B. and Djaldetti, M. (1979): Two abnormal clones in the bone marrow cells of a patient with paroxysmal nocturnal hemoglobinuria. Clin. Genet. 16:178-182

Como, R.M. and Graze, P.R. (1979): Emergence of a cell line with extreme hypodiploidy in blast crisis of chronic myelocytic leukemia. Blood 53:707-711

Contrafatto, G. (1977): Marker chromosome of macroglobulinemia identified by G-banding. Cytogenet. Cell Genet. 18:370-373

Cooperman, B.S. and Klinger, H.P. (1981): Double minute chromosomes in a case of acute myelogenous leukemia resistant to chemotherapy. Cytogenet. Cell Genet. 30:25-30

Couturier-Turpin, M.H., Couturier, D., Nepveux, P., Louvel, A., Chapuis, Y. and Guerre, J. (1982): Human chromosome analysis in 24 cases of primary carcinoma of the large intestine: Contribution of the G-banding technique. Br. J. Cancer 46:856-869

Dallapiccola, B. and Alimena, G. (1979): Inactive normal X in a female leukaemic patient with an acquired X/autosome translocation. Hum. Genet. 48:169-177

Daniel, A., Francis, S.E., Stewart, L.A. and Barber, S. (1978): A near haploid clone: 25,XY,t(9;22)(q34;q11) from a patient in blast crisis of chronic myeloid leukaemia. Scand. J. Haematol. 21:99-103

Darbyshire, P.J. and Eden, O.B. (1983): Monosomy 7 in a girl with Ph1 positive chronic granulocytic leukemia. Cancer Genet. Cytogenet. 9:233-238

Davidson, W.M. and Knight, L.A. (1973): Acquired trisomy 9. Lancet (1):1510

Davis, E., Rowley, J.D., Miller, W. and Hoffman, P.C. (1981): Undetected chromosome abnormalities in leukemia: a cautionary note. New Engl. J. Med. 304:1109

De Rossi, G. and Alimena, G. (1981): Membrane markers and 14q+ acute lymphocytic leukemia. Acta Haematol. 65:138-139

Debiec-Rychter, M., Kaluzewski, B., Zajaczkowska, D. and Pokuszynska, K. (1982): Mosaicism 48,XX,+8,+21/47,XX,+21 in Down syndrome and rapid progression from preleukaemia to acute leukaemia. Lancet (2):448-449

Dewald, G., Dines, D.E., Weiland, L.H. and Gordon, H. (1978): Usefulness of chromosome examination in the diagnosis of malignant pleural effusions. New Engl. J. Med. 295:1494-1500

Dewald, G.W., Pierre, R.V. and Phyliky, R.L. (1982): Three patients with structurally abnormal X chromosomes, each with Xq13 breakpoints and a history of idiopathic acquired sideroblastic anemia. Blood 59:100-105

Dewald, G.W., Morrison-DeLap, S.J., Schuchard, K.A., Spurbeck, J.L. and Pierre, R.V. (1983): A possible specific chromosome marker for monocytic leukemia: Three more patients with t(9;11)(p22;q24) and another with t(11;17)(q24;q21), each with acute monoblastic leukemia. Cancer Genet. Cytogenet. 8:203-212

DiBenedetto, J.Jr., Padre-Mendoza, T. and Albala, M.M. (1979): Pure red cell hypoplasia associated with long-arm deletion of chromosome 5. Hum. Genet. 46:345-348

Diamond, L.W., Bearman, R.M., Berry, P.K., Mills, B.J., Nathwani, B.N., Weisenburger, D.D., Winberg, C.D., Teplitz, R.L. and Rappaport, H. (1980): Prolymphocytic leukemia: flow microfluorometric, immunologic, and cytogenetic observations. Am. J. Hematol. 9:319-330

Dileo, P.E., Muller, H.J., Obrecht, J.-P., Speck, B., Buhler, E.M. and Stalder, G.R. (1977): Loss of the Y chromosome from bone marrow cells of males with myeloproliferative disorders. Acta Haematol. 57:310-320

Dor, J.F., Mattei, J.F., Mattei, M.G., Giraud, F. and Mongin, M. (1977): Subacute myelocytic leukemia associated with the Philadelphia chromosome and supplementary translocation: 9-12. Biomedicine 27:131-134

Douglass, E.C. and Freeman, D.L. (1983): Hypotetraploidy in erythroleukemia. Cancer Genet. Cytogenet. 8:231-234

Douglass, E.C., Magrath, I.T., Lee, E.C. and Whang-Peng, J. (1980a): Cytogenetic studies in non-African Burkitt lymphoma. Blood 55:148-155

Douglass, E.C., Poplack, D.G. and Whang-Peng, J. (1980b): Involvement of chromosome No. 22 in neuroblastoma. Cancer Genet. Cytogenet. 2:287-291

Douglass, E.C., Magrath, I.T. and Terebelo, H. (1982): Burkitt cell leukemia without abnormalities of chromosomes No. 8 and 14. Cancer Genet. Cytogenet. 5:181-185

Drew, M. and Hulten, M. (1983): Personal communication.

Edelson, R.L., Berger, C.L., Raafat, J. and Warburton, D. (1979): Karyotype studies of cutaneous T cell lymphoma: evidence for clonal origin. J. Invest. Dermatol. 73:548-550

Ekblom, M., Elonen, E., Vuopio, P., Heinonen, K., Knuutila, S., Gahmberg, C.G. and Andersson, L.E. (1982): Acute erythroleukaemia with L3 morphology and the 14q+ chromosome. Scand. J. Haematol. 29:75-82

Elfenbein, G.J., Borgaonkar, D.S., Bias, W.B., Burns, W.H., Saral, R., Sensenbrenner, L.L., Tutschka, P.J., Zaczek, B.S., Zander, A.R., Epstein, R.B., Rowley, J.D. and Santos, G.W. (1978): Cytogenetic evidence for recurrence of acute myelogenous leukemia after allogenic bone marrow transplantation in donor hematopoietic cells. Blood 52:627-636

Ellman, L., Hammond, D. and Atkins, L. (1979): Eosinophilia, chloromas and a chromosome abnormality in a patient with a myeloproliferative syndrome. Cancer 43:2410-2413

Engel, E., McGee, B.J., Flexner, J.M., Russell, M.T. and Myers, B.J. (1974): Philadelphia chromosome (Ph1) translocation in an apparently Ph1-negative, minus G22, case of chronic myeloid leukemia. New Engl. J. Med. 291:154

Engel, E., McGee, B.J., Flexner, J.M. and Krantz, S.B. (1975a): Translocation of the Philadelphia chromosome onto the 17 short arm in chronic myeloid leukemia: A second example. New Engl. J. Med. 293:666

Engel, E., McKee, L.C., Flexner, J.M. and McGee, B.J. (1975b): 17 long arm isochromosome. A common anomaly in malignant blood disorders. Ann. Genet. 18:56-60

Engel, E., McGee, B.J., Flexner, J.M. and Krantz, S.B. (1975c): Chromosome band analysis in 19 cases of chronic myeloid leukemia: 9 chronic, 10 blastic, two with Ph1 (22q-) translocation on 17 short arm. Ann. Genet. 18:239-240

Engel, E., McGee, B.J., Myers, B.J., Flexner, J.M. and Krantz, S.B. (1977): Chromosome banding patterns of 49 cases of chronic myelocytic leukemia. New Engl. J. Med. 296:1295

Esseltine, D.W., Vekemans, M., Seemayer, T. Reece, E., Gordon, J. and Whitehead, V.M. (1982): Significance of a (4;11) translocation in acute lymphoblastic leukemia. Cancer 50:503-506

Ferro, M.T., San Roman, C., Ranada, J.M.F. and Mayayo, M. (1982): 15;17 translocation in acute promyelocytic leukemia. Cancer Genet. Cytogenet. 7:89-92

Fialkow, P.J., Jacobson, R.J., Singer, J.W., Sacher, R.A., McGuffin, R.W. and Neefe, J.R. (1980): Philadelphia chromosome (Ph1)-negative chronic myelogenous leukemia (CML): a clonal disease with origin in a multipotent stem cell. Blood 56:70-73

Finan, J., Daniele, R., Rowlands Jr, D. and Nowell, P. (1978): Cytogenetics of chronic cell leukemia, including two patients with a 14q+ translocation. Virchows Arch. B Cell Pathol. 29:121-127

Findley, L., Kurnick, J.E., Peakman, D.C. and Robinson, A. (1979): Chromosome deletion (46,XX,del(20)(q11)) in agnogenic myeloid metaplasia. Hum. Genet. 47:207-211

First International Workshop on Chromosomes in Leukemia. (1978a): Chromosomes in Ph1-positive chronic granulocytic leukaemia. Brit. J. Haematol. 39:305-309

First International Workshop on Chromosomes in Leukemia. (1978b): Chromosomes in acute non-lymphocytic leukaemia. Brit. J. Haematol. 39:311-316

Fitzgerald, P.H., McEwan, C.M., Hamer, J.W. and Beard, M.E.J. (1980): Richter's syndrome with identification of marker chromosomes. Cancer 46:135-138

Fitzgerald, P.H., McEwan, C., Fraser, J. and Beard, M.E.J. (1981): A complex Ph1 translocation in a patient with primary thrombocythaemia. Brit. J. Haematol. 47:571-575

Fitzgerald, P.H., Morris, C.M. and Giles, L.M. (1982a): Direct versus cultured preparation of bone marrow cells from 22 patients with acute myeloid leukemia. Hum. Genet. 60:281-283

Fitzgerald, P.H., Giles, L.M., Morris, C.M., Heaton, D.C. and Beard, M.E.J. (1982b): Cytogenetic studies of acute promyelocytic leukemia. Cancer Genet. Cytogenet. 7:299-305

Fitzgerald, P.H., Morris, C.M., Fraser, G.J., Giles, L.M., Hamer, J.W., Heaton, D.C. and Beard, M.E.J. (1983): Nonrandom cytogenetic changes in New Zealand patients with acute myeloid leukemia. Cancer Genet. Cytogenet. 8:51-66

Fleischman, E.W. and Prigogina, E.L. (1975): G banding in cytogenetic study of hemoblastoses. Humangenetik 26:335-342

Fleischman, E.W. and Prigogina, E.L. (1977): Karyotype peculiarities of malignant lymphomas. Hum. Genet. 35:269-279

Fleischman, E.W., Prigogina, E.L., Volkova, M.A. and Petkovitch, I. (1977): Unusual translocation (10;22) in chronic myelogenous leukemia. Hum. Genet. 39:127-129

Fleischman, E.W., Prigogina, E.L., Volkova, M.A. and Kulagina, O.E. (1979): Chromosomal marker 20q- in cases of osteomyelosclerosis and CML. Hum. Genet. 50:101-104

Fleischman, E.W., Prigogina, E.L., Volkova, M.A., Frenkel, M.A., Zakhartchenko, N.A., Konstantinova, L.N., Puchkova, G.P. and Balakirev, S.A. (1981): Correlations between the clinical course, characteristics of blast cells, and karyotype patterns in chronic myeloid leukemia. Hum. Genet. 58:285-293

Fleischmann, T. and Krizsa, F. (1975): Marker chromosome in myeloproliferative syndrome. Acta Haematol. 54:59-63

Fleischmann, T. and Krizsa, F. (1977): Chromosomes in malignant lymphomas (Study on short-term lymph node cultures). Haematologia 11:47-55

Fleischmann, T., Håkansson, C.H., Levan, A. and Möller, T. (1972): Multiple chromosome aberrations in a lymphosarcomatous tumor. Hereditas 70:243-258

Fleischmann, T., Håkansson, C.H. and Levan, A. (1976): Chromosomes of malignant lymphomas. Studies in short-term cultures from lymph nodes of twenty cases. Hereditas 83:47-56

Foerster, W., Medau, H.J. and Löffler, H. (1974): Chronische myeloische Leukämie mit Philadelphia-Chromosom und Tandem-Translokation am 2. Chromosom Nr. 22; 46,XX,tan(22q+;22q-). Klin. Wochenschr. 52:123-126

Fonatsch, C., Burrichter, H., Schaadt, M., Kirchner, H.H. and Diehl, V. (1982): Translocation t(8;22) in peripheral lymphocytes and established lymphoid cell lines from a patient with Hodgkin's disease followed by acute lymphatic leukemia. Int. J. Cancer 30:321-327

Forman, E.N., Padre-Mendoza, T., Smith, P.S., Barker, B.E. and Farnes, P. (1977): Ph1- positive childhood leukemias: spectrum of lymphoid-myeloid expressions. Blood 49:549-558

Fraisse, J., Jaubert, J., Vasselon, C. and Brizard, C.P. (1980): Etude cytogenetique de 44 cas de LMC en bandes R. Nouv. Rev. Fr. Hematol. Suppl. 22:86

Fraisse, J., Jaubert, J., Vasselon, C. et Brizard, C.P. (1981a): Trois observations de leucémie aigue promyélocytaire. Nouv. Rev. Fr. Hematol. 23:163-164

Fraisse, J., Lenoir, G., Vasselon, C., Jaubert, J., and Brizard, C.P. (1981b): Variant tranclocation in Burkitt's lymphoma:8;22 translocation in a French patient with an Epstein-Barr virus-associated tumor. Cancer Genet. Cytogenet. 3:149-153

Francesconi, D. and Pasquali, F. (1978a): 8/21 translocation, loss of the Y chromosome and Philadelphia chromosome. Brit. J. Haematol. 38:149-150

Francesconi, D. and Pasquali, F. (1978b): Three chromosomes (7;9;22) rearrangement and the origin of the Philadelphia chromosome. Hum. Genet. 43:133-137

Fraser, J., Hollings, P.E., Fitzgerald, P.H., Day, W.A., Clark, V., Heaton, D.C., Hamer, J.W. and Beard, M.E.J. (1981): Acute promyelocytic leukemia: cytogenetics and bone-marrow culture. Int. J. Cancer 27:167-173

Freycon, F., Fraisse, J., Lepetit, J.L., Pouyau, G., Bertheas, M.F. and Gourre, D. (1982): Cytogenetics follow-up of two cases of chronic myelogenous leukemia (CML) with BMT. Exp. Hematol. 10, suppl. 10:84

Fukuhara, S. and Rowley, J.D. (1978): Chromosome 14 translocations in non-Burkitt lymphomas. Int. J. Cancer 22:14-21

Fukuhara, S., Shirakawa, S. and Uchino, H. (1976): Specific marker chromosome 14 in malignant lymphomas. Nature 259:210-211

Fukuhara, S., Rowley, J.D. and Variakojis, D. (1978a): Banding studies of chromosomes in a patient with mycosis fungoides. Cancer 42:2262-2268

Fukuhara, S., Rowley, J.D., Variakojis, D. and Sweet Jr, D.L. (1978b): Banding studies on chromosomes in diffuse "histiocytic" lymphomas: Correlation of 14q+ marker chromosome with cytology. Blood 52:989-1002

Fukuhara, S., Rowley, J.D., Variakojis, D. and Golomb, H.M. (1979a): Chromosome abnormalities in poorly differentiated lymphocytic lymphoma. Cancer Res. 39:3119-3128

Fukuhara, S., Ueshima, Y., Shirakawa, S., Uchino, H. and Morikawa, S. (1979b): 14q translocations, having a break point at 14q13, in lymphoid malignancy. Int. J. Cancer 23:739-743

Gaeke, M.E., Vardiman, J.W., Miller, W., Medenica, M., Hopper, J.E. and Rowley, J.D. (1981): Human T-cell lymphoma with suppressor effects on the mixed lymphocyte reaction (MLR). I. Morphological and cytogenetic analysis. Blood 57:634-641

Gahrton, G., Zech, L. and Lindsten, J. (1974a): A new variant translocation (19q+;22q-) in chronic myelocytic leukemia. Exp. Cell Res. 86:214-216

Gahrton, G., Lindsten, J. and Zech, L. (1974b): Involvement of chromosomes 8, 9, 19 and 22 in Ph1 positive and Ph1 negative chronic myelocytic leukemia in the chronic or blastic stage. Acta Med. Scand. 196:355-360

Gahrton, G., Friberg, K. and Zech, L. (1977): A new translocation involving three chromosomes in chronic myelocytic leukemia, 46,XY,t(9;11;22). Cytogenet. Cell Genet. 18:75-81

Gahrton, G., Friberg, K., Lindsten, J. and Zech, L. (1978a): Duplication of part of the long arm of chromosome 1 in myelofibrosis terminating in acute myeloblastic leukemia. Hereditas 88:1-5

Gahrton, G., Friberg, K. and Zech, L. (1978b): Constitutional chromosomal aberration, t(4;12)(q23;p11), in a patient with Ph1 positive chronic myelocytic leukemia. Hereditas 89:169-173

Gahrton, G., Friberg, K. and Zech, L. (1979): Translocation between chromosome 7 and chromosome 22, t(7;22)(p22;q12), in a patient with chronic myelocytic leukemia. Hum. Genet. 49:225-227

Gahrton, G., Robert, K.-H., Friberg, K., Zech, L. and Bird, A.G. (1980a): Extra chromosome 12 in chronic lymphocytic leukaemia. Lancet (1):146-147

Gahrton, G., Zech, L., Nilsson, K., Lönnqvist, B. and Carlström, A. (1980b): 2 translocations, t(11;14) and t(1;6), in a patient with plasma cell leukaemia and 2 populations of plasma cells. Scand. J. Haematol. 24:42-46

Gahrton, G., Robert, K.-H., Friberg, K., Zech, L. and Bird, A.G. (1980c): Nonrandom chromosomal aberrations in chronic lymphocytic leukemia revealed by polyclonal B-cell-mitogen stimulation. Blood 56:640-647

Gahrton, G., Robert, K.-H. Friberg, K., Juliusson, G., Biberfeld, P. and Zech, L. (1982): Cytogenetic mapping of the duplicated segment of chromosome 12 in lymphoproliferative disorders. Nature 297:513-514

Gall, J.A., Boggs, D.R., Chervenick, P.A., Pan, S. and Fleming, R.B. (1976): Discordant patterns of chromosome changes and myeloblast proliferation during the terminal phase of chronic myeloid leukemia. Blood 47:347-353

Gallagher, R., Collins, S., Trujillo, J., McCredie, K., Ahearn, M., Tsai, S., Metzgar, R., Aulakh, G., Ting, R., Ruscetti, F. and Gallo, R. (1979): Characterization of the continuous, differentiating myeloid cell line (HL-60) from a patient with acute promyelocytic leukemia. Blood 54:713-733

Gardner, H.A., Gallie, B.L., Knight, L.A. and Phillips, R.A. (1982): Multiple karyotypic changes in retinoblastoma tumor cells: presence of normal chromosome No. 13 in most tumors. Cancer Genet. Cytogenet. 6:201-211

Garson, O.M. (1980): Acute non-lymphocytic leukaemia. Clin. Haematol. 9:39-54

Geary, C.G., Catovsky, D., Wiltshaw, E., Milner, G.R., Scholes, M.C., Noorden, S. van, Wadsworth, L.D., Muldal, S., MacUver, J.E. and Galton, D.A.G. (1975): Chronic myelomonocytic leukaemia. Brit. J. Haematol. 30:289-302

Geraedts, J.P.M. and Haak, H.L. (1976): Trisomy 6 associated with aplastic anemia. Hum. Genet. 35:113-115

Geraedts, J.P.M., Mol, A., Ottolander, G.J. den, Van Der Ploeg, M. and Pearson, P.L. (1977): Variation in the chromosomes of CML patients. Helsinki Chromosome Conference 1977. Abstr. Book p. 194

Geraedts, J.P.M., Weber, R.F.A., Kerkhofs, H. and Leeksma, C.H.W. (1980a): The preleukemic syndrome. II. Cytogenetic findings. Acta Med. Scand. 207:447-454

Geraedts, J.P.M., Ottolander, G.J. den, Ploem, J.E. and Muntinghe, O.G. (1980b): An identical translocation between chromosome 1 and 7 in three patients with myelofibrosis and myeloid metaplasia. Brit. J. Haematol. 44:569-575

Geurts van Kessel, A.H.M., Agthoven, A.J. van, Groot, P.G. de and Hagemeijer, A. (1981): Characterization of a complex Philadelphia translocation (1p-;9q+;22q-) by gene mapping. Hum. Genet. 58:162-165

Gibbs, T.J., Wheeler, M.V., Bellingham, A.J. and Walker, S. (1977): The significance of the Philadelphia chromosome in acute lymphoblastic leukaemia: a report of two cases. Brit. J. Haematol. 37:447-453

Gilbert, F., Balaban, G., Moorhead, P., Bianchi, D. and Schlesinger, H. (1982): Abnormalities of chromosome 1p in human neuroblastoma tumors and cell lines. Cancer Genet. Cytogenet. 7:33-42

Goffman, T.E., Mulvihill, J.J., Carney, D.N., Triche, T.J. and Whang-Peng, J. (1983): Fatal hypereosinophilia with chromosome 15q- in a patient with multiple primary and familial neoplasms. Cancer Genet. Cytogenet. 8:197-202

Goh, K., Jacox, R.F. and Anderson, F.W. (1980): Chromosomal abnormalities. Findings in a patient with lymphoma and rheumatoid arthritis treated with intra-articular Gold Au 198. Arch. Pathol. Lab. Med. 104:473-475

Goldman, J.M., Najfeld, V. and Th'ng, K.H. (1975): Agar culture and chromosome analysis of eosinophilic leukaemia. J. Clin. Pathol. 28:956-961

Goldstone, A.H., McVerry, B.A., Janossy, G. and Walker, H. (1979): Clonal identification in acute lymphoblastic leukemia. Blood 53:892-898

Golomb, H.M., Rowley, J., Vardiman, J., Baron, J., Locker, G. and Krasnow, S. (1976): Partial deletion of long arm of chromosome 17. Arch. Intern. Med. 136:825-828

Golomb, H.M., Lindgren, V. and Rowley, J.D. (1978a): Hairy cell leukemia: An analysis of the chromosomes of 26 patients. Virchows Arch. B. Cell. Path. 29:113-120

Golomb, H.M., Lindgren, V. and Rowley, J.D. (1978b): Chromosome abnormalities in patients with hairy cell leukemia. Cancer 41:1374-1380

Golomb, H.M., Vardiman, J.W., Rowley, J.D., Testa, J.R. and Mintz, U. (1978c): Correlation of clinical findings with quinacrine-banded chromosomes in 90 adults with acute nonlymphocytic leukemia. An eight-year study (1970-1977). New Engl. J. Med. 299:613-619

Golomb, H.M., Testa, J.R., Vardiman, J.W., Butler, A.E. and Rowley, J.D. (1979): Cytogenetic and ultrastructural features of de novo acute promyelocytic leukemia; The University of Chicago Experience (1973-1978). Cancer Genet. Cytogenet. 1:69-78

Golomb, H.M., Rowley, J.D., Vardiman, J.W., Testa, J.R. and Butler, A. (1980): "Microgranular" acute promyelocytic leukemia: A distinct clinical, ultrastructural, and cytogenetic entity. Blood 55:253-259

Granberg, I., Gupta, S. and Zech, L. (1973): Chromosome analyses of a metastatic gastric carcinoma including quinacrine fluorescence. Hereditas 75:189-194

Greef, I. de, Geraedts, J.P.M. and Leeksma, C.H.W. (1982): Malignant hematologic disorders in two Robertsonian 13:14 translocation carriers. Cancer Genet. Cytogenet. 7:181-184

Green, R.J., Findley, H.W. Jr., Chen, A.T.L. and Ragab, A.H. (1982): Characterization of a new chromosomal marker for acute lymphoblastic leukemia from a long-term cell line. Cancer Genet. Cytogenet. 7:257-269

Greenberg, B.R., Wilson, F.D., Woo, L. and Jenks, H.M. (1978): Cytogenetics of fibroblastic colonies in Ph1-positive chronic myelogenous leukemia. Blood 51:1039-1044

Gustavsson, A., Mitelman, F. and Olsson, I. (1977): Acute myeloid leukaemia with the Philadelphia chromosome. Scand. J. Haematol. 19:449-452

Gyger, M. and Bonny, Y. (1981): Monosomy 7 syndrome. New Engl. J. Med. 305:1155-1156

Gyger, M., Bonny, Y. and Forest, L. (1982): Childhood monosomy 7 syndrome. Am. J. Hematol. 13:329-334

Gödde, E. and Zelewski, E. von (1980): A clone with loss of the Y chromosome in Ph1-positive chronic myelocytic leukemia (CML). Cancer Genet. Cytogenet. 2:7-11

Gödde-Salz, E., Schwarze, E.-W., Stein, H., Lennert, K. and Grote, W. (1981a): Cytogenetic findings in T-zone lymphoma. J. Cancer Res. Clin. Oncol. 101:81-89

Gödde-Salz, E., Schwarze, E-W., Lennert, K. and Grote, W. (1981b): T(Y;14) - a new type of 14q+ marker chromosome. Cancer Genet. Cytogenet. 3:89-90

Hagemeijer, A., Smit, E.M.E. and Bootsma, D. (1979a): Improved identification of chromosomes of leukemic cells in methotrexate-treated cultures. Cytogenet. Cell Genet. 23:208-212

Hagemeijer, A., Zanen, G.E. van, Smit, E.M.E. and Hählen, K. (1979b): Bone marrow karyotypes of children with nonlymphocytic leukemia. Pediat. Res. 13:1247-1254

Hagemeijer, A., Hählen, K., Smit, E.M.E. and Zanen, G.E. van (1980a): C-group chromosome abnormalities in bone marrow cells of three children with dyshaematopoiesis of unknown origin. Brit. J. Haematol. 46:377-385

Hagemeijer, A., Stenfert Kroeze, W.F. and Abels, J. (1980b): Cytogenetic follow-up of patients with nonlymphocytic leukemia I. Philadelphia chromosome-positve chronic myeloid leukemia. Cancer Genet Cytogenet 2:317-326

Hagemeijer, A., Hählen, K. and Abels, J. (1981a): Cytogenetic follow-up of patients with nonlymphocytic leukemia. II. Acute nonlymphocytic leukemia. Cancer Genet. Cytogenet. 3:109-124

Hagemeijer, A., Sizoo, W., Smit, E.M.E. and Abels, J. (1981b): Translocation (5p;17q) in blast crisis of chronic myeloid leukemia. Cytogenet. Cell Genet. 30:205-210

Hagemeijer, A., Löwenberg, B. and Abels, J. (1982a): Analysis of the breakpoints in translocation (15;17) observed in 4 patients with acute promyelocytic leukemia. Hum. Genet. 61:223-227

Hagemeijer, A., Hählen, K., Sizoo, W. and Abels, J. (1982b): Translocation (9;11)(p21;q23) in three cases of acute monoblastic leukemia. Cancer Genet. Cytogenet. 5:95-105

Hansson, A. and Korsgaard, R. (1974): Cytogenetical diagnosis of malignant pleural effusions. Scand. J. resp. Dis. 55:301-308

Hartley, S.E. (1983): Personal communication.

Hartley, S.E. and Cook, M.K. (1979): Near-haploidy in a case of chronic myeloid leukemia. Cancer Genet. Cytogenet. 1:169-176

Hartley, S.E. and McBeath, S. (1981): Cytogenetic follow-up in chronic myeloid leukemia. Cancer Genet. Cytogenet. 3:37-46

Hartley, S.E. and McCallum, C.J. (1981): The 5q- chromosome in a case of erythroid hypoplasia. Cancer Genet. Cytogenet. 3:33-36

Hartley, S.E. and Sainsbury, C. (1981): Acute leukaemia and same chromosome abnormality in monozygotic twins. Hum. Genet. 58:408-410

Hartley, S.E. and Toolis, F. (1980): Double minute chromosomes in a case of acute myeloblastic leukemia. Cancer Genet. Cytogenet. 2:275-280

Hayata, I. and Sasaki, M. (1976): A case of Ph1-positive chronic myelocytic leukemia associated with complex translocations. Proc. Jap. Acad. 52:29-32

Hayata, I., Kakati, S. and Sandberg, A.A. (1973): A new translocation related to the Philadelphia chromosome. Lancet (2):1385

Hayata, I., Kakati, S. and Sandberg, A.A. (1974): On the monoclonal origin of chronic myelocytic leukemia. Proc. Jap. Acad. 50:381-385

Hayata, I., Kakati, S. and Sandberg, A.A. (1975a): Another translocation related to the Ph1 chromosome. Lancet (1):1300

Hayata, I., Sakurai, M., Kakati, S. and Sandberg, A.A. (1975b): Chromosomes and causation of human cancer and leukemia. XVI. Banding studies of chronic myelocytic leukemia, including five unusual Ph1 translocations. Cancer 36:1177-1191

Hays, T., Morse, H.G. and Robinson, A. (1981): 9;22;15 complex translocation in Ph1 chromosome positive CML revealed by Giemsa-11 procedure in apparent lymphoid cells of blastic crisis. Cancer Genet. Cytogenet. 4:283-292

Heaton, D.C., Fitzgerald, P.H., Fraser, G.J. and Abbott, G.D. (1981): Transient leukemoid proliferation of the cytogenetically unbalanced +21 cell line of a constitutional mosaic boy. Blood 57:883-887

Hecht, F., Kuban, D.J., Berger, C., Kaiser-McCaw Hecht, B. and Sandberg, A.A. (1983): Adenocarcinoma of the gallbladder: Chromosome abnormalities in a genetic form of cancer. Cancer Genet. Cytogenet. 8:185-190

Hellström, K., Hagenfeldt, L., Larsson, A., Lindsten, J., Sundelin, P. and Tiepolo, L. (1971): An extra C chromosome and various metabolic abnormalities in the bone marrow from a patient with refractory sideroblastic anaemia. Scand. J. Haematol. 8:293-306

Hjorth, M., Mark, J. and Tibblin, E. (1980): A hypotriploid stemline with 4 Ph1 chromosomes in erythroleukemic blast crisis of a CML-patient with a long survival time. Hereditas 93:333-336

Hoeltge, G.A., Dyment, P.G. and Slovak, M.L. (1982): Acute lymphocytic leukemia with microblastosis and near haploidy (26 chromosomes): a case report. Med. Ped. Oncol. 10:53-59

Honoré, L.H., Dill, F.J. and Poland, B.J. (1974): The association of hydatidiform mole and trisomy 2. Obstet. Gynecol. 43:232-237

Horland, A.A., Wolman, S.R., Distenfeld, A. and Cohen, T. (1976): Another variant translocation in chronic myelogenous leukemia. New Engl. J. Med. 294:164-165

Hossfeld, D.K. (1974a): Identification of chromosome anomalies in the blastic phase of chronic myelocytic leukemia (CML) by Giemsa- and quinacrine-banding techniques. Humangenetik 23:111-118

Hossfeld, D.K. (1974b): No chromosome 9q+ in Ph1-negative CML. Nature 249:864

Hossfeld, D.K. (1975a): Additional chromosomal indication for the unicellular origin of chronic myelocytic leukemia. Z. Krebsforsch 83:269-273

Hossfeld, D.K. (1975b): Chronic myelocytic leukemia: Cytogenetic findings and their relations to pathogenesis and clinic. Ser. Haematol. 8:53-72

Hossfeld, D.K. (1978): Chromosome 14q+ in a retinoblastoma. Int. J. Cancer 21:720-723

Hossfeld, D.K. (1983): Personal communication.

Hossfeld, D.K. and Köhler, S. (1979): New translocations in chronic granulocytic leukaemia: t(X;22)(p22;q11) and t(15;22)(q26;q11). Brit. J. Haematol. 41:185-191

Hossfeld, D.K. and Schmidt, C.G. (1978): Chromosome findings in effusions from patients with Hodgkin's disease. Int. J. Cancer 21:147-156

Hossfeld, D.K. and Wendehorst, E. (1974): Ph1-negative chronic myelocytic leukemia with a missing Y chromosome. Acta Haematol. 52:232-237

Hossfeld, D.K., Bremer, K., Meusers, P., Wendehorst, E. and Reis, H.E. (1975): Extramedullary manifestation of the blastic phase of chronic myelocytic leukemia: A chromosome study. Z. Krebsforsch. 84:49-57

Hossfeld, D.K., Höffken, K., Schmidt, C.G. and Diedrichs, H. (1976): Chromosome abnormalities in angioimmunoblastic lymphadenopathy. Lancet (1):198

Hossfeld, D.K., Higi, M., Köhler, S., Miller, A. and Zschaber, R. (1980): A subtype of the prototypic karyotype in acute myeloid leukemia t(8;21)(q22;q22), del(9)(q13;q23). Blut 40:27-32

Hsu, L.Y.F., Papenhausen, P., Greenberg, M.L. and Hirschhorn, K. (1974a): Trisomy D in bone marrow cells in a patient with chronic myelogenous leukemia. Acta Haematol. 52:61-64

Hsu, L.Y.F., Alter, A.V. and Hirschhorn, K. (1974b): Trisomy 8 in bone marrow cells of patients with polycythemia vera and myelogenous leukemia. Clin. Genet. 6:258- 264

Hsu, L.Y.F., Pinchiaroli, D., Gilbert, H.S., Wittman, R. and Hirschhorn, K. (1977): Partial trisomy of the long arm of chromosome 1 in myelofibrosis and polycythemia vera. Am. J. Hematol. 2:375-383

Hsu, L.Y.F., Greenberg, M.L., Kohan, S. and Wittman, R. (1979): Trisomy 13 in bone marrow cells in acute myelocytic leukemia and myelofibrosis. Clin. Genet. 15:327-331

Huang, C.S., Gomez, G.A., Kohno, S.-I., Sokal, J.E. and Sandberg, A.A. (1979): Chromosomes and causation of human cancer and leukemia. XXXIV. A case of "hypereosinophilic syndrome" with unusual cytogenetic findings in a chloroma, terminating in blastic transformation and CNS leukemia. Cancer 44:1284-1289

Humbert, J.R., Morse, H.G., Hutter Jr, J.J., Rose, B. and Robinson, A. (1978): Non-leukemic dividing cells in the blood of leukemic patients. Am. J. Hematol. 4:217-224

Humphrey, M.J., Hutter, J.J. Jr. and Tom, W.W. (1981): Hypereosinophilia in a monosomy 7 myeloproliferative disorder in childhood. Am. J. Hematol. 11:107-110

Hurd, D.D., Vukelich, M., Arthur, D.C., Lindquist, L.L., McKenna, R.W., Peterson, B.A. and Bloomfield, C.D. (1982): 15;17 translocation in acute promyelocytic leukemia. Cancer Genet. Cytogenet. 6:331-337

Hurley, J.N., Fu, S.M., Kunkel, H.G., Chaganti, R.S.K. and German, J. (1980): Chromosome abnormalities of leukaemic B lymphocytes in chronic lymphocytic leukaemia. Nature 283:76-78

Hustinx, T.W.J. and Rutten, F.J. (1983): Personal communication.

Hustinx, T.W.J., Burghouts, J.T.M., Scheres, J.M.J.C. and Smits, A.P.T. (1980): A case of AMMoL with 8/21 translocation and loss of the Y as probably secondary events. Cancer 45:285-288

Hyder, D.M., Bottomley, S.S. and Bottomley, R.H. (1978): A new chromosome abnormality in idiopathic sideroblastic anemia: 46,XY,del11q23. Am. J. Hematol. 5:239-245

Högstedt, B., Nilsson, P.G. and Mitelman, F. (1981): Micronuclei in erythropoietic bone marrow cells: Relation to cytogenetic pattern and prognosis in acute nonlymphocytic leukemia. Cancer Genet. Cytogenet. 3:185-193

ISCN (1978): . - Birth Defects: Original Article Series, Vol. XIV, No. 8, The National Foundation, New York, 1978

Imamura, N., Kamada, N. and Kuramoto, A. (1981): Pre-B-cell leukemia with the 14q+ chromosome. New Engl. J. Med. 305:701

Inoue, S., Ravindranath, Y., Thompson, R.I., Zuelzer, W.W. and Ottenbreit, M.J. (1977): Cytogenetics of juvenile type chronic granulocytic leukemia. Cancer 39:2017-2024

Ishihara, T. (1983): Personal communication.

Ishihara, T. and Minamihisamatsu, M. (1981): A possible mechanism of the formation of unusual Ph1 translocations in chronic myelocytic leukemia. Natl. Inst. Radiol. Sci. Ann. Rep. 21:51

Ishihara, T., Kohno, S.-I. and Kumatori, T. (1974): Ph1 translocation involving chromosomes 21 and 22. Brit. J. Cancer 29:340-342

Ishihara, T., Minamihisamatsu, M. and Kohno, S-I. (1982): Near-haploid conversion in Ph1-positive chronic myelocytic leukemia. Proc. Jap. Acad. 58:165-168

Ishihara, T., Sasaki, M., Oshimura, M., Kamada, N., Yamada, K., Okada, M., Sakurai, M., Sugiyama, T., Shiraishi, Y. and Kohno, S.-I. (1983): A summary of cytogenetic studies on 534 cases of chronic myelocytic leukemia in Japan. Cancer Genet. Cytogenet. 9:81-92

Izakovic, V., Horak, I. and Krizan, P. (1982): Clonal evolution of karyotype in blastic phase of CML. Neoplasma 29:613-623

Jacobson, R.J., Salo, A. and Fialkow, P.J. (1978): Agnogenic myeloid metaplasia: A clonal proliferation of hematopoietic stem cells with secondary myelofibrosis. Blood 51:189-194

Jonasson, J., Gahrton, G., Lindsten, J., Simonsson-Lindemalm, C. and Zech, L. (1974): Trisomy 8 in acute myeloblastic leukemia and sideroachrestic anemia. Blood 43:557-563

Jones Cruciger, Q.V., Pathak, S. and Cailleau, R. (1976): Human breast carcinomas: marker chromosomes involving 1q in seven cases. Cytogenet. Cell Genet. 17:231-235

Jotterand-Bellomo, M. (1978): One case of a Ph1 chromosome resulting from translocation of the distal end of 22q onto the short arm of chromosome 15. Cytogenet. Cell Genet. 21:168-169

Kaffe, S., Hsu, L.Y.F. and Hirschhorn, K. (1974): Acquired trisomies 12 and 7. Lancet (1):261-262

Kaffe, S., Hsu, L.Y.F., Hoffman, R. and Hirschhorn, K. (1978): Association of 5q- and refractory anemia. Am. J. Hematol. 4:269-272

Kaiser McCaw, B., Hecht, F., Harnden, D.G. and Teplitz, R.L. (1975): Somatic rearrangement of chromosome 14 in human lymphocytes. Proc. Natl. Acad. Sci. 72:2071-2075

Kaiser McCaw, B., Epstein, A.L., Kaplan, H.S. and Hecht, F. (1977): Chromosome 14 translocation in African and North American Burkitt's lymphomas. Int. J. Cancer 19:482-486

Kakati, S., Hayata, I., Oshimura, M. and Sandberg, A.A. (1975): Chromosomes and causation of human cancer and leukemia. X. Banding patterns in cancerous effusions. Cancer 36:1729-1738

Kakati, S., Hayata, I. and Sandberg, A.A. (1976a): Chromosomes and causation of human cancer and leukemia. XIV. Origin of a large number of markers in a cancer. Cancer 37:776-782

Kakati, S., Oshimura, M. and Sandberg, A.A. (1976b): Chromosomes and causation of human cancer and leukemia. XIX. Common markers in various tumors. Cancer 38:770-777

Kakati, S., Song, S.Y. and Sandberg, A.A. (1977): Chromosomes and causation of human cancer and leukemia. XXII. Karyotypic changes in malignant melanoma. Cancer 40:1173-1181

Kakati, S., Barcos, M. and Sandberg, A.A. (1979): Chromosomes and causation of human cancer and leukemia. XXXVI. The 14q+ anomaly in an American Burkitt lymphoma and its value in the definition of lymphoproliferative disorders. Med. Ped. Oncol. 6:121-129

Kakati, S., Barcos, M. and Sandberg, A.A. (1980): Chromosomes and causation of human cancer and leukemia. XLI. Cytogenetic experience with non-Hodgkin, non-Burkitt lymphomas. Cancer Genet. Cytogenet. 2:199-220

Kamada, N., Okada, K., Oguma, N., Tanaka, R., Mikami, M. and Uchino, H. (1976): C-G translocation in acute myelocytic leukemia with low neutrophil alkaline phosphatase activity. Cancer 37:2380-2387

Kamada, N., Dohy, H., Okada, K., Oguma, N., Kuramoto, A., Tanaka, K. and Uchino, H. (1981): In vivo and in vitro activity of neutrophil alkaline phosphatase in acute myelocytic leukemia with 8;21 translocation. Blood 58:1213-1217

Kamihira, S., Tomonaga, Y., Tagawa, M., Sadamori, N., Nonaka, M., Kinoshita, K., Itchimaru, M. and Tomonaga, A. (1977): An autopsy case of sideroblastic anemia associated with cytogenetic aberrations (t(2q+;5q-)) in bone marrow cells. Jap. J. Clin. Hematol. 18:1370-1377

Kaneko, Y. and Sakurai, M. (1977): 15/17 translocation in acute promyelocytic leukaemia. Lancet (1):961

Kaneko, Y. and Sakurai, M. (1980): Acute lymphocytic leukemia (ALL) with near-haploidy - a unique subgroup of ALL?. Cancer Genet. Cytogenet. 2:13-18

Kaneko, Y., Sakurai, M. and Hattori, M. (1978a): A case of acute myelogenous leukemia with an 8-21 translocation, missing Y, and additional karyotypic abnormalities. Am. J. Hematol. 4:273-280

Kaneko, Y., Sakurai, M. and Hattori, M. (1978b): Childhood acute myelogenous leukemia with an 8-21 chromosome translocation. J. Pediatr. 93:1066-1067

Kaneko, Y., Rowley, J.D., Check, I., Variakojis, D. and Moohr, J.W. (1980): The 14q+ chromosome in pre-B-ALL. Blood 56:782-785

Kaneko, Y., Rowley, J.D., Variakojis, D., Chilcote, R.R., Moohr, J.W. and Patel, D. (1981a): Chromosome abnormalities in Down's syndrome patients with acute leukemia. Blood 58:459-466

Kaneko, Y., Hayashi, Y. and Sakurai, M. (1981b): Chromosomal findings and their correlation to prognosis in acute lymphocytic leukemia. Cancer Genet. Cytogenet. 4:227-235

Kaneko, Y., Abe, R., Sampi, K. and Sakurai, M. (1982a): An analysis of chromosome findings in non-Hodgkin's lymphomas. Cancer Genet. Cytogenet. 5:107-121

Kaneko, Y., Rowley, J.D., Maurer, H.S., Variakojis, D. and Moohr, J.W. (1982b): Chromosome pattern in childhood acute nonlymphocytic leukemia (ANLL). Blood 60:389-399

Kaneko, Y., Larson, R.A., Variakojis, D., Haren, J.M. and Rowley, J.D. (1982c): Nonrandom chromosome abnormalities in angioimmunoblastic lymphadenopathy. Blood 60:877-887

Kaneko, Y., Variakojis, D., Kluskens, L. and Rowley, J.D. (1982d): Lymphoblastic lymphoma: cytogenetic, pathologic, and immunologic studies. Int. J. Cancer. 30:273-279

Kaneko, Y., Rowley, J.D., Variakojis, D., Chilcote, R.R., Check, I. and Sakurai, M. (1982e): Correlation of karyotype with clinical features in acute lymphoblastic leukemia. Cancer Res. 42:2918-2929

Kaneko, Y., Kimpara, H., Kawai, S. and Fujimoto, T. (1983): 8;21 chromosome translocation in eosinophilic leukemia. Cancer Genet. Cytogenet. 9:181-183

Kardon, N., Schulman, P., Degnan, T.J., Budman, D.R., Davis, J. and Vinciguerra, V. (1982a): Cytogenetic findings in the dysmyelopoietic syndrome. Cancer 50:2834-2838

Kardon, N.B., Slepowitz, G. and Kochen, J.A. (1982b): Childhood acute lymphoblastic leukemia associated with an unusual 8;14 translocation. Cancer Genet. Cytogenet. 6:339-343

Karpas, A., Fischer, P. and Swirsky, D. (1982): Human plasmacytoma with an unusual karyotype growing in vitro and producing light-chain immunoglobulin. Lancet (1):931-933

Kaufmann, U., Löffler, H., Foerster, W., Desaga, J.F. and Koch, F. (1974): Fehlendes Chromosom Nr. 7 in der präleukämischen Phase einer Myeloblastenleukose bei einem Kind. Blut 29:50-61

Kerkhofs, H., Hagemeijer, A., Leeksma, C.H.W., Abels, J., Den Ottolander, G.J., Somers, R., Gerrits, W.B.J., Langenhuiyen, M.M.A.C., Von Dem Borne, A.E.G.Kr., Van Hemel, J.O. and Geraedts, J.P.M. (1982): The 5q- chromosome abnormality in haematological disorders: a collaborative study of 34 cases from the Netherlands. Brit. J. Haematol. 52:365-381

Kessous, A., Colombies, P., Sudries, M., Bourrouillou, G., Pris, J. and Clement, D. (1980): Complex Ph1 translocation in chronic myeloid leukemia. Cancer Genet. Cytogenet. 2:335-337

Khalid, G., Li, Y-S., Flemans, R.J. and Hayhoe, F.G.J. (1981): Chromosomal abnormalities in a case of hairy cell leukaemia. Leukemia Res. 5:431-435

Kirkland, D.J., Welch, S.G., Povey, S., Najfeld, V., Price, D.J. and Lawler, S.D. (1980): Glutamic pyruvate transaminase phenotypes in polycythaemia rubra vera. Brit. J. Haematol. 44:407-413

Knight, L.A., Davidson, W.M. and Cuddigan, B.J. (1974): Acquired trisomy 9. Lancet (1):688

Knoerr-Gaertner, H., Schuhmann, R., Kraus, H., and Uebele-Kallhardt, B. (1977): Comparative cytogenetic and histologic studies on early malignant transformation in mesothelial tumors of the ovary. Hum. Genet. 35:281-297

Knuutila, S., Vuopio, P., Borgström, G.H. and Chapelle, A. de la (1980): Higher frequency of 5q- clone in bone marrow mitoses after culture than by a direct method. Scand. J. Haematol. 25:358-362

Knuutila, S., Vuopio, P., Elonen, E., Siimes, M., Kovanen, R., Borgström, G.H. and Chapelle, A. de la (1981): Culture of bone marrow reveals more cells with chromosomal abnormalities than the direct method in patients with hematologic disorders. Blood 58:369-375

Knuutila, S., Ruutu, T., Partanen, S. and Vuopio, P. (1983): Chromosome 1q+ in erythroid and granulocyte-monocyte precursors in a patient with essential thrombocythemia. Cancer Genet. Cytogenet. 9:245-249

Kohn, G., Manny, N., Eldor, A. and Cohen, M.M. (1975): De novo appearance of the Ph1 chromosome in a previously monosomic bone marrow (45,XX,-6): Conversion of a myeloproliferative disorder to acute myelogenous leukemia. Blood 45:653-657

Kohno, S-I. and Sandberg, A.A. (1980): Chromosomes and causation of human cancer and leukemia: XXXIX. Usual and unusual findings in Ph1-positive CML. Cancer 46:2227-2237

Kohno, S-I, Abe, S. and Sandberg, A.A. (1979a): The chromosomes and causation of human cancer and leukemia: XXXVIII. Cytogenetic experience in Ph1-negative chronic myelocytic leukemia (CML). Am. J. Hematol. 7:281-291

Kohno, S.-I., Van Den Berghe, H. and Sandberg, A.A. (1979b): Chromosomes and causation of human cancer and leukemia. XXXI. Dq- deletions and their significance in proliferative disorders. Cancer 43:1350-1357

Kolitz, J.E., Schulman, P., Kardon, N., Budman, D.R., Vinciguerra, V.P., Broekman, A. and Degnan, T.J. (1981): A complex variant Philadelphia (Ph1) chromosome translocation involving chromosomes No. 11, 14, and 22 in a case of chronic myelogenous leukemia. Cancer Genet. Cytogenet. 4:185-188

Kondo, K. and Sasaki, M. (1979): Cytogenetic studies in four cases of acute promyelocytic leukemia (APL). Cancer Genet. Cytogenet. 1:131-138

Kondo, K. and Sasaki, M. (1982): Further cytogenetic studies on acute promyelocytic leukemia. Cancer Genet. Cytogenet. 6:39-46

Kovacs, G. (1978a): Abnormalities of chromosome No. 1 in human solid malignant tumours. Int. J. Cancer 21:688-694

Kovacs, G. (1978b): Banding analysis of three primary cancers. Acta Cytol. 22:538-541

Kovacs, G. (1981): Preferential involvement of chromosome 1q in a primary breast carcinoma. Cancer Genet. Cytogenet. 3:125-129

Kristoffersson, U (1983): Personal communication.

Kristoffersson, U., Olsson, H., Mark-Vendel, E. and Mitelman, F. (1981): Fine needle aspiration biopsy: a useful tool in tumor cytogenetics with special reference to malignant lymphomas. Cancer Genet. Cytogenet. 4:53-60

Kross, J., Schulman, P., Kardon, N., Budman, D., Vinciguerra, V. and Degnan, T. (1981): Association of monosomy 7 with myelodysplasia following chemotherapy for Hodgkin's disease: serial observations. Cancer Genet. Cytogenet. 3:155-159

Kusnetsova, L.E., Prigogina, E.L., Pogosianz, H.E. and Belkina, B.M. (1982): Similar chromosomal abnormalities in several retinoblastomas. Hum. Genet. 61:201-204

Kusyk, C.J., Seski, J.C., Medlin, W.V. and Edwards, C.L. (1981): Progressive chromosome changes associated with different sites of one ovarian carcinoma. J. Natl. Cancer Inst. 66:1021-1025

Kusyk, C.J., Turpening, E.L., Edwards, C.L., Wharton, J.T. and Copeland, L.J. (1982): Karyotype analysis of four solid gynecologic tumors. Gynecol. Oncol. 14:324-338

Kwan, Y.-L., Sing, S., Vincent, P.C. and Gunz, F.W. (1977): Metamorphosis of chronic granulocytic leukaemia arising in an extramedullary site. Leukemia Res. 1:301-307

Köpf, I., Swolin, B. and Weinfeld, A. (1982): Cytogenetic studies on primary thrombocythaemia. Hereditas 97:217-220

Labal de Vinuesa, M., Slavutsky, I., Dupont, J., Bianchi de Di Risio, C. and Brieux de Salum, S. (1981): Ph1-negative chronic myelocytic leukemia (CML) with an unusual karyotype. Cancer Genet. Cytogenet. 3:347-351

Lai, J.L., Jouet, J.P., Bauters, F. and Deminatti, M. (1982): Anomalie chromosomique novelle par translocation t(20;22) au cours d'une leucémie myéloide chronique. Nouv. Presse Med. 11:3270

Lampert, F., Phebus, C.K., Huhn, D., Meyer, G. and Greifenegger, M. (1972): Leukemic xanthomatosis with a missing No. 9 chromosome. Z. Kinderheilk. 112:251-260

Larson, R.A., Sweet, D.L., Golomb, H.M., Testa, J.R. and Rowley, J.D. (1982): Response to 5-azacytidine in patients with refractory acute nonlymphocytic leukemia and association with chromosome findings. Cancer 49:2222-2225

Lawler, S.D., Lobb, D.S. and Wiltshaw, E. (1974): Philadelphia-chromosome positive bone-marrow cells showing loss of the Y in males with chronic myeloid leukaemia. Brit. J. Haematol. 27:247-252

Lawler, S.D., Secker-Walker, L.M., Summersgill, B.M., Reeves, B.R., Lewis, J., Kay, H.E.M. and Hardisty, R.M. (1975): Chromosome banding studies in acute leukaemia at diagnosis. Scand. J. Haematol. 15:312-320

Lawler, S.D., O'Malley, F. and Lobb, D.S. (1976): Chromosome banding studies in Philadelphia chromosome positive myeloid leukaemia. Scand. J. Haematol. 17:17-28

Lawler, S.D., Summersgill, B., Clink, H. MacD. and McElwain, T.J. (1980): Cytogenetic follow-up study of acute non-lymphocytic leukaemia. Brit. J. Haematol. 44:395-405

Lele, K.P., Filippa, D.A. and Chaganti, R.S.K. (1981): Cytogenetic studies of hairy cell leukemia. Cancer Genet. Cytogenet. 4:325-330

Lessard, M. and Le Prise, P.-Y. (1982): Cytogenetic studies in 56 cases with Ph1-positive hematologic disorders. Cancer Genet. Cytogenet. 5:37-49

Lessard, M. and Le Prise, P.-Y. (1983): Personal communication.

Lessard, M., Le Prise, P.-Y., Gandhour, C. and Richter, J.L. (1980): Etude cytogenetique de cinquante leucemies myeloides chroniques. Nouv. Rev. Fr. Hematol. Suppl. 22:88

Lessard, M., Duval, S. and Fritz, A. (1981): Unusual translocation and chronic myelocytic leukemia: "Masked" Philadelphia chromosome (Ph1). Cancer Genet. Cytogenet. 4:237-244

Levan, G. and Mitelman, F. (1975): Clustering of aberrations to specific chromosomes in human neoplasms. Hereditas 79:156-160

Levan, G. and Mitelman, F. (1977): Chromosomes and the etiology of cancer. In Chromosomes Today (Eds. A. de la Chapelle and M. Sorsa), Elsevier/North Holland Biomedical Press, Amsterdam, pp.363-371

Levan, G. and Mitelman, F. (1979): Absence of late-replicating X-chromosome in a female patient with acute myeloid leukemia and the 8;21 translocation. J. Natl. Cancer Inst. 62:273-275

Levitt, R., Pierre, R.V., White, W.L. and Siekert, R.G. (1978): Atypical lymphoid leukemia in ataxia telangiectasia. Blood 52:1003-1011

Li, Y.-S. and Hayhoe, F.G.J. (1982): Cytogenetic study in acute myeloid leukaemia using peripheral blood samples sent by post. J. Clin. Pathol. 35:861-865

Li, F.P., Hecht, F., Kaiser-McCaw, B., Baranko, P.V. and Upp Potter, N. (1981a): Ataxia-pancytopenia: syndrome of cerebellar ataxia, hypoplastic anemia, monosomy 7, and acute myelogenous leukemia. Cancer Genet. Cytogenet. 4:189-196

Li, Y.S., Khalid, G., Flemans, R.J. and Hayhoe, F.G.J. (1981b): A case of acute myelomonocytic leukaemia with Philadelphia and small F group chromosomes. Brit. J. Haematol. 48:175-176

Li, Y.S., Khalid, G. and Hayhoe, F.G.J. (1983): Correlation between chromosomal pattern, cytological subtypes, response to therapy, and survival in acute myeloid leukaemia. Scand. J. Haematol. 30:265-277

Liang, W. and Rowley, J.D. (1978): 14q+ marker chromosomes in multiple myeloma and plasma-cell leukaemia. Lancet (1):96

Liang, W., Hopper, J.E. and Rowley, J.D. (1979): Karyotypic abnormalities and clinical aspects of patients with multiple myeloma and related paraproteinemic disorders. Cancer 44:630-644

Liang, J.C., Gaulden, M.E. and Herndon, J.H. (1980): Chromosome markers and evidence for clone formation in lymphocytes of a patient with Sézary syndrome. Cancer Res. 40:3426-3429

Lilleyman, J.S., Potter, A.M., Watmore, A.E., Cooke, P., Sokol, R.J. and Wood, J.K. (1978): Myeloid karyotype and the malignant phase of chronic granulocytic leukaemia. Brit. J. Haematol. 39:317-323

Linch, D.C., Walker, H., Roberts, P., McKinnon, J., Goldstone, A.H. and Huehns, E.R. (1982): A chronic myeloproliferative disorder associated with monosomy 7 in the bone marrow cells; normal karyotype in acute transformation. Brit. J. Haematol. 51:439-444

Linder, D., Kaiser McCaw, B. and Hecht, F. (1975a): Parthenogenic origin of benign ovarian teratomas. New Engl. J. Med. 292:63-66

Linder, D., Hecht, F., Kaiser McCaw, B. and Campbell, J.R. (1975b): Origin of extragonadal teratomas and endodermal sinus tumours. Nature 254:597-598

Lindgren, V. and Rowley, J.D. (1977): Comparable complex rearrangements involving 8;21 and 9;22 translocations in leukaemia. Nature 266:744-745

Lindquist, R., Gahrton, G., Friberg, K. and Zech, L. (1978): Trisomy 8 in the chronic phase of Philadelphia negative chronic myelocytic leukaemia. Scand. J. Haematol. 21:109-114

Lobb, D.S., Reeves, B.R. and Lawler, S.D. (1972): Identification of isochromosome 17 in myeloid leukaemia. Lancet (1):849-850

Lundh, B., Mitelman, F., Nilsson, P.G., Stenstam, M. and Söderström, N. (1975): Chromosome abnormalities identified by banding technique in a patient with acute myeloid leukaemia complicating Hodgkin's disease. Scand. J. Haematol. 14:303-307

Lyall, J.M. and Garson, O.M. (1978): Non-random chromosome changes in the blastic transformation stage of Ph1-positive chronic granulocytic leukaemia. Leukemia Res. 2:213-222

Lyall, J.M., Brodie, G.N. and Garson, O.M. (1978): A variant chromosomal translocation found in a series of 24 patients with Philadelphia positive chronic granulocytic leukaemia. Aust. N.Z. J. Med. 8:288-289

Lönnqvist, B., Gahrton, G., Eriksson, P., Friberg, K. and Zech, L. (1979): Isochromosome 17 in a patient with a myeloproliferative disorder terminating in eosinophilic leukemia. Acta Med. Scand. 206:321-325

Löwenberg, B., Hagemeijer, A. and Swart, K. (1982): Karyotypically distinct subpopulations in acute leukemia with specific growth requirements. Blood 59:641-645

MacDougall, L.G., Brown, J.A., Cohen, M.M. and Judisch, J.M. (1974): C-monosomy myeloproliferative syndrome: a case of 7-monosomy. J. Pediat. 84:256-259

Mahmood, T., Robinson, W.A., Hamstra, R.D. and Wallner, S.F. (1979): Macrocytic anemia, thrombocytosis and nonlobulated megakaryocytes. The 5q- syndrome, a distinct entity. . Am. J. Med. 66:946-950

Mamaeva, S.E., Mamaev, N.N., Jartseva, N.M., Belyaeva, L.V. and Scherbakova, E.G. (1983): Complete or partial trisomy for the long arm of chromosome 1 in patients with various hematologic malignancies. Hum. Genet. 63:107-112

Mammon, Z., Grinblat, J. and Joshua, H. (1976): Philadelphia chromosome with t(6;22)(p25;q12). New Engl. J. Med. 294:827-828

Mandel, E.M., Shabtai, F., Gafter, U., Klein, B., Halbrecht, I. and Djaldetti, M. (1977): Ph1-positive acute lymphocytic leukemia with chromosome 7 abnormalities. Blood 49:281-287

Mann, B.D., Sparkes, R.S., Kern, D.H. and Morton, D.L. (1983): Chromosomal abnormalities of a mediastinal embryonal cell carcinoma in a patient with 47,XXY Klinefelter syndrome: Evidence for the premeiotic origin of a germ cell tumor. Cancer Genet. Cytogenet. 8:191-196

Manolov, G. and Manolova, Y. (1972): Marker band in one chromosome 14 from Burkitt lymphomas. Nature 237:33-34

Manolova, Y., Manolov, G., Apostolov, P. and Levan, A. (1979a): The same marker chromosome, mar17p+, in four consecutive cases of multiple myeloma. Hereditas 90:307-310

Manolova, Y., Manolov, G., Kieler, J., Levan, A. and Klein, G. (1979b): Genesis of the 14q+ marker in Burkitt's lymphoma. Hereditas 90:5-10

Marinello, M.J. and Levan, A. (1982): Ring-shaped double minutes in human acute myelocytic leukemia and in the murine SEWA sarcoma. A comparison. Hereditas 96:39-48

Marinello, M.J., Morita, M. and Sandberg, A.A. (1981): "Masked" Philadelphia chromosome (Ph1) due to an unusual translocation. Cancer Genet. Cytogenet. 3:227-232

Mark, J. (1973a): Karyotype patterns in human meningiomas. A comparison between studies with G- and Q-banding techniques. Hereditas 75:213-220

Mark, J. (1973b): Origin of the ring chromosome in a human recurrent meningioma studied with G-band technique. Acta Pathol. Microbiol. Scand. Sect. A 81:588-590

Mark, J. (1973c): The fluorescence karyotypes of three human meningiomas with hyper-diploid-hypotriploid stemlines. Acta Neuropathol. 25:46-53

Mark, J. (1975a): Histiocytic lymphomas with the marker chromosome 14q+. Hereditas 81:289-292

Mark, J. (1975b): Two pseudodiploid human breast carcinomas studied with G-band technique. Europ. J. Cancer 11:815-819

Mark, J. (1976): G-band analyses of a human intestinal leiomyosarcoma. Acta Pathol. Microbiol. Scand. Sect. A 84:538-540

Mark, J. (1977): On the specificity of medium-sized isomarker chromosomes in non-Burkitt lymphomas. Acta Pathol. Microbiol. Scand. Sect. A 85:557-558

Mark, J. (1978): Monosomy 14, monosomy 22 and 13q-. Three chromosomal abnormalities observed in cells of two malignant mesotheliomas studied by banding techniques. Acta Cytol. 22:398-401

Mark, J., Levan, G. and Mitelman, F. (1972a): Identification by fluorescence of the G chromosome lost in human meningomas. Hereditas 71:163-168

Mark, J., Mitelman, F. and Levan, G. (1972b): On the specificity of the G abnormality in human meningomas studied by the fluorescence technique. Acta Pathol. Microbiol. Scand. Sect. A 80:812-820

Mark, J., Ekedahl, C. and Arenander, E. (1976): Cytogenetical observations in two cases of clinicopathologically suspected malignant lymphomas. Hereditas 84:225-230

Mark, J., Ekedahl, C. and Hagman, A. (1977): Origin of the translocated segment of the 14q+ marker in non-Burkitt lymphomas. Hum. Genet. 36:277-282

Mark, J., Ekedahl, C. and Dahlenfors, R. (1978): Characteristics of the banding patterns in non-Hodgkin and non-Burkitt lymphomas. Hereditas 88:229-242

Mark, J., Dahlenfors, R. and Ekedahl, C. (1979): Recurrent chromosomal aberrations in non-Hodgkin and non-Burkitt lymphomas. Cancer Genet. Cytogenet. 1:39-56

Mark, J., Dahlenfors, R., Ekedahl, C. and Stenman, G. (1980): The mixed salivary gland tumor - a normally benign human neoplasm frequently showing specific chromosomal abnormalities. Cancer Genet. Cytogenet. 2:231-241

Mark, J., Dahlenfors, R. and Ekedahl, C. (1981a): Chromosomal deviations and their specificity in human mixed salivary gland tumours. Anticancer Res. 1:49-57

Mark, J., Ekedahl, C. and Dahlenfors, R. (1981b): Polyclonal chromosomal evolution in a cultured human acinic cell tumour. Anticancer Res. 1:45-48

Mark, J., Dahlenfors, R., Ekedahl, C. and Stenman, G. (1982a): Chromosomal patterns in a benign human neoplasm, the mixed salivary gland tumour. Hereditas 96:141-148

Mark, J., Dahlenfors, R. and Ekedahl, C. (1982b): On double-minutes and their origin in a benign human neoplasm, a mixed salivary gland tumour. Anticancer Res. 2:261-264

Marsh, W.L., Chaganti, R.S.K., Gardner, F.H., Mayer, K., Nowell, P.C. and German, J. (1974): Mapping human autosomes: Evidence supporting assignment of Rhesus to the short arm of chromosomes No. 1. Science 184:966-968

Martin, P., Levin, B., Golomb, H.M. and Riddell, R.H. (1979): Chromosome analysis of primary large bowel tumors. A new method for improving the yield of analyzable metaphases. Cancer 44:1656-1664

Martin, P.J., Najfeld, V., Hansen, J.A., Penfold, G.K., Jacobson, R.J. and Fialkow, P.J. (1980): Involvement of the B-lymphoid system in chronic myelogenous leukaemia. Nature 287:49-50

Matsunaga, M., Sadamori, N., Tomanaga, Y., Tagawa, M. and Ichimaru, M. (1976): Chronic myelogenous leukemia with an unusual karyotype: 46,XY,t(17q+;22q-). New Engl. J. Med. 295:1537

Mazur, E.M., Lovett, D.H., Enriquez, R.E., Breg, W.R. and Papac, R.J. (1979): Angioimmunoblastic lymphadenopathy.Evolution to a Burkitt-like lymphoma. Am. J. Med. 67:317-324

McCulloch, P.B., Dent, P.B., Hayes, P.R. and Liao, S.-K. (1976): Common and individually specific chromosomal charactertistics of cultured human melanoma. Cancer Res. 36:398-404

McDermott, A., Romain, D., Fraser, I.D. and Scott, G.L. (1978): Isochromosome 17q in two cases of acute blast transformation in myeloproliferative disorders. Hum. Genet. 45:215-218

McGlave, P.B., Kim, T.H., Hurd, D.D., Arthur, D.C., Ramsay, N.K.C. and Kersey, J. (1982): Successful allogeneic bone-marrow transplantation for patients in the accelerated phase of chronic granulocytic leukaemia. Lancet (2):625-627

McKenna, R.W., Parkin, J., Bloomfield, C.D., Sundberg, R.D. and Brunning, R.D. (1982): Acute promyelocytic leukaemia: a study of 39 cases with identification of a hyperbasophilic microgranular variant. Brit. J. Haematol. 50:201-214

Mecucci, C., Donti, E., Bocchini, V., Tabilio, A. and Martelli, M.F. (1982): 5q-syndrome in a patient with chronic exposure to ionizing radiation. Cancer Genet. Cytogenet. 5:75-80

Michalski, K.A., Miles, J.H. and Perry, M.C. (1982): Unusual Ph1 translocation in a preleukemia. Cancer Genet. Cytogenet. 6:89-90

Michalski, K., Meyers, L., Hakami, N., Miles, J. and Germain, C. (1983): The translocation 9;11 in acute monoblastic leukemia. Cancer Genet. Cytogenet. 9:307-308

Minowada, J., Oshimura, M., Tsubota, T., Higby, D.J. and Sandberg, A.A. (1977): Cytogenetic and immunoglobulin markers of human leukemic B-cell lines. Cancer Res. 37:3096-3099

Mintz, U., Vardiman, J., Golomb, H.M. and Rowley, J.D. (1979): Evolution of karyotypes in Philadelphia (Ph1) chromosome-negative chronic myelogenous leukemia. Cancer 43:411-416

Misawa, S. (1978): Cytogenetic studies of chronic myelocytic leukemia. J. Kyoto Pref. Univ. Med. 87:807-831

Mitelman, F. (1974): Heterogeneity of Ph1 in chronic myeloid leukaemia. Hereditas 76:315-316

Mitelman, F. (1975): Comparative cytogenetic studies of bone marrow and extramedullary tissues in chronic myeloid leukemia. Ser. Haematol. 8:113-117

Mitelman, F. (1980a): Tumor etiology and chromosome pattern - evidence from human and experimental neoplasms. In Genes, Chromosomes and Neoplasia (Eds. F.E. Arrighi, P.N. Rau and E. Stubblefield) Raven Press, New York, pp. 335-350

Mitelman, F. (1980b): Cytogenetics of experimental neoplasms and non-random chromosome correlations in man. Clin. Haematol. 9:195-219

Mitelman, F. (1981): Marker chromosome 14q+ in human cancer and leukemia. Adv. Cancer Res. 34:141-170

Mitelman, F. (1983): Unpublished observations.

Mitelman, F. and Brandt, L. (1974): Chromosome banding pattern in acute myeloid leukaemia. Scand. J. Haematol. 13:321-330

Mitelman, F. and Levan, G. (1976a): Clustering of aberrations to specific chromosomes in human neoplasms. II. A survey of 287 neoplasms. Hereditas 82:167-174

Mitelman, F. and Levan, G. (1976b): Do only a few chromosomes carry genes of prime importance for malignant transformation?. Lancet (2):264

Mitelman, F. and Levan, G. (1978): Clustering of aberrations to specific chromosomes in human neoplasms. III. Incidence and geographic distribution of chromosome aberrations in 856 cases. Hereditas 89:207-232

Mitelman, F. and Levan, G. (1979): Chromosomes in neoplasia: an appeal for unpublished data. Cancer Genet. Cytogenet. 1:29-32

Mitelman, F. and Levan, G. (1981): Clustering of aberrations to specific chromosomes in human neoplasms. IV. A survey of 1 871 cases. Hereditas 95:79-139

Mitelman, F., Brandt, L. and Levan, G. (1973): Identification of isochromosome 17 in acute myeloid leukaemia. Lancet (2):972

Mitelman, F., Brandt, L. and Nilsson, P.G. (1974a): Cytogenetic evidence for splenic origin of blastic transformation in chronic myeloid leukaemia. Scand. J. Haematol. 13:87-92

Mitelman, F., Mark, J., Nilsson, P.G., Dencker, H., Norryd, C. and Tranberg, K.-G. (1974b): Chromosome banding pattern in human colonic polyps. Hereditas 78:63-68

Mitelman, F., Brandt, L. and Nilsson, P.G. (1974c): The banding pattern in Philadelphia-chromosome negative chronic myeloid leukemia. Hereditas 78:302-304

Mitelman, F., Nilsson, P.G. and Brandt, L. (1975a): Abnormal clones resembling those seen in blast crisis arising in the spleen in chronic myelocytic leukemia. J. Natl. Cancer Inst. 54:1319-1321

Mitelman, F., Panani, A. and Brandt, L. (1975b): Isochromosome 17 in a case of eosinophilic leukaemia. An abnormality common to eosinophilic and neutrophilic cells. Scand. J. Haematol. 14:308-312

Mitelman, F., Levan, G. and Brandt, L. (1975c): Highly malignant cells with normal karyotype in G-banding. Hereditas 80:291-293

Mitelman, F., Levan, G., Nilsson, P.G. and Brandt, L. (1976a): Non-random karyotypic evolution in chronic myeloid leukemia. Int. J. Cancer 18:24-30

Mitelman, F., Nilsson, P.G., Levan, G. and Brandt, L. (1976b): Non-random chromosome changes in acute myeloid leukemia. Chromosome banding examination of 30 cases at diagnosis. Int. J. Cancer 18:31-38

Mitelman, F., Brandt, L. and Nilsson, P.G. (1978): Relation among occupational exposure to potential mutagenic/carcinogenic agents, clinical findings, and bone marrow chromosomes in acute nonlymphocytic leukemia. Blood 52:1229-1237

Mitelman, F., Klein, G., Andersson-Anvret, M., Forsby, N. and Johansson, B. (1979a): 14q+ marker chromosome in an EBV-genome-negative lymph node without signs of malignancy in a patient with EBV-genome-positive nasopharyngeal carcinoma. Int. J. Cancer 23:32-36

Mitelman, F., Andersson-Anvret, M., Brandt, L., Catovsky, D., Klein, G., Manolov, G., Manolova, Y., Mark-Vendel, E. and Nilsson, P.G. (1979b): Reciprocal 8;14 translocation in EBV-negative B-cell acute lymphocytic leukemia with Burkitt-type cells. Int. J. Cancer 24:27-33

Mitelman, F., Nilsson, P.G., Brandt, L., Alimena, G., Montuoro, A. and Dallapiccola, B. (1979c): Chromosomes, leukaemia, and occupational exposure to leukaemogenic agents. Lancet (2):1195-1196

Mitelman, F., Nilsson, P.G., Brandt, L., Alimena, G., Gastaldi, R. and Dallapiccola, B. (1981): Chromosome pattern, occupation, and clinical features in patients with acute non-lymphocytic leukemia. Cancer Genet. Cytogenet. 4:197-214

Miyamoto, K. (1980): Chromosome abnormalities in patients with chronic myelocytic leukemia. Acta Med. Okayama 34:367-382

Miyamoto, K., Miyano, K., Miyoshi, I., Hamasaki, K., Nishihara, R., Terao, S., Kimura, I., Maeda, K., Matsumura, K., Nishijima, K. and Tanaka, T. (1980a): Chromosome 14q+ in a Japanese patient with Burkitt's lymphoma. Acta Med. Okayama 34:61-65

Miyamoto, K., Sato, J., Miyoshi, I., Nishihara, R., Terao, S., Hara, M. and Kimura, I. (1980b): 8-14 translocation in a Japanese Burkitt's lymphoma. Acta Med. Okayama 34:139-142

Miyamoto, K., Hayashi, K., Tsubota, T. and Tanaka, T. (1981a): 14q12 translocation in a non-Burkitt lymphoma. Acta Med. Okayama 35:285-287

Miyamoto, K., Hamasaki, K., Kitajima, K., Adachi, T., Tanaka, T. and Sato, J. (1981b): Abnormalities of chromosome no. 1 related to blood dyscrasias: study of 10 cases. Acta Med. Okayama 35:137-141

Miyamoto, K., Sato, J., Kitajima, K., Hiraki, S., Mori, K. and Tanaka, T. (1982a): Chromosome 8-14 translocation in a non-african Burkitt's lymphoma with leukemic conversion. Acta Med. Okayama 36:157-160

Miyamoto, K., Kitajima, K., Suemaru, S. and Tanaka, T. (1982b): Karyotypic findings and prognosis of adult T-cell leukemia patients. Gann 73:854-856

Miyoshi, I., Kubonishi, I., Uchida, H., Hiraki, S., Toki, H., Tanaka, T., Masuji, H. and Hiraki, K. (1976): Direct implantation of Ph1 chromosome-positive myeloblasts into newborn hamsters. Blood 47:355-361

Miyoshi, I., Sumita, M., Sano, K., Nishihara, R., Miyamoto, K., Kimura, I. and Sato, J. (1979a): Marker chromosome 14q+ in adult T-cell leukemia. New Engl. J. Med. 300:921

Miyoshi, I., Hiraki, S., Kimura, I., Miyamoto, K. and Sato, J. (1979b): 2/8 translocation in a Japanese Burkitt's lymphoma. Experientia 35:742-743

Miyoshi, I., Miyamoto, K., Sumida, M., Nishihara, R., Lai, M., Yoshimoto, S., Sato, J. and Kimura, I. (1981a): Chromosome 14q+ in adult T-cell leukemia. Cancer Genet. Cytogenet. 3:251-259

Miyoshi, I., Hamazaki, K., Kubonishi, I., Yoshimoto, S., Kitajima, K., Kimura, I., Miyamoto, K., Sato, J., Yorimitsu, S., Tao, S., Ishibashi, K. and Tokuda, M. (1981b): A variant translocation (8;22) in a Japanese patient with Burkitt lymphoma. Gann 72:176-177

Miyoshi, I., Hamasaki, K., Miyamoto, K., Nagase, K., Narahara, K., Kitajima, K., Kimura, I. and Sato, J. (1981c): Chromosome translocations in Burkitt's lymphoma. New Engl. J. Med. 304:734

Mohandas, T., Anderson, C. and Okun, D. (1980): Chronic myelogenous leukemia with a complex translocation. Cancer Genet. Cytogenet. 2:19-21

Morita, M., Minowada, J. and Sandberg, A.A. (1981): Chromosomes and causation of human cancer and leukemia. XLV. Chromosome patterns in stimulted lymphocytes of chronic lymphocytic leukemia. Cancer Genet. Cytogenet. 3:293-306

Morse, H.G., Humbert, J.R., Hutter, J.J. and Robinson, A. (1977): Karyotyping of bone-marrow cells in hematologic disaeses. Hum. Genet. 37:33-39

Morse, H.G., Ducore, J.M., Hays, T., Peakman, D. and Robinson, A. (1978): Multiple leukemic clones in acute leukemia of childhood. Hum. Genet. 40:269-278

Morse, H., Hays, T., Rose, B. and Robinson, A. (1979a): Chromosome 1 abnormalities in relapse and terminal stages in childhood leukemia. Med. Ped. Oncol. 7:9-16

Morse, H.G., Hays, T. and Odom, L.F. (1979b): Unusual cytogenetics in a case of acute lymphoblastic leukemia. Med. Ped. Oncol. 7:257-262

Morse, H.G., Hays, T., Patterson, D., Robinson, A. (1982a): Giemsa-11 technique. Applications in the chromosomal characterization of hematologic specimens. Hum. Genet. 61:141-144

Morse, H.G., Heideman, R., Hays, T. and Robinson, A. (1982b): 4;11 translocation in acute lymphoblastic leukemia: a specific syndrome. Cancer Genet. Cytogenet. 7:165-172

Mufti, G.J., Hamblin, T.J. and Seabright, M. (1982): Acute transformation of a myeloproliferative state in sideroblastic anaemia with abnormal karyotype. J. Med. Genet. 19:478

Muir, P.D., Occomore, M.A., Thornley, B., Singh, S. and Gunz, F.W. (1977): The value of chromosome banding methods in the study of adult acute leukaemia. Pathology 9:323-330

Muldal, S., Mir, M.A., Freeman, C.B. and Geary, C.G. (1975): A new translocation associated with the Ph1 chromosome and an acute course of chronic granulocytic leukaemia. Brit. J. Cancer 31:364-368

Musilova, J., Michalova, K., Dvorak, O., Slavik, S., Mericka, O. and Novotna, J. (1981): Cytogenetic study of malignant and benign effusions. Neoplasma 28:463-471

Nacheva, E., Fischer, P., Haas, O., Manolova, Y., Manolov, G. and Levan, A. (1982): Acute myelogenous leukemia in a child with primary involvement of chromosomes 11 and X. Hereditas 97:273-288

Nagao, K., Yonemitsu, H., Yamaguchi, K. and Okuda, K. (1977): A case of acute myeloblastic leukemia with Ph1 chromosome showing translocation 9q+;22q-. Blood 50:259-262

Najfeld, V. (1976): Isochromosome 17 in a case of chronic erythroleukaemia. Scand. J. Haematol. 17:101-104

Najfeld, V. (1983): Personal communication.

Najfeld, V., Singer, J.V., James, M.C. and Fialkow, P.J. (1978a): Trisomy of 1q in preleukaemia with progression to acute leukaemia. Scand. J. Haematol. 21:24-28

Najfeld, V., Price, T.H., Adamson, J.W. and Fialkow, P.J. (1978b): Myelofibrosis with complex chromosome abnormality in a patient with erythrocytosis due to hemoglobin Rainier and treated with 32P. Am. J. Hematol. 5:63-69

Najfeld, V., Fialkow, P.J., Karande, A., Nilsson, K., Klein, G. and Penfold, G. (1980): Chromosome analyses of lymphoid cell lines derived from patients with chronic lymphocytic leukemia. Int. J. Cancer 26:543-549

Najfeld, V., Thorning, D., Doney, K.C. and Fialkow, P.J. (1981): Acquired trisomy 12 and absent Y chromosome in a patient with acute undifferentiated leukaemia. Scand. J. Haematol. 26:130-136

Najfeld, V., Tobe, R. and Fialkow, P.J. (1983): "Masked" Ph1 chromosome in a complex three-way translocation. Cancer Genet. Cytogenet. 8:19-26

Nicoara, S., Mutoianu, E. and Gingold, N. (1977): Ph1-positive polycythemia vera. Rev. Roum. Med. - Med. Int. 15:57-62

Nilsson, P.G., Brandt, L. and Mitelman, F. (1977a): Prognostic implications of chromosome analysis in acute non-lymphocytic leukemia. Leukemia Res. 1:31-34

Nilsson, P.G., Brandt, L. and Mitelman, F. (1977b): Relation between age and chromosomal aberrations at diagnosis of adult non-lymphocytic leukemia. Leukemia Res. 1:385-386

Nora, A., Heideman, R., Peakman, D. and Morse, H. (1979): Cytogenetic studies in an acute leukemia following cerebellar astrocytoma. Hum. Genet. 50:157-161

Nordenson, I. (1983): Personal communication.

Nordenson, I., Lenner, P. and Roos, G. (1983): Patient with B-cell neoplasia (immunoblastic sarcoma) and the Philadelphia chromosome. Cancer Genet. Cytogenet. 9:37-44

Norman, C.S. and Boucher, B.J. (1978): Atypical chronic myelogenous leukemia with Philadelphia (Ph1) chromosome and an additional translocation. Cancer 41:1123-1127

Norrby, A., Ridell, B., Swolin, B. and Westin, J. (1982): Rearrangement of chromosome No. 3 in a case of preleukemia with thrombocytosis. Cancer Genet. Cytogenet. 5:257-26

Nowell, P.C. (1983): Personal communication.

Nowell, P.C. and Finan, J.B. (1977): Isochromosome 17 in atypical myeloproliferative and lymphoproliferative disorders. J. Natl. Cancer Inst. 59:329-333

Nowell, P.C. and Finan, J. (1978): Chromosome studies in preleukemic states. IV. Myeloproliferative versus cytopenic disorders. Cancer 42:2254-2261

Nowell, P.C., Jensen, J. and Gardner, F. (1975): Two complex translocations in chronic granulocytic leukemia involving chromosomes 22, 9 and a third chromosome. Humangenetik 30:13-21

Nowell, P.C., Jensen, J., Winger, L., Danilele, R. and Growney, P. (1976a): T cell variant of chronic lymphocytic leukaemia with chromosome abnormality and defective response to mitogens. Brit. J. Haematol. 33:459-468

Nowell, P.C., Jensen, J., Gardner, F., Murphy, S., Chaganti, R.S.K. and German, J. (1976b): Chromosome studies in "Preleukemia". III. Myelofibrosis. Cancer 38:1873-1881

Nowell, P.C., Rowlands, Jr, D.T., Daniele, R.P., Berger, B.M. and Dupont, G. (1979): Changes in membrane markers and chromosome patterns in chronic T-cell leukemia. Clin. Immunol. Immunopath. 12:323-330

Nowell, P.C., Shankey, T.V., Finan, J., Guerry, D. and Besa, E. (1981): Proliferation, differentiation, and cytogenetics of chronic leukemic B lymphocytes cultured with mitomycin-treated normal cells. Blood 57:444-451

Nowell, P.C., Finan, J.B. and Vonderheid, E.C. (1982): Clonal characteristics of cutaneous T cell lymphomas: cytogenetic evidence from blood, lymph nodes, and skin. J. Invest. Dermatol. 78:69-75

Okada, M., Miyazaki, T. and Kumota, K. (1977): 15/17 translocation in acute promyelocytic leukaemia. Lancet (1):961

Okada, M., Katahira, J. and Mizoguchi, H. (1982): Missing Y clones in a patient with lymphoblastic leukemia and paroxysmal nocturnal hemoglobinuria. Cancer Genet. Cytogenet. 7:307-311

Olah, É. and Rak, K. (1981): Prognostic value of chromosomal findings in the blast phase of Ph1-positive chronic myeloid leukaemia (CML). Int. J. Cancer 27:287-295

Olah, É., Kiss, A. and Jako, J. (1980): Chromosome abnormalities, clinical and morphological manifestations in metamorphosis of chronic myeloid leukemia. Int. J. Cancer 26:37-45

Olah, É., Stenszky, V., Kiss, A., Kovacs, I., Balogh, E. and Karmazsin, L. (1981a): Familial leukemia: Ph1 positive acute lymphoid leukemia of a mother and her infant. Blut 43:265-272

Olah, É., Kiss, A., Balogh, E. and Rak, K. (1981b): Cytogenetic observations in chronic myeloid leukemia (CML).Correlations of the cytogenetic findings with morphological cytochemical features of blast cells in various types of blastic phase. Neoplasma 28:325-332

Olinici, C.D., Marinca, E., Macavei, I. and Dobay, O. (1978a): Missing X chromosome and ring chromosome 21 in a case of acute myelomonocytic leukemia. Arch. Geschwulstforsch. 48:202-204

Olinici, C.D., Petrov, L., Macavei, I. and Dobay, O. (1978b): Different cell clones in bone marrow and spleen of a patient with chronic myelocytic leukemia (CML) in blastic phase. Cancer 42:2707-2709

Ondreyco, S.M., Kjeldsberg, C.R., Fineman, R.M., Vaninetti, S. and Kushner, J.P. (1981): Monoblastic transformation in chronic myelogenous leukemia: presentation with massive hepatic involvement. Cancer 48:957-963

Oshimura, M. and Sandberg, A.A. (1975): Isochromosome 17 in prostatic cancer. J. Urol. 114:249-250

Oshimura, M. and Sandberg, A.A. (1977): Chromosomes and causation of human cancer and leukemia. XXV. Significance of the Ph1 (including unusual translocations) in various acute leukemias. Cancer 40:1149-1160

Oshimura, M., Sonta, S. and Sandberg, A.A. (1976a): Trisomy of the long arm of chromosome No. 1 in human leukemia. J. Natl. Cancer Inst. 56:183-184

Oshimura, M., Hayata, I., Kakati, S. and Sandberg, A.A. (1976b): Chromosomes and causation of human cancer and leukemia. XVII. Banding studies in acute myeloblastic leukemia (AML). Cancer 38:748-761

Oshimura, M., Freeman, A.I. and Sandberg, A.A. (1977a): Chromosomes and causation of human cancer and leukemia. XXVI. Banding studies in acute lymphoblastic leukemia (ALL). Cancer 40:1161-1172

Oshimura, M., Freeman, A.I. and Sandberg, A.A. (1977b): Chromosomes and causation of human cancer and leukemia XXIII. Near-haploidy in acute leukemia. Cancer 40:1143-1148

Oshimura, M., Ohyashiki, K., Vehara, M., Miyasaka, Y., Osamura, S. and Tonomura, A. (1981a): Chronic myelogenous leukemia with translocations (3q-;9q+) and (17q-;22q+). Possible crucial cytogenetic events in the genesis of CML. Hum. Genet. 57:48-51

Oshimura, M., Ohyashiki, K., Tonomura, A. and Terada, H. (1981b): A 14q+ chromosome in a malignant lymphoma in a patient with Down's syndrome. Cancer Genet. Cytogenet. 4:245-250

Oshimura, M., Ohyashiki, K., Terada, H., Takaku, F. and Tonomura, A. (1982a): Variant Ph1 translocations in CML and their incidence, including two cases with sequential lymphoid and myeloid crises. Cancer Genet. Cytogenet. 5:187-201

Oshimura, M., Ohyashiki, K., Mori, M., Terada, H. and Takaku, F. (1982b): Cytogenetic and hematologic findings in acute myelogenous leukemia, M2 according to the FAB classification. Gann 73:212-216

Padre-Mendoza, T., Forman, E.N., Farnes, P., Barker, B.E. and Smith, P.S. (1978): Short Y chromosome and Ph1 chromosome in acute monomyelocytic leukaemia. Lancet (1):667

Page, B.M., Watt, J.L., Reid, I.N., Davidson, R.J.L. and Walker, W. (1979): Clonal evolution of marker chromosomes in a case of myelofibrosis with myeloid metaplasia and myeloblastic transformation. Acta Haematol. 61:301-309

Panani, A., Papayannis, A.G., Kyrkou, K. and Gardikas, C. (1977): Cytogenetic studies in preleukaemia using the G-banding staining technique. Scand. J. Haematol. 18:301-308

Panani, A., Papayannis, A.G. and Gardikas, C. (1979): Chronic myelogenous leukemia with an unusual karyotype. Hum. Genet. 48:109-111

Panani, A., Papayannis, A.G. and Sioula, E. (1980): Chromosome aberrations and prognosis in preleukaemia. Scand. J. Haematol. 24:97-100

Papa, G., Alimena, G., Annino, L., Anselmo, A.P., Ciccone, F., De Luca, A.M., Granati, L., Petti, N. and Mandelli, F. (1979): Acute nonlymphoid leukaemia following Hodgkin's disease. Scand. J. Haematol. 23:339-347

Parkin, J.L., Arthur, D.C., Abramson, C.S., McKenna, R.W., Kersey, J.H., Heideman, R.L. and Brunning, R.D. (1982): Acute leukemia associated with the t(4;11) chromosome rearrangement: Ultrastructural and immunologic characteristics. Blood 60:1321-1331

Pasquali, F. (1983): Personal communication.

Pasquali, F. and Casalone, R. (1981): Rearrangement of three chromosomes (Nos. 2, 8 and 21) in acute myeloblastic leukemia. Evidence for more than one specific event. Cancer Genet. Cytogenet. 3:335-339

Pasquali, F., Francesconi, D., Casalone, R. and Ippoliti, G. (1979a): Partial trisomy 1 due to 1/17 translocation in Ph1-positive chronic myelocytic leukemia. Hum. Genet. 49:277-282

Pasquali, F., Casalone, R., Francesconi, D., Peretti, D., Fraccaro, M., Bernasconi, C. and Lazzarino, M. (1979b): Transposition of 9q34 and 22 (q11-qter) regions has a specific role in chronic myelocytic leukemia. Hum. Genet. 52:55-67

Pasquali, F., Bernasconi, P., Casalone, R., Fraccaro, M., Bernasconi, C., Lazzarino, M., Morra, E., Alessandrino, E.P., Marchi, M.A. and Sanger, R. (1982a): Pathogenetic significance of "pure" monosomy 7 in myeloproliferative disorders. Analysis of 14 cases. Hum. Genet. 62:40-51

Pasquali, F., Panarello, C., Bernasconi, P. and Casalone, R. (1982b): The isochromosome (17q) in chronic myelocytic leukaemia: mechanism of origin, centromeric function and clonal evolution. Hum. Genet. 62:89-90

Pathak, S., Siciliano, M.J., Cailleau, R., Wiseman, C.L. and Hsu, T.C. (1979): A human breast adenocarcinoma with chromosome and isoenzyme markers similar to those of the HeLa line. J. Natl. Cancer Inst. 62:263-271

Pathak, S., Strong, L.C., Ferrell, R.E. and Trindade, A. (1982): Familial renal cell carcinoma with a 3;11 chromosome translocation limited to tumor cells. Science 217:939-941

Pearson, J., Ilgren, E.B. and Spriggs, A.I. (1982): Lymphoma cells in cerebrospinal fluid confirmed by chromosome analysis. J. Clin. Pathol. 35:1307-1311

Pedersen, B. and Boesen, A.M. (1983): Extreme hypodiploidy in a case of myelomonocytic crisis of chronic myelogenous leukemia. Cancer Genet. Cytogenet. 9:101-112

Pedersen-Bjergaard, J., Haahr, S., Philip, P., Thomsen, M., Jensen, G., Ersböll, J. and Nissen, N.I. (1980): Abolished production of interferon by leucocytes of patients with the acquired cytogenetic abnormalities 5q- or -5 in secondary and de-novo acute nonlymphocytic leukaemia. Brit. J. Haematol. 46:211-223

Pedersen-Bjergaard, J., Philip, P., Thing Mortensen, B., Ersböll, J., Jensen, G., Panduro, J. and Thomsen, M. (1981): Acute nonlymphocytic leukemia, preleukemia, and acute myeloproliferative syndrome secondary to treatment of other malignant diseases. Clinical and cytogenetic characteristics and results of in vitro culture of bone marrow and HLA typing. Blood 57:712-723

Pedersen-Bjergaard, J., Vindelöv, L., Philip, P., Ruutu, P., Elmgreen, J., Repo, H., Christensen, I.J., Killmann, S.A. and Jensen, G. (1982): Varying involvement of peripheral granulocytes in the clonal abnormality -7 in bone marrow cells in preleukemia secondary to treatment of other malignant tumors:cytogenetic results compared with results of cytometric DNA analysis and neutrophil chemotaxis. Blood 60:172-179

Penchansky, L., Wollman, M., Whiteside, T. and Malatack, J. (1981): Philadelphia chromosome - positive pre-B-lymphocytic leukemia in a child. Am. J. Pediat. Hematol. Oncol. 3:265-268

Petit, P. (1983): Personal communication.

Petit, P. and Van Den Berghe, H. (1979a): A chromosomal abnormality (21q-) in primary thrombocytosis. Hum. Genet. 50:105-106

Petit, P. and Van Den Berghe, H. (1979b): The 5q- and additional chromosome anomalies in two patients with acute myeloid leukemia. Ann. Genet. 22:103-105

Petit, P. and Van Den Berghe, H. (1981): Particular secondary chromosome changes in chronic leukemia t(8:17): report of two cases. Ann. Génét. 24:25-27

Petit, P., Alexander, M. and Fondu, P. (1973): Monosomy 7 in erythroleukaemia. Lancet (2):1326

Petit, P., Gottlob, R. and Germeau, A. (1978): Report of an atypical Ph1 translocation in CML and 5q- in two AML cases. Clin. Genet. 14:307

Petit, P., Fryns, J.P., Masure, R. and Van Den Berghe, H. (1982): Isodicentric (X)(q13): a new characteristic chromosomal anomaly in myeloproliferative syndrome? Cancer Genet. Cytogenet. 7:339-341

Philip, P. (1975a): Trisomy 11 in acute phase of chronic myeloid leukemia. Acta Haematol. 54:188-191

Philip, P. (1975b): Trisomy 8 in acute myeloid leukaemia. Scand. J. Haematol. 14:140-147

Philip, P. (1975c): Marker chromosome 14q+ in multiple myeloma. Hereditas 80:155-156

Philip, P. (1976): G banding analysis of complex aneuploidy in a case of erythroleukaemia. Scand. J. Haematol. 16:365-368

Philip, P. and Drivsholm, A. (1976): G-banding analysis of complex aneuploidy in multiple myeloma bone marrow cells. Blood 47:69-77

Philip, P., Muller-Berat, N. and Killmann, S.-A. (1976): Philadelphia chromosome in acute lymphocytic leukaemia. Hereditas 84:231-232

Philip, P., Lange Wantzin, G., Krogh Jensen, M. and Drivsholm, A. (1977a): Trisomy 8 in acute myeloid leukaemia: A non-random event. Scand. J. Haematol. 18:163-169

Philip, P., Krogh Jensen, M. and Pallesen, G. (1977b): Marker chromosome 14q+ in non-endemic Burkitt's lymphoma. Cancer 39:1495-1499

Philip, P., Krogh Jensen, M., Killmann, S.-A., Drivsholm, A. and Hansen, N.E. (1978a): Chromosomal banding patterns in 88 cases of acute nonlymphocytic leukemia. Leukemia Res. 2:201-212

Philip, P., Slater, R.M. and Sandberg, A.A. (1978b): Burkitt type 14+ marker chromosome in B-cell type acute lymphocytic leukaemia. Hereditas 89:268

Philip, P., Drivsholm, A., Hansen, N.E., Jensen, M.K. and Killmann, S-A. (1980): Chromosomes and survival in multiple myeloma. A banding study of 25 cases. Cancer Genet. Cytogenet. 2:243-257

Philip, T., Lenoir, G.M., Fraisse, J., Philip, I., Bertoglio, J., Ladjaj, S., Bertrand, S. and Brunat-Mentigny, M. (1981): EBV-positive Burkitt's lymphoma from Algeria, with a three-way rearrangement involving chromosomes 2, 8 and 9. Int. J. Cancer 28:417-420

Pickthall, V.J. (1976): Detailed cytogenetic study of a metastatic bronchial carcinoma. Brit. J. Cancer 34:272-278

Pittman, S., Catovsky, D. and Morilla, R. (1979): Unusual translocations in Ph1-positive acute lymphoblastic leukaemia. Lancet (1):986

Pittman, S., Morilla, R. and Catovsky, D. (1982): Chronic T-cell leukemias. II. Cytogenetic studies. Leukemia Res. 6:33-42

Potter, A.M., Sharp, J.C., Brown, M.J. and Sokol, R.J. (1975): Structural rearrangements associated with the Ph1 chromosome in chronic granulocytic leukaemia. Humangenetik 29:223-228

Potter, A.M., Watmore, A.E., Cooke, P., Lilleyman, J.S. and Sokol, R.J. (1981): Significance of non-standard Philadelphia chromosomes in chronic granulocytic leukaemia. Brit. J. Cancer 44:51-54

Pravtcheva, D., Andreeva, P. and Tsaneva, R. (1976): A new translocation in chronic myelogenous leukemia. Hum. Genet. 32:229-232

Priest, J.R., Robison, L.L., McKenna, R.W., Lindquist, L.L., Warkentin, P.I., LeBien, T.W., Woods, W.G., Kersey, J.H., Coccia, P.F. and Nesbit, M.E. (1980): Philadelphia chromosome positive childhood acute lymphoblastic leukemia. Blood 56:15-22

Prieto, F., Badia, L., Calabuig, J.R., Mayans, J., Perez-Sirvent, M.L., Martinez, J., Luno, E. and Marty, M.L. (1976): Alteraciones cromosomicas en las anemias refractarias con mieloblastosis parcial. Sangre 21:701-712

Prieto, F., Badia, L., Herranz, C., Redon, J., Caballero, M., Artiges, E. and Baguena, J. (1978a): Chromosoma marcador (14q+) y linfomas. Rev. Esp. Oncol. 25:399-408

Prieto, F., Badia, L., Mayans, J., Besalduch, J. and Marty, M.L. (1978b): Trisomia de los brazos largos del cromosoma 1 e inversion del cromosoma 9 en celulas leucemicas. Sangre 23:64-68

Prieto, F., Badia, L., Castell, V., Perez-Sirvent, M.L. and Marty, M.L. (1981): Citogenética de las leucemias agudas. Sangre 26:738-755

Prigogina, E.L. and Fleischman, E.W. (1975a): Certain patterns of karyotype evolution in chronic myelogeneous leukaemia. Chromosome abnormalities in CML. Humangenetik 30:113-119

Prigogina, E.L. and Fleischman, E.W. (1975b): Marker chromosome 14q+ in two non-Burkitt lymphomas. Humangenetik 30:109-112

Prigogina, E.L., Fleischman, E.W., Volkova, M.A. and Frenkel, M.A. (1978): Chromosome abnormalities and clinical and morphologic manifestations of chronic myeloid leukemia. Hum. Genet. 41:143-156

Prigogina, E.L., Fleischman, E.W., Puchkova, G.P., Kulagina, O.E., Majakova, S.A., Balakirev, S.A., Frenkel, M.A., Khvatova, N.V. and Peterson, I.S. (1979): Chromosomes in acute leukemia. Hum. Genet. 53:5-16

Pui, C-H., Williams, D.L., Scarborough, V., Jackson, C.W., Price, R. and Murphy, S. (1982): Acute megakaryoblastic leukaemia associated with intrinsic platelet dysfunction and constitutional ring 21 chromosome in a young boy. Brit. J. Haematol. 50:191-200

Rajasekariah, P., Illes, I. and Garson, O.M. (1982): Double 9;22 translocation with hyperdiploidy appearing in blastic transformation of chronic granulocytic leukemia. Cancer Genet. Cytogenet. 7:85-88

Rajendra, B.R., Lee, M-1, Nissenblatt, M.J., Gartenberg, G., Rose, D.V. and Sciorra, L.J. (1981): The occurrence of the Philadelphia chromosome in essential thrombocytosis. Hum. Genet. 56:287-291

Raposa, T., Natarajan, A.T. and Granberg, I. (1974): Identification of Ph1 chromosome and associated translocation in chronic myelogenous leukemia by Hoechst 33258. J. Natl. Cancer Inst. 52:1935-1938

Rausen, A.R., Kim, H.J., Burstein, Y., Rand, S., McCaffrey, R.M. and Kung, P.C. (1977): Philadelphia chromosome in acute lymphatic leukaemia of childhood. Lancet (1):432

Reeves, B.R. (1973): Cytogenetics of malignant lymphomas. Studies utilising a Giemsa-banding technique. Humangenetik 20:231-250

Reeves, B.R. and Pickup, V.L. (1980): The chromosome changes in non-Burkitt lymphomas. Hum. Genet. 53:349-355

Reeves, B.R. and Stathopoulos, G. (1976): Cytogenetic and cell-surface marker studies in two non-Hodgkin's lymphomata of T-cell origin. Hum. Genet. 31:203-210

Reeves, B.R., Lobb, D.S. and Lawler, S.D. (1972): Identity of the abnormal F-group chromosome associated with polycythaemia vera. Humangenetik 14:159-161

Reichmann, A., Riddell, R.H., Martin, P. and Levin, B. (1980): Double minutes in human large bowel cancer. Gastroenterol. 79:334-339

Reichmann, A., Martin, P. and Levin, B. (1981): Chromosomal banding patterns in human large bowel cancer. Int. J. Cancer 28:431-440

Reichmann, A., Levin, B. and Martin, P. (1982a): Human large-bowel cancer: correlation of clinical and histophathological features with banded chromosomes. Int. J. Cancer 29:625-629

Reichmann, A., Martin, P. and Levin, B. (1982b): Karyotypic findings in a colonic villous adenoma. Cancer Genet. Cytogenet. 7:51-57

Riccardi, V.M., Humbert, J.R. and Peakman, D. (1978): Acute leukemia associated with trisomy 8 mosaicism and a familial translocation 46,XY,t(7;20)(p13;p12). Am. J. Med. Genet. 2:15-21

Riet-Fox, M.F. van der, Retief, A.E. and Niekerk, W.A. (1979): Chromosome changes in 17 human neoplasms studied with banding. Cancer 44:2108-2119

Robert, K.H., Gahrton, G., Friberg, K., Zech, L. and Nilsson, B. (1982): Extra chromosome 12 and prognosis in chronic lymphocytic leukaemia. Scand. J. Haematol. 28:163-168

Rochon, M. and Vaillancourt, L. (1980): Nouvelle translocation complexe réalisant un chromosome Philadelphie masqué. Union Med. Canada 109:1337-1340

Rochon, M. and Vaillancourt, L. (1981): Translocation t(15;17), leucémie aigue promyélocytaire: a propos d'une observation. Union Méd. Canada 110:138-141

Rolovic, Z., Ciric, M. and Mijovic, A. (1983): A case of chronic myeloid leukemia without Ph1 translocation. Cancer Genet. Cytogenet. 8:75-79

Roos, G., Nordenson, I., Osterman, B., Jorpes, P. and Rudolphi, O. (1982): Patient with acute B-cell leukemia of Burkitt's type (L3) and marker chromosomes including an (8;14) translocation. Leukemia Res. 6:27-31

Roozendaal, K.J., Van Der Reijden, H.J. and Geraedts, J.P.M. (1981): Philadelphia chromosome positive acute lymphoblastic leukaemia with T-cell characteristics. Brit. J. Haematol. 47:145-147

Rossi, G. de, Alimena, G., Guglielmi, C., Gastaldi, R., Lopez, M., Martelli, M. and Mandelli, F. (1982): B-cell acute lymphoid leukemia (ALL) with lymphoblasts expressing surface immunoglobulins only at relapse. Leukemia Res. 6:855-859

Roth, D.G., Cimino, M.C., Variakojis, D., Golomb, H.M. and Rowley, J.D. (1979): B cell acute lymphoblastic leukemia (ALL) with a 14q+ chromosome abnormality. Blood 53:235-243

Roth, D.G., Tichman, C.M. and Rowley, J.D. (1980): Chronic myelodysplastic syndrome (preleukemia) with the Philadelphia chromosome. Blood 56:262-264

Roush, G.C. (1981): Unusual chromosomal tranclocation, t(1p+,7q-), and familial leukemia in a patient with chronic myelomonocytic leukemia. Conn. Med. 45:767-771

Rowley, J.D. (1973a): A new consistent chromosomal abnormality in chronic myelogenous leukemia identified by quinacrine fluorescence and Giemsa staining. Nature 243:290-293

Rowley, J.D. (1973b): Identification of a translocation with quinacrine fluorescence in a patient with acute leukemia. Ann. Génét. 16:109-112

Rowley, J.D. (1973c): Acquired trisomy 9. Lancet (2):390

Rowley, J.D. (1973d): Deletions of chromosome 7 in haematological disorders. Lancet (2):1385-1386

Rowley, J.D. (1974a): Do human tumors show a chromosome pattern specific for each etiologic agent?. J. Natl. Cancer Inst. 52:315-320

Rowley, J.D. (1974b): Missing sex chromosomes and translocations in acute leukaemia. Lancet (2):835-836

Rowley, J.D. (1974c): Absence of the 9q+ chromosome in Ph1 negative chronic myelogenous leukaemia. J. Med. Genet. 11:166-170

Rowley, J.D. (1975a): Nonrandom chromosomal abnormalities in hematologic disorders of man. Proc. Natl. Acad. Sci. 72:152-156

Rowley, J.D. (1975b): Abnormalities of chromosome 1 in myeloproliferative disorders. Cancer 36:1748-1757

Rowley, J.D. (1976): 5q- acute myelogenous leukemia: Reply. Blood 48:626

Rowley, J.D. (1983): Personal communication.

Rowley, J.D. and Potter, D. (1976): Chromosomal banding patterns in acute nonlymphocytic leukemia. Blood 47:705-721

Rowley, J.D., Wolman, S.R. and Horland, A.A. (1976): Another variant translocation in chronic myelogenous leukemia - revisited. New Engl. J. Med. 295:900-901

Rowley, J.D., Golomb, H.M. and Vardiman, J. (1977a): Nonrandom chromosomal abnormalities in acute non-lymphocytic leukemia in patients treated for Hodgkin disease and non-Hodgkin lymphomas. Blood 50:759-770

Rowley, J.D., Golomb, H.M. and Dougherty, C. (1977b): 15/17 translocation, a consistent chromosomal change in acute promyelocytic leukaemia. Lancet (1):549-550

Rowley, J.D., Golomb, H.M., Vardiman, J., Fukuhara, S., Dougherty, C. and Potter, D. (1977c): Further evidence for a non-random chromosomal abnormality in acute promyelocytic leukemia. Int. J. Cancer 20:869-872

Rowley, J.D., Resnick, G.D., Whitman, S.L., Senterfit, L. and Silver, R.T. (1979): Variant Ph1 translocation in chronic myeloid leukemia. Blood 54:294-295

Rowley, J.D., Golomb, H.M. and Vardiman, J.W. (1981a): Nonrandom chromosome abnormalities in acute leukemia and dysmyelopoietic syndromes in patients with previously treated malignant disease. Blood 58:759-767

Rowley, J.D., Variakojis, D., Kaneko, Y. and Cimino, M. (1981b): A Burkitt-lymphoma variant translocation (2p-;8q+) in a patient with ALL, L3 (Burkitt type). Hum. Genet. 58:166-167

Rowley, J.D., Alimena, G., Garson, O.M., Hagemeijer, A., Mitelman, F. and Prigogina, E.L. (1982): A collaborative study of the relationship of the morphological type of acute nonlymphocytic leukemia with patient age and karyotype. Blood 59:1013-1022

Rozynkowa, D. (1983): Personal communication.

Rozynkowa, D., Stepien, J., Kowalewski, J. and Nowakowski, A. (1977): Nonrandom chromosome rearrangements in 27 cases of human myeloid leukemia. Hum. Genet. 39:293-301

Rutten, F.J., Hustinx, T.W.J., Scheres, J.M.J.C. and Wagener, D.J.T. (1974): Trisomy-9 in the bone marrow of a patient with acute myelomonoblastic leukaemia. Brit. J. Haematol. 26:391-394

Ruutu, P., Ruutu, T., Vuopio, P., Kosunen, T.U. and Chapelle, A. de la. (1977a): Defective chemotaxis in monosomy 7. Nature 265:146-147

Ruutu, P., Ruutu, T., Vuopio, P., Kosunen, T.U. and Chapelle, A. de la. (1977b): Function of neutrophils in preleukaemia. Scand. J. Haematol. 18:317-325

Sadamori, N. and Sandberg, A.A. (1983a): 14q+ and 6q- anomalies in a case with hairy cell leukemia. Cancer Genet. Cytogenet. 8:89-90

Sadamori, N. and Sandberg, A.A. (1983b): The clinical and cytogenetic significance of C-banding on chromosome #9 in patients with Ph1-positive chronic myeloid leukemia. Cancer Genet. Cytogenet. 8:235-241

Sadamori, N., Matsunaga, M., Yao, E., Nishino, K., Tomonaga, Y., Tagawa, M., Kusano, M. and Ichimary, M. (1980): Chromosomes in the chronic phase of chronic granulocytic leukemia. Cancer Genet. Cytogenet. 1:299-310

Sadamori, N., Tomonaga, Y., Tagawa, M. and Ichimaru, M. (1981a): A case of atypical acute myelogenous leukemia with the 5q- chromosome. Acta Haematol. Jpn. 44:961-965

Sadamori, N., Ikeda, S., Muta, T., Ichimaru, M. and Matsunaga, M. (1981b): Erythroblastic transformation of Philadelphia chromosome (Ph1)-positive chronic myelogenous leukemia associated with marked chromosomal rearrangements. Cancer Genet. Cytogenet. 3:353-357

Sakurai, M., Oshimura, M., Kakati, S. and Sandberg, A.A. (1974): 8-21 translocation and missing sex chromosomes in acute leukaemia. Lancet (2):227-228

Sakurai, M., Sasaki, M., Kamada, N., Okada, M., Oshimura, M., Ishihara, T. and Shiraishi, Y. (1982a): A summary of cytogenetic, morphologic, and clinical data on t(8q-;21q+) and t(15q+;17q-) translocation leukemias in Japan. Cancer Genet. Cytogenet. 7:59-65

Sakurai, M., Kaneko, Y. and Abe, R. (1982b): Further characterization of acute myelogenous leukemia with t(8;21) chromosome translocation. Cancer Genet. Cytogenet. 6:143-152

Sakurai, M., Hayashi, Y., Abe, R. and Nakazawa, S. (1983): Chromosome translocations, surface immunoglobulins, and Epstein-Barr virus in Japanese Burkitt's lymphoma. Cancer Genet. Cytogenet. 8:275-276

San Roman, C., Ferro, M.T., Fernandez Ranada, J.M. and Steegman, J.L. (1982): Translocation (11;14) in B-cell lymphoproliferative disorders. Cancer Genet. Cytogenet. 7:279-286

Sandberg, A.A. (1977): Chromosome markers and progression in bladder cancer. Cancer Res. 37:2950-2956

Sandberg, A.A. (1980): The chromosomes in human cancer and leukemia. Elsevier North-Holland, New York, 776 pp.

Sandberg, A.A., Kohno, S-I., Wake, N. and Minowada, J. (1980): Chromosomes and causation of human cancer and leukemia. XLII. Ph1-positive ALL: an entity within myeloproliferative disorders?. Cancer Genet. Cytogenet. 2:145-174

Sandberg, A.A., Abe, S., Kowalczyk, J.R., Zedgenidze, A., Takeuchi, J. and Kakati, S. (1982a): Chromosomes and causation of human cancer and leukemia. L. Cytogenetics of leukemias complicating other diseases. Cancer Genet. Cytogenet. 7:95-136

Sandberg, A.A., Hecht, B.K., Ondreyco, S.M., Prieto, F. and Hecht, F. (1982b): Translocations involving chromosomes #3 and #12: hematologic diseases associated with abnormalities of these chromosomes. Cancer Genet. Cytogenet. 7:1-17

Sandberg, A.A., Wake, N. and Kohno, S. (1982c): Chromosomes and causation of human cancer and leukemia. XLVII. Severe hypodiploidy and chromosome conglomerations in ALL. Cancer Genet. Cytogenet. 5:293-307

Sasaki, M., Muramoto, J., Makino, S., Hara, Y., Okada, M. and Tanaka, E. (1975): Two cases of acute myeloblastic leukemia associated with a 9/22 translocation. Proc. Jap. Acad. 51:193-197

Sasaki, M., Okada, M., Kondo, I. and Muramoto, J. (1976): Chromosome banding patterns in 27 cases of acute myeloblastic leukemia. Proc. Jap. Acad. 52:505-508

Sasaki, M., Kondo, K. and Tomiyasu, T. (1983): Cytogenetic characterization of ten cases of Ph1-positive acute myelogenous leukemia. Cancer Genet. Cytogenet. 9:119-128

Saxon, A., Stevens, R.H. and Golde, D.W. (1979): Helper and suppressor T-lymphocyte leukemia in ataxia telangiectasia. New Engl. J. Med. 300:700-704

Scheres, J.M.J.C., Hustinx, T.W.J., Vaan, G.A.M. de and Rutten, F.J. (1978): 15/17 translocation in acute promyelocytic leukaemia. Hum. Genet. 43:115-117

Schmidt, R., Dar, H., Santorineou, M. and Sekine, I. (1975): Ph1 chromosome and loss and reappearance of the Y chromosome in acute lymphocytic leukaemia. Lancet (1):1145

Schröder, J., Vuopio, P. and Autio, K. (1981): Chromosome changes in human chronic lymphocytic leukemia. Cancer Genet. Cytogenet. 4:11-21

Seabright, M. (1983): Personal communication.

Secker Walker, L.M. and Hardy, J.D. (1976): Philadelphia chromosome in acute leukemia. Cancer 38:1619-1624

Secker Walker, L.M. and Sandler, R.M. (1978): Acute myeloid leukaemia with monosomy-7 follows acute lymphoblastic leukaemia. Brit. J. Haematol. 38:359-366

Secker-Walker, L.M., Summersgill, B.M., Swansbury, G.J., Lawler, S.D., Chessels, J.M. and Hardisty, R.M. (1976): Philadelphia-positive blast crisis masquerading as acute lymphoblastic leukaemia in children. Lancet (2):1405

Secker-Walker, L.M., Swansbury, G.J., Lawler, S.D. and Hardisty, R.M. (1979): Bone marrow chromosomes in acute lymphoblastic leukaemia: a long-term study. Med. Ped. Oncol. 7:371-385

Second International Workshop on Chromosomes in Leukemia. (1980a): General report. Cancer Genet. Cytogenet. 2:93-96

Second International Workshop on Chromosomes in Leukemia. (1980b): Morphological analysis of acute promyelocytic leukemia (M3) and t(8;21) cases. Cancer Genet. Cytogenet. 2:97-98

Second International Workshop on Chromosomes in Leukemia. (1980c): Cytogenetic, morphologic, and clinical correlations in acute non-lymphocytic leukemia with t(8q-;21q+). Cancer Genet. Cytogenet. 2:99-102

Second International Workshop on Chromosomes in Leukemia. (1980d): Chromosomes in acute promyelocytic leukemia. Cancer Genet. Cytogenet. 2:103-107

Second International Workshop on Chromosomes in Leukemia. (1980e): Chromosomes in preleukemia. Cancer Genet. Cytogenet. 2:108-113

Seidal, T., Mark, J., Hagmar, B. and Angervall, L. (1982): Alveolar rhabdomyosarcoma: a cytogenetic and correlated cytological and histological study. Acta path. microbiol. immunol. scand. Sect. A. 90:345-354

Sekhon, G.S., Taysi, K. and Moedjono, S. (1978): An unusual chromosome translocation in a case with CML. Birth Defects: Original Article Series, XIV 6c, 1978. The National Foundation, New York

Sessarego, M., Grammenu, S., Bianchi Scarra, G. and Ajmar, F. (1979): A case of chronic myelogenous leukemia with unusual chromosomal abnormality. Leukemia Res. 3:271-275

Sessarego, M., Canepa, L., Grammenu, S., Bianchi Scarra, G.L. and Ajmar, F. (1981a): Marked karyotype abnormalities in two cases of acute myelogenous leukemia. Cancer Genet. Cytogenet. 4:303-309

Sessarego, M., Bianchi Scarra, G.L., Ajmar, F. and Salvidio, E. (1981b): Karyotype evolution in a case of chronic myelogenous leukemia with an unusual Philadelphia chromosome translocation, t(4;22), and an additional translocation, t(3;5). Cancer Genet. Cytogenet. 3:47-51

Sessarego, M., Pasquali, F., Scarra, G.L.B. and Ajmar, F. (1982): Masked Philadelphia chromosome caused by translocation (9;11;22). Cancer Genet. Cytogenet. 8:319-323

Shabtai, F. (1983): Personal communication.

Shabtai, F. and Halbrecht, I. (1981a): Interpretation of a marker chromosome 17p in multiple myeloma. Hereditas 95:11-14

Shabtai, F. and Halbrecht, I. (1981b): Studies of banded chromosomes in patients with acute lymphocytic leukemia, including one patient with the Burkitt-type (L3). Cancer Genet. Cytogenet. 3:11-18

Shabtai, F., Weiss, S., Van Der Lijn, E., Lewinski, U., Djaldetti, M. and Halbrecht, I. (1978): A new cytogenetic aspect of polycythemia vera. Hum. Genet. 41:281-287

Shabtai, F., Lewinski, U.H., Har-Zahav, L., Gafter, U., Halbrecht, I. and Djaldetti, M. (1979): A hypodiploid clone and its duplicate in acute lymphoblastic leukemia. Am. J. Clin. Pathol. 72:1018-1024

Shabtai, F., Gafter, U., Weiss, S., Djaldetti, M. and Halbrecht, I. (1980): New complex Ph1 translocation t(10;14;22) in bone marrow cells and in PHA-stimulated peripheral blood cultures in chronic myelocytic leukaemia. J. Cancer Res. Clin. Oncol. 96:287-294

Shah-Reddy, I., Mayeda, K., Mirchandani, I. and Koppitch, F.C. (1982): Sézary syndrome with a 14:14(q12:q31) translocation. Cancer 49:75-79

Sharp, J.C., Potter, A.M. and Guyer, R.J. (1975): Karyotypic abnormalities in transformed chronic granulocytic leukaemia. Brit. J. Haematol. 29:587-591

Sharp, J.C., Potter, A.M. and Wood, J.K. (1976): Non-random and random chromosomal abnormalities in transformed chronic granulocytic leukaemia. Scand. J. Haematol. 16:5-12

Shashaty, G.G. and Baumiller, R.C. (1980): Philadelphia chromosome-negative chronic myelogenous leukemia with trisomy D. Arch. Pathol. Lab. Med. 104:376-378

Sheer, D., Solomon, E., Greaves, M.F. and Lister, T.A. (1982): 15/17 chromosome translocation in acute promyelocytic leukemia. Cancer Genet. Cytogenet. 5:353-354

Shiloh, Y., Naparstek, E. and Cohen, M.M. (1979): Cytogenetic investigation of leukemic and preleukemic disorders. Isr. J. Med. Sci. 15:500-506

Shiraishi, Y., Hayata, I., Sakurai, M. and Sandberg, A.A. (1975): Chromosomes and causation of human cancer and leukemia. XII. Banding analysis of abnormal chromosomes in polycythemia vera. Cancer 36:199-202

Shiraishi, Y., Yamamoto, K., Taguchi, H., Ueda, N. and Shiomi, F. (1980): Philadelphia chromosome in pure red cell aplasia: a preleukemic state?. Cancer Genet. Cytogenet. 2:1-5

Shiraishi, Y., Taguchi, H., Niiya, K., Shiomi, F., Kikukawa, K., Kubonishi, S., Ohmura, T., Hamawaki, M. and Ueda, N. (1982): Diagnostic and prognostic significance of chromosome abnormalities in marrow and mitogen response of lymphocytes of acute nonlymphocytic leukemia. Cancer Genet. Cytogenet. 5:1-24

Sieff, C.A., Chessells, J.M., Harvey, B.A.M., Pickthall, V.J. and Lawler, S.D. (1981): Monosomy 7 in childhood: a myeloproliferative disorder. Brit. J. Haematol. 49:235-249

Sikand, G.S., Taysi, K., Strandjord, S.E., Griffith, R. and Vietti, T.J. (1980): Trisomy 21 in bone marrow cells of a patient with a prolonged preleukemic phase. Med. Ped. Oncol. 8:237-242

Slater, R.M. and Kraker, J. de (1982): Chromosome number 11 and Wilms' tumor. Cancer Genet. Cytogenet. 5:237-245

Slater, R.M., Philip. P., Badsberg, E., Behrendt, H., Hansen, N.E. and Heerde, P. van. (1979): A 14q+ chromosome in a B-cell acute lymphocytic leukemia and in a leukemic non-endemic Burkitt lymphoma. Int. J. Cancer 23:639-647

Slater, R.M., Behrendt, H. and Heerde, P. van (1982): Cytogenetic studies on four cases of non-endemic Burkitt lymphoma. Med. Ped. Oncol. 10:71-84

Slavutsky, I., Labal de Vinuesa, M., Dupont, J., Mondini, N. and Brieux de Salum, S. (1981): Abnormalities of chromosome No. 1: two cases with lymphocytic lymphomas. Cancer Genet. Cytogenet. 3:341-346

Slee, P.H.T.J., Everdingen, J.J.E. van, Geraedts, J.O.M., Velde, J. te and Ottolander, G.J. den (1981): Familial myeloproliferative disease. Acta Med. Scand. 210:321-327

Smadja, N., James, J., Zittoun, R. and Debray, J. (1980): LMC avec chromosome Philadelphie et translocation complexe (2;9;22) "secondaire" a une la indifferenciee. Nouv. Rev. Fr. Hematol. Suppl. 22:88

Smadja, N., James, J., Krulik, M., Zittoun, R. and Debray, J. (1983a): Chronic myelogenous leukemia with a Philadelphia chromosome resulting from a complex translocation (2;9;22), following an undifferentiated acute leukemia. Cancer Genet. Cytogenet. 8:1-8

Smadja, N., Krulik, M., Genot, J.-Y., Audebert, A.A. and Debray, J. (1983b): Hematologic and cytogenetic study of two cases of acute leukemia associated with breast cancer. Cancer Genet. Cytogenet. 9:185-196

Sokal, G., Michaux, J.L., Van Den Berghe, H., Cordier, A., Rodhain, J., Ferrant, A., Moriau, M., Bruyéere, M. de and Sonnet, J. (1975): A new hematologic syndrome with a distinct karyotype: The 5q- chromosome. Blood 46:519-533

Sonta, S. and Sandberg, A.A. (1977a): A new complex Ph1 translocation involving three chromosomes. J. Natl. Cancer Inst. 58:1583-1585

Sonta, S.-I. and Sandberg, A.A. (1977b): Chromosomes and causation of human cancer and leukemia. XXVIII. Value of detailed chromosome studies on large numbers of cells in CML. Am. J. Hematol. 3:121-126

Sonta, S.-I. and Sandberg, A.A. (1978a): Chromosomes and causation of human cancer and leukemia. XXX. Banding studies of primary intestinal tumors. Cancer 41:164-173

Sonta, S.-I. and Sandberg, A.A. (1978b): Chromosomes and causation of human cancer and leukemia. XXIX. Further studies on karyotypic progression in CML. Cancer 41:153-163

Sonta, S., Oshimura, M. and Sandberg, A.A. (1976): Chromosomes and causation of human cancer and leukemia XXI. Cytogenetically unusual cases of leukemia. Blood 48:697-705

Sonta, S., Oshimura, M., Evans, J.T. and Sandberg, A.A. (1977): Chromosomes and causation of human cancer and leukemia. XX. Banding patterns of primary tumors. J. Natl. Cancer Inst. 58:49-53

Soukup, S.W. and Neely, J.E. (1981): Chromosome studies in a case of monoblastic leukemia. Cancer Genet. Cytogenet. 4:331-335

Sparkes, R.S., Como, R. and Golde, D.W. (1980): Cytogenetic abnormalities in ataxia telangiectasia with T-cell chronic lymphocytic leukemia. Cancer Genet. Cytogenet. 1:329-336

Spriggs, A.I., Holt, M.M. and Bedford, J. (1976): Duplication of part of the long arm of chromosome 1 in marrow cells of a treated case of myelomatosis. Blood 48:595-599

Srivastava, A.K., Gruppo, R.A. and Siegrist, C.W. (1981): Chromosomal anomaly of 6q in chronic myelogenous leukemia (CML). Cancer Genet. Cytogenet. 3:131-136

Stenman, G., Dahlenfors, R., Mark, J. and Sandberg, N. (1982): Adenoid cystic carcinoma: a third type of human salivary gland neoplasm characterized cytogenetically by reciprocal translocations. Anticancer Res. 2:11-16

Stern, R., Sorenson, J., Wurster-Hill, D.H., Cornwell III, G.G. and McIntyre, O.R. (1979a): Chromosome changes in a patient achieving complete remission in the acute phase of chronic myelogenous leukemia. Am. J. Hematol. 6:155-161

Stern, R., Widirstky, S., Wurster-Hill, D.H., Allen, R.D., Smith, K.A., Cornwell III, G.G. and Cornell, C.J. Jr. (1979b): Hand-mirror cell leukemia associated with mental retardation: Immunologic, chromosome, and morphological studies. Blood 54:703-712

Stoll, C. (1983): Personal communication.

Stoll, C. and Oberling, F. (1978): Non-random clonal evolution in 45 cases of chronic myeloid leukemia. Leukemia Res. 3:61-66

Stoll, C. and Oberling, F. (1982): Y chromosome duplication in chronic myeloid leukemia. Nouv. Rev. Fr. Hematol. 24:9-12

Stoll, C., Oberling, F. and Flori, E. (1978): Chromosome analysis of spleen and/or lymph nodes of patients with chronic myeloid leukemia (CML). Blood 52:828-838

Streuli, R.A., Testa, J.R., Vardiman, J.W., Mintz, U., Golomb, H.M. and Rowley, J.D. (1980): Dysmyelopoietic syndrome: sequential clinical and cytogenetic studies. Blood 55:636-644

Streuli, R.A., Kaneko, Y., Variakojis, D., Kinnealey, A., Golomb, H.M. and Rowley, J.D. (1981): Lymphoblastic lymphoma in adults. Cancer 47:2510-2516

Sudries, M., Kessous, A., Bouroullou, G., Colombies, P. and Clement, D. (1980): Frequence des translocations inhabituelles du chromosome Ph1 impliquant trois chromosomes dans une etude de 68 cas de leucemie myeloide chronique. Nouv. Rev. Fr. Hematol. Suppl. 22:89

Swansbury, G.J. and Lawler, S.D. (1980): Chromosomes and prognosis in preleukaemia: four cases of 5q- with other karyotypic abnormality. Leukemia Res. 4:611-618

Sweet, D.L., Golomb, H.M., Rowley, J.D. and Vardiman, J.M. (1979): Acute myelogenous leukemia and throbocythemia associated with an abnormality of chromosome No. 3. Cancer Genet. Cytogenet. 1:33-37

Swolin, B., Weinfeld, A., Ridell, B., Waldenström, J. and Westin, J. (1981): On the 5q- deletion: clinical and cytogenetic observations in ten patients and review of the literature. Blood 58:986-993

Swolin, B., Weinfeld, A., Waldenström, J. and Westin, J. (1983): Cytogenetic studies of bone marrow and extramedullary tissues and clinical course during metamorphosis of chronic myelocytic leukemia. Cancer Genet. Cytogenet. 9:197-209

Symann, M., Montpellier, C. de, Ninane, J. and Van Den Berghe, H. (1982): "Spontaneous" erythroid progenitor cells in the circulation and monosomy 7 in juvenile chronic myelogenous leukemia. Cancer Genet. Cytogenet. 6:183-185

Tabachnik, E., Garty, R., Zaizov, R. and Chemke, J. (1979): Myeloproliferative disease of childhood associated with a trisomy 21 clone. Acta Haematol. 62:90-93

Tadano, J., Niwa, M., Nagumo, F., Watanabe, K. and Arimori, S. (1980): A case of acute myeloblastic leukemia with chromosomes showing 46,XY = 11/46,XY,21q- = 39. Toaki J. Exp. Clin. Med. 5:45-50

Takeuchi, J., Ohshima, T. and Amaki, I. (1981): Cytogenetic studies in adult acute leukemias. Cancer Genet. Cytogenet. 4:293-302

Tanzer, J., Najean, Y., Frocrain, C. and Bernheim, A. (1977): Chronic myelocytic leukemia with a masked Ph1 chromosome. New Engl. J. Med. 296:571-572

Tanzer, J., Frocrain, C. and Gardais, J. (1980a): Chromosome Philadelphie inhabituels au cours de la leucemie myeloide chronique (LMC). Nouv. Rev. Fr. Hematol. Suppl. 22:90

Tanzer, J., Frocrain, C., Alcalay, D. and Desmarest, M.C. (1980b): Pleuropericardite a cellules de Burkitt chez un poitevin. Chromosome 14q+ par translocation (1;14), association a une translocation (8;22), a la perte de l'y et a de nombreuses autres anomalies. Nouv. Rev. Fr. Hematol. Suppl. 22:107

Teerenhovi, L., Borgström, G.H., Mitelman, F., Brandt, L., Vuopio, P., Timonen, T., Almqvist, A. and Chapelle, A. de la. (1978): Uneven geographical distribution of 15;17-translocation in acute promyelocytic leukaemia. Lancet (2):797

Teerenhovi, L., Borgström, G.H., Lintula, R., Ruutu, T., Lahtinen, R., Chapelle, A. de la and Vuopio, P. (1981): The 5q- chromosome in preleukaemia and acute leukaemia. Scand. J. Haematol. 27:119-129

Testa, J.R., Kinnealey, A., Rowley, J.D., Golde, D.W. and Potter, D. (1978a): Deletion of the long arm of chromosome 20 (del(20)(q11)) in myeloid disorders. Blood 52:868-877

Testa, J.R., Golomb, H.M., Rowley, J.D., Vardiman, J.W. and Sweet Jr, D.L. (1978b): Hypergranular promyelocytic leukemia (APL): Cytogenetic and ultrastructural specificity. Blood 52:272-280

Testa, J.R., Mintz, U., Rowley, J.D., Vardiman, J.W. and Golomb, H.M. (1979): Evolution of karyotypes in acute nonlymphocytic leukemia. Cancer Res. 39:3619-3627

Testa, J.R., Kanofsky, J.R., Rowley, J.D. and Baron, J.M. (1981a): Multiple cytogenetically abnormal clones in two polycythemia vera patients. Hum. Genet. 57:165-168

Testa, J.R., Kanofsky, J.R., Rowley, J.D., Baron, J.M. and Vardiman, J.W. (1981b): Karyotypic patterns and their clinical significance in polycythemia vera. Am. J. Hematol. 11:29-45

Testa, J.R., Jumamoy, L.M. and Wiernik, P.H. (1982): Variant Philadelphia chromosome translocations in two patients with chronic myelogenous leukemia. Cancer Genet. Cytogenet. 7:79-84

Testa, J.R., Oguma, N., Pollak, A. and Wiernik, P.H. (1983): Near-tetraploid clones in acute leukemia. Blood 61:71-78

Third International Workshop on Chromosomes in Leukemia. (1981a): Chromosomal abnormalities in acute lymphoblastic leukemia. Cancer Genet. Cytogenet. 4:101-110

Third International Workshop on Chromosomes in Leukemia. (1981b): Clinical significance of chromosomal abnormalities in acute lymphoblastic leukemia. Cancer Genet. Cytogenet. 4:111-137

Third International Workshop on Chromosomes in Leukemia. (1981c): Report on essential thrombocythemia. Cancer Genet. Cytogenet. 4:138-142

Tiepolo, L. and Zuffardi, O. (1973): Identification of normal and abnormal chromosomes in tumor cells. Cytogenet. Cell Genet. 12:8-16

Togawa, A., Hasegawa, K., Mitake, T., Mannoji, M., Takemoto, Y., Yamada, O., Miyajima, K., Yoshimoto, M., Yawata, Y. and Yamada, K. (1981): A chromosome analysis in a patient with chronic myelogenous leukemia and gastric cancer. Acta Haematol. Jpn. 44:590-594

Tomiyasu, T., Sasaki, M. and Abe, S. (1980): Long arm deletion of chromosome No. 5 in a case of Philadelphia chromosome-positive chronic myelocytic leukemia. Cancer Genet. Cytogenet. 2:309-315

Tomiyasu, T., Sasaki, M., Kondo, K. and Okada, M. (1982): Chromosome banding studies in 106 cases of chronic myelogenous leukemia. Jpn. J. Human Genet. 27:243-258

Tosato, G., Whang-Peng, J., Levine, A.S. and Poplack, D.G. (1978): Acute lymphoblastic leukemia followed by chronic myelocytic leukemia. Blood 52:1033-1036

Trent, J. (1983): Personal communication.

Trent, J.M. and Davis, J.R. (1979): D-group chromosome abnormalities in endometrial cancer and hyperplasia. Lancet (2):361

Trent, J.M. and Salmon, S.E. (1980): Human tumour karyology: marked analytic improvement by short-term agar culture. Brit. J. Cancer 41:867-874

Trent, J.M. and Salmon, S.E. (1981): Karyotypic analysis of human ovarian carcinoma cells cloned in short term agar culture. Cancer Genet. Cytogenet. 3:279-291

Trent, J.M., Davis, J.R. and Durie, B.G.M. (1983a): Cytogenetic analysis of leukaemic colonies from acute and chronic myelogenous leukaemia. Brit. J. Cancer 47:103-109

Trent, J.M., Rosenfeld, S.B. and Meyskens, F.L. (1983b): Chromosome 6q involvement in human malignant melanoma. Cancer Genet. Cytogenet. 9:177-180

Tricot, G. and Van Den Berghe, H. (1982): Isolated peripheral thrombocytopenia as presenting symptom in preleukemia: a report of two cases with 11q+. Cancer Genet. Cytogenet. 5:147-151

Tricot, G., Broeckaert-van Orshoven, A., Casteels-van Daele, M. and Van Den Berghe, H. (1981): 8/21 translocation in acute myeloid leukaemia. Scand. J. Haematol. 26:168-176

Tricot, G., Criel, A. and Verwilghen, R.L. (1982): Thrombocytopenia as presenting symptom of preleukaemia in 3 patients. Scand. J. Haematol. 28:243-250

Tricot, G., Fryns, J.-P., Thomas, J., Moerman, P., Broeckaert-Van Orshoven, A., Vermaelen, K. and Van Den Berghe, H. (1983): 5q- anomaly in a patient with disseminated teratoma. Cancer Genet. Cytogenet. 9:239-244

Trujillo, J.M. and Cork, A. (1981): Clinical implications of cytogenetic abnormalities in acute nonlymphocytic leukemia. Am. J. Pediat. Hematol. Oncol. 3:425-431

Trujillo, J.M., Cork, A., Ahearn, M.J., Youness, E.L. and McCredie, K.B. (1979): Hematologic and cytologic characterization of 8/21 translocation acute granulocytic leukemia. Blood 53:695-706

Tsuchimoto, T., Buhler, E.M., Stalder, G.R., Mayr, A.C. and Obrecht, J.P. (1974): Deletion of chromosome 7 in polycythaemia vera. Lancet (1):566

Ueda, K., Kawaguchi, Y., Kodama, M., Tanaka, Y., Usui, T. and Kamada, N. (1981): Primary myelofibrosis with myeloid metaplasia and cytogenetically abnormal clones in 2 children with Down's syndrome. Scand. J. Haematol. 27:152-158

Ueshima, Y., Fukuhara, S., Hattori, T., Uchiyama, T., Takatsuki, K. and Uchino, H. (1981): Chromosome studies in adult T-cell leukemia in Japan: significance of trisomy 7. Blood 58:420-425

Ueshima, Y., Fukuhara, S., Nagai, K., Takatsuki, K. and Uchino, H. (1983): Cytogenetic studies and clinical aspects of patients with plasma cell leukemia and leukemic macroglobulinemia. Cancer Res. 43:905-912

Vahdati, M., Graafland, H. and Emberger, J.M. (1983a): Karyotype analysis of B-lymphocytes transformed by Epstein-Barr virus in 21 patients with B cell chronic lymphocytic leukemia. Hum.Genet. 63:327-331

Vahdati, M., Graafland, H. and Emberger, J.M. (1983b): Isochromosome 17q in cell lines of two cases of B cell chronic lymphocytic leukemia. Cancer Genet. Cytogenet. 9:227-232

Van Den Akker, J., Taillemite, J.L., Portnoi, M.F., Porrier, N. le and Najman, A. (1980): Trois remaniements chromosomiques (2;9;22) a l'origin d'un Ph1. Nouv. Rev. Fr. Hematol. Suppl. 22:90

Van Den Berghe, H. (1983): Personal communication.

Van Den Berghe, H., David, G., Michaux, J.-L., Sokal, G. and Verwilghen, R. (1976): 5q- acute myelogenous leukemia. Blood 48:624-625

Van Den Berghe, H., David, G., Broeckaert-Van Orshoven, A., Louwagie, A. and Verwilghen, R. (1978): Unusual Ph1 translocation in acute myeloblastic leukemia. New Engl. J. Med. 299:360

Van Den Berghe, H., Louwagie, A., Broeckaert-Van Orshoven, A., David, G. and Verwilghen, R. (1979a): Chromosome analysis in two unusual malignant blood disorders presumably induced by benzene. Blood 53:558-566

Van Den Berghe, H., David, G., Broeckaert-Van Orshoven, A., Louwagie, A., Verwilghen, R., Casteels-Van Daele, M., Eggermont, E. and Eeckels, R. (1979b): A new chromosome anomaly in acute lymphoblastic leukemia (ALL). Hum. Genet. 46:172-180

Van Den Berghe, H., Louwagie, A., Broeckaert-Van Orshoven, A., David, G., Verwilghen, R., Michaux, J.L. and Sokal, G. (1979c): Chromosome abnormalities in acute promyelocytic leukemia (APL). Cancer 43:558-562

Van Den Berghe, H., Parloir, C., Gosseye, S., Englebienne, V., Cornu, G. and Sokal, G. (1979d): Variant translocation in Burkitt lymphoma. Cancer Genet. Cytogenet. 1:9-14

Van Den Berghe, H., Petit, P., Broeckaert-Van Orshoven, A., Louwagie, A., Baere, H. de and Verwilghen, R. (1979e): Simultaneous occurrence of 5q- and 21q- in refractory anemia with thrombocytosis. Cancer Genet. Cytogenet. 1:63-68

Van Den Berghe, H., Louwagie, A., Broeckaert-Van Orshoven, A., David, G., Verwilghen, R., Michaux, J.L. and Sokal, G. (1979f): Philadelphia chromosome in human multiple myeloma. J. Natl. Cancer Inst. 63:11-16

Van Den Berghe, H., Broeckaert-Van Orshoven, A., Louwagie, A., Verwilghen, R., Michaux, J.-L. and Sokal, G. (1979g): Transformation of polycythemia vera to myelofibrosis and late appearance of a 5q- chromosome anomaly. Cancer Genet. Cytogenet. 1:157-167

Van Den Berghe, H., Vermaelen, K., Broeckaert-Van Orshoven, A., Delbeke, M-J., Benoit, Y., Orye, E., Van Eygen, M. and Logghe, N. (1983): Pentasomy 21 characterizing spontaneously regressing congenital acute leukemia. Cancer Genet. Cytogenet. 9:19-24

Van Der Blij-Philipsen, M., Breed, W.P.M. and Hustinx, T.W.J. (1977): A case of chronic myeloid leukemia with a translocation (12;22)(p13;q11). Hum. Genet. 39:229-231

Van Der Weyden, M.B., Bailey, L. and Garson, O.M. (1978): Human adenosine deaminase and chromosome 20. Experientia 34:531

Vang Nielsen, K., Rönne, M. and Jakobsen, A. (1982): A unique case of cervical carcinoma. Comparison of chromosome analysis and flow cytometric measurements. Hereditas 97:65-72

Vercherat, M., Tous, J. and Pochat, D. (1980): Un nouveau cas de leucémie aigue lympholastique de l'enfant avec chromosome Philadelphie. Ann. Péd. 27:53-56

Verhest, A. (1983): Personal communication.

Verhest, A., Schoubroeck, F. van, Wittek, M., Naets, J.P. and Denolin-Reubens, R. (1976): Specificity of the 5q- chromosome in a distinct type of refractory anemia. J. Natl. Cancer Inst. 56:1053-1054

Verhest, A., Lustman, F., Wittek, M., Schoubroeck, F. van and Naets, J.P. (1977): Cytogenetic evidence of clonal evolution in 5q- anemia. Biomedicine 27:211-212

Verhest, A., Lustman, F. and Debusscher, L. (1980): 9q+ marker chromosome in chronic myelogenous leukaemia without the Ph1. Brit. J. Haematol. 46:493-494

Verma, R.S. and Dosik, H. (1977): The value of reverse banding in detecting bone marrow chromosomal abnormalities: Translocation between chromosomes 1, 9, and 22 in a case of chronic myelogenous leukemia (CML). Am. J. Hematol. 3:171-175

Verma, R.S. and Dosik, H. (1979): A case of chronic myelogenous leukemia (CML) with a translocation between chromosomes 12 and 22, t(12;22)(p13;q13), resulting in a Philadelphia (Ph1) chromosome. Cytogenet. Cell Genet. 23:274-276

Verma, R.S. and Dosik, H. (1981): 17p+ in a patient with chronic myelogenous leukemia (CML): its differentiation from an isochromosome 17,(i(17q)). Cancer Genet. Cytogenet. 3:55-60

Vilpo, J.A., Klemi, P., Lassila, O., Schröder, J. and Chapelle, A. de la (1979): Transformation in chronic granulocytic leukaemia. Different blast cell clones in different anatomical sites. Acta Haematol. 62:247-250

Vilpo, J.A., Dryzun, B., Klemi, P., Lassila, O. and Chapelle, A. de la (1980): Extramedullary pleural blast crisis during otherwise chronic phase in chronic granulocytic leukaemia. Europ. J. Cancer 16:885-891

Vloten, W.A. van, Pet, E.A. and Geraedts, J.P.M. (1979): Chromosome studies in mycosis fungoides. Brit. J. Dermatol. 102:507-513

Vykoupil, K.F., Thiele, J., Stangel, W., Krmpotic, E. and Georgii, A. (1980): Polycythemia Vera. I. Histopathology, ultrastructure and cytogenetics of the bone marrow in comparison with secondary polycythemia. Virchows Arch. A Path. Anat. Histol. 389:307-324

Waddell, C.C., Brown, J.A. and Zelnick, P.W. (1980): Chronic myelogenous leukemia. Association with 45 XO Philadelphia chromosome karyotype and prolonged survival. Arch. Int. Med. 140:270-271

Wake, N., Hreshchyshyn, M.M., Piver, S.M., Matsui, S-i. and Sandberg, A.A. (1980): Specific cytogenetic changes in ovarian cancer involving chromosomes 6 and 14. Cancer Res. 40:4512-4518

Wake, N., Slocum, H.K., Rustum, Y.M., Matsui, S-i. and Sandberg, A.A. (1981): Chromosomes and causation of human cancer and leukemia. XLIV. A method for chromosome analysis of solid tumors. Cancer Genet. Cytogenet. 3:1-10

Warburton, D. and Bluming, A. (1973): A "Philadelphia-like" chromosome derived from the Y in a patient with refractory dysplastic anemia. Blood 42:799-804

Warburton, D. and Shah, N. (1976): A 9/11 translocation in a child with Ph1-negative chronic myelogenous leukemia. J. Pediat. 88:599-601

Watt, J.L. (1983): Personal communication.

Watt, J.L., Hamilton, P.J. and Page, B.M. (1977): Variation in the Philadelphia chromosome. Hum. Genet. 37:141-148

Watt, J.L., Khaund, R.R., Allan, S.G., Smith, C.C. and Stephen, G.S. (1982): An unusual karyotype in preleukemia. Cancer Genet. Cytogenet. 7:67-72

Watt, J.L., King, D.J., Palmer, J.B.D. and Davidson, R.J.L. (1983): The heterogeneity of the 5q- chromosome marker in refractory anemia. Cancer Genet. Cytogenet. 9:113-118

Wayne, A.W., Sharp, J.C., Joyner, M.V., Sterndale, H. and Pulford, K.A.F. (1979): The significance of Ph1 mosaicism: a report of six cases of chronic granulocytic leukaemia and two cases of acute myeloid leukaemia. Brit. J. Haematol. 43:353-360

Webber, L.M. and Garson, O.M. (1983): Fluorodeoxyuridine synchronization of bone marrow cultures. Cancer Genet. Cytogenet. 8:123-132

Weber, T.H., Wegelius, R., Borgström, G.H., Gahmberg, C.G. and Andersson, L.C. (1979): A case of pure monocytic leukaemia in a child.Characterization of cellular morphology, membrane markers, surface glycoproteins and karyotype. Scand. J. Haematol. 22:47-52

Weh, H.J. and Hossfeld, D.K. (1982): Translocation t(4;11) in acute lymphocytic leukemia (ALL). Blut 44:271-274

Weh, H.J., Zschaber, R. and Hossfeld, D.K. (1982): Double minute chromosomes: a frequent marker in leukemic patients with a previous history of malignant disease?. Cancer Genet. Cytogenet. 5:279-280

Weinfeld, A., Westin, J., Ridell, B. and Swolin, B. (1977a): Polycythemia vera terminating in acute leukaemia. A clinical, cytogenetic and morphologic study in 8 patients treated with alkylating agents. Scand. J. Haematol. 19:255-272

Weinfeld, A., Westin, J. and Swolin, B. (1977b): Ph1-negative eosinophilic leukaemia with trisomy 8. Scand. J. Haematol. 18:413-420

Weiss, A.F., Portmann, R., Fischer, H., Simon, J. and Zang, K.D. (1975): Simian virus 40-related antigens in three human meningiomas with defined chromosome loss. Proc. Natl. Acad. Sci. 72:609-613

Westin, J. (1976): Chromosome abnormalities after chlorambucil therapy of polycythaemia vera. Scand. J. Haematol. 17:197-204

Westin, J., Wahlström, J. and Swolin, B. (1976): Chromosome studies in untreated polycythaemia vera. Scand. J. Haematol. 17:183-196

Wetter, O., Linder, K-H., Hossfeld, D.K., Lunscken, C. and Schmitt-Gräff, A. (1980): Plasma cell dyscrasias - a comparative study of cell surface properties in plasma cell leukemia and myeloma. Leukemia Res. 4:249-259

Whang-Peng, J., Knutsen, T.A. and Lee, E.C. (1973): Dicentric Ph1 chromosome. J. Natl. Cancer Inst. 51:2009-2012

Whang-Peng, J., Knutsen, T., Lee, E.C. and Leventhal, B. (1976a): Acquired XO/XY clones in bone marrow of a patient with paroxysmal nocturnal hemoglobinuria (PNH). Blood 47:611-619

Whang-Peng, J., Broder, S., Lee, E. and Young, R.C. (1976b): Unusual clonal evolution in a case of chronic myelogenous leukemia. Acta Haematol. 56:345-354

Whang-Peng, J., Knutsen, T., Ziegler, J. and Leventhal, B. (1976c): Cytogenetic studies in acute lymphocytic leukemia: Special emphasis in long-term survival. Med. Ped. Oncol. 2:333-351

Whang-Peng, J., Gralnick, H.R., Knutsen, T., Brereton, H., Chang, P., Schechter, G.P. and Lessin, L. (1977): Small F chromosome in myelo- and lymphoproliferative diseases. Leukemia Res. 1:19-30

Whang-Peng, J., Lee, E., Knutsen, T., Chang, P. and Nienhuis, A. (1978): Cytogenetic studies in patients with myelofibrosis and myeloid metaplasia. Leukemia Res. 2:41-56

Whang-Peng, J., Knutsen, T., O'Donnell, J.F. and Brereton, H.D. (1979): Acute non-lymphocytic leukemia and acute myeloproliferative syndrome following radiation therapy for non-Hodgkin's lymphoma and chronic lymphocytic leukemia. Cancer 44:1592-1600

Whang-Peng, J., Lee, E., Knutsen, T. and Solanki, D. (1981): Dicentric isochromosome for the long arm of chromosome #17, dic i(17q), in a patient with chronic myelogenous leukemia (CML). Cancer Genet. Cytogenet. 3:233-236

Whang-Peng, J., Bunn, P.A. Jr, Knutsen, T., Matthews, M.J., Schechter, G. and Minna, J.D. (1982a): Clinical implications of cytogenetic studies in cutaneous T-cell lymphoma (CTCL). Cancer 50:1539-1553

Whang-Peng, J., Bunn Jr., P.A., Kao-Shan, C.S., Lee, E.C., Carney, D.N., Gazdar, A. and Minna, J.D. (1982b): A nonrandom chromosomal abnormality, del 3p(14-23), in human small cell lung cancer (SCLC). Cancer Genet. Cytogenet. 6:119-134

Whaun, J.M., Lin, C.C., Biederman, B., Cornish, S.J. and Dundas, J.B. (1981): Myeloproliferative disorder with unusual marrow chromosome constitution. Cancer 48:1164-1169

Whittaker, J.A., Davies, P. and Khurshid, M. (1975): Absence of the Y chromosome in patients with chronic granulocytic leukaemia. Acta Haematol. 54:350-357

Wick, M.R., Li, C.-Y. and Pierre, R.V. (1982): Acute nonlymphocytic leukemia with basophilic differentiation. Blood 60:38-45

Williams, W.C. and Weiss, G.B. (1982): Megakaryoblastic transformation of chronic myelogenous leukemia. Cancer 49:921-926

Wurster-Hill, D. (1983): Personal communication.

Wurster-Hill, D.H. and Maurer, L.H. (1978): Cytogenetic diagnosis of cancer: Abnormalities of chromosomes and polyploid levels in the bone marrow of patients with small cell anaplastic carcinoma of the lung. J. Natl. Cancer Inst. 61:1065-1075

Wurster-Hill, D.H., McIntyre, O.R., Cornwell, G.G. and Maurer, L.H. (1973): Marker-chromosome 14 in multiple myeloma and plasma-cell leukaemia. Lancet (2):1031

Wurster-Hill, D., Whang-Peng, J., McIntyre, O.R., Hsu, L.Y.F., Hirschhorn, K., Modan, B., Pisciotta, A.V., Pierre, R., Balcerzak, S.P., Weinfeld, A. and Murphy, S. (1976): Cytogenetic studies in polycythemia vera. Semin. Hematol. 13:13-32

Yamada, Y. and Furusawa, S. (1976): Preferential involvement of chromosomes No. 8 and No. 21 in acute leukemia and preleukemia. Blood 47:679-686

Yamada, K., Yoshioka, M. and Oami, H. (1977): A 14q+ marker and a late replicating chromosome No. 22 in a brain tumor. J. Natl. Cancer Inst. 59:1193-1195

Yamada, K., Kondo, T., Yoshioka, M. and Oami, H. (1980): Cytogenetic studies in twenty human brain tumors: association of No. 22 chromosome abnormalities with tumors of the brain. Cancer Genet. Cytogenet. 2:293-307

Yamada, K., Sugimoto, E., Amano, M., Imamura, Y., Kubota, T. and Matsumoto, M. (1983a): Two cases of acute promyelocytic leukemia with variant translocations: The importance of chromosome No. 17 abnormality. Cancer Genet. Cytogenet. 9:93-99

Yamada, K., Shionoya, S., Amano, M. and Imamura, Y. (1983b): A Burkitt-type 8;14 translocation in a case of plasma cell leukemia. Cancer Genet. Cytogenet. 9:67-70

Yunis, J.J. (1982): Comparative analysis of high-resolution chromosome techniques for leukemic bone marrows. Cancer Genet. Cytogenet. 7:43-50

Yunis, J.J., Bloomfield, C.D. and Ensrud, K. (1981): All patients with acute nonlymphocytic leukemia may have a chromosomal defect. New Engl. J. Med. 305:135-139

Yunis, J.J., Oken, M.M., Kaplan, M.E., Ensrud, K.M., Howe, R.R. and Theologides, A. (1982): Distinctive chromosomal abnormalities in histologic subtypes of non-Hodgkin's lymphoma. New Engl. J. Med. 307:1231-1236

Zaccaria, A., Baccarani, M., Gugliotta, L., Guarini, A., Betti, S. and Tura, S. (1980): 21q- in primary thrombocythemia. Cancer Genet. Cytogenet. 1:337-344

Zaccaria, A., Rosti, G. and Testoni, N. (1982): Reciprocal translocation (11q+;17q-) in a patient with acute monoblastic leukemia. Nouv. Rev. Fr. Hematol. 24:389-390

Zaccaria, A., Barbieri, B., Castoldi, G.L., Ferraresi, P., Finelli, C., Hossfeld, D.K., Mitelman, F., Rosti, G., Testoni, N. and Van Den Berghe, H. (1983a): Normal bone marrow karyotype in paroxysmal nocturnal hemoglobinuria - a cooperative European study. Cancer Genet. Cytogenet. 9:211-215

Zaccaria, A., Rosti, G., Testoni, N., Mazza, P., Cantore, M. and Tura, S. (1983b): Acute nonlymphocytic leukemias and dysmyelopoietic syndromes in patients treated for Hodgkin's lymphoma. Cancer Genet. Cytogenet. 9:217-226

Zahavi, I., Shabtai, F., Appel, S., Rudniki, C. and Djaldetti, M. (1982): Ultrastructural and chromosomal studies in a patient with hypergranular (M3) promyelocytic leukemia with two abnormal clones. Acta Haematol. 68:52-57

Zang, K. and Zankl, H. (1983): Personal communication.

Zankl, H. and Zang, K.D. (1972): Cytological and cytogenetical studies on brain tumors. IV. Identification of the missing G chromosome in human meningiomas as No. 22 by fluorescence technique. Humangenetik 14:167-169

Zankl, H. and Zang, K.D. (1981): Marker chromosome 20q- does not arise only in bone marrow disorders. Cancer Genet. Cytogenet. 3:85-87

Zankl, H., Weiss, A.F. and Zang, K.D. (1975a): Cytological and cytogenetical studies on brain tumors. VI. No evidence for a translocation in 22-monosomic meningiomas. Humangenetik 30:343-348

Zankl, H., Seidel, H. and Zang, K.D. (1975b): Cytological and cytogenetical studies on brain tumors V. Preferential loss of sex chromosomes in human meningiomas. Humangenetik 27:119-128

Zankl, H., Ludwig, B, May, G. and Zang, K.D. (1979): Karyotypic variations in human meningioma cell cultures under different in vitro conditions. J. Cancer Res. Clin. Oncol. 93:165-172

Zankl, H., Huwer, H. and Zang, K.D. (1982): Cytogenetic studies on the nucleolar organizer region (NOR) activity in meningioma cells with normal and hypodiploid karyotypes. Cancer Genet. Cytogenet. 6:47-53

Zech, L., Lindsten, J., Uden, A.-M. and Gahrton, G. (1975): Monosomy 7 in two adult patients with acute myeloblastic leukaemia. Scand. J. Haematol. 15:251-255

Zech, L., Haglund, U., Nilsson, K. and Klein, G. (1976a): Characteristic chromosomal abnormalities in biopsies and lymphoid-cell lines from patients with Burkitt and non-Burkitt lymphomas. Int. J. Cancer 17:47-56

Zech, L., Gahrton, G., Killander, D., Franzen, S. and Haglund, U. (1976b): Specific chromosomal aberrations in polycythemia vera. Blood 48:687-696

Zhang, S., Zech, L. and Klein, G. (1982): High-resolution analysis of chromosome markers in Burkitt lymphoma cell lines. Int. J. Cancer 29:153-157

Zuelzer, W.W., Inoue, S., Thompson, R.I. and Ottenbreit, M.I. (1976): Long-term cytogenetic studies in acute leukemia of children; The nature of relapse. Am. J. Hematol. 1:143-190